Handbook of GC-MS

Hans-Joachim Hübschmann

Related Titles

Anderson, J., Berthod, A., Pino, V., Stalcup, A.M. (eds.)

Analytical Separation Science

2015
Print ISBN: 978-3-527-33374-5

Weiss, J., Weiss, T.

Handbook of Ion Chromatography
4 Edition

2014
Print ISBN: 978-3-527-32928-1

Maurer, H.H., Pfleger, K., Weber, A.A.

Mass Spectral and GC Data of Drugs, Poisons, Pesticides, Pollutants and Their Metabolites
4 Edition

2011
Print ISBN: 978-3-527-32992-2

Blumberg, L.M.

Temperature-Programmed Gas Chromatography

2010
Print ISBN: 978-3-527-32642-6

Encyclopedia of Analytical Chemistry, Supplementary VS1-S3

2010
Print ISBN: 978-0-470-97333-2

Hans-Joachim Hübschmann

Handbook of GC-MS

Fundamentals and Applications

Third Edition

WILEY-VCH
Verlag GmbH & Co. KGaA

Author

Dr. Hans-Joachim Hübschmann
Thermo Fisher Scientific
Blk 33, Marsiling Industrial Estate Road 3
#07-06
Singapore 739256
Singapore

■ All books published by **Wiley-VCH** are carefully produced. Nevertheless, authors, editors, and publisher do not warrant the information contained in these books, including this book, to be free of errors. Readers are advised to keep in mind that statements, data, illustrations, procedural details or other items may inadvertently be inaccurate.

Library of Congress Card No.: applied for

British Library Cataloguing-in-Publication Data
A catalogue record for this book is available from the British Library.

Bibliographic information published by the Deutsche Nationalbibliothek
The Deutsche Nationalbibliothek lists this publication in the Deutsche Nationalbibliografie; detailed bibliographic data are available on the Internet at <http://dnb.d-nb.de>.

© 2015 Wiley-VCH Verlag GmbH & Co. KGaA, Boschstr. 12, 69469 Weinheim, Germany

Print ISBN: 978-3-527-33474-2
ePDF ISBN: 978-3-527-67433-6
ePub ISBN: 978-3-527-67432-9
Mobi ISBN: 978-3-527-67431-2
oBook ISBN: 978-3-527-67430-5

Cover-Design Adam-Design, Weinheim, Germany
Typesetting Laserwords Private Limited, Chennai, India
Printing and Binding Markono Print Media Pte Ltd, Singapore

Printed on acid-free paper

Für Lotta und Rosa

Contents

Foreword

I consider the *Handbook of GC-MS* as a great contribution in the field of analytical chemistry. Dr. Hans-Joachim Hübschmann's vast hands-on knowledge and experience in handling with the GC-MS technology is well reflected throughout the book, and I strongly feel that it will offer a comprehensive support to the students, researchers, and practitioners of GC and GC-MS to deal with various routine and innovative applications in the field of food and environmental chemistry. Importantly, it endeavors to bridge the gap between theory and practice.

In food and environmental analyses, the chemists routinely face numerous challenges starting from sample preparation to optimization of the instrumentation method. Every stakeholder demands a rapid turnaround time for analysis with an expectation that a sample be tested across the laboratories with comparable proficiency. With every new set of sample matrix, the analytical issues and challenges seem to get compounded. Furthermore, analysis of any food or environmental samples for contaminant residues is a challenging task because the residues are often present in minute trace levels, typically at sub-ppm concentrations. Hence, a laboratory chemist needs to have a thorough understanding of the facets of the instrumentation technology so that it can be effectively utilized for measurements with satisfactory precision and accuracy. This consideration places a great importance on scientific reliability of analysts' technical skills, and this fundamental aspect is well tackled throughout this Handbook.

While operating a GC-MS instrument, the chemists routinely face diverse kinds of problems, which include lack of repeatability in analyte response, matrix effect, and so on. As a ready manual for the users of GC and GC-MS, this Handbook provides an answer to all such problems. The book has explained the fundamental concepts of sample preparation and provides an in-depth description of sample introduction to the GC system, column selection, and various advanced aspects, for example, 2D GC separations. The detection techniques described include the relatively simple detectors such as FID with a gradual transition to the complicated mass spectrometers, thus covering almost every single aspect of the GC and GC-MS technology with befitting explanatory examples.

It is a pleasure to commend this well-conceived book that is topical and contributes to the body of analytical methodologies involving GC and GC-MS technology. This is the kind of one-stop reference book that should be on the shelf of every laboratory that works with GC or GC-MS.

January 2015

Kaushik Banerjee, PhD, FRSC, FNAAS
ICAR National Fellow and Principal Scientist
National Referral Laboratory
ICAR-National Research Centre for Grapes
Pune 412 307, India

Preface to the Third Edition

Analytical sciences do not stand still, much technical advancement became routine for new solutions and increased productivity. GC-MS instrumentation, the undisputed workhorses in the analytical laboratories, made significant steps ahead with higher integrated, automated sample preparation and extraction methods, and improved matrix robustness, selectivity for real-life samples and the growing potential for multi-compound methods.

Major impact on standardized methods came from the wide distribution of the QuEChERS methodology and from extraction micro-methods such as the solid-phase microextraction (SPME). QuEChERS is on its way beyond the successful utilization for pesticides into environmental and drug applications. New micro-methods allow an easy handling by robotic autosamplers. Not unexpected, also mass spectrometric detection advanced, not in sensitivity, but significantly in selectivity, which translates for sure into sensitivity when analysing "dirty" real-life samples from less cleaned multi-method extracts. In particular, triple quadrupole mass spectrometry became a new standard for routine target compound analysis for many hundreds of compounds in one analytical run, providing unprecedented productivity for trace analyses. With this high-target analyte capacity, triple quadrupole GC-MS/MS instruments replace more and more, the initially highly successful but capacity limited, ion-trap technology. Accurate mass measurements are on the verge in GC-MS, extending the targeted analysis approach into the realm of untargeted trace compound analysis.

This third edition of the *Handbook of GC-MS* reflects the changing analytical requirements in GC-MS analysis with significant technology updates, additions of new fundamenttal topics and new applications based on current best practice methods.

Special focus has been set to the widely used and popular sample preparation methods as there are the pressurized liquid extraction (PLE), the thermal extraction of materials and food (outgassing), and in detail on the QuEChERS pesticide sample preparation used for GC-MS and LC-MS. As a consequence of the higher matrix load of these extracts and the recommended GC injection techniques, concurrent backflush, preventive maintenance and inlet deactivation became current topics of discussion. Olfactometry solutions have been added for applications in flavour analysis.

The applications section was updated with best practice solutions as of present demand from many laboratories. The sections describing the analytical conditions have been standardized and provide the complete method details for reproduction of the application in the reader's laboratory.

The selection of applications, due to the limited capacity of the Handbook, puts special focus on the following:

- Volatile analysis with static and dynamic headspace, multiple headspace extraction (MHE), as well as thermal desorption
- Pesticide analysis with multi-methods using single and triple quadrupole instruments and QuEChERS sample preparation
- Food safety and environmental analysis including most recent developments of the versatile SPME method
- Metabolomics analyses workflow using GC-MS/MS for identification and quantitation
- Persistent organic pollutants (POPs) analysis covering dioxins, polychlorinated biphenyls (PCBs) and the brominated flame retardants (BFRs), also featuring "Fast GC" for increased sample throughput
- Drugs of abuse screening from hair or urine matrices
- Extractables, leachables and outgassing analysis from leather, textiles, car interior materials or food

This third edition of the *Handbook of GC-MS* is a comprehensive update of current best practice GC-MS methodology compiled from practical laboratory work. It would not be possible without the contribution and support of many colleagues from analytical laboratories or from the analytical instruments industry, driving innovation and implementing new solutions for successful analytical methods. In particular, I am greatly indebted to the contributions, discussions and comments provided by Mike Buchanan from Sigma-Aldrich/Supelco; Dave Hope and Pat Pond from the Pacific Rim Lab in Vancouver, Canada; Chris Llewellynn and Elizabeth Woolfenden from Markes International in Llantrissant, UK; Cindy Llorente, Rosario Jimenez and Nese Sreenivasulu from the International Rice Research Institute (IRRI) in Los Banos, Philippines; Professor Hans-Ulrich Melchert, retired from the Robert Koch Institute in Berlin, Germany; Professor Janusz Pawliszyn from the University of Waterloo in Canada; Peter Pichler from the Brechbuehler AG in Schlieren, Switzerland. I am also grateful to my directly collaborating colleagues at Thermo Fisher Scientific Alex Chen in Melbourne, Australia; Benedicte Desroy in Villebon-sur-Yvette, France; Inge de Dobbeleer in Breda, Netherlands; Silvia Gemme and Jeremy Matthews in Singapore; Joachim Gummersbach in Dreieich, Germany; Aarti Karkhanis and Soma Dasgupta in Mumbai and Kolkata, India; Dirk Krumwiede and Heinz Mehlmann in Bremen, Germany; Paolo Magni and Massimo Santoro in Rodano, Italy; and Chongtian Yu in Shanghai. Also, I would like to thank as well the many communicating readers of previous editions providing valuable comments and feedback.

Singapore   *Hans-Joachim Hübschmann*
January 2015

1
Introduction

Detailed knowledge of the chemical processes in plants, animals and in our environment with air water and soil, about the safety of food and products has been made possible only through the power of modern instrumental analysis. In an increasingly short time span, more and more data are being collected. The absolute detection limits for organic substances are down in the attomole region, and counting individual molecules per unit time has already become a reality. In food safety and environmental analysis we are making measurements at the level of background contamination. Most samples subjected to chemical trace analysis carry high matrix, as are even blank samples. With the demand for decreasing detection limits by legal regulations, in the future effective sample preparation and separation procedures in association with highly selective detection techniques will be of critical importance for analysis. In addition, the number of substances requiring detection is increasing and with the broadening possibilities for analysis, so is the number of samples. The increase in analytical sensitivity is exemplified in the case of dioxins with 2,3,7,8-TCDD (tetrachlorodibenzodioxin), the most toxic food and feed contamination known today (Table 1.1).

Capillary gas chromatography (GC) is today the most important analytical method in organic chemical analysis for the determination of individual low molecular substances in complex mixtures. Mass spectrometry (MS) as the detection method gives the most meaningful data, arising from the direct determination of the substance molecule or of fragments. The results of mass spectrometry are therefore used as a reference for other indirect detection processes and finally for confirmation of the facts. The complete integration of MS and GC into a single GC-MS system has shown itself to be synergistic in every respect. While at the beginning of the 1980s MS was considered to be expensive, complicated and time-consuming or personnel-intensive, there is now hardly a GC laboratory which is not equipped with a GC-MS system. At the beginning of the 1990s MS became more widely recognized and furthermore an indispensable detection method for GC. The simple construction, clear function and an operating procedure, which has become easy because of modern computer systems, have resulted in the fact that GC-MS is widely used alongside traditional spectroscopic methods. The universal detection technique, together with high selectivity and very high sensitivity, has made GC-MS important for a broad

Handbook of GC-MS: Fundamentals and Applications, Third Edition. Hans-Joachim Hübschmann.
© 2015 Wiley-VCH Verlag GmbH & Co. KGaA. Published 2015 by Wiley-VCH Verlag GmbH & Co. KGaA.

Table 1.1 Sensitivity progress in mass spectrometry.

Year	Instrumental technique	Limit of detection (pg)
1967	GC-FID (packed column)	500
1973	GC-MS (quadrupole, packed column)	300
1976	GC-MS-SIM (magnetic instrument, capillary column)	200
1977	GC-MS (magnetic sector instrument)	5
1983	GC-HRMS (double focussing magnetic sector instrument)	0.15
1984	GC-MSD/SIM (quadrupole mass 'selective detector')	2
1986	GC-HRMS (double focussing magnetic sector instrument)	0.025
1989	GC-HRMS (double focussing magnetic sector instrument)	0.010
1992	GC-HRMS (double focussing magnetic sector instrument)	0.005
2006	GC×GC-HRMS (using comprehensive GC)	0.0003
2013	Cryogenic zone compression (t-CZC) GC-HRMS	0.0002

GC, gas chromatography; FID, flame ionization detector; MS, mass spectrometry; SIM, selected ion monitoring; and HRMS, high-resolution mass spectrometry.

spectrum of applications. Even higher selectivity is provided by the structure selective MS/MS and elemental formula providing accurate mass technologies for modern multi-residue methods with short sample preparation and clean-up steps. Benchtop GC-MS systems have completely replaced in many applications the stand-alone GC with selective detectors. Even GC-MS/MS has found its way to routine replacing many single quadrupole systems today.

The control of the chromatographic separation process still contributes significantly to the exploitation of the analytical performance of the GC-MS system (or according to Konrad Grob: "Chromatography takes place in the column!"). The analytical prediction capabilities of a GC-MS system are, however, dependent upon mastering the spectrometry. The evaluation and assessment of the data is leading to increasingly greater challenges with decreasing detection limits and the increasing number of compounds sought or found. While quantitation today is the main application for GC-MS, trace analysis methods and the appropriate data processing require additional measures for confirmation of results by mass spectrometric methods.

The high performance of GC lies in separation of substance mixtures and providing the transient signal for data deconvolution. With the introduction of fused silica columns, GC has become the most important and powerful separation method of analysing complex mixtures of products. GC-MS accommodates the current trend towards multi-methods or multi-component analyses (e.g. of pesticides, solvents, etc.) in an ideal way. Even isomeric compounds, which are present, for example, in essential oils, metabolic profiling, in polychlorinated biphenyls (PCBs) or dioxins, are separated by GC, while in many cases their mass spectra are almost indistinguishable. The high efficiency as a routine process is achieved through the high speed of analysis and the short turnaround time and thus guarantees high productivity with a high sample throughput. Adaptation and optimization for different tasks only requires a quick change of column. In many cases, however, and here one is relying on the explanatory power of the mass spectrometer, one type of a medium polar column can be used for different applications by adapting the sample injection technique and modifying the method parameters.

The area of application of GC and GC-MS is limited to substances which are volatile enough to be analysed by GC. The further development of column technology in recent years has been very important for application to the analysis of high-boiling compounds. Temperature-stable phases now allow elution temperatures of up to 500 °C for stable compounds. A pyrolyser in the form of a stand-alone sample injection system extends the area of application to involatile substances by separation and detection of thermal decomposition products. A typical example of current interest for GC-MS analysis of high-boiling compounds is the determination of polyaromatic hydrocarbons, which has become a routine process using the most modern column material.

The coupling of GC with MS using fused silica capillaries has played an important role in achieving a high level of chemical analysis. In particular in the areas of environmental analysis, analysis of residues and forensic science the high information content of GC-MS analyses has brought chemical analysis into focus through sometimes sensational results. For example, it has been used for the determination of anabolic steroids in cough mixture and the accumulation of persistent organic pollutants in the food chain. With the current state of knowledge, GC-MS is an important method for monitoring the introduction, the location and fate of man-made substances in the environment, foodstuffs, chemical processes and biochemical processes in the human body. GC-MS has also made its contribution in areas such as the ozone problem, the safeguarding of quality standards in foodstuffs production, in the study of the metabolism of pharmaceuticals or plant protection agents or in the investigation of polychlorinated dioxins and furans produced in certain chemical processes, to name but a few.

The technical realization of GC-MS coupling occupies a very special position in instrumental analysis. Fused silica columns are easy to handle, can be changed rapidly and are available in many high-quality forms. New microfluidic switching technologies extend the application without compromising performance for flow switching or parallel detection solutions. The optimized carrier gas streams show good compatibility with mass spectrometers, which is true today for both carrier gases, helium and hydrogen. Coupling can therefore take place easily by directly connecting the GC column to the ion source of the mass spectrometer.

The obvious challenges of GC and GC-MS lie where actual samples contain involatile components (matrix). In this case the sample must be processed before the analysis appropriately, or suitable column-switching devices need to be considered for backflushing of high-boiling matrix components. The clean-up is generally associated with enrichment of trace components. In many methods, there is a trend towards integrating sample preparation and enrichment in a single instrument. Headspace and purge and trap techniques, thermodesorption or SPME (solid phase microextraction) are coupled online with GC-MS and got integrated into the data systems for seamless control.

Future development will continue for a highly productive multi-compound trace analysis for the quantitation of mostly regulated target compounds. In addition, especially to comply with the aspects of food safety and product-safety requirements also non-targeted analytical techniques for the identification of potentially hazardous contaminants will evolve applying combined full scan and accurate mass capabilities.

1.1
The Historical Development of the GC-MS Technique

The foundation work in both GC and MS, which led to the current realization, was published at the end of the 1950s. At the end of the 1970s and the beginning of the 1980s, a rapid increase in the use of GC-MS in all areas of organic analysis began. The instrumental technique has now achieved a mature level for the once much specialized technique to become an indispensable routine procedure.

1910 : The physicist J.J. Thompson developed the first mass spectrometer and proved for the first time the existence of isotopes (^{20}Ne and ^{22}Ne). He wrote in his book 'Rays of Positive Electricity and their Application to Chemical Analysis': 'I have described at some length the application of positive rays to chemical analysis: one of the main reasons for writing this book was the hope that it might induce others, and especially chemists, to try this method of analysis. I feel sure that there are many problems in chemistry which could be solved with far greater ease by this than any other method'. Cambridge 1913. In fact, Thompson developed the first isotope ratio mass spectrometer (IRMS).

1910 : In the same year, M.S. Tswett published his book in Warsaw on 'Chromophores in the Plant and Animal World'. With this he may be considered to be the discoverer of chromatography.

1918 : Dempster used electron impact ionization for the first time.

1920 : Aston continued the work of Thompson with his own mass spectrometer equipped with a photoplate as detector. The results verified the existence of isotopes of stable elements (e.g. ^{35}Cl and ^{37}Cl) and confirmed the results of Thompson.

1929 : Bartky and Dempster developed the theory for a double-focussing mass spectrometer with electrostat and magnetic sector.

1934 : Mattauch and Herzog published the calculations for an ion optics system with perfect focussing over the whole length of a photoplate.

1935 : Dempster published the latest elements to be measured by MS, Pt and Ir. Aston thus regarded MS to have come to the end of its development.

1936 : Bainbridge and Jordan determined the mass of nuclides to six significant figures, the first accurate mass application.

1937 : Smith determined the ionization potential of methane (as the first organic molecule).

1938 : Hustrulid published the first spectrum of benzene.

1941 : Martin and Synge published a paper on the principle of gas liquid chromatography, GLC.

1946 : Stephens proposed a time of flight (TOF) mass spectrometer: *Velocitron*.

1947 : The US National Bureau of standards (NBS) began the collection of mass spectra as a result of the use of MS in the petroleum industry.

1948 : Hipple described the ion cyclotron principle, known as the *Omegatron* which now forms the basis of the current ion cyclotron resonance (ICR) instruments.

1950 : Gohlke published for the first time the coupling of a gas chromatograph (packed column) with a mass spectrometer (Bendix TOF).

1950 : The Nobel Prize for chemistry was awarded to Martin and Synge for their work on GLC (1941).

1950 : McLafferty, Biemann and Beynon applied MS to organic substances (natural products) and transferred the principles of organic chemical reactions to the formation of mass spectra.

1952 : Cremer and co-workers presented an experimental gas chromatograph to the ACHEMA in Frankfurt; parallel work was carried out by Janák in Czechoslovakia.

1952 : Martin and James published the first applications of GLC.

1953 : Johnson and Nier published an ion optic with a 90° electric and 60° magnetic sector, which, because of the outstanding focussing properties, was to become the basis for many high-resolution, organic mass spectrometers (Nier/Johnson analyser).

1954 : Paul published his fundamental work on the quadrupole analyser.

1955 : Wiley and McLaren developed a prototype of the present TOF mass spectrometer.

1955 : Desty presented the first GC of the present construction type with a syringe injector and thermal conductivity detector. The first commercial instruments were supplied by Burrell Corp., Perkin Elmer and Podbielniak Corp.

1956 : A German patent was granted for the QUISTOR (quadrupole ion storage device) together with the quadrupole mass spectrometer.

1958 : Paul published about his research on the quadrupole mass filter as
 – a filter for individual ions
 – a scanning device for the production of mass spectra
 – a filter for the exclusion of individual ions.

1958 : Ken Shoulders manufactured the first 12 quadrupole mass spectrometers at Stanford Research Institute, California.

1958 : Golay reported for the first time on the use of open tubular columns for GC.

1958 : Lovelock developed the argon ionization detector as a forerunner of the electron capture detector (ECD, Lovelock and Lipsky).

1962 : U. von Zahn designed the first hyperbolic quadrupole mass filter.

1964 : The first commercial quadrupole mass spectrometers were developed as residual gas analysers (Quad 200 RGAs) by Bob Finnigan and P.M. Uthe at EAI (Electronic Associates Inc., Paolo Alto, California).

1966 : Munson and Field published the principle of chemical ionization (CI).

1968 : The first commercial quadrupole GC-MS system for organic analysis was supplied by Finnigan Instruments Corporation to the Stanford Medical School Genetics Department.

1978 : Dandenau and Zerenner introduced the technique of fused silica capillary columns.

1978 : Yost and Enke introduced the triple-quadrupole technique.

1982 : Finnigan obtained the first patents on ion trap technology for the mode of selective mass instability and presented the ion trap detector as the first universal MS detector with a PC data system (IBM XT).

1989 : Prof. Wolfgang Paul, Bonn University, Germany, received the Nobel Prize for physics for work on ion traps, together with Prof. Hans G. Dehmelt, University of Washington in Seattle, and Prof. Norman F. Ramsay, Harvard University, USA.

2000 : Alexander Makarov published a completely new mass analyser concept called *Orbitrap* suitable for accurate mass measurements of low ion beams.

2005 : Introduction of a new type of hybrid Orbitrap mass spectrometer by Thermo Electron Corporation, Bremen, Germany, for MS/MS, very high mass resolution and accurate mass measurement on the chromatographic time scale.

2009 : Amelia Peterson *et al.*, University Wisconsin, Prof. Josh Coon group, published first results on the implementation of an EI/CI interface on a hybrid Orbitrap system for ultra-high resolution GC-MS using a GC-Quadrupole-Orbitrap configuration for full scan, SIM, MS/MS and SRM (selected reaction monitoring) at the ASMS conference.

2015 : Market introduction of the first high resolution accurate mass GC-MS system using Orbitrap technology by Thermo Fisher Scientific, Austin, TX, USA.

2
Fundamentals

2.1
Sample Preparation

In the analytical process of sampling, sample preparation, clean-up, separation, detection and data processing the sample preparation is certainly the most critical one for residue analysis. Low-level analytes of different chemical nature are embedded in complex matrices and need to become isolated and concentrated. In many cases today, the sample preparation can be an integral part of practical GC-MS analysis, as we know this from headspace, solid phase micro extraction (SPME) or thermal desorption (TD), only to name a few of the many options. The current trend is clearly directed to automated instrumental technique limiting expensive and error prone manual work to the essential. The concentration processes in this development are of particular importance for coupling with capillary GC-MS, as in trace analysis the limited sample capacity of capillary columns must be compensated for. It is therefore necessary both that overloading of the stationary phase by matrix is avoided and that the detection limits of mass spectrometric detection are taken into consideration. To optimize separation on a capillary column, strongly interfering components of the matrix must be removed before applying an extract. The primarily universal character of the mass spectrometer poses conditions on the preparation of a sample which are to some extent more demanding than those of an element-specific detector, such as ECD (electron capture detector) or NPD (nitrogen/phosphorus detector) unless highly selective techniques as MS/MS or high resolution accurate measurements are applied. The clean-up and analyte concentration, which forms part of sample preparation, must therefore in principle always be regarded as a necessary preparative step for GC-MS analysis. The differences in the concentration range between various samples, differences between the volatility of the analytes and that of the matrix and the varying chemical nature of the substances are important for the choice of a suitable sample preparation procedure.

Off-line techniques (as opposed to online coupling or hyphenated techniques) have the particular advantage that samples can be processed in parallel and the extracts can be subjected to other analytical processes besides GC-MS. Online techniques have the special advantage of sequential processing of the samples without intermediate manual steps. The online clean-up allows an optimal time overlap which gives the sample preparation the same amount of time as the

Handbook of GC-MS: Fundamentals and Applications, Third Edition. Hans-Joachim Hübschmann.
© 2015 Wiley-VCH Verlag GmbH & Co. KGaA. Published 2015 by Wiley-VCH Verlag GmbH & Co. KGaA.

analysis of the preceding sample. This permits maximum use of the instrument and automatic operation.

Online processes generally offer potential for higher analytical quality through lower contamination from the laboratory environment and, for smaller sample sizes, lower detection limits with lower material losses. Frequently, total sample transfer is possible without taking aliquots or diluting e.g. by large volume injection techniques. Volatility differences between the sample and the matrix allow, for example, the use of extraction techniques such as the static or dynamic (purge and trap, P&T) headspace techniques as typical GC-MS coupling techniques. These are already used as online techniques in many laboratories. Where the volatility of the analytes is insufficient, other extraction procedures, for example, thermal extraction, pyrolysis or online SPE (solid phase extraction) techniques are being increasingly used online. SPE in the form of microextraction, gel permeation chromatography (GPC), liquid chromatography (LC)-GC coupling or extraction with pressurized or supercritical fluids show high analytical potential here.

Future developments in sample preparation are expected from the application of nanomaterials. Due to their specific properties and the flexibility in tailored surface modifications carbon-based nanomaterials have already found a wide range of applications in different sample preparation procedures. They can be used as selective adsorbents by direct interaction between the analyte and the nanoparticles (Zhang, 2013) (Table 2.1).

2.1.1
Solid Phase Extraction

From the middle of the 1980s, a new extraction technique called solid phase extraction (SPE) began to revolutionize the enrichment, extraction and clean-up of analytical samples. Following the motto 'The separating funnel is a museum piece', the time-consuming and arduous liquid/liquid extraction has increasingly been displaced from the analytical laboratory (Bundt *et al.*, 1991). Today the euphoria of the rapid and simple preparation with disposable columns has lessened as a result of a realistic consideration of their performance levels and limitations. A particular advantage over the classical liquid/liquid partition is the low consumption of expensive and sometimes harmful solvents. The amount of apparatus and space required is low for SPE. Parallel processing of several samples is therefore quite possible. Besides an efficient clean-up, the necessary concentration of the analyte frequently required for GC-MS is achieved by SPE.

In SPE, strong retention of the analyte is required, which prevents migration through the carrier bed during sample application and washing. Specific interactions between the substances being analysed and the chosen adsorption material are exploited to achieve retention of the analytes and removal of the matrix. An extract which is ready for analysis is obtained by changing the eluents. The extract can then be used directly for GC and GC-MS in most cases. The choice of column materials permits the exploitation of different separating mechanisms of adsorption chromatography, normal-phase and reversed-phase chromatography, and also ion exchange and size exclusion chromatography (Figure 2.1).

Table 2.1 Comparison of carbon-based nanomaterials in sample preparation (Zhang, 2013).

Materials	Special material characteristics	Derivatization methods	Sample preparation potential	Current status*	Perspective*
Graphene	1. Both available sides for adsorption 2. Easily synthesized in lab 3. Easily modified with functional groups	Modified via graphene oxide	SPE, MSPE, SPME, LDI substrate, μ-SPE, PMME, MSDP, SRSE	*****	*****
Carbon nanotubes	1. Mostly used carbon nanomaterials 2. Be easily covalently or non-covalently functionalized	Oxidized CNTs can be grafted via creation of amide bonds, N2-plasma and radical addition	SPE, MSPE, SPME, LDI substrate, μ-SPE, SRSE	*****	*****
Carbon nanofibers	1. Easily available on a large scale 2. Larger dimensions without coating or functionalization	Polar groups can be introduced by treating with the concentrated nitric acid	SPE, SPME, on-line μ-SPE	***	****
Fullerenes	Firstly used carbon nanomaterials	Covalently bonded to other reagents	SPE, on-line SPE, MSPE, SPME, LLE, SFE, ion-pair precipitation	****	
Nanodiamonds	1. Chemical inertness and hardness 2. More expensive than others	Functionalized with H/D-terminated, halogenated, aminated, hydroxylated and carboxylated surfaces	SPE, LDI substrate	***	***
Carbon nanocones, disks and nanohorns	1. Multiple structures 2. Lower aggregation tendency	Functionalized by microwave (for SWNHs)	SPE, SPME	**	**

*The star number represents the research status and perspective which is mainly based on properties of materials, the number of published articles and publication year.
LDI - laser desorption ionization; MSPE- Magnetic solid-phase extraction; PMME- Poly (methyl methacrylate); Matrix solid-phase dispersion (MSDP); Stir rod sorptive extraction (SRSE); Liquid/liquid extraction (LLE); Singlewall carbon nanohorn (SWNH)

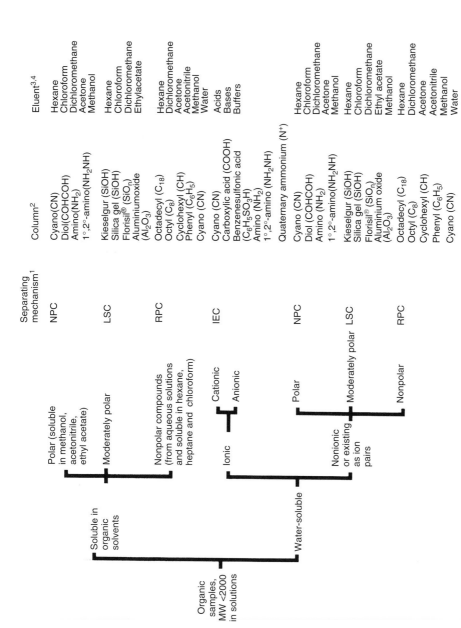

Figure 2.1 Key to choosing SPE columns and eluents. The choice of the SPE phase depends on the molecular solubility of the sample in a particular medium and on its polarity. The sample matrix is not considered (J. T. Baker).

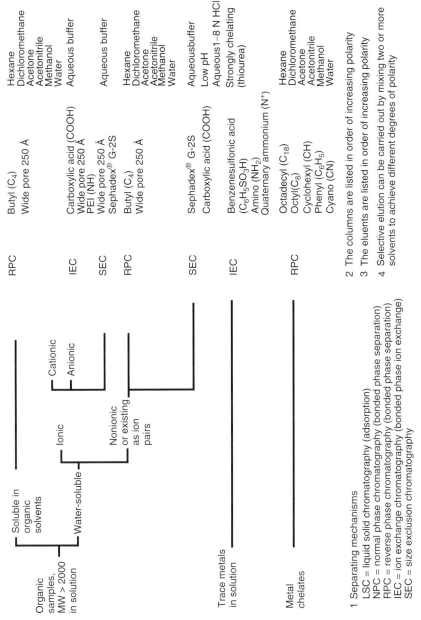

Figure 2.1 (continued)

The chart reads as follows:

Organic samples, MW > 2000 in solution
- Soluble in organic solvents
  - RPC — Butyl (C₄) Wide pore 250 Å — Hexane, Dichloromethane, Acetone, Acetonitrile, Methanol, Water
- Water-soluble
  - Ionic
    - Cationic
      - IEC — Carboxylic acid (COOH) Wide pore 250 Å, PEI (NH) — Aqueous buffer
    - Anionic
      - SEC — Wide pore 250 Å, Sephadex® G-2S — Aqueous buffer
  - Nonionic or existing as ion pairs
    - RPC — Butyl (C₄) Wide pore 250 Å — Hexane, Dichloromethane, Acetone, Acetonitrile, Methanol, Water

Trace metals in solution
- SEC — Sephadex® G-2S, Carboxylic acid (COOH) — Aqueous buffer, Low pH, Aqueous 1–8 N HCl
- IEC — Benzenesulfonic acid (C₆H₅SO₃H), Amino (NH₂), Quaternary ammonium (N⁺) — Strongly chelating (thiourea)

Metal chelates
- RPC — Octadecyl (C₁₈), Octyl (C₈), Cyclohexyl (CH), Phenyl (C₆H₅), Cyano (CN) — Hexane, Dichloromethane, Acetone, Acetonitrile, Methanol, Water

1 Separating mechanisms
LSC = liquid solid chromatography (adsorption)
NPC = normal phase chromatography (bonded phase separation)
RPC = reverse phase chromatography (bonded phase separation)
IEC = ion exchange chromatography (bonded phase ion exchange)
SEC = size exclusion chromatography

2 The columns are listed in order of increasing polarity
3 The eluents are listed in order of increasing polarity
4 Selective elution can be carried out by mixing two or more solvents to achieve different degrees of polarity

The physical extraction process, which takes place between the liquid phase (the liquid sample containing the dissolved analytes) and the solid phase (the adsorption material) is common to all SPEs. The analytes are usually extracted successfully because the interactions between them and the solid phase are stronger than those with the solvent or the matrix components. After the sample solution has been applied to the solid phase bed, the analytes become enriched on the surface of the SPE material. All other sample components pass unhindered through the bed and can be washed out. The maximum sample volume that can be applied is limited by the breakthrough volume (BTV) of the analyte. Elution is achieved by changing the solvent. For this, there must be a stronger interaction between the elution solvent and the analyte than between the latter and the solid phase. The elution volume should be as small as possible to prevent subsequent solvent evaporation.

In analytical practice, two SPE processes have become established. Cartridges are mostly preferred for liquid samples (Figures 2.2 and 2.3). If the GC-MS analysis reveals high contents of plasticizers, the plastic material of the packed columns must first be considered and in special cases a change to glass columns must be made. For sample preparation using slurries or turbid water, which rapidly lead to deposits on the packed columns, SPE disks should be used. Their use is similar to that of cartridges. Additional contamination, for example, by plasticizers, can be ruled out for residue analysis in this case (Figure 2.4)

A large number of different interactions are exploited for SPE (Figure 2.2). Selective extractions can be achieved by a suitable choice of adsorption materials. If the eluate is used for GC-MS the detection characteristics of the mass spectrometer in particular must be taken into account. Unlike an ECD, which is still widely used as a selective detector for substances with high halogen content in environmental analysis, a GC-MS system can be used to detect a wide range of non-halogenated substances. The use of a selected ion monitoring technique (SIM, MID, multiple ion detection) therefore requires better purification of the

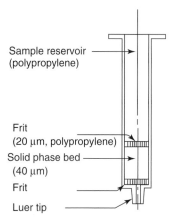

Sample reservoir
(polypropylene)

Frit
(20 μm, polypropylene)
Solid phase bed
(40 μm)
Frit
Luer tip

Figure 2.2 Construction of a packed column for solid phase extraction (J. T. Baker).

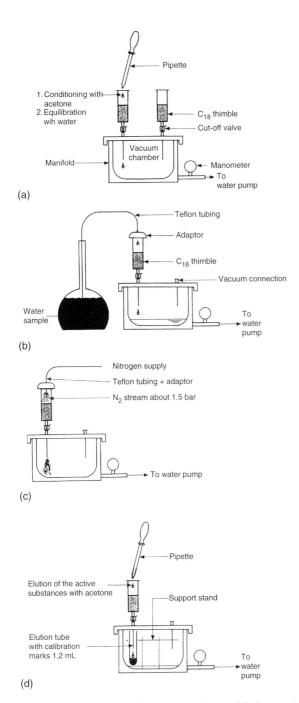

Figure 2.3 Enrichment of a water sample on solid phase cartridges with C_{18}-material (Hein and Kunze, 1994): (a) conditioning, (b) loading (extraction) and washing, (c) drying and (d) elution of active substances.

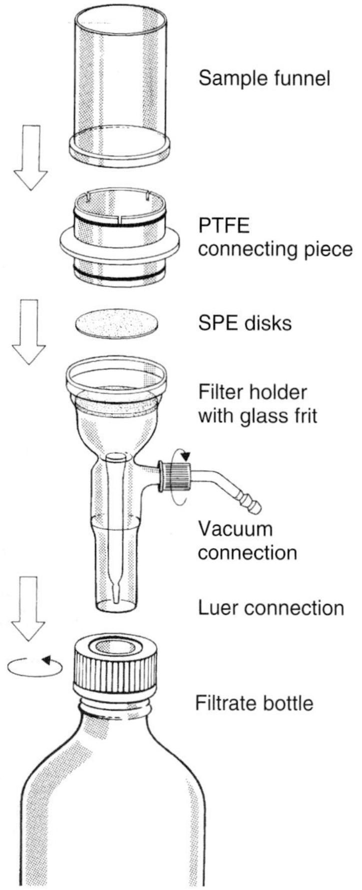

Sample funnel

PTFE
connecting piece

SPE disks

Filter holder
with glass frit

Vacuum
connection

Luer connection

Filtrate bottle

Figure 2.4 Apparatus for solid phase extraction with SPE disks (J. T. Baker).

extracts obtained by SPE. In the case of the processing of dioxins and PCBs (polychlorinated biphenyls) from waste oil, for example, a silica gel column charged with sulfuric acid is also necessary. Extensive oxidation of the non-specific hydrocarbon matrix is thus achieved. The quality and reproducibility of SPE depends on criteria comparable to those which apply to column materials used in high-pressure liquid chromatography (HPLC).

2.1.2
Solid Phase Microextraction

'The simplification of sample preparation and its integration with both sampling and the convenient introduction of extracted components to analytical instruments are a significant challenge and an opportunity for the contemporary analytical chemist' writes the inventor Prof. Janusz Pawliszyn in his preface to the Handbook of SPME' (Pawliszyn, 2012).

The solvent-free micro extraction technique SPME is an important step towards the instrumentation and automation of the SPE technique for online sample preparation and introduction to GC-MS (Zhang *et al.*, 1994; Eisert and Pawliszyn, 1997; Lord and Pawliszyn, 1998). It involves exposing a fused silica fibre coated with a liquid polymeric material to a sample containing the analyte. As an extraction and enrichment technique it compares to P&T methods (MacGillivray *et al.*, 1994). Also derviatization steps can be coupled to the extraction process for polar compounds and improved efficiencies (Pan *et al.*, 1997). The typical dimensions of the active fibre surface are $1\,cm \times 100\,\mu m$. The analyte diffuses from the gaseous, headspace or liquid sample phase into the fibre surface and partitions into the coating material according to the first partition coefficient. The agitated sample is typically incubated at a constant temperature before and during the sampling process to achieve maximum recovery and precision for quantitative assays (Gorecki and Pawliszyn, 1997). The application of internal standards or the use of the standard addition method for quantitation is strongly recommended for achieving low relative standard deviations (Nielsson *et al.*, 1994; Poerschmann *et al.*, 1997). After the equilibrium is established, the fibre with the collected analytes is withdrawn from the sample and transferred into a GC injector, either manually or more convenient via an autosampler. The analyte is desorbed thermally in the hot injector from the coating. The fibre material is used for a large number of samples in automated serial analysis. Modern autosampler are capable to exchange the fibre holder to provide automated access to different fibre characteristics according to the analyte requirements.

SPME offers distinct advantages for automated sample preparation including reduced time per sample, less manual sample manipulation resulting in an increased sample throughput and, in addition, a significant reduction of organic solvent use in the routine laboratory (Pawliszyn, 1997; Berlardi and Pawliszyn, 1989; Arthur *et al.*, 1992).

An SPME unit consists of a length of fused silica fibre coated with a phase similar to those used inside of chromatography columns. The phase can be mixed with solid adsorbents, for example, divinylbenzene (DVB) polymers, templated resins or porous carbons. The fibre is attached to a stainless steel plunger in a protective syringe like holder, used manually or in a specially prepared autosamplers. The plunger on the syringe retracts the fibre for storage and piercing septa, and exposes the fibre for extraction and desorption of the sample. A spring in the assembly keeps the fibre retracted, reducing the chance of it being damaged (Figure 2.5a,b).

The design of a portable, disposable SPME holder with a sealing mechanism gives flexibility and ease of use for on-site sampling. The sample can be extracted and stored by placing the tip of the fibre needle in a septum. Lightweight disposable holders can be used for the lifetime of the fibre. SPME could also be used as an indoor air-sampling device for GC and GC-MS analysis. For this technique to be successful, the sample must be stable to storage.

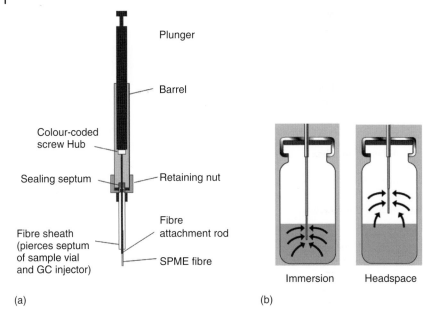

Plunger

Barrel

Colour-coded
screw Hub

Sealing septum

Retaining nut

Fibre
attachment rod

Fibre sheath
(pierces septum
of sample vial
and GC injector)

SPME fibre

Immersion   Headspace

(a)   (b)

Figure 2.5 (a) SPME plunger for autosampler use and (b) SPME principle for liquid immersion and headspace applications.

A major shortcoming of SPME is the lack of fibres that are polar enough to extract very polar or ionic species from aqueous solutions without first changing the nature of the species before derivatization. Ionic, polar and involatile species have to be derivatized to GC amenable species before SPME extraction.

Since the introduction in 1989 SPME became quickly popular for a variety of applications in trace residue analysis, also including the analysis of volatile analytes and gases consisting of small molecules (Potter and Pawliszyn, 1992; Boyd-Boland and Pawliszyn, 1995; Boyd-Boland *et al.*, 1996; Magdic *et al.*, 1996; Martos *et al.*, 1997; Gorecki *et al.*, 1998). The amount of analyte extracted is dependent upon the distribution constant (Zhang and Pawliszyn, 1993). The higher the distribution constant, the higher is the quantity of analyte extracted. As described in static headspace the addition of salt to the sample matrix enhances the amount of analyte extracted by the SPME fiber coating (Steffen and Pawliszyn, 1996). Generally, a thicker film is required to retain small molecules and a thinner film is used for larger molecules with high distribution constants. The polarity of the fibre and the type of coating can also increase the sampling efficiency (Table 2.2). The initially introduced fibre coatings had been used preferably for extractions from sample headspace solid phase micro extraction (HS-SPME) (Ligor and Buszewski, 2008), and found limited applications using the sample immersion technique due to low recoveries by matrix interaction (Luo *et al.*, 1998). New multilayer coatings have recently been developed in particular for closing the gap of applications for low volatile polar compounds using the direct immersion solid phase micro extraction technique (DI-SPME) (Risticevic *et al.*, 2009; Souza Silva and Pawliszyn, 2012).

Table 2.2 Types of SPME fibres available.

Non-polar fibres
- Cellulose acetate/polyvinylchloride (PVC), alkane selective (Farajzadeh and Hatami, 2003)
- Powdered activated carbon (PAC), for BTEX and halocarbons (Shutao *et al.*, 2006)
- Polydimethylsiloxane (PDMS), 7, 30, 100 µm coating (Saraullo *et al.*, 1997)

Bipolar fibres
- Carboxen/polydimethylsiloxane (CAR/PDMS), 75, 85 µm coating
- Divinylbenzene/Carboxen/polydimethylsiloxane (DVB/CAR/PDMS), 30, 50 µm coating
- Divinylbenzene/Carboxen/polydimethylsiloxane (DVB/CAR/PDMS), 30, 50 µm coating
- Polydimethylsiloxane/divinylbenzene (PDMS/DVB), 60, 65 µm coating

Polar fibres
- Carbowax/Divinylbenzene (CW/DVB), 65, 70 µm coating
- Carbowax/templated resin (CW/TPR), 50 µm coating
- Polyacrylate (PA), 85 µm coating

Blended fibre coatings contain porous material, such as divinylbenzene (DVB) and carboxen, blended in either PDMS or carbowax.
Benzene, toluene, ethylbenzene, and xylenes (BTEX).

New capabilities in SPME are expected from the use of nanomaterials. Graphene was immobilized on the SPME fibre. The nanomaterial showed improved extraction efficiency, higher mechanical and thermal stability, and with more than 250 stable extractions a significantly longer life span than commercial PDMS (polydimethylsiloxane) or PDMS/DVB-coated fibres (Arthur and Pawliszyn, 1990; Huang *et al.*, 2012). A ionic-liquid-mediated multi-walled carbon nanotube (CNT)-poly(dimethylsiloxane) hybrid coating was used as solid-phase microextraction adsorbent for methyl-t-butyl ether (MTBE) in water. This innovative fibre as reported has high thermal stability of >320 °C and a long lifespan with over 210 analyses (Vatani and Yazdi, 2014).

Basic Rules for SPME Fibre Choice (Supelco):

- Low molecular weight or volatile compounds usually require a 100 µm PDMS-coated fibre.
- Larger molecular weight or semi-volatile compounds are more effectively extracted with a 30 µm PDMS fibre or a 7 µm PDMS fibre.
- To extract very polar analytes from polar samples, use an 85 µm polyacrylate-coated fibre.
- More volatile polar analytes, such as alcohols or amines, are adsorbed more efficiently and released faster with a 65 µm PDMS/DVB-coated fibre.

- A 60 μm PDMS/DVB fibre is a general purpose fibre for HPLC.
- For trace-level volatiles analysis, use a 75 μm PDMS/Carboxen fibre.
- For an expanded range of analytes (C3-C20), use a 50/30 DVB/Carboxen on PDMS fibre (Tables 2.3 and 2.4).

Table 2.3 Fibre selection guide (Supelco, 2013).

Analyte type	Recommended fibre
Gases and low molecular weight compounds (M 30-225)	75 μm/85 μm Carboxen/poly-dimethylsiloxane
Volatiles (M 60-275)	100 μm polydimethylsiloxane
Volatiles, amines and nitro-aromatic compounds (M 50-300)	65 μm polydimethylsiloxane/divinylbenzene
Polar semi-volatiles (M 80-300)	85 μm polyacrylate
Non-polar high molecular weight compounds (M 125-600)	7 μm polydimethylsiloxane
Non-polar semi-volatiles (M 80-500)	30 μm polydimethylsiloxane
Alcohols and polar compounds (M 40-275)	60 μm Carbowax (PEG)
Flavour compounds: volatiles and semi-volatiles, C3-C20 (M 40-275)	50/30 μm divinylbenzene/Carboxen on polydimethylsiloxane on a StableFlex fibre
Trace compound analysis (M 40-275)	50/30 μm divinylbenzene/Carboxen on polydimethylsiloxane on a 2 cm StableFlex fibre
Amines and polar compounds (HPLC use only)	60 μm polydimethylsiloxane/divinylbenzene

Table 2.4 Summary of method performance for several analysis examples.

Parameter	Formaldehyde	Triton-X-100	Phenylurea pesticides	Amphetamines
LOD	2–40 ppb	1.57 μg/L	<5 μg/L	1.5 ng/mL
Precision (%)	2–7	2–15	1.6–5.6	9–15
Linear range	50–3250 ppb	0.1–100 mg/L	10–10 000 μg/L	5–2000 ng/mL
Linearity (r_2)	0.9995	0.990	0.990	0.998
Sampling time	10–300 s	50 min	8 min	15 min

The principle of SPME found its extension in the Stir Bar Sorptive Extraction tool SBSE (Sandra *et al.*, 2000, Hoffmann *et al.*, 2000). The extraction is performed with a glass-covered, magnetic stir bar, which is coated with PDMS and other extraction phases. The difference to SPME is the capacity of the sorptive PDMS phase (Bicchi *et al.*, 2009). While in the SPME fibre an extraction phase volume of about 15 µL is used, with SBSE a significantly enlarged volume of up to 125 µL is available. The larger volume of sorption phase provides a better phase ratio with increased recovery resulting in up to 250-fold lowered detection limits (Deußing, 2005). Many samples can be extracted at the same time. Extraction times run are significantly higher than SPME with typically up to 60 min. Transfer to the GC is achieved by thermal desorption of the stir bar in a suitable injection system. Thermolabile compounds can be alternatively dissolved by liquids (SBSE/LD). The excellent sensitivity and reproducibility have been demonstrated, for instance, in the low-level multi-residue detection of endocrine disrupting chemicals in drinking water (Serodio and Nogueira, 2004).

2.1.3
Pressurized Liquid Extraction

The trend to fully automated instrumental extraction methods has also reached the time- and solvent-consuming Soxhlet extraction process. A comparison of modern extraction methods such as microwave assisted extraction (MAE), supercritical fluid extraction (SFE), SPME and pressurized liquid extraction (PLE) or pressurized fluid extraction (PFE) with the traditional Soxhlet method, finds that PLE has found widespread application due to a number of advantageous features, including the most practical ones: ease-of-use, versatility and reproducibility (Table 2.5). There are both strong analytical and economical advantages with PLE: The high extraction efficiency, savings in solvent consumption (and waste generation) and the short extraction times led to a steadily growing number of applications (Richter *et al.*, 1996; Li, 2003; Oleszek-Kudlak *et al.*, 2007). PLE also limits glassware handling, and facilitates a safe laboratory working environment and has become a popular green extraction technology for different classes of organic contaminants extracted from numerous matrices in environmental, food, feed, biological and technical samples.

Table 2.5 Comparison of extraction methods for solid samples soil, sediment, fly ash, sludge and solid wastes (Oleszek-Kudlak *et al.*, 2007).

Extraction method	US EPA method	Avg. volume of solvent (mL)	Avg. extraction time	Acquisition cost	Operating cost/sample
Soxhlet extraction	3540C	300	16–24 h	Very low	Very high
Automated Soxhlet extraction	3541	50	2 h	Moderate	Low to moderate
Pressurized fluid extraction	3545A	10–30	10–15 min	High	Low
Microwave extraction	3546	25–40	10–20 min	Moderate	Low
UltraSonic extraction	3550C	300	~30 min	Low	High
Supercritical fluid extraction	3560/3561/ 3562	10	20–50 min	Moderate to high	Moderate to high

PFE or PLE also known as *accelerated solvent extraction (ASE)* is a technique used for extracting the analytes of interest from a solid, semisolid or liquid sample using an organic solvent at elevated temperatures up to 200 °C (390 F) and pressures up to 1500 psi (100 bar) in a combination of static and dynamic extraction steps. The elevated pressure increases the boiling temperature of the solvent and decreases its viscosity, thereby allowing faster extraction to be conducted at relatively higher temperatures. The benefit of higher temperature extraction is primarily speed. The PFE extraction process is significantly faster than traditional methods such as Soxhlet extraction. It also covers a wide and flexible sample range from 1 to 100 g. In most cases typically, a 10 g sample is used for trace analysis methods. PFE is the general term for the automated technique of extracting solid and semisolid samples with liquid, organic or aqueous solvents used in the official EPA method 3445A. The PFE method is also marketed under the term *Accelerated Solvent Extraction (ASE)* by former Dionex Corp. (today part of Thermo Fisher Scientific), USA, covered by a number of international patents (Dionex, 2006). Automated PFE instrumentation allows the independent control of temperature and pressure conditions for each sample cell. This control is critical for a high analyte recovery and reproducibility.

At the typical PFE conditions, the solvent characteristics change compared to the traditional Soxhlet technique and allow an improved separation of target analytes from the matrix (Hubert *et al.*, 2001). For instance, the extraction of relatively polar phenols from soil samples is possible using the non-polar solvent *n*-hexane. PFE extractions typically use the known organic solvents n-hexane, cyclohexane, toluene, dichloromethane, methanol, acetone as applied in Soxhlet methods. For biological samples and for the extraction of water soluble components, water-based media in mixtures with organic solvents are applied (Curren, 2002). This variation gives access to polar compounds for subsequent LC-MS analysis. Due to a mostly sharp matrix/analyte separation by PFE, the collected extracts require less time consuming clean-up typically employing SPE or gel permeation chromatography (GPC) steps (Figure 2.6).

Thermal degradation does not occur during PFE extractions. Using PFE techniques, the sample is completely and constantly surrounded by solvent; thus, oxidative losses are minimized if the extraction solvent is degassed and oxygen is excluded. Since the analytes are in the heated zone for only short periods of time, thermal losses do not occur if appropriate temperatures are used for the extractions.

Traditional techniques, such as Soxhlet, can take 4−48 h. With PFE, analyte recoveries equivalent to those obtained using traditional extraction methods can be achieved in only 15−30 min. Although PFE uses the same aqueous and organic solvents as traditional extraction methods, it uses them more efficiently. A typical PFE extraction is done using 50−150 mL of solvent, see Table 2.6 for comparison.

Approximately 75% of all PFE extractions can be completed in less than 20 min using the standard PFE extraction conditions (100 °C, 1500 psi). But, PFE

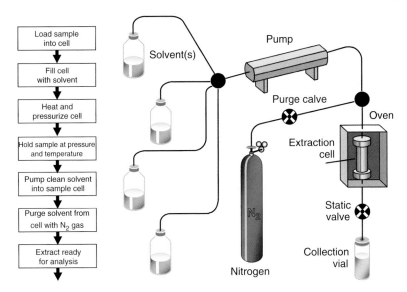

Figure 2.6 Schematics of a pressurized extraction unit. (Courtesy Dionex Corp.)

Table 2.6 Comparison of standard laboratory extraction techniques.

Techniques	Solvent volume per sample (mL)	Extraction time per sample (h)	Cost per sample[a] (in US $)
Manual Soxhlet	200–500	4–48	23.50
Automated Soxhlet	50–100	1–4	16.25
Sonication	100–300	0.5–1	20.75
SFE	8–50	0.5–2	17.60
Microwave	50–100	0.5–1	15.25
ASE	15–40	0.2–0.3	11.20

a) Cost per sample in US $ based on 2000 samples per year, average values, based on US comparison (Dionex, 2006).

extractions are matrix dependent and require further optimization when applied to different matrices. Method development starts with standard conditions and evaluates both the recovery in the first step and the result of a second extraction of the given new sample. Also, the initial application of a standard matrix, for example, sea sand is a helpful step. If these initial parameters do not provide the recoveries desired, the temperature is increased to improve the efficiency of the extraction.

Adding static cycles, increasing static time and selecting a different solvent are additional variables that can be used to optimize a method.

Recommended Extraction Conditions (US EPA, 1998)

For semi-volatiles, organophosphorus pesticides, organochlorine pesticides, herbicides and PCBs:

Oven temperature	100 °C
Pressure	1500–2000 psi
Static time	5 min (after 5 min pre-heat equilibration)
Flush volume	60% of the cell volume
Nitrogen purge	60 s at 150 psi (purge time may be extended for larger cells)
Static cycles	1

For PCDDs (polychlorinated dioxins)/PCDFs (polychlorinated furans):

Oven temperature	150–175 °C
Pressure	1500–2000 psi
Static time	5–10 min (after 5 min pre-heat equilibration)
Flush volume	60–75% of the cell volume
Nitrogen purge	60 s at 150 psi (purge time may be extended for larger cells)
Static cycles	2 or 3

PFE/PLE extraction is internationally accepted by the regulatory bodies, among others by the US EPA SW-846 Method 3545A, which can be used in place of Methods 3540, 3541, 3550 and 8151, by the US Contract Laboratory Program (CLP) statement of work (SOW) OLM04.2, the Chinese method GB/T 19649-2005 and German Method L00.00-34 both for pesticides, also the ASTM Standard Practice D-7210 for additives in polymers and D-7567 for gel content of polyolefins.

Method 3545A can be applied to the extraction of base/neutrals and acids (BNAs), chlorinated pesticides and herbicides, PCBs, organophosphorus pesticides, dioxins and furans and total petroleum hydrocarbons (TPHs).

Scope and Application of EPA Method 3545A

1.1 Method 3545A is a procedure for extracting water-insoluble or slightly water-soluble organic compounds from soils, clays, sediments, sludges and waste solids. The method uses elevated temperature (100–180 °C) and pressure (1500–2000 psi) to achieve analyte recoveries equivalent to those from Soxhlet extraction, using less solvent and taking significantly less time than the Soxhlet procedure. This procedure was developed and validated on a commercially available, automated extraction system.

1.2 This method is applicable to the extraction of semi-volatile organic compounds, organophosphorus pesticides, organochlorine pesticides, chlorinated herbicides, PCBs and PCDDs/PCDFs, which may then be analysed by a variety of chromatographic procedures.

(US EPA Method 3545A, January 1998)

2.1.3.1 In-Cell Sample Preparation

Selective PFE can be achieved with specific analyte or matrix-related procedures even derivatization steps (Sanchez-Prado *et al.*, 2010) taking place during the extraction step, saving additional manipulation of the extract for increased productivity (Figure 2.7). Numerous PFE in-cell procedures are known for various applications (Haglund, 2010) also covering trace analysis by GC-MS. The most typical ones with water removal and hydrocarbon matrix oxidation are highlighted in the following in more detail.

2.1.3.2 In-Cell Moisture Removal

Sample drying can be accomplished in several ways such as air drying and oven drying before extraction. However, these approaches are not suited when analysing volatile or semi-volatile components as they would be removed from the sample before extraction or analysis. The in-cell method can remove moisture when the wet sample is mixed with a water absorbing polymer (Ullah *et al.*, 2012).

Another common method for moisture removal is by using salts such as sodium sulfate, calcium chloride, magnesium sulfate, calcium sulfate and the like. These salts tend to associate to water molecules to form hydrated salts. Sodium sulfate, for example, tends to clump together when water is present. Sodium sulfate is not suitable for in-cell moisture removal and extraction in ASE. Sodium sulfate can dissolve in hot solvent to a certain extent and can precipitate downstream in

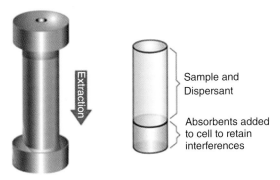

Figure 2.7 PFE in-cell sample preparation procedure with additional adsorbent layer.

some instances clogging the outlet frit, tubes and valves. Moreover, sodium sulfate becomes an aggregate hard lump upon water absorption.

Typically, a water-absorbent polymer is applied comprising a copolymer of a basic monomer with an acidic monomer. This combination is suitable for moisture removal under high ionic strength conditions. Results also demonstrate that when the polymer is mixed with diatomaceous earth (DE), absorbent the water-removal efficiency increases significantly. Water-removal efficiency increases significantly due to improved flow characteristics.

The amount of water-absorbing polymer needed is ~0.20 g for absorbing 1 g of water at room temperature. In PFE, the water-absorbing ability of the polymer decreases with increasing temperature. For example, at 100 °C a 4 g polymer and 4 g DE can remove roughly 10 g of water completely. Hexane is used as the extraction solvent. Recoveries for nitro-, alkyl- or chloro-substituted phenols range from 81% to 98%. New developments of improved water-absorbing polymers can remove up to 5 g of water per gram of the polymer at room temperature.

2.1.3.3 In-Cell Hydrocarbon Oxidation

Spinnel *et al.* further developed the modular PFE approach for dioxin analysis. A matrix retainer module filled with sulfuric acid impregnated silica gel was introduced between the sample. In this case, salmon muscle was mixed with sodium sulfate and the carbon trap. This procedure provided extracts clean enough to be analysed by GC–HRMS (high-resolution mass spectrometry) analysis after solvent evaporation to an appropriate final volume.

The results obtained were in good agreement with those of a Soxhlet reference method. A slightly higher total dioxin equivalency concentration was obtained for the modular PFE (89.3 ± 8.3 pg/g fresh weight) than for Soxhlet extraction (81.3 ± 7.8 pg/g fresh weight), which indicate an improved extraction efficiency of the PFE method (Spinner, 2008).

A miniaturized PFE device has been designed and used for selective fast determination of the endogenous environmentally relevant PCB congeners in fatty foodstuffs (Ramos *et al.*, 2007). Compared with conventional PFE procedures, the developed analytical method reduces sample amounts to about 100 mg, solvent consumption to 3.5 mL and allows complete sample preparation in a single step (reported 17 min per sample), which results in a significant reduction of the analysis cost. The use of $SiO_2 - H_2SO_4$ has been found to be a distinct advantage as compared to previously reported selective PFE protocols, allowing an efficient preliminary fat removal. The miniaturized PFE device allows the quantitative extraction and purification of extracts for instance of PCBs in a single step, requiring lower solvent, sorbent and sample consumption and a reduced analysis cycle time. The achieved relative standard deviation (RSD) was found in the range of 4–23%, which is similar to those for reference methods.

2.1.4
Online Liquid Chromatography Clean-Up

The necessary effort of sample preparation in trace analysis is driving cost and limiting productivity not only in the GC-MS laboratory today. Especially a series of manual steps for extraction, evaporation and transfer are responsible for long analysis times, high cost and also the reason for contaminations and analyte losses. Coupling LC clean-up online with the GC separation opens a great potential for a significant reduction in cost per sample, full automation and high productivity for standardized analyses (Noij and van der Kooi, 1995). An online coupling of GPC or HPLC as clean-up procedure is a mature technique first described by Majors (1980). Commercial solutions are available since more than 20 years and established especially for routine analysis in many laboratories covering a wide range of applications, for example, for pesticides, PCBs, dioxins or plasticizers (Munari and Grob, 1990; De Paoli *et al.*, 1992; Jongenotter *et al.*, 1999; Che *et al.*, 2014). A specific advantage for trace analyses is the concentrated eluate allowing lower detection limits and reproducibility. High-extract dilutions introducing additional contaminations with a subsequent need for evaporation as known from Soxhlett or SPE extractions are avoided.

The instrumental set-up is built by an additional LC system with autosampler, HPLC separation column and switching valve in front of the GC (see Figure 2.8).

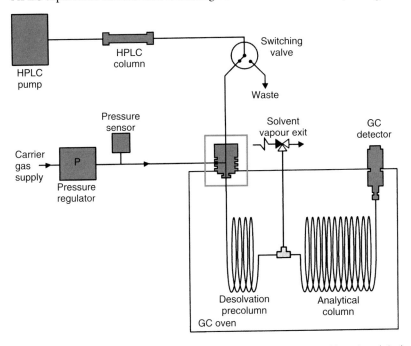

Figure 2.8 Schematic of an on-line LC clean-up GC system (Trestianu, Munari, and Grob, 1996).

An online LC detector is necessary during method development with spiked samples. The GC is equipped with an injector suitable for large volume injections (LVI). A regular programmed temperature vaporizer (PTV) with a packed insert liner and a solvent vapor vent valve, or an on-column injector with retention gap and solvent vapour exit can be used. Conventional liquid extraction techniques can be replaced with this setup completely and run fully automated for a large sample series.

The significantly higher separation efficiency in GPC or regular HPLC columns (number of available theoretical plates) can be exploited for group type separations (Kerkdijk *et al.*, 2007). Eluents applied for the online clean-up need to be compatible with GC as typical for normal phase LC (reversed phase separations need fraction collection and solvent exchange). The analyte containing fractions of the LC eluate are cut via a sample loop, or more flexible for optimization, using a time-controlled transfer to the GC injector. Typically, injection volumes of 100–1000 µL are employed using concurrent solvent evaporation or retention gap LVI techniques as described in Section 2.2.6 GC-MS Inlet Systems.

2.1.5
Headspace Techniques

One of the most elegant possibilities for instrumental sample preparation and sample transfer for GC-MS systems is the use of the headspace technique (Figure 2.9). Here all the frequently expensive steps, such as extraction of the sample, clean-up and concentration are dispensed with. Using the headspace technique, the volatile substances in the sample are separated from the matrix. The latter is not volatile under the conditions of the analysis. The tightly closed sample vessels, which, for example, are used for the static headspace procedure,

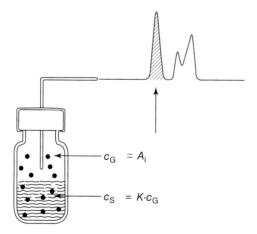

Figure 2.9 Principle of headspace analysis (Kolb, 1980) A_i area of the GC signal of the ith component. c_G, c_S concentrations in the gas and sample phases, respectively, K partition coefficient.

can frequently even be filled at the sampling location. The risk of false results (loss of analyte) as a result of transportation and further processing is thus reduced.

The extraction of the analytes is based on the partition of the very and moderately volatile substances between the matrix and the gas phase above the sample. After the partition equilibrium has been set up, the gas phase contains a qualitatively and quantitatively representative cross-section of the sample and is therefore used for analysing the components to be determined. All involatile components remain in the headspace vial and are not analysed. For this reason, the coupling of headspace instruments and GC-MS systems is particularly favourable (Pigozzo *et al.*, 1991). Since the interfering organic matrix is not involved, a longer duty cycle of the instrument and outstanding sensitivity are achieved. Furthermore, headspace analyses are easily and reliably automated for this reason and achieve a higher sample throughput in a 24 h operating period.

There are limitations to the coupling of headspace analysis with GC-MS systems if moisture is driven out of the sample as well. In certain cases, water can impair the focussing of volatile components at the beginning of the GC column. Impairment of the GC resolution can be counteracted by choosing suitable polar and thick film GC columns, by removing the water before injection of the sample.

It is also known that water affects the stability of the ion source of the mass spectrometer detector, which nowadays is becoming ever smaller. In the case of repeated analyses, the effects are manifested by a marked response loss and poor reproducibility. In such cases, special precautions must be taken, in particular in the choice of ion source parameters.

The headspace technique is very flexible and can be applied to the most widely differing sample qualities. Liquid or solid sample matrices are generally used, but gaseous samples can also be analysed readily and precisely using this method. Even the water content in food or pharmaceutical products can be determined by headspace GC (Kolb, 1993). Both qualitative and, in particular, quantitative determinations are carried out coupled to GC-MS systems.

There are two methods of analysing the volatiles of a sample which have very different requirements concerning the instrumentation: the static and dynamic (purge and trap, P&T) headspace techniques. The areas of use overlap partially but the strengths of the two methods are demonstrated in the different types of applications.

2.1.5.1 Static Headspace Technique

The term *headspace analysis* was coined in the early 1960s when the analysis of substances with odours and aromas in the headspace of tins of food was developed (Hachenberg and Schmidt, 1977; Kolb, 1980). The equilibrium of volatile substances between a sample matrix and the gas phase above it in a closed static system is the basis for static headspace gas chromatography (HSGC). The term *headspace* is often used without the word static, for example, headspace analyses, headspace sampler, and so on.

An equilibrium is set up in the distribution of the substances being analysed between the sample and the gas phase. The concentration of the substances in the

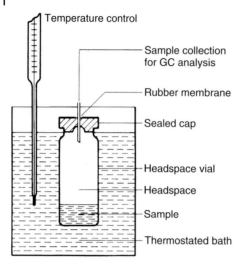

Figure 2.10 Sample handling in the static headspace method (Hachenberg, 1988).

gas phase then remains constant. An aliquot is taken from the gas phase and is fed into the GC-MS system via a transfer line (Figure 2.10).

The static headspace method is therefore an indirect analysis procedure, requiring special care in performing quantitative determinations. The position of the equilibrium depends on the analysis parameters (e.g. temperature) and also on the sample matrix itself. The matrix dependence of the procedure can be counteracted in various ways. The matrix can be standardized, for example, by addition of Na_2SO_4 or Na_2CO_3. Other possibilities include the standard addition method, internal standardization or the multiple headspace extraction procedure (MHE) as published by (Kolb and Ettre, 1991; Zhu *et al.*, 2005) (Figure 2.11).

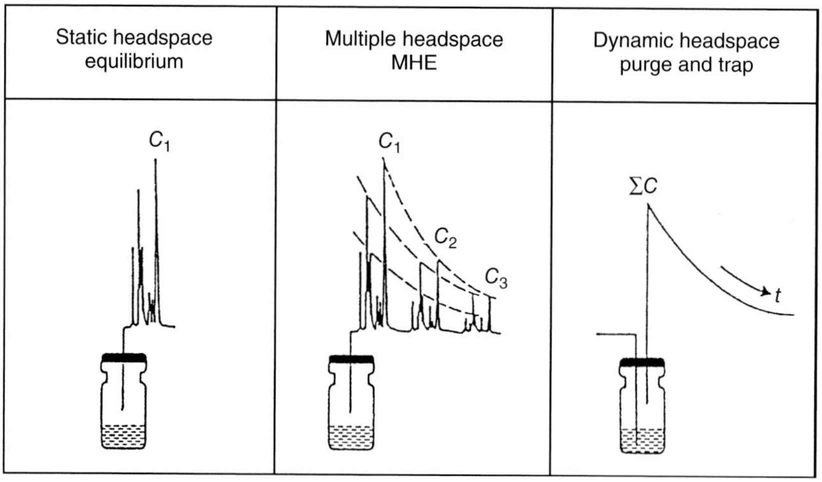

Static headspace equilibrium	Multiple headspace MHE	Dynamic headspace purge and trap
C_1	C_1 C_2 C_3	ΣC t

Figure 2.11 Gas extraction techniques (Kolb, 1992).

Static Headspace Analysis

In static headspace analysis, the samples are taken from a closed static system (closed headspace bottle) after the thermodynamic equilibrium (partition) between the liquid/solid matrix and the headspace above it has been established.

For coupling of HSGC with MS, the internal standard procedure has proved particularly successful for quantitative analyses. Besides the headspace-specific effects, possible variations in the MS detection are also compensated for. The best possible precision is thus achieved for the whole procedure. The MHE procedure can be used in the same way.

In static headspace analysis, the partition coefficient of the analytes is used to assess and plan the method (Kolb and Ettre, 2006). For the partition coefficient K of a volatile compound, the following equation is valid:

$$K = \frac{c_S}{c_G} = \frac{\text{concentration in the sample}}{\text{concentration in the gas phase}} \tag{2.1}$$

The partition coefficient depends on the temperature at equilibrium. This must therefore be kept constant for all measurements.

Rearranging the equation gives:

$$c_S = K \cdot c_G \tag{2.2}$$

As the peak area A determined in the GC-MS analysis is proportional to the concentration of the substance in the gas phase c_G, the following is valid:

$$A \approx c_G = \frac{1}{K} \cdot c_S \tag{2.3}$$

To be able to calculate the concentration of a substance c_0 in the original sample, the mass equilibrium must be referred to:

$$M_0 = M_S + M_G \tag{2.4}$$

The quantity M_0 of the volatile substance in the original sample has been divided at equilibrium into a portion in the gas phase M_G and another in the sample matrix M_S. Replacing M by the product of the concentration c and volume V gives:

$$c_0 \cdot V_0 = c_S \cdot V_S + c_G \cdot V_G \tag{2.5}$$

By using the definition of the partition coefficient given in Eq. (2.2), the unknown parameter c_S can be replaced by $K \cdot c_G$:

$$c_0 \cdot V_0 = K \cdot c_G \cdot V_S + c_G \cdot V_G \tag{2.6a}$$

$$c_0 = c_G \cdot \frac{V_S}{V_0} \cdot \left(K + \frac{V_G}{V_S} \right) \tag{2.6b}$$

The starting concentration c_0 of the sample, assuming $V_0 = V_S$, is given by:

$$c_0 = c_G \cdot \left(K + \frac{V_G}{V_S} \right) \tag{2.7}$$

As the peak area determined is proportional to the concentration of the volatile substance in the gas phase c_G, the following equation is valid, corresponding to the proportionality between the gas phase and the peak area:

$$c_0 \approx A \cdot (K + \beta) \tag{2.8}$$

whereby β is the phase ratio V_G/V_S (see Eq. (2.7)) and therefore describes the degree of filling of the headspace vessel (Ettre and Kolb, 1991).

The effects on the sensitivity of a static headspace analysis can be easily derived from the ratio given in Eq. (2.8):

$$A \approx c_0 \cdot \frac{1}{K + \beta} \quad \text{with } \beta = \frac{V_G}{V_S} \tag{2.9}$$

The peak area determined from a given concentration c_0 of a component depends on the partition coefficient K and the sample volume V_S (through the phase ratio β).

For substances with high-partition coefficients (e.g. ethanol, isopropyl alcohol, dioxane in water), the sample volume has no effect on the peak area determined. However, for substances with small partition coefficients (e.g. cyclohexane, trichloroethylene and xylene in water), the phase ratio β determines the headspace sensitivity. Doubling the quantity of sample leads in this case to a doubling of the peak area (Figure 2.12). For all quantitative determinations of compounds with small partition coefficients, an exact filling volume must be maintained.

The sensitivity of detection in static headspace analysis can also be increased through the lowering the partition coefficient K by setting up the equilibrium at a higher temperature. Substances with high partition coefficients benefit particularly from higher equilibration temperatures (Table 2.7, e.g. alcohols in water). Raising the temperature is, however, limited by increased matrix evaporation (water) and the danger of bursting the sample vessel or the seal.

Changes in the sample matrix can also affect the partition coefficient. For samples with high water contents, an electrolyte can be added to effect a salting out process (Figure 2.13). For example, for the determination of ethanol in water the addition of NH_4Cl gives a twofold and addition of K_2CO_3 an eightfold increase in the sensitivity of detection. The headspace sensitivity can also be raised by the addition of nonelectrolytes. In the determination of residual monomers in polystyrene dissolved in DMF, adding increasing quantities of water, for example, in the determination of styrene and butanol, leads to an increase in peak area for styrene by a factor of 160 and for n-butyl alcohol by a factor of only 25 (Figure 2.14).

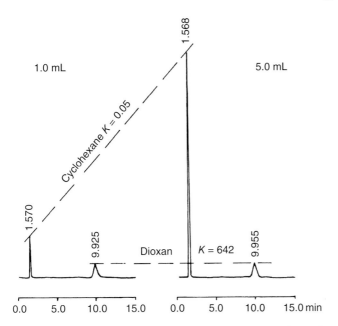

Figure 2.12 Sample volume and sensitivity in the static headspace for cyclohexane ($K = 0.05$) and dioxan ($K = 642$) with sample volumes of 1 and 5 mL, respectively (Kolb, 1992).

Table 2.7 Partition coefficients K of selected compounds in water.

Substance	40 °C	60 °C	80 °C
Tetrachloroethane	1.5	1.3	0.9
1,1,1-Trichloroethane	1.6	1.5	1.2
Toluene	2.8	1.6	1.3
o-Xylene	2.4	1.3	1.0
Cyclohexane	0.07	0.05	0.02
n-Hexane	0.14	0.04	< 0.01
Ethyl acetate	62.4	29.3	17.5
n-Butyl acetate	31.4	13.6	7.6
Isopropyl alcohol	825	286	117
Methyl isobutyl ketone	54.3	22.8	11.8
Dioxan	1618	642	288
n-Butyl alcohol	647	238	99

(Kolb, 1992).

Quantitation by Multiple Headspace Extraction

The total concentration of an analyte in a sample can be determined by MHE from one vial (Maggio *et. al.*, 1991). This is possible because the decline in peak area follows a first-order kinetics according to Eq. (2.10). At any time t, the concentration of the analyte depends on the initial concentration c_0 and the

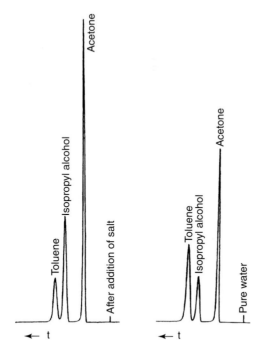

Figure 2.13 Matrix effects in static headspace. The effect of salting out on polar substances (Kolb, 1992).

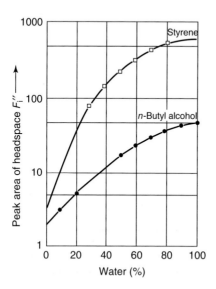

Figure 2.14 Increasing water concentrations in the determination of styrene as a residual monomer lead to a sharp increase in response (Hachenberg, 1988).

constant q in the exponent, describing the decline of the measured peak areas (Kolb and Ettre, 1991).

$$c = c_0 \times e^{-qt} \tag{2.10}$$

The MHE procedure is carried out in a series of consecutive runs from one and the same sample vial. Immediately after the first headspace analysis, the vial is depressurized to the atmosphere and incubated with the same method again for the next analysis. This step can also be accomplished by many types of headspace autosamplers. In case of a headspace syringe autosampler, an optional 'MHE device' is required, which is used to depressurize the vial (Vulpius and Baltensperger, 2009). With this stepwise procedure Eq. (2.10) can be rewritten in Eq. (2.11) by replacing the concentration with peak areas (with i the number of injections, A_i the peak area after i injections and A_1 the peak area of the first injection), and for the time t with the number of analysis from the same vial. The measurements of consecutive extractions show an exponential decrease of the peak areas.

$$A_i = A_1 \times e^{-q(i-1)} \tag{2.11}$$

Plotting the peak areas logarithmic against the number of extractions, a linear relationship can be expressed.

$$\ln A_i = \ln A_1 - q(i-1) \tag{2.12}$$

In practice, we do not need a large number of injections. Due to the linear relationship in Eq. (2.12), the total area can be calculated from the area of the first run. The slope q, the decline in peak areas, is taken from the slope expressed in the regression calculation of the logarithmic graph. The total peak area, which is equivalent to the total concentration of the analyte in the sample, is given by the summation of all potential extractions A_i in Eq. (2.13).

$$\sum_{i=1}^{\infty} A_i = \frac{A_1}{1 - e^q} \tag{2.13}$$

An example from the MHE quantitation of ethylenoxide in PVC (polyvinylchloride) demonstrates the calculation steps (Petersen, 2008). A series of six measurements is performed providing the peak area data given in Table 2.8 (Figure 2.15).

Table 2.8 MHE peak area results of six consecutive headspace runs.

Run#	Area (cts)	ln (Area)
1	151 909	11.93
2	63 127	11.05
3	26 802	10.20
4	10 963	9.30
5	5768	8.66
6	2240	7.71

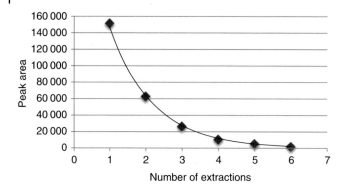

Figure 2.15 Graphical representation of the MHE measurements of Table 2.8, peak area versus run number.

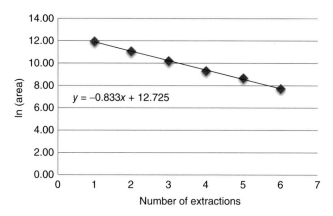

Figure 2.16 Logarithmic plot of the MHE area results of Table 2.8, ln (peak area) versus run number, with regression formula.

The calculation for the total analyte peak area in the sample as of Eq. (2.13) with $A_1 = 151\,909$ from Run 1 from Table 2.8 and the slope $q = -0.833$ from the regression calculation in Figure 2.16, and $e^{-0.833} = 0.4347$, gives:

$$\sum_{i=1}^{6} A_i = \frac{A_1}{1 - e^q} = \frac{151\,909}{1 - 0.4347} = 268\,743\,\text{cts} \tag{2.14}$$

This result is calculated using all six data points. With only two measurements, the slope can be calculated with a good and acceptable precision compared to the detailed experiment as shown in Table 2.9. The difference in the calculated result is low with only 3.3%. For the final determination of the analyte concentration in

Table 2.9 Comparison of the MHE experiment using six or first two measurements.

	6 Runs	2 Runs	Delta
Area A_1 (cts)	151 909	151 909	0
Slope q	−0.8330	−0.8781	−0.0451
exp(q)	0.4347	0.4156	0.0192
Total area A_i (cts)	268 743	259 928	3.3%

the sample, a regular response calibration with external or internal standards is required.

Headspace Analysis Operation

To shorten the equilibration time the most up-to-date headspace samplers have devices available for mixing the samples. For liquid samples, vigorous mixing strongly increases the phase boundary area and guarantees rapid delivery of analyte-rich sample material to the phase boundary.

Teflon-coated stirring rods have not proved successful because of losses through adsorption and crosscontamination. However, shaking devices which mix the sample in the headspace bottle through vertical or rotational movement have proved effective (Table 2.10). A particularly refined shaker uses changing excitation frequencies of 2–10 Hz to achieve optimal mixing of the contents at each degree of filling (Figure 2.17).

The resulting equilibration times are in the region of 10 min on using shaking devices compared with 45–60 min without mixing.

In addition, quantitative measurements are more reliable. The use of shaking devices in static headspace techniques can reduce RSDs to less than 2%.

Table 2.10 Comparison of the analyses with/without shaking of the headspace sample (volatile halogenated hydrocarbons), average values in parts per billion and precision in standard deviations.

Substance	Without shaking			With shaking		
Ethylbenzene	353	18	(5.2%)	472	8	(1.7%)
Toluene	336	20	(5.9%)	411	4	(1.0%)
o-Xylene	324	13	(4.1%)	400	7	(1.8%)
Benzene	326	18	(5.4%)	372	5	(1.3%)
1,3-Dichlorobenzene	225	13	(5.6%)	255	5	(2.1%)
1,2,4-Trichlorobenzene	207	9	(4.2%)	225	6	(2.5%)
Bromobenzene	213	11	(5.2%)	220	5	(2.1%)

After Tekmar.

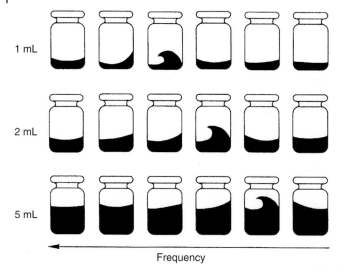

1 mL

2 mL

5 mL

Frequency

Figure 2.17 Effect of variable shaking frequencies (2–10 Hz) on different depths of filling in headspace vessels (Perkin-Elmer).

Static Headspace Injection Techniques

- Pressure Balanced Injection (Figure 2.18):
  Variable quantity injected: can be controlled (programmed) by the length of the injection process, injection volume = injection time × flow rate of the GC column, no drop in pressure to atmospheric pressure, depressurizing to initial pressure of column. Change in pressure in the headspace bottle: reproducible pressure build-up with carrier gas, mixing, initial pressure in column maintained during sample injection. Sample losses: none known.
- Syringe Injection (Figure 2.19):
  Variable injection volume through pump action: easily controlled, drop in pressure to atmospheric on transferring the syringe to the injector.
  Pressure change in the headspace bottle: varies with the action of the syringe plunger (volume removed) on sample removal from the sealed bottle, compensation by injection of carrier gas necessary.
  Sample losses: possible through condensation on depressurizing to atmospheric pressure, losses through evaporation from the syringe after it has been removed from the septum cap of the bottle, in comparison the largest surface contact with the sample.
- Sample Loop (Figure 2.20):
  Quantity injected on the instrument side: fixed volume by loop, changes in the volume injected requires change of the sample loop, drop in pressure to atmospheric or variable backpressure on filling the sample loop. Pressure change in the headspace bottle: compensation by pressurization with carrier

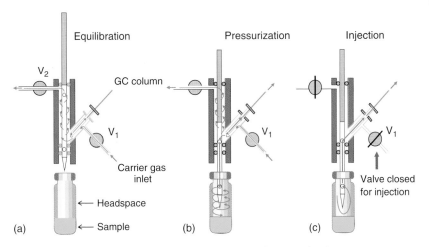

Figure 2.18 Injection techniques for static headspace: the principle of pressure balanced injection. (a) equilibration, (b) pressurization and (c) injection phase (Perkin-Elmer).

> gas, complete depressurizing of the pressure built up in the bottle through thermostating, the expansion volume must be a multiple of the sample loop to be able to fill it reproducibly.
>
> Sample losses: for larger sample loops attention must be paid to dilution with carrier gas during the pressure build-up phase, possible condensation on depressurizing to atmospheric pressure.

2.1.5.2 Dynamic Headspace Technique (Purge and Trap)

The trace analysis of volatile organic compounds (VOCs) is of continual interest, for example, in environmental monitoring, in outgassing studies on packaging material, in the analysis of flavours and fragrances, and also in occupational and public health screening (Ashley *et al.*, 1992). The dynamic headspace technique, known as *purge and trap (PAT, P&T)*, is a process in which highly and moderately volatile organic compounds are continuously extracted from a matrix and concentrated in an adsorption trap (Wylie, 1987, 1988; Lin *et al.*, 1993). The substances evaporated from the internal trap by thermal desorption reach the gas chromatograph as a concentrated sample plug where they are finally separated and detected (Westendorf, 1985, 1987, 1989a).

Early P&T applications in the 1960s already involved the analysis of body fluids. In the 1970s, P&T became known because of the increasing requirement for the testing of drinking water for volatile halogenated hydrocarbons. The analysis of drinking water for the determination of a large number of these compounds in ppq quantities (concentrations of less than 1 µg/L) became possible only through the concentration process, which was part of purge and trap gas chromatography

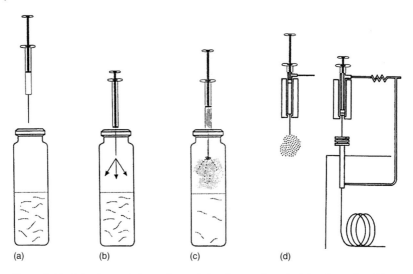

Figure 2.19 Injection techniques for static headspace: the principle of transfer with a gas-tight syringe. (a) Sample heating, (b) pressurisation, (c) sampling and (d) inject.

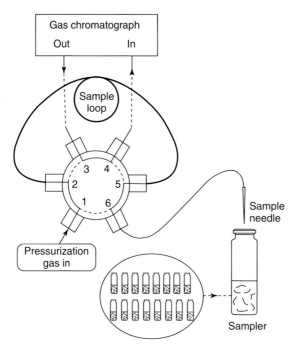

Figure 2.20 Injection techniques for static headspace: the principle of application with a sample loop (Tekmar). Sequence: 1. Heat sample at a precise temperature over a set period of time. 2. Pressurize with carrier gas. 3. Fill the sample loop. 4. Inject to GC by switching the six-port valve.

(P&T-GC). This technique is now frequently used to detect residues of volatile organic compounds in the environment and in the analysis of liquid and solid foodstuffs. In particular, the coupling with GC-MS systems allows its use as a multi-component process for the automatic analysis of a large series of samples.

An analytically interesting variant of the procedure is the fine dispersion of liquid samples in to a carrier gas stream. This so-called *spray and trap* process uses the high surface area of the sample droplets for an effective gas extraction. The procedure also works very well with foaming samples. A detergent content of up to 0.1% does not affect the extraction result. The spray and trap process is therefore particularly suitable for use with mobile GC-MS analyses (Matz 1993).

Modes of Operation of Purge and Trap Systems

A P&T analysis procedure consists of three main steps:

1) Purge phase with simultaneous concentration
2) Desorption phase
3) Baking out phase.

The Purge Phase During the purge phase the volatile organic components are stripped from the matrix. The purge gas (He or N_2) is finely dispersed in the case of liquid samples (drinking water, waste water) by passing through a special frit in the base of a U-tube (frit sparger, Figure 2.21a). The surface of the liquid can be greatly increased by the presence of very small gas bubbles and the contact between the liquid sample and the purge gas maximized.

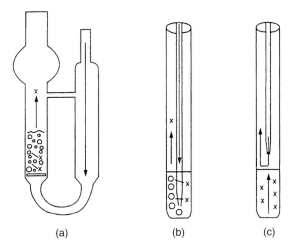

(a) (b) (c)

Figure 2.21 Possibilities for sample introduction in purge and trap. (a) U tube with/without frits (fritless/frit sparger) for water samples. (b) Sample vessel (needle sparger) for water and soil samples (solids). (c) Sample vessel (needle sparger) for foaming samples for determination of the headspace sweep.

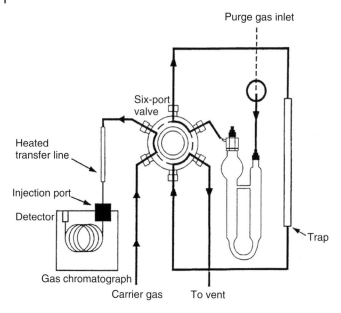

Figure 2.22 Gas flow schematics of the purge and trap-GC coupling, and switching of phases at the six-port valve. Bold line: purge and baking out phase and dotted line: desorption phase (Tekmar).

The purge gas extracts the analytes from the sample and transports them to a internal trap. The analytes are retained in this trap and concentrated while the purge gas passes out through the vent. The desorption gas, which is provided by the carrier gas regulation of the GC, enters in this phase via the 6-port switching valve and transfer line the GC injector and maintains the constant gas flow for the column (Figure 2.22).

Solid samples are analysed in special vessels (needle sparger, see Figure 2.21b,c) into which a needle with side openings is dipped. For foaming samples, the headspace sweep technique can be used.

The total quantity of volatile organic compounds removed from the sample depends on the purge volume. The purge volume is the product of the purge flow rate and the purge time. Many environmental samples are analysed at a purge volume of 440 mL. This value is achieved using a flow rate of 40 mL/min and a purge time of 11 min. A purge flow rate of 40 mL/min gives optimal purge efficiency. Changes in the purge volume should consequently be made only after adjusting the purge time. Although a purge volume of 440 mL is optimal in most cases, some samples may require larger purge volumes for adequate sensitivity to be reached (Figure 2.23).

The purge efficiency is defined as that quantity of the analytes, which is purged from a sample with a defined quantity of gas. It depends upon various factors. Among them are: purge volume, sample temperature and nature of the sparger (needle or frit), the nature of the substances to be analysed and that of the matrix.

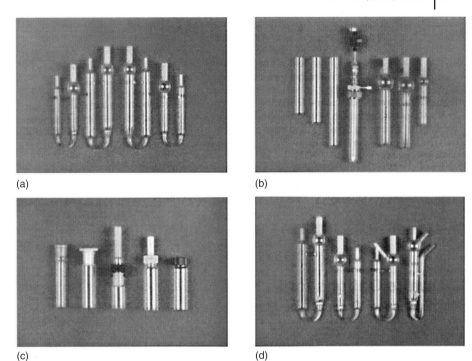

(a)

(b)

(c)

(d)

Figure 2.23 Glass apparatus for the purge and trap technique (Tekmar). (a) U-tube with/without frit (5 and 25 mL sizes). (b) Needle sparger: left: single use vessels, middle: glass needle with frits, right: vessels with foam retention, 5, 20 and 25 mL volumes. (c) Special glass vessels: 25 mL flat-bottomed flask, 40 mL flask with seal, 20 mL glass (two parts) with connector, 40 mL glass with flange, 40 mL screw top glass. (d) U-tube for connection to automatic sample dispensers: left: 25 and 5 mL vessel with side inlet, right: 5 mL vessel with upper inlet for sample heating, 25 mL special model with side inlet.

The purge efficiency has a direct effect on the percentage recovery (the quantity of analyte reaching the detector).

Control of Purge Gas Pressure during the Purge Phase

The adsorption and chromatographic separation of volatile halogenated hydrocarbons is improved significantly by regulating the pressure of the purge gas during the purge phase. By additional back pressure control during this phase, a very sharp adsorption band is formed in the trap, from which, in particular, the highly volatile components profit, since a broader distribution does not occur. The risk of the analytes passing through the trap is almost completely excluded under the given conditions.

During the desorption phase, the narrow adsorption band determines the quality of the sample transfer to the capillary column. The result is clearly improved peak symmetry, and thus better GC resolution and an improvement in the sensitivity of the whole procedure.

The Dry Purge Phase

To remove water from a hydrophobic adsorption trap (e.g. with a Tenax filling), a dry purge phase is introduced. During this step, most of the water condensed in the trap is blown out by dry carrier gas. Purge times of about 6 min are typical.

The Desorption Phase

During the desorption phase, the trap is heated and subjected to a backflush (BKF) with carrier gas. The reversal of the direction of the gas flow is important in order to desorb the analytes in the opposite direction to the concentration by the trap. In this way, narrow peak bands are obtained.

The time and temperature of the desorption phase affect the chromatography of the substances to be analysed. The desorption time should be as short as possible but sufficient to transfer the components quantitatively on to the GC column. Most of the analytes are transferred to the GC column during the first minute of the desorption step. The desorption time is generally 4 min.

The temperature of the desorption step depends on the type of adsorbent in the trap (Table 2.11). The most widely used adsorbent, Tenax, desorbs very efficiently at 180 °C without forming decomposition products. The peak shape of compounds eluting early can be improved by inserting a desorb-preheat step. Here the trap is preheated to a temperature near the desorption temperature before the valve is switched for desorption and before the gas flows freely through the trap. Gas is not passed through the trap during the preheating step, but the analytes are nevertheless desorbed from the carrier material. When the gas stream is passed through the trap after switching the valve, it purges the substances from the trap in a concentrated carrier gas cloud. Highly volatile compounds which are not focussed at the beginning of the column thus give rise to a narrower peak shape. A preheating temperature of 5 °C below the desorption temperature has been found to be favourable.

VOCARB material, which is used in current applications involving volatile halogenated hydrocarbons, can be desorbed at higher temperatures than Tenax up to 290 °C (Table 2.12). At higher desorption temperatures, however, the possibility of catalytic decomposition of some substances must be taken into account (see also Section 2.1.5).

Water Removal

Water is evaporated and trapped from aqueous or moist samples as well. Most of which is disposed of by the dry purge step, particularly when using Tenax adsorption traps with low water retention. Residual moisture can still be transferred to the GC column during the desorption step (Madden and Lehan, 1991). As the resolution of highly volatile substances on capillary columns would be impaired and the detection by the mass spectrometer would be affected, additional devices are used to remove water. In particular, where the P&T technique is used with ECD or MS as detectors, reliable water removal is necessary. Different technical solutions working automated during the desorption phase are in use with the P&T instruments of different manufacturers.

Table 2.11 List of trap materials used in the purge and trap procedure with details of applications and recommended analysis parameters.

Trap number	Adsorbent	Application	Drying possible?	Drying time (min)	Desorb pre-heat (°C)	Desorb temperature (°C)	Bake-out temperature (°C)	Bake-out time (min)	Conditioning temperature (°C) time (min)	Remarks
1	Tenax	All substances down to CH_2Cl_2	Yes	2–6	220	225	230	7–10	230 10	Low response with brominated substances, high back pressure, background with benzene, toluene and ethylbenzene
2	Tenax Silica gel	All substances except freons	No	—	220	225	230	10–12	230 10	Low response with brominated substances, high back pressure, background with benzene, toluene and ethylbenzene
3	Tenax Silica gel Activated charcoal	All substances including freons	No	—	220	225	230	10–12	230 10	Low response with brominated substances, high back pressure, background with benzene, toluene and ethylbenzene

(continued overleaf)

Table 2.11 (*Continued*)

Trap number	Adsorbent	Application	Drying possible?	Drying time (min)	Desorb pre-heat temperature (°C)	Desorb temperature (°C)	Bake-out temperature (°C)	Bake-out time (min)	Conditioning temperature (°C) time (min)	Remarks
4	Tenax Activated charcoal	All substances down to CH_2Cl_2 and gases	No	—	220	225	230	7–10	230 10	Low response with brominated substances, high back pressure, background with benzene, toluene and ethylbenzene
5	OV-1 Tenax Silica gel Activated Charcoal	All substances including freons	No	—	220	225	230	10–12	230 10	Low response with brominated substances, high back pressure, background with benzene, toluene, ethylbenzene
6	OV-1 Tenax Silica gel	All substances except freons	No	—	220	225	230	10–12	230 10	Low response with brominated substances, high back pressure, background with benzene, toluene and ethylbenzene

									Remarks
7	OV-1 Tenax	Yes	2–6	220	225	230	7–10	230 10	Low response with brominated substances, high back pressure, background with benzene, toluene and ethylbenzene
8	Carbopak B Carbosieve SIII	Yes	11	245	250	260	4–10	260 20–30	Losses of CCl_4
9	VOCARB 3000	Yes	1–3	245	250	260	4	290 4h	Response factors see Table 2.12
10	VOCARB 4000	Yes	1–3	245	250	260	4	270 4h	Low response with chlorinated compounds, high back pressure, quantitative losses of chloroethyl vinyl ethers
11	BTEXTRAP Carbopak B Carbopak C	Yes	1–3	245	250	260	4	270	—
12	Tenax GR Graphpac-D	Yes	1–4	245	250	260	12	—	—

All substances down to CH_2Cl_2 (row 7); All substances including freons (rows 8–12).

Trap numbers 1–8: Tekmar; trap numbers 9–11: SUPELCO and trap number 12: Alltech.

Table 2.12 Response factors and standard deviations for components with EPA method 624 using a VOCARB 3000 trap and 5 mL of sample (250 °C desorption temperature, reference bromochloromethane/difluorobenzene, concentrations in ppb, Supelco 1988).

Compound	20	50	100	150	200	Average	Standard deviation	% Relative standard deviation
Methyl chloride	0.933	0.962	0.984	0.906	1.287	1.014	0.139	13.7
Vinyl chloride	1.037	1.167	1.466	1.536	1.333	1.308	0.185	14.1
Methyl bromide	1.018	1.257	1.079	1.077	1.290	1.144	0.108	9.5
Ethyl chloride	0.419	0.582	0.503	0.557	0.536	0.519	0.056	10.9
Trichlorofluoromethane	0.972	1.533	1.266	1.491	1.507	1.354	0.213	15.8
1,1-Dichloroethylene	0.873	1.207	1.203	1.264	1.180	1.145	0.139	12.1
Dichloromethane	1.550	1.134	1.231	1.325	1.11	1.270	0.159	12.5
1,2-Dichloroethylene	0.973	0.991	0.944	1.124	1.007	1.008	0.062	6.1
1,1-Dichloroethane	2.064	2.093	2.013	2.186	2.175	2.106	0.066	3.1
Chloroform		2.368	2.193	2.419	2.373	2.373	0.104	4.4
Tetrachloroethane	1.205	1.472	1.302	1.325	1.476	1.356	0.105	7.7
Carbon tetrachloride	1.067	1.426	1.350	1.240	1.335	1.284	0.124	9.6
Benzene	0.947	0.949	0.906	0.973	1.024	0.960	0.039	4.0
1,2-Dichloroethane	0.046	0.045	0.047	0.050	0.049	0.047	0.002	4.2
Trichloroethylene	0.412	0.495	0.485	0.504	0.540	0.487	0.042	8.6
1,2-Dichloropropane	0.535	0.530	0.513	0.537	0.549	0.533	0.012	2.2
Bromodichloromethane	0.082	0.080	0.084	0.088	0.087	0.084	0.003	3.4
2-Chloroethyl vinyl ether	0.043	0.039	0.043	0.047	0.041	0.043	0.003	6.2

cis-1,3-Dichloropropene	0.898	0.977	1.001	1.086	1.082	1.009	0.070	6.9
Toluene	0.861	1.208	1.141	1.250	1.283	1.149	0.151	13.2
trans-1,3-Dichloropropene	0.312	0.300	0.280	0.299	0.289	0.296	0.011	3.7
1,1,2-Trichloroethane	0.593	0.486	0.488	0.526	0.507	0.520	0.039	7.5
Tetrachloroethylene	0.364	0.488	0.492	0.497	0.558	0.480	0.063	13.2
Dibromochloromethane	0.818	0.712	0.731	0.787	0.785	0.767	0.039	5.1
Chlorobenzene	1.135	1.116	1.117	1.183	1.242	1.159	0.048	4.2
Ethylbenzene	0.561	0.508	0.476	0.504	0.543	0.518	0.030	5.8
Bromoform	1.070	0.882	0.920	1.013	0.918	0.961	0.070	7.3
1,1,2,2-Tetrachloroethane	0.082	0.069	0.070	0.078	0.065	0.073	0.006	8.4
1,3-Dichlorobenzene	1.277	1.215	1.237	1.338	1.451	1.304	0.085	6.5
1,4-Dichlorobenzene	1.397	1.279	1.291	1.391	1.509	1.374	0.084	6.1
1,2-Dichlorobenzene	1.351	1.233	1.188	1.284	1.350	1.281	0.064	5.0

The water-removal system used in several technical evolutions by Tekmar (Teledyne Tekmar, Mason, OH, USA) is shown in principle in Figure 2.24. The Tekmar Moisture Control System (MCS) uses a temperature-controlled steel capillary loop. If the dew point for water is reached in the MCS, a stationary water phase is formed. BTEX (benzene/toluene/ethylbenzene/xylene) and volatile halogenated hydrocarbons analytes can pass through unaffected. Polar components such as alcohols can also pass through the MCS system at moderate temperatures but with a retardation by the water phase. When the desorption phase is finished, the tubing of the MCS is dried by baking out in a carrier gas backflush. An example for VOC analysis showing good efficiency and retention of the residual moisture during the desorption process is shown in Figure 2.25 (Johnson and Madden, 1990).

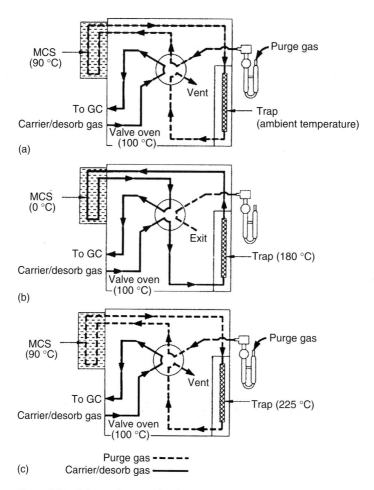

Figure 2.24 Scheme showing the three phases of the purge and trap cycle with water removal (MCS, Tekmar). (a) Purge phase, (b) desorption phase and (c) baking out phase.

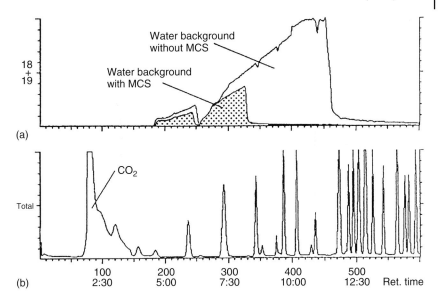

(a)

(b)

Figure 2.25 GC-MS analysis using purge and trap (a) Mass chromatogram for water (*m/z* 18 + 19) with and without removal (without MCS) of the moisture driven out. (b) Total ion chromatogram with volatile substances (volatile halogenated hydrocarbons) after removal of the water (with MCS).

Another approach for water removal is patented by O.I. Analytical (Xylem Inc., OI Analytical, College Station, TX, USA). A Cyclone Water Management™ system takes advantage of a cyclonic effect exploiting the propensity of heavy molecules to travel nearer the centre of a gas tube and lighter ones to travel farther out (a similar effect was used in former glass jet separators coupling packed columns to mass spectrometer ion sources). 'Cooling' the H_2O molecules by interaction with carrier gas allows to 'collect' in the bottom of the fan cooled device by gravity, while the analyte gas stream continues to the GC for measurement with about 96% less of water. The principle of operation is demonstrated in Figure 2.26. The response of polar molecules is improved as there is no passage of a stationary water phase required.

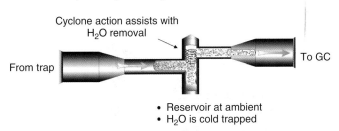

Figure 2.26 Cyclonic water removal during the purge and trap desorption phase. (Courtesy O.I. Analytical.)

Bake Out Phase

When required, the trap can be subjected to a baking out phase (trap-bake mode) after the desorption of the analytes. During desorption, involatile organic compounds are released from the trap at elevated temperatures and driven off. Bake out conditions the trap material for the next analysis.

2.1.5.3 Coupling of Purge and Trap with GC-MS Systems

There are two fundamentally different possibilities for the installation of a P&T instrument coupled to a GC-MS system, which depend on the areas of use and the number of P&T samples to be processed per day.

In many laboratories, the most flexible arrangement involves connecting a P&T system to the gas supply of the GC injector. The P&T concentrator is connected in such a way that either manual or automatic syringe injection can still be carried out. The carrier gas is passed from the GC carrier gas regulation to the central 6-port valve of the P&T system (see Figure 2.22). From here the carrier gas flows back through the continuously heated transfer line to the injector of the gas chromatograph. A particular advantage of this type of installation lies in the fact that the injector can still be used for liquid samples. This allows the manual injection of control samples or the operation of liquid autosamplers. In addition, it guarantees that the whole additional tubing system remains free from contamination even when the P&T instrument is on standby.

When the sample is transferred to the GC column, it should be ensured that adequate focussing of the analytes is achieved. For this purpose, GC columns are available with film thicknesses of about 1.8 μm or more, which are used as well for the analysis of volatile halogenated hydrocarbons with GC-MS systems (types 502, 624 or volatiles). The injector system is operated using a split ratio, which can be selected depending on the detection limit required.

A complete splitless injection of the analytes on to the GC column can only be achieved with injector or on-column cryofocussing (see also Section 2.2.5.7). For the on-column cryofocussing, the column is connected to the transfer line or the central 6-port valve of the P&T unit in such a way that the beginning of the column passes through a zone about 10 cm long which can be cooled by an external cooling agent, such as liquid CO_2 or N_2, to -30 to $-150\,°C$. All components which reach the GC column after desorption from the trap can be frozen out in a narrow band at the beginning of the analytical column. This enables the highest sensitivities to be achieved, in particular with mass spectrometers (Westendorf, 1989b).

As large quantities of water are also focussed together with the analytes in the case of moist or aqueous samples, removal of moisture in the desorption phase is particularly important with cryofocussing techniques. Insufficient removal of water leads to the deposition of ice and blockage of the capillaries. The consequences are poor focussing or complete failure of the analysis.

2.1.5.4 Headspace versus Purge and Trap

Both instrumental extraction techniques have specific advantages and limitations when coupled to GC and GC-MS. This should be taken into consideration when choosing an analysis procedure. In particular, the nature of the sample material, the concentration range for the measurement and the effort required to automate the analyses for large numbers of samples play a significant role. The recovery and the partition coefficient, and thus the sensitivity which can be achieved, are relevant to the analytical assessment of the procedure. For both procedures, it must be possible to vaporize the substances being analysed below 150 °C and then to partition them in the gas phase. The vapour pressure and solubility of the analytes in the sample matrix, as well as the extraction temperature, affect both procedures (Figure 2.27).

How then do the techniques differ? For this, the terms *recovery* and *sensitivity* must be defined. For both methods, the recovery depends on the vapour pressure, the solubility and the temperature. The effects of temperature can be dealt with because it is easy to increase the vapour pressure of a compound by raising the temperature during the vaporization step. With the P&T technique, the term *percentage recovery* is used. This is the amount of a compound which reaches the gas chromatograph for analysis relative to the amount which was originally present in the sample. If a sample contains 100 pg benzene and 90 pg reach the GC column, the percentage recovery is 90%. In the static headspace technique, a simple expression like this cannot be used because it is possible to use a large

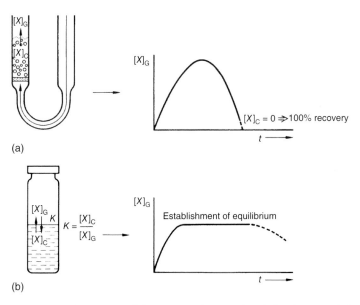

Figure 2.27 Comparison of the purge and trap and static headspace techniques. (a) Purge and trap and (b) static headspace.

number of types of vial, injection techniques and injection volumes, which always apply aliquots to the analysis.

The commonly used term connected with the static headspace method is the partition coefficient, as mentioned earlier. The partition coefficient K is defined as the quantity of a compound in the sample divided by the quantity in the vapour phase. Therefore, the smaller the partition coefficient is, the higher the sensitivity. It should be noted that a partition coefficient is valid only for the analysis parameters at which it has been determined. These include the temperature, the size of the sample vial, the quantity of sample (weight and volume), the nature of the matrix and the size of the headspace. After the partition coefficient, the quantity injected is the next parameter affecting headspace sensitivity. The quantity injected is limited by a range of factors. For example, only a limited quantity can be removed from the headspace of a closed vessel. Attempts to remove a larger quantity of sample vapour would lead to a partial vacuum in the sample vessel as a rule of thumb the same amount of carrier gas is injected into the headspace vial, e.g. by a syringe sampler, that is programmed to be withdrawn (Figure 2.19). Loop type sampler pressurize the vial before loop filling (Figure 2.20). This is difficult to reproduce. Furthermore, only a limited quantity can be injected on to a GC column without causing peak broadening. For larger quantities of sample, cryofocussing is necessary. Capillary columns require cold trapping at injection quantities of more than about 200 µL.

An alternative injection system involves pressure-balanced injection (Figure 2.18). This technique uses a time controlled injection directly from the vial into the GC carrier gas stream. The headspace sample is introduced onto the column without using a gas syringe, thus avoiding fractionation due to pressure changes in the syringe. Since the needle is sealed, there are no losses of headspace gas during transfer. The injection needle is passed through the septum into the headspace in order to pressurize the headspace vial to at least the column head pressure. After equilibrium has been established, this pressure is released during injection on to the GC column over a short programmable time interval. This allows larger quantities to be injected. Due to the fact that this technique is controlled by 'injection time' to inject the sample, the absolute volume of the sample is unknown. It is impossible to measure the exact quantity injected; however, the reproducibility of this method is very high. Limitations typically arise from the necessary sample transfer line. This includes a potential for sample carry-over and the fact that the injection port is always occupied by the transfer line (or the column is connected as transfer line directly to the HS unit bypassing the injector), and therefore not available for manual or liquid autosampler use.

First Example: Volatile Halogenated Hydrocarbons

A comparison with actual concentration values makes the differences between the static headspace and P&T techniques very clear. The percentage recovery for the P&T technique is, for example, for an environmental sample of volatile halo-genated hydrocarbons: 95% for chloroform, 92% for bromodichloromethane, 87% for chlorodibromomethane and 71% for bromoform. For a sample of 5 mL, which contains 1 ppb of each substance (i.e. a total quantity of 5 ng of each compound), 4.75, 4.6, 4.35 and 3.55 ng are recovered. In a typical static headspace system with a sample vessel of 21 mL containing 15 mL of sample at 70 °C, the partition coef-ficients for the corresponding volatile halogenated hydrocarbons are: 0.3, 0.9, 1.5 and 3.0. This means that the quantities in the 5 mL of headspace are: 11.5, 7.9, 6.0 and 3.8 ng. On injection of 20 µL of the headspace gas mixture on to a stan-dard capillary column, the quantities injected are 0.05, 0.03, 0.02 and 0.015 ng. For a larger injection (0.5 mL) using cryofocussing, the quantities injected are 1.2, 0.8, 0.6 and 0.4 ng. The P&T technique is therefore more sensitive than the static headspace procedure for these volatile halogenated hydrocarbons by factors of 4.1, 5.8, 7.2 and 9.3 (Table 2.13).

Table 2.13 Lower application limits (µg/L) for headspace and purge and trap techniques in the analysis of water from the river Rhine.

Substance	Lower application limits (µg/L) for headspace	Lower application limits (µg/L) for purge and trap
Dichloromethane	1	0.5
Chloroform	0.1	0.05
1,1,1-Trichloroethane	0.1	0.02
1,2-Dichloroethane	5	0.5
Carbon tetrachloride	0.1	0.02
Trichloroethylene	0.1	0.05
Bromodichloromethane	0.1	0.02
1,1,2-Trichloroethane	0.5	0.5
Dibromochloromethane	0.1	0.05
Tetrachloroethylene	0.1	0.02
1,1,2,2-Tetrachloroethane	0.1	0.02
Bromoform	0.1	0.2
1,1,2,2-Tetrachloroethane	0.1	0.02
Benzene	5	0.1
Toluene	5	0.05
Chlorobenzene	5	0.02
Ethylbenzene	5	0.1
m-Xylene	5	0.05
p-Xylene	5	0.05
o-Xylene	5	0.1
Triethylamine	—	0.5
Tetrahydrofuran	—	0.5
1,3,5-Trioxan	—	0.5

After Willemsen, Gerke and Krabbe (1993).

Static headspace	Dynamic headspace
Headspace	Purge and trap
Extraction:	
Waiting for establishment of equilibrium	Continuous disturbance of the equilibrium by the purge gas
Intermediate steps:	
Closed vessel at constant temperature	Enrichment of the substances driven off in a trap
Sample injection:	
Removal of a preselected volume from the headspace	Thermal desorption from the trap

Second Example: Cooking Oils

For compounds with lower recoveries, for example, in the analysis of free aldehydes in cooking oils, the difference is even clearer. At 150 °C, the P&T analysis gives recoveries of 47%, 59% and 55% for butanal, 2-hexenal and nonane. For a sample of 0.5 mL containing 100 ppb of the compounds, 24, 30 and 28 ng are recovered. In the static headspace analysis, the partition coefficients for these compounds at 200 °C are all higher than 200. Assuming the value of 200, the quantity of each of these compounds in 5 mL of headspace is 0.7 ng. For an injection of 0.5 mL, the quantity injected is therefore only 0.07 ng. The differences in sensitivity favouring P&T analysis are therefore 343, 428 and 400!

Third Example: Residual Solvents in Plastic Sheeting

Comparable ratios are obtained in the analysis of a solid sample, for example, the analysis of residual solvents in a technical product. A run using the P&T technique and 10 mL of sample at 150 °C gave a recovery of 63% for toluene. The sample contained 1.6 ppm, which corresponds to a quantity of 101 ng. The partition coefficient in the static headspace technique at 150 °C (for a sample of 1 g) is 95. The quantity of residual solvent in 19 mL of headspace is therefore 17 ng. For an injection of 0.5 mL, 0.4 ng are injected. The quantity injected is therefore smaller by a factor of 250 than that in the P&T analysis. Furthermore, the reproducibility of this analysis was 7% for the P&T technique and 32% for the static headspace analysis (RSD).

The theoretically achievable or effectively necessary sensitivity is not the only factor deciding the choice of procedure. The specific interactions between the analytes and the matrix, the nature of the sample with for instance the potential of foaming, the performance of the detectors available and the legally required detection limits play a more important role. US EPA methods are mainly based on P&T use while European methods usually apply static headspace methods.

Besides the sensitivity, there are other aspects which must be taken into account when comparing the P&T and static headspace techniques.

The static headspace technique is very simple and quick. The procedure is well documented in the literature, and for many applications the sensitivity is more than adequate, so that its use is usually favoured over that of the P&T technique. There are areas of application where good results are obtained with the static headspace technique which cannot be improved upon by the P&T method. These include the forensic determination of alcohol in blood, of free fatty acids in cell cultures, of ethanol in fermentation units or drinks and residual water in polymers. This also applies to studies on the determination of ionization constants of acids and bases and the investigation of gas phase equilibria.

However, for many other samples, specific limitations besides sensitivity arise on use of the static headspace technique, which can be overcome using the P&T technique. In every static headspace system, all the compounds present in the headspace are injected, not only the organic analytes. This means that an air peak is also obtained. All air contaminants (a widely occurring problem in many laboratories) are visible and oxygen can impair the service life of the capillary column at high temperatures. In addition, a larger quantity of water is injected in this method than in the P&T method. For drinks containing carbonic acid, CO_2 can lead to the build-up of excess pressure. Even at room temperature, an undesirably high pressure can build up in the sample vials, which can lead to flooding of the GC column during the analysis. In addition, a very large quantity of CO_2 is injected. Heating the sample enhances these effects. For safety reasons, sample vials with safety caps to guard against excess pressure have to be used in these cases. Dust particles lead to further problems in the case of powder samples. To achieve the necessary sensitivity for an analysis, the sample often has to be heated to a temperature which is higher than that necessary in the P&T technique. This can lead to thermal decomposition, which is frequently observed with foodstuffs. In addition, oxygen cannot be eliminated from the sample before heating in a static headspace system. It is therefore impossible to prevent oxidation of the sample contents. During the thermostatting phase, additional problems arise as a result of the septum used. Substances emitted from the septum (e.g. CS_2 from butyl rubber septa) falsify the chromatogram. The permeability of the septum to oxygen presents a further hazard.

Advantages of the Static Headspace Technique

- The static headspace can be easily automated. All commercial headspace samplers operate automatically for only a few or large numbers of samples.
- For manual qualitative preliminary samples, a gas-tight syringe heated-up in the GC oven is satisfactory.
- Samples, which tend to foam or contain unexpectedly high concentrations of analyte, do not generally lead to faults or cross-contamination.
- All sample matrices (solid, liquid or gaseous) can be used directly usually without expensive sample preparation.
- Headspace samplers are often readily portable and, when required, can be rapidly connected to different types of GC instruments.

- The sample vessels (special headspace vials with caps) are intended only for single use. There is no additional workload of cleaning glass equipment and hence cross-contamination does not occur.
- Headspace vials can be filled and sealed at the sampling point outside the laboratory in certain cases. This dispenses with transfer of the sample material and eliminates the possibility of loss of sample. Moreover, the risk of inclusion of contamination from the laboratory environment is reduced.
- Because of the high degree of automation, the cost of analysing an individual sample is kept low.

Limitations of the Static Headspace Technique

- On filling headspace vials, a corresponding quantity of air is enclosed in the vial unless filling is carried out under an inert gas atmosphere. However, this is expensive. During thermostatting, undesirable side reactions can occur as a result of atmospheric oxygen in the sample.
- During injection from headspace vials, air regularly gets into the GC system and can affect sensitive column materials.
- In the case of moist or aqueous samples, a considerable quantity of water vapour gets on to the column. This requires special measures to ensure the integrity of the early eluting peaks. Headspace samplers do not provide devices for moisture removal as is standard on P&T systems.
- When mass spectrometers are used, special attention must be paid to the stability of ion sources. On insufficient heating, the surface of the source and the lens system become increasingly coated with moisture in the course of the work (caused by the injection of water vapour). As a result, the focussing of the mass spectrometer can change and this can impair quantitative work.
- Quantitation is matrix-dependent. Standardization measures are necessary, such as addition of carbonate, internal standards and MHE procedures.
- The ability of substances to be analysed is limited by the maximum possible filling of the headspace vials and by the partition coefficients. For compounds with low partition coefficients, larger quantities of sample do not lead to an increase in sensitivity. If the maximum possible equilibration temperature is being used, it is almost impossible to increase the sensitivity further.
- The quantity injected is limited and where the sensitivity is insufficient, multiple extraction with cryofocussing is necessary.
- Undesired blank values can be obtained through contaminated air (laboratory air) which gets into the headspace vial or through bleeding of the septum caps.
- Excess vial pressure can cause headspace vials to burst (e.g. with drinks containing carbonic acid or high equilibration temperatures). This always puts instruments out of operation for long periods and results in considerable clean-up costs. Only the use of special vial caps (with spring rings) together with special sealed vials which can release excess pressure into the atmosphere prevents bursting.

Advantages of the Purge and Trap Technique

- The sample quantity (maximum 25–50 mL) can easily be adapted to give the required sensitivity.
- A pre-purge step can remove atmospheric oxygen from the sample, even at room temperature if required.
- No septum or other permeable material is placed in the way of the sample.
- Substances with high-partition coefficients in the static headspace can be determined with good yields.
- There are no excess pressure problems. The entire gas stream is passed pressure-free through the trap to the vent.
- Water vapour can be kept out of the GC-MS system by means of a dry purge step and a moisture removal system.
- In automatic operations adding an internal standard without manual measures is quite straightforward.

Limitations of the Purge and Trap Technique

- Foaming samples require special treatment, the headspace sweep technique.
- The purge vessels (made of glass) are reused and must be cleaned carefully. Economical single-use vessels are currently available only in polymer materials.
- Larger quantities of sample require longer purge times.
- For highly contaminated samples, the BTV (break through volume) of the trap must be taken into consideration.
- If the baking out step is inadequate, the danger of a carry-over exists.
- Coupling with capillary GC necessitates the use of small split ratios or cryofocussing.

2.1.6
Adsorptive Enrichment and Thermal Desorption

In the analysis of air or gas samples, an extraordinarily large number of components of the most widely differing classes of compounds and over a very wide concentration range have to be considered. GC-MS is the method of choice for the determination of such volatile organic compounds. Usually, the concentration of the substances of interest is too low for the direct measurement of an air sample, and therefore enrichment on suitable sorbent materials is necessary. The concentration on solid adsorption material allows the accumulation of organic components from large volumes of gas. Typical areas of use include soil gas, workplace air monitoring, gases from landfill sites, urban air pollution and indoor air analysis. Compounds to be analyzed cover the wide range of GC amenable compounds with a boiling range up to typically C40. Most inorganic permanent gases like O_2, O_3, CO, CO_2, SO_2, NO_2 and also methane cannot be accumulated. Exceptions hereto are H_2S, CS_2, N_2O and SF_6 which are well absorbed.

Besides needing sufficient storage capacity to eliminate risk of analyte break-through during sampling (breakthrough volumes (BTV) must be greater than sampling volumes), the adsorption material is expected to have a low affinity for water (air moisture). This is not only important for GC-MS analysis. Neither water nor CO_2 should have a negative effect on the BTV of the organic components (Knobloch and Engewald, 1994). The ambient air generally has high moisture content. However, particular precautions must be taken in the case of combustion gases. Desorption of the enriched components from the carrier materials should be complete and without thermal changes.

The selection of the most suitable sorbent material or a combination of adsorbents for sampling and release of the analytes is of highest importance when developing a thermal desorption (TD) method. The choice of sorbent principally depends on the vapour pressure of the analyte. The more volatile the analyte to be trapped, the stronger the sorbent must be. As a rule of thumb, the boiling point of the analytes can be used as a guideline. The sorbent must quantitatively retain the compounds of interest from the volume of air sampled. For the desorption process, the sorbent material must then release those compounds efficiently and must provide a high thermal stability not to contribute to the background of the analysis. The expected adsorption and stability properties should not change for a large number of analyses, even on repeated use of the sampling tube (Brown and Shirey, 2002; Markes, 2012a).

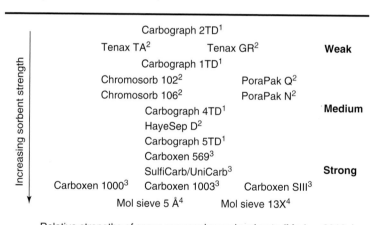

Relative strengths of some commonly used sorbents (Markes 2012a).

[1] = Graphitized carbon blacks.
[2] = Porous polymers.
[3] = Carbonized molecular sieves.
[4] = Zeolite molecular sieves

In air analysis adsorption materials, such as Tenax, Carbon blacks, car-bonised molecular sieves and XAD resins, are generally used (see Figure 2.28 and Table 2.14) (Sigma-Aldrich, 2014). Pure activated charcoal is indeed an

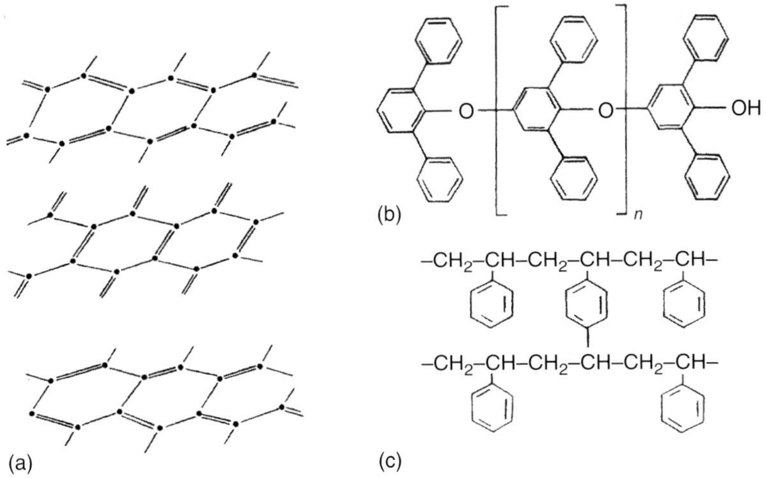

(a) (b) (c)

Figure 2.28 Surface model for common sorbent materials (Supelco). (a) Carbotrap, surface area about 100 m²/g, uniform charge distribution over all carbon atom centres. (b) Tenax (2,6-diphenyl-p-phenylene oxide), surface area about 24 m²/g, non-uniform charge distribution, the charge is essentially localized on the oxygen atoms. (Tenax-TA has replaced Tenax-GC as a new material of higher purity; Tenax-GR is a graphitized modification) (c) Amberlite XAD-2, surface area about 300 m²/g, non-uniform charge distribution, less polar than Tenax (XAD-4, about 800 m²/g).

Table 2.14 Adsorption materials and frequently described areas of use.

XAD-4	C_1/C_2-chlorinated hydrocarbons R11 Halogenated narcotics Vinyl chloride Ethylene oxide Styrene
Tenax TA 35-60 mesh	For boiling points 80–200 °C C_1/C_2-chlorinated hydrocarbons General solvents, volatile halogenated Hydrocarbons BTEX Phenols
Porapak	High-boiling, non-polar substances Halogenated narcotics Ethylene oxide
Carbotrap/VOCARB (various)	For boiling points −15 to −120 °C C_1/C_2-chlorinated hydrocarbons volatile Halogenated hydrocarbons BTEX Styrene
Molecular sieve 5 Å	N_2O

outstanding adsorbent for many organic compounds; however, these can only be sufficiently recovered using liquid solvents (Krebs *et al.*, 1991). Complete thermal desorption of organic analytes from charcoal of charcoal requires extremely high temperatures (>600 °C). This can lead to pyrolytic decomposition of the organic compounds, which are then no longer detected in the residue analysis.

Thermally stable materials offering a wide range of sorbent strengths have been developed over the last 30 or 40 years for use in TD applications. These include standard carbon blacks (Carbographs and Carbopacks/traps), carbonised molecular sieves such as the Carboxen series and modified carbon blacks (Carbograph 5 TD and Carbopack X) which offer much of the strength of carbonized molecular sieves but with minimal water retention. All are now in widespread use as sorbents in thermal desorption tubes (Betz, *et al.*, 1989).

Like Tenax, GCBs also have a low affinity for water. These adsorption materials can be dried with a dry gas stream, for example, the GC carrier gas (in the direction of adsorption!), without significant loss of material (dry purge).

VOCARB traps are special combinations of the adsorption materials Carbopack (GCB) and Carboxen (carbon molecular sieve). These combinations have been optimized in the form of VOCARB 3000 and VOCARB 4000 for volatile and lessvolatile compounds respectively, corresponding to the EPA methods 624/1624 and 542.2 (Supelco, 1992). VOCARB 4000 exhibits higher adsorptivity for less volatile components, such as naphthalenes and trichlorobenzenes. However, it shows catalytic activity towards 2-chloroethyl vinyl ether (complete degradation!), 2,2-dichloropropane, bromoform and methyl bromide.

Volatile polar components are enriched on highly polar absorbent materials. The combination of Tenax TA and silica gel has proved particularly successful for the enrichment of polar compounds. Water absorption by silica gel need to be taken care of in the application for air analysis.

To cover a wider range of molecular sizes, various adsorption materials are combined with one another. For example, the Carbotrap multibed adsorption tube (Figure 2.29) consists of three materials: Carbopack C with its low surface area of 12 m^2/g is used to enrich high molecular weight components, such as

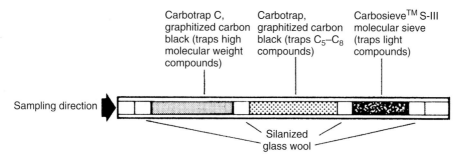

Figure 2.29 Multibed adsorption/desorption tube Carbotrap 300 (Supelco). Carbotrap, Carbotrap C: graphitized carbon black, GCB, surface area about 12 m^2/g. Carbosieve S-III: carbon molecular sieve, surface area about 800 m^2/g and pore size 15–40 Å.

Table 2.15 Breakthrough volumes [L] for Carbotrap 300 adsorption/desorption tubes (Supelco, 2013).

Substance	Carbosieve S-III	Carbotrap	Carbopack C
Vinylchloride	158	1.1 [a)]	—
Chloroform	—	—	—
1,2.Dichloroethane	—	0.4	27.8[a)]
1,1,1-Trichloroethane	—	2.7 [a)]	—
Carbon tetrachloride	—	4.7 [a)]	—
1,2.Dichloropropane	—	6.8	—
Trichloroethylene	—	2.5	—
Bromoform	—	1.7	—
Tetrachloroethylene	—	2.2	—
Chlorobenzene	—	316	—
n-Heptane	—	262	—
1-Heptene	—	284[a)]	—
Benzene	—	2.3	—
Toluene	—	130	—
Ethylbenzene	—	—	12.9
p-Xylene	—	—	11.2
m-Xylene	—	—	11.0 [a)]
o-Xylene	—	—	11.0 [a)]
Cumene	—	—	27.8[a)]

a) Theoretical value.

alkylbenzenes, polyaromatic hydrocarbons or PCBs, directly at the inlet. All the more volatile substances pass through to the subsequent layers. A layer of Carbotrap particles (Carbopack B) separated by glass wool is characterized by the higher particle size of 20/40 mesh. Volatile organic substances, in particular, are excellently adsorbed on Carbotrap and thermally desorbed with high recoveries (Supelco, 1986). To adsorb C_2-hydrocarbons Carbosieve S-III material with a particularly high surface area of $800 \, m^2/g$ and a pore size of $15-40 \, \text{Å}$ is placed at the end of the adsorption tube (patent, BASF AG, Ludwigshafen, Germany) (Table 2.15). Carbon molecular sieves have lower water retention that regular molecular sieves allowing them to be used in atmospheres with high moisture contents.

Many regulated methods recommend the application of certain approved adsorbent materials. A selection of internationally used methods is provided in Table 2.16.

2.1.6.1 Sample Collection

Sample collection can be either passive or active (Figure 2.30). For passive collection, diffusion tubes with special dimensions are used. The content of substances in the surrounding air is integrated, taking the collection time into account.

Table 2.16 Thermal desorption tube selection by regulated method (Supelco, 2013).

Method	Thermal desorption tube
ASTM D6196	Carbotrap™ 100, Carbopack™ B, Chromosorb® 106
EPA TO-1	Tenax® TA
EPA TO-2	Carbosieve™ SIII
EPA TO-14	Air toxics
EPA TO-17	Carbotrap™ 217, Carbotrap™ 300, Carbotrap™ 317
EPA 1P-1B	Tenax TA, Carbotrap™ 349
EPA 0030, EPA 0031, EPA SW-846	VOST stack sampling tubes: Tenax TA (35/60 mesh), Tenax TA: Petroleum Charcoal (2 : 1)
MDHS 72	Chromosorb® 106
NIOSH 2549	Carbotrap™ 349

Active collection devices require a calibrated pump with which a predetermined volume is drawn through the adsorption tube. Having estimated the expected concentrations, for example, for indoor air 100 mL/min and for outdoor air 1000 mL/min, the air is drawn through the prepared adsorption tube over a period of 4 h. After sample collection, the adsorption tube must be closed tightly to exclude additional uncontrolled contamination. The storage performance of thermo desorption sampling tubes were tested to be stable over several weeks using Tenax TA and other sorbent materials under different storage conditions (Brown 2014).

Brown and Purnell carried out thorough investigations on the determination of BTVs. The latter generally vary widely with the collection rate. On use of Tenax as the sorbent the ideal collection rate is 50 mL/min, in any case, however, performance remains optimum (i.e. breakthrough volumes remain constant) provided flow rates remain in the range 20 to 200 mL/min. Flow rates up to 500 mL/min can also be used for short duration sampling (up to 15 minutes). Note that pump flow rates below 5 or 10 mL/min are subject to unacceptable error due to diffusive ingress and should not be used. Moisture does not affect the BTVs with Tenax (unlike other porous materials). Furthermore, sample collection is greatly affected by temperature. An increase in temperature increases the BTV (about every 10 K doubles the volume) (Brown and Purnell, 1979; Figg, *et al.*, 1987; Manura, 1995).

2.1.6.2 Calibration

In the calibration of thermal desorption tubes, the same conditions should predominate as in sample collection. Methods such as the liquid application of a calibration solution to the adsorption materials or the comparison with direct injections have been shown to be unsatisfactory. Alternatively, Certified Reference Standards (CRSs) are available (e.g. Markes Int., UK). CRS tubes are recommended in many key standard methods (e.g. US EPA Method TO-17) for auditing purposes and as a means of establishing analytical quality control. CRS tubes are often certified traceable to primary standards, have a minimum shelf life of typically 6 months. They are available ready for use with concentrations

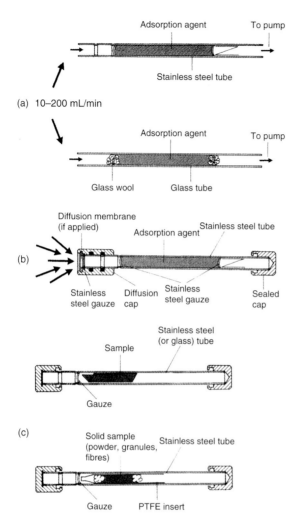

Figure 2.30 Sample collection with standard 3.5-inch (89 mm) long × 1/4-inch (6.4 mm O.D.) × 5 mm I.D. stainless steel thermal desorption tubes (Perkin-Elmer). (a) Active sample collection with pump (e.g. personal air sampler), (b) passive sample collection by diffusion and (c) direct introduction of solid samples.

varying from 10 ng to 100 µg per component. Chromatograms from a shipping blank, and an example analysis of a CRS tube should be supplied along with the CRS certificate.

CRS tubes are available loaded with benzene, toluene and xylene (BTX) at levels of 100 ng or 1 µg per component; TO-17 standards at 25 ng per component of benzene, toluene, xylene, dichloromethane, 1,1,1-trichloroethane, 1,2,4-trimethylbenzene, methyl-*t*-butyl ether, butanol, ethyl acetate and methylethyl ketone. Custom CRS tubes are also available on a variety of sorbents for a wide range of compounds from different vendors.

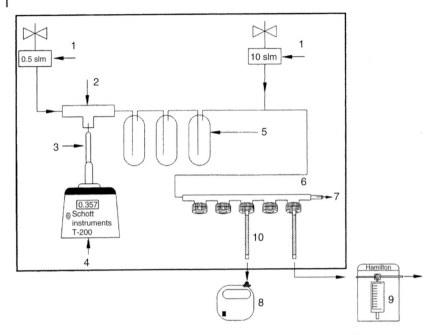

Figure 2.31 Principle of introducing measured volumes of standard atmosphere onto thermal desorption tubes. (Tschickard, 1993).

Customized calibration tubes can be prepared by using standard atmospheres. The process for the preparation of standard atmospheres by continuous injection into a regulated air stream has been described in standardized guidelines (Figure 2.31) (Kommission Reinhaltung der Luft im VDI, 1981; HSL, 1990). In this process, an individual component or a mixture of the substances to be determined is continuously charged to an injector through which air is passed (complementary gas), using a thermostatted syringe burette. The air quantity (up to 500 mL/min) is adjusted using a mass flow meter. The complementary gas can be diluted by mixing with a second air stream (dilution gas up to 10 L/min). Moistening the gas can be carried out inside or outside the apparatus. In this way, concentrations in the ppm range can be generated. For further dilution, for example, for calibration of pollution measurements, a separate dilution stage is necessary. The gas samples to be tested are drawn out of the calibration station into a glass tube with several outlets. In this case, active or passive sample collection is possible. Continuous injection has the advantage that the preparation of mixtures is very flexible (Tables 2.17 and 2.18). However, accurate standard atmospheres are notoriously difficult to generate and maintain. For this reason many modern thermal desorption standards specify calibration of thermal desorption methods by introduction of small volumes of gas- or liquid phase standards onto the sampling end of thermal desorption tubes in a stream of inert carrier gas. For example see EN ISO 16017 or ASTM D6196.

Table 2.17 Evaluation of test tubes which were prepared using the calibration unit by continuous injection ($n = 10$).

Component	Mean value	Standard deviation	Relative standard deviation (%)
1,1,1-Trichlorethane	562.6	4.95	0.88
Dichloromethane	538.6	7.93	1.47
Benzene	753.3	9.48	1.25
Trichloroethylene	627.3	16.7	2.67
Chloroform	626.6	12.1	1.94
Tetrachloroethylene	698.2	6.11	0.88
Toluene	1074	6.88	0.64
Ethylbenzene	358.1	2.91	0.81
p-Xylene	736.3	4.85	0.66
m-Xylene	731.6	4.77	0.65
Styrene	755.8	4.60	0.61
o-Xylene	389.5	3.49	0.89

Table 2.18 Analysis results of BTEX determination of certified samples after calibration with the calibration unit.

Mass (µg)	Benzene	Toluene	m-Xylene
Measured value	1.071	1.136	1.042
Required value (certified)	1.053	1.125	1.043
Standard deviation (certified)	0.014	0.015	0.015

2.1.6.3 Desorption

The elution of the organic compounds collected involves extraction by a solvent (displacement) or thermal desorption. Pentane, CS_2 and benzyl alcohol are generally used as extraction solvents. CS_2 is very suitable for activated charcoal, but cannot be used with polymeric materials, such as Tenax or Amberlite XAD, because decomposition occurs. As a result of displacement with solvents, the sample is extensively diluted, which can lead to problems with the detection limits on mass spectrometric detection. With solvents additional contamination can occur. The extracts are usually applied as solutions. The readily automated static headspace technique can also be used for sample injection. This procedure has also proved to be effective for desorption using polar solvents, such as benzyl alcohol or ethylene glycol monophenyl ether (1% solution in water, Krebs, 1991).

In thermal desorption, the concentrated volatile components are released by rapid heating of the adsorption tube and after preliminary focussing, usually within the instrument, are injected into the GC-MS system for analysis (Table 2.19). Automated thermal desorption gives better sensitivity, precision and accuracy in the analysis. The number of manual steps in sample processing is reduced. Through frequent reuse of the adsorption tube and complete elimination

Table 2.19 Desorption temperatures for common adsorption materials and possible interfering components which can be detected by GC-MS.

Adsorbent	Desorption (°C)	Maximum temperature (°C)	Interfering components
Carbotrap	up to 330	>400	Not determined
Tenax	150–250	375	Benzene, toluene and trichloroethylene
Molecular sieve	250	350	Not determined
Porapak	200	250	Benzene, xylene and styrene
VOCARB 3000	250	>400	Not determined
VOCARB 4000	250	>400	Not determined
XAD-2/4	150	230	Benzene, xylene and styrene

of solvents from the analysis procedure, a significant lowering of cost per sample is achieved.

Thermal desorption has become a routine procedure because of program-controlled samplers. The individual steps are method controlled and monitored internally by the instrument. For the sequential processing of a large number of samples, autosamplers with capacities of up to 100 adsorption tubes are commercially available.

Before and after sample collection, the adsorption tubes need to be hermetically sealed. This can be achieved with long term storage caps comprising 2-piece, 1/4-inch metal compression fittings and combined PTFE fittings. These have been evaluated for long term storage by independent reference laboratories and found to ensure sample stability for many months (even years) at room temperature. Sorbent tubes also have to be kept sealed throughout automated thermal desorption analysis – both to preserve the integrity of sampled tubes and to prevent the ingress of contaminants onto desorbed/analysed tubes. During automated TD, tubes are fitted with temporary Teflon caps. Sampling tubes also can be sealed by use of the DiffLok™ end-cap technique at either end of the tube. Diffusion-locking is a patented technology that keeps the sample tubes sealed at ambient pressures, but allows gas to flow through the tubes whenever pressure is applied (Woolfenden, 2003). Diffusion-locking does not involve any kind of valve or other moving parts, and is thus inherently simple and robust. With the inlet and outlet tube of a sampler being sufficiently narrow and long, the process of diffusion of vapours into or out of an attached sorbent tube can be reduced almost to zero. The sealing of the tubes protects the sample from ingress or loss of volatiles at all stages of the monitoring process (Markes, 2012b) (Figure 2.32).

Tubes using check valves are not suitable for diffusive monitoring. In case Teflon caps are used, they are removed inside the instrument before measurement and the adsorption tube is inserted into the desorption oven. Before desorption, the tightness of the seal to the instrument is tested as the initial step of the automated desorption process by monitoring an appropriate carrier gas pressure for a short time. Carrier gas is passed through the adsorption tubes in the desorption oven at temperatures of up to 400 °C (in the reverse direction to the adsorption!). The

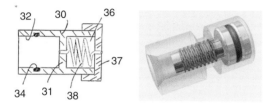

Figure 2.32 Schematic of a DiffLok™ analytical end-cap for TD sample tubes (Markes, 2003) 30 cap body, 31 wall, 32 socket, 34 O-ring seal to fit over a sample tube, 36 hollow compartment, 37 closure and 38 wound capillary tube.

components released are collected in a cold trap inside the apparatus. The intermediately trapped sample then is flushed to the GC column by rapidly heating the cold trap. This two-stage desorption and the use of the multiple split technique enable the measuring range to be adapted to a wide range of substance concentrations. The sample quantity can be adapted to the capacity of the capillary column used through suitable split ratios both before and after the internal cold trap (Figure 2.33).

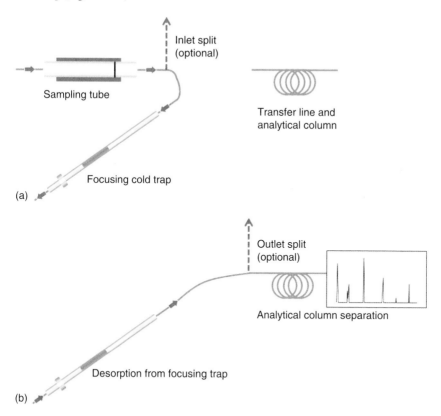

Figure 2.33 Multiple split technique for thermal desorption (Markes).

The thermal desorption process is usually carried out automatically. The high sensitivity and reliability of thermal desorption does not mean that it has no disadvantages. Until recently, TD was invariably a one-shot process with no sample remaining after desorption to repeat the process if anything went wrong. However, with a recent development by Markes International (SecureTD-Q) it has become possible to quantitatively and routinely re-collect the split portion of the sample for repeat analysis. This overcomes the historical one-shot limitation of thermal desorption methods and considerably simplifies method and considerably data validation.

For high throughput applications, an electronic tube tagging using radio frequency identification (RFID) tube tags is available for industry standard sorbent tubes and 4.5-in. (DAAMS, Depot Area Air Monitoring System) tubes.

Possible and, for certain carrier materials, already known decomposition reactions have been mentioned. For this reason, another method is favoured by the EPA for air analysis. Electropolished and passivated canisters, called SUMMA canister expressing the highest inertness, of about 2 L capacity, maximum up to 15 L are evacuated for sample collection. The whole air samples collected onsite on opening the canister can be measured several times in the laboratory following EPA methods such as TO-14 or TO-15. Suitable samplers are used, which are connected online with GC-MS. Cryofocussing is used to concentrate the analytes from the volumes collected. If required, the sample can be dried with a semipermeable membrane (Nafion drier) or by condensation of the water (MCS, see also section 2.1.5.2). Adsorption materials are not used in these processes.

2.1.7
Pyrolysis

The use of pyrolysis extends the area of use of GC-MS coupling to samples which cannot be separated by GC because they cannot be desorbed from a matrix or evaporated without decomposition (Wampler, 1995; Brodda *et al.*, 1993). In analytical pyrolysis, a large quantity of energy is passed into a sample so that fragments are formed which can be gas chromatographed reproducibly. The pyrolysis reaction initially involves thermal cleavage of C−C bonds, for example in the case of polymers. Thermally induced chemical reactions within the pyrolysis product are undesired side reactions and can be prevented by reaching the pyrolysis temperature as rapidly as possible (Ericsson, 1980). The reactions initiated by the pyrolysis are temperature-dependent. To produce a reproducible and quantifiable mixture of pyrolysis products, the heating rates, and the pyrolysis temperatures in particular, should be kept constant (Ericsson, 1985). The sample and its contents can be characterized using the chromatographic sample trace (pyrogram) or by the mass spectroscopic identification of individual pyrolysis products (Irwin, 1982).

The use of pyrolysis apparatus with GC-MS systems imposes particular requirements on them. The sample quantity applied must correspond to the capacity of commercial fused silica capillary columns. It is usually in the microgram range

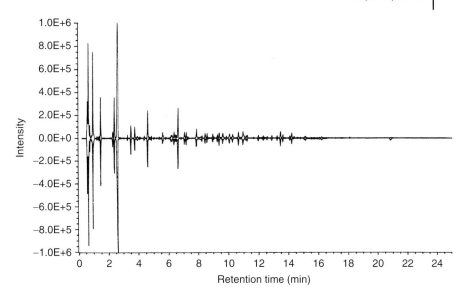

Figure 2.34 Reproducibility of the pyrolysis of an automotive paint on two consecutive days using a foil pyrolyser coupled to GC-MS, mirrored representation (Steger, Audi AG).

or less (Hancox *et al.*, 1991). By selecting a suitable split ratio, peak equivalents for the pyrolysis products can be achieved in the middle nanogram range. These lie well within the range of modern GC-MS systems. The small sample quantities are also favourable for analytical pyrolysis in another aspect. The reproducibility of the procedure increases as the sample quantity is lowered as a more rapid heat transport through the sample is possible (Figure 2.34). Side reactions in the sample itself and as a result of reactive pyrolysis products are increasingly eliminated.

Because of the small sample quantities and the high reproducibility of the results, analytical pyrolysis has experienced a renaissance in recent years. Both in the analysis of polymers with regard to quality, composition and stability (Ericsson, 1978, 1990), and in the areas of environmental analysis, foodstuffs analysis and forensic science, pyrolysis has become an important analytical tool, the significance of which has been increased immensely by coupling with GC-MS (Klusmeier *et al.*, 1988; Zaikin *et al.*, 1990; Galletti and Reeves, 1991; Hardell, 1992; Matney *et al.*, 1994).

Analytical pyrolysis is available using different processes: high-frequency pyrolysis (Curie point pyrolysis), furnace and foil pyrolysis. The processes differ principally through the different means of energy input and the different temperature rise times (TRTs). Pyrolysis instruments can easily be connected to GC and GC-MS systems. The reactors currently used are constructed so they can be placed on top of GC injectors (split operation) and can be installed for short-term use only if required. Also automated pyrolysis units are available providing high sample throughput (Snelling, *et al.*, 1994)

Analytical Pyrolysis Procedures

Procedure	Foil	High frequency	Micro-furnace
Carrier	Pt foil	Fe/Ni alloys	Cup
Pyrolysis temperature	Variable, resistively heated up to 1400 °C	Fixed, alloy Curie temperatures up to 1040 °C	Variable, resistively heated furnace up to 1000 °C
TRT	<8 ms	up to 200 ms	up to 3 °C/s

2.1.7.1 Foil Pyrolysis

In foil pyrolysis, the sample is applied to a thin platinum foil (Figure 2.35). The thermal mass of this device is extremely low. After application of a heating current, any desired temperature up to about 1400 °C can be achieved within milliseconds. The extremely high heating rate results in high reproducibility. The temperature of the Pt foil can be controlled by its resistance. However, the temperature can be measured and controlled more precisely and rapidly from the radiation emitted by the Pt foil. An exact calibration of the pyrolysis temperature can be carried out and the course of the pyrolysis is recorded by this feedback alone. Besides endothermic pyrolysis, exothermic processes can also be detected and recorded.

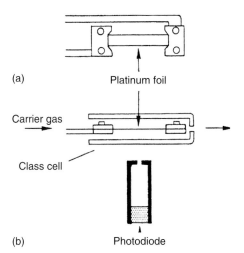

(a)   Platinum foil

Carrier gas

Class cell

(b)   Photodiode

Figure 2.35 Scheme of a Pt foil pyrolyser with temperature control by means of fibre optic cable (Pyrola). (a) View of the Pt foil with carrier gas and current inlets. (b) Side view of the pyrolysis cell with the glass cell (about 2 mL volume, Pyrex) and photodiode under the Pt foil for calibration and monitoring of the pyrolysis temperature.

Pyrolysis Nomenclature
(Uden, 1983; Ericsson and Lattimer, 1989)

- *Pyrolysis*:
  A chemical degradation reaction initiated by thermal energy alone.
- *Oxidative pyrolysis*:
  A pyrolysis which is carried out in an oxidative atmosphere (e.g. O_2).
- *Pyrolysate*:
  The total products of a pyrolysis.
- *Analytical pyrolysis*:
  The characterization of materials or of a chemical process by instrumental analysis of the pyrolysate.
- *Applied pyrolysis*:
  The production of commercially usable materials by pyrolysis.
- *Temperature/time profile (TTP)*:
  The graph of temperature against time for an individual pyrolysis experiment.
- *Temperature rise time (TRT)*:
  The time required by a pyrolyser to reach the pre-set pyrolysis temperature from the start time.
- *Flash pyrolysis*:
  A pyrolysis which is carried out with a short TRT to achieve a constant final temperature.
- *Continuous pyrolyser*:
  A pyrolyser where the sample is placed in a preheated reactor.
- *Pulse pyrolyser*:
  A pyrolyser where the sample is placed in a cold reactor and then rapidly heated.
- *Foil pyrolyser*:
  A pyrolyser where the sample is applied to metal foil or band which is directly heated as a result of its resistance.
- *Curie point pyrolyser*:
  A pyrolyser with a ferromagnetic sample carrier which is heated inductively to its Curie point.
- *Temperature-programmed pyrolysis*:
  A pyrolysis where the sample is heated at a controlled rate over a range of temperatures at which pyrolysis occurs.
- *Sequential pyrolysis*:
  Pyrolysis where the sample is repeatedly pyrolysed for a short time under identical conditions (kinetic studies).
- *Fractionated pyrolysis*:
  A pyrolysis where the sample is pyrolysed under different conditions in order to investigate different sample fractions.
- *Pyrogram*:
  The chromatogram (GC, GC-MS) or spectrum (MS) of a pyrolysate.

2.1.7.2 Curie Point Pyrolysis

High-frequency pyrolysis uses the known property of ferromagnetic alloys of losing their magnetism spontaneously above the Curie temperature (Curie point). At this temperature, a large number of properties change, such as the electrical resistance or the specific heat. Above the Curie temperature, ferromagnetic substances exhibit paramagnetic properties. The possibility of reaching a defined and constant temperature using the Curie point was first realized by Simon and Giacobbo (1965). In a high frequency field, a ferromagnetic alloy does not absorb any more energy above its Curie point and remains at this temperature. As the Curie temperature is alloy-dependent, another Curie temperature can be used by changing the material. If a sample is applied to a ferromagnetic material and is heated in an energy rich high frequency field (Figure 2.36), the pyrolysis takes place at the temperature determined by the choice of alloy and thus its Curie point (Schulten *et al.*, 1987; Oguri and Kirn, 2005). The temperature rise profiles for various metals and alloys are shown in Figure 2.37.

In practice, sample carriers in the form of loops, coils or simple wires made of different alloys at fixed temperature intervals are used. Ferromagnetic metal foils (Pyrofoil™) are available for 21 different pyrolysis temperatures ranging from 160 °C to 1040 °C (JAI Ltd., Tokyo, Japan). In this case, the sample gets crimped into the metal foil using a dedicated tool before analysis (Oguri and Kirn, 2005).

A potential disadvantage of Curie point pyrolysis is the longer TRT compared with the direct resistively heated foil pyrolysis. The temperature rise time of up to 200 ms to reach the Curie point is significantly slower and depends on the

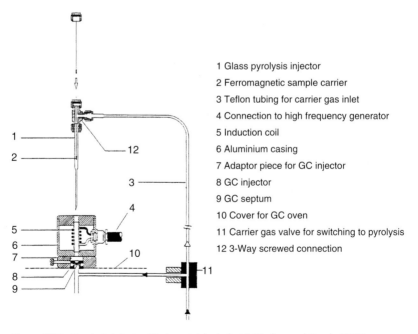

1 Glass pyrolysis injector
2 Ferromagnetic sample carrier
3 Teflon tubing for carrier gas inlet
4 Connection to high frequency generator
5 Induction coil
6 Aluminium casing
7 Adaptor piece for GC injector
8 GC injector
9 GC septum
10 Cover for GC oven
11 Carrier gas valve for switching to pyrolysis
12 3-Way screwed connection

Figure 2.36 Pyrolysis injector (Curie point hydrolysis) (Fischer and Kusch, 1993).

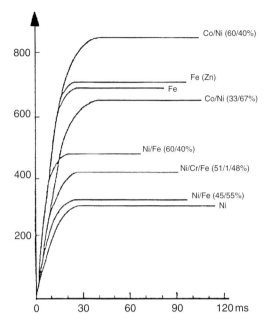

Figure 2.37 Curie temperatures and temperature/time profiles of various ferromagnetic materials (Simon and Giacobbo, 1965).

device used. There are also effects due to the not completely inert surface of the sample carrier. They can manifest themselves in the inadequate reproducibility of the pyrolysis process for analytical purposes. Copolymers with thermally reactive functional groups (e.g., free OH, NH_2 or COOH groups) cannot be analysed by Curie point pyrolysis.

For the analytical assessment of the coupling of pyrolyser with GC and GC-MS systems, a high-boiling mixture of cholesterol with n-alkanes has been proposed (Gassiot-Matas). Figure 2.38 shows the evaporation of cholesterol (500 ng) with the C_{34} and C_{36}-n-alkanes (50 ng of each). Before the intact cholesterol is detected, its dehydration product appears. The intensity of this peak increases with small sample quantities and increasing temperatures in the region of substance transfer in the GC injector. The high peak symmetries of the signals of the alkanes, the cholesterol and its dehydration product indicate that the coupling is functioning well.

There are various possibilities for the evaluation of pyrograms obtained with a GC-MS system. With classical FID (flame-ionization detector) detection, the pyrogram pattern is compared only with known standards. However, with GC-MS systems, the mass spectra of the individual pyrolysis products can be evaluated. By using libraries of spectra, substances and substance groups can be identified. GC-MS pyrograms can be selectively investigated for trace components even in complex separating situations by using the characteristic mass fragments of minor components. The comparison of sample patterns becomes meaningful through

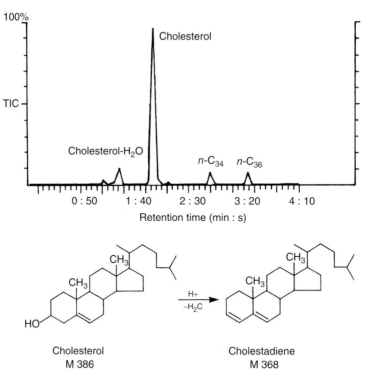

Figure 2.38 Analysis of cholesterol and *n*-alkanes (C_{34}, C_{36}) for testing the pyrolysis coupling to the GC injector (Richards 1988). Conditions: GC column 4 m × 0.22 mm ID × 0.25 μm CP-Sil5, 200–320 °C, 30 °C/min, GC-ITD.

the choice of mass chromatograms of substance-specific fragment ions. GC-MS programs, with its full pattern information, can be stored and compared for identity or similarity. Suitable software systems are commercially available providing a numerical measure on similarity instead of only a visual inspection (Chromsearch, Axel Semrau GmbH, Germany; GC Concordance, Spectrochrom, Bouc Bel Air, France).

In the library search for the spectra of pyrolysis products, special care must be taken. Commercial spectral libraries consist of spectra of particular substances that have been fully characterized, which is not the case for the majority of pyrolysis products. Classical MS fragmentation rules apply for interpretation of search results. Depending on the sample material, however, an extremely large number of reaction products are formed on pyrolysis, which cannot be completely separated even under the best GC conditions. If the detection is sensitive enough, this situation is shown clearly by the mass chromatogram. Also, for polymers, there is a typical appearance of homologous fragments, which must also be taken into account. These series can also be shown easily using mass chromatograms through the choice of suitable fragment ions.

Quantitative determinations using pyrolysis benefit particularly from the selectivity of GC-MS detection. The precision is comparable to that of liquid injections.

2.1.7.3 Micro-furnace Pyrolysis

Another very flexible solution for numerous applications in pyrolysis, outgassing or thermal extraction experiments is the use of a micro-furnace, which can be resistively heated to temperatures above 1000 °C (Roussis and Fedora, 1998). The sample is placed in a small, low, thermal mass cup of 50–80 µL volume, which is dropped into the heated pyrolysis furnace consisting of a vertical quartz tube. The furnace tube size is comparable to a GC inlet liner with 4.5 mm inner diameter and 12 mm length and allows programmable temperature profiles with heating rates of up to 200 °C/min (Frontier Laboratories Ltd., Japan).

Typical sample sizes are in the 0.1 mg range, allowing a fast heat transfer to the set desorption or pyrolysis temperature. The transfer of the pyrolysis products is achieved by a carrier gas flow through the heated zone and a needle tip on the bottom of the furnace reaching into the injector of the GC.

Besides the flash pyrolysis of solid samples, the furnace solution allows different sampling devices for a wider bandwidth of analytical experiments (see Figure 2.39). This solution can provide access to additional analytical applications with the introduction of liquid samples using a regular micro syringe, the online pyrolysis for the analysis of high-pressure reactions in glass capsules, the thermal desorption of small amounts of solid materials, or the combination with a subsequent second reactor for the reaction with catalyst materials. Unique is the

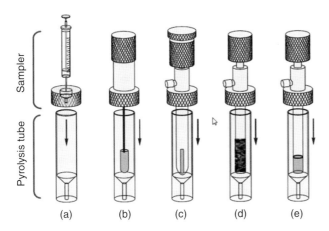

(a)   (b)   (c)   (d)   (e)

Figure 2.39 Five micro-furnace pyrolysis tubes for different pyrolysis and thermal desorption applications. (Frontier Laboratories Ltd., Japan.) (a) Liquid sampler for direct liquid sample injection using a micro syringe. (b) Single-shot sampler for flash pyrolysis using sample cup. (c) On-line micro reaction sampler for the analysis of high-pressure reactions in glass capsules. (d) Micro TD sampler for the analysis of enriched compounds with absorbent. (e) On-line micro UV sampler for the analysis of photo, thermal and oxidative degradation products by UV irradiation.

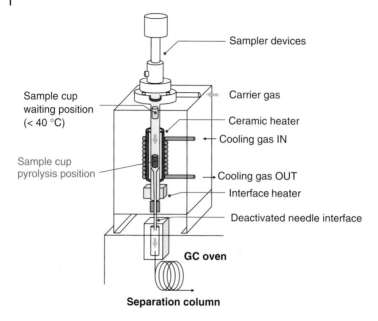

Sampler devices

Sample cup
waiting position
(< 40 °C)

Carrier gas

Ceramic heater

Cooling gas IN

Sample cup
pyrolysis position

Cooling gas OUT

Interface heater

Deactivated needle interface

GC oven

Separation column

Figure 2.40 Schematic diagram of the micro furnace pyrolyzer EGA/PY-3030D. (Frontier Laboratories Ltd.)

online UV radiation of samples for the analysis of photo, thermal and oxidative degradation products (Figure 2.40).

2.1.8
Thermal Extraction (Outgassing)

In a thermal extraction device, the sample is placed in a temperature controlled chamber, which can be operated at a constant temperature for material emissions testing (Figure 2.41). Extraction chambers for outgassing are commercially available in very different dimensions to accommodate even large sample sizes (Markes, 2009a, 2011). This allows both the thermal extraction of volatile (VOCs) and semi-volatile organic carbons (SVOCs) components from liquid and solid samples or even technical components (Markes, 2009b, 2012; US Environmental Protection Agency, 1996). Conditioned Tenax or multi-sorbent tubes are attached to the constantly heated desorption chamber. A controlled air flow of typically 10–500 mL/min or inert gas (He, N$_2$) is passed through. Any vapors emitted by the sample under the selected conditions are swept onto an attached sorbent tube by the flow of gas. After vapour collection the tube is transferred to a TD unit and thermally desorbed for GC or GC-MS analysis.

The difference between thermal extraction and the analytical pyrolysis systems described earlier lies in the consideration of the sample quantity and the applied temperatures. Even inhomogeneous materials, bulk materials or complete assemblies can be investigated at temperatures below pyrolysis, releasing

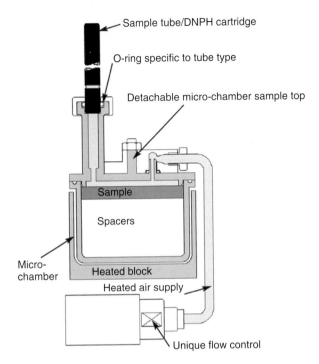

Sample tube/DNPH cartridge

O-ring specific to tube type

Detachable micro-chamber sample top

Sample

Spacers

Micro-
chamber

Heated block

Heated air supply

Unique flow control

Figure 2.41 Thermal extractor schematic (Micro-Chamber/Thermal Extractor™). (Markes Int.)

embedded volatile components. The applications are numerous and range from food produce (Figure 2.42) to semiconductor components (Figure 2.43). They cover, for instance, the emissions of volatile and semi-volatile organic compounds from materials that can adversely impact indoor or in-vehicle air quality. National and international reference methods for regulations such as the European Construction Products Regulation, the German protocol for fire-resistant floorings (AgBB, Ausschuss zur gesundheitlichen Bewertung von Bauprodukten) and the Californian CHPS protocol for public school building programs (CHPS) specify the determination of emissions from construction materials using conventional test chambers/operated at ambient temperature (Methods ISO/EN 16000-6/-9/-10/-11, ASTM D5116-97, ASTM D7143-05, etc.) in order to certify product performance. However, in this case, smaller micro-chamber/thermal extractor devices provide a useful and complementary quick screening method for factory product control, inhouse comparative checks and tests on raw materials etc. (Reference ASTM D7706 and ISO 12219-3). Direct thermal extraction may be used to measure SVOCs in a wide range of solid, resinous and liquid materials, and eliminates complex liquid extraction steps. It is also used by the paint industry (US EPA Method 311 for paints) for evaluating 'low VOC' products and the German automotive industry for testing car interior components (Method VDA 278, 2001) (Markes, 2009). Outgassing devices are commercially available

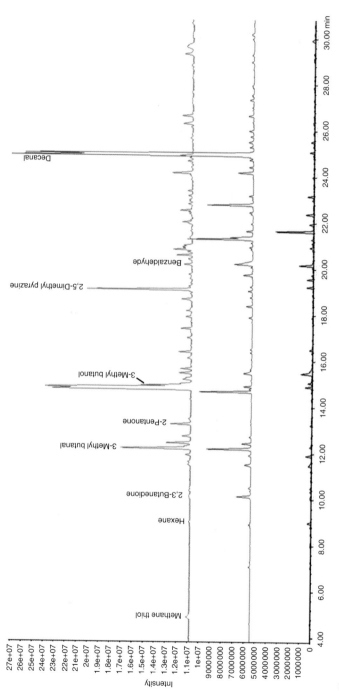

Figure 2.42 Thermal extraction chromatograms of a cheese sample at $t = 0$ days (black) and $t = 1$ day (blue) and $t = 10$ days (red). (Markes Int.)

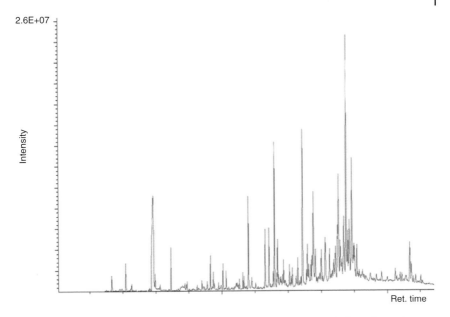

2.6E+07

Intensity

Ret. time

Figure 2.43 Bulk emissions from a sample of printed circuit board (including electronic components) tested at 120 °C. (Markes Int.)

in different chamber sizes up to several litres of volume and capacity from one to multiple channels for parallel thermal extraction. Maximum temperatures are typically in the range of 200–400 °C. In particular for electronic components, such devices can range in geometry to receive electronic components, silicon wafers or complete hard discs, provided by companies such as Markes Int., UK or JAI Ltd., Japan.

In the field of environmental analysis, thermal extraction is proposed by an EPA method for the quantitative analysis of semi-volatile compounds from solid sample materials. The US EPA method 8275 is a thermal extraction capillary GC-MS procedure for the rapid and quantitative determination of targeted PCBs and PAHs (polynuclear aromatic hydrocarbons) in soils, sludges and solid wastes. This method requires extraction temperatures of 340 °C for 3 min for the quantitative desorption of the PCBs (EPA, 1996).

2.1.9
QuEChERS Sample Preparation

> *The QuEChERS method is a streamlined approach that makes it easier and less expensive for analytical chemists to examine pesticide residues in food.*

> Steven Lehotay, 2013

The sample preparation and clean-up processes for pesticide analysis have not been standardized, but a clear trend over the recent years indicates a strong preference of labs using the QuEChERS method. QuEChERS is a dispersive SPE technique for extracting multi-residue pesticides from fruits and vegetables, in general, low-fat food commodities. The name is formed as an acronym from 'Quick, Easy, Cheap, Effective, Rugged and Safe' (Anastassiades, 2011). This new sample preparation method was first presented for the extraction of pesticides from fruit and vegetables during the fourth European Pesticide Residue Workshop in Rome followed by a publication in the journal of the AOAC (Anastassiades *et al.*, 2002, 2003). As the sample preparation methods for pesticides in the past have been optimized for different matrices and mirrored the individual experience and knowledge of many trace chemical laboratories, QuEChERS today is taking the lead for an increasingly standardized methodology, creating an international pool of experience for the growing number of pesticides to analyse on residue level (Cunha *et al.*, 2007). This trend is impressively pictured with the feedback from EU proficiency tests indicating the used sample preparation method. Figure 2.44 shows the statistics starting from 2006 over the following 6 years. While the traditional methods such as Luke, Specht and solvent extractions steadily decline, the number of labs using QuEChERS is strongly increasing and dominating the types of sample preparation methods applied in pesticide residue labs.

The steadily growing international use of the QuEChERS method is based on what the method wants to convey: speed, ease-of-use, minimized use of solvents and low cost per sample (Khan, 2012). It combines several sample preparation steps, and replaces older, tedious extraction methods with a comprehensive approach of good performance for an increasing number of analytes. Reported recoveries are typically high in the range of 90–110%, as shown in Figure 2.45 (Lehotay, 2007, 2013). Another driving aspect for further widespread use is the compatibility with the requirements of high throughput laboratories (Mol *et al.*, 2007; Pinto *et al.*, 2010). Although performed mostly manual there are already

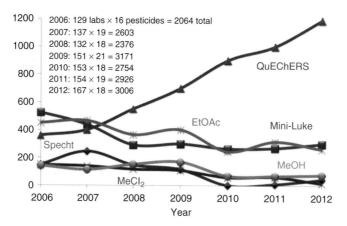

Figure 2.44 Number of results reported in EU PT samples by method. (Lehotay, 2013, adapted from Paula Medina Pastor, EURL-FV.)

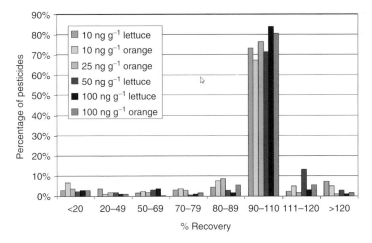

Figure 2.45 Recoveries in the QuEChERS method, 229 pesticides analysed by GC-MS and LC-MS/MS (Lehotay, 2013).

automated instrumental solutions available (Settle, 2010; Teledyne Tekmar, 2013). QuEChERS further expands currently also outside of pesticides analysis. More applications including POPs (persistent organic pollutants) and environmental and veterinary drug residue methods are reported to significantly widen the scope of applications in the near future (Luetjohann *et al.*, 2009; Brondi *et al.*, 2011; Usui *et al.*, 2012). A wide variety of matrices that QuEChERS was applied to are reported, including animal products such as meat, fish, kidney, chicken, or milk, cereals, honey, wines, juices and more.

QuEChERS Stands for

- Quick: Less time is required to process samples, compared to other techniques allowing a higher sample throughput, typically 15 samples can be manually prepared within 1 h.
- Easy: Requires less handling of extracts than other techniques, fewer steps are required.
- Cheap: Less sorbent material is needed for smaller sample sizes.
- Effective: The technique gives high and accurate recovery levels for a wide range of different compound types.
- Rugged: The method can detect a large number of pesticides from different food commodities, including charged and polar pesticides.
- Safe: Unlike other techniques, it does not require the use of chlorinated solvents. The extraction is typically carried out using acetonitrile, which is both GC and LC compatible.

Pesticides Sample Preparation Using QuEChERS:

1) Sample comminution and homogenization (liquid N_2, dry ice)
2) Extraction: Shake $10-15$ g sample with solvent and salts
3) Centrifuge for 1 min
4) Clean-up: Mix a portion with sorbent
5) Centrifuge for 1 min
6) Analyse pesticides via GC-MS/MS and LC-MS/MS

What are current limitations of QuEChERS? Steve Lehotay summarized the status in a recent presentation 'revisiting the advantages of the QuEChERS' (Lehotay, 2013). Although there is one name of the methodology used in common, there are different modified versions in use, partially customized for specific purposes. There are as well distinct differences in the original, and the currently approved AOAC and the EN methods that need to be recognized for their effects on recoveries (see Table 2.20) (European Standard EN15662, 2008). In general, spices, tea, cereals and fatty matrices give problems and require special treatments. Figure 2.46 shows the drop in recovery with increasing fat content of the sample. While polar compounds appear less affected, the apolar lipophilic compounds drop in recovery significantly above 10% fat content. Concerning individual pesticides problems are reported with captan, folpet and captafol.

Another distinct limitation is the high matrix content in QuEChERS extracts. This requires special care on the inlet systems and columns for GC, for the LC ionisation, and raises the requirements for more selective MS analysers, working best with triple quadrupole and accurate mass MS systems. Acetonitrile as solvent

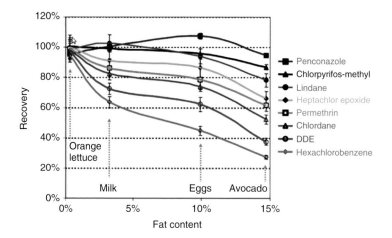

Figure 2.46 Pesticide recoveries versus fat content (Lehotay, 2013).

Table 2.20 QuEChERS methods comparison (Lehotay, 2013).

Original	AOAC 2007.01	CEN 15662
Reference Anastassiades (2003)	Reference Lehotay *et al.* (2005)	Reference Anastassiades (2007)
10–15 g sub sample	10–15 g sub sample	10–15 g sub sample
10–15 mL MeCN	10–15 mL 1% HOAc in MeCN	10–15 mL MeCN
Shake	Shake	Shake
0.4 g/mL anh.MgSO$_4$	0.4 g/mL anh.MgSO$_4$	0.4 g/mL anh.MgSO$_4$
0.1 g/mL NaCl	0.1 g/mL NaOAc	0.1 g/mL NaCl
		0.1 g/mL Na$_3$Cit·2 H$_2$O
		0.05 g/mL Na$_2$Cit·1.5 H$_2$O
Shake	Shake	Shake
Centrifuge	Centrifuge	Centrifuge
150 mg/mL anh.MgSO$_4$	150 mg/mL anh.MgSO$_4$	150 mg/mL anh.MgSO$_4$
25 mg/mL PSA	50 mg/mL PSA	
Shake and centrifuge	Shake and centrifuge	Shake and centrifuge
Option	Option	—
+50 mg C18	+50 mg C18	
+7.5 mg GCB	+2.5–7.5 mg GCB	
Comments	Comments	Comments
• Sodium chloride is used to reduce polar interferences.	• Employs 1% acetic acid in acetonitrile and sodium acetate buffer to protect base sensitive analytes from degradation.	• The European method includes sodium chloride to limit polar interferences and several buffering reagents to preserve base sensitive analytes.
• Provides the cleanest extraction because it uses fewer reagents.	• A USDA study has demonstrated that this method provides superior recovery for pH sensitive compounds when compared to the other two QuEChERS methods.	• Sodium hydroxide used in the citrus step should be avoided as it can add impurities to the extract as well as damage the sorbent used in the clean-up step.
• Does not use acetic acid which may be problematic in GC-MS analysis.		
• Uses dispersive clean-up procedures.	• The approach uses acetic acid in the extraction step. The acetic acid can overload the PSA sorbent used in the clean-up step making it ineffective and possibly causing GC resolution issues.	Sample preparation and extraction: Freeze samples to −20 °C. Homogenize with dry ice until a free flowing powder is formed. The sample is then extracted into solvent. Dispersive or cartridge SPE is used for clean-up.

for GC shows limitations due to the popular use of less polar phases in GC-MS such as the very common apolar 5% phenyl phase. The initial publication was using a mid-polar 35% phenyl-65% dimethyl arylene column phase with higher affinity to acetonitrile as solvent (Anastassiades, 2003). Solvent exchange or simply the dilution of the extracts with a more compatible GC solvent such as ethylacetate is a proven solution (Hetmanski *et al.*, 2010). Large volume GC injectors (SSL-CSR, PTV) allow the application of higher diluted extract volumes compensating for the dilution effect (Hoh and Mastovska, 2007). On the LC-MS side, the high matrix load gives increased rise to matrix effects resulting in potential analyte quenching, which can be addressed by dilution of the extract (Stahnke *et al.*, 2012).

When acetonitrile is poured into the extraction tube containing the homogenized sample, an exothermic reaction occurs between the magnesium sulfate and water. This step may lead to reduced recoveries of the volatile pesticides. To overcome this effect, the sample can be weighed directly into an empty centrifuge tube followed by the addition of acetonitrile. The tube can then be immersed in an ice bath with slow addition of salts.

The QuEChERS extract can be applied directly for LC-MS analysis. Exchanging the acetonitrile solvent for GC-MS analysis is recommended for the use of apolar GC columns. Hexane/acetone or ethylacetate are used instead as more GC column compatible solvents (Thermo Scientific, 2010): To exchange solvents, evaporate a 5 mL aliquot of cleaned up QuEChERS extract almost to dryness under a gentle stream of nitrogen at 40 °C for about 1 h. Immediately remove the tube to prevent over-drying. Add 900 µL aliquot of hexane/acetone (9:1). Cap the tube and vortex it for 15 s. Transfer 1 mL of extract to a 2 mL clean-up tube, cap it tightly and vortex for 30 s. Centrifuge the extract for 5 min at 3000 rpm and then transfer 200 µL of the upper light coloured clear extract to an autosampler vial, preferably with a small glass insert for injection. For a matrix spike calibration, spike the individual calibration levels into each extract for the calibration curve in matrix.

The QuEChERS Reagents and Their Function

Extraction

Magnesium sulfate, anhydrous	Facilitates solvent partitioning and improves recovery of polar analytes.
Acetic acid	Is used to adjust pH, see buffers.
Acetonitrile	Is an organic solvent providing the best characteristics for extracting the broadest range of pesticides with the least number of co-extractables. Amenable for both LC and GC analysis.
Buffers	Prevents degradation of pH sensitive analytes by maintaining optimal pH.
Sodium chloride	Reduces the amount of polar interferences.

Clean-up	
Aminopropyl	Removes sugars and fatty acids. Serves the same function as PSA (primary secondary amine), but is less likely to catalyse degradation of base sensitive analytes. Aminopropyl has a lower capacity for clean-up than PSA.
ChloroFiltr®	A polymeric sorbent for selective removal of chlorophyll from acetonitrile extracts without loss of polar aromatic pesticides. It is designed to replace GCB for the efficient removal of chlorophyll without loss of planar analytes.
C18	Removes long chain fatty compounds, sterols and other non-polar interferences.
GCB	Is a strong sorbent for removing pigments, polyphenols and other polar compounds. Examples of planar (polar aromatic) pesticides which may be removed: chlorothalonil, coumaphos, hexachlorobenzene, thiabendazole, terbufos and quintozene.
Magnesium sulfate, anhydrous	Removes water from organic phase.
PSA	Is used in the removal of sugars and fatty acids, organic acids, lipids and some pigments. When used in combination with C18, additional lipids and sterols can be removed.

(UTC 2011)

2.2
Gas Chromatography

Chromatography, in all its forms, undoubtedly represents the most outstanding development to date in physical chemistry for the separation of molecules. During my lifetime it has revolutionized nearly every field of analytical chemistry.

Denis H. Desty, 1991

2.2.1
Sample Inlet Systems

Much less attention was paid to this area at the time when packed columns were used. On-column injection was the state of the art technique and was in no way a limiting factor for the quality of the chromatographic separation. With the introduction of capillary techniques in the form of glass or fused silica capillaries, high resolution gas chromatography (HRGC) maintained its presence in laboratories and GC and GC-MS made a great technological advance. Many well-known names in chromatography are associated with important contributions to GC sample introduction: Desty, Ettre, Grob, Halasz, Poy, Schomburg and Vogt among others. The exploitation of the high separating capacity of capillary columns now requires perfect control of a problem-orientated sample injection technique.

Sample injection should satisfy the following requirements (Schomburg, 1983, 1987):

- Achieving the optimal efficiency of the column.
- Achieving a high signal/noise (S/N) ratio through peaks which are as steep as possible in order to be certain of the detection and quantitative determination of trace components at sufficient resolution (no band broadening).
- Avoidance of any change in the quantitative composition of the original sample (systematic errors, accuracy).
- Avoidance of statistical errors which are too high for the absolute and relative peak areas (precision).
- Avoidance of thermal and/or catalytic decomposition or chemical reaction of sample components.
- Sample components which cannot be evaporated must not reach the column or must be removed easily (pre-column backflush). Involatile sample components lead to decreases in separating capacity through peak broadening and shortening of the service life of the capillary column.
- In the area of trace analysis, it is necessary to transfer the substances to be analysed to the separating system with as little loss as possible. Here the injection of larger sample volumes (up to 100 µL) is desirable.
- Simple handling, service and preventive maintenance of the sample injection system play an important role in routine applications.
- The possibility of automation of the injection is important, not only for large numbers of samples, but also as automatic injection is superior to manual injection for achieving a low standard deviation.

The sample injection is of fundamental importance for the quality of the chromatographic analysis with all GC and GC-MS systems. Careless injection of a sample extract frequently overlooks the outstanding possibilities of the capillary technique. Poor injection cannot be compensated for even by the choice of the best column material, or the best choice of detectors. The use of a mass spectrometer as a detector can be much more powerful if the chromatography is of the best quality. In fact, the use of GC-MS systems shows very often that as soon as GC is coupled to MS, the chromatography is rapidly degraded to an inlet route. Effort is put too quickly into the optimization of the parameters of the MS detector without exploiting the much wider potential of the GC side. Each GC-MS system is only as good as the chromatography allows!

The starting point for the discussion of sample inlet systems is the target of creating a sample zone at the top of the capillary column for the start of the chromatography which is as narrow as possible. This narrow sample band principally determines the quality of the chromatography as the peak shape at the end of the separation cannot be better (narrower, more symmetrical etc.) than at the beginning. As an explanatory model, chromatography can be described as a chain of distillation plates. The number of separation steps (number of plates) of a column is used by many column producers as a measure of the separating capacity of a column. In this sense, sample injection means the application to the first plate of the column. In capillary GC, the volume of such a plate is less than 0.01 µL. The sample extracts used in trace analysis are generally very dilute, making larger quantities of solvent (>1 µL) necessary. This limited sample capacity of capillary columns shows the importance of the split and splitless sample injection techniques for HRGC. In this context, Victor Pretorius and Wolfgang Bertsch need to be cited:

If the column is described as the heart of chromatography, then sample introduction may, with some justification, be referred to as the *Achilles heel*. It is the least understood and the most confusing aspect of modern GC (Pretorius, 1983).

In practice, different types of injectors are used (see Table 2.18). The sample injection systems are classified as hot or cold according to their function. A separate section is dedicated to direct on-column injection.

2.2.2
Carrier Gas Regulation

The carrier gas pressure regulation of GC injectors found in commercial instrumentation follows two different principles which influence injection modes and the adaptation of automated sample preparation devices. Most of the modern GC instruments are equipped with electronic pressure and flow control units (EPC, electronic pressure control). The pressure sensor control loop must be taken special care of when modifying the carrier gas routes through external devices such as headspace, P&T or thermal desorption units. In these installations, the carrier gas flow is often directed from the GC regulation to the external device and returned again via the capillary transfer line to the GC injector. Also, when applying large volume injection (LVI) techniques, knowledge of the individual pressure regulation scheme is required for successful operation.

2.2.2.1 Forward Pressure Regulation
The classical control of column flow rates is achieved by a 'forward pressure' regulation. The inlet carrier gas flow is regulated by the EPC valve in front of the injector. The pressure sensor is installed close to the injector body and hence to the column head for fast feedback on the regulation either in the carrier gas supply line to the injector, the septum purge or split exit line. The split exit usually is regulated further down the line by a separate flow-regulating unit (see Figure 2.47). An in-line filter cartridge is typically used to prevent the lines from accumulating trace contaminations in the carrier gas, which needs to be exchanged frequently.

The forward injector pressure regulation is the most simple and versatile regulation design used in all split/splitless injection modes as well as the cold on-column injector. It uniquely allows the LVI in the splitless mode using the concurrent solvent recondensation (CSR) and the closure of the septum vent during the splitless phase (Magni, 2003). Also, split and splitless injections for narrow bore columns, as used in Fast GC, benefit from the forward pressure regulation with a stable and very reproducible regulation at low carrier gas flow that is not usually possible with mechanical or electronic mass flow controllers. Also, the switch of injection techniques in one injector, for example, the PTV body between regular split/splitless

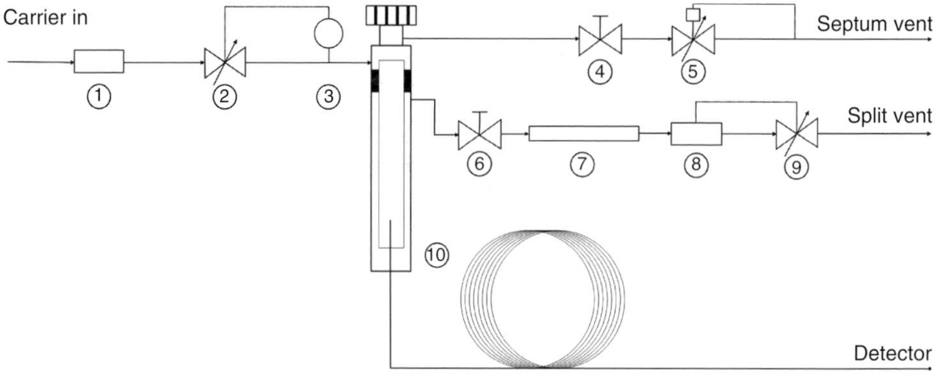

Figure 2.47 Injector forward pressure regulation. 1. Carrier gas inlet filter, 2. proportional control valve, 3. electronic pressure sensor, 4. septum purge on/off valve, 5. septum purge regulator, 6. split on/off valve, 7. split outlet filter cartridge, 8. electronic flow sensor, 9. proportional control valve and 10. injector with column installed.

and on-column injection with a retention gap is possible without any modification of the injector gas flow regulation.

The adaptation of external devices such as headspace, P&T or thermal desorption is straightforward, with the pressure sensor installed close to the injector head which is generally the case with commercial forward-regulated GC instruments. If the pressure sensor is installed in the split line of the injector, it has to be moved close to the inlet to provide a short carrier gas regulation loop. Carrier gas flow and pressure in the external device is controlled by the EPC module of the GC. The position of the pressure sensor in the flow path of an inlet is important for accurate measurement of the inlet pressure to get rapid feedback control.

LVIs are also straightforward using forward pressure-regulated injectors. Typically, the regular split/splitless device can be used for larger volume injections of up to 10 or 50 μL without any hardware modification, exploiting the concurrent solvent recondensation effect. For injections in the range of up to 30 μL, only the injection volume programming of the autosampler and the first oven isothermal phase have to be adjusted accordingly letting elute the wider solvent peak before ramping the GC oven.

2.2.2.2 Back Pressure Regulation

The term *back pressure* describes the position of the EPC valve behind the injector in the split exit line, in combination with a mass flow controller in front of the injector (see Figure 2.48). In this widely used carrier gas regulation scheme for split/splitless injectors, the pressure sensor is typically found in the septum purge line close to the injector body to ensure a pressure measurement close to the column head. A filter cartridge may need to be used to protect the regulation unit from any carrier gas contamination.

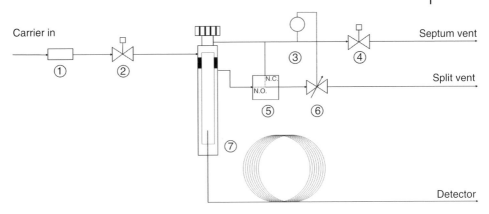

Figure 2.48 Injector back pressure regulation. 1. Carrier gas inlet filter, 2. mass flow regulator, 3. electronic pressure sensor, 4. septum purge regulator, 5. solenoid valve, 6. electronic pressure control valve and 7. injector with column installed.

The total carrier gas flow to the back pressure-regulated injector is set by the mass flow controller and is typically held constant independently of the column head pressure. The total flow into the injector is split into three flow paths the split, septum purge and column flow. The column flow is then adjusted and kept constant by varying the split exit flow according to the required column flow conditions. This allows the exact and independent setting of split flows from column flows. However, when the back pressure regulated inlet is used in split mode, a large amount of sample passes through the regulation in the split exit line. This mode of operation requires special care using a filter cartridge with frequent exchange to prevent the system from flow restrictions, chemical attack and finally clogging.

When adapting an external inlet device, the outlet of the mass flow controller is directed on the shortest route to the external device. A sample transfer line from the external device is returned to the injector either using the regular carrier inlet or by piercing a needle through the regular septum. The pressure sensor must be relocated from the split line close to the EPC valve to achieve correct flows and sufficient pressure control stability within a short feedback loop due to the additional flow restrictions in the external device. The carrier gas flow through the external device is now regulated by the mass flow controller. The GC inlet pressure at the injector and the effective column flow is regulated independently by the EPC module through the EPC valve.

Alternatively, an external auxiliary EPC module for feeding the external device is recommended.

2.2.2.3 Carrier Gas Saving
The supply for helium carrier gas for GC-MS became increasingly difficult and expensive in many regions. Saving helium carrier gas is especially important with

the expected further increasing cost for the near future. Several options to save or replace helium as carrier gas have been discussed for years. While hydrogen is the classical carrier gas in GC, there are limitations for use with GC-MS due to its reactivity with analyte hydrogenations and technical requirements for appropriate vacuum systems to cope with the increased flow rates. The replacement by nitrogen for chromatography is not an option due to the reduced separation power. Switching to nitrogen can be an option during the standby operation, but a considerable time needs to be calculated for flushing the GC system before operation with helium can start again. Two practical options have proven to be practically useful to replace or significantly reduce the helium carrier gas consumption.

Most of the modern GC systems are equipped with a device for electronic carrier gas control (EPC). The EPC function allows the convenient setting of the optimum gas flow parameters during the injection process, but also allows a different setting during the separation phase after injection. This applies in particular to the split flow. The split flows are typically high during a split injection or directly after a splitless injection phase in the range of 30–100 mL/min. The split flow consumes most of the carrier gas, but without any analytical function after injection other than keeping the injector clean for the next injection. After completing the injection phase, it is not necessary to maintain such high split flows, and it is advisable to save carrier gas.

A carrier gas saver function of the EPC can set split flow to a lower value of 10–20 mL/min depending on the injector design. The low split flow is activated about 2–3 min after the injection at the start of the oven temperature program and remains during the complete development of the chromatogram. This still keeps the injector clean from any potential deposit and saves a significant amount of carrier gas throughout the much longer chromatography time.

As the split flow with up to 100 mL/min, is the major load of a GC system compared to the typical column flow range of 1–2 mL/min, there is a consequent solution available which replaces helium completely during the injection process, and maintains the low consuming column separation flow with helium. This helium saver injector requires a different design with two separate gas inlets and electronic gas controls for helium carrier and nitrogen as auxiliary gas (patent Thermo Fisher Scientific). Figure 2.49 demonstrates the basic operation. A diffusion tube at the bottom of the injector body feeds the helium carrier to the analytical column while separating the carrier gas from the auxiliary nitrogen gas. The injection and sample vaporization processes is performed in nitrogen atmosphere, also serving the septum purge and split lines. All gas flows are EPC controlled to move the sample vapour cloud into the analytical column in either split or splitless operation. The column separation is maintained as usual with helium as carrier gas resulting in a significant saving in helium carrier gas. No modifications to the chromatographic methods are required.

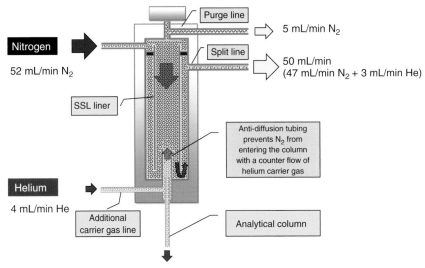

Figure 2.49 Helium saver injector operation principle. (Thermo Fisher Scientific).

2.2.3
Injection Port Septa

The injection port septum is an unsolved weak point in GC. It is arguably the most important but often overlooked functional part of the injector. The septum is used to seal the injection port, and withstands the inlet pressure and provides smooth and repeated access for the syringe needle to inject the sample. The syringe needle punches the septum repeatedly in a small area causing only gradual small leaks, which are detectable with an electronic leak detector already before a noticeable analyte response drop occurs. The risk of leaks increases with use, older septa take longer to re-seal than new ones (Hinshaw, 2008).

The septum material is based on silicon rubber, made of PDMS, a similar polymer that is used for the stationary phases of the fused silica columns. The ideal injection port septum should exhibit low bleed, resist leaks, is easy to penetrate for the syringe needle and re-seal itself after the injection, reliably for a large number of analyses. The septum performance is vital to the reproducibility of results, sensitivity, peak shape and constant flow conditions within the chromatographic system. It is widely accepted that silicon septa degrade relative to the operating temperature of the injector, the operation at higher pressures and the physical interaction with the syringe needle, especially with modern fast injection autosamplers. This mechanical stress causes shedding of particles from the septum but also abrasion from the outside of the metal needle which can accumulate in the inlet liner (de Zeeuw, 2013). A difference in septa materials deposited in the insert liner may cause differences in what types of compounds become reactive in the GC inlet.

Septum bleed can become a serious problem during trace analysis, as the leached compounds elute in the middle of the GC run with high intensity. Persistent bleed causes low sensitivity and poor quantification. Once a septum is punctured, a small amount of silicon can be transferred into the injector with each injection. This is quickly transferred to the top of the column, and if the column is at a low temperature, focused here. Once the column is heated, the bleed components elute. The longer the column is held at a low temperature, the more intense the presence of bleed peaks in the chromatogram. The major component of septum bleed is a series of siloxanes with increasing molecular weight (see Table 3.16 and Figure 3.168). Other observed bleed components are phthalates and also hydrocarbons (Warden, 2007).

Hot split/splitless injectors of different manufacturers can have very different temperatures at the septum base. For a given inlet temperature set point, the temperature at the septum is typically lower than the set point, but varies among instruments. Inlets with a larger gradient over the injector length and a cooler septum base typically experience fewer problems with septa sticking, but may be prone to increased high boiler contamination. Operators of instruments with inlets which are more evenly heated with a smaller temperature gradient and a hotter septum base need to consider using septa that are rated for the highest operation temperature and setting the inlet at the lowest permissible temperature (Grossman, 2013). Septa brands, however, specify a single, maximum operating temperature which corresponds to the inlet set point, not the actual septum base temperature the septum can withstand and still function properly. All commercial septa today withstand a nominal injector temperature of 350 °C without melting or hardening. Special care is advised when installing a new septum to avoid a typical inherent overtightening, as the silicon septa tend to swell significantly with increasing temperature.

Bleed and temperature optimized septa (BTO™, Chromatography Research Supplies, Inc., Louisville, KY, USA) are well suited for trace GC and GC-MS applications which require low bleed and extended sealing capabilities for high injection numbers. As an additional benefit, many septa provide a pre-drilled needle guide for increased septum lifetime, which makes them ideal for the use with liquid autosamplers as well as with SPME. BTO septa are specified for use with up to 400 °C inlet temperature, are preconditioned and require very little additional conditioning before use. It is recommended to change septa at the end of a day, so the septa can condition overnight to ensure the lowest bleed for the next day analyses.

All GC septa may adsorb contaminants when handled without gloves or from volatile organics present in the air. Any such contaminants picked up by a septum may elute during analysis and appear as septa bleed or additional ghost peaks. To avoid contamination, septa generally need to be stored in their original glass containers with the lid tightly closed. Forceps are recommended for handling septa, and if the septa need to be handled with fingers, then only the edge should be touched and not the sealing surface.

The reported average lifetime in terms of the number of injections ranges from 150 to 450 injections for silicon rubber septa and up to 2000 injections for the Merlin seal, depending on the septum type and conditions used (Westland, 2012). Due to the necessary frequent exchange of the inlet liner, it is recommended to replace both, the septum and insert liner regularly on a preventive maintenance schedule.

2.2.3.1 Septum Purge

All modern GC instruments offer the injector function called *septum purge*, introduced 1972 by Kurt Grob. A constant flow in the range of 0.5–5 mL/min carrier gas to purge any volatiles from the bottom of the septum has proven to be reliable. This small carrier gas flow prevents leaching contaminations of the septum material from entering into the insert liner. It purges the so-called septum bleed to a dedicated septum purge exit line (see Figures 2.58 and 2.65).

The septum purge flow is recommended to stay on as well during the injection phase. With properly chosen injection conditions keeping the solvent/sample vapour cloud inside of the insert liner, there will be no loss of sample analytes via the septum purge outlet (see Section 2.2.5.1 Hot Sample Injection).

2.2.3.2 The MicroSeal Septum

The Merlin MicroSeal™ Septum is a unique long-life replacement for the conventional silicon septa on split/splitless and PTV injectors. MicroSeal is a trademark of the Merlin Instrument Company, Half Moon Bay, CA, USA. Functionally, the MicroSeal provides a micro valve two-step sealing mechanism (see Figure 2.50):

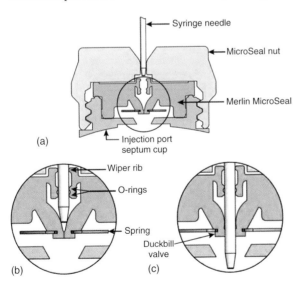

Figure 2.50 MicroSeal injection sequence – a 2-step injection process. (a) MicroSeal installed under regular injector cap, (b) the needle is first penetrating the dust wiper rib and (c) the needle is penetrating the spring loaded duck bill seal into the injector liner (Merlin Instrument Company).

A double O-ring type and a top wiper rib around the syringe needle improve resistance to particulate contamination, and a spring-assisted duckbill seals the injection port. The MicroSeal can be easily installed on many injector types by removing the existing septa nut without any modifications. It can be used at injection port temperatures up to 400 °C. It is commercially available for operation up to 45 psi as low pressure MicroSeal or up to 100 psi with the high pressure series. A special design is available also for SPME injections (Merlin, 2012).

The seal is made from Viton material, a high temperature-resistant fluorocarbon elastomer also providing high resistance to wear. It greatly reduces shedding of septum particles into the injection port liner. Because the syringe needle does not pierce a septum layer, it is eliminating a major source of silicon septum bleed and ghost peaks. Due to this sealing mechanism, it is especially suited for trace analysis applications requiring particular low background conditions (Figure 2.51). The low syringe insertion force makes it also ideal for manual injections.

The MicroSeal is usable in either manual or autosampler applications with 0.63 mm diameter (23 gauge) blunt cone syringes (see Figure 2.52). The 23/26

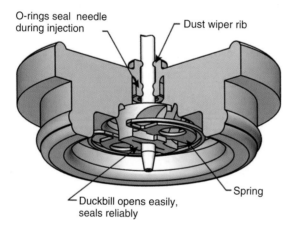

Figure 2.51 MicroSeal cutaway view (Merlin Instrument Company).

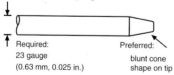

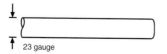

Figure 2.52 MicroSeal compatible needle and SPME probe styles.

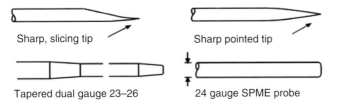

Sharp, slicing tip

Sharp pointed tip

Tapered dual gauge 23–26

24 gauge SPME probe

Figure 2.53 Do not use these needle and probe styles with the MicroSeal. Note: Sharp-edged or sharp-pointed needles can pierce and damage the duckbill seal.

dual gauge tapered needles and 24 gauge SPME probes will not seal properly in the MicroSeal, (see Figure 2.53). The longer lifetime for many thousands of injections reduces the chances for septum leaks especially during extended autosampler runs.

2.2.4
Injection Port Liner

The inlet liners of the injection port are the point of sample evaporation and determine to a high degree the GC performance in terms of sensitivity, reproducibility and sample integrity. Inlet liners are retaining nonvolatile matrix components and protect the analytical column from performance degradation. They are consumables necessary to be replaced on a regular basis depending on the sample matrix load on a daily or weekly basis according to a preventive maintenance plan.

The liner choice need to fit the analytical method setup if split or splitless injections are planned, or the basic injection techniques hot needle with thermo spray or the cold needle with liquid band formation are preferred. Different liner geometries are available to support the individual injection and vaporization process. There is not a single liner that serves all injection modes. So, a few basic rules help identifying the liner of choice for the chosen injection technique.

2.2.4.1 Split Injection
Split injections are used to perform a sample 'dilution' in the injector with carrier gas for a reduction of the amount of sample which enters the column. For concentrated (undiluted) samples, this is necessary due to the limited sample capacity of fused silica columns. The amount of sample wasted via the split line relative to the column carrier gas flow is adjusted by the split flow rate. Split liners provide high flow speeds inside the liner for a short sample band at the beginning of the column. Room is required between the needle exit and column entry for the evaporation of solvent and analytes, and the homogeneous mixing with the carrier gas before the sample cloud 'flies' across the column entry. Straight liners with and without glass wool are used for split injections (Schomburg *et al.*, 1977). The inside diameter should be low to guarantee high-split flows through the injector body out to the split line (see Figure 2.55).

The residence times of samples in the liner are short compared to a splitless injection. With low split ratios labile compounds can also be analysed with good results, reducing the residence time and degradation in the hot liner significantly.

2.2.4.2 Splitless Injection

Splitless injections are typical for trace analysis with a total sample transfer into the analytical column. Inlet liners for splitless injection need to provide room for the solvent cloud to expand during evaporation (see Table 2.21), and typically, liners with larger inner diameters are used (see Figure 2.55) (Hinshaw, 1992). The knowledge of liner volume and solvent vapour volume is essential for the optimum parameter setting. Typically, the 'goose neck' liners with a tapered bottom end are preferred for splitless injections preventing analytes getting in touch with metal surfaces at the bottom of the liner with potential memory and degradation. Overfilling of the liner need to be avoided by proper selection of the solvent and the maximum tolerated 'at once' injection volume.

A known disadvantage of goose neck liners can be caused by septum particles, which can get in contact with the sample vapours at the bottom of the tapered liner. In this case, the use of straight liners for splitless injections can be advantageous as septum particles accumulate at the base of the injector and do less interfere with the sample.

The transfer time of analytes from injection and evaporation into the analytical column is the residence time of the solvent/analyte cloud in the inlet liner. The residence time depends on the flow conditions within the inlet. The sample vapours are diluted exponentially as they are transferred into the analytical column increasing the residence time for a certain part of the analytes. Labile compounds require a short residence time in the liner for reduced surface contacts.

Using a surge pressure (pressure pulse) during the transfer phase with a two to three times higher pressure than the regular column flow head pressure allows emptying the liner more quickly with shorter splitless times. The surge pressure needs to be released through the analytical column after switch off for about 2 min before the split valve can be opened to avoid a back stream and loss of sample from the analytical column.

2.2.4.3 Liner Activity and Deactivation

The activity of inlet liners can be a number one limitation for the analytical performance and is still a constant challenge in GC. Liner activity can be caused by the material itself, an insufficient or degrading deactivation, but is also built up during regular use by the deposit of nonvolatile matrix components. The liner activity changes with the first matrix sample injected. While fulfilling a major task of preventing low or nonvolatile matrix from entering the analytical column, the matrix deposit in the liner is causing adsorption and decomposition effects for subsequent injections. A quality control and preventive maintenance plan is required with regular liner exchange according to the sample matrix burden. The same applies for glass wool. A glass wool plug can become rapidly active and needs to be replaced or even removed with the inlet liner for the analysis of sensitive compounds.

Significant progress is noticeable in recent years in reducing the liner activity by chemically modifying the glass surface or applying high temperature coatings. Commercially deactivated liners are available from different manufacturers using proprietary processes, for instance, with phenylmethyl surface deactivation. These treatments offer a highly efficient deactivation with stability even to high temperature ranges above 400 °C, and cover a wide polarity range. While there is the expectation for a general-purpose liner deactivation, there will be compromises. Specific deactivations are available, for example, for basic compounds such as amines. The same deactivation procedures are used for the widely used borosilicate glass liner and wool. It is recommended to test different liners for a specific application for inertness, recovery and robustness, done with a typical matrix and analyte concentrations close to the method detection limits (MDLs).

The liner activity is usually measured and compared by using injections of endrin and dichloro-diphenyl-trichloroethane (DDT). The breakdown products are endrin ketone and endrin aldehyde, respectively dichloro-diphenyl-dichloroethane (DDD). With breakdown percentages less than 15% each, an insert liner is accepted to be inert (e.g. EPA method 8081b).

Glass wool in the liner, although required for the fast cold needle 'liquid band' injection (see 2.2.5.2), can be a root cause of inlet liner activity. Approximately 50 mg of deactivated glass wool is typically used in a liner providing a large surface for interaction with sensitive analytes. There are different types of glass wool: regular or base-deactivated borosilicate glass wool and also deactivated fused silica quartz wool with low alkali content. If the glass wool in the liner becomes the major source of activity by decomposition or analyte absorption due to insufficient deactivation, the only remedy is to change liners more often, or work, if possible, using baffled liners without the wool.

Straight liners can be mechanically cleaned with solvents, sonication or a soft brush (e.g. a pipe cleaner), and reused, but require a new deactivation. Do not scratch or use acids to leach the liner surface; this will increase the liner activity significantly. Several liner deactivation methods are reported in the literature typically offering a gas phase or coated deactivation (Figure 2.54). For a simple and efficient 'in-house' coated deactivation, the treatment with Surfasil™ or Aquasil™ has proven to be very efficient and durable, and is used and recommended, for instance, for the multi-residue analysis of pesticides. The immersion procedure for the precleaned liners is easy to perform with only a few steps (Thermo, 2008). Liners with a glass wool plug can also be deactivated using the SurfaSil or AquaSil procedure. Several liners can be treated at once, and stored for later use:

1) Dilute the SurfaSil siliconizing fluid in a non-polar organic solvent such as acetone, toluene, carbon tetrachloride, methylene chloride, chloroform, xylene or hexane. Typical working concentrations are 1–10% mass to volume.

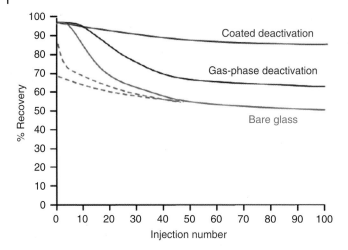

Figure 2.54 Injection port liner activity dependence from the injection number (Klee, 2013)

2) Completely immerse or flood the dry inlet liner to be coated in the diluted SurfaSil solution for at least 5–10 s. Agitate the solution to ensure a uniform coat. A thin film will immediately coat the liner's surface.
3) Rinse the liner with the same solvent in which the reagent was diluted.
4) Rinse the liner with methanol. This rinse is required to prevent interaction of the SurfaSil coating with water and thus, reversing siliconization.
5) Air-dry the liner for 24 h or heat at 100 °C for 20–60 min.

Other inlet liner deactivation procedures are using gas phase or solution silylation procedures. The gas phase silylation provides the more stable liner deactivation for high temperature use than a liquid process, but is a more laborious procedure (Rood, 2007):

1) Place the liner in a glass test tube that can be easily flame sealed.
2) Heat the neck of the tube until a 2–3 mm opening remains.
3) Add two to three drops of diphenyltetramethyldisilazane.
4) Immediately flush the tube with a stream of dry nitrogen or argon.
5) Immediately flame seal the glass tube.
6) Heat the test tube at 300 °C for 3 h.
7) Allow to cool to room temperature.
8) Open the tube and rinse the liner with pentane or hexane.
9) Dry the liner at 75–100 °C

2.2.4.4 Liner Geometry

Split liners are typically narrow and straight tubes, while splitless liners are wider and typically show a tapered end, see Figure 2.55. These bottom tapers keep the sample cloud in the centre of the liner for an optimized transfer into the column. Tapers prevent a possible exposure of sample components to the metal (gold) seal

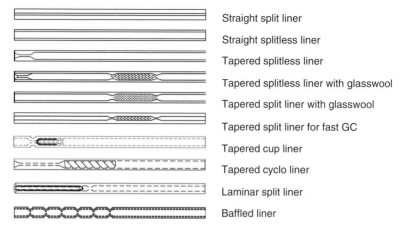

Figure 2.55 Injection port liner types for split and splitless injections.

at the bottom of the inlet body. With tapered liners, the tip of the column is best positioned just above of the tapered section. A thin glass or metal rod pushed upwards in the liner by the column during installation gives valuable assistance in finding the correct position of the column tip.

Other liner types introduce mechanical obstacles for improved vaporization and mixing with carrier gas such as the baffled, cup or cyclo liners. Although those inlet liners offer improved injection conditions, they can hardly be cleaned and reused; so they are excluded from routine use with matrix samples.

Special liner types are used for headspace applications. In many cases, the transfer line of a headspace sampler ends in a syringe type needle. Additional liner dead volume and turbulence must be avoided for the transfer of the sample plug into the analytical column. Special straight 'headspace liners' with narrow diameters are available for many GC models.

Direct on-column injections can also be done using a regular syringe using a temperature programmable injector (PTV). This requires a special inlet liner type, which centres a 0.53 mm ID pre-column in the liner, while the top part of the liner serves as a needle guide. Regular needle diameters of 0.45 mm OD are used. This allows the syringe needle to enter, centred and deep enough into the pre-column for a direct liquid injection into the pre-column. By this way on-column injections can be automated using regular GC autosampler units.

2.2.5
Vaporizing Sample Injection Techniques

In case liquid samples are applied to GC or GC-MS, the most widely used injection technique evaporates the liquid sample in the inlet liner of the injector in order to transfer the analytes into the analytical column. Classical injection techniques involve applying the sample solution in constantly heated injectors. Both the solvent and the dissolved analytes evaporate in an evaporation tube specially

Peak area normalised
for C_8 (= 100%)

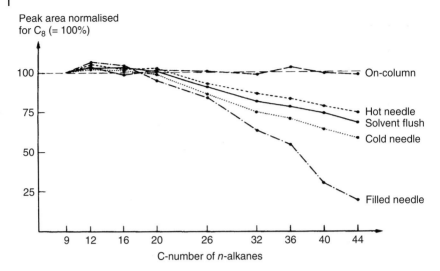

Figure 2.56 Discrimination among alkanes using different injection techniques (Grob, 2001). The peak areas are normalized to $C_8 = 100$%.

fitted for the purpose (inlet liner) and the sample vapor is mixed with the carrier gas. Injector temperatures of ~200 °C to above 300 °C are typically applied for evaporation.

The transfer can only be partial for concentrated extracts (split mode), or a total transfer of the sample into the column for trace analysis is performed (splitless mode). Both injection methods require a different parameter setting, choice of inlet liners and oven program start temperature to achieve the optimum performance. Also the possible injection volumes need to be considered. The operating procedures of split injection and total sample transfer (splitless) differ according to whether there is partial or complete transfer of the solvent/sample on to the column.

The problem of discrimination of high boiling analytes on injection into hot injectors arises with the question of what is the best injection technique. Figure 2.56 shows the effects of various injection techniques on the discrimination between various alkanes. The reference technique is the direct cold on-column injection avoiding the evaporation steps. While in the boiling range similar to a hydrocarbon chain length of up to about C_{16}, hardly any differences are observed; discrimination can be avoided for higher boiling compounds by a suitable choice of the injection technique. Two vaporizing injection modes can be distinguished: The *hot needle thermo spray injection* and the *cold needle liquid band injection*, both with particular analytical advantages and limitations.

2.2.5.1 Hot Needle Thermo Spray Injection Technique

The hot needle injection can be used with hot SSL injectors both in split and splitless modes, using an empty inlet liner. Common to the hot needle techniques is that the sample plug is pulled up into the barrel of the syringe. A short delay of 3−5 s is used after the syringe needle entered the injector for heating up the

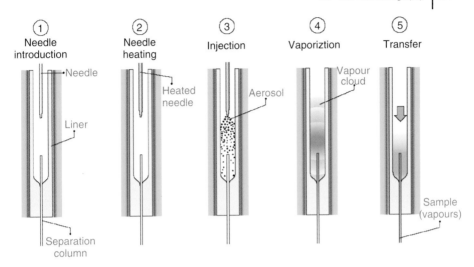

Figure 2.57 Hot needle thermo spray injection technique (in splitless mode).

complete needle length in the hot injector. The injection itself is then carried out quickly pushing the plunger rapidly down and removing the needle immediately (see Figure 2.57). This injection technique is preferred for all manual injections, and can also be used for autosampler injections. The thermo spray technique requires a hot injector head for an even heating of the syringe needle.

The important benefit of the thermo spray injection is the evaporation of substances in a cloud of solvent vapour. Due to the hot needle, a spray of droplets and vapour leaves the needle. Wall contacts of the analyte to the insert liner are minimized protecting polar and sensitive compounds from adsorption to the glass surface with potential degradation reactions. The thermo spray injection is the most gentle injection technique for labile and reactive compounds (Grob, 2001). The sample volume needs to be adjusted considering the liner dimensions that the generated vapour cloud stays safely inside the inlet liner and overfilling is avoided.

Solvent Flush Hot Needle Injection
This is the injection procedure of choice for hot injectors and provides lowest discrimination of high boiling compounds. Approximately $0.5 - 1\,\mu L$ solvent or a derivatizing agent are first taken up in the syringe, then a small plug of $0.5\,\mu L$ air and finally the sample in the desired volume. The procedure is also known as *sandwich injection* technique. Before the injection the liquid sample plug is drawn up into the body of the syringe. The volumes can thus be controlled better on the scale for a manual injection. The injection involves first inserting the needle into the injector, waiting for the needle to heat up for $3-5\,s$ and then rapidly injecting the sample. The solvent flush hot needle injection can be carried out very well manually or by using a programmable autosampler.

Understanding the relationships between inlet liner geometry, carrier gas flow, injection volume and solvent vapour volume is essential for optimizing splitless injections. Especially the expansion volume of the solvent used must be taken into account. The maximum liquid injection volume needs to be adjusted in the hot

Table 2.21 Typical solvent expansion volumes.

Injection volume (μL)	H$_2$O (μL)	CS$_2$ (μL)	CH$_2$Cl$_2$ (μL)	Hexane (μL)	Iso-Octane (μL)
0.1	142	42	40	20	16
0.5	710	212	200	98	78
1.0	1420	423	401	195	155
2.0	2840	846	802	390	310
3.0	4260	1270	1200	585	465
4.0	5680	1690	1600	780	620
5.0	7100	2120	2000	975	775

The expansion volumes given here refer to an injection temperature of 250 °C and a column pressure of 0.7 bar (10 psi). The values for other temperatures and pressures can be calculated according to $V_{Exp.} = 1/P \cdot n \cdot R \cdot T$.

injection mode to the capacity of the installed inlet liner (volume) which can vary between 0.3 and 2 mL depending on its type. Table 2.21 provides the expansion volumes of typical GC solvents for often used injection volumes at standard inlet conditions. For proper operation the solvent expansion volume has to stay safely within the inlet liner capacity.

To determine the time required to flush the contents of the injector completely on to the column at least twice the insert volume is applied, corresponding to a yield of 90 - 95 %. During transfer of the sample and solvent vapor cloud into the column the re-condensation on the column head exploiting the *solvent effect* is required. The minimum isothermal time of the oven temperature to stay below the effective solvent boiling is calculated from twice the inlet liner volume by the set carrier gas flow rate, see also Table 2.22.

2.2.5.2 Cold Needle Liquid Band Injection Technique

The cold needle injection technique avoids the heating of the syringe needle in the injector. This prevents potential discrimination caused by the evaporation of the sample in the needle. A band of liquid enters the inlet liner by a rapid injection process. This requires special precaution to prevent the liquid to reach the bottom of the liner, and finally rinsing to the bottom of the injector getting lost for injection. The solution for a liquid band injection is usually a plug of deactivated fused silica glass wool at the bottom of the liner. Other absorbent materials, even solid materials like Chromosorb™, can also be used as well. Typically a plug of 1-2 cm of deactivated fused silica quartz wool is used. The position of the plug depends on the split mode used. A high position of the plug for split injections allows enough mixing with the carrier gas, while a low position of the glass wool plug in splitless operation allows a short distance to the column entry with minimum wall contacts for the evaporating analytes. Using the glass wool plug at the top of the liner allows the syringe needle tip to penetrate the glass wool plug during injection for extracts with high load of non-volatile matrix material. The injected solvent is kept by the adsorbent plug without rinsing down the liner. Discrimination effects are

Table 2.22 Effective solvent boiling points for different inlet pressures in °C, sorted by BP.

Solvent	Ambient	100 kPa	200 kPa	300 kPa	400 kPa	500 kPa
iso-Pentane	28	49	65	77	87	98
Diethylether	35	54	72	84	93	101
n-Pentane	35	57	72	84	93	102
Dichloromethane	40	60	75	87	96	105
MTBE	55	72	91	104	118	124
Acetone	56	77	90	100	109	116
Methylacetate	57	72	91	104	115	124
Chloroform	61	81	97	109	121	130
Methanol	65	82	97	107	116	124
n-Hexane	69	95	111	124	136	145
Ethylacetate	77	97	114	126	136	145
Cyclohexane	81	106	122	137	149	159
Acetonitrile	82	105	120	131	140	148
Toluene	110	134	149	161	170	178

minimized as the samples reach the glass wool as aspirated from the sample vial. Evaporation starts from the glass wool plug, but cannot be supported by the low thermal mass glass wool alone. Evaporating solvent from the adsorbent leads to a local temperature decrease keeping the analytes focussed in a small region for the transfer to the column. As the glass wool plug with the sample extract cools down during the evaporation phase, it allows a gentle evaporation process, thereby preventing a sudden flash evaporation of the complete sample with the fatal risk of overfilling the liner volume.

The liquid band injection requires fast autosamplers completing the injection phase within at best 100 ms only, preventing the needle to become hot. It cannot be performed manually. The needle penetration depth into the injector liner is only very short, just penetrating the septum and remaining with the needle tip at the liner entry for injection. Analytically, the liquid band injection is typically used with sample extracts carrying a high matrix load. It provides the highest integrity of the injected sample, preventing matrix deposit and potential discrimination during evaporation from a hot syringe needle. A known limitation is caused by a potentially increasing activity of the glass wool losing its deactivation after a number of injections, and by the accumulating deposit of unevaporated matrix in the glass wool plug. A more frequent liner exchange becomes necessary.

2.2.5.3 Filled Needle Injections

This injection procedure is no longer up-to-date and should be avoided with hot vaporizing injectors. It is associated with certain types of syringes which, on measuring out volumes below 1 μL of the sample extracts, can only allow the liquid plug into the injection needle. High discrimination of high boiling components is observed, see Figure 2.56.

2.2.5.4 Split Injection

After evaporation of the liquid sample in the insert, with the split technique, the sample/carrier gas stream is divided. The larger, variable portion leaves the injector via the split exit and the smaller portion passes on to the column. The split ratios can be adapted within wide limits to the sample concentration, the sensitivity of the detector and the capacity of the capillary column used. Typical split ratios are mostly used in the range 1 : 10 to 1 : 100 or more. The start values for the temperature program of a GC oven are independent of the injection procedure using the split technique. If the oven temperature is kept below the boiling point of the solvent at a given pressure, the reconcentration of the solvent into the column needs to be considered when calculating the split ratio. In this case, more sample enters into the column and consequently, the split ratio is not as calculated by the measured flow ratios. For this reason, it is good practice to keep the column temperature sufficiently high to avoid any recondensation during the split injection, if the split ratio is important.

For concentrated samples, the variation of the split ratio and the volume applied represents the simplest method of matching the quantity of substance to the column load and to the linearity of the detector. This is of great practical benefit in all quality control applications. Even in residue analysis, the split technique is not unimportant. By increasing the split stream, the carrier gas velocity in the injector is increased and this allows the highly accelerated transport of the sample cloud past the orifice of the column. This permits a very narrow sample zone to be applied to the column. To optimize the process, the possibility of a split injection at a split ratio of less than 1 : 10 should also be considered. In particular, on coupling with static headspace or P&T techniques, better peak profiles and shorter analysis times are achieved. The smallest split ratio that can be used depends on the internal volume of the insert. For liquid injection in hot split mode in general a high carrier gas flow in the liner is required, but also a sufficient evaporation and mixing with carrier gas before the sample cloud passes the column inlet as the split point. Narrow diameter liner increase the flow speed for sharp injection bands, wider liners give a reduced carrier gas speed best suitable for wide bore columns. Too small diameter liners cause insufficient evaporation especially with the higher flow of widebore columns and can induce a partial splitting with non-repeatable data. Liners with wool, frit or other flow 'obstacles' like baffles or cups improve the mixing process, see Figure 2.55.

A disadvantage of the split injection technique is the uncontrollable discrimination with regard to the sample composition. This applies particularly to samples with a wide boiling point range. Quantitation using external standardisation is particularly badly affected by this. Because of the deviation of the effective split ratio from that set up, this value should not be used in the calculation. Quantitation with an internal standard or alternatively the standard addition procedure should be used.

2.2.5.5 Splitless Injection (Total Sample Transfer)

With the total sample transfer technique, the sample is injected into the hot injector with the split valve closed (Figure 2.58). The volume of the injector insert must

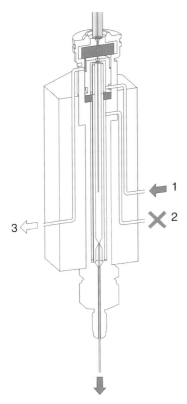

Figure 2.58 Hot split/splitless injector with a single tapered splitless liner and analytical column installed (Thermo Fisher Scientific; Milan). 1. Carrier gas inlet, 2. split exit line and 3. septum purge line.

be able to hold the solvent/sample vapour cloud completely. Because of this, special insert liners (vaporizer) are recommended for splitless operations. Depending on the insert used and the solvent, there is a maximum injection volume which allows the vapour to be held in the insert. Inserts that are too small, on explosive evaporation of the solvent, lead to the expansion of the sample vapour beyond the geometry of the inserts into the cold regions of the injector, and cause a probable loss by the septum purge as well as memory effects. Pressure waves of subsequent injections bring back deposited material as carry over from the split line into the next analysis. Inserts that are too wide lead to significant dilution of the sample cloud and thus to prolonged transfer times and losses through diffusion. A recent compromise made by many manufacturers involves insert volumes of 1 mL for injection volumes of about $1-2\,\mu L$. Splitless injections require an optimized needle length as stated by the injector manufacturer that defines the point of evaporation in the center of the liner. The septum purge should not be closed even with a splitless injection. With many commercial SSL injectors the closure of the septum purge is optional and can be used to favour the auto-pressure surge. For application of surge pressures during injection, the septum purge will be closed.

With the correct choice of insert, the sample cloud does not reach the septum so that the low purge flow does not have any effect on the injection itself.

The carrier gas flushes the sample cloud continuously from the inlet liner on to the column. This process generally takes about 30–90 s for complete transfer depending on the insert volume, sample volume and carrier gas flow. There is an exponential decrease in concentration caused by mixing and dilution with the carrier gas. Longer transfer times are generally not advisable because the sample band becomes broadened in the column. Ideal transfer times allow three to five times the volume of the insert to be transferred to the column. This process is favoured by high-carrier gas flow rates and an increased head pressure during the transfer process (surge pressure). Hydrogen is preferred to helium as carrier gas with respect to sample injection. For the same reason, a column diameter of 0.32 mm is preferred for the splitless technique compared to narrower diameters, in order to be able to use optimal flow rates in the injector. In the same way, a pressure surge step (pressure pulse) is available in modern GC instrumentation using electronic pressure regulation to improve the sample transfer.

For hot injectors capillary columns with an internal diameter of 0.32 mm are more suitable, as the transfer of the sample from the inlet liner to the column takes place more rapidly with less diffusion. However, on coupling with MS detectors, the higher column flow rate can exceed the maximum compatible flow rate in some quadrupole MS systems, and need to be checked. Columns with an internal diameter of 0.25 mm are the popular alternative and should be used in hot injectors with 2 mm internal diameter inserts and low injection volumes to obtain optimal results. If inserts with a wide internal diameter of 4 mm are used for larger injection volumes, the solvent effect must be particularly exploited for focusing (Figure 2.67).

As the transfer times are long compared with the split injection and lead to a considerable distribution of the sample cloud on the column, the resulting band broadening must be counteracted by a suitable temperature of the column oven. Splitless injection requires the oven temperature at injection below the boiling point of the solvent for sufficient refocussing of the sample (Grob and Grob, 1978a). This is achieved by the so-called *solvent effect*. A rule of thumb is that the solvent effect operates best at an oven temperature of 10–15 °C below the effective boiling point of the solvent at the set inlet pressure. A table of the effective boiling points at different inlet pressures for solvents typically used in GC is given in Table 2.22. The effective boiling point at a certain pressure depends on the specific heat of evaporation of the solvent and can be calculated for solvents using the Clausius–Clapeyron relation.

The solvent condensing on the column walls acts temporarily as an auxiliary phase, accelerates the transfer of the sample cloud on to the column by volume contraction due to re-condensation. The solvent phase holds the sample components and focusses them at the beginning of the column with increasing evaporation of solvent into the carrier gas stream (solvent peak). Here the use of 'GC compatible solvents' that are miscible with the stationary phase becomes important. 'Incompatible' solvents do not generate a suitable solvent effect and need to be avoided, mixed with a compatible solvent, or applied in split mode. The splitless technique therefore requires working with temperature programs. Because of the almost complete transfer of the sample on to the column, total sample transfer is the method of choice for residue analysis (Grob, 1994, 1995). After the sample transfer into the analytical column is completed, the split valve is opened until the end of the analysis to prevent the further entry of sample material or contaminants on to the column.

Because of the longer residence times in the injector, with the splitless technique, there is an increased risk of thermal or catalytic decomposition of labile components. There are losses through adsorption on the surface of the insert, which can usually be counteracted by suitable deactivation. Much more frequently there is an (often intended) deposit of involatile sample residues in the insert or septum particles get collected at the bottom of the liner. This makes it necessary to regularly check and clean the insert according to a preventive maintenance procedure.

2.2.5.6 Concurrent Solvent Recondensation

The large volume injection possible on regular splitless injectors did not garner much attention until the transfer mechanism of the sample vapour into the column involved was further investigated and fully understood. The concurrent solvent recondensation technique (CSR) technique permits the injection of increased sample amounts up to 50 µL by using a conventional split/splitless injector with forward pressure regulation. This can be achieved in a very simple and straightforward way, since all processes are self-regulating. Moreover, the technique is robust towards contaminants and therefore it is suitable when complex samples have to be injected in larger amounts (Magni and Porzano, 2003; patent Thermo Fisher Scientific).

The recondensation of the solvent inside of the capillary column is causing a volume contraction and hence a pressure drop at the beginning of the column. A strong pressure difference occurs between column and injector liner that significantly speeds up the sample transfer. The sample and solvent vapour is drawn down from the insert by a strong suction effect due to the lower pressure inside of the column and continuously condensed to form the liquid band at the beginning of the column (see Figure 2.59). For proper operation of the CSR technique, the SSL injector has to be operated in the forward pressure regulation. A back pressure regulation would compensate for the occurring inlet pressures causing sample losses.

Sample volumes of up to 10 µL can be injected in liquid band mode by using the regular 0.25 mm capillary columns, while sample volumes of up to 50 µL require the wider diameter of an empty retention gap with 0.32 or 0.53 mm ID to accept the complete amount of liquid injected, and some silanized glass wool in the insert liner (focus liner).

The practical benefit of CSR is the flexibility for the injection of a wider range of diluted sample volumes reducing significantly the time for extract preconcentration. As the CSR injection is performed with the regular SSL injector hardware without any modifications, only the autosampler injection volume,

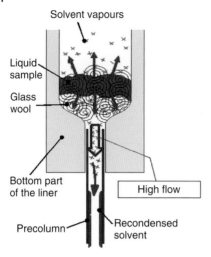

Figure 2.59 Principle of concurrent solvent recondensation in splitless injection (Magni and Porzano, 2003).

when using regular 10 μL syringes, has to be programmed in the methods. For increased volumes, the initial isothermal time needs to be extended accordingly up to the end of the solvent evaporation from the pre-column. The increased solvent vapour cloud needs to leave the condensation region (band formation) before starting the oven program. The remaining oven temperature program remains unchanged.

Concurrent Solvent Recondensation

Key steps (see Figure 2.59)

- Restricted evaporation rate:
  Injection with liquid band formation: The collection of the liquid on the glass wool allows a slow evaporation from a 'single' droplet.

- Increased transfer rate:
  Auto-pressure surge due to the large amount of solvent: A temporary increase of the inlet pressure rapidly drives the vapours into the column where the recondensation process starts. The concurrent solvent recondenzation in the pre-column is generating a strong suction effect.

- Solvent evaporation in the pre-column:
  The oven temperature is kept below the boiling point up to the end of solvent evaporation from the pre-column.

2.2.5.7 Concurrent Backflush

The popular multi-component methods with short sample preparation and extract clean-up such as QuEChERS require special care for the chromatographic system in GC and GC-MS. Extraction methods using low polar solvents, for example, ethylacetate, cyclohexane, acetone or any blend of these, are used for instance in pesticide extractions (DIN EN 12393-2; Anastassiades, 2003). While there is a high recovery reported even for a large number of pesticide components, medium and less polar compounds of high molecular weight get into the extracts. Fruits and vegetables from fresh as well as from processed food give rise to high amounts of a large variety of lipid components as a matrix of high boiling compounds. Although present in the final extracts, the sample extracts often look clear and almost colourless, and no visual quality control is possible.

Once injected into the chromatographic system, high boiling substances persist in the inlet liner of the GC and on the analytical column. While inlet liners typically are exchanged after a certain number of injections, the analytical column gets incrementally contaminated. The elution of high boiling compounds at high oven temperatures used for baking does not occur sufficiently. Matrix compounds accumulate and deliver an increasingly high background level with routine analysis of a large number of samples. Bake out procedures add additional time between samples, do not work efficiently and finally lead to reduced column lifetime. The situation is even worse with more polar columns with limited temperature range that cannot be baked out efficiently without damage to the column film. One solution can be a reversed column flow after the elution of the last analyte of interest, but this does not prevent the high boiling matrix from entering the column, and adds additional analysis time.

An optimum solution would be the separation of the analytes during the injection process from the high boiling matrix material. The more volatile low molecular weight analytes, for example, pesticides, are able to travel quickly into the analytical column, while high boilers move slowly and can be kept in the insert liner and a pre-column. Analytes that reach the analytical column stop motion at the beginning of the column and get focused by the column film. At this point, the pre-column can be swept backwards during the complete analysis run. For maintenance purposes, the pre-column as a cheap consumable can be replaced easily like the injector insert liner (Munari, 2000).

Concurrent Backflush Operation

The concurrent BKF system consists of four elements: A three-way solenoid valve (BKF valve) in the carrier gas line, a wide-bore pre-column, a purged T-connector in the GC oven and a flow restrictor (see Figures 2.60 and 2.61). The analytical column is pushed about 3–5 cm into the wide-bore pre-column to ensure full chromatographic integrity avoiding analyte contact to the surface of the T-piece. The flow restriction connects the carrier gas line of the injector with the BKF line to purge the T-piece. The flow of the restriction is usually factory set.

Injection   During injection, the BKF valve is off, see Figure 2.60. The carrier gas is directed in the regular way to the GC injector. A small flow provided by the restrictor grants sufficient purging of the T-connection to avoid dead volumes. The flow restrictor is designed to provide a purge flow of about 5% of the main

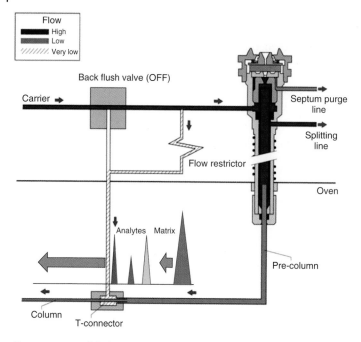

Figure 2.60 Backflush OFF – Injection and analyte transfer to pre-column and analytical column.

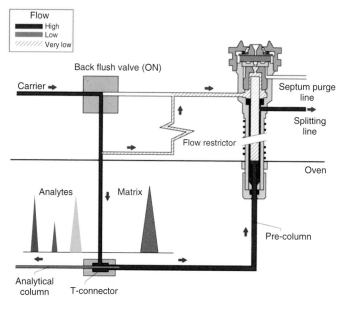

Figure 2.61 Backflush ON – Cleaning pre-column and injector from matrix during the analysis run.

flow when the split valve is closed. In the standard configuration, the pre-column consists of a 2 m × 0.53 mm ID uncoated fused silica tubing. Coated pre-columns improve inertness but lead to increased transfer times.

Backflush Activated
When the BKF valve is activated, the carrier gas flow is diverted directly to the T-connection and the analytical column to proceed with the regular chromatographic separation, see Figure 2.61. At the same time, the flow through the pre-column is reversed. The majority of the carrier gas now enters the inlet from the bottom and is vented through the split line cleaning the pre-column. All high-boiling sample matrix residing at this time in the pre-column gets flushed backwards. The pre-column and inlet liner of the injector get cleaned from high-boiling matrix components of no analytical interest concurrent to the regular separation process and analysis. A small carrier gas stream is provided by the restrictor to the top of the injector, purging the insert liner and septum during the cleaning phase (see Figure 2.61).

The timing of switching the BKF valve is set such that all analytes pass the pre-column to the analytical column. This time is called the *transfer time*.

Concurrent Backflush Setup
In a first step, a standard chromatogram using the optimized column temperature program without the use of BKF is acquired as reference. For BKF operation, no changes to the standard oven temperature program are required. In the following step, the BKF operation is optimized with standards starting with a long transfer time.

1) First deactivate the BKF mode.
2) Inject a standard with the optimized oven temperature program required for the complete elution of the analytes of interest.
3) Enable BKF and program a long transfer time. For reference, use the elution time of the last component of interest. For pesticides analysis, the last eluting analyte usually is e.g. deltamethrin.
4) On a PTV injector enable the 'Cleaning Phase' and set the cleaning time to match the total analysis time, and the split flow to about 50–100 mL/min (Figure 2.62). Using a hot SSL injector, the injector temperature remains unchanged during BKF operation.
5) Inject the sample and compare the chromatogram with the reference obtained at point 2.
6) If the last component of interest is present in the chromatogram, this will be the correct transfer time to set for the BKF. The transfer time can be shortened until the peak of the last component has a lower area or is not present. Then increase the transfer time in 1 min steps (Figure 2.63).
7) The BKF will be activated reproducibly for all subsequent analyses at the same time. If the temperature program and carrier flow conditions are modified, the optimum performance needs to be checked with the injection of a standard solution again.

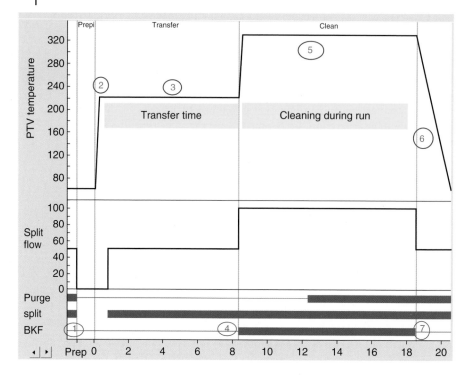

Figure 2.62 PTV operation with backflush and cleaning phase. 1. Sample injection; 2. PTV heating to transfer temperature; 3. transfer time from pre-column into analytical column; 4. switch BKF valve, reverse flow in the pre-column; 5. cleaning phase at elevated temperature on a PTV; 6. return PTV temperature for next injection and 7. switch BKF off for next injection.

The analytical benefits provided by the use of a concurrent BKF for sample extracts with matrix content are multiple (Hetmanski, 2009; Hildmann, 2013). All of them support the increased productivity of a GC and GC-MS system:

- No change to the oven temperature program in use
- Chromatographic integrity is maintained even with difficult matrix samples
- Time saving by avoiding additional bake-out phases of the column
- Shorter cycle times – stop after last eluting analyte
- Lower final oven temperatures improve cool down times to the next analysis
- The possibility to better select the column phase for the separation of the lighter components
- Increased column lifetime – no high bake-out temperatures
- No clipping of the analytical column – stable retention times
- Thinner column films possible for less matrix and faster separation
- Compatible with cold PTV and hot SSL injectors
- Compatible with LVI methods
- Easy maintenance of the injection system for pre-column exchange with BKF on (Figures 2.62–2.64).

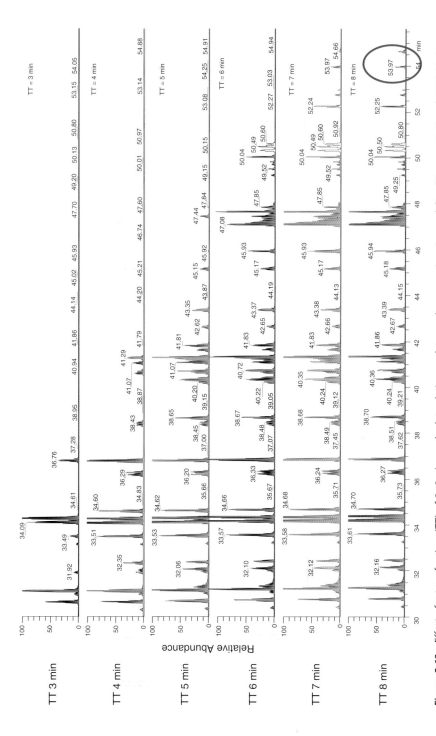

Figure 2.63 Effect of a transfer time (TT) of 3–8 min on the late eluting analytes and on retention time stability. (Courtesy R. Fussel, FERA York, UK.)

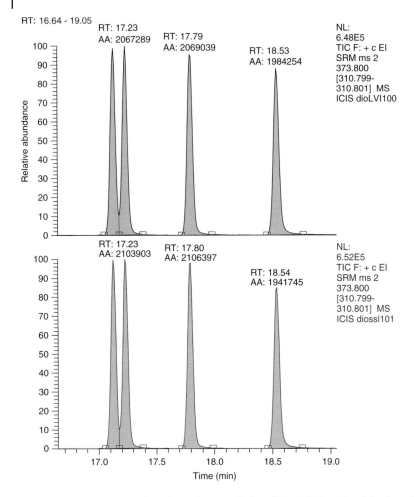

Figure 2.64 Comparison of regular and large volume injection. EPA 1613 CS3 standard, mass chromatograms of the hexachlorodibenzofuran congener peaks with retention time and peak area. (a) 80 μL PTV-LVI injection of the 1 : 80 diluted standard solution. (b) 1 μL PTV splitless injection of the undiluted standard solution. Conditions: Thermo Scientific TSQ Quantum GC, column: TR-5MS, 30 m × 0.25 mm ID × 0.1 μm film, TRACE GC with PTV, packed liner with deactivated glass wool.

2.2.6
Temperature Programmable Injection Systems

These increasingly popular injection systems are particularly designed for a rapid change of temperature between the injection and transfer phase. The sample extract is ejected from the syringe needle in liquid form into a specially designed 'cold' vaporizer. The insert temperature during injection is well below the boiling point of the solvent. Heating only begins after the syringe needle has been removed from the injection zone. For trace residue analysis PTV cold injection systems for split and splitless injection are becoming more widely used. The cold

Table 2.23 Choice of a suitable injector system.

Characteristics of the sample	Hot split	Hot splitless	PTV split	PTV splitless	PTV solvent split	On column
Concentrated samples	+	−	+	−	−	−
Trace analysis	≈	+	−	+	+	+
Diluted extracts	−	−	−	−	+	+
Narrow boiling range	+	+	+	+	+	+
Wide boiling range	−	+	+	+	−	+
Volatile substances	+	+	+	+	−	−
Low-volatile substances	−	≈	+	+	+	+
With involatile matrix	+	+	+	+	+	−
Thermolabile substances	≈	−	≈	≈	≈	+
Automated injection	+	+	+	+	+	+

+ recommended, ≈ can be used and − not recommended.

injection of a liquid sample eliminates a potentially selective evaporation from the syringe needle, which, in the case of hot injection procedures, can lead to discrimination against high-boiling components. Discrimination is also avoided as a result of explosive evaporation of the solvent in hot injectors. Colder regions of hot injection system, such as the septum area or the tubing leading to the split valve, can serve as expansion areas if the permitted injection volume is exceeded and can retain individual sample components through adsorption. The individual concepts of cold injection differ in the transfer of the sample to the column and in the possibility for using the split exit. These exist both between different cold injection techniques and in comparison with hot injection techniques.

In spite of the many advantages of the cold injection system, in practice there are also limits to its use. These exist both between different cold injection techniques and in comparison with hot injection techniques (Table 2.23). These include the analysis of thermally labile substances. Because of the low injection temperature a cold injection is expected to be particularly suitable for labile substances. However, during the heating phase, the residence times of the substances in the insert are long enough to initiate thermal decomposition. In this case, only on-column injection can be used because it completely avoids external evaporation of the sample for transfer to the column (see Section 2.2.6.3). A test for thermal decomposition was suggested by Donike with the injection of a mixture of the same quantities of fatty acid TMS esters (C10 to C22 thermolabile) and n-alkanes (C12 to C32 thermally stable). If no thermal decomposition takes place, all the substances appear with the same peak intensity.

2.2.6.1 The PTV Cold Injection System

The PTV injector (programmable temperature vaporizer, also called multimode inlet) with split or splitless operation is based on the systematic work of Vogt (1979), Poy (1981) and Schomburg (1981) (Figures 2.65 and 2.66). In particular, emphasis was placed on the precise and accurate execution of quantitative analyses of complex mixtures with a wide boiling point range (Poy, *et al.*, 1981,

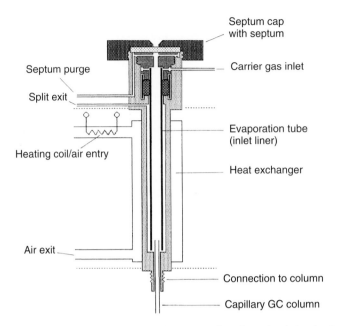

Septum cap
with septum

Septum purge

Split exit

Carrier gas inlet

Heating coil/air entry

Evaporation tube
(inlet liner)

Heat exchanger

Air exit

Connection to column

Capillary GC column

Figure 2.65 PTV with air as the heating medium from the design by Poy.

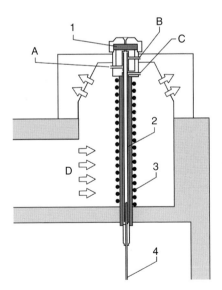

Figure 2.66 PTV split/splitless injector with direct heating (Thermo Fisher Scientific).
(A) Carrier gas inlet, (B) septum purge line, (C) split exit line, (D) active cooling. 1. septum,
2. injector body, 3. direct heating, and 4. analytical column.

1982; Grob, 1986). Particularly at the beginning of the experiments, absence of discrimination for substances up to above C_{60} was documented and its suitability for involatile substances, such as polyaromatic hydrocarbons, was stressed. Examples of the analysis of triglycerides demonstrate the injection of samples up to above C_{50} and the analysis of crude oil fractions up to C_{90}! The PTV process combines the advantages of the hot and on-column injection techniques (Vogt *et al.*, 1979a/b; Poy *et al.*, 1981; Saravalle *et al.*, 1987; Müller and Stan, 1987, 1989; Stan and Müller, 1987).

There are many advantages of cold split and splitless sample injection:

- Discrimination as a result of fractionated evaporation effects from the syringe needle does not occur. The extracts enters the inlet liner as a liquid.
- A defined volume of liquid can be injected reproducibly.
- The sample components evaporate as a result of controlled heating of the injection area in the order of their boiling points. The solvent evaporates first and leaves the sample components in the injection area without causing a spatial distribution of the analytes in the injector as a result of explosive evaporation.
- Aerosol and droplet formation is avoided.
- As the evaporator does not have to take up the complete expansion volume of the injected sample solution, smaller inserts with smaller internal volumes can be used. The consequently more rapid transfer to the column lowers band broadening of the peaks and thus improves the S/N ratio.
- If the boiling points of the sample components and the solvent differ by more than 100 °C, larger sample volumes (up to more than 100 µL) can be injected using the solvent split mode.
- Impurities and residues which cannot be evaporated do not get on to the analytical column.
- With concentrated samples the possibility can be used of adapting the injection to the capacity of the column and the dynamic range of the detector by selecting a suitable split ratio.

The PTV cold injection technique is mainly used for residue analysis in the areas of pesticides, pharmaceuticals, polyaromatic hydrocarbons, brominated flame retardants, dioxins and PCBs. However, the enrichment of volatile halogenated hydrocarbons, the direct analysis of water and the formation of derivatives in the injector also demonstrate the broad versatility of the PTV type injector (Bergna *et al.*, 1989; Mol *et al.*, 1993; Blanch *et al.*, 1994).

2.2.6.2 The PTV Injection Procedures

PTV Total Sample Transfer

The splitless injection for the injection of a maximum sample equivalent on to the column is the standard requirement for residue analysis. The sample is taken up in a suitable solvent and injected at a low temperature. The split valve is closed. The PTV temperature at injection should correspond to the boiling point of the solvent at ambient pressure. During the injection phase, the oven temperature is kept below the PTV temperature. It must be below the boiling point of the

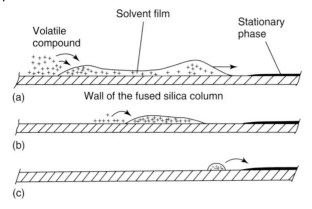

(a)

(b)

(c)

Volatile compound

Solvent film

Stationary phase

Wall of the fused silica column

Figure 2.67 Refocussing by means of the solvent effect in splitless injection (after Grob). (a) Condensation of solvent at the beginning of the capillary column by lowering the oven temperature below the boiling point. The condensed solvent acts as a stationary phase and dissolves the analytes. At the same time, this front migrates and evaporates in the carrier gas flow. (b) The continuous evaporation of the solvent film concentrates the analytes on to a narrow ring (band) in the column. For this process, no stationary phase at the beginning of the column is necessary. (c) The substances concentrated in a narrow band meet the stationary phase. The separation begins with a sharp band with the raising oven temperature.

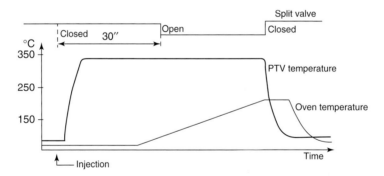

Figure 2.68 PTV total sample transfer (splitless injection).

solvent (Table 2.22) in order to exploit the necessary solvent effect for focussing the substances at the beginning of the column (Figure 2.67). If the focussing of the substances is unsuccessful or insufficient, the peaks of the components eluting early are broad and are detected with a low S/N ratio.

The injector is heated a few seconds after the injection when the solvent has already evaporated and has reached the column (Figure 2.68). Typically, this time interval is between 5 and 30 s. For high-boiling substances in particular, longer residence times have been found to be favourable. The heating rate should be moderate in order to achieve a smooth evaporation of the sample components required for transfer to the column. Heating rates of about 10–14 °C/s have proved to be suitable. The optimal heating rate depends on the dimensions of the insert liner and the flow rate of the carrier gas in the insert.

PTV Total Sample Transfer

- Split valve closed
- PTV at the ambient boiling point of the solvent
- Oven temperature below the boiling point of the solvent
- Start of PTV heating about 5–10 s after injection
- Start of the GC temperature program about 30–120 s after injection
- PTV remains hot until the end of the analysis.

Filling the insert with silanized glass wool has proved effective for the absorption of the sample liquid and for rapid heat exchange. However, the glass wool clearly contributes to enlarging the active surface area of the injector and should therefore be used only in the analysis of noncritical compounds (alkanes, chlorinated hydrocarbons, etc.). For the injection of polar or basic components special care is required keeping the glass wool deactivated by frequent replacements, an empty deactivated insert is recommended.

The use of capillary columns of 25 m in length and 0.25 mm internal diameter can be recommended without limitations for cold injection systems. The coupling to a mass spectrometer is particularly favourable because of the generally low flow rates. Also, the injection of larger sample volumes is straightforward and does not impair the operation of the mass spectrometer (Figure 2.64). For these reasons the use of cold injection systems for GC-MS is particularly recommended.

The transfer of the analytes to the analytical column for a 1–2 µL injection is completed after about 30–120 s, depending on the liner dimensions and carrier gas flow. Larger injection volumes require increased transfer times. The temperature program of the GC oven stays isothermal during this transfer phase, and gets started right after completion. At the same time, at the start of the GC oven programming, the injector gets purged of any remaining residues by opening the split valve. If required, a second PTV heating ramp and plateau can follow as cleaning phase to bake out the insert at elevated temperatures. The high injection or cleaning temperature of the PTV is retained until the end of the analysis to keep the injector free from possible adsorptions and the accumulation of impurities from the carrier gas inlet tubing. The start of cooling of the PTV is adjusted so that both the PTV and the oven are ready for the next analysis at the same time.

A special form of the PTV cold injection system consists of the column and insert connected via a press-fit attachment. Total sample transfer is possible in the same way as the classical PTV injection, but no split is present to allow flushing of the injection area after evaporation. Injectors of this type can be used only for total sample transfer (e.g. a SPI (septum-equipped programmable injector), Figure 2.69).

PTV Split Injection

In this mode of operation, the split valve is open throughout (Figure 2.70). In classical sample application, this mode of injection is suitable for concentrated solutions, through which the column loading can be adapted to its capacity and

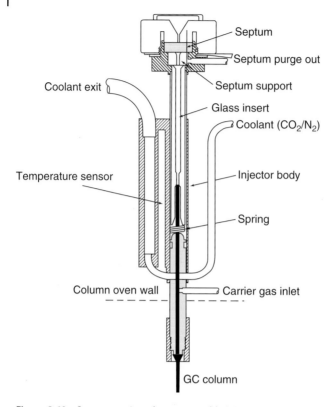

Figure 2.69 Septum-equipped programmable injector, SPI (Finnigan/Varian).

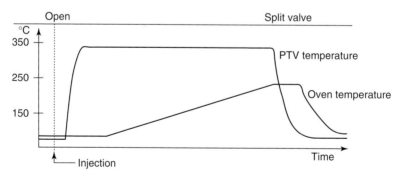

Figure 2.70 PTV split injection.

the nature of the detector by regulating the split flow (Poy, 1982). Cold injection systems are characterized by the particular dimensions of the inserts. This allows a high carrier gas flow rate at the split point. Compared to hot split injectors, smaller split ratios of about 1 : 5 are possible. In an individual case, it is possible to work in the split mode and thus achieve better S/N ratios and shorter analysis times, compared with total sample transfer.

PTV Split Injection

- Split valve open
- PTV below the ambient boiling point of the solvent
- Start of PTV heating about 5-10 s after injection
- Oven temperature to be chosen freely, typically above solvent BP.
- PTV remains hot until the end of the analysis.

All other PTV adjustments of time and temperature are completely unchanged compared with total sample transfer! Of particular importance for the split injection using the PTV cold injection system is the fact that the sample is injected into the cold injector (see also Section 2.2.7.5). The choice of start temperature of the GC oven is now no longer coupled to the boiling point of the solvent because of the solvent effect and is chosen according to the retention conditions (Tipler and Johnson, 1989).

PTV Solvent Split Injection

This technique is particularly suitable for residue analysis for the injection of solutions of diluted extracts with low analyte concentrations of high boiling analytes, (Figure 2.71). An overview of current large volume injection methods is given in Table 2.24. The solvent split mode allows the most efficient sample throughput shortening time-consuming pre-concentration steps. With suitable parameter settings, very large solvent quantities, also for a fraction collection from an online sample preparation and being limited only by practical considerations, can be applied (Staniewski and Rijks, 1991; Mol et al., 1994).

To inject larger sample volumes (from about $2\,\mu L$ to well over $100\,\mu L$), it is advantageous for the insert to be filled at least with silanized glass wool or packing of Tenax, Chromosorb, Supelcoport or another inert carrier material about $0.5-1\,cm$ wide. The split valve is open during the injection phase. The PTV is kept at a temperature that corresponds to the boiling point of

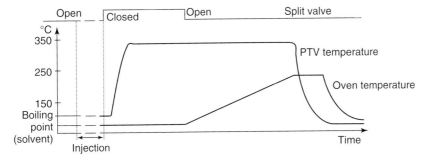

Figure 2.71 PTV solvent split injection.

Table 2.24 Comparison of large volume injection methods.

	LV-on-column with solvent vapour exit	LVI-PTV with solvent split	LV-split/splitless with concurrent solvent recondensation
Typical injection volume	150 μL	100 μL	30 μL
Robust versus complex matrix	No	Yes	Yes
Volatiles analysis	Yes	Need optimization	Yes
Suitability for thermolabile and actives compounds	High	Medium	Medium
Solvent vented	Yes	Yes	No
Requires uncoated pre-column	Yes	No[a]	Yes
Number of parameters	Medium	Large	Small[b]
Software assisted set-up	Yes	No	Yes

a) It can be necessary if large amount of solvent is retained for improving volatiles recovery.
b) Up to 10 μL with regular 0.25 ID columns.

the solvent. The oven temperature is below the PTV temperature and thus below the boiling point of the solvent at the carrier gas head pressure. The maximum injection rate depends upon how much solvent per unit time can be evaporated in the insert and eliminated through the split tubing by the carrier gas. The at-once injection of large volumes can best be tested easily by applying the desired amount of solvent with the liner outside of the injector. During injection, a high split flow of >100 mL/min is recommended to focus the analytes on the packing material by local consumption of evaporation heat. For the ease-of-method development of the injection, the data recording of the GC or GC-MS system is started so that the course of the solvent peak can be followed (Figure 2.72).

When quantities of more than 10 μL are applied, the split valve is closed only when the decline of the solvent peak begins to show on the display, meaning that the injector is free of most of the solvent. After the split valve has been closed, the PTV injector can be heated up as usual. After about 30–120 s the transfer of the enriched sample from the insert to the column is complete and the GC temperature program can be started. The split valve then opens and the PTV is held at the injection temperature until the end of the analysis.

The PTV solvent split mode also allows the derivatization of substances in the injector. This procedure simplifies sample preparation considerably. The sample extract is treated with the derivatizing agent (e.g. TMSH (trimethylsulfonium hydroxide) methylation) and injected into the PTV. The excess solvent is blown out during the solvent split phase. The derivatization reaction takes place during the heating phase and the derivatized substances pass on to the column (Färber *et al.*, 1991, 1992).

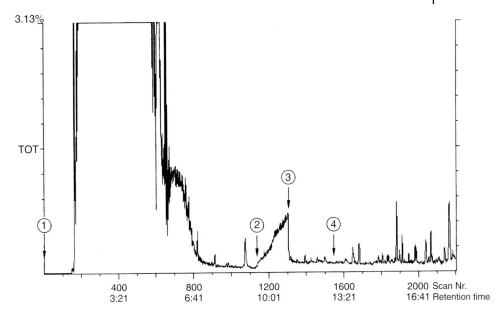

Figure 2.72 The course of an analysis with large quantities of solvent using the PTV cold injection system filled with a 1 cm Chromosorb plug 100 µL PCB solution (200 pg/µL). After the dead time, a broad solvent peak is registered. After the solvent peak is eluted, the split valve is closed and the PTV heated up. A smaller roughly triangular solvent peak is produced from adsorbed material. After injection, the split valve is opened again and the oven temperature program started. The peaks eluting show good resolution and are free from tailing. (1). Start of the injection: split open. (2). Start of PTV heating: split closed. (3). Baking out the PTV: split open. (4). Start of the GC temperature program: split open, and typically the start of the data acquisition.

PTV Large Volume Injection

The PTV large volume injection mode (LVI-PTV) allows the repeated automated injection of sample volumes in the range up to 100 µL and even more (Müller *et al.*, 1993; Mol *et al.* 1994; Hoh and Mastovska, 2008). The large volume mode requires that the sample components are less volatile than the used solvent. The LVI-PTV operation needs the additional installation of a heated solvent split valve to prevent solvent vapour to condensate and plug the split exhaust line. The operation is further facilitated by a widebore pre-column and BKF valve which is recommended to prevent large amounts of solvent vapour entering the analytical column and for cleaning the injector during the heat-off step after injection.

As described in the previous section for PTV solvent split injections, a wide liner needs to be filled either by silanized quartz wool or other suitable packing material to retain the sample and prevent solvent from rinsing through the liner. The solvent capacity of the liner can easily be tested outside of the injector by adding the intended amount of solvent. No rinsing of the solvent may be observed when holding the insert liner upright on a sheet of paper.

2.2.6.3 On-Column Injection

Another powerful injection technique is completely avoiding the vaporization of the sample. The nonvaporizing sample injection involves injecting the liquid 'cold' sample directly on to the column (Grob and Grob, 1978b). In the era of packed columns an on-column injection was the state of the art (although not known as such at the time). The difference between that and the present procedure lies in coping with the small diameters of capillary columns which have caused the term *filigree* to be applied to the on-column technique. Schomburg introduced the first on-column injector for capillary columns in 1977 under the designation of direct injection and described the process of sample injection very precisely. A year later, a variant using syringe injection was developed by Grob. Today's commercial on-column injection involves injecting the sample directly on to the column (internal diameter 0.32 mm) in liquid form using a standard syringe with a 75 mm long steel needle of 0.23 mm external diameter. More favourable dimensioning is possible on use of a retention gap with an internal diameter of 0.53 mm, which can even be used with standard needles of 0.45 mm OD. The use of retention gaps allows autosamplers to be employed for on-column sample injection.

The injector itself does not have a complicated construction and can be serviced easily and safely. The carrier gas feed is situated and the capillary fixed centrally in the lower section. The middle section carries a rotating valve as a seal and in the upper section there is the needle entry point which allows central introduction of the syringe needle. In all types of operation, the whole injector block remains cold and warming by the oven is also prevented by a surrounding air stream.

Small sample volumes of less than 10 µL can be injected directly on to a capillary column. Larger volumes require a deactivated retention gap of appropriate dimensions (Figure 2.73). The retention gap is connected to the column with zero dead volume connectors (e.g. press fit connectors). For injection, the column should be operated at a high flow rate (about 2–3 mL/min for He). During the injection the oven temperature must be below the effective boiling point of the solvent. This corresponds to a temperature of about 10–15 °C above the boiling point of the solvent under ambient conditions. As a rule of thumb, the boiling point increases by 1 °C for every 0.1 bar of head pressure. The injection is carried out by pressing the syringe plunger down rapidly (about 15–20 µL/s) to avoid the liquid being sucked up between the needle and the wall of the column through capillary effects, otherwise the subsequent removal of the syringe would result in loss of sample. For larger sample volumes, from about 20–30 µL, the injection rate must, however, be lowered (to about 5 µL/s) as the liquid plug causes increasing resistance and reversed movement of the liquid must be avoided (Figure 2.74). The oven temperature program with the heating ramp can then begin when the solvent peak is clearly decreasing (Grob, 1983).

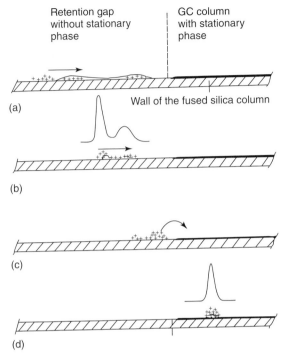

Retention gap
without stationary
phase

GC column
with stationary
phase

Wall of the fused silica column

(a)

(b)

(c)

(d)

Figure 2.73 Formation of the start band on use of a retention gap (after Grob). (a) Starting situation after injection: the retention gap is wetted by the sample extract. (b) The solvent evaporates and leaves an undefined substance distribution in the retention gap. (c) The substance reaches the stationary phase of the GC column. Considerable retention corresponding to the capacity factor of the column begins. (d) The analytes injected with the sample have been concentrated in a narrow band at the beginning of the column; this is the starting point for the chromatographic separation.

On-Column Injection

- Flow rate about 2–3 mL/min
- Oven temperature at the effective boiling point of the solvent (normal boiling point plus 1 °C per 0.1 bar pre-pressure)
- Rapid injection of small volumes
- Oven temperature to be kept at the evaporation temperature
- Only start the heating program after elution of the solvent peak.

The on-column technique has the following advantages over a cold injection system:

- The risk of thermal decomposition during the injection is practically eliminated. A substance evaporates at the lowest possible temperature and is heated to its elution temperature at the maximum (Cavagna *et al.*, 1994).

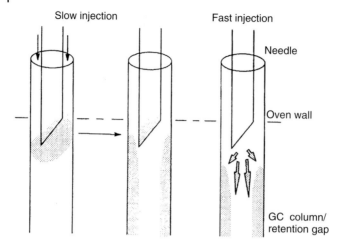

Figure 2.74 Slow and fast on-column injections. For the slow injection, the exterior of the needle is contaminated and part of the sample is lost.

- Only on-column injection allows the quantitative and reproducible sample injection on to the column without losses. Low-volatile substances, in particular, are transferred totally without discrimination.
- Defined volumes can be injected with high reproducibility. Standard deviations of about 1% can be achieved.
- The injection volumes can be varied within wide ranges without additional optimization. At the beginning of the heating program, only the duration of elution of the solvent peak needs to be taken into account for the initial isothermal phase.

For use in GC-MS systems, attention should be paid to the injection at the necessary high flow rates in connection with the maximum permitted carrier gas loading of the mass spectrometer. As the injection system is always cold, moisture can accumulate from samples or insufficiently purified carrier gas (Figure 2.75). Because of this, the sensitivity of the mass spectrometer could be compromized. The effect is easy to detect because the mass spectrometer continually registers the water background and because the on-column injector can easily be opened at a particular time so that the carrier gas can pass out via the splitter.

If the water background decreases after the dead time, this effect must be taken into consideration during planning of the analysis, as the response behaviour of substances can change during the working day.

Other system-related limitations to the on-column technique for certain types of sample are:

- Concentrated samples may not be injected on-column. In these cases, preliminary dilution is necessary or the use of the split injector is recommended (minimum on-column injection volume 0.2 μL).

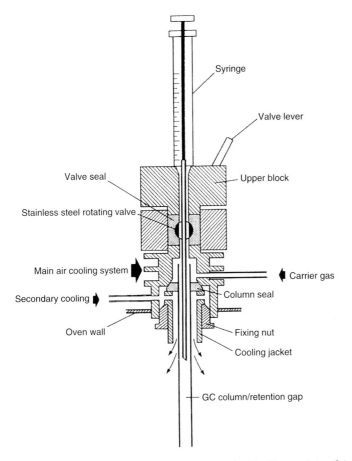

Figure 2.75 On-column injector schematics (Carlo Erba/Thermo Scientific).

- Samples containing high amounts of matrix rapidly lead to quality depletion. Here a retention gap is absolutely necessary. However, impurities in the retention gap quickly give rise to adsorption and peak broadening because of the small capacity, so that changing it regularly is necessary. For dirty samples, hot or cold PTV sample injection is preferred.
- Samples containing volatile components are not easy to control, as focussing by the solvent effect is inadequate under certain circumstances. If changing the solvent does not help, changing to a hot or cold split or splitless system is recommended.

2.2.6.4 PTV On-Column Injection

A regular PTV injector can also be used successfully for on-column injections. This method allows the straightforward automation of on-column injections using regular autosamplers with standard syringes. The insert liner used needs to be equipped with a restrictor at the top. This restrictor functions as a needle guide into the column. Syringes with the regular 0.47 mm OD needles can be used allowing the direct injection into wide-bore columns or retention gaps (pre-columns) of equivalent dimensions, for example, 0.53 mm ID. For this purpose, the column inlet is pushed up until it gets positioned and centered at the bottom of the restrictor site of the insert liner.

In PTV on-column mode, the injector body and the column oven is set for injection to a temperature below the solvent boiling point (see Table 2.23). After a short injection time of up to 20 s the injector is heated up for sample transfer. The oven program is started right after the completed sample transfer into the analytical column. The split valve remains opened with a low flow rate of only a few millilitre per minute. The PTV temperature is maintained high throughout the chromatographic run as usual.

2.2.6.5 Cryofocussing

Cryofocussing should not be regarded as an independent injection system, but nevertheless should be treated individually in the list of injectors because the static headspace and P&T systems can be coupled directly to a cryofocusing unit as a GC injector. Furthermore, some thermodesorption systems already contain a cryo focus unit. Many simpler designed instruments, however, do not, and require external focussing to ensure their proper function. Generally, for the direct analysis of air or gases from indoor rooms or at the workplace, or of emissions, also for large volumes of headspace gases, a cryofocussing unit is required for concentration and injection on to the GC column (Poy and Cobelli, 1985; Klemp *et al.*, 1993).

Packed columns can effectively concentrate the substances contained in gaseous samples at the beginning of the column because of their high sample capacity. Capillary columns under normal working conditions cannot form sufficiently narrow initial bands for the substances to be detected from the gas volumes being handled in direct air analysis or on heating up traps (P&T, thermodesorption) because of their comparatively low sample capacity. Additional effective cooling is necessary. The entire oven space can be cooled, but this has a major disadvantage: the requirement in terms of time and cooling agents is immense.

A special cryofocussing unit cools down the beginning of the GC column and allows on column focussing of the analytes without requiring a retention gap. The column film present improves the efficiency at the same time by acting as an adsorbent. For this purpose, the capillary column is inserted into a 1/16 in. stainless steel tube via an opening in the oven lid of the chromatograph and is connected via a connecting piece to the transfer line of the sampler (e.g. headspace, P&T, canister). This stainless steel tube is firmly welded to a cold finger and a thermoelement (Figure 2.76). The tube is also surrounded by a differential heating coil which heats the inlet side more rapidly than the exit to the GC oven. In this way, a uniform temperature gradient is guaranteed in all the phases of the operation. Because of the gradient, on cooling, for example, with liquid nitrogen, all the analytes are focussed into a narrow band at the beginning of the capillary column, as the substances migrate more slowly at the start of the band than at the end.

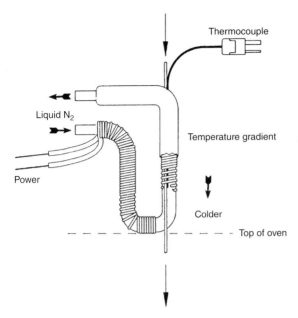

Figure 2.76 Schematics of a cryofocussing unit for use with fused silica capillary columns (Tekmar).

After concentration and focussing, the chromatographic separation starts with the heating of this region. On careful control of the gradient, the heating rate does not affect the efficiency of the column (heating rates between 100 and 2000 °C/min).

Because of on-column focussing, the cryotrap has the same sample capacity as the column has for these analytes. A breakthrough of the cryotrap as a result of too high an analyte concentration can be prevented by changing the column. Overloading the column would, in any case, result in poor separation. The diagram in Figure 2.77 gives the temperatures at which a breakthrough of the analytes must be reckoned with in cryofocussing. Larger film thicknesses and internal diameters permit higher focussing temperatures, which favour the mobilization of the analytes from the column film.

2.2.6.6 PTV Cryo-Enrichment

This injection technique is suitable for trace analysis and for concentrating volatile compounds, which are present in large volumes of gas, for example, in gas sampling, thermodesorption or the headspace technique. Sample transfer into the injector is carried out while the PTV is in trapping mode. For this, the insert is filled with a small quantity (about 1 – 2 mg, about 1 – 3 cm wide) of Tenax, Carbosieve or another thermally stable adsorbent. The PTV injector is cooled with liquid CO_2 or nitrogen until injection. During the injection, the split valve is open as in the solvent split mode. Like the total sample transfer method, the oven temperature is kept correspondingly low in order to focus the components at the beginning of the column.

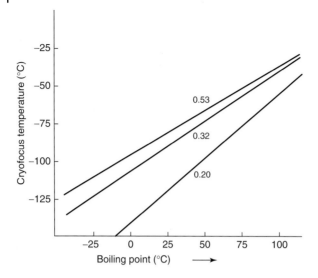

Figure 2.77 Breakthrough temperatures in cryofocussing for various column internal diameters as a function of the boiling point of the analyte (Teledyne Tekmar, 2013). Film thicknesses: 0.25 μm at 0.20 mm internal diameter, 1.0 μm at 0.32 mm internal diameter and 3.0 μm at 0.53 mm internal diameter.

The gaseous sample is passed slowly through the injector and the organic substances retained on the adsorbent. Before injection the PTV can be heated to a low temperature with the split still open to dry the Tenax material, if required. For transfer of the sample to the GC column, the split valve is closed and the PTV is heated to effect the total sample transfer of the concentrated components. After about 30–120 s, the temperature program of the oven is started and the split valve is opened (Figure 2.78). The PTV remains at the same temperature until the end of the analysis. This last step is particularly important in cryo-enrichment because the cooled adsorbent could become enriched with residual impurities from the carrier gas, which would lead to ghost peaks in the subsequent analysis.

2.2.7
Capillary Column Choice and Separation Optimization

There are no hard-and-fast rules for the choice of columns for GC-MS coupling. The choice of the correct phase is made on the typical criterion: 'like dissolves like'. If substances exhibit no interaction with the stationary phase, there will be no retention and the substances leave the column at the dead time or get hardly separated from each other. The polarity of the stationary phase should correspond to the polarity of the substances being separated (see Table 2.25). Less polar substances are better separated on non-polar phases and vice versa.

An exception to this general rule of thumb is given with the newly developed ionic liquid stationary phases. The characteristics and guide to potential applications are discussed in a separate section at the end of this chapter.

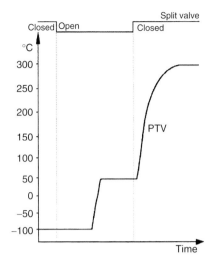

Figure 2.78 PTV cryo-enrichment at −100 °C (with an optional fractionation step here at 50 °C) and injection with total sample transfer.

Chose More Polar Stationary Phases

- Weaker retention of non-polar compounds
- Stronger retention of polar compounds
- Shift of compounds with specific interactions
- Exceptions with ionic liquid phases, see Section 2.2.6.6

For coupling with MS the carrier gas flow and the specific noise of the column (column bleed) are included in the criteria governing the choice of column. When considering the optimal carrier gas flow, the maximum carrier gas load of the mass spectrometer must be taken into account. The limit for small benchtop mass spectrometers with quadrupole analysers is about 2–4 mL/min. Larger instruments with high pumping capacity can generally tolerate higher loads. These conditions limit the column diameter which can be used.

There can be no compromises concerning column bleed in GC-MS. Column bleed generally contributes to chemical noise where MS is used as the mass-dependent detector, and curtails the detection limits. The optimization of a particular S/N ratio can also be effected in GC-MS by selecting particularly thermally stable stationary phases with a low tendency to bleed. For use in trace analysis, stationary phases for high-temperature applications have proved particularly useful (Figures 2.79 and 2.80). Besides the phase itself, the film thickness also plays an important role. Thinner films and shorter columns exhibit lower column bleed.

Table 2.25 Composition of stationary phases for fused silica capillary columns with column designations for comparable phases from different manufacturers, arranged in order of increasing polarity and columns with special phases (selection).

Phase composition	Agilent (J&W)	Agilent (Varian/Chrompack)	Alltech	Macherey-Nagel	PerkinElmer
100% dimethyl polysiloxane	HP-1, DB-1	CP Sil 5 CB	AT-1MS	OPTIMA 1	Elite -1
100% dimethyl polysiloxane (low bleed)	DB-1ms UI, Ultra-1, HP-1ms, HP-1ms UI	VF-1ms, CP-Sil 5, CP Sil 5 CB Low Bleed/MS	AT-1ht, AT-1, EC-1	OPTIMA-1ms, OPTIMA-1ms Accent	–
100% dimethyl polysiloxane (high temp.)	DB-1HT	VF-1HT	–	–	–
50% n-octyl 50% methyl siloxane	–	–	–	–	–
5% phenyl 95% dimethyl polysiloxane	DB-5, HP-5, HP-5ms, Ultra-2	VF-5MS, CP Sil 8 CB, CP Sil 8 CB MS	AT-5, AT-5ms, EC-5	OPTIMA 5	Elite-5, Elite-5 ms, Elite-5ht
5% phenyl 95% dimethyl polysiloxane (base modified)	–	–	–	–	–
5% phenyl 95% dimethyl polysilarylene (low bleed)	DB-5ms	VF-XMS	–	OPTIMA-5ms	–
5% phenyl 95% dimethyl polysilarylene (high temp.)	–	–	–	OPTIMA-5ms Accent, OPTIMA-5 HT	–
5% phenyl 95% dimethyl polysilarylene (high temp.)	DB-5HT	VF-5HT	–	–	–
5% phenyl 95% dimethyl polysilarylene	DB-XLB	–	–	OPTIMA XLB	–
5% phenyl polycarborane-siloxane	–	–	–	–	–
8% phenyl polycarborane-siloxane	–	–	–	–	–
10% phenyl 90% dimethyl polysiloxane	–	–	–	–	–
20% phenyl 80% dimethyl polysiloxane	–	–	AT-20, EC-20	–	–
35% phenyl 65% dimethyl polysiloxane	DB-35, HP-35	VF-35MS	AT-35, AT-35ms	–	Elite-35, Elite-35ms
35% phenyl 65% dimethyl polysiloxane (base modified)	–	–	–	–	–
35% phenyl 65% dimethyl arylene siloxane	DB-35ms	–	–	OPTIMA 35 MS	–
50% phenyl 50% methyl polysiloxane	DB-17ms, HP-50	–	AT-50	OPTIMA 17	Elite-17, Elite-17ht, Elite-17ms
50% phenyl 50% dimethyl polysiloxane	–	VF-17ms, CP-Sil 24 CB	–	OPTIMA 17 MS	–
50% diphenyl 50% dimethyl polysiloxane	DB-17, HP-50+, HP-17	CP-Sil 24 CB	AT-50	OPTIMA 17	–
50% phenyl 50% dimethyl arylene siloxane	–	–	AT-50ms	–	–
65% phenyl 35% dimethyl polysiloxane	–	TAP-CB	–	–	–
65% diphenyl 35% dimethyl polysiloxane	–	–	–	–	–
75% phenyl 25% dimethyl polysiloxane	DB-17, HP-50+	VF-17ms	–	–	–
100% methylphenyl polysiloxane	–	CP Sil 24 CB	–	OPTIMA 17	–
35% trifluoropropyl 65% dimethyl polysiloxane	DB-200	VF-200MS	–	OPTIMA 210	–
50% trifluoropropyl 50% dimethyl polysiloxane	DB-210	–	AT-210	OPTIMA 210	–
–	–	–	–	–	–
6% cyanopropylphenyl 94% dimethyl polysiloxane	DB-1301, HP-1301, DB-624, HP-624, HP-Fast GC	VF-624ms, CP-1301, VF-1301ms	AT-624, AT-1301	OPTIMA 1301, OPTIMA 624, OPTIMA 624 LB	Elite-1301
14% cyanopropylphenyl 86% dimethyl polysiloxane	DB-1701, HP-1701, PAS-1701, DB-1701P	CP Sil 19 CB, VF-1701ms	AT-1701	OPTIMA 1701	Elite-1701

Phenomenex	Quadrex	Restek	SGE	Supelco	Thermo Scientific	US Pharmacopeia Designation
ZB-1	007-1	Rtx-1	BP1	SPB-HAP, SPB-1, SPB-1 Sulfur	TG-1MS, TG-1MT, TR-1MS	G1, G2, G38
ZB-1HT Inferno, ZB-1ms	–	Rtxi-1ms	SolGel-1ms	Equity-1	–	G1, G2, G38
ZB-1HT	–	Rxi-1HT	BPX1	–	–	–
–	–	–	–	SPB-Octyl, Petrocol DH Octyl	–	–
ZB-5	007-5	Rtx-5	BP5, BP5MS	SAC-5, Equity-5, SPB-5	TG-5MS, TG-5 SILMS, TG-5MT, TR-5, TR-5MS,	G27; G36, G41
–	–	–	–	PTA-5	TG-5MS AMINE	–
ZB -5MS, ZB-5MSi, ZB-Semivolatiles	007-5ms	Rxi-5ms	BPX5	SLB-5ms	–	G27, G36
ZB-5HT	–	Rxi-5Sil MS	–	–	TG-5HT, TR-5 HT, TR-8270	–
ZB-5HT Inferno	–	Rxi-5HT	–	–	–	–
ZB-XLB-HT Inferno	–	Rtx-XLB	–	–	TG-XLBMS, TR-527	–
–	–	–	HT5	HT-5 (aluminum clad)	–	–
–	–	–	HT8	–	TR-PCB 8MS	–
–	007-10, 007-20	–	–	–	–	G41
–	–	Rtx-20	–	SPB-20	–	G28, G32
ZB-35, ZB-35HT Inferno	007-35	Rtx-35, Rtx-35SilMS	–	SPB-35	TG-35MS, TR-35MS	G42
–	–	–	–	–	TG-35MS AMINE	–
–	–	–	BPX35, BPX608	–	–	–
–	–	Rtx-50	–	–	TG-17MS	G3
ZB-50	–	Rxi-17SilMS	–	–	–	–
ZB-50	–	Rxi-17	–	SPB-50	–	G3
–	–	–	BPX50	–	TR-50MS	–
–	007-65HT	Rtx-65, Rtx-65TG	–	–	–	G17
–	–	Rtx-65	–	–	–	G17
–	–	–	–	–	–	G17
–	007-17	Rtx-50	–	SP-2250	–	G3
–	007-210	Rtx-200, Rtx-200MS	–	–	TG-200MS	G6
–	–	–	–	–	–	G6
–	–	–	–	–	–	–
–	–	–	–	–	–	–
ZB-624	007-1301	Rtx-1301, Rtx-624	BP624	SPB-624, OVI-G43	TG-1301MS, TG-624, TR-V1	G43
ZB-1701, ZB-1701P	007-1701	Rtx-1701	BP10(1701)	Equity-1701	TG-1701MS, TR-1701	G46

Phase composition	Agilent (J&W)	Agilent (Varian/Chrompack)	Alltech	Macherey-Nagel	PerkinElmer
33% cyanopropylphenyl 67% dimethyl polysiloxane	–	–	–	OPTIMA 240	–
50% cyanopropylphenyl 50% dimethyl polysiloxane	DB-225, HP-225, DB-225ms	CP Sil 43 CB	AT-225	OPTIMA 225	Elite-225
25% cyanopropyl 25% dimethyl polysiloxane	–	–	–	–	–
50% cyanopropyl 50% dimethyl polysiloxane	DB-23	–	AT-SILAR	–	–
50% cyanopropyl 50% phenyl-methyl polysiloxane	DB-225, HP-225	–	–	OPTIMA 225	–
70% cyanopropyl polysilphenylene-siloxane	–	–	–	–	–
70% biscyanopropyl/ 30% cyano-propylphenyl polysiloxane	–	–	–	–	–
78% cyanopropyl 22% dimethyl polysiloxane	–	–	–	–	–
80% biscyanopropyl/ 20% cyano-propylphenyl polysiloxane	–	VF-23ms	AT-SILAR-90	–	–
88% cyanopropylaryl polysiloxane	HP-88	CP-Sil 88	–	–	–
90% cyanopropyl polysilphenylene-siloxane	–	–	–	–	–
90% biscyanopropyl/ 10% cyano-propylphenyl polysiloxane	–	CP Sil 84	–	–	–
biscyanopropyl polysiloxane	–	–	AT-SILAR-100	–	–
1,2,3-tris(2-cyanoethoxy)propane	–	–	–	–	–
Divinylbenzene-ethylene glycol-dimethacrylate	HP-PLOT U	–	–	–	–
DVB 4-vinylpyridine polymer	–	CP-PoraPLOT S	–	–	–
PEG polyethylene glycol	HP-INNOWax	CP Wax 52 CB	AT-Wax, AT-WAXms, AT-AquaWax, EC-WAX	Permabond CW 20M, OPTIMA WAX, OPTIMA WAXplus	Elite-WAX
PEG polyethylene glycol	DB-WAX, DB-WaxFF, DB-WAXetr	VF-WAXms	–	–	–
PEG polyethylene glycol	–	–	–	–	–
PEG - base modified, for amines and basic compounds	CAM	CP Wax 51	AT-CAM	–	–
PEG - acid modified, for acidic compounds	DB-FFAP, HP-FFAP	CP Wax 58 CB	AT-1000, EC-1000, AT-AquaWax-DA	Permabond FFAP, OPTIMA FFAP, OPTIMA FFAPplus	Elite-FFAP
1,12-di(tripropylphosphonium) dodecane bis (trifluoro-methylsulfonyl)imide	–	–	–	–	–
1,12-di(tripropylphosphonium) dodecane bis (trifluoro-methylsulfonyl)imide trifluoromethylsulfonate	–	–	–	–	–
tri(tripropylphosphonium-hexanamido) triethylamine bis (trifluoromethylsulfonyl)imide	–	–	–	–	–
1,9-di(3-vinylimidazolium)nonane bis(trifluoromethylsulfonyl)imide	–	–	–	–	–
1,5-di(2,3-dimethylimidazolium) pentane bis (trifluoro-methylsulfonyl)imide	–	–	–	–	–
PLOT Columns:					
PLOT phase packed polystyrene-divinylbenzene	HP-PLOT Q	CP-PoraPLOT Q, CP-PoraBond Q	–	–	Elite-Q
PLOT phase packed divinylbenzene 4-vinylpyridine	–	–	–	–	–
PLOT phase packed divinylbenzene ethyleneglycol dimethylacrylate	–	–	–	–	–
PLOT phase packed Alumina	HP-PLOT Al2O3, GS-Alumina	CP-AL2O3	–	–	Elite-Alumina

Phenomenex	Quadrex	Restek	SGE	Supelco	Thermo Scientific	US Pharmacopeia Designation
–	–	–	–	–	–	–
–	007-225	–	–	–	TG-225MS, TR-225	G7
–	–	–	–	–	–	G19
–	–	–	–	–	–	G5
–	–	Rtx-225	–	SPB-225	–	G7, G19
–	–	–	BPX70	–	TR-FAME	–
–	–	–	–	–	–	–
–	007-23	–	–	–	–	–
–	–	–	–	SP-2330	–	G8
–	–	–	–	–	–	–
–	–	–	BPX90	–	–	–
–	007-23	Rtx-2330	–	SP-2331, SP-2380	TG-POLAR	G48
–	–	Rt-2560	–	SP-2560, SP-2340	–	G5
–	–	–	–	TCEP	–	–
–	–	Rt-U-BOND	–	–	–	G45
–	–	Rt-S-BOND, MXT-S-BOND	–	–	–	–
ZB-WAX, ZB-WAXplus	007-CW, BTR-CW	Stabilwax, Rtx-Wax	SolGel-WAX	SPB-PUFA, Omegawax, SU-PELCOWAX 10	TG-WaxMS,TG-WaxMT, TR-Wax, TR-WaxMS	G14, G15, G16, G20, G39
–	–	–	BP20(Wax)	–	–	G20
–	–	–	–	PAG	–	G18
–	–	Stabilwax-DB	–	Carbowax-Amine	TG-WaxMS B	–
–	007-FFAP	Stabilwax-DA	BP21(FFAP)	Nukol, SP-1000	TG-WaxMS A, TR-FFAP	G25, G35
–	–	–	–	SLB-IL59, SLB-IL60, SLB-IL82	–	–
–	–	–	–	SLB-IL61	–	–
–	–	–	–	SLB-IL76	–	–
–	–	–	–	SLB-IL100	–	–
–	–	–	–	SLB-IL111	–	–
–	PLT-Q	Rt-Q-BOND, MXT-Q-BOND	–	Supel-Q-PLOT	TR-Bond Q/Q+	–
–	–	–	–	–	TR-Bond S	–
–	–	–	–	–	TR-Bond U	–
–	PLT-AL203	Rt-Alumina BOND	–	Alumina sulfate PLOT, Alumina chloride PLOT	TR-Bond Alumina (Na$_2$SO$_4$/KCl)	–

Phase composition	Agilent (J&W)	Agilent (Varian/Chrompack)	Alltech	Macherey-Nagel	PerkinElmer
PLOT phase packed carbon molecular sieve	–	–	–	–	–
PLOT phase packed Molsieve 5A	HP-PLOT Molesieve	CP-Molsieve 5A	–	–	Elite-Molsieve
Chiral Phases:					
permethylated ß-cyclodextrin	–	CP-Cyclodextrin ß	–	–	Elite-Betacydex
beta-Cyclodextrin in phenyl-based stationary phase	HP-Chiral β	–	–	–	–
30%-heptakis (2,3-di-O-methyl -6-O-t-butyl dimethylsilyl)-B-cyclodextrin in 1701	CycloSil-B	–	–	–	–
1,2,3-tris(cyanoethoxy)propane	–	CP-TCEP	–	–	–
Proprietary Phases:	DB-VRX	–	–	–	–
	DB-XLB	–	–	–	–
	DB-TPH	–	–	–	–
	DB-MTBE	–	–	–	–
	GC-GasPro	–	–	–	–
	GS-CarbonPLOT	–	–	–	–
	GS-OxyPLOT	–	–	–	–
				OPTIMA δ-6	
				OPTIMA δ-3	
Application Specific Columns					
Volatile organic analysis-EPA methods 502.2, 524.2, 601, 602, 624, 8010, 8020, 8240, 8260	DB-VRX, DB-624	–	AT-502	–	–
Volatile organic analysis-EPA methods 502.2, 524.2, 601, 602, 624, 8010, 8020, 8240, 8260	–	CP Sil 13 CB	–	–	Elite-624, Elite-Volatiles
Herbicides EPA method 507	–	–	–	–	–
Semivolatile organic analysis EPA methods 625, 1625, 8270, CLP	DB-5.625	–	–	–	–
Organochlorine pesticides-EPA Methods 8081, 608, and CLP Pesticides.	–	–	–	–	Elite-608
Organophosphorus pesticides-EPA Method 8141A	–	–	–	–	–
Organochlorine pesticides-EPA Methods 8081, 608, and CLP Pesticides	–	CP Sil 8 CB, CP Sil 19 CB	AT-Pesticide	–	PE- -2, PE -608, PE-1701
ASTM Method D2887	DB-2887	CP-SimDist-CB	AT-2887	–	–
ASTM Method 3710	–	–	AT-3710	–	–
PONA Analysis	–	CP Sil PONA CB	AT- Petro	–	Elite-PONA
Simulated Distillation	DB-HT SimDis	–	–	–	–
Amines and Basic Compounds	–	–	–	–	–
Fatty Acid Methyl Esters (FAMEs) (70% cyanopropyl polysilphenyl-siloxane)	–	–	AT-FAME	–	–
Blood Alcohol Analysis	DB-ALC1, DB-ALC2, HP-Blood Alcohol	–	–	–	–
Residual Solvents in Pharmaceuticals	–	–	–	–	–
Residual Solvents in Pharmaceuticals	–	–	–	–	–
Fragrances and flavors	–	–	–	–	–
Explosives (8% phenyl polycarbonate-siloxane)	–	–	–	–	–
Dioxins and furans	–	–	–	–	–
PCB Congeners	–	–	–	–	–
Sulfur Analysis	–	–	AT-Sulfur	–	–

Phenomenex	Quadrex	Restek	SGE	Supelco	Thermo Scientific	US Pharmacopeia Designation
–	–	–	–	Carboxen-1010 PLOT, Carboxen-1006 PLOT	–	–
–	PLT-5 A	Rt-Msieve 5A, MXT-Msieve 5A	–	Mol Sieve 5A PLOT	TR-Bond Msieve 5A	–
–	–	Rt-ßDEXm	Cydex-B	CHIRALDEX, Supelco DEX	–	–
–	–	–	–	–	–	–
–	–	–	–	–	–	–
–	–	Rt-TCEP	–	TCEP	–	–
ZB-MultiResidue-1	–	Rtx-CLPesticides, Stx-CLPesticides, Rtx-440	–	MET-Biodiesel	TG-SQC	–
ZB-MultiResidue-2	–	Rtx-CLPesticides2, Stx-CLPesticides2	–	–	TG-VRX	–
ZB-Drug-1	–	Stx-500	–	–	TG-VMS	–
ZB-Bioethanol	–	–	–	–	–	–
–	–	–	–	–	–	–
–	–	–	–	–	–	–
–	–	Rtx-VMS, Rtx-VGC	BPX-VOLATILES	VOCOL	TR-524	–
OV-624	007-624, 007-502	Rtx-VRX, Rtx-502.2, Rtx-624, Rtx-Volatiles	BP624	SPB-624	TR-525	–
–	–	–	–	Sub-Herb	–	–
–	–	–	–	–	TR-8270	–
–	007-608	Rtx-CLPesticides, Rtx-CLPesticides2, Stx-CLPesticides, Stx-CLPesticides2	–	–	TG-OCP I/II	–
–	–	Rtx-OPPesticides, Rtx-OPPesticides2	–	–	TG- OPP I/II	–
ZB-5, ZB-35, ZB-1701, ZB-50	007-2, 007-608, 007-17, 007-1701	Rtx-5, Rtx-35, Rtx-50, Rtx-1701	BP-5, BP-10, BP-608	SPB-5, SPB-608, SPB-1701	–	G3
–	007-1-10V-1.0F	Rtx-2887	–	Petrocol 2887, Petrocol EX2887	TR-SimDist	–
–	–	–	–	–	–	–
–	007-1-10V-0.5F	Rtx-1PONA	BP1 PONA	Petrocol DH	–	–
ZB-1XT SimDist	–	MXT-500 Sim Dist	HT 5	Petrocol 2887. EX2887	TR-SimDist	–
–	–	Rtx-5 Amine	–	PTA-5	–	G50
–	–	FAMEWAX	–	Omegawax	TR-FAME	–
ZB-BAC-1, ZB-BAC-2	–	Rtx-BAC1, Rtx-BAC2	–	–	TG ALC I/II	–
–	–	Rtx-G27	–	–	–	G27
–	–	Rtx-G43	–	OVI-G43	–	G43
–	007-CW	Rt-CW20M F&F	BP20M	–	–	–
–	–	Rtx-TNT, Rtx-TNT2	–	–	TR-8095	–
–	–	Rtx-Dioxin, Rtx-Dioxin2	–	–	TR-Dioxin5MS	–
–	–	Rtx-PCB	–	–	TR-PCB8MS	–
–	–	–	–	–	–	–

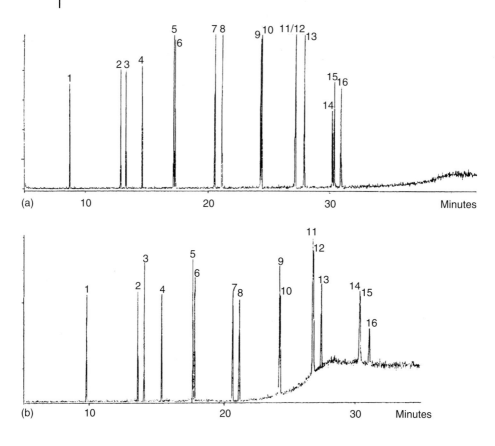

(a)   10   20   30   Minutes

(b)   10   20   30   Minutes

Polynuclear aromatic hydrocarbons

(a) Phase:	HT5, 0.10 µm film thickness	(b) Phase:	BP1, 0.25 1 µm film thickness
Column :	25 m × 0.22 mm ID	Column :	25 m × 0.22 mm ID
Start temperature:	50 °C, 2 min	Start temperature:	50 °C, 2 min
Program rate:	10 °C/min	Program rate:	10 °C/min
Final temperature:	420 °C, 5 min	Final temperature:	300 °C, 10 min
Detector :	HP 5971 MS	Detector :	HP 5971 MS
Scan range:	35 – 550 Da	Scan range:	35 – 550 Da
Injection:	Split 50 : 1	Injection:	Split 50 : 1

1. Naphthalene
2. Acenaphthylene
3. Acenaphthene
4. Fluorene
5. Phenanthrene
6. Anthracene

7. Pyrene
8. Fluoranthene
9. Chrysene
10. Benzo(a)anthracene
11. Benzo(b)fluoranthene
12. Benzo(k)fluoranthene

13. Benzo(a)pyrene
14. Indeno(1,2,3,-cd)pyrene
15. Dibenzo(ah)anthracene
16. Benzo(ghi)perylene (0.5 ng per component)

Figure 2.79 Comparison of a high temperature phase (a) SGE HT5 (siloxane-carborane) with a conventional silicone phase (b) SGE BP1 (dimethylsiloxane) using a polyaromatic hydrocarbon standard (SGE). The long temperature program for the high temperature phase up to 420 °C allows the elution of components 14, 15 and 16 during the oven temperature ramp at an elution temperature of about 350 °C. Sharp narrow peaks with low column bleed give better detection conditions for high-boiling substances (GC-MS system HP-MSD 5971).

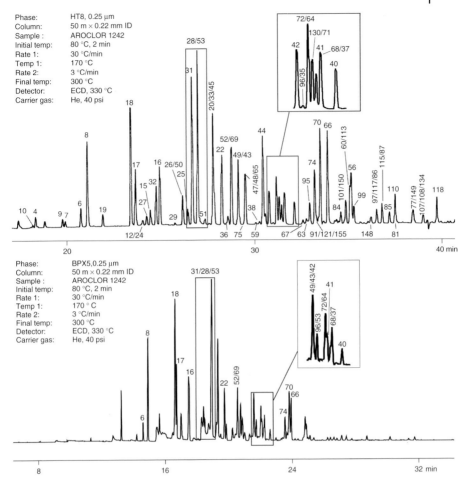

Figure 2.80 An example of the different selectivities of stationary phases for the Aroclor mixture 1242 (see in particular the separation of the critical congeners 31, 28 and 53. (Courtesy SGE.) (a) Carborane phase – HT8 50 m × 0.22 mm × 0.25 µm and (b) 5%-Phenyl phase – BPX5 50 m × 0.22 mm × 0.25 µm.

2.2.7.1   Sample Capacity

The sample capacity is the maximum quantity of an analyte with which the phase can be loaded (Table 2.26). An overloaded column exhibits peak fronting, an asymmetrical peak which has a gradient on the front and a sharp slope on the back side. This effect can increase until a triangular peak shape is obtained, a so-called shark fin. Overloading occurs rapidly if a column of the wrong polarity is chosen. The capacity of a column depends on the internal diameter, the film thickness, polarity and the solubility of a substance in the phase.

Table 2.26 Sample capacities and optimum flow rates for common column diameters.

Internal diameter (mm)	0.18	0.25	0.32	0.53
Film thickness (µm)	0.20	0.25	0.25	1.00
Sample capacity (ng)	<50	50–100	400–500	1000–2000
Theoretical plates per metre of column	5300	3300	2700	1600
Optimum flow rate at				
20 cm/s helium (mL/min)	0.3	0.7	1.2	2.6
40 cm/s hydrogen (mL/min)	0.6	1.4	2.4	5.2

Sample Capacity

- Increases with internal diameter
- Increases with film thickness
- Increases with solubility

2.2.7.2 Internal Diameter

The internal diameter of columns used in capillary GC varies from 0.1 mm (micro-bore capillary) via 0.15 and 0.18 mm (narrow-bore) and the standard columns with 0.25 (normal) and 0.32 mm (wide-bore) to 0.53 mm. For the direct coupling with mass spectrometers, in practice only the columns up to 0.32 mm internal diameter are used. Mega-bore columns with a diameter of 0.53 mm (halfmil) are mainly used for pre-columns (retention gap, deactivated, no stationary phase) or to replace packed 1/8″ steel columns in specially engineered GC instruments.

The internal diameter affects the resolving power and the analysis time (Figure 2.81). At constant film thicknesses, lower internal diameters are preferred in order to achieve higher chromatographic resolution. As the flow per unit time decreases at a particular carrier gas velocity, the analysis time increases. In the case of complex mixtures, changing to a column with a smaller internal diameter gives better separation of critical pairs of compounds (Table 2.27). In practice, it has been shown that even changing from 0.25 mm internal diameter to 0.20 mm allows an improvement in the separation of, for example, PCBs. Another advantage of using a narrow-bore column is that the optimal linear velocity of the carrier gas also increases, which allows shorter analysis time. The optimal linear velocity that provides the highest efficiency is 32.7 cm/s for 0.15 mm ID compared to 28.5 cm/s for the conventional 0.25 mm ID GC column (van Ysacker, 1993; Khan, 2013) as can be read from the HETP plots (height equivalent to a theoretical plate) for capillary GC columns of different diameter in Figure 2.82.

Smaller Internal Diameter (at identical film thickness)

- Increases the resolution
- Decreases the analysis time.

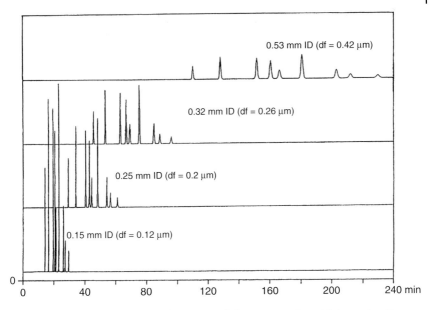

Figure 2.81 Effect of increasing column internal diameter together with increasing film thickness on peak height and retention time. The columns have the same phase ratio! (Chrompack).

Table 2.27 Effect of column diameter and linear carrier gas velocity on the flow rate.

Internal diameter (mm)	Linear carrier gas velocity		Flow rate	
	He (cm/s)	H$_2$ (cm/s)	He (mL/min)	H$_2$ (mL/min)
0.18	30−45	45−60	0.5−0.7	0.7−0.9
0.25	30−45	45−60	0.9−1.3	1.3−1.8
0.32	30−45	45−60	1.4−2.2	2.2−2.8
0.54	30−45	45−60	4.0−6.0	6.0−7.9

2.2.7.3 Film Thickness

The variation in the film thickness at a given internal diameter and column length gives the user the possibility for optimization of special separation tasks. As a rule, thick films are used for volatile compounds and thin films for high-boilers and trace analysis.

Thick film columns with coatings of more than 1.0 µm can separate very low boiling compounds well, for example, volatile halogenated hydrocarbons. Through the large increase in capacity with thicker films, it is even possible to dispense with additional oven cooling during injection in the analysis of volatile halogenated hydrocarbons using headspace or P&T. Start temperatures above

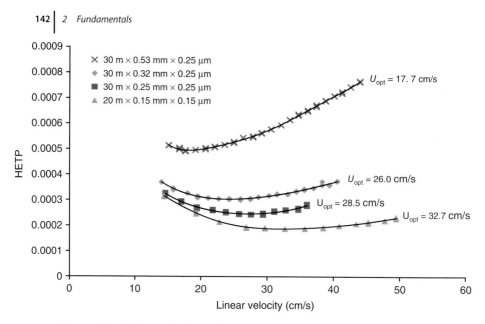

Figure 2.82 Influence of column diameter on optimal gas velocity and separation power (HETP).

room temperature are usual. However, thick film columns exhibit increased column bleed at elevated temperatures.

For the trace residue analysis of all other substances, thin film columns with coating thicknesses of about 0.1 μm to 0.25 μm have proved to be very effective in GC-MS. Thin film columns give narrow rapid peaks and can be used in higher temperature ranges without significant column bleed. The elution temperatures of the compounds decrease with thin films and, at the same program duration, the analysis can be extended to compounds with higher molecular weights and thermolabile compounds, for example, the decabromodiphenylether (PBDE 209). The duration of the analysis for a given compound becomes shorter, but the capacity of the column decreases also limiting the load for matrix samples (Figure 2.83).

For the analysis of polar compounds it needs to be considered that thicker films provide higher inertness and less residual column activity by better shielding from any polar remaining free glass silanol groups.

Increasing Film Thickness

- Improves the resolution of volatile compounds
- Increases the analysis time
- Increases the analyte elution temperatures

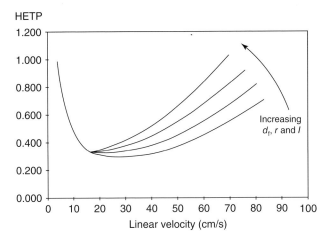

HETP

Linear velocity (cm/s)

Increasing
d_f, r and l

Figure 2.83 Effect of film thickness (d_f), internal diameter (r) and length (l) on the optimum carrier gas flow (van Deemter curves for helium as the carrier gas).

The Relationship between Film Thickness and Internal Diameter

The phase ratio of a capillary column is determined by the ratio of the volume of the gaseous mobile phase (internal volume) to the volume of the stationary phase (coating). From Table 2.28, the phase ratio can be read for each combination of film thickness and internal diameter (assuming the same film type and the same column length). High values mean high separation power. The same values show combinations with the same separating capacity. For GC-MS, optimal separations can be planned and also other conditions, such as carrier gas flow and column bleed can be taken into consideration. To achieve better separation, it is possible to change to a smaller film thickness at the same internal diameter or to keep the film thickness and choose a higher internal diameter. For example, a fast GC column of 0.18 mm ID and 0.10 μm film has almost the double phase ratio than the commonly used 0.25 mm ID column with 0.25 μm film. The phase ratio is tripled when switching to a 0.25 mm ID column with 0.1 μm film which is typical for trace analysis applications. Using Table 2.28, the separation efficiency can easily be optimized to the required conditions.

2.2.7.4 Column Length

The analytical column should be as short as possible. The most common lengths for standard columns are 30 or 60 m. Fast GC separations go with 10 or 15 m lengths. Greater lengths are usually not necessary in general residue analysis with GC-MS systems, and are reserved for special well-documented separation purposes as known for instance for FAMEs, or the complex PCB analyses.

Shorter columns would be desirable for simpler separations, but they are with the same diameter at the limit of the maximum flow for the mass spectrometer

Table 2.28 Effect of column diameter and film thickness on the phase ratio (separation power).

Internal diameter (mm)	Film thickness							
	0.10 μm	0.25 μm	0.50 μm	1.0 μm	1.50 μm	2.0 μm	3.0 μm	5.0 μm
0.18	450	180	90	45	30	23	15	9
0.25	625	250	125	63	42	31	21	13
0.32	800	320	160	80	53	40	27	16
0.53	1325	530	265	128	88	66	43	27

used. Here the switch to fast GC applications using smaller diameters should be considered. Doubling the column length only results in an improvement in the separation by a factor of 1.4 ($\sqrt{2}$) while the analysis time is almost doubled (and the cost of the column also!). For isothermal chromatography, the retention time is directly proportional to the column length. With programmed operations the increased heating rates lead to higher elution temperatures of the compounds. On changing to a longer column, the temperature program should always be optimized again to achieve optimal retention times.

Doubling the Column Length

- Resolution only increases by a factor of 1.4
- Costs are doubled
- Retention times are doubled
- Sample throughput (productivity) is cut by half
- The temperature program must be optimized again

2.2.7.5 Setting the Carrier Gas Flow

The maximum separating capacity of a capillary column can be exploited only with an optimized carrier gas flow. With direct GC-MS coupling, the carrier gas flow is affected slightly by the vacuum on the detector side. In practice, the adjustment is no different from that in classical GC systems. The separating efficiency of a capillary column is given as theoretical plates in chromatographic terminology (see Section 2.2.8). A high analytical separating capacity is always accompanied by a large number of plates (number of separation steps) so the height equivalent HETP decreases. In van Deemter curves, the HETP is plotted against the carrier gas velocity (not flow!). The minimum of one of these curves gives the optimal adjustment for a particular carrier gas for isothermal and programmed constant flow operations.

For helium, the optimum carrier gas velocity for standard columns is about 24 cm/s. As the viscosity of the carrier gas increases in the course of the oven

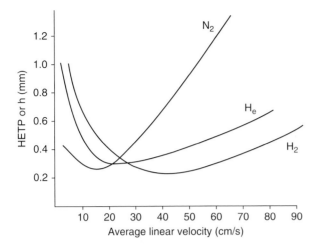

Figure 2.84 van Deemter curves for nitrogen, helium and hydrogen as the carrier gas. The values refer to a standard column of 30 m length, 0.25 mm internal diameter and 0.25 µm film thickness (Restek Corporation, 1993).

temperature program, in practice, at the lower start temperature a higher velocity is used (about 30 cm/s) with constant pressure setups. Velocities which are too high lower the efficiency only marginally. The use of constant flow conditions with EPC is recommended to maintain the carrier gas flow in optimum conditions with temperature ramped analyses.

With hydrogen as the carrier gas a much higher velocity (>40 cm/s) can be used, which leads to significant shortening of the analysis time. Furthermore, with hydrogen, the right hand branch of the van Deemter curve is more flat, so that a further increase in the gas velocity is possible without impairing the efficiency (Figures 2.84 and 2.85). The separating efficiency is retained and time is gained. Because of the often very limited pumping capacity in commercial GC-MS systems (see Section 2.3.6), only limited use can be made of these advantages with hydrogen. Caused by the expected shortage of the natural helium supply and the already increasing prices hydrogen becomes an economical alternative. Hydrogen generators are available for a safe and on-site premium quality supply.

2.2.7.6 Properties of Column Phases

GC columns with polysiloxane stationary phases are most commonly used (Figures 2.86 to 2.100). The polysiloxane backbone is either 100% methyl substituted, or in large variety modified with other functional groups defining the phase polarity and interaction. Modifications of the backbone as polysilarylene or siloxancarborane improves the thermal stability, a criterion of special importance for GC-MS trace analysis applications. The polar polyethylene glycol stationary

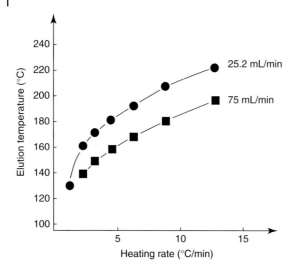

Figure 2.85 The effect of heating rate and carrier gas flow on the elution temperature (retention temperature) (Karasek and Clement, 1988).

phases (PEG) are not substituted. As the PEG film is less stable they are operated at lower temperature limits.

Chemically bonded stationary phases are covalently bonded to the fused silica surface, are highly temperature stable, and can be solvent rinsed for cleaning from matrix contaminations. In contrast, cross-linked stationary phases are not bonded to the surface, only use chemical links between the polymer chains.

Oxygen degrades the stationary phases. For column storage it is recommended to seal the ends with a septum, and store in a dark place, typically the column box, to prevent stationary phase damage by oxidation, moisture and UV light. Not used installed GC columns should get a low carrier gas flow with split open at low flow as well (prepare a standby method at low oven temperature). All column manufacturers provide upper and lower temperature limits for use of the individual columns and stationary phases:

- Low temperature limit: Below this temperature only limited phase interaction occurs with loss of peak shape and efficiency.
- Isothermal upper limit: The column film is stable for operation up to this temperature for long time with low bleed. The isothermal temperature limit need to be considered when setting the transfer line temperature to the MS.
- Temperature program upper limit: The column should be used for up to 15 min only with GC oven temperature programs up to this maximum temperature above the isothermal limit. Column bleed increases visibly with a rising baseline, but no damage to the phase will occur. Extended operation at or above the maximum temperature increases bleed and shortens the column lifetime.

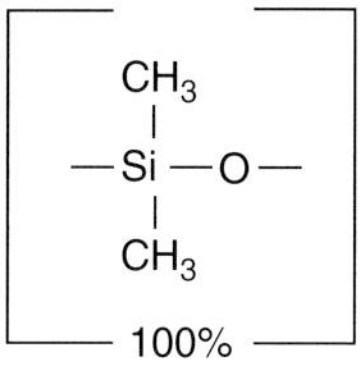

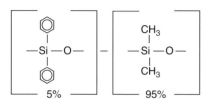

Figure 2.86 100% dimethyl-polysiloxane.

Polarity	Least polar bonded phase
Use	Boiling point separations for solvents, petroleum products, pharmaceuticals
Properties	Minimum temperature −60 °C Maximum temperature 340–430 °C Helix structure

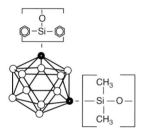

Figure 2.87 5% diphenyl-95% dimethyl-polysiloxane.

Polarity	Nonpolar, bonded phase
Use	Boiling point, point separations for aromatic compounds, environmental samples, flavours, aromatic hydrocarbons
Properties	Minimum temperature −60 °C Maximum temperature 340 °C

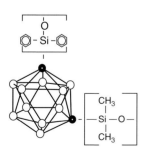

Figure 2.88 Siloxane-carborane, comparable to 5% phenyl.

| Use | • Ideal for GC-MS coupling because of very low bleeding • All environmental samples • All medium and high molecular weight substances • Polyaromatic hydrocarbons, PCBs, waxes, triglycerides |
| Properties | Minimum temperature −10 °C Maximum temperature 480 °C (highest operating temperature of all stationary phases, aluminium coated, 370 °C polyimide coated), high temperature phase |

Figure 2.89 Siloxane-carborane, comparable to 8% phenyl.

Polarity	Weakly polar, similar to 8% phenylsiloxane
Use	• Ideal for GC-MS coupling because of very low bleeding • All environmental samples, can be used universally • Volatile halogenated hydrocarbons, solvents – polyaromatic hydrocarbons, pesticides, only column that separates all PCB congeners
Properties	Minimum temperature −20 °C Maximum temperature 370 °C

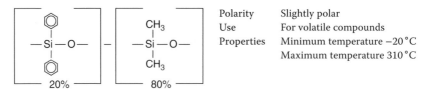

Polarity	Slightly polar
Use	For volatile compounds
Properties	Minimum temperature −20 °C
	Maximum temperature 310 °C

Figure 2.90   20% diphenyl-80% dimethyl-polysiloxane.

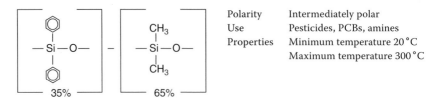

Polarity	Intermediately polar
Use	Pesticides, PCBs, amines
Properties	Minimum temperature 20 °C
	Maximum temperature 300 °C

Figure 2.91   35% diphenyl-65% dimethyl-polysiloxane.

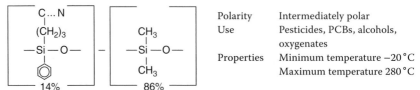

Polarity	Intermediately polar
Use	Pesticides, PCBs, alcohols,
	oxygenates
Properties	Minimum temperature −20 °C
	Maximum temperature 280 °C

Figure 2.92   14% cyanopropylphenyl-86% dimethyl-polysiloxane.

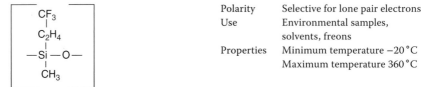

Polarity	Selective for lone pair electrons
Use	Environmental samples,
	solvents, freons
Properties	Minimum temperature −20 °C
	Maximum temperature 360 °C

Figure 2.93   100% trifluoropropylmethyl-polysiloxane.

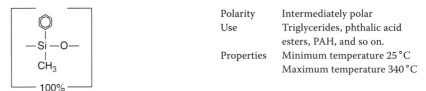

Polarity	Intermediately polar
Use	Triglycerides, phthalic acid
	esters, PAH, and so on.
Properties	Minimum temperature 25 °C
	Maximum temperature 340 °C

Figure 2.94   100% phenyl-methyl-polysiloxane.

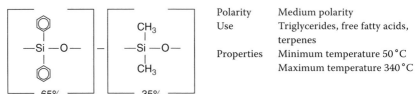

Polarity	Medium polarity
Use	Triglycerides, free fatty acids,
	terpenes
Properties	Minimum temperature 50 °C
	Maximum temperature 340 °C

Figure 2.95   65% diphenyl-35% dimethyl-polysiloxane.

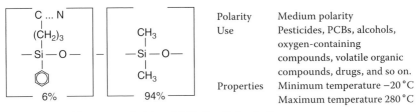

Polarity	Medium polarity	
Use	Pesticides, PCBs, alcohols, oxygen-containing compounds, volatile organic compounds, drugs, and so on.	
Properties	Minimum temperature $-20\,°C$ Maximum temperature $280\,°C$	

Figure 2.96   6% cyanoprophylphenyl-94% dimethyl-polysiloxane.

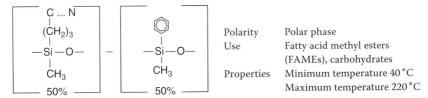

Polarity	Polar phase
Use	Fatty acid methyl esters (FAMEs), carbohydrates
Properties	Minimum temperature $40\,°C$ Maximum temperature $220\,°C$

Figure 2.97   50% cyanopropylmethyl-50% phenylmethyl-polysiloxane.

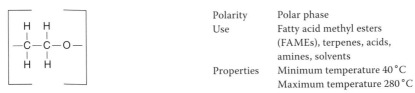

Polarity	Polar phase
Use	Fatty acid methyl esters (FAMEs), terpenes, acids, amines, solvents
Properties	Minimum temperature $40\,°C$ Maximum temperature $280\,°C$

Figure 2.98   100% carbowax polyethyleneglycol 20M.

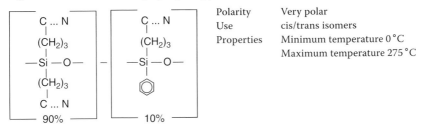

Polarity	Very polar
Use	cis/trans isomers
Properties	Minimum temperature $0\,°C$ Maximum temperature $275\,°C$

Figure 2.99   90% biscyanopropyl-10% phenylcyanopropyl-polysiloxane.

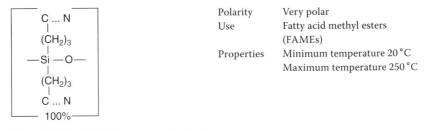

Polarity	Very polar
Use	Fatty acid methyl esters (FAMEs)
Properties	Minimum temperature $20\,°C$ Maximum temperature $250\,°C$

Figure 2.100   100% biscyanopropyl-polysiloxane.

2.2.7.7 Properties of Ionic Liquid Phases

Ionic liquids are a new class of stationary films in capillary GC with remarkable properties and benefits for GC-MS applications as well. Stable room-temperature ionic liquids (RTILs) are organic salts with melting points at or below room temperature (Berthod *et al.*, 2003). They can be applied for fused silica coating and form a liquid stationary film in which ions are present, but do not contain neutral molecules. The current phases are not (yet) chemically bonded. They are made of the equal number of positive and negative ions so that the liquid is electrically neutral. Although not new and already known for various chemical applications, a first evaluation for use in GC was performed by Armstrong in 1999. The high viscosity, high thermal stability and high polarity make them very suitable as alternative polar phases for GC separations (Vidal *et al.*, 2012). A first commercial ionic liquid GC column was introduced by Supelco in 2008 and further development has yielded additional phases of differing polarity. A range of columns is now available encompassing different phase types, from a polyethylene glycol (PEG) equivalent polarity with improved thermal stability SLB-IL59/60 to phases with even higher polarity SLB-IL111 (Supelco, 2013).

The main physicochemical properties of RTILs are very compatible with the requirements of capillary GC (Berthod, 2003). They remain liquid over a temperature range of up to 300 °C, are of high chemical stability and have practically no vapour pressure, promising a low bleed at high elution temperatures. The properties of RTILs depend on the nature and size of both their cation and anion constituents, and offer a high potential for customized phases to special applications. A general structure is shown in Figure 2.101. The combination of high polarity with high temperature stability along with a new selectivity mechanism offers a great potential for future GC and GC-MS applications (Armstrong *et al.*, 2009).

The retention mechanism is different from the traditional silicon phase systems. It appears from the available application reports that the phase solubility is of low importance (as of 'like separates like'). Retention of the analytes is mainly caused by electrostatic and dipole/dipole interactions. With increased polarities, peak widths of low polar analytes increase, for example, of petroleum hydrocarbons, due to a poor wetting of the phase surface, while polar analytes are increasingly retained (Whitmarsh, 2012). Ionic liquid stationary phases seem to have a

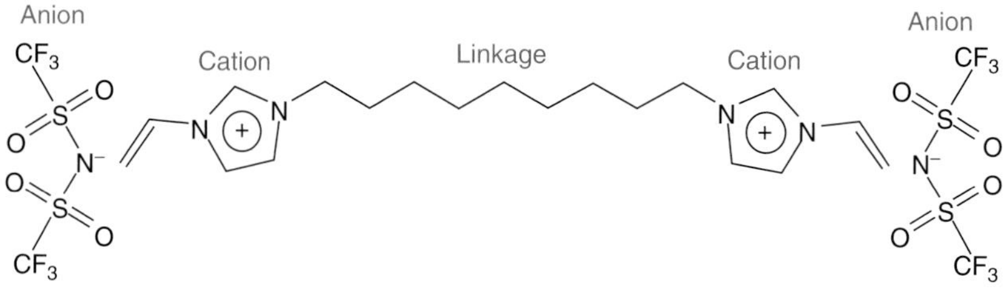

Figure 2.101 General anion/cation structure of ionic liquid phases, 1,9-di(3-vinyl-imidazolium) nonane bis(trifluoromethyl) sulfonyl imidate phase of the SLB-IL100 column (Supelco).

dual nature. They appear to act as a low-polarity stationary phase to non-polar compounds. However, molecules with strong proton donor groups, are strongly retained (Armstrong, 1999).

Analytical Benefits of Ionic Liquid Phases

- Greater thermal stability compared to polysiloxane polymers and PEGs. The traditional polar PEG phases show intense bleed in GC-MS applications, and are not useful for high temperature trace analysis applications.
- Lower column bleed, longer life and higher thermal limits, especially for polar column applications in GC-MS.
- More resistant to damage from moisture and oxygen.
- Numerous combinations of cations and anions are possible, allowing the design of 'tailored' selectivities, applications or functions.

As expected from the ionic liquid phase properties, increasing column polarity reduces the retention time for non-polar hydrocarbons, as shown in Figure 2.102 for fatty acid methyl esters (FAMEs), comparing a traditional polar poly(biscyanopropyl) siloxane phase with the most polar ionic liquid phase in SLB-IL111. Significant is the reduction in analysis time while maintaining the peak separation and the reduced column bleed. With the same temperature program, the compounds are eluting at lower oven temperature. The improvement in productivity occurs to such an extent that a C31 n-alkane elutes at almost half of the retention time on a polar ionic liquid phase (SLB-IL111) compared to a traditional 5% phenyl phase, using identical oven temperature programs (Whitmarsh, 2012).

In food safety FAMEs analysis, the separation of the fatty acid cis/trans compounds is a routine analysis of high importance. For a comparison, a 38 component FAME mix was analysed on the traditional polar wax and an ionic liquid column under identical conditions. The ion liquid column SLB-IL60 was chosen for the complimentary selectivity to PEG phases, but offers higher maximum temperatures and a significantly lower bleed. The resulting chromatograms are shown in Figure 2.103.

Analytical conditions for Figure 2.103 FAMEs separation: (a) Omegawax column, bonded phase poly(ethylene glycol), temperature limit 280 °C programmed. (b) SLB-IL60 column (bottom), non-bonded phase 1,12-di(tripropylphosphonium)dodecane bis(trifluoromethylsulfonyl)imide, temperature limit 300 °C programmed. GC conditions: Omegawax, 30 m × 0.25 mm ID, 0.25 µm. SLB-IL60: 30 m × 0.25 mm ID, 0.20 µm. Oven: 170 °C, 1 °C/min to 225 °C, injection temperature: 250 °C, carrier gas: helium, 1.2 mL/min, detector: FID, 260 °C. Injection: 1 µL, 100:1 split, liner: 4 mm ID, split/splitless type, single

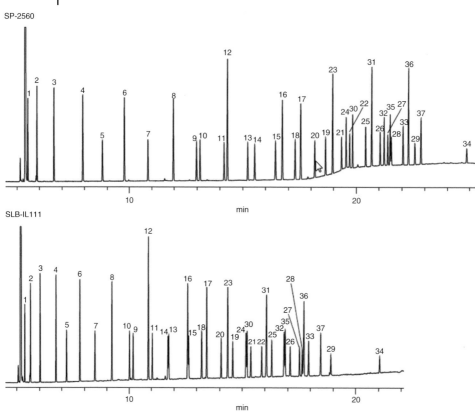

Figure 2.102 Comparison of an ionic liquid phase with classical highly polar phase system (poly(biscyanopropyl siloxane phase). (Supelco.)

tapered wool packed FocusLiner™. Sample: Supelco 37-Component FAME Mix + C22:5n3, in methylene chloride.

The separation of the FAMEs compounds on the ionic liquid SLB-IL60 column is achieved in a much shorter runtime of 36 versus 52 min on a standard Omegawax column, which is a 30% gain in analysis time with even improved separation quality (Figure 2.103). The resolution of the *cis/trans* oleic acids C18:1n9c (peak 17) and C18:1n9t (peak 18) is well achieved by the differing selectivity of the ionic liquid column. The *trans* oleic acid C18:2n6t (peak 20) also elutes before the cis compound C18:2n6c (peak 19) and shows a very clear baseline separation (Stenerson, *et al.*, 2013, 2014).

Ionic liquids provide a different selectivity from traditional silicon polymer GC phases. They are not susceptible to same stability issues as siloxane and PEG-based phases, allow higher maximum temperatures than traditional phases of comparable polarity. The extended temperature operating range is best seen in comparison with the known PEG phases. GC-MS applications benefit from a lower bleed and achieve the compound elution at lower oven temperatures.

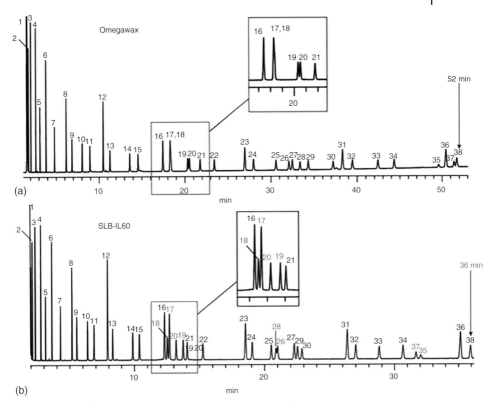

Figure 2.103 FAMEs separation for oleic and linolenic acid cis/trans analysis (#17/18, #19/20) on a traditional wax (a) and ionic liquid phase (b) column. Legend of the FAMEs peak numbers of Figure 2.103: 1. C4:0, 2. C6:0, 3. C8:0, 4. C10:0, 5. C11:0, 6. C12:0, 7. C13:0, 8. C14:0, 9. C14:1, 10. C15:0, 11. C15:1, 12. C16:0, 13. C16:1, 14. C17:0, 15. C17:1, 16. C18:0, 17. C18:1n9c, 18. C18:1n9t, 19. C18:2n6c, 20. C18:2n6t, 21. C18:3n6, 22. C18:3n3, 23. C20:0, 24. C20:1n9, 25. C20:2, 26. C20:3n6, 27. C21:0, 28. C20:3n3, 29. C20:4n6, 30. C20:5n3, 31. C22:0, 32. C22:1n9, 33. C22:2, 34. C23:0, 35. C22:5n3, 36. C24:0, 37. C22:6n3 and 38. C24:1n9. (Supelco.)

Ionic liquid column phases extend the polarity range of GC phases, improve lab productivity by shorter run times and can be used for a wide variety of applications.

2.2.8
Chromatography Parameters

All chromatography processes are based on the multiple repetition of minute separation steps, such as the continuous dynamic partition of the components between two phases (Martin and Synge, 1941; Bock, 1974).

In a model, chromatography can be regarded as a continuous repetition of partition steps. The starting point is the partition of a substance between two phases

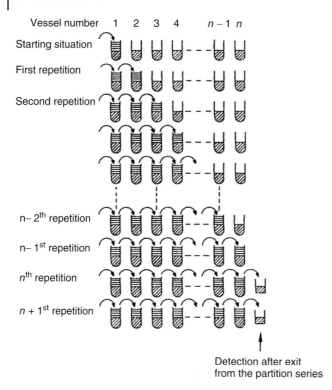

Vessel number   1 2 3 4 $n-1$ n

Starting situation

First repetition

Second repetition

$n-2^{th}$ repetition

$n-1^{st}$ repetition

n^{th} repetition

$n+1^{st}$ repetition

Detection after exit
from the partition series

Figure 2.104 Partition series: mode of operation with two auxiliary phases.

in a separating funnel. Suppose a series of separating funnels is set up which all contain the same quantity of phase

1) As this phase remains in the separating funnels, it is called the *stationary phase*. The sample is placed in the first separating funnel dissolved in a second auxiliary phase.
2) After establishing equilibrium through shaking, phase 2 is transferred to the next separating funnel. The auxiliary phase thereby becomes the mobile phase. Fresh mobile phase is placed into the first separating funnel, and the process goes on for a large number of n separation funnels (Figure 2.104).

The results of this type of partition with 100 vessels and two analytes A and B are shown in Figure 2.105. The prerequisite for this is the validity of the Nernst equation. For detection, the concentrations of A and B in the vessels are determined.

With the model described, so many separating steps are carried out that the mobile phase leaves the system of 100 vessels and the individual components A and B are removed, one after the other, from the series of vessels. This process is known as *elution*.

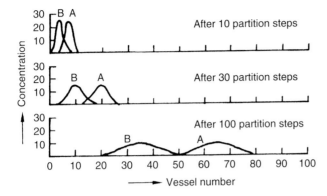

Figure 2.105 Partition of substances A and B after 10, 30 and 100 separating steps. After 10 steps A and B are hardly separated, after 30 steps quite well and after 100 steps practically completely. The two substances are partitioned among an increasing number of vessels and the concentrations decrease more and more ($a_A = 2 : a_B = 0.5$).

2.2.8.1 The Chromatogram and Its Meaning

The substances eluted are transported by the mobile phase to the detector and are registered as Gaussian curves (peaks). The peaks give qualitative and quantitative information on the mixture investigated.

- *Qualitative*: The retention time is the time elapsing between injection of the sample and the appearance of the maximum of the signal at the detector. The retention time of a component is always constant under the same chromatographic conditions. A peak can therefore be identified by a comparison of the retention time with a standard (pure substance).
- *Quantitative* : The height and area of a peak is proportional to the quantity of substance injected. Unknown quantities of substance can be determined by a comparison of the peak areas (or heights) with known concentrations.

In the ideal case, the peaks eluting are in the shape of a Gaussian distribution (bell-shaped curve, Figure 2.106). A very simple explanation of this shape is the different paths taken by the molecules through the separating system (multipath effect), which is caused by diffusion processes (Eddy diffusion) (Figure 2.107).

Under defined conditions, the time required for elution of a substance A or B at the end of the separating system, the retention time t_R, is characteristic of the substance. It is measured from the start (sample injection) to the peak maximum (Figure 2.108).

At a constant flow rate, t_R is directly proportional to the retention volume V_R.

$$V_R = t_R \cdot F \tag{2.15}$$

where $F =$ flow rate in mL/min.

The retention volume shows how much mobile phase has passed through the separating system until half of the substance has eluted (peak maximum!).

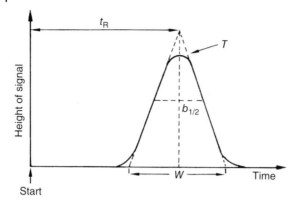

Figure 2.106 Parameters determined for an elution peak t_R retention time, $b_{1/2}$ half width and W base width.

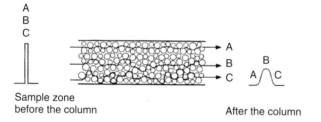

Figure 2.107 Eddy diffusion in packed columns (multipath effect) (Meyer, 2010).

2.2.8.2 Capacity Factor k′

The retention time t_R depends on the interaction of the analyte with the stationary phase, but also on the flow rate of the mobile phase and the length of the column. If the mobile phase moves slowly or the column is long, t_0 is large and so is t_R. Thus, t_R is not suitable for the comparative characterization of a substance, for example, between two laboratories. It is better to use the capacity factor, also known as the k' value, which relates the net retention time t_R' to the dead time:

$$k' = \frac{t_R'}{t_0} = \frac{t_R - t_0}{t_0} \tag{2.16}$$

Thus, the k' value is independent of the column length and the flow rate of the mobile phase and represents the molar ratio of a particular component in the stationary and mobile phases. Large k' values mean long analysis times.

The k' value is related to the partition coefficient K as follows:

$$k' = K \cdot \frac{V_1}{V_g} \tag{2.17}$$

where V_1 = volume of the liquid stationary phase and V_g = volume of the mobile phase.

The capacity factor is, therefore, directly proportional to the volume of the stationary phase (or for adsorbents, their specific surface area in m^2/g).

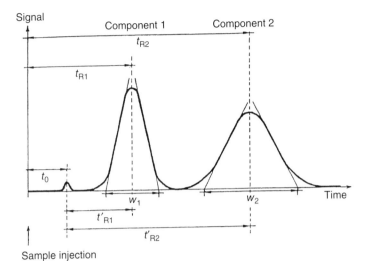

Figure 2.108 The chromatogram and its parameters (Meyer, 2010).

W Peak width of a peak. $W = 4\sigma$ with σ = standard deviation of the Gaussian peak.

t_0 *Dead time* of the column; the time which the mobile phase requires to pass through the column. The linear velocity u of the solvent is calculated from

$$u = \frac{L}{t_0} \quad \text{with} \quad L = \text{length of the column}$$

A substance which is not retarded, that is, a substance which does not interfere with the stationary phase, appears at t_0 at the detector.

t_R *Retention time:* the time between the injection of a substance and the recording of a peak maximum.

t_R' *Net retention time.* From the diagram it can be seen that $t_R = t_0 + t_R'$. t_0 is the residence time in the *mobile* phase. The substances separated differ in their residence times in the *stationary* phase t_R'. The longer a substance stays dissolved in the stationary phase, the later it is eluted.

α is a measure of the relative retention and is given by:

$$\alpha = \frac{k_2'}{k_1'} = \frac{K_2}{K_1} \quad (k_2' > k_1') \tag{2.18}$$

In the case where $\alpha = 1$, the two components 1 and 2 are not separated because they have the same k' values.

The relative retention α is thus a measure of the selectivity of a column and can be manipulated by choice of a suitable stationary phase. (In principle, this is also true for the choice of the mobile phase, but in GC-MS helium or hydrogen are, in fact, always used.)

2.2.8.3 Chromatographic Resolution

A second model, the theory of plates, was developed by Martin and Synge in 1941. This is based on the functioning of a fractionating column, then as now a widely used separation technique. It is assumed that the equilibrium between two phases on each plate of the column has been fully established. Using the plate theory, mathematical relationships can be derived from the chromatogram, which are a practical measure of the sharpness of the separation and the resolving power.

The chromatography column is divided up into theoretical plates, that is, into column sections in the flow direction, the separating capacity of each one corresponding to a theoretical plate. The length of each section of column is called the *height equivalent to a theoretical plate (HETP)*. The HETP value is calculated from the length of the column L divided by the number of theoretical plates N:

$$\text{HETP} = \frac{L}{N} \text{ [mm]} \tag{2.19}$$

The number of theoretical plates is calculated from the shape of the eluted peak. In the separating funnel model, it is shown that with an increasing number of partition steps the substance partitions itself between a larger number of vessels. A separation system giving sharp separation concentrates the substance band into a few vessels or plates. The more plates there are per length of a separation system, the sharper the eluted peaks. A column is more effective, the more theoretical plates it has (Figure 2.109).

The number of theoretical plates N is calculated from the peak profile. The retention time t_R at the peak maximum and the width at the base of the peak measured as the distance between the cutting points of the tangents to the inflexion points with the base line are determined from the chromatogram (see Figure 2.106).

$$N = 16 \cdot \left(\frac{t_R}{W}\right)^2 \tag{2.20}$$

where t_R = retention time and W = peak width.

For asymmetric peaks, the half width (the peak width at half height) is used:

$$N = 8 \ln 2 \cdot \left(\frac{t_R}{W_h}\right)^2 \tag{2.21}$$

where t_R = retention time and W_h = peak width at half height.

The width of a peak in the chromatogram determines the resolution of two components at a given distance between the peak maxima (Figure 2.110). The resolution R is used to assess the quality of the separation:

$$R \approx \frac{\text{retention difference}}{\text{peak width}}$$

The resolution R of two neighbouring peaks is defined as the quotient of the distance between the two peak maxima, that is, the difference between the two retention times t_R and the arithmetic mean of the two peak widths:

$$R = 2 \cdot \frac{t_{R2} - t_{R1}}{W_1 + W_2} = 1.198 \cdot \frac{t_{R2} - t_{R1}}{W_{h1} + W_{h2}}$$

where W_h = peak width at half height.

Figure 2.111 shows what one can expect optically from a value for R calculated in this way. At a resolution of 1.0, the peaks are not completely separated, but it

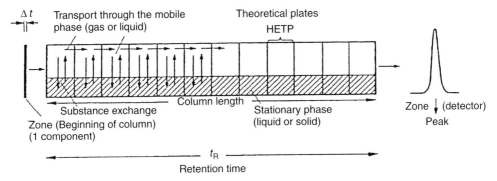

Figure 2.109 Substance exchange and transport in a chromatography column are optimal when there are as many phase transfers as possible with the smallest possible expansion of the given zones (Schomburg, 1981).

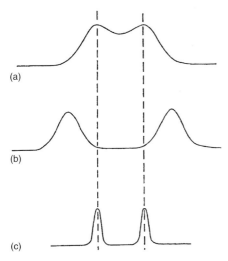

(a)

(b)

(c)

Figure 2.110 Resolution. (a/c) peaks with the same retention time, (a/b) Peaks with the same peak width, and (b/c) separation with the same resolution (Meyer, 2010).

can definitely be seen that there are two components. The tangents to the inflexion points just touch each other and the peak areas only overlap by 2%.

For the precise determination of the peak width, the tangents to the inflexion points can be drawn in manually (Figure 2.112). For a critical pair, for example, stearic acid (C_{18-0}) and oleic acid (C_{18-1}) the construction of the tangents is shown in Figure 2.113.

2.2.8.4 Factors Affecting the Resolution

Rearranging the resolution equation and putting in the capacity factor $k' = (t_R - t_0)/t_0$, the selectivity factor $\alpha = k'_2/k'_1$ and the number of theoretical plates N gives an important basic equation for all chromatographic elution processes.

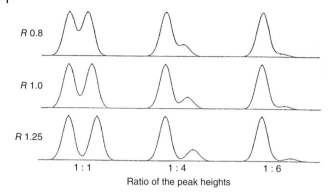

Figure 2.111 Resolution of two neighbouring peaks (Snyder and Kirkland, 1979).

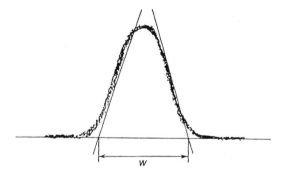

Figure 2.112 Manual determination of the peak width using tangents to the inflexion points (Meyer, 2010).

The resolution R is related to the selectivity α (relative retention), the number of theoretical plates N and the capacity factor k' by:

$$R = \frac{1}{4}(\alpha - 1) \times \frac{k'}{1 + k'} \times \sqrt{N}$$

$$\underbrace{\qquad}_{I} \quad \underbrace{\qquad}_{II} \quad \underbrace{\quad}_{III}$$

(2.22)

The Selectivity Term

R is directly proportional to $(\alpha - 1)$. An increase in the ratio of the partition coefficients leads to a sharp improvement in the resolution, which can be achieved, for example, by changing the polarity of the stationary phase for substances of different polarities.

As the selectivity generally decreases with increasing temperature, difficult separations must be carried out at as low a temperature as possible.

The change in the selectivity is the most effective of the possible measures for improving the resolution. As shown in Table 2.29, more plates are required to achieve the desired resolution when α is small.

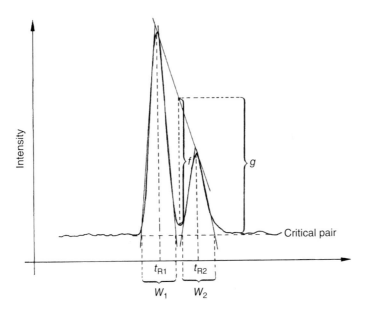

Figure 2.113 Determination of the resolution and peak widths for a critical pair.

Table 2.29 Relationship between relative retention α and the chromatographic resolution R.

Relative retention n	$R = 1.0$	$R = 1.5$
1.005	650 000 plates	1 450 000 plates
1.01	163 000	367 000
1.05	7 100	16 000
1.10	3 700	8 400
1.25	400	900
1.50	140	320
2.0	65	145

Figure 2.114 shows the effect of relative retention and number of plates on the separation of two neighbouring peaks:

- At high relative retention, the number of theoretical plates in the column does not need to be large to achieve satisfactory resolution (Figure 2.114a). The column is poor but the system is selective.
- A high relative retention and large number of theoretical plates give a resolution which is higher than the optimum. The analysis is unnecessarily long (Figure 2.114b).
- At the same (small) number of theoretical plates as in Figure 2.114a, but at a smaller relative retention, the resolution is strongly reduced (Figure 2.114c).

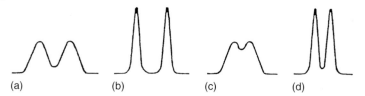

Figure 2.114 (a–d) Relative retention, number of plates and resolution.

- If the relative retention is small, a large number of theoretical plates are required to give a satisfactory resolution (Figure 2.114d).

The Retardation Term

Here the resolution is directly proportional to the residence time of a component in the stationary phase based on the total retention time. If the components stayed only in the mobile phase ($k' = 0$!), there would be no separation.

For very volatile or low molecular weight non-polar substances, there are only weak interactions with the stationary phase. Thus, at a low k' value the denominator $(1 + k')$ of the term is large compared with k' and R is therefore small.

This also applies to columns with a small quantity of stationary phase and column temperatures which are too high. To improve the resolution, a larger content of stationary phase can be chosen (greater film thickness).

The Dispersion Term

The number of plates N characterizes the performance of a column (Figure 2.115). However, the resolution R only increases with the square root of N. As N is directly

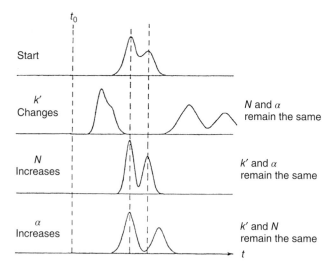

Figure 2.115 Effect of capacity factor, number of plates and relative retention on the chromatogram. (Snyder and Kirkland, 1979).

proportional to the column length L, the performance is only proportional to the square root of the column length.

Doubling the column length therefore only increases the resolution by a factor of 1.4. Since the retention time t_R is proportional to the column length, for an improvement in the resolution by a factor of 1.4, the analysis time is doubled.

2.2.8.5 Maximum Sample Capacity

The maximum sample capacity can be derived from the equations concerning the resolution. Under ideal conditions,

$$\frac{f}{g} = 100\% \tag{2.23}$$

where f = the area under the line connecting the peak maxima and g = the height of the connecting line above the base line, measured in the valley between the peaks, (see Figure 2.113).

The maximum sample capacity of a column is reached if f/g falls below 90% for a critical pair. If too much sample material is applied to a column, the k' value and the peak width are no longer independent of the size of the sample, which ultimately affects the identification and the quantitation of the results (Figure 2.116).

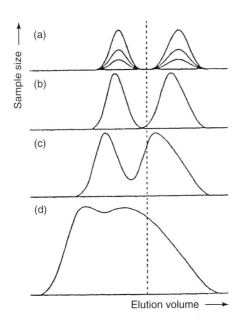

Figure 2.116 Change in the chromatogram with increasing sample size (Snyder and Kirkland, 1979). (a) Constant k' values, (b–d) Increasing changes to the retention behaviour through overloading.

2.2.8.6 Peak Symmetry

In exact quantitative work (integration of the peak areas), a maximum asymmetry must not be exceeded; otherwise, there will be errors in determining the cut-off point of the peak with the base line.

For practical reasons, the peak symmetry T is determined at a height of 10% of the total peak height (Figure 2.117):

$$T = \frac{b_{0.1}}{a_{0.1}} \tag{2.24}$$

where $a_{0.1}$ = peak width distance from the peak front to the maximum measured at 0.1 h. $b_{0.1}$ = peak width distance from the maximum to the end of the peak measured at 0.1 h.

T should ideally be 1.0 for a symmetrical peak, but for a practical quality measure, it should not be greater than 2.5. If the tailing exceeds higher values, there will be errors in the quantitative area measurement because the point where the peak reaches the base line is very difficult to determine and has large impact on the final peak area.

2.2.8.7 Optimization of Carrier Gas Flow

The flow of the mobile phase affects the rate of substance transport through the stationary phase. High-flow rates allow rapid separation. However, the efficiency is reduced because of the reduced exchange time of substances between the stationary and mobile phases (as of Figure 2.109) and leads to peak broadening. On the other hand, peak broadening caused by diffusion of the components within the mobile phase is only held up by increasing the flow rate.

The aim of flow optimization for given column properties at a given temperature and with a given carrier gas is to find the flow rate which gives either the maximum number of separation steps or at adequate efficiency the shortest possible analysis time.

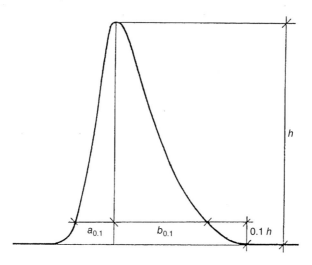

Figure 2.117 Asymmetric peak.

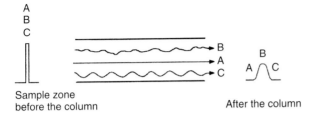

Sample zone
before the column

After the column

Figure 2.118   Path differences for laminar flow in capillary columns (multipath effect, caused by turbulence at high flow rates).

The HETP and the number of theoretical plates N depends on the flow rate of the mobile phase v according to van Deemter. The linear flow rate v of the mobile phase is calculated from the chromatogram:

$$v = \frac{L}{t_0} \qquad (2.25)$$

with L = length of the column in cm and t_0 = dead time in s.

The following affect the optimum flow rate:

1) The diffusion on the peak broadening. This effect is independent of flow and naturally. The Eddy diffusion has an impact on packed columns only (Figure 2.107), dependent on the nature of the packing material and the density of packing. For open capillary (tabular) columns (Figure 2.118), Eddy diffusion does not occur, but flow turbulences need to be taken into account.

2) The axial diffusion on peak broadening. This diffusion occurs in and against the direction of flow and decreases with increasing flow rate. Axial diffusion is also increased with isothermal oven programming sections, reduced with temperature programming.

3) Incomplete partition equilibrium. The transfer of analyte between the stationary and mobile phases only has a finite rate relative to that of the mobile phase, corresponding to the diffusion rates. The contribution to peak broadening increases with increasing flow rate of the mobile phase.

For the maximum efficiency of the separation (Figure 2.119), a flow rate u_{min} must be chosen as a compromise between these opposing effects. The position of the minimum is affected by:

- the quantity of the stationary phase (e.g., film thickness),
- the particle size of the packing material (for packed columns),
- the diameter of the column and
- the nature of the mobile phase (diffusion coefficient, viscosity).

Modern GC systems equipped with electronic pressure control (EPC) allow the setting of the optimum carrier gas flow independent of the oven temperature programming in *constant flow* mode. For practical reasons, the effective flow rates in *constant pressure* modes are set above the optimum flow rates as the increased carrier gas viscosity at higher oven temperature reduce the column flow during temperature programming. In this case, the right branch of the van Deempter

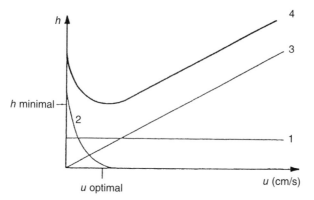

Figure 2.119 The van Deemter curve. (1). Proportion of Eddy diffusion and flow distribution at band broadening. (2). Proportion of longitudinal diffusion. (3). Proportion of substance exchange phenomena. (4). Resulting curve H(u), called the *van Deemter curve*.

graph needs to be considered where the slope of the curve is low (Figure 2.119). Also for increased speed and productivity a small loss in column resolution by increased flow rates is more than compensated by the advantage of a short analysis time which holds true for helium and especially for hydrogen.

In particular, in GC × GC, the flow rate of the second short column is accepted to be significantly above the ideal flow rate using the direct connection to the modulator. A split device would be needed between the modulator and second column for the independent adjustment of the flow rate. Most GC × GC applications do not use a flow adjustment for the second column because of the cut in substance concentration and hence sensitivity of the method. This is probably the most important factor, the overwhelming increase in peak separation and S/N by using a high-flow compromise.

Definition of Chromatographic Parameters

Carrier gas velocity	$v = L/t_0$	The average linear carrier gas velocity has an optimal value for each column with the lowest possible height equivalent to a theoretical plate (see van Deemter); v is independent of temperature, L = length of column and t_0 = dead time.
Partition coefficient	$K = c_1/c_g$	Concentration of the substance in the stationary phase (liquid) divided by the concentration in the mobile phase (gas). K is constant for a particular substance in a given chromatographic system.

	$K = k' \cdot \beta$	K is also expressed as the product of the capacity ratio (k') and the phase ratio (β).
Capacity ratio (partition ratio)	$k' = K \cdot V_1 / V_g$	Determines the retention time of a compound. V_1 = volume of the liquid stationary phase and V_g = volume of the gaseous mobile phase,
	$k' = (t_R - t_0)/t_0$	or by the degree of retention of an analyte relative to an unretained peak. t_R = retention time of the analyte and t_0 = dead time.
Phase ratio	$\beta = r/2d_f$	r = internal column radius and d_f = film thickness
Number of theoretical plates	$N = 5.54 \, (t_R/W_h)^2$	The number of theoretical plates is a measure of the efficiency of a column. The value depends on the nature of the substance and is valid for isothermal work. N = number of theoretical plates, t_R = retention time of the substance and W_h = peak width at half height.
HETP	$h = L/N$	Is a measure of the efficiency of a column independent of its length. L = length of the column and N = number of theoretical plates.
Resolution	$R = 2 \, (t_j - t_i)/(W_j + W_i)$	Gives the resolving power of a column with regard to the separation of components i and j (isothermally). t_i = retention time of substance i; t_j = retention time of substance j and $W_{i,j}$ = peak width at half height of substances i, j.
Separation factor	$\alpha = k'_j / k'_i$	Measure of the separation of the substances i, j.
Trennzahl number	$TZ = \frac{t_{R(x+1)} - t_{R(x)}}{W_{h(x+1)} + W_{h(x)}} - 1$	The Trennzahl number is, like the resolution, a means of assessing the efficiency of a column and is also used for temperature-programmed work. TZ gives the number of components which can be resolved between two homologous n-alkanes.

Effective plates	$N_{\text{eff.}} = 5.54 \left((t_{R(i)} - t_0) / W_{h(i)} \right)$	The effective number of theoretical plates takes the dead volume of the column into account.
Retention volume	$V_R = t_R \cdot F$	Gives the carrier gas volume required for elution of a given component. $F = $ carrier gas flow.
Kovats index	$KI = $ $100 \cdot c + 100 \dfrac{\log(t'_R)_x - \log(t'_R)_c}{\log(t'_R)_{c+1} - \log(t'_R)_c}$	The Kovats index is used for isothermal work. $t'_R = $ corrected retention times for standards and substances $t'_R = t_R - t_0$.
Modified Kovats index	$RI = 100 \cdot c + 100 \dfrac{(t'_R)_x - (t'_R)_c}{(t'_R)_{c+1} - (t'_R)_c}$	The modified Kovats index according to van den Dool and Kratz is used with temperature programming.

2.2.8.8 Effect of Oven Temperature Ramp Rate

The oven temperature ramp rates show a significant impact on the separation performance of a column. Increasing oven ramp rates reduces the column separation power and leads to higher elution temperature of compounds. In this case, efficiency cannot be used as a measure of column performance, instead peak width or peak capacity are generally used. The peak capacity (P_c) decreases by 22% as a temperature ramp rate increases from 10 to 20 °C/min. The practical implication is that the effective elution temperature of late-eluting analytes is also increased. The maximum operating temperature of the column may not be high enough to elute compounds. Hence, high oven temperature ramp rates increase the risk that high boiling compounds elute in the final isothermal phase with significant peak broadening, increased column bleed, and reduced S/N ratios. Also isothermal phases between oven temperature rates should be avoided, as this reduces separation power by diffusion. Alternatively reduced heating rates should be considered. Optimized moderate oven ramp rates reduce the effective compound elution temperatures and maintain high separation performance and the elution within the oven temperature ramp. Thermo labile compounds benefit in response from lower ramp rates with lower elution temperatures.

The example in Figure 2.120 displays the separation of a phenol mix with different GC oven heating rates. The speed of analysis increases with increasing oven temperature ramp rate at the expense of separation power and an increased analyte elution temperature. Each 16 °C/min increase in temperature ramp rate reduces the retention factor by 50%, but at reduced peak resolution. With a ramp rate of 20 °C/min pentachlorophenol (compound 11) elutes at 238 °C, while a ramp of 10 °C/min leads to a 30 °C less elution temperature of 208 °C, but with the trade-off of a 50% increased analysis time. However, if resolution is sufficient then temperature ramp rates should be optimized for increased productivity. In splitless injection for trace analysis, a quick jump from the low oven temperature below the solvent boiling point to a moderately high oven

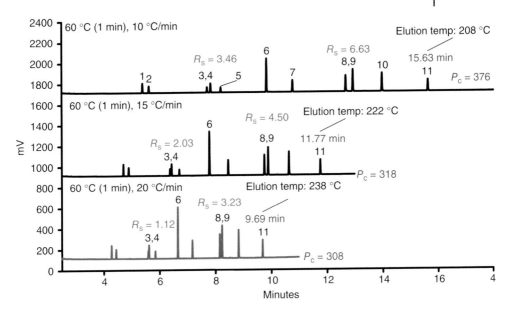

Figure 2.120 Effect of oven temperature ramp rate on analysis time, resolution and elution temperature (Khan, 2013). Experimental conditions: Column type TG-5MS, 30 m × 0.25 mm × 0.25 µm; inlet SSL at 250 °C; carrier gas 1.2 mL/min helium, constant flow; split injection 80 : 1; injection volume 1.0 µL; detector FID at 280 °C; oven temperature program as indicated with each chromatogram, Rs peak resolution #3,4 and #8,9. P_c peak capacity. Analytes: 1. phenol, 2. 2-chlorophenol, 3. 2-nitrophenol, 4. 2,4-dimethylphenol, 5. 2,4-dichlorophenol, 6. 4-chloro-3-methylphenol, 7. 2,4,6-trichlorophenol, 8. 2,4-dinitrophenol, 9. 4-nitrophenol, 10. 2-methyl-4,6-dinitrophenol and 11. pentachlorophenol.

temperature is preferred, with a following slow heating ramp for compound separation.

Further, optimization of the oven temperature ramp versus the carrier gas flow establishes the desired final separation conditions of short analysis time with still good peak resolution and a reduced elution temperature for the analytes. The example in Figure 2.121 demonstrates the win in analysis time of more than 50% from 16:08 to 7:36 min for pentachlorophenol at a low elution temperature of 208 °C.

2.2.9
Fast Gas Chromatography Solutions

The pressure of workload in modern analytical laboratories necessitates an increased throughput of samples, and hence method development is largely focussed on productivity (Figure 2.122). However, speed of chromatographic analysis and enhancement of peak separation generally require analytical method optimization in opposite directions. Basically, two adjacent approaches are

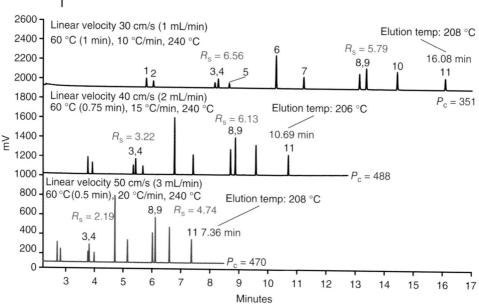

Figure 2.121 Effect of adjusting the linear velocity combined with increased oven temperature ramp rates (Khan, 2013). Experimental conditions and analytes as of Figure 2.120, Rs peak resolution #3,4 and #8,9.

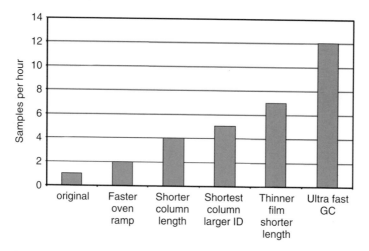

Figure 2.122 Increased sample throughput by progress in column technology.

increasingly used to combine advancements in speed and peak separation. In the existing conventional GC ovens, the use of narrow bore columns below the standard 0.25 mm ID offers a viable practical 'Fast GC' solution for almost every GC and GC-MS with electronic pressure regulation. Alternatively, the installation of cartridges for direct column heating are today commonly used even for 'Ultra-Fast' GC separations.

A recent publication has studied the performance of different columns featuring different lengths and internal diameters at different heating rates and with different carrier gas types and flow rates (Facchetti *et al.*, 2002).

2.2.9.1 Fast Chromatography

The term *Fast GC* compares to the conventionally used fused silica columns of typical dimensions with lengths of 30 m or longer and inner diameters of 0.25 mm or higher. These column types became standard since the introduction of the fused silica capillaries used for most applications in GC-MS. With improved and partially automated sample preparation methods, GC analysis time is becoming the rate determining step for productivity in many laboratories. A reduction of GC-MS analysis time is possible with the utilization of alternatively available techniques using a reduced column length and inner diameter with appropriate film thicknesses and oven temperature programming (Donato *et al.*, 2007; Khan, 2013). See Table 2.30 for an equivalence list for a 1:1 column replacement with fast GC capillary columns.

A very good example describing the progress in GC separation technology and advancement in GC productivity is given with an optimization strategy starting from a packed column separation (Facchetti *et al.*, 2002). The sample used for this comparison is the 16 component standard mixture used in EPA 610 analysis of PAHs (Table 2.31). The EPA method is antiquated and describes a 45 min analysis using a packed column, 1.8 m long × 2 mm ID glass, packed with 3% OV-17 on Chromosorb (EPA, 1997). All fused silica capillary columns used in this example were 5% phenyl-polysilphenylene-siloxane coated TRACE TR-5MS columns (Thermo Fisher Scientific, Bellefonte, PA, USA) with dimensions as specified with each chromatogram (Warden *et al.*, 2004). Helium is the carrier gas in each case. Flow rates and oven programs were optimized to the column dimensions.

Table 2.30 Capillary column dimensions that can be replaced 1 : 1 in order to achieve fast GC analysis.

Present column	Fast GC column
15 m × 0.25 mm × 0.25 μm	10 m × 0.15 mm × 0.15 μm
30 m × 0.25 mm × 0.25 μm	20 m × 0.15 mm × 0.15 μm
60 m × 0.25 mm × 0.25 μm	40 m × 0.15 mm × 0.15 μm
15 m × 0.32 mm × 0.25 μm	10 m × 0.15 mm × 0.15 μm
30 m × 0.32 mm × 0.25 μm	15 m × 0.15 mm × 0.15 μm
60 m × 0.32 mm × 0.25 μm	30 m × 0.15 mm × 0.15 μm

Table 2.31 Compounds in the EPA 610 standard on *packed* column.

Peak	Component	Retention time (min)
1	Naphthalene	4.50
2	Acenaphthylene	10.40
3	Acenaphthene	10.80
4	Fluorene	12.60
5	Phenanthrene	15.90*
6	Anthracene	15.90*
7	Fluoranthene	19.80
8	Pyrene	20.60
9	Benzo(*a*)anthracene	24.70*
10	Chrysene	24.70*
11	Benzo(*b*)fluoranthene	28.00*
12	Benzo(*k*)fluoranthene	28.00*
13	Benzo(*a*)pyrene	29.40
14	Dibenzo(*a*,*h*)anthracene	36.20*
15	Indeno(1,2,3-*cd*)pyrene	36.20*
16	Benzo(*ghi*)perylene	38.60

Note: Four pairs of peaks remain unresolved using the EPA 610 packed column technique, marked with *.

The chromatograms were run in either constant flow or constant pressure mode as indicated using a TRACE Ultra GC with a TriPlus autosampler (Thermo Fisher Scientific, Milan, Italy) with a FID detector.

Initially, capillary GC methods attempted to simulate the packed column method with a run time of 45 min. The example in Figure 2.123 shows a capillary separation where the oven program is slowed down to keep the run time similar to the packed column method, and helium has been substituted for the original nitrogen carrier gas. However, with the increasing pressures on analysis time, and the flexibility within the EPA 610 method, it has been possible to greatly reduce run time by altering simple GC variables.

A number of variables within the GC setup can be manipulated to shorten the run time, bearing in mind the need to maintain resolution, elution order and sensitivity as outlined in Table 2.32. The first strategy for reducing run time should be to modify the oven ramp by increasing the ramp where the peaks are well separated with large retention time spaces and applying a slow ramp where extra separation is required (see Figure 2.124).

Reducing column length is the second strategy for reducing run time. Note the need to increase the oven ramp rate in order to elute the high boiling materials, while retaining peak shape (see Figure 2.125 with the switch to a 15 m, Figure 2.126 to a 7 m column, and the finally optimized separation in Figure 2.127.)

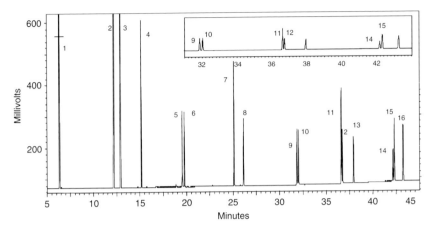

Figure 2.123 Simulation of a packed column separation performed on a 30 m capillary column. Carrier gas has been changed from nitrogen to helium and compared with a 3% OV17 packed column, the 5% phenyl-methylpolysiloxane capillary column separates four previously unresolved peak pairs in a similar run time. Column: 30 m TRACE TR-5MS; ID 0.25 mm, Film 0.25 μm; Initial Temp: 100 °C; Rate 1: 5 °C/min to 300 °C; Carrier Gas: Helium; Gas Flow: 1.5 mL/min (constant flow); Detector: FID; Split Ratio: 50:1; Injection vol.: 1.0 μL by TriPlus autosampler. Sample run rate is 1 per h (includes oven cooling).

Table 2.32 Variables affecting GC run time.

Parameter	Run time effects
Temperature ramp rate	Higher starting temperature and steeper heating rate → peaks elute faster
Column length	Shorter column → shorter run time
Carrier gas flow rate	Each gas type has an optimum linear flow range within which speed can be adjusted. Higher than optimum flow rates are acceptable for productivity increase if critical separations remain compliant.
Film thickness	Thicker film → more interaction of solutes → longer run time
	Thinner film → 'Fast Chromatography'
Column internal diameter	Reducing internal diameter increases column efficiency → shorter runtimes using 'Fast Chromatography'
Ultra-fast technology	Combination of short column, small internal diameter, thin film, special ultra-fast heating coil for ultra-fast separations. Requires optional Ultra-Fast (UF) or Low-Thermal-Mass (LTM) GC modules.

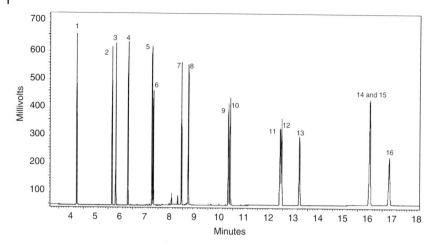

Figure 2.124 Optimization of the oven program has cut the run time by more than half without changing any other parameters. There is some loss of resolution between peaks 14 and 15 compared with the 45 min run; however, the extra specificity of an MS detector would resolve these peaks. Column: 30 m TRACE TR-5MS; ID 0.25 mm, Film 0.25 μm; Initial Temp: 90 °C hold 1 min; Rate 1: 25 °C/min to 290 °C; Rate 2: 4 °C/min to 320 °C hold 5 min; Carrier Gas: Helium; Gas Flow: 1.5 mL/min. (constant flow); Detector: FID; Split Ratio: 50 : 1; Injection vol.: 0.5 μL manual injection. Sample run rate is 2 per h (includes oven cooling).

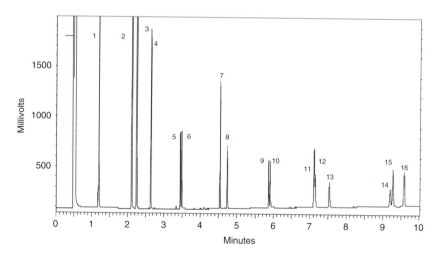

Figure 2.125 Illustration of the effect of simply reducing column length by half (15 m) using constant flow reducing the run time to 10 min. Column: 15 m TRACE TR-5MS; ID 0.25 mm, Film 0.25 μm; Initial Temp: 120 °C hold 0.2 min; Rate 1: 25 °C/min to 260 °C; Rate 2: 7 °C/min to 300 °C hold 3 min; Carrier Gas: Helium; Gas Flow: 1.5 mL/min (constant flow); Detector: FID; Split Ratio: 50 : 1; Injection vol.: 1.0 μL by TriPlus autosampler. Sample run rate is 4 per h (includes oven cooling).

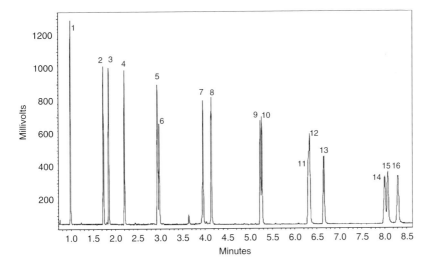

Figure 2.126 Further reduction of the column length by half to 7 m but with larger ID, illustrates the effect of film thickness on run time. The larger ID makes the column less efficient. Therefore, to maintain the resolution, a similar oven program is needed, making the analysis time only slightly shorter than on the 15 m column. Column: 7 m TRACE TR-5MS; ID 0.32 mm, Film 0.25 μm; Initial Temp: 120 °C hold 1 min; Rate 1: 25 °C/min to 250 °C; Rate 2: 10 °C/min to 300 °C hold 5 min; Gas Flow: 1.5 mL/min (constant flow mode) Carrier Gas: Helium; Detector: FID; Split Ratio: 50 : 1; Injection vol.: 0.5 μL manual injection; Sample run rate is 5 per h (includes oven cooling).

2.2.9.2 Ultra-Fast Chromatography

The term *Ultra-Fast Gas Chromatography* (UF-GC) describes an advanced column-heating technique that is taking advantage of short, narrow bore capillary columns, and very high temperature programming rates (Facchetti and Cadoppi, 2005). This technique offers to shorten the analysis time by a factor up to 30 compared to conventional capillary GC. Ultra-fast GC conditions typically apply the use of short 0.1 mm ID narrow bore capillary columns of 2.5–5 m in length with elevated heating rates of 100–1200 °C/min providing peak widths of 100 ms or less, see Figure 2.128. At this point, the limiting factor for sample throughput becomes the ability of the column to cool quickly enough between runs. In addition to a fast heating/cooling regime, a fast conventional or MS detection with an acquisition rate better than 25 Hz for reliable peak identification is required in, for example, a restricted scan, SIM/MID or TOF (time-of-flight) mode, to achieve the necessary sampling rate across individual peaks. Currently, the fast scanning quadrupole technologies and time-of-flight analyzer (TOF) are the MS technologies that are able to provide the required speed of detection for an ultra-fast GC-MS at data rates of 50 Hz or better with partial or full mass spectral information.

The special instrumentation required for ultra-fast chromatography comprises a dedicated ultra-fast column module (UFM), see Figure 2.129, or low-thermal-mass device (LTM), comprising a specially assembled fused silica column for

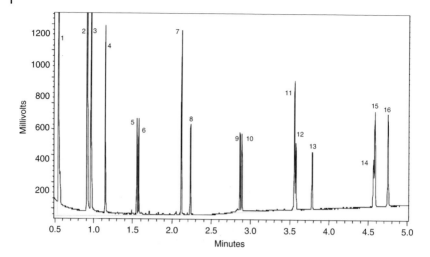

Figure 2.127 The combination of using a narrow 0.1 mm column coated with a thinner film is the strategy for fast analysis. In comparison with the 7 m 0.25 µm film column, the longer 10 m × 0.1 µm film is faster. In addition, using constant pressure has enhanced the analysis speed down to under 5 min. The effect of running at constant pressure is to greatly speed up the linear velocity at the beginning of the run where the peaks are well separated. Later in the run, when the temperature is high, the velocity is lower allowing more dwell time on the column for the later peaks which elute close together.

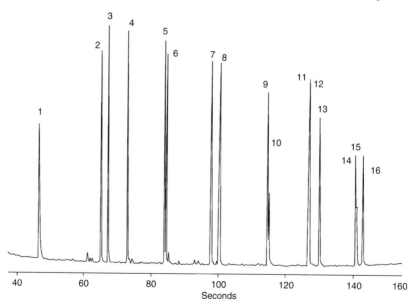

Figure 2.128 Ultra-fast column and detector can run the PAH sample in just 160 s. Column: 5 m Ultra-fast TR5-MS; ID 0.1 mm, Film 0.1 µm; Flow Rate: 1.0 mL/min; Carrier gas: Hydrogen; Oven: 40 °C hold 0.3 min; Rate 1: 2 °C/S to 330 °C. Sample run rate is 12 per h (includes column cooling).

Figure 2.129 UFM column module installed in a regular GC oven (Mega, Legnano, Italy; TRACE GC Ultra, Thermo Fisher Scientific, Milan, Italy).

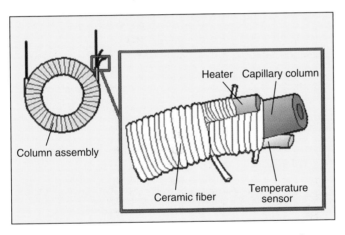

Figure 2.130 Column wrapping detail with heating element and temperature sensor for ultra-fast capillary chromatography.

direct resistive heating, wrapped with a heating element and a temperature sensor, see Figure 2.130 (Facchetti *et al.*, 2002). The assembly is held in a compact metal cage to be installed inside the regular GC oven (which is not active during ultra-fast operation). Temperature programming rates can be achieved as high as 1200 °C/min. UFM/LTM modules are commercially available providing different column lengths and diameters (Magni *et al.*, 2002).

Another compelling example in Figure 2.131 shows the ultra-fast analysis of a mineral oil performed in only 2 min still maintaining enough resolution to separate pristane and phytane from *n*-C17 and *n*-C18.

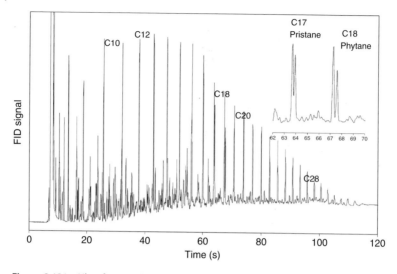

Figure 2.131 Ultra-fast GC chromatogram of a mineral oil sample. The chromatogram was obtained using a 5 m, 0.1 mm ID, 0.1 μm film thickness SE54 column with a temperature program from 40 (12 s) to 350 °C (6 s) at 180 °C/min.

2.2.10
Multi-Dimensional Gas Chromatography

The search for innovative solutions to increase the limiting peak capacities in GC came up many years ago with two-dimensional separations. The development of instrumental techniques focussed in the beginning on the transfer of specific peak regions of interest, the unresolved 'humps', to a second separation column. This technique, widely known as *heart cutting*, found many successful applications in solving critical research oriented projects but its use was not widespread in routine laboratories.

The objective to increase the peak capacity for the resolution of entire complex chromatograms and not only of few discrete congested areas could be satisfied first with the introduction of a comprehensive two-dimensional GC (GC×GC), the most powerful two-dimensional solution today (Liu and Phillips, 1991). While the initial ideas of multi-dimensional gas chromatography (MDGC) relied on the use of two regular capillary columns (MDGC, selectivity tuning), current comprehensive two-dimensional chromatography benefits from the alignment with fast GC on the second separation dimension (Beens *et al.*, 2000; Bertsch, 2000). The result is a significantly improved separation power. With the commercial availability of GC×GC instrumentation this technique found avid utilization for a widespread range of applications within the first few years (Marriott *et al.*, 2003).

Particularly, for the separation of complex mixtures, GC×GC showed far greater resolution and a significant boost in signal to noise, with no increase in analysis times. Further, the hyphenation of GC×GC with MS detection providing three independent analytical dimensions made this technique ideal for the measurement of organic components within complex samples such as those from environmental, petrochemical and biological analysis. The significantly increased chromatographic resolution in GC × GC allows separation of many previously undetectable components. Fast scanning mass spectrometers are required to maintain the excellent chromatographic resolution, provided today with full scan data by TOF, quadrupole or high resolution MS systems (Hoh *et al.*, 2008). More than 15 000 peaks could be detected in an ambient air sample from the city of Augsburg, Germany, by thermal desorption comprehensive two-dimensional GC with TOF-MS detection (Welthagen *et al.*, 2003).

Multidimensional GC

The great advantage of the combination of multiple chromatographic separation steps is the increase in peak capacity. Peak capacity is the maximum number of peaks that can be resolved in a given retention time frame. The more peaks a combination of techniques is able to resolve, the more complex samples can be analysed. When a sample is separated using two dissimilar columns, the maximum peak capacity Φ_{max} will be the product of the individual column's peak capacity Φ_n.

$$\Phi_{max} = \Phi_1 \times \Phi_2$$

For example, if each separation mode generates peak capacities of 1100 in the first dimension and 30 in the second, the theoretical peak capacity of the 2D experiment will be 33 000, a huge gain in separation space, which would theoretically compare to the separation power of a 12 000 m column in the normal single-dimension analysis. To achieve this gain, however, the two techniques should be totally orthogonal, that is, based upon completely different separation mechanisms.

Comparison of peak capacities in normal, heart-cut and comprehensive two-dimensional chromatography

Normal chromatography

Heart-cut 2D chromatography

Comprehensive 2D chromatography

2.2.10.1 Heart Cutting

The goal of heart cutting is the increase of peak capacity for target substances in congested regions of unresolved compounds. Two capillary columns are connected in series, typically by means of a valveless flow switching system. One or more short retention time slices are sampled onto a second separation column typically of different separation characteristic. A valveless switching device is ideally suited for high-speed and inert flow switching. It is based on the principle of pressure balancing and was introduced by Deans already in 1968. This technique found a wider recognition first with state-of-the-art EPC units, which allowed the integration of column switching as part of regular GC methods for routine applications (Deans, 1968; McNamara *et al.*, 2003).

Heart cutting requires a sample specific individual setup of the analysis strategy. The sections of interest have to be previously identified and selected by retention time from the first chromatogram to be 'cut' into the second column. Several heart-cuts can take place within a chromatogram as of analytical need. The choice of column length is application specific and does not interfere with the 'cutting' process. The second dimension column can be of the same but recommended of a different film polarity. A monitor detector is required, typically a universal FID, which continuously observes the separation on the first dimension. A second detector, often an MS detector, acquires data from the second dimension in now 'higher' chromatographic resolution (De Alencastro *et al.*, 2003).

2.2.10.2 Comprehensive GC - GC × GC

True multi-dimensional chromatography requires two independent (orthogonal) separations mechanisms and the conservation of the first separation into the second dimension. Comprehensive GC × GC today is the most developed and most powerful multi-dimensional chromatographic technique. The technique has been widely accepted and applied to the analysis of complex mixtures. Commercial instrumentation is available at a mature technological standard for routine application (Figure 2.132).

In contrast to heart cutting, the complete first dimension effluent is continuously separated in slices (which can be interpreted as a continuous sequence of cuts) and forms a three-dimensional data space adding a second separation dimension with retention times in both dimensions and the peak intensities by classical or MS detection. No monitor detector for the first dimension or prior identification of an individual retention time window is required. No increase in total analysis time is involved (see Figure 2.133).

The timing regime of GC × GC should be relatively slow with a regular capillary column for the first dimension and a short fast GC column for the second-dimension separation. Usually both columns connected by a modulator interface are installed in the same GC oven. The second dimension column should be of a different polarity than the first one to provide different kind of

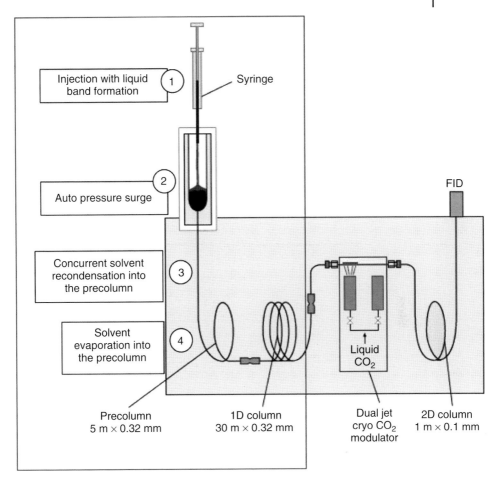

Figure 2.132 Comprehensive GC schematics using LVSI and dual jet cryo modulator (Cavagnino et al., 2003).

interactions in substance class separations. Typically, non-polar columns are used in the first dimension for a separation along the substance boiling points. Polar film interactions characterize the further enhanced separation on the second dimension, which takes place at a low temperature gradient, almost isothermal, due to the speed of elution and only small oven temperature change in this short time. Column lengths in the second dimension are as short as 1–2 m for fast GC conditions (see also Section 2.2). In total, the duration of analysis in GC×GC is not increased compared to conventional techniques and the same sample throughput can be achieved.

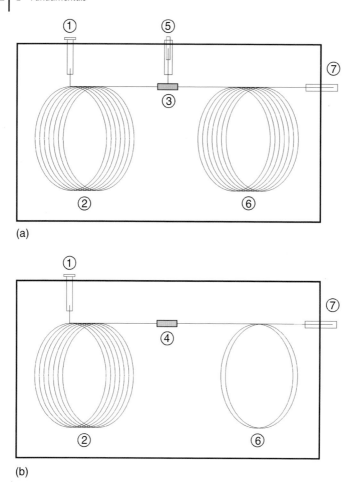

Figure 2.133 Two-dimensional GC Principles. (a) Heart cutting with monitor detector and (b) comprehensive GC × GC. 1. Injector; 2. primary separation column, first dimension; 3. flow switching device; 4. modulator device; 5. monitor detector; 6. secondary separation column, second dimension and 7. MS transfer line or conventional detector, second dimension.

The operation of the interface between the both columns, called the *modulator*, is key to GC × GC. The modulator operates in an alternating trap and inject mode with a modulation frequency related to the fast second-dimension separation. Short elution sections from the first column are integrated by cold trapping and injected for the second-dimension separation. Each injection pulse generates a high speed chromatogram (Bertsch, 1999; Beens *et al.*, 2001).

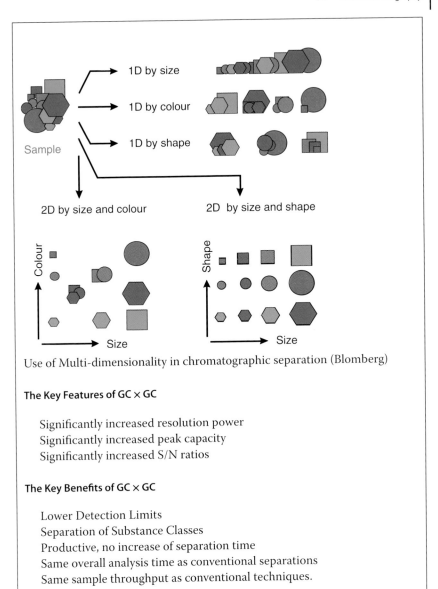

Use of Multi-dimensionality in chromatographic separation (Blomberg)

The Key Features of GC × GC

Significantly increased resolution power
Significantly increased peak capacity
Significantly increased S/N ratios

The Key Benefits of GC × GC

Lower Detection Limits
Separation of Substance Classes
Productive, no increase of separation time
Same overall analysis time as conventional separations
Same sample throughput as conventional techniques.

The collection of a large series of second-dimension chromatograms results in a large amplification of the column peak capacities as the first-dimension peaks are distributed on a large series of fast second-dimension chromatograms. The effect of increased separation can be visualized best with 3D contour or peak apex plots, illustrating the commonly used term *comprehensive* chromatography, see Figure 2.134a (Cavagnino *et al.*, 2003). The very high separation power offered by the comprehensive two-dimensional GC allows, even in the case of complex

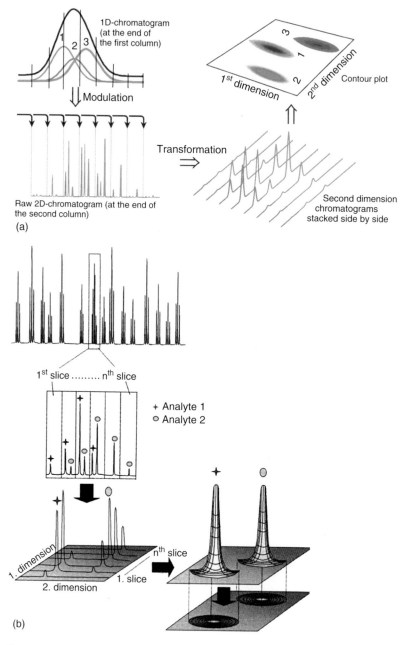

1D-chromatogram
(at the end of
the first column)

Modulation

Raw 2D-chromatogram (at the end of
the second column)

(a)

Transformation

Second dimension
chromatograms
stacked side by side

Contour plot

1st dimension

2nd dimension

1st slice .... nth slice

+ Analyte 1
○ Analyte 2

1. dimension

2. dimension   1. slice

nth slice

(b)

Figure 2.134 (a) Visualization of the principle of multi-dimensional GC separation with the final generation of the contour plot of an unresolved co-elution of three substances in the first dimension (Dallüge *et al.*, 2003); (b) Construction of a three-dimensional contour plot for analytes 1 and 2 from a modulated chromatogram (Kellner *et al.*, 2004).

Figure 2.135   Dual jet cryo modulator, using a stretched column region (Cavagnino *et al.*, 2003).

matrices, a good separation of target compounds from the interferences with a significant increase in signal to noise. This emerging technology has substantially enhanced the chromatographic resolution of complex samples and proven its great potential to further expand the capabilities of modern GC-MS systems.

2.2.10.3   Modulation

The key to successful application of GC × GC is the ability to trap and modulate efficiently the first column effluent into the second dimension. It is important to modulate faster than the peak width of first dimension peaks, so that multiple second dimension peaks (slices) are obtained (see Figure 2.134a). The moderator unit has to provide minimized bandwidths in sample transfer to retain the first dimension resolution and also allow the rapid remobilization. Practically, a high peak compression is achieved by trapping the effluent from the first column on a short band inside of the modulator, being responsible for the significant increase of S/N values (Patterson *et al.*, 2006). This requires a high frequency handling of the column effluent with a reproducible trapping and controlled substance release. Practically applied modulation frequencies range from 4 to 6 s duration, depending on the second dimension column parameters.

The thermal modulator solutions for substance trapping are:

- Trapping by thick stationary film and heat pulse to release substances.
- Cryogenic zone placed over a modulator capillary. Substances are released when removed and resume movement by oven temperature. Also, dual jet cryo modulators have been introduced and commercially available to decouple the processes of collection from the first column and sampling into the second dimension, see Figures 2.135 and 2.136a (Beens *et al.*, 1998). The operation principle of a dual cryo jet modulator is illustrated in Figure 2.136b (Kellner *et al.*, 2004).
- Jet-pulsed modulators, synchronizing hot and cold jet pulses.
- Flow switching valve solutions, slow, not for fast continuous operation, used mainly for heart cutting purposes.

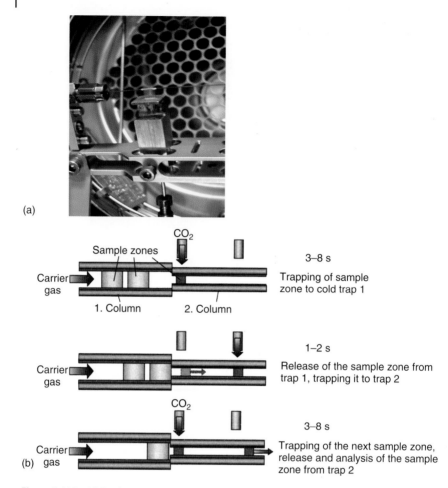

Figure 2.136 (a) Dual jet cryo modulator in, trapping operation on the first column (Cavagnino *et al.*, 2003). (b) Principle of dual jet cryo modulation (Kellner *et al.*, 2004).

The availability of commercial instrumentation providing method-controlled modulators, ability to cool via cryojets, chilled air, liquid CO_2 or liquid N_2 and rapidly heat the trapping areas for injection has made $GC \times GC$ straightforwardly applicable today to perform routinely.

2.2.10.4 Detection

The high speed conditions of the secondary column deliver very sharp peaks of typically 200 ms base width or below and hence require fast detectors with an appropriate high detection rate. Due to the strong peak band compression obtained during the modulation step, low detection limits are reached. Fast scanning quadrupole MS (scan speed >10 000 Da/s) have been applied by scanning a restricted mass range with an acquisition rate of 25 Hz as a cost effective alternative to TOF-MS for compound identification and quantitation (Mondello *et al.*, 2008). Even sector mass spectrometers have been used

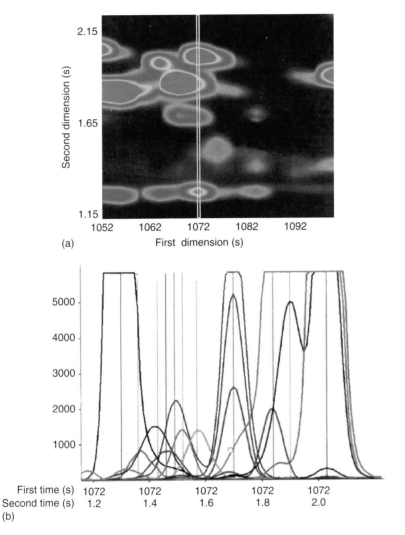

Figure 2.137 Peak deconvolution in full scan GC×GC/TOF-MS. (a) Three-dimensional contour plot, scan 1072 highlighted and (b) scan 1072, each vertical indicates a deconvoluted compound.(Dimandja, 2004, reprinted with permission from Analytical Chemistry, Copyright American Chemical Society.)

successfully in the fast SIM/MID technique, with excellent sensitivity in environmental trace analysis (Patterson *et al.*, 2006). TOF detection and fast scanning quadrupole instruments in a restricted mass range offer full scan analysis capabilities allowing a detailed peak deconvolution for the generation of pure mass spectra also for unresolved trace components, see Figure 2.137 (also see Section 2.2.10.5 Data Handling).

With regard to sensitivity and selectivity, GC×GC/TOF-MS has proven to compare well with the classical HRGC-HRMS methods. Isotope ratio measurements of the most intense ions for both natives and isotopically labelled internal standards ensured the required selectivity (Horii *et al.*, 2004). Potentially interfering matrix compounds are kept separated from the compounds to be measured in the two-dimensional chromatographic space (Focant *et al.*, 2004).

2.2.10.5 Data Handling

Data systems for comprehensive GC require special tools extending the available features of most current GC-MS software suites. Because of the 3D matrix of multiple chromatograms and conventional or MS detection, special tools are required to display and evaluate comprehensive GC separations. The time/response data streams are converted to a matrix format for the two-dimensional contour plot providing a colour code of peak intensities (Figure 2.137), or the three-dimensional surface plot generation (Figure 2.138), for visualization.

Integrated software programs are commercially available providing total peak areas also for quantification purposes (Cavagnino *et al.*, 2003; Reichenbach *et al.*, 2004; GC Image 2015). Dedicated to the comprehensive qualitative sample characterization with GC×GC/TOF-MS analyses, as well as the quantitative

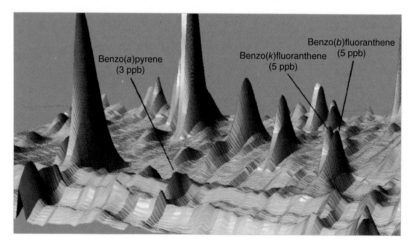

Figure 2.138 Three-dimensional peak view of a GC×GC analysis of a cigarette smoke extract (Cavagnino *et al.*, 2003).

Columns:	
Pre-column	5 m deactivated retention gap, 0.32 mm ID
First dimension column	Rtx-5 30 m, 0.32 mm ID, 0.25 µm df
Second dimension column	BPX-50 1 m, 0.1 mm ID, 0.1 µm df
Carrier	Helium
Operative conditions:	
Oven	90 °C (8min) to 310 °C @ 3 °C/min
Programmed flow	2.5 mL/min (4min) to 0.8 mL/min @ 5mL/min/min
Injection volume	30 µL splitless
Splitless time	0.8 min
Moderator	CO_2 dual-jet modulator
Modulation time	6 s
Data system	HyperChrom for acquistion and data processing

analysis of specific mixture analytes, a GC-MS software package is commercially available (Leco Corp, 2005).

Advanced chemometric processing options are an active area of current research and will advance the GC×GC technology in the near future. In comprehensive GC using TOF detection (GC×GC-MS), data can also be visualized and processed as an image. Each resolved compound produces a two-dimensional peak and can be visually distinguished through pseudo colour mapping, generating a three-dimensional surface. Digital image processing methods, such as creating a difference image (by subtraction of chromatograms) or addition (by addition of chromatograms in different colours) will provide insights previously not available to analytical samples. For GC×GC-MS, mass spectral searching can be used in conjunction with pattern matching (Hollingsworth *et al.*, 2006). Those features demonstrate the great potential of GC×GC-MS with its large peak capacity at a largely improved speed of analysis.

2.2.10.6 Moving Capillary Stream Switching

Another often used valveless flow switching device for heart-cutting purposes or switching the flow direction to different detectors is the Moving Capillary Stream Switching system (MCSS). The MCSS is a miniaturized and very effective flow switching system based on the position of capillary column ends in a small transfer tube. The column effluent is not directed by pressure variations but 'delivered' to the required column by moving the end of the delivering column close to the column outlet of choice into an auxiliary carrier gas stream, see Figure 2.139 (Sulzbach, 1991; Brechbuehler 2007).

Up to five capillary columns can be installed to a small (a few centimetre long) hollow tip of glass, the 'glass dome' (see Figure 2.139), which is located inside the GC oven. The column ends are positioned at different locations inside the glass dome. The delivering pre-column inlet is movable in position over a length of about 1 cm. One of the installed columns is used for a variable make up flow of carrier gas from a second independent regulation. This make-up gas facilitates the parallel coupling of MS detectors in parallel to classical detectors or IRMS (isotope ratio mass spectrometry) to prevent the access of the ions source vacuum to the split region. Instead of an additional detector, the additional line can also be used for monitoring the middle pressure between the two columns. Commercial products are available for MDGC covering a variable split range.

The analyte carrier stream fed into the dome by the pre-column can be split into a variable stream upwards, serving a second analytical column and another stream downwards depending on the position of the pre-column end, for example, to another column outlet, which can lead to a monitor detector. If the delivering column is pushed fully up, the eluate enters the second column and quantitatively transfers the analytes eluted in that position. There is no change of the pressure conditions in the whole switching system during the transfer period. Therefore, the carrier gas flow rates through both of the columns are always kept stable. Hence, there is no influence of the number or duration of switching processes on the retention times on either column. Analytes can only contact inert glass

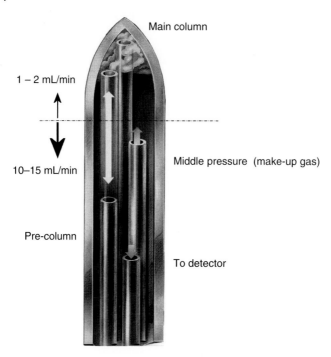

Main column

1 – 2 mL/min

10–15 mL/min

Middle pressure (make-up gas)

Pre-column

To detector

Figure 2.139 Visualization of MCSS glass dome operation – the flow of the movable pre-column is directed to the main column. (Brechbuehler AG, Schlieren, Switzerland.)

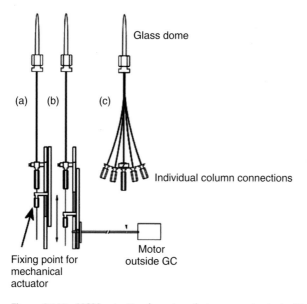

Glass dome

(a)   (b)   (c)

Individual column connections

Fixing point for
mechanical
actuator

Motor
outside GC

Figure 2.140 MCSS actuation by external stepper motor (a, b side view, c top view).

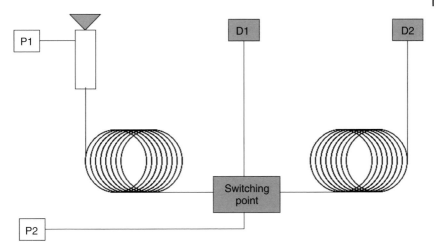

Figure 2.141 Column switching schematics with pre-column, MCSS flow switch at the switch point and main separation column (P1, P2 carrier gas regulations, D1 monitor detector and D2 main detector).

surfaces, fused silica and the column coating; likewise when the sample is injected in any heated injector. Furthermore, the system is free of diffusion and maintains the chromatographic resolution of the pre-column. No influences from ambient atmosphere could be observed using ECD, MS or in IRMS detection (Guth, 1996).

The movement of the pre-column is electrically actuated from outside of the GC oven, completely automated and sequence controlled through the time functions of the GC (see Figure 2.140). The MCSS can be used for a number of tasks affording the increased power of multi-dimensional heart-cutting GC separation (Figure 2.141). The range of applications covers the classical MDGC in a single or double oven concept, as well as detector switching, backflush (BKF), preparative GC and more.

MCSS Applications

Chiral separations of essential oils (FID/MS)
Classification of fuels (FID)
Matrix exclusion (FID/PND/ECD/MS)
Sniffing devices in parallel to detectors (MS)
Coplanar PCBs (double ECD/MS/IRMS)
Pesticides (ECD, NPD, FPD)
Preparative sampling (Microprep Trap)
All kinds of GC–IRMS applications
Parallel GC-MS/IRMS detection
and many others.

2.2.11
Classical Detectors for GC-MS Systems

Classical detectors are important for the consideration of GC-MS coupling if an additional specific means of detection is to be introduced parallel to MS. The classical detectors can be grouped by the response characteristics into the concentration dependant detectors (analyte concentration in the carrier gas) such as the thermal conductivity detector (TCD), and the mass-flow-dependent detectors such as FID, NPD as is the MS (Hill, 1992). A change of make-up gases do not affect the signal in mass flow dependant detectors.

A parallel detection to MS with a thermal conductivity detector is not found often in practice as the mass spectrometric analysis of gases is mostly carried out with specially configured MS systems (RGA, residual gas analyser, mass range <100 Da). Also, the parallel coupling of an FID does not lead to results which are complementary to those of MS, as both detection processes give practically identical total ion chromatograms. The response factors for most of the organic substances are comparable. A parallel MS detection to the FID can be used for a routine FID method development (see applications section).

A more application-oriented classification is possible with selective and non-selective detectors. Additional information to MS can be obtained with parallel element-specific detectors ECD, FPD or NPD. The detection limits which can be achieved with an ECD or a NPD are usually comparable to those attainable using a mass spectrometer in SIM mode, but different mechanism of selectivity apply. On splitting the carrier gas flow, the ratio of the two parts must be considered when planning such a setup. Normally a larger proportion is passed into the mass spectrometer so that in residue analysis low concentrations of substances do not fall below the detection limit. The use of such flow dividers for quantitative determinations must be checked in an individual case, as a constant split cannot be expected for all boiling point ranges. A constant split ratio can be attained by using the constant pressure mode. The constant flow mode will change the flow conditions at the split point due to the pressure variation with increasing oven temperature.

Applications can cover the rapid screening, for example, on the intensity of halogenated compounds using an ECD with an intelligent decision for a subsequent MS analysis of the same positively screened sample for a mass selective quantitation.

2.2.11.1 Flame-Ionization Detector (FID)
With the FID, the substances to be detected are burned (oxydized) in a hydrogen flame and are thus partially ionized (Table 2.33). The FID is a mass-flow-dependent detector. The signal is dependent on the rate at which analyte molecules enter the detector from the analytical column. The response of the FID is not affected by the makeup gas flow rate.

For operation, the FID jet is set to a negative potential, hence the produced positive ions are neutralized. The corresponding electrons are captured at the

Table 2.33 Reactions in the FID.

Pyrolysis	CH_3^0, CH_2^0, CH^0, C^0
Exited radicals	O_2^*, OH^*
Ionization	$CH_2^0 + OH^* \rightarrow CH_3O^+ + e^-$
	$CH^0 + OH^* \rightarrow CH_2O^+ + e^-$
	$CH^0 + O_2^* \rightarrow CHO_2^+ + e^-$
	$C^0 + OH^* \rightarrow CHO^+ + e^-$

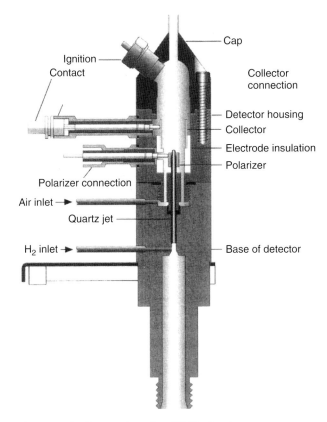

Figure 2.142 Construction of an FID (Finnigan).

ring-shaped collector electrode to give a signal current (Figure 2.142). The electrode is at a potential which is about 200 V more positive than the jet.

Provided that only hydrogen burns in the flame, only radical reactions occur. No ions are formed. If organic substances with C–H and C–C bonds get into the flame, they are first pyrolysed. The carbon-containing radicals are oxidized by oxygen and OH radicals formed in the flame. The excitation energy leads to ionization of the oxidation products. Only substances with at least one C–H or C–C bond are detected, but not permanent gases, carbon tetrachloride or water.

If a reactor (hydrogenator, methanizer) is connected before the FID, the latter can be converted into an extremely sensitive detector for CO and CO_2.

The oxygen-specific detector (O-FID) uses two reactors. At first, hydrocarbons are decomposed into carbon, hydrogen and carbon monoxide at above 1300 °C. CO is then converted into methane in the hydrogenation reactor and detected with the FID. With the O-FID, for example, oxygen-containing components in fuels can be selectively detected (Schneider *et al.*, 1982).

FID

Universal detector
Mass-flow-dependent detector

Advantages	High dynamics high sensitivity
	Robust
Use	Hydrocarbons, for example, fuels, odor substances, BTEX, polyaromatic hydrocarbons **(PAHs)**, and so on.
	Comparison of diesel with petrol.
	Important all round detector.
Limitations	As it is universal, its performance is poor for trace analysis in complex matrices.
	Low or no response for fully oxidized, highly chlorinated or brominated substances.
	Not suitable for permanent gas, except for CO and CO_2 with methanizer.

2.2.11.2 Nitrogen-Phosphorous Detector (NPD)

An NPD basically is a modified FID detector with a rubidium or cesium chloride bead on a Pt wire inside a heater coil situated close to the hydrogen jet and the collector electrode. It is also called the *alkali flame-ionization detector (AFID)*, see Figure 2.143. Nitrogen and phosphorus can be selectively detected. The sensitivity for the specific detection of nitrogen or phosphorus can be four orders of magnitude greater than that for carbon.

The alkali beads are heated to red heat both electrically and in the flame and are excited to alkali emission. They are always at a negative potential compared with the collector electrode. The heated alkali bead emits electrons by thermionic emissions which are collected at the anode providing the background current of the detector.

In the nitrogen/phosphorous (NP) mode, the hydrogen flow is set just below the minimum required for ignition. The rubidium or cesium bead ignites the hydrogen catalytically, and forms cold plasma. Excitation of the alkali metal results in ejection of electrons, which are detected as a background current between anode and cathode. Nitrogen or phosphorus containing analytes increase the current. For the detection of phosphorus, the jet is grounded. The electrons emitted by the

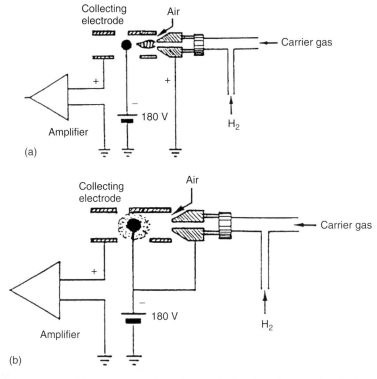

Figure 2.143 (a) Diagram of an NPD (P operation) and (b) diagram of an NPD (N operation).

hydrocarbon parts of the molecule cannot exceed the negative potential of the beads and do not reach the collector electrode. The electrons from the specific alkali reaction reach the collector electrode unhindered (Table 2.34). Since the alkali metal bead is consumed over time, it must be replaced on a regular basis.

Phosphorus-containing substances are first converted in the flame into phosphorus oxides with an uneven number of electrons. Anions formed in the alkali reaction by the addition of an electron are oxidized by OH radicals. The electrons added are released and produce a signal current.

Table 2.34 Reactions with P compounds in the NPD.

$$\overline{O} = \underline{\underline{P}} + A^* \rightarrow \left[\overline{O} = \underline{\underline{P}}\right]^- + A^+$$

$$\overline{O} = \dot{P} = \overline{O} + A^* \rightarrow \left[\overline{O} = P = \overline{O}\right]^- + A^+$$

$$\left[\overline{O} = \underline{\underline{P}}\right]^- + OH^\circ \rightarrow HPO_2 + e^-$$

$$\left[\overline{O} = P = \overline{O}\right]^- + OH^\circ \rightarrow HPO_3 + e^-$$

$$HPO_3 + H_2O \rightarrow H_3PO_4$$

A = alkali.

Table 2.35 Reactions with N compounds in the NPD.

Pyrolysis of CNC compounds $\rightarrow$ C$\equiv$N$|^{\circ}$

CN$^{\circ}$ + A* $\rightarrow$ CN$^-$ + A*
CN$^-$ + H$^{\circ}$ $\rightarrow$ HCN + e$^-$
CN$^-$ + OH$^{\circ}$ $\rightarrow$ HCNO + e$^-$

A = alkali.

Like phosphorus, nitrogen has an uneven number of electrons. Under the reducing conditions of the flame cyanide and cyanate radicals are formed, which can undergo the alkali reaction (Table 2.35). For this, the input of hydrogen and air are reduced. Instead of the flame the hydrogen burns in the form of cold plasma around the electrically heated alkali beads.

In order to form the required cyanide and cyanate radicals, the C−N structure must already be present in the molecule. Nitro compounds are detected, but not nitrate esters, ammonia or nitrogen oxides. By taking part in the alkali reaction the cyanide radical receives an electron. Cyanide ions are formed, which react with other radicals to give neutral species. The electron released provides the detector signal.

NPD

Specific detector
Mass-flow-dependent detector

Advantages	High selectivity and sensitivity.
	Ideal for trace analyses.
Use	Only for CN and P-containing compounds.
	Plant protection agents.
	Chemical warfare gases, explosives.
	Pharmaceuticals.
Limitations	Additional detector for ECD or MS.
	Quantitative measurements with an internal standard are recommended.
	To some extent, time-consuming optimization of the Rb beads.

2.2.11.3 Electron Capture Detector (ECD)

The operation of the ECD, invented already 1957 by James Lovelock, is based on the effect that thermal electrons can be 'captured' by electronegative analytes eluting from the analytical column (Wentworth and Chen, 1981). The ECD consists of an ionization chamber, which contains a nickel plate, on the surface of which a thin layer of the radioactive isotope ^{63}Ni has been applied (about 10−15 mC, Figure 2.144). The makeup gas (N_2 or Ar/10% methane) is ionized by the ß-radiation of the ^{63}Ni foil (typically 10 mCurie (370 MBq), low energy beta particle spectrum, 100 yrs half life). The free electrons migrate towards the collector electrode and provide the background current of the detector. Substances with electronegative groups reduce the background current by capturing electrons and forming negative molecular ions. The reduction of the background current is proportional to the analyte quantity in the sample. The main reactions in the ECD are dissociative electron capture (Table 2.36) and electron capture (Table 2.37, see also Section 2.3.1.3 Chemical Ionization). Negative molecular ions can recombine with positive carrier gas ions.

Electron capture is more effective, the slower the electrons move. For this reason, sensitive ECDs are operated using a pulsed DC voltage. By changing the pulse frequency, the current generated by the electrons is kept constant. The pulse frequency thus becomes the actual detector signal.

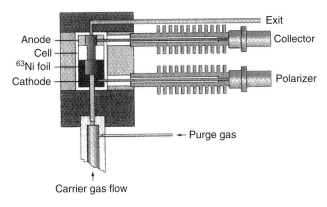

Figure 2.144 Construction of an ECD (Finnigan).

Table 2.36 Substance reactions in the ECD.

---

1. Dissociative electron capture
$$AB + e^- \rightarrow A^0 + B^-$$
2. Addition of an electron
$$AB + e^- \rightarrow (AB^-)^*$$

Table 2.37 Basic reactions in the ECD.

$$CG \xrightarrow{\beta} CG^+ + e^-$$
Electron capture
$$M + e^- \rightarrow M^-$$
Recombination
$$M^- + CG^+ \rightarrow M^° + CG$$

CG, carrier gas, makeup gas N_2 or Ar/10% methane.

ECD

Selective, mass-flow-dependent-detector

Non-destructive, can be stacked with FID, NPD

The ECD reacts with all electronegative elements and functional groups, such as -F, -Cl, -Br, $-OCH_3$ and $-NO_2$ with a high response. All hydrocarbons (generally the matrix) remain transparent, although present.

Advantages	Selectivity for Cl, Br, methoxy and nitro groups.
	Transparency of all hydrocarbons (= matrix).
	High sensitivity.
Disadvantages	Radioactive radiator, therefore local handling authorization necessary.
	Sensitive to misuse.
	Mobile use only under limited conditions.
Use	Typical detector for environmental analysis.
	Ideal for trace analysis.
	Pesticides, OCPs, PCBs, dioxins.
	Volatile halogenated hydrocarbons, freons, nitrous oxides.
Limitations	Substances with low halogen contents (Cl_1, Cl_2) only have a low response (Table 2.38).
	Volatile halogenated hydrocarbons with low chlorine levels are better detected with FID or MS.
	Limited dynamic range.
	Multipoint calibration necessary.

Table 2.38 Relative response as of conversion rates in the ECD for molecules with different degrees of halogenation.

Molecule	Conversion rate (cm^3/mol·s)	Main product
CH_2Cl_2	1×10^{-11}	Cl^-
$CHCl_3$	4×10^{-9}	Cl^-
CCl_4	4×10^{-7}	Cl^-
CH_3CCl_3	1×10^{-8}	Cl^-
$CH_2ClCHCl_2$	1×10^{-10}	Cl^-
CF_4	7×10^{-13}	M^-
CF_3Cl	4×10^{-10}	Cl^-
CF_3Br	1×10^{-8}	Br^-
CF_2Br_2	2×10^{-7}	Br^-
C_6F_6	9×10^{-8}	M^-
$C_6F_5CF_3$	2×10^{-7}	M^-
$C_6F_{11}CF_3$	2×10^{-7}	M^-
SF_6	4×10^{-7}	M^-
Azulene	3×10^{-8}	M^-
Nitrobenzene	1×10^{-9}	M^-
1,4-Naphthoquinone	7×10^{-9}	M^-

2.2.11.4 Photo Ionization Detector (PID)

The photo ionization detector (PID) operates on the principle of energy absorption of photons emitted by a UV lamp in the detector housing. The absorbed energy leads to ionization of the molecule by release of an electron from the excited molecule M^* according to Eq. (2.26).

$$M + h \cdot \nu \rightarrow M^* \rightarrow M^+ + e^- \tag{2.26}$$

The high selectivity of the PID is based on the different energy levels of the emitted emission lines by the used UV lamps. The different types of lamps are filled either with argon, hydrogen, krypton or with other gases to cover an ionization energy range from 8.4 to 11.8 eV. Ionization occurs only if the ionization potential (IP) of the molecules M is lower than the energy level of the emitted UV bands.

The number of ions produced is proportional to the absorption coefficient of the molecule and the intensity of the lamp.

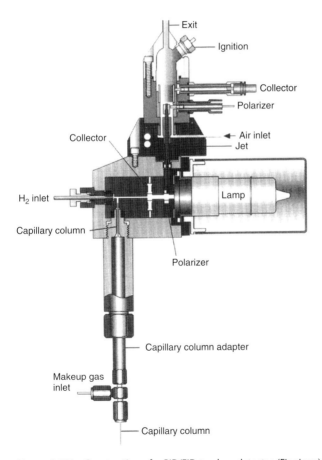

Figure 2.145 Construction of a PID/FID tandem detector (Finnigan).

Lamp type (eV)	Typical applications
8.4	Amines, PAH
9.6	Volatile aromatics, BTEX
10.2	General application
11.8	Aldehydes, ketones

The UV lamp is usually removable for exchange to applications requiring different selectivity. While the lamp housing is kept relatively cool (<100 °C), the quartz ionization chamber containing the two measurement electrodes is the heated part of the PID (>300 °C). It is closed by a quartz or alkalifluoride window for entry of the UV light. Due to the non-destructive nature of the PID, a second detector can be put in series or mounted on top, see Figure 2.145 with a nonselective FID.

The PID is mainly used for the analysis of aromatic pollutants, for example, BTEX or PAH, or halogenated compounds in environmental applications. Many EPA methods, for example, 602, 502 or 503.1, cover priority pollutants in surface or drinking water. Typical for the PID is the detection of impurities air, also with mobile detectors. This is due to the fact that the energy of the UV lamp is sufficient to ionize the majority of organic air contaminants, but insufficient to ionize the air components oxygen, nitrogen, water, argon and carbon dioxide.

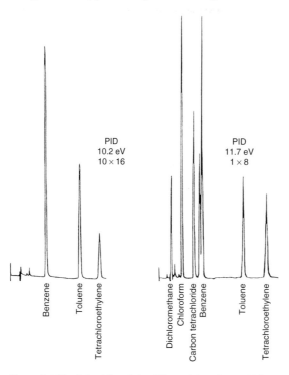

Figure 2.146 Selectivity of the PID at 10.2 and 11.7 eV for a mixture of aromatic and chlorinated hydrocarbons (HNU).

PID

Universal and selective, non-destructive, concentration-dependent detector
The energy-rich radiation of a UV lamp (Figure 2.145) ionizes the substances
to be analysed more or less selectively, depending on the energy content
(Table 2.39); measurement of the overall ion flow.

Advantages	The selectivity can be chosen, see Figure 2.146. High sensitivity. No gas supply is required. Robust, no maintenance required. Ideal for mobile and field analyses.
Use	Field analysis of volatile halogenated hydrocarbons, BTEX, polyaromatic hydrocarbons (PAHs), and so on.
Limitations	Only for substances with low IPs ($\leq 10\,eV$), consult manufacturers' data.

Table 2.39 Ionization potentials of selected analytes.

Substance	Ionization potential (eV)
Helium	24.59
Argon	15.76
Nitrogen	15.58
Hydrogen	15.43
Methane	12.98
Ethane	11.65
Ethylene	11.41
2-Chlorobutane	10.65
Acetylene	10.52
n-Hexane	10.18
2-Bromobutane	9.98
n-Butyl acetate	9.97
Isobutyraldehyde	9.74
Propene	9.73
Acetone	9.69
Benzene	9.25
Methyl isothiocyanate	9.25
N,N-Dimethylformamide	9.12
2-Iodobutane	9.09
Toluene	8.82
n-Butylamine	8.72
o-Xylene	8.56
Phenol	8.50

2.2.11.5 Electrolytical Conductivity Detector (ELCD)

In the electrolytical conductivity detector (ELCD) the eluate from the GC column passes into a Ni reactor in which all substances are completely oxidized or reduced at an elevated temperature (about 1000 °C).

The products dissociate in circulating water which flows through a cell where the conductivity is measured between two Pt spirals (Figure 2.147). The change in conductivity is the measurement signal. Ionic compounds can be measured without using the reactor (Piringer and Wolff, 1984; Ewender and Piringer, 1991).

The Hall detector has a comparable function to the ELCD and is a typical detector for packed or halfmil columns (0.53 mm internal diameter) because of the large volume of its measuring cell. Because of the latter significant peak

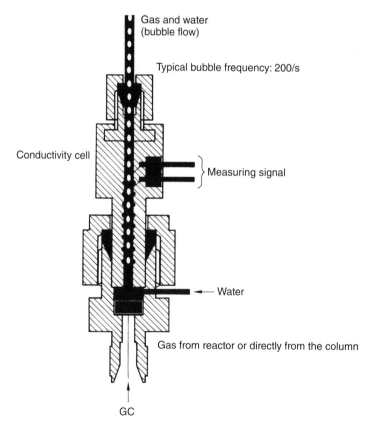

Figure 2.147 Construction of an ELCD (patent Fraunhofer Gesellschaft, Munich, Germany).

tailing occurs with capillary columns. Special constructions for use with normal bore columns (0.25, 0.32 mm internal diameter) are available with special instructions.

ELCD

Selective, concentration-dependent detector

Advantages	Can be used for capillary chromatography. Selectivity can be chosen for halogens, amines, nitrogen and sulfur. High sensitivity at high selectivity. Can be used for halogen detection without a source of radioactivity, therefore no authorization necessary. Simple calibration, as the response is directly proportional to the number of heteroatoms in the analyte, for example, the proportion of Cl in the molecule.
Use	Environmental analysis, for example, volatile halogenated hydrocarbons, PCBs. Selective detection of amines, for example, in packaging or foodstuffs. Determination of sulfur-containing components.
Limitations	The sensitivity in the halogen mode just reaches that of ECD, so that its use is particularly favourable in association with concentration procedures, such as P&T or thermodesorption.

2.2.11.6 Flamephotometric Detector (FPD)

Flamephotometric detectors (FPDs) are used as one- or two-flame detectors. In a hydrogen-rich flame, P- and S-containing radicals are in an excited transition state. On passing to the ground state, a characteristic band spectrum is emitted (S: 394 nm, P: 526 nm). The flame emissions initiated by the eluting analytes (chemiluminescence) are determined using an optical filter and amplified by a photomultiplier (Figure 2.148).

The analytical capability of the FPD can be expanded by connecting a second photomultiplier tube with a different optical filter on the same detector base, for example, to monitor P and S containing substances in parallel.

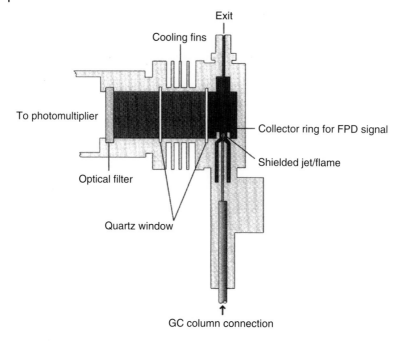

Figure 2.148   Construction of an FPD (Finnigan).

FPD

Selective, mass-flow-dependent detector

Advantages	In phosphorus mode, comparable selectivity to FID with high dynamics.
	High selectivity in sulfur mode.
Use	Mostly selective for sulfur and phosphorus compounds (e.g., pesticides).
	Detection of sulfur compounds in a complex matrix.
Limitations	Adjustment of the combustion gases important for reproducibility and selectivity.
	In sulfur mode, quenching effect is possible because of too high hydrocarbon matrix (double flame necessary).
	Sensitivity in sulfur mode is not always sufficient for trace analysis.

2.2.11.7   Pulsed Discharge Detector (PDD)

The pulsed discharge detector (PDD) is a universal and highly sensitive nonradioactive and non-destructive detector, also known as a *helium photoionization detector*. It is based on the principle of the photoionization by radiation

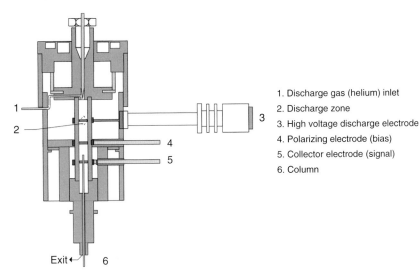

1. Discharge gas (helium) inlet
2. Discharge zone
3. High voltage discharge electrode
4. Polarizing electrode (bias)
5. Collector electrode (signal)
6. Column

Figure 2.149 Pulsed discharge detector (PDD) in cutaway view (Thermo Scientific).

arising from the transition of diatomic helium to the dissociative ground state (Figure 2.149). PDD chromatograms compare very well with FID or MS in full scan mode.

The response to organic compounds is linear over five orders of magnitude with minimum detectable quantities in the low picogram or femtogram range. The response to permanent gases is positive with minimum detectable quantities in the low parts per billion range. The performance of the detector is negatively affected by the presence of any impurities in the gas flows (carrier, discharge), therefore, the use of a high quality grade of helium (99.999% pure or better) as carrier and discharge gases is strongly recommended. Because even the highest quality carrier gas may contain some water vapour and fixed gas impurities, a helium purifier is typically included as part of the detector system.

The PDD detector consists of a quartz cell supplied from the top with ultra pure helium as discharge gas that reaches the discharge zone consisting of two of electrodes connected to a high voltage pulses generator (pulsed discharge module). The eluates from the column, flowing counter to the flow of helium from the discharge zone, are ionized by photons at high energy arising from metastable helium generated into the discharge zone. The resulting electrons are accelerated and measured as electrical signal by the collector electrode. The discharge and carrier gas flows are opposite. For this reason, it is necessary that the discharge gas flow is greater than the carrier gas flow to avoid that the eluates from the column reach the discharge zone with consequent discharge electrodes contamination.

The discharge and carrier gas flow out together from the bottom of the cell where it is possible to measure the sum of both at the outlet on the back of the instrument.

The PDD chromatograms show a great similarity to the classical FID detector and offers comparable performance without the use of a flame, radioactive emitter or combustible gases. The PDD in helium photoionization mode is an excellent replacement for FIDs in petrochemical or refinery environments, where the flame and use of hydrogen can be problematic. In addition, when the helium discharge gas is doped with a suitable noble gas, such as argon, krypton or xenon (depending on the desired cut-off point), the PDD can function as a specific photoionization detector for selective determination of aliphatics, aromatics, amines, as well as other species.

Another very typical application is the area of pure gas analysis. Due to the very high sensitivity, the PDD is able to perform the analysis of impurities in several pure industrial gas mixtures. The availability of a highly sensitive detector (pulse discharge) makes the configured GC system suitable for determination of compounds in a range of concentration ranging from hundreds of ppm down to the ppb trace level. The extremely high flexibility and versatility of the system permits, for instance, the characterization of impurities in pure noble gases such as Xe and Kr.

Some PDD detectors also offer an electron capture mode being selective for monitoring high electron affinity compounds such as freons, chlorinated pesticides and other halogen compounds. For such type of compounds, the minimum detectable quantity (MDQ) is at the femtogram or low picogram level. The PDD is similar in sensitivity and response characteristics to a conventional radioactive ECD, and can be operated at temperatures up to $400\,°C$. For operation in this mode, He and CH_4 are introduced just upstream from the column exit.

2.2.11.8 Olfactometry

GC–Olfactometry (GC–O) or 'sniffing' describes techniques that use the human nose to detect and evaluate volatile compounds eluting from a GC separation (Delahunty, 2006). Assessors sniff the eluate from a specifically designed odour port parallel to FID or MS detection. GC–O applications have become common not limited to the food and flavour industry to assign specific flavour characteristics to each of the volatile compounds identified. The human nose plays the role of the detector. However, the human nose is often more sensitive than any physical detector, and GC–O exhibits supplementary capabilities that can be applied to any fragrant product.

Olfactometry techniques can be classified into two categories: dilution methods, which are based on successive dilutions of an aroma extract until no odour is perceived at the sniffing port of the GC, and the intensity methods, in which the aroma extract is injected and the assessor records the odour intensity and perception as a function of time. The technical solution is straightforward with a split at the end of a chromatographic column and a heated transfer line to a GC external sniffer port. The eluting compounds are splitted, for example, 1:50 to an FID or MS detector and the sniffing port. The column effluent is combined at the sniffer port with a laminar stream of inert make-up gas, which is heated to a constant temperature and additionally humidified.

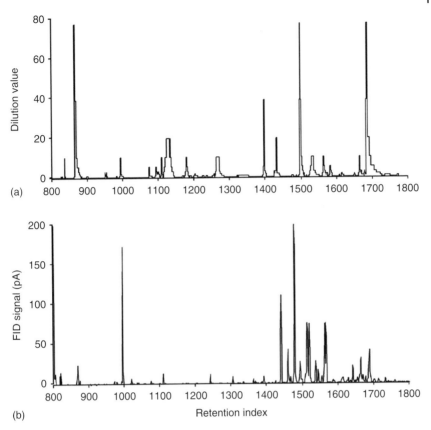

Figure 2.150 Comparison of (a) the GC-O aromagram with (b) the FID chromatogram of a hop essential oil sample (Delahunty, 2006).

For data recording, the peak to odour correlation is documented by specialized fragrance chemists. Different solutions are available to note the odour intensity and characteristics, usually connected as an additional channel on an existing data system, or integrator. Figure 2.150 shows as example the aromagram of a hop essential oil sample with the olfactometry response compared to the FID detection. A strong odour perception can be noted with peaks of only low classical detector response. Compound identification starts from here using retention index information and full scan MS with library search and classical spectrum interpretation tools.

2.2.11.9 Classical Detectors Parallel to the Mass Spectrometer

In principle, there are two possibilities for operating another detector parallel to the mass spectrometer. The sample can be already divided in the injector and passed through two identical columns. The second more difficult but controllable solution is the split of the eluate at the end of the column.

The division of the sample on to two identical columns can be realized easily and carried out very reliably, but different retention times need to be considered due to the vacuum impact on the MS side. Since the capacities of the two columns are additive, the quantity of sample injected can be adjusted in order to make use of the operating range of the mass spectrometer. Two standard columns can be installed for most injectors without further adaptation being necessary. In the simplest case, the connection can be made using ferrules with two holes in them. It must be ensured that there is a good seal. A better connection involves an adaptor piece with a separate screw-in joint for each column. With this construction the independent positioning of the columns in the injector is possible. Suitable adaptors can be obtained for all common injectors.

In GC-MS, the division of the flow at the end of the column requires a considerable higher effort because the direct coupling of the branch to the mass spectrometer causes reduced pressure at the split point. For this reason, the use of a simple Y piece is seldom possible with standard columns. The consequence would be a reversed flow through the detector connected in parallel caused by the wide vacuum impact of the MS into the column. The effect is equivalent to a leakage of air into the mass spectrometer. A split at the end of the column must therefore be carried out with high flow rates, using a makeup gas, and calculate length and diameter of the transfer capillaries as required. Precise split of the eluent is possible, for example, using the glass cap cross divider (Figure 2.151). Here, the column, the transfer capillaries to the mass spectrometer, the parallel detector and a makeup gas inlet all meet. By choosing the internal diameter and the position of the end of the column in the glass cap (also known as the *glass dome*), the ratio of the split can easily be adjusted. The advantage of this solution lies in the free choice of column so that small internal diameters with comparatively low flow rates can also be used (Bretschneider and Werkhoff, 1988a/b).

The calculation of gas flow rates through outlet splitters with fixed restrictors is desirable to have a means of estimating the rate of gas flow through a length of fixed restrictor tubing of specified dimensions. Conversely, for other applications it is necessary to estimate the length and ID of restriction tubing required to yield a desired flow rate at a specified head pressure. Within limits, the following calculation can be used for this purpose.

$$V = \frac{\pi \cdot P \cdot r^4 \cdot (7.5)}{L \cdot \eta}$$

with:

V	=	volumetric gas flow rate (cm^3/min)
P	=	pressure differential across the tube (dyne/cm^2)
	=	× 68 947.6 (PSI)
R	=	tube radius (cm)
L	=	tube length (cm)
η	=	gas viscosity (poise [dyne-s/cm^2])

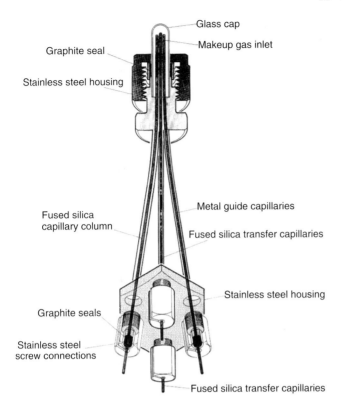

Figure 2.151 Flow divider after Kaiser, Rieder (Seekamp Company, Werkhoff Splitter, Glass Cap Cross).

One of the obvious, limitations to this calculation is the critical nature of the radius measurement. Since this is a fourth power term, small errors in the ID measurement can result in relatively large errors in the flow rate. This becomes more critical with smaller diameter tubes (i.e., less than 100 µm). A 100× microscope (if available) is a convenient tool for determining the exact ID of a particular length of restriction tubing.

The gas viscosity values can be determined from the graph of gas viscosity against temperature for hydrogen, helium and nitrogen (Figure 2.152), or internet based gas viscosity calculators.

2.2.11.10 Microchannel Devices

The microchannel or microfluidic devices used for flow switching in GC are new developments which became available due to progresses made in precise laser machining capabilities of metal films and appropriate metal surface deactivation solutions. The features required for these microchannels devices are laser cut into metal sheet (shims) with thicknesses from 20 to 500 µm. The resulting channel dimensions are similar to the conventionally fused silica

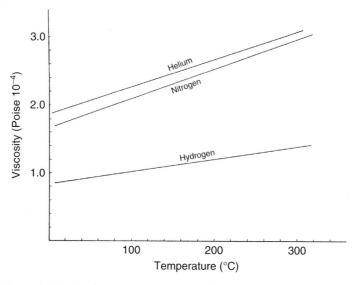

Figure 2.152 Graphs to determine viscosity for hydrogen, helium and nitrogen.

capillary columns, avoiding dead volumes and peak broadening. Shim stacks are bonded together with top and bottom plates for column connections, having low thermal mass for fast GC oven ramp rates. The outer plates are thicker (usually 500 μm) and carry the connection sockets for fused silica analytical or transfer columns. This microchannel concept based on shim stacks allows devices to be designed for instance ideally suitable for flow splitting to different detectors (see Figure 2.153), flow switching such as heart-cutting or the Deans

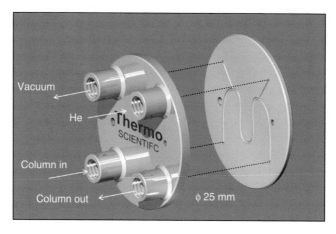

Figure 2.153 Microchannel switching device used for GC-MS detector flow switching (DFS, Dual Data flow switch, Thermo Fisher Scientific).

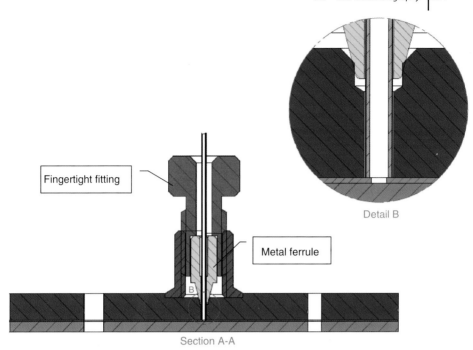

Detail B

Section A-A

Figure 2.154   Microchannel disk finger tight connection system using SilTite™ metal ferrules. (SGE.)

switch applications, also enabling custom designs to be developed for special analytical tasks.

The metal surface deactivation provides flow channels that are inert even to reactive analytes at trace levels. The inertness achieved is substantially better than can be achieved on conventional inlet liner glass surfaces and is similar to the inertness of deactivated fused silica columns.

The column connection system is special. The microchannels can be blocked if particulate material is allowed to enter the flow path. Graphite ferrules as well as polyimide (Vespel™) ferrules must never be used with microchannel devices. Metal ferrules (SilTite™) must be used, see Figure 2.154. The metal ferrules have been demonstrated to be extremely reliable with gas-tight seals suitable for mass spectrometers. Because the thermal coefficient of expansion of the metal ferrules matches the coefficient of thermal expansion of the housing, the seal does not leak with the GC oven cycling unlike polymer ferrules do. The miniature metal ferrules are designed to be used with finger force only on a knurled fitting. Wrenches should never be used. If the finger force only requirement is followed, the ferrules cannot be overtightened avoiding damage to the fitting or tubing.

2.3
Mass Spectrometry

> *Looking back on my work in MS over the last 40 years, I believe that my major contribution has been to help convince myself, as well as other mass spectrometrists and chemists in general, that the things that happen to a molecule in the mass spectrometer are in fact chemistry, not voodoo; and that mass spectrometrists are, in fact, chemists and not shamans.*
>
> *Seymour Meyerson, Research Department, Amoco Corporation*

Mass spectrometers are instruments for producing and analysing mixtures of ions for components of differing mass, for exact mass determination and quantitation. The substances to be analysed are introduced to an ion source and ionized. In GC-MS systems, there is continuous transport of substances by the carrier gas into the ion source. Mass spectrometers basically differ in the construction of the analyser as a beam or ion storage instrument. Mass separation is achieved on the basis of the ion mass over charge value (m/z). In GC-MS ions are generally singly charged (few known exceptions for instance with PAHs), so the m/z value directly gives the mass value with m/1. The mass units used are 'u' or the more common 'Da' (Dalton). The performance of a mass spectrometric analyser is determined by the resolving power for differentiation between masses with small differences, the mass range and the ion transmission required to achieve high detection sensitivity. Depending on the type of mass analyser employed the mass of an ion can be determined as unit mass (low mass resolution) or accurate mass (high mass resolution).

2.3.1
Ionization Processes

2.3.1.1 Reading Mass Spectra
What does a mass spectrum mean? In the graphical representation, the mass to charge ratio m/z of ions measured is plotted along the horizontal line. As ions with

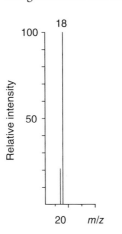

Figure 2.155 McLafferty's unknown spectrum.

unit charge are generally involved in GC-MS with a few exceptions (e.g., polyaromatic hydrocarbons), this axis is generally taken as the mass scale and gives the molar mass of an ion (see Figure 2.155). The intensity scale shows the frequency of occurrence of an ion under the chosen ionization conditions. The scale is usually given both in percentages relative to the base peak (100% intensity) and in measured intensity values (counts). As neutral particles are lost in the fragmentation or rearrangement of a molecular ion M^+ and cannot be detected by the analyser, the mass of these neutral particles is deduced from the difference between the fragment ions and the molecular ion (or precursor fragments) (Figure 2.156).

Introduction

Learning how to identify a simple molecule from its electron ionization (EI) mass spectrum is much easier than from other types of spectra. The mass spectrum shows the mass of the molecule and the masses of pieces from it. Thus, the chemist does not have to learn anything new – the approach is similar to an arithmetic brain-teaser. Try one and see.

In the bar-graph form of a spectrum (see Figure 2.155), the abscissa indicates the mass (actually m/z, the ratio of mass to the number of charges on the ions employed) and the ordinate indicates the relative intensity. If you need a hint, remember the atomic weigths of hydrogen and oxygen are 1 and 16, respectively.

Prof. McLafferty, Interpretation of Mass Spectra (1993)

In EI at an ionisation energy of 70 eV, the extent of the fragmentation reactions observed for most organic compounds is independent of the design of the ion source. For building up libraries of spectra, the comparability of the mass spectra produced is thus ensured. All commercially available libraries of mass spectra are run under these standard conditions and allow the fragmentation pattern of an unknown substance to be compared with the spectra available in the library (see Section 3.2.4).

The Time Aspect in the Formation and Determination of Ions in MS

Ion generation
- Flight time of an electron through an organic molecule (70 eV) 10^{-16} s
- Formation of the molecular ion M^+ (EI) 10^{-12} s
- Fragmentation reactions finished 10^{-9} s
- Rearrangement reactions finished $10^{-6} - 10^{-7}$ s
- Lifetime of metastable ions $10^{-3} - 10^{-6}$ s

Flight times of ions
- Magnetic sector analyser 10^{-5} s
- Quadrupole analyser 10^{-4} s
- Ion trap analyser (storage times) $10^{-2} - 10^{-6}$ s
- Orbitrap analyser $10^{-3} - 10^{0}$ s

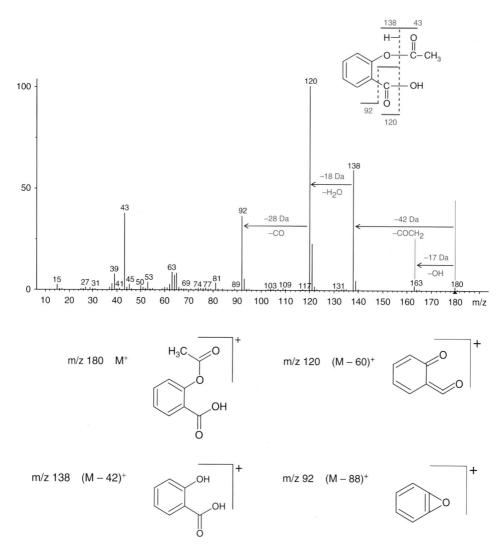

Figure 2.156 A typical line spectrum in organic mass spectrometry with a molecular ion, fragment ions and the loss of neutral particles (acetylsalicylic acid, $C_9H_8O_4$, M 180).

Types of Ions in Mass Spectrometry

- *Molecular ion* :
  The unfragmented positive or negatively charged ion with a mass equal to the molecular mass and of a radical nature because of the unpaired electron.
- *Quasimolecular ion* :
  Ions associated with the molecular mass which are formed through CI, for example, as $(M + H)^+$, $(M - H)^+$ or $(M - H)^-$ and are not radicals.
- *Adduct ions* :
  Ions which are formed through addition of charged species, for example, $(M + NH_4)^+$ through CI with ammonia as the CI gas.
- *Fragment ions* :
  Ions formed by cleavage of one or more bonds.
- *Rearrangement products*:
  Ions which are formed following bond cleavage and migration of an atom (see McLafferty rearrangement).
- *Metastable ions* :
  Ions (m_1) which lose neutral species (m_2) during the flight time through the analyser and are detected on a magnetic sector instrument with mass $m^* = (m_2)^2/m_1$.
- *Base ion* :
  This ion has the highest intensity (100%) (base peak) in a mass spectrum.

2.3.1.2 Electron Ionization

Electron ionization (EI) is the standard process in all GC-MS instruments. An ionization energy of 70 eV is used in all commercial instruments. Only a few benchtop instruments still allow the user to adjust the ionization energy for specific purposes. In particular, for magnetic sector instruments the ionization energy can be reduced to about 15 eV (Gross, 2004). This allows EI spectra with a high proportion of molecular information to be obtained (low voltage ionization) (Figure 2.157). The technique has decreased in importance due to its inherently low sensitivity since the introduction of chemical ionization (CI).

The process of EI ionization can be explained by a wave or a particle model. The amount of energy transferred in this process is called the *ionization energy*. The current theory is based on the interaction between the energy-rich electron beam with the outer electrons of a molecule. At around 70 eV, the de Broglie wavelength of the electrons matches the length of typical bonds in organic molecules (about 0.14 nm) and the energy transfer to an organic analyte molecule is maximized by resonance, leading to the strongest possible ionization and fragmentation reactions. Energy absorption initially leads to the formation of a molecular ion M^+ by loss of an electron. The excess energy causes excitation in the rotational

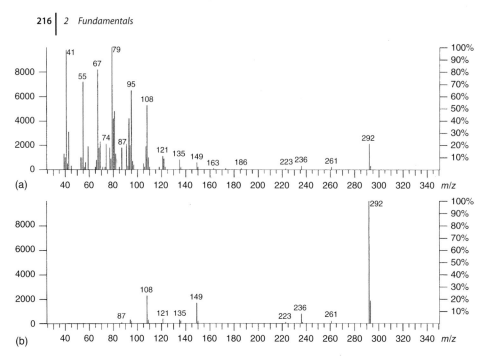

Figure 2.157 EI spectra of methyl linolenate, $C_{17}H_{29}COOCH_3$ (Spiteller and Spiteller, 1973). Recording conditions: direct inlet, (a) 70 eV, (b) 17 eV.

and vibrational energy levels of this radical cation. The subsequent processes of fragmentation depend on the amount of excess energy and the capacity of the molecule for internal stabilization.

The generation of EI mass spectra is well understood with a rich documentation of the fragmentation processes. But in contrast to NMR, mass spectra still cannot be calculated from a theoretical standpoint. A general method to compute EI mass spectra based on a combination of fast quantum chemical methods, molecular dynamics and stochastic preparation of 'hot' ionized species was published by Grimme. It provides mass spectra that compare well with their experimental counterparts, even in subtle details (Grimme, 2013).

The concept of localized charge empirically describes the fragmentation reactions (Budzikiewicz, 2005). The concept was developed from the observation that molecular ions preferentially fragment bonds near heteroatoms (N, O, S Figure 2.158) or π electron systems. This is attributed to the fact that a positive or negative charge is stabilized by an electronegative structure element in the molecule or one favouring mesomerism. Bond breaking can be predicted by subsequent electron migrations or rearrangement. These types of processes include a-cleavage, allyl cleavage, benzyl cleavage and the McLafferty rearrangement (see also Section 3.2.5).

The energy necessary for ionization of organic molecules is lower than the effective applied energy of 70 eV and is usually less than 15 eV (Table 2.40). The EI operation of all MS instruments at the high ionization energy of 70 eV was

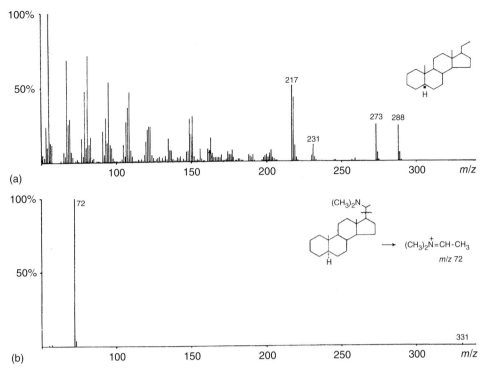

(a)

(b)

Figure 2.158 Effect of structural features on the appearance of mass spectra – concept of localized charge (Budzikiewicz, 2005). (a) Mass spectrum of 5α-pregnane and (b) mass spectrum of 20-dimethylamino-5α-pregnane. The α-cleavage of the amino group dominates in the spectrum; information on the structure of the sterane unit is completely absent!

Table 2.40 First ionization potentials (eV) of selected substances.

Helium	24.6
Nitrogen	15.3
Carbon dioxide	13.8
Oxygen	12.5
Propane	11.07
1-Chloropropane	10.82
Butane	10.63
Pentane	10.34
Nitrobenzene	10.18
Benzene	9.56
Toluene	9.18
Chlorobenzene	9.07
Propylamine	8.78
Aniline	8.32

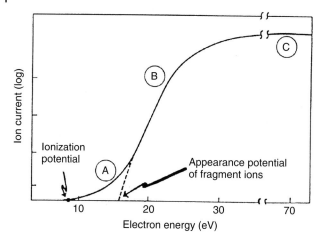

Figure 2.159 Increase in the ion current with increasing electron energy (Frigerio, 1974). (A) Threshold region after reaching the appearance potential; molecular ions are mainly produced here. (B) Build-up region with increasing production of fragment ions. (C) Routine operation, stable formation of fragment ions.

established and is kept constant with regard to sensitivity (ion yield) and the comparability of the mass spectra obtained, in particular for library generation and search purposes.

Assuming a constant substance stream into an ion source, Figure 2.159 shows the change in intensity of the signal with increasing ionization energy. The steep rise of the signal intensity only begins when the IP is reached. Low measurable intensities just below the IP are produced as a result of the inhomogeneous composition of the electron beam. Generally, the increase in signal intensity continues with increasing ionization energy until a plateau is reached. A further increase in the ionization energy is now indicated by a slight decrease in the signal intensity. An electron with an energy of 50 eV has a velocity of 4.2×10^6 m/s and crosses a molecular diameter of a few Ångstroms in about 10^{-16} s!

Further increases in the signal intensity are, therefore, not achieved via the ionization energy with beam instruments, but by using measures to increase the density and the dispersion of the electron beam. The application of pairs of magnets to the ion source can be used, for example.

The standard ionization energy of 70 eV which has been established for many years is also aimed at making mass spectra comparable. At an ionization energy of 70 eV, which is in excess of that required for ionization to M^+, remains as excess energy in the molecule, assuming maximum energy transfer. As a result, fragmentation reactions occur and lead to an immediate decrease in the concentration of M^+ ions in the ion source. At the same time, stable fragment ions are increasingly formed.

Each mass spectrum is the quantitative analysis by the analyser system of the processes occurring during ionization. It is recorded as a line diagram. The involved fragmentation and rearrangement processes are extensively known today. They serve as fragmentation rules for the manual interpretation of mass spectra and thus for the identification of unknown substances.

2.3.1.3 Chemical Ionization

In electron ionisation (EI) molecular ions M^+ are first produced by bombardment of the molecule M with high energy electrons (70 eV). The high excess energy in M^+ (the IP of organic molecules is below 15 eV) leads to unimolecular fragmentation into fragment ions ($F1^+$, $F2^+$, …) and uncharged species.

The EI mass spectrum shows the fragmentation pattern. The nature (*m/z* value) and frequency (intensity %) of the fragmentation can be read directly from the line spectrum. The loss of neutral particles is shown by the difference between the molecular ion and the fragments formed from it.

Which line in the EI spectrum is the molecular ion? Only a few stable molecules give dominant M^+ ions, for example, aromatics and their derivatives, such as PCBs and dioxins. The molecular ion is frequently only present with a low intensity. With the small quantities of analytes applied, as is the case with GC-MS, the molecular information can be identified only with difficulty among the noise (matrix), or it fragments completely and does not appear in the spectrum (Howe *et al.*, 1981).

Figure 2.160, which shows the EI/CI spectra of the phosphoric acid ester Tolclofos-methyl, is an example of this. The base peak in the EI spectrum shows a Cl atom, noticeable by the typical isotope pattern of m/z 265/267. Loss of a methyl group $(M - 15)^+$ gives *m/z* 250. Is *m/z* 265 the nominal molecular mass? The CI spectrum shows *m/z* 301 for a protonated ion so the nominal molecular mass could be 300 Da. A chlorine isotope pattern of two Cl atoms is also visible.

How can both EI and CI spectra be completed? Obviously, Tolclofos fragments completely in EI by loss of a Cl atom to *m/z* 265 as $(M - 35)^+$. With CI, this fragmentation does not occur. The attachment of a proton retains the complete molecule with all components of the elemental formula by formation of the quasimolecular ion $(M + H)^+$.

The importance of EI spectra for identification and structure confirmation is due to the fragmentation pattern. Searches through libraries of spectra are typically based on EI spectra. With the introduction of the CI capabilities for internal ionization ion trap systems, a commercial CI library of spectra with more than 300 pesticides was introduced only at that time by Finnigan.

The term *chemical ionization*, unlike EI, covers all soft ionization techniques which involve an exothermic chemical reaction in the gas phase mediated by a reagent gas and its reagent ions. Stable positive or negative ions are formed as products. Unlike the molecular ions of EI ionization, the quasimolecular ions of CI are not radicals (Harrison, 1992).

The principle of CI was first described by Munson and Field (Munson and Field, 1966). CI has now developed into a widely used technique for structural determination and quantitation in GC-MS (Stan and Kellner, 1981). Instead of an open, easily evacuated ion source, a closed ion volume is necessary for carrying out CI. In the high vacuum environment of the ion source, a reagent gas pressure of about

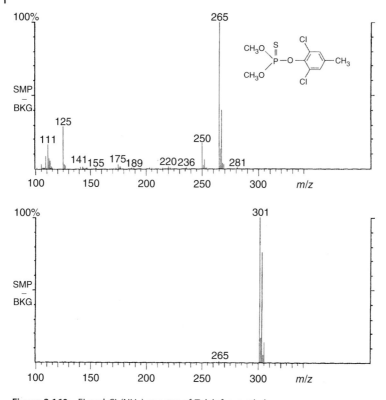

Figure 2.160 EI and CI (NH$_3$) spectra of Tolclofos-methyl.

1 Torr must be maintained to achieve the desired CI reactions. Depending on the construction of the instrument, either changing the ion volume, a change of the whole ion source, or only a software switch is necessary. Through the straightforward technical realization, in the case of combination ion sources, and through the use of ion trap mass spectrometers, CI has become established in residue and environmental analysis, even for routine methods.

CI uses considerably less energy for ionizing the molecule M. CI spectra, therefore, have fewer or no fragments and thus generally give important information on the molecule itself.

The use of CI is helpful in structure determination, confirmation or determination of molecular weights, and also in the determination of significant substructures (Keller *et al.*, 1989). Additional selectivity can be introduced into mass spectrometric detection by using the CI reaction of certain reagent gases, for example, the detection of active substances with a transparent hydrocarbon matrix. Analyses can be quantified selectively, with high sensitivity and unaffected by the low molecular weight matrix, by the choice of a quantitation mass in the upper molecular weight range. The spectrum of analytical possibilities with CI is not limited to the basic reactions described briefly here. Furthermore, it opens up the whole field of chemical reactions in the gas phase.

The Principle of Chemical Ionization

In CI, two reaction steps are always necessary. In the primary reaction, a stable cluster of reagent ions is produced from the reagent gas through electron bombardment. The composition of the reagent gas cluster is typical for the gas used. The cluster formed usually shows up on the screen for adjustment.

In the secondary reaction, the molecule M in the GC eluate reacts with the ions in the reagent gas cluster. The ionic reaction products are detected and recorded as the CI spectrum. It is the secondary reaction which determines the appearance of the spectrum. Only exothermic reactions give CI spectra. In the case of protonation, this means that the proton affinity PA of M must be higher than that of the reagent gas PA(R) (Figure 2.161 and Table 2.41).

Through the choice of the reagent gas R, the quantity of energy transferred to the molecule M and thus the degree of possible fragmentation and the question of selectivity can be controlled. If PA(R) is higher than PA(M), no protonation occurs. When a non-specific hydrocarbon matrix is present, this leads to transparency of the background, while active substances, such as plant protection agents, appear with high S/N ratios.

For CI, many types of reaction can be used analytically. In gas phase reactions, not only positive, but negative ions can also be formed (Budzikiewicz, 1981). Commercial GC-MS systems usually detect positive ions (positive chemical ionization, PCI) and negative ions (negative chemical ionization, NCI) separately. Specially equipped instruments also allow the simultaneous detection of positive and negative ions produced by CI (Bowadt *et al.*, 1993). The alternating reversal of polarity during scanning (pulsed positive ion negative ion chemical ionization, PPINICI) produces two complementary data files from one analysis (Hunt *et al.*, 1976).

Positive Chemical Ionization   Essentially, four types of reaction mainly contribute to the formation of positive ions. As in all CI reactions, reaction partners meet in the gas phase and form a transfer complex $M \cdot R^+$. In the following types of reaction the transfer complex is either retained or reacts further.

Protonation   Protonation is the most frequently used reaction in PCI. Protonation leads to the formation of the quasimolecular ion $(M + H)^+$, which can then

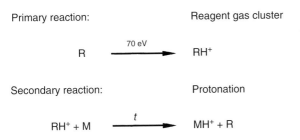

Primary reaction: Reagent gas cluster

$$R \xrightarrow{\text{70 eV}} RH^+$$

Secondary reaction: Protonation

$$RH^+ + M \xrightarrow{\;t\;} MH^+ + R$$

Figure 2.161 Primary and secondary reactions in protonation.

Table 2.41 Proton affinities of some simple compounds.

Aliphatic amines

NH_3	857	$n\text{-}Pr_2NH$	951
$MeNH_2$	895	$i\text{-}Pr_2NH$	957
$EtNH_2$	907	$n\text{-}Bu_2NH$	955
$n\text{-}PrNH_2$	913	$i\text{-}Bu_2NH$	956
$i\text{-}PrNH_2$	917	$s\text{-}Bu_2NH$	965
$n\text{-}BuNH_2$	915	Me_3N	938
$i\text{-}BuNH_2$	918	Et_3N	966
$s\text{-}BuNH_2$	922	$n\text{-}Pr_3N$	976
$t\text{-}BuNH_2$	925	$n\text{-}Bu_3N$	981
$n\text{-}Amyl\text{-}NH_2$	918	Me_2EtN	947
$Neopentyl\text{-}NH_2$	920	$MeEt_2N$	957
$t\text{-}Amyl\text{-}NH_2$	929	$Et_2\text{-}n\text{-}PrN$	970
$n\text{-}Hexyl\text{-}NH_2$	920	Pyrrolidine	938
$Cyclohexyl\text{-}NH_2$	925	Piperidine	942
Me_2NH	922	N-methylpyrrolidine	952
MeEtNH	930	N-methylpiperidine	956
Et_2NH	941	$Me_3Si(CH_2)_3NMe_2$	966

Oxides and sulfides

H_2O	723	$n\text{-}Bu_2O$	852
MeOH	773	$i\text{-}PrO\text{-}t\text{-}Bu$	873
EtOH	795	$n\text{-}Pentyl_2O$	858
$n\text{-}PrOH$	800	Tetrahydrofuran	834
$t\text{-}BuOH$	815	Tetrahydropyran	839
Me_2O	807	H_2S	738
MeOEt	844	MeSH	788
Et_2O	838	Me_2S	839
$i\text{-}PrOEt$	850	MeSEt	851
$n\text{-}Pr_2O$	848	Et_2S	859
$i\text{-}Pr_2O$	861	$i\text{-}Pr_2S$	875
$t\text{-}BuOMe$	852	H_2Se	742

Disubstituted alkanes

$NH_2CH_2CH_2NH_2$	947	$Me_2N(CH_2)_6NMe_2$	1041
$NH_2CH_2CH_2CH_2NH_2$	973	$NH_2CH_2CH_2OMe$	933
$NH_2CH_2CH_2CH_2CH_2NH_2$	995	$NH_2CH_2CH_2CH_2OH$	952
$NH_2(CH_2)_5NH_2$	986	$NH_2(CH_2)_6OH$	966
$NH_2(CH_2)_2NH_2$	989	$NH_2CH_2CH_2CH_2F$	928
$Me_2NCH_2CH_2NMe_2$	996	$NH_2CH_2CH_2CH_2Cl$	928
$Me_2NCH_2CH_2CH_2NH_2$	1002	$OHCH_2CH_2CH_2CH_2OH$	886
$Me_2NCH_2CH_2CH_2NMe_2$	1012	2,4-Pentanedione	886
$Me_2N(CH_2)_4NMe_2$	1028	—	—

Substituted alkylamines and alcohols

$CH_2FCH_2NH_2$	890	CF_3NMe_2	815
$CHF_2CH_2NH_2$	871	CHF_2CH_2OH	755
$CF_3CH_2NH_2$	850	CF_3CH_2OH	731

Table 2.41 *(Continued)*

$(CF_3)_3CNH_2$	800	CCl_3CH_2OH	760
$CF_3CH_2NHCH_3$	880	Piperazine	936
$CF_3CH_2CH_2NH_2$	885	1,4-Dioxan	811
$CF_3CH_2CH_2CH_2NH_2$	900	Morpholine	915
$CH_2FCH_2CH_2NH_2$	914	—	—

Unsaturated amines and anilines

$CH_2=CHCH_2NH_2$	905	$m\text{-}CH_3OC_6H_4NH_2$	906
Cyclo-$C_3H_5NH_2$	899	$p\text{-}CH_3OC_6H_4NH_2$	899
$CH_2=C(CH_3)CH_2NH_2$	912	$p\text{-}ClC_6H_4NH_2$	876
$HC\equiv CCH_2NH_2$	884	$m\text{-}ClC_6H_4NH_2$	872
$(CH_2=CCH_3CH_2)_3N$	964	$m\text{-}FC_6H_4NH_2$	870
$(CH_2=CHCH_2)_3N$	958	C_6H_5NHMe	912
$(HC\equiv CCH_2)_3N$	916	$C_6H_5NMe_2$	935
$C_6H_5NH_2$	884	$C_6H_5CH_2NH_2$	918
$m\text{-}CH_3C_6H_4NH_2$	896	$C_6H_5CH_2NMe_2$	953
$p\text{-}CH_3C_6H_4NH_2$	896	—	—

Other N, O, P, S compounds

Aziridine	902	$CH_2=CHCN$	802
Me_2NH	922	ClCN	759
N-Methylaziridine	926	BrCN	770
Me_3N	938	CCl_3CN	760
MeNHEt	930	CH_2ClCN	773
2-Methylaziridine	916	CH_2ClCH_2CN	795
Pyridine	921	$NCCH_2CN$	757
Piperidine	942	i-PrCN	819
MeCH=NEt	931	n-BuCN	818
n-PrCH=NEt	942	cyclo-C_3H_6CN	824
$Me_2C=NEt$	959	Ethylene oxide	793
HCN	748	Oxetane	823
CH_3NH_2	895	CH_2O	741
MeCN	798	MeCHO	790
EtCN	806	$Me_2C=O$	824
n-PrCN	810	Thiirane	818

Carbonyl compounds, iminoethers and hydrazines

EtCHO	800	CF_3CO_2-n-Bu	782
n-PrCHO	809	$HCO_2CH_2CF_3$	767
n-BuCHO	808	$NCCO_2Et$	767
i-PrCHO	808	$CF_3CO_2CH_2CH_2F$	764
Cyclopentanone	835	$(MeO)_2C=O$	837
HCO_2Me	796	HCO_2H	764
HCO_2Et	812	$MeCO_2H$	797
HCO_2-n-Pr	816	$EtCO_2H$	808
HCO_2-n-Bu	818	FCH_2CO_2H	781
$MeCO_2Me$	828	$ClCH_2CO_2H$	779
$MeCO_2Et$	841	CF_3CO_2H	736

(continued overleaf)

Table 2.41 *(Continued)*

MeCO$_2$-n-Pr	844	n-PrNHCHO	878
CF$_3$CO$_2$Me	765	Me$_2$NCHO	888
CF$_3$CO$_2$Et	777	MeNHNH$_2$	895
CF$_3$CO$_2$-n-Pr	781	—	—
Substituted pyridines			
Pyridine	921	4-CF$_3$-pyridine	890
4-Me-pyridine	935	4-CN-pyridine	880
4-Et-pyridine	939	4-CHO-pyridine	900
4-t-Bu-pyridine	945	4-COCH$_3$-pyridine	909
2,4-diMe-pyridine	948	4-Cl-pyridine	910
2,4-di-t-Bu-pyridine	967	4-MeO-pyridine	947
4-Vinylpyridine	933	4-NH$_2$ pyridine	961
Bases weaker than water			
HF	468	CO$_2$	530
H$_2$	422	CH$_2$	536
O$_2$	423	N$_2$O	567
Kr	424	CO	581
N$_2$	475	C$_2$H$_6$	551
Xe	477	—	—

All data in (kJ mol^{-1}) calculated for proton affinities PA(M) at 25 °C corresponding to the reaction M + H$^+$ − MH$^+$ (Aue and Bowers, 1979).

Table 2.42 Reagent gases for proton transfer.

Gas	Reagent ion	m/z	PA (kJ mol^{-1})
H$_2$	H$_3^+$	3	422
CH$_4$	CH$_5^+$	17	527
H$_2$O	H$_3$O$^+$	19	706
CH$_3$OH	CH$_3$OH$_2^+$	33	761
i-C$_4$H$_{10}$	t-C$_4$H$_9^+$	57	807
NH$_3$	NH$_4^+$	18	840

undergo fragmentation:

$$M + RH^+ \rightarrow MH^+ + R$$

Normally, methane, water, methanol, isobutane or ammonia are used as a protonating reagent gas (Table 2.42). Methanol occupies a middle position with regard to fragmentation and selectivity. Methane is less selective and is designated a hard CI gas. Isobutane and ammonia are typical soft CI gases often used besides other more selective protonating reagent gas systems (Dorey *et al.*, 1994).

The CI spectra formed through protonation show the quasimolecular ions $(M + H)^+$. Fragmentations start with this ion. For example, loss of water shows up as $M - 17$ in the spectrum, formed through $(M + H)^+ - H_2O$!

The existence of the quasimolecular ion is often indicated by low signals from addition products of the reagent gas. In the case of methane, besides $(M + H)^+$, $(M + 29)^+$ and $(M + 41)^+$ appear (see methane), and for ammonia, besides $(M + H)^+$, $(M + 18)^+$ appears with varying intensity (see ammonia).

Hydride Abstraction In this reaction, a hydride ion (H^-) is transferred from the substance molecule to the reagent ion:

$$M + R^+ \rightarrow RH + (M - H)^+$$

This process is observed, for example, in the use of methane when the C_2H_5 ion (m/z 29) contained in the methane cluster abstracts hydride ions from alkyl chains.

With methane as the reagent gas, both protonation and hydride abstraction can occur, depending on the reaction partner M. The quasimolecular ion obtained is either $(M - H)^+$ or $(M + H)^+$. In charge exchange reactions M^+ is also formed.

Charge Exchange The charge exchange reaction gives a radical molecular ion with an odd number of electrons as in EI. Accordingly, the quality of the fragmentation is comparable to that of an EI spectrum. The extent of fragmentation is determined by the IP of the reagent gas.

$$M + R^+ \rightarrow R + M^{\cdot+}$$

The ionisation potential (IP) of most organic compounds are below 15 eV. Through the choice of reagent gas it can be controlled whether a dominant molecular ion appears in the spectrum or whether and how extensively fragmentation occurs. In the extreme case, spectra similar to those with EI are obtained. Common reagent gases for ionization by charge exchange are benzene, nitrogen, carbon monoxide, nitric oxide or argon (Table 2.43).

Table 2.43 Reagent gases for charge exchange reactions.

Gas	Reagent ion	m/z	IP (eV)
C_6H_6	$C_6H_6^+$	78	9.3
Xe	Xe^+	131	12.1
CO_2	CO_2^+	44	13.8
CO	CO^+	28	14.0
N_2	N_2^+	28	15.3
Ar	Ar^+	40	15.8
He	He^+	4	24.6

In the use of methane, charge exchange reactions as well as protonations can be observed, in particular for molecules with low proton affinities.

Adduct Formation   If the transition complex described earier does not dissociate, the adduct is visible in the spectrum:

$$M + R^+ \rightarrow (M + R)^+$$

This effect is seldom made use of in GC-MS analysis in contrast to being the prevailing reaction in ESI-LC-MS (electrospray ionization), but must be taken into account on evaluating CI spectra. The enhanced formation of adducts is always observed with intentional protonation reactions where differences in the proton affinity of the participating species are small. High reagent gas pressure in the ion source favours the effect by stabilizing collisions.

Frequently, an $(M + R)^+$ ion is not immediately recognized, but can give information which is as valuable as that from the quasimolecular ion formed by protonation (see Methane). Cluster ions of this type, nevertheless sometimes, make interpretation of spectra more difficult, particularly when the transition complex does not lose immediately recognizable **neut**ral species.

Negative Chemical Ionization   Negative ions are also formed during ionization in MS even under EI conditions, but their yield is so extremely low that it is of no use analytically. The intentional production of negative ions can take place by attachment of thermal electrons (analogous to an ECD), by charge exchange or by extraction of acidic hydrogen atoms (Dougherty, 1981; DePuy *et al.*, 1982; Stout and Steller, 1984; Horning *et al.*, 1991).

Charge Transfer   The ionization of the sample molecule is achieved by the transfer of an electron between the reagent ion and the molecule M

$$M + R^- \rightarrow M^- + R$$

The reaction can only take place if the electron affinity (EA) of the analyte M is greater than that of the electron donor R.

$$EA(M) > EA(R)$$

In practical analysis, charge transfer to form negative ions is less important unlike the formation of positive ions by charge exchange.

Proton Abstraction   Proton transfer in NCI can be understood as proton abstraction from the sample molecule. In this way, all substances with acidic hydrogens, for example, alcohols, phenols or ketones, can undergo soft ionization.

$$M + R^- \rightarrow (M-H)^- + RH$$

Proton abstraction only occurs if the proton affinity of the reagent gas ion is higher than that of the conjugate base of the analyte molecule. A strong base, for example, OH^-, is used as the reagent for ionization. Substances which are more basic than the reagent are not ionized (Smit and Field, 1977).

Reagent gases and organic compounds were arranged in order of gas phase acidity by Bartmess and McIver in (1979). The order corresponds to the reaction enthalpy of the dissociation of their functional groups into a proton and the corresponding base. Table 2.44 can be used for controlling the selectivity via proton abstraction by choosing suitable reagent gases.

Extensive fragmentation reactions can be excluded by proton abstraction. The energy released in the exothermic reaction is essentially localized in the new compound RH. The anion formed does not contain excess energy for extensive fragmentation.

Reagent Ion Capture   The capture of negative reagent gas ions was described in the early 1970s by Manfred von Ardenne and coworkers for the analysis of long-chain aliphatic hydrocarbons with hydroxyl ions (Ardenne *et al.*, 1971).

$$M + R^- \rightarrow M \cdot R^-$$

Besides associative addition with weak bases, adduct formation can lead to a new covalent bond. The ions formed are more stable than the comparable association products.

Substitution reactions, analogous to an S_N2 substitution in solution, occur more frequently in the gas phase because of poor solvation and low activation energy. Many aromatics give a peak at $(M + 15)^-$, which can be attributed to substitution of H by an O^- radical. Substitution reactions are also known for fluorides and chlorides. Fluoride is the stronger nucleophile and displaces chloride from alkyl halides.

Electron Capture   Electron capture with formation of negative ions is the NCI ionization process that is most frequently used in the GC-MS analysis. There is a direct analogy with the behaviour of substances in ECD and the areas of use are similar (Class, 1991). The commonly used term *ECD-MS* indicates the parallel mechanisms and applications. With NCI, the lowest detection limits in organic MS have been reached (Figure 2.162). The detection of 100 ag octafluoronaphthalene corresponds to the detection of about 200 000 molecules!

At the same energy, electrons in the CI plasma have a much higher velocity (mobility) than those of the heavier positive reagent ions.

$$E = m/2 \cdot v^2 \qquad m(e^-) = 9.12 \cdot 10^{-28} g$$
$$m(CH_5^+) = 2.83 \cdot 10^{-22} g$$

Electron capture as an ionization method is 100 – 1000 times more sensitive than ion/molecule reactions which are limited by diffusion. For substances with high electron affinities, higher sensitivities can be achieved than with PCI. Substances

Table 2.44 Scale of gas phase acidities.

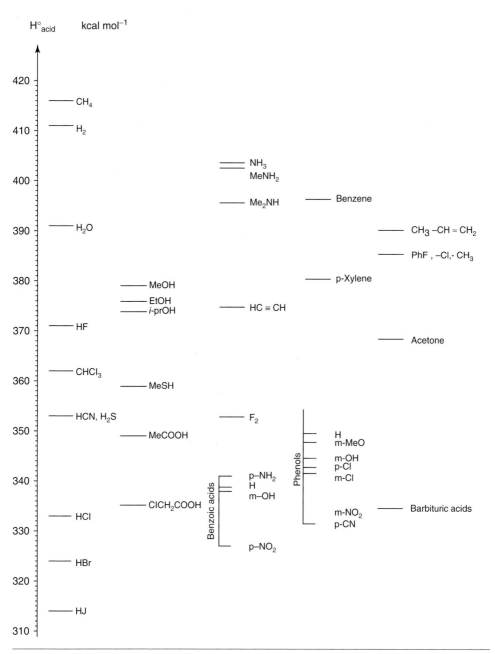

(Bartmess and McIver, 1979).

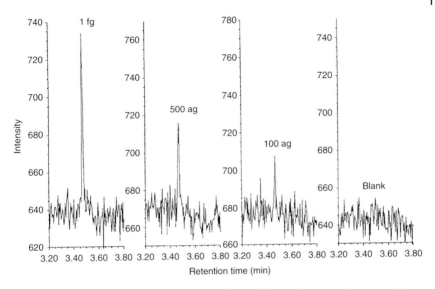

Figure 2.162 GC-MS detection of traces of $10^{-15}-10^{-16}$ g octafluoronaphthalene by NCI detection of the molecular ion m/z 272 (McLafferty and Michnowicz, 1992).

which have a high NCI response typically have a high proportion of halogen or nitro groups (Figure 2.163). NCI permits the detection of trace components in complex biological matrices (Figure 2.164). In practice, it has been shown that about five to six halogen atoms in the molecule detection with NCI give a higher specific response than that using EI (Figure 2.165). For this reason, in analysis of polychlorinated dioxins and furans, although chlorinated, EI is the predominant method, in particular in the detection of 2,3,7,8-TCDD in trace analysis (Crow, et al., 1981; Buser and Müller, 1994; Chernetsova, 2002).

Another feature of NCI measurements is the fact that, like ECD, the response depends not only on the number of halogen atoms, but also on their position in the molecule. Precise quantitative determinations are therefore only possible with defined reference systems via the determination of specific response factors.

The key to sensitive detection of negative ions lies in the production of a sufficiently high population of thermal electrons. The extent of formation of M^- at sufficient electron density depends on the electron affinity of the sample molecule, the energy spectrum of the electron population and the frequency with which molecular anions collide with neutral particles and become stabilized (collision stabilization of the radical ion). Also with an ion trap analyser, using external ionization, NCI can be utilized analytically. The storage of electrons in the ion trap itself is not possible because of their low mass (internal ionization).

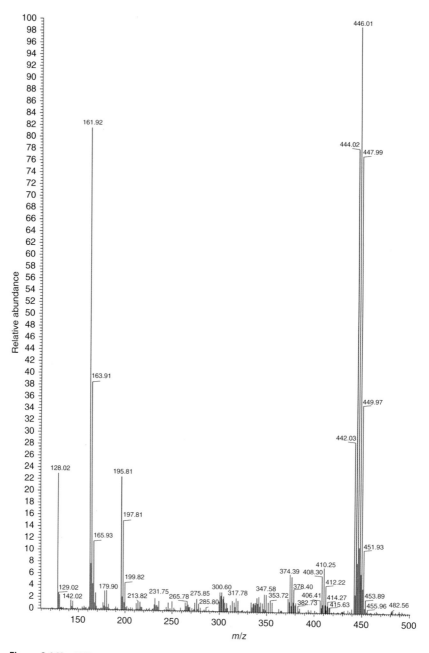

Figure 2.163 NCI spectrum of a Toxaphen component (Parlar 69, (2,2,5,5,6-exo,8,9,9,10,10-Decachlorobornane)). The typical fragmentation known from EI ionization is absent, the total ion current is concentrated on the molecular ion range (Hainzl, 1994; Theobald, 2007).

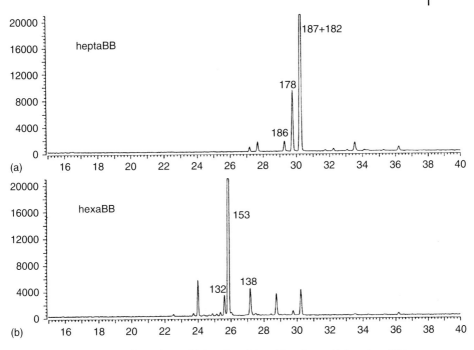

Figure 2.164 Detection of heptabromobiphenylene (a) and hexabromobiphenylene (b) in human milk by NCI (Fürst, 1994).

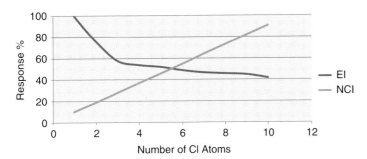

Figure 2.165 Response dependence in the EI and NCI modes of the increasing number of Cl atoms in the molecule. The decrease in the response in the EI mode with increasing Cl content is caused by splitting of the overall signal by individual isotopic masses.

In the NCI mode, the response increases with increasing Cl content through the increase in the electronegativity for electrons capture (analogous to an ECD, compound dependent).

For residue analysis, derivatization with perfluorinated reagents (e.g., heptafluorobutyric anhydride, perfluorobenzoyl chloride) in association with NCI is important (Knapp, 1979; Pierce, 1982). Besides being easier to chromatograph, the derivatized analytes show high electron affinities due to the high number of halogens and allow a very sensitive detection.

Reagent Gas Systems

Methane   Methane is one of the longest known and best-studied reagent gases. It is known and used as a hard reagent gas for general CI applications and has been replaced by softer ones in many areas of analysis.

The reagent gas cluster of methane is formed by a multistep reaction, which gives two dominant reagent gas ions with m/z 17 and 29, and in lower intensity an ion with m/z 41.

$$CH_4 \text{ at } 70\,eV \rightarrow CH_4^+, CH_3^+, CH_2, CH^+ \text{ and others}$$
$$CH_4^+ + CH_4 \rightarrow CH_5^+ + CH_3 \qquad m/z\,17 \quad 50\%$$
$$CH_3^+ + CH_4 \rightarrow C_2H_5^+ + H_2 \qquad m/z\,29 \quad 48\%$$
$$CH_2^+ + CH_4 \rightarrow C_2H_3^+ + H_2 + H$$
$$C_2H_3^+ + CH_4 \rightarrow C_3H_5^+ + H_2 \qquad m/z\,41 \quad 2\%$$

Good CI conditions are achieved if a ratio of m/z 17 to m/z 16 of 10:1 is achieved. Experience shows that the correct methane pressure is that at which the ions m/z 17 and 29 dominate in the reagent gas cluster and have approximately the same height with good resolution. The ion m/z 41 should also be recognizable with lower intensity (Figure 2.166).

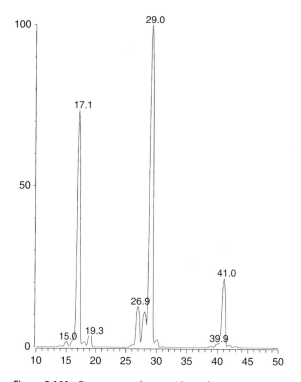

Figure 2.166   Reagent gas cluster with methane.

Methane is mainly used as the reagent gas in protonation reactions, charge exchange processes (PCI) and in pure form or as a mixture with N_2O in the formation of negative ions (NCI). In protonation, methane is a hard reagent gas, causing addition fragmentation of the molecular ion. For substances with lower proton affinity, methane frequently provides the final possibility of obtaining CI spectra. The unique adduct ions $(M + C_2H_5)^+ = (M + 29)^+$ and $(M + C_3H_5)^+ = (M + 41)^+$ formed by the methane cluster confirm the molecular mass interpretation. These adduct ions are easily seen as a mass difference of 28 resp. 40 from the protonated molecular ion $(M+H)^+$.

Ammonia

To supply the CI system with ammonia, a steel cylinder with a special reducing valve is necessary. Because of the aggressive properties of the gas, the entire tubing system must be made of stainless steel. In ion trap instruments, only the ammonium ion NH_4^+ with mass m/z 18 is formed in the reagent gas cluster. At higher pressures, adducts of the ammonium ion with ammonia $(NH_3)_n NH_4^+$ can be formed in the ion source.

Attention: Very often the ammonium ion is confused with water. Freshly installed reagent gas tubing generally has an intense water background with high intensities at m/z 18 and 19 as H_2O^+/H_3O^+. Clean tubing and correctly adjusted NH_3 CI gas shows no intensity at mass m/z 19 (Figure 2.167)!

Ammonia is a very soft reagent gas for protonation. The selectivity is correspondingly high, which is made use of in the residue analysis of many active substances. Fragmentation reactions only occur to a small extent with ammonia CI.

Adduct formation with NH_4^+ occurs with substances where the proton affinity differs little from that of NH_3 and can be used to confirm the molecular mass interpretation. In these cases the formation and addition of higher $(NH_3)_n NH_4^+$ clusters is observed with instruments with threshold pressures of about 1 Torr. Interpretation and quantitation can thus be impaired with such compounds.

Isobutane

Like methane, CI with isobutane has been known and is well documented for years. The *t*-butyl cation (m/z 57) is formed in the reagent gas cluster and is responsible for the soft character of the reagent gas (Figure 2.168).

Isobutane is used for protonation reactions of multifunctional and polar compounds. Its selectivity is high and there is very little fragmentation. In practice, with continued use significant coating of the ion source through soot formation has been reported, which can even lead to dousing of the filament. This effect depends on the adjustment and on the instrument. In such cases, ammonia can be used instead.

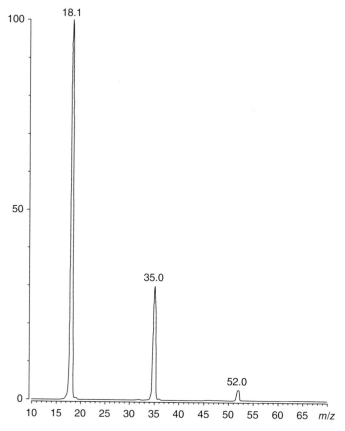

Figure 2.167 Ammonia as reagent gas.

Liquid CI Gases

Methanol Because of its low vapour pressure, methanol is ideal for CI in ion trap instruments with internal ionization. Neither pressure regulators nor cylinders or a long tubing system are required. The connection of a glass flask or a closed tube containing methanol directly on to the CI inlet is sufficient. In addition, every laboratory has methanol available. Also, for ion traps using external ionization and quadrupole instruments, liquid Cl devices are commercially available.

$(CH_3OH \cdot H)^+$ is formed as the reagent ion, which is adjusted to high intensity with good resolution (Figure 2.169). The appearance of a peak at *m/z* 47 shows the dimer formed by loss of water (dimethyl ether), which is only produced at sufficiently high methanol concentrations. It does not function as a protonating reagent ion, but its appearance shows that the pressure adjustment is correct.

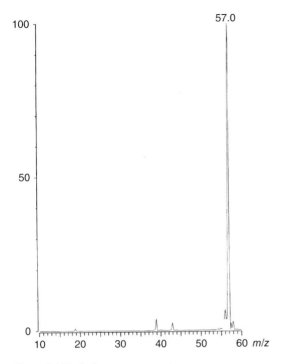

Figure 2.168 Isobutane as reagent gas.

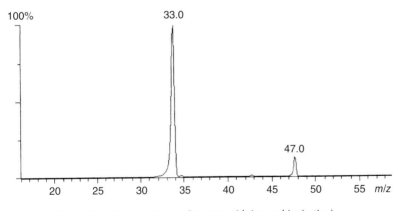

Figure 2.169 Methanol as reagent gas (ion trap with internal ionization).

Methanol is used exclusively for protonation. Because of its medium proton affinity, methanol allows a broad spectrum of classes of compounds to be determined. It is therefore suitable for a preliminary CI measurement of compounds not previously investigated. The medium proton affinity does not give any pronounced selectivity. However, substances with predominantly alkyl character remain transparent. Fragments have low intensities.

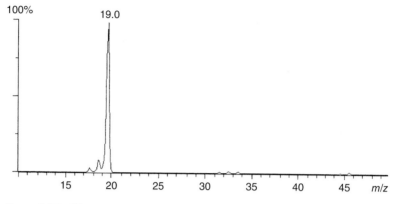

Figure 2.170 Water as reagent gas (ion trap with internal ionization).

Water For most mass spectroscopists, water is a problematic substance. However, as a reagent gas, water has extraordinary properties. Because of the high conversion rate into H_3O^+ ions, and due to its low proton affinity, water achieves a high response for many compounds when used as the reagent gas. The spectra obtained usually have few fragments and concentrate the ion beam on a dominant ion.

When water is used as the reagent gas (Figure 2.170), the intensity of the H_3O^+ ion should be as high as possible. With ion trap instruments with internal ionization, no additional equipment is required. However, a short tube length should be used for good adjustment.

For instruments with high pressure CI ion sources, the use of a heated reservoir and completely heated inlet tubing are imperative. Quadrupole instruments and ion traps using external ionization require a special liquid CI device with heated CI gas reservoir and inlet lines.

The use of water is universal. In the determination of polyaromatic hydrocarbons, considerable increases in response compared with EI detectors are found. Analytical procedures have even been published for nitroaromatics. Water can also be used successfully as a reagent gas for screening small molecules, for example, volatile halogenated hydrocarbons (industrial solvents), as it does not interfere with the low scan range for these substances.

Aspects of Switching between EI and CI

Quadrupole and Magnetic Sector Instruments To initiate CI reactions and to guarantee a sufficient conversion rate, an ion source pressure of about 1 Torr in an analyser environment of $10^{-5} - 10^{-7}$ Torr is necessary for beam instruments. For this, the EI ion source is replaced by a special CI source, which must have a gastight connections to the GC column, the electron beam and the ion exit in order to maintain the pressure in these areas.

Combination sources with mechanical devices for sealing the EI to the CI source have so far only proved successful with magnetic sector instruments. With quadrupole instruments combination ion sources (or ion volumes) are available. The performance in both ionisation modes is compromised. The response is below optimized sources, and EI/CI mixed spectra may be produced.

Increased effort is required for conversion, pumping and calibrating the CI source in beam instruments. Because of the high pressure, the reagent gas also leads to rapid contamination of the ion source and thus to additional cleaning measures in order to restore the original sensitivity of the EI system. Readily exchangeable ion volumes have been shown to be ideal for CI applications. This permits a high CI quality to be attained and, after a rapid exchange, unaffected EI conditions to be restored.

Ion Trap Instruments Ion trap mass spectrometers with internal ionization can be used for CI without hardware conversion. Because of their mode of operation as storage mass spectrometers, only a very low reagent gas pressure is necessary for instruments with internal ionization. The pressure is adjusted by means of a special needle valve which is operated at low leak rates and maintains a partial pressure of only about 10^{-5} Torr in the analyser. The overall pressure of the ion trap analyser of about $10^{-4} - 10^{-3}$ Torr remains unaffected by it. CI conditions thus set up give rise to the term *low pressure CI*. Compared to the conventional ion source used in high pressure CI, in protonation reactions, for example, a clear dependence of the CI reaction on the proton affinities of the reaction partners is observed. Collision stabilization of the products formed does not occur with low pressure CI. This explains why "high pressure" CI-typical adduct ions are not formed here, which would confirm the identification of the (quasi)molecular ion (e.g., with methane besides $(M + H)^+$, also $M + 29$ and $M + 41$ are expected). The determination of ECD-active substances by electron capture (NCI) is not possible with low pressure CI (Yost, 1988).

Switching between EI and CI modes in an ion trap analyser with internal ionization takes place with a keyboard command or through the scheduled data acquisition sequence in automatic operations. All mechanical devices necessary in beam instruments are dispensed with completely. The ion trap analyser is switched to a CI scan function internally without effecting mechanical changes to the analyser itself.

The CI reaction is initiated when the reagent ions are generated by changing the operating parameters and a short reaction phase has taken place in the ion trap analyser. The scan function used in the CI mode with ion trap instruments (see Figure 2.171) clearly shows two plateaus which directly correspond to the primary and secondary reactions. After the end of the secondary reaction, the product ions, which have been produced and stored, are determined by the mass scan and the CI spectrum registered. In spite of the presence of the reagent gas, typical EI spectra can therefore be registered in the EI mode. The desired CI is made possible by simply switching to the CI operating parameters.

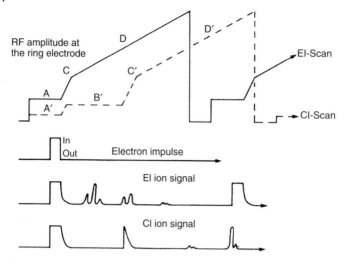

Figure 2.171 Switching between the EI and CI scan functions in the case of an ion trap analyser with internal ionization (Finnigan). EI scan: A, Ionization and storage of ions; C, Starting mass; D, Recording of an EI mass spectrum. CI scan: A', Ionization and storage of reagent gas ions; B', Reaction of reagent gas ions with neutral substance molecules; C', Starting mass and D', Recording of a CI mass spectrum.

On using autosamplers, it is therefore possible to switch alternately between EI and CI data acquisition and thus use both ionization processes routinely in automatic operations. The risk of additional contamination by CI gas does not occur with ion trap instruments because of the very small reagent gas input. It allows this mode of operation to be run without impairing the analytical quality.

Ion trap instruments with external ionization comprise an ion source with a conventional design. Changing between EI and CI ionization takes place by changing the ion volume. CI in the classical manner of high pressure CI is carried out and thus the formation of negative ions by electron capture (NCI) in association with an ion trap analyser is made possible.

2.3.2
Resolution Power

2.3.2.1 Resolving Power and Resolution in Mass Spectrometry
The resolving power of a mass spectrometer describes the smallest mass differences which can be separated by the mass analyser (Webb, 2004b). Resolution and resolving power in MS today are defined differently depending on the analyser or instrument type, and is often stated without the indication of the definition employed. The new IUPAC definitions of terms used in MS provide a precise definition.

Mass Resolution

The smallest mass difference Δm between two equal magnitude peaks so that the valley between them is a specified fraction of the peak height.

Mass Resolving Power

In a mass spectrum, the observed mass divided by the difference between two masses that can be separated: $m/\Delta m$.

The procedure by which Δm was obtained and the mass at which the measurement was made shall be reported as full width at half maximum (FWHM) or 10% valley of two adjacent peaks of similar height (equals 5% peak height of a single isolated mass peak).

The mass resolving power R can be calculated by the comparison of two adjacent mass peaks m_1 and m_2 of about equal height. R is defined as

$$\frac{m_1}{\Delta m}$$

with m_1 the lower value of two adjacent mass peaks. Δm of the two peaks is taken at an overlay (valley) of 10%, or at 50% in case of the FWHM definition. In the 10% valley definition, the height from the baseline to the junction point of the two peaks is 10% of the full height of the two peaks. Each peak at this point is contributing 5% to the height of the valley.

Because it is difficult to get two mass spectral peaks of equal height adjacent to one another the practical method of calculating Δm as typically done by the instruments software is using a single mass peak of a reference compound. The peak width is measured in 5% height for the 10% valley definition and at 50% peak height for the FWHM definition as shown in Figure 2.172. The resolving power calculated using the FWHM method gives values for R that are about twice of that which is determined by the 10% valley method. This can be checked using the intercept

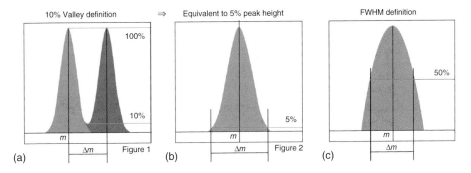

Figure 2.172 (a–c) Comparison of different mass resolution definitions 10% valley (equal to 5% peak height) and FWHM (Münster and Taylor, 2009).

theorems calculation in a triangle in which the ratios of height to width are equal:

$$\frac{h_1}{w_1} = \frac{h_2}{w_2}$$

$$w_2 = \left(\frac{h_2}{h_1}\right) \times w_1$$

and as the peak height above 5% intensity compares to the half peak height very close by a factor of 2:

$$h_2 \sim 2 \times h_1$$

following

$$w_2 \approx \left(2 \times \frac{h_1}{h_1}\right) \times w_1$$

$$\approx 2 \times w_1$$

for

h_1	=	Half peak height
w_1	=	Peak width at 50% height
h_2	=	Peak height above 5% to top
w_2	=	Peak width at 5%

following for the mass peak width that

$$\Delta m\,(5\% \text{ peak height}) \approx 2 \times \Delta m\,(50\% \text{ peak height})$$

consequently

$$R(10\% \text{ valley}) \times 2 \approx R(\text{FWHM})$$

In practice, a resolving power R 60 000 at 10% valley compares directly to a specification of R 120 000 at FWHM.

The mass resolving power for magnetic sector instruments is historically given with a 10% valley definition. A mass peak in high resolution magnetic sector MS is typically triangular with this analyser type. The peak width in 5% height is of valuable diagnostic use for non-optimal analyser conditions. Therefore, this method provides an excellent measure for the quality of the peak shape together with the resolving power which would not be available at the half maximum condition. The typical broad peak base was initially observed in time of flight mass spectrometers and caused the 10% valley definition would only calculate poor resolution values for this type of analyser. The more practical approach to describe the resolving power of TOF analysers is the measurement at half peak height, or as it is usually stated, at the *full* peak *width* at *half maximum* (FWHM). Both resolving power conditions compare by a factor of 2 as outlined above.

Constant Resolving Power over the Mass Range

Double focussing mass spectrometers, using both electric and magnetic fields to separate ions operate at constant mass resolving power. At a resolving power of 10 000, these instruments separate ions of m/z 1000 and m/z 1000.1. In this example, Δm is 0.1 and M is 1000, therefore $R = 1000/0.1$ or 10 000. For practical use, this property of constant resolving power over the entire mass range means, that with a resolving power of 1000, values of 0.01 Da can be separated at m/z 100, that is $R = 10\,000 = m/\Delta m = 100/0.01$. This implies that the visible distance of mass peaks on the m/z scale decreases with increasing m/z, see Figure 2.173 and 2.181.

High-resolution mass spectrometers are using data system controlled split systems at the entrance and exit of the ion beam to and from the analyser for resolution adjustment. Practically, this can be done either manually or under software control. The adjustment is performed first by closing the entrance slit to get a flat top peak. Next, the exit slit is closed accordingly until the intensity starts to decrease and a triangular peak is formed (see Figure 2.174). No other adjustments in the focussing conditions are necessary to achieve the resolution setting. Figure 2.175 show the example of the separation and peak widths of the two ions Ar^+ and $C_3H_4^+$. With increasing resolution power from R 1000 to 9500 the peak width becomes significantly narrower allowing the detection of additional ion species with small mass differences.

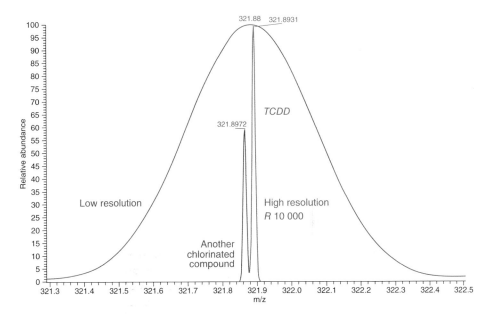

Figure 2.173 Low and high resolved 2,3,7,8-tetrachlorodibenzodioxin mass peak, m/z 321.8931 at $R = 10\,000$ (10% valley). The background interference of m/z 321.8672 cannot be resolved at low resolution.

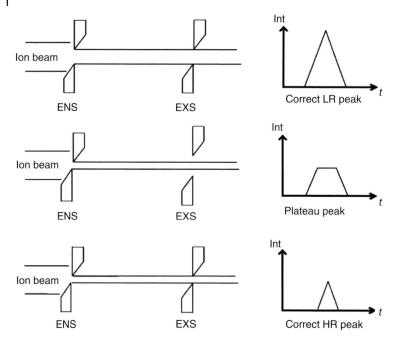

Figure 2.174 Principle of the high resolution adjustment (ENS entrance slit, EXS exit slit).

Constant Resolution Over the Mass Range

Quadrupole and ion trap mass analysers show constant peak widths and mass differences over the entire mass range. Hence, both analyser types operate at constant mass resolution with increasing resolving power (see Figure 2.181). This means, that the ability to separate ions at m/z 100 and m/z 1000 is the same. If Δm is 1 mass unit at m/z 100, the resolution at m/z 100 is 1 and the resolving power R is 100/1 or 100. If Δm is 1 at m/z 1000, the resolution at m/z 1000 is also 1, but the resolving power R at m/z 1000 is 1000/1 or 1000. Consequently, these types of analysers operate at increasing resolving power with increasing m/z value. Accordingly, the maximum resolving power, which is usually specified with a commercial quadrupole system, is dependent on the maximum specified mass range of the employed quadrupole hardware.

The maximum resolution obtained with commercial quadrupole systems is constant throughout the entire mass range with possible peak widths in the range of 0.4–5.0 Da (FWHM). Only hyperbolic shaped quadrupole rods of a special length and precision machining allow the operation mode for narrower mass peak widths down to 0.1 Da without losing significant ion transmission, see Figure 2.176.

Enhanced Mass Resolution with Quadrupole Analysers

Already in 1968 a publication by Dawson and Whetten dealt with the resolution capabilities of quadrupoles (Dawson, 1968). The theoretical investigation covered round rods as well as hyperbolic rods and indicated the higher resolution

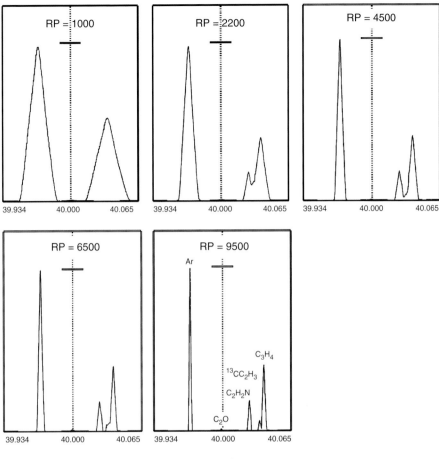

Figure 2.175 Example of the separation and peak widths of the two ions Ar^+ and $C_3H_4^+$ with increasing resolution power (RP).

Ar^+:	39.9624 Da
C_2O^+:	39.9949 Da
$C_2H_2N^+$:	40.0187 Da
$^{13}C_1C_2H_3^+$:	40.0268 Da
$C_3H_4^+$:	40.0313 Da

potential of quadrupole analysers using hyperbolic rods (see Figure 2.177). Especially with precisely machined hyperbolic rods, sufficient ion transmission is achieved even at higher resolving power, making this technology especially useful for target compound analysis with increased selectivity. The maximum resolving power of quadrupole analysers depends on the number of cycles an ion is exposed to the electromagnetic fields inside of the quadrupole assembly, hence on operation frequency and the length of the rods.

A recent development for the improved mass resolution of quadrupole analysers has been published and commercialized for a SIM technique in MS/MS named 'H-SRM' (< 0.4 Da) or 'U-SRM' (< 0.2 Da), selected reaction monitoring by using enhanced mass resolution. Hyperbolic quadrupole rods of a true hyperbolic pole face, high-precision four-section design with a rod length of 25 cm length

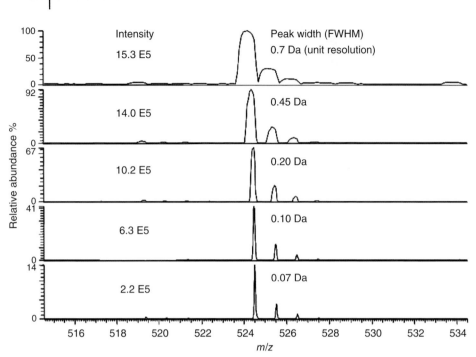

Figure 2.176 Increased peak resolution of hyperbolic quadrupoles see Figure 2.178. Note the decrease in peak height of factor 2 at 0.1 Da peak width compared to unit resolution. (Thermo Fisher Scientific.)

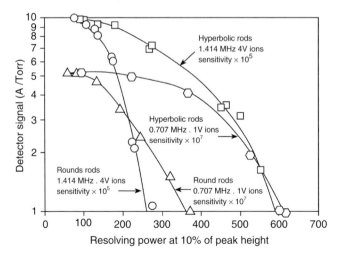

Figure 2.177 Ion transmission of round and hyperbolic quadrupole systems (resolving power ag. ion transmission), first documented investigation by Brubaker (1968) (two curves left: round rods; two curves right: hyperbolic rods).

Figure 2.178 Low resolution (left, TSQ 7000) and enhanced mass resolution (right, TSQ Quantum HyperQuad^(TM)) quadrupole analysers. (Thermo Fisher Scientific.)

are employed in triple quadrupole GC- and LC-MS/MS systems for the selective quantification of target compounds. The increased field radius of 6 mm provides a significantly increased ion transmission for trace analysis (HyperQuad technology, Thermo Fisher Scientific) (Figure 2.178).

By operating the first quadrupole in higher resolution mode, the selectivity for MS/MS analysis of a parent target ion within a complex matrix is significantly increased. Higher S/N values for the product ion peak lead to significantly lowered method detection limits (MDLs). Using the hyperbolic quadrupole rods, the reduction from 0.7 Da peak width to 0.1 Da reduces overall signal intensity only by half but S/N increases by a factor better than a factor of 10. In many cases, at low levels it even first allows to detect the target peak well above the background. By using round rods, the drop in signal intensity would be close to 100%, and would lose all sample signals. Hence, round rods can only be applied for a higher resolved and higher selective detection mode compromising ion transmission, which translates directly to less detection sensitivity.

To eliminate the uncertainty associated with possible mass drifts in H-SRM and U-SRM acquisitions, algorithms that correct the calibration table for mass position and peak width have been developed (Jemal, 2003; Liu, 2006). Data are obtained from U-SRM quantitation at a peak width setting of 0.1 Da at FWHM for the precursor ion. Deviations detected for mass position or peak width in internal standards is used to internally adjust the mass calibration. The data is then used to determine the deviations if any for all target calibrant ions. The calibration correction method (CCM) is submitted and executed within the sequence of data acquisition. A scan-by-scan CCM has been applied to accommodate a variety of factors that influence precursor ion mass drift and ensure performance eliminating signal roll-off. With real-life samples exhibiting strong matrix backgrounds, a significant increase in S/N is achieved by using the U-SRM technique (Figure 2.179).

Analytical benefits of highly resolving quadrupole systems in GC-MS/MS analysis are the separation of ions with the same nominal m/z value for increased selectivity, in particular the high resolution precursor ion selection for MS/MS mode (Table 2.45).

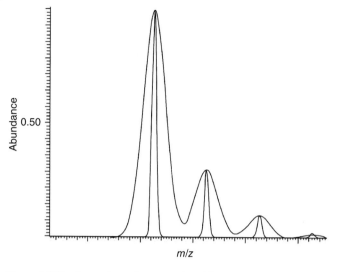

Figure 2.179 Comparison of mass peak widths at 0.7 and 0.1 Da at FWHM.

Table 2.45 Ion transmission for low and enhanced resolution quadrupoles at 0.2 Da mass peak width FWHM, 100% equals 0.7 Da FWHM (Jemal, 2003).

Quadrupole type	SIM (%)	SRM (%)
Enhanced resolution	70	53
Low resolution	11	0.15

Mass Analysers for Accurate Mass Measurement

Mass analyser	Maximum resolving power (FWHM)	Mass accuracy [ppm]	Dynamic range	Inlet methods	MS/MS capability
Time-of-flight	5000 – 50 000	5	$10^2 - 10^4$	GC-MS, LC-MS	In hybrid systems
Hyperbolic triple quadrupole	10 000	5	$10^4 - 10^6$	GC-MS, LC-MS	Typical application
Magnetic sector	120 000 – 150 000	2	$10^5 - 10^6$	GC-MS, LC-MS, Direct Inlets	Limited
FT orbitrap	>500 000	<1	$10^4 - 10^6$	LC-MS	Extensive use
FT ion cyclotron resonance	>1 000 000	<0.2	$10^3 - 10^4$	LC-MS	Extensive use

2.3.2.2 Unit Mass Resolution

A mass spectrum obtained with an ion trap or quadrupole mass spectrometer (Figures 2.180 and 2.183) shows the well-known characteristic with the constant distance between two mass signals and their mass peak widths over the whole mass range! How far the instrument can scan to higher masses does not change the peak width, hence it does not change the available mass resolution (Miller and Denton, 1986). The quadrupole/ion trap peaks of water ($m/z = 17/18$) are of the same width and have the same separation as the masses in the upper mass range (Figure 2.181).

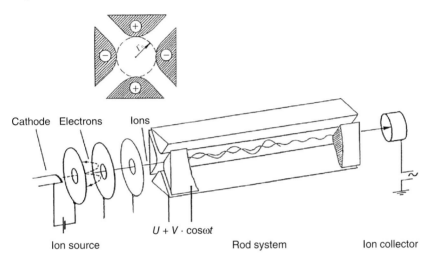

Cathode Electrons   Ions

$U + V \cdot \cos\omega t$

Ion source   Rod system   Ion collector

Figure 2.180   Diagram of a quadrupole mass spectrometer.

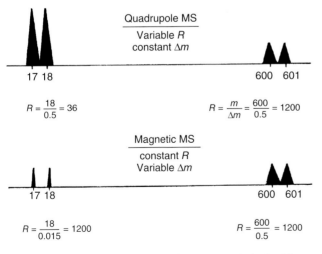

Quadrupole MS

Variable R
constant Δm

17 18   600   601

$$R = \frac{18}{0.5} = 36$$ $$R = \frac{m}{\Delta m} = \frac{600}{0.5} = 1200$$

Magnetic MS

constant R
Variable Δm

17 18   600   601

$$R = \frac{18}{0.015} = 1200$$ $$R = \frac{600}{0.5} = 1200$$

Figure 2.181   Comparison of the resolution power R obtained from quadrupole/ion trap and magnetic sector analysers.

Since the distance between two signals, the resolution Δm, is constant over the whole mass range for these instrument types, the formula $R = m/\Delta m$ would have the result that the resolving power R would be directly proportional to the highest possible mass m (Figure 2.182). At a peak separation of one mass unit, the resolving power in the lower mass range would be small using the formula for the resolution (e.g., for water $R \approx 18$) and in the upper mass range higher (e.g., $R \approx 614$ for FC43).

The following conclusions may be drawn from these facts concerning the assessment of quadrupole and ion trap instruments:

1) The formula $R = m/\Delta m$ does not give any meaningful figures for quadrupole and ion trap instruments and therefore cannot be used without specifying the mass used for calculating R.

2) Because all quadrupole and ion trap instruments provide the same quality of nominal mass peak resolution they are said to work at unit mass resolution (see Figure 2.182). The terms 'unit mass resolution' or 'nominal mass resolution' should be used instead of calculating the resolution power R.

3) The visible optical resolution and peak width is the same in the upper and lower mass ranges. It can easily be seen that the signals corresponding to whole numbers are well separated. This corresponds in practice to the typically used resolution for quadrupole and ion trap instruments.

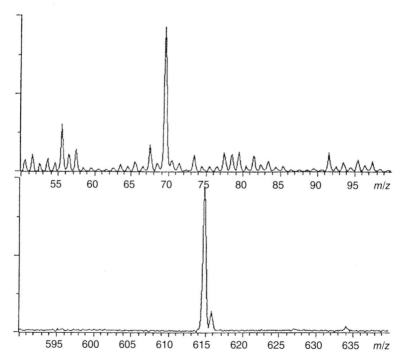

Figure 2.182 Diagram of the mass signals obtained with constant peak widths using an ion trap analyser in the lower and upper mass ranges (calibration standard FC43).

4) The mass resolution, which is constant over the whole mass range, is set up by the manufacturer in the electronics of the instrument and is the same for all types and manufacturers. The peak width is chosen in such a way that the distance between two neighbouring nominal mass signals corresponds to one mass unit (1 u = 1000 mu, the unit Da is used alternatively). High mass resolution, as in a magnetic sector or Orbitrap instrument, is not possible for quadrupole and ion trap instruments within the framework of the scan technique used.

5) The mass range of quadrupole and ion trap instruments varies but still has no effect on the mass resolution.

Frequently, the terms *mass range* and *unit mass resolution* are mixed up when giving a quality criterion for a mass range above 1000 Da for a quadrupole instrument (it is not obvious that a mass range up to 4000 Da, for example, is always accompanied by unit mass resolution). The effective attainable resolution for a real measurable signal of a reference compound is accurate and meaningful.

The different types of analyser for quadrupole rod and quadrupole ion trap instruments function on the same mathematical basis (Paul and Steinwedel, 1953; Paul *et al.*, 1958; and patent specification 1956) and, therefore, show the same resolution properties (Figure 2.183).

The display of line spectra on screen must be considered completely separately from the resolution of the analyser. By definition, a mass peak with unit mass resolution has a base width of one mass unit or 1000 mDa. On the other hand, the position of the top of the mass peak (centroid) can be calculated exactly. Data sometimes given with one or several digits of a mass unit lead to the false impression of a resolution higher than unit mass resolution. Components appearing at the same time at the ion source with signals of the same nominal mass, but slightly differing exact mass, which can naturally occur in GC-MS (as a result of co-eluates, the matrix, column bleed, etc.) cannot be separated at unit mass

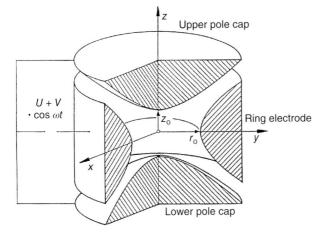

Figure 2.183 Diagram of an ion trap analyser (Finnigan).

resolving power of the quadrupole and ion trap analysers (see Section 2.3.1.3). The position of the centroid, therefore, cannot be used for any sensible evaluation. In no case is this the basis for the calculation of a possible empirical formula! Depending on the manufacturer, the labelling of the spectra can be found with pure nominal masses to several decimal places and can usually be altered by the user. It requires special software tools for internal mass calibration and profile type mass acquisition instead of centroiding to exploit the excellent quadrupole mass stability for accurate mass calculations (e.g. Cerno Bioscience, Danbury, CT, USA, MassWorks software)

2.3.2.3 High Mass Resolution

The flight paths of ions with different m/z values follow a different course as a result of the magnetic and electric fields in a magnetic sector instrument (Figure 2.184). The ion entrance and exit slit systems mask the ion beam and determine with their settings the mass resolution during operation (see Figures 2.174 and 2.184). Spectra can be recorded by continually changing the operational parameters of the instrument, for example, the acceleration voltage or strength of the magnetic field. The width of the ion beam is determined by the source slit at the ion source. The beams must not overlap, or only to a very small extent so that ions of different masses can be registered consecutively (Figure 2.185).

The resolution power R of neighbouring signals (Figure 2.186) for magnetic sector instruments is calculated according to

$$R = \frac{m}{\Delta m} \tag{2.27}$$

with m = mass and Δm = distance between neighbouring masses.

According to this formula, the value of resolution power is dimensionless.

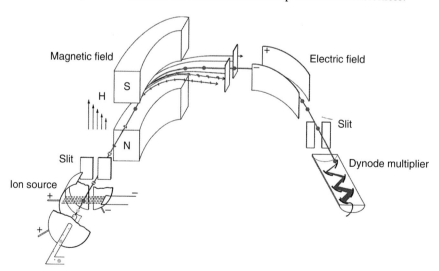

Figure 2.184 Principle of the double focussing magnetic sector mass spectrometer (reversed Nier Johnson geometry: BE).

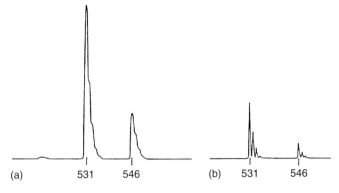

(a)   531   546   (b)   531   546

Figure 2.185 Section of a poorly resolved spectrum (a) and of a better, more highly resolved one (b) which, however, results in a lowering of intensity of the signals (Budzikiewicz, 2005).

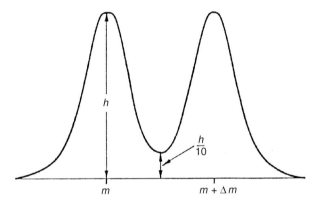

Figure 2.186 General resolution conditions for two mass signals −10% valley definition (Budzikiewicz, 2005).

Using high mass resolution the accurate mass of a molecular ion or of fragment ions can be determined. If the precision of the measurement is high enough candidates for the empirical formula can be calculated (Figure 2.187). The higher the masses, the more interferences are possible by meaningful elemental compositions. A selection of the large number of realistic chemical formulae for the mass 310 ($C_{22}H_{46}$, M 310.3599) is shown in Figure 2.188. To differentiate between the individual signals, a precision in the mass determination of 2 ppm or better would be necessary for this example. The maximum resolution and precision that can be achieved characterizes the slit system and the quality of the ion optics of the magnetic sector instrument.

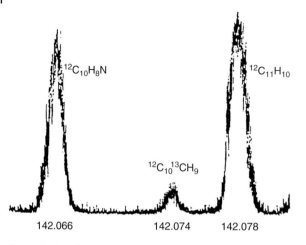

$^{12}C_{10}H_8N$

$^{12}C_{11}H_{10}$

$^{12}C_{10}{}^{13}CH_9$

142.066   142.074   142.078

Figure 2.187 Diagram showing mass signals in a high resolution spectrum at $R = 61\,000$ (Finnigan MAT 90).

Example Calculation of the Necessary Minimum Resolution

What mass spectroscopic resolution is required to obtain the signals of carbon monoxide (CO), nitrogen (N_2) and ethylene (C_2H_4) which are passed through suitable tubing into the ion source of a mass spectrometer?

Substance	Nominal mass	Exact mass
CO	m/z 28	m/z 27.994910
C_2H_4	m/z 28	m/z 28.006148
N_2	m/z 28	m/z 28.031296

For MS separation of CO and C_2H_4, which appear at the same time in accordance with the formula Eq. (2.27) $R = m/\Delta m$ given above, a resolving power of at least 2500 is necessary (see also Figure 2.175).

All MS systems with low resolution need preliminary GC separation of the components (CO, C_2H_4 and N_2 would then arrive one after the other in the ion source of an MS). This is the case for all ion trap and quadrupole instruments.

2.3.2.4 The Orbitrap Analyser

The 'Orbitrap' is a new high-resolution mass analyser invented by Alexander Makarov (2000). An inner spindle electrode is enclosed by two electrodes in a small device of only a few centimetre length. Ions are injected into the Orbitrap

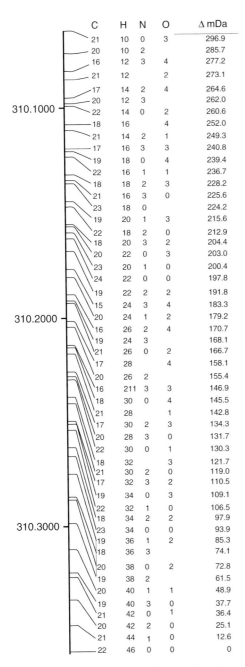

C	H	N	O	Δ mDa
21	10	0	3	296.9
20	10	2		285.7
16	12	3	4	277.2
21	12		2	273.1
17	14	2	4	264.6
20	12	3		262.0
22	14	0	2	260.6
18	16		4	252.0
21	14	2	1	249.3
17	16	3	3	240.8
19	18	0	4	239.4
22	16	1	1	236.7
18	18	2	3	228.2
21	16	3	0	225.6
23	18	0		224.2
19	20	1	3	215.6
22	18	2	0	212.9
18	20	3	2	204.4
20	22	0	3	203.0
23	20	1	0	200.4
24	22	0	0	197.8
19	22	2	2	191.8
15	24	3	4	183.3
20	24	1	2	179.2
16	26	2	4	170.7
19	24	3		168.1
21	26	0	2	166.7
17	28		4	158.1
20	26	2		155.4
16	211	3	3	146.9
18	30	0	4	145.5
21	28		1	142.8
17	30	2	3	134.3
20	28	3	0	131.7
22	30	0	1	130.3
18	32		3	121.7
21	30	2	0	119.0
17	32	3	2	110.5
19	34	0	3	109.1
22	32	1	0	106.5
18	34	2	2	97.9
23	34	0	0	93.9
19	36	1	2	85.3
18	36	3		74.1
20	38	0	2	72.8
19	38	2		61.5
20	40	1	1	48.9
19	40	3	0	37.7
21	42	0	1	36.4
20	42	2	0	25.1
21	44	1	0	12.6
22	46	0	0	0

Left-axis reference marks: 310.1000, 310.2000, 310.3000

Figure 2.188   Exact masses of chemically realistic empirical formulae consisting of C, H, N ($\leq$3), O ($\leq$5) given in deviations (Δ mDa) from the molecular mass of $C_{22}H_{46}$ *m/z* 310 (McLafferty and Turecek, 1993).

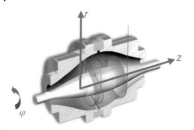

Figure 2.189 Schematics of the Orbitrap analyser.

and creating an image current by oscillation of the trapped ions around the spindle and between the outer electrodes on the z-axis, see Figure 2.189. The signal taken at the electrodes is digitized and converted by Fourier transformation to obtain mass spectra. The frequency of the oscillation between the electrodes is directly proportional to mass as of Eq. (2.28) for the motion in z direction (with k and q instrument constants).

$$\omega_z = \sqrt{\frac{k}{m/q}} \tag{2.28}$$

As the frequency of ion motion can be determined with high precision, the mass determination results with high accuracy below 1 ppm and very high mass resolution above 500 000 (Figure 2.190). The mass precision and resolution depends on the measurement time of the trapped ions. Orbitrap MS systems are applied in LC-MS since many years for proteomics and metabolomics especially in life sciences, but also show its high capabilities in the small molecular domain for multi-residue trace analyses of, for example, pesticides or drugs, both in LC-MS (Fürst and Bernsmann, 2010; Kaklamanos *et al.*, 2013) and GC-MS applications (Peterson *et al.*, 2009, 2010).

With its capabilities in high mass resolution combined with the uncompromized detection sensitivity and speed, the analytical advantages of the Orbitrap analyser for GC-MS are developing quickly into a combined targeted and non-targeted trace analysis.

2.3.2.5 High and Low Mass Resolution in the Case of Dioxin Analysis
Can dioxin analysis be carried out with quadrupole and ion trap instruments?

The main limitation is the analytical selectivity, precision and sensitivity in matrix (Beck *et al.*, 1988). The sensitivity of single quadrupole and ion trap instruments appears to be adequate with clean standard samples, however, the selectivity is not, and requires triple quadrupole MS/MS methods (Malavia *et al.*, 2008; Slayback and Taylor, 1983).

Errors potentially created when using the SIM technique are smaller for full scan data collection with ion traps and on using MS/MS techniques, as usually a complete spectrum or a product ion spectrum are available for further confirmation. Errors caused by false positive peak detection and poor detection limits as a result of matrix overlap, which cannot be totally excluded, generally limit the use of low resolution mass spectrometers. For screening tests, positive results should

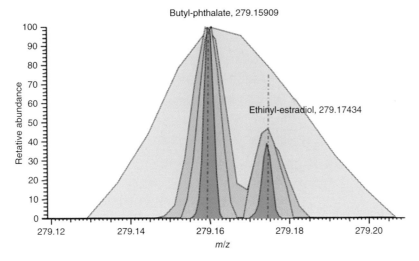

Figure 2.190 Separation by accurate mass of LC coeluting buthyl-phthalate and ethinyl-estradiol, with R 10 000 no separation (yellow), R 30 000 valley separation (grey), R 50 000 baseline separation (blue) and R 100 000 fully separated (green).

be followed by high resolution GC-MS confirmation (Eljarrat and Barcelo, 2002; exact masses see Thermo Electron Corp, 2004).

The high resolution and mass precision of the magnetic sector instrument allows the accurate mass to be recorded, e.g. 2,3,7,8-TCDD (tetrachlorodibenzodioxin) at m/z 321.8937 instead of a nominal mass of 322 and thus masks the known interference effects (Table 2.46). As a result, very high selectivity with very low detection limits of <10 fg are achieved, which gives the necessary assurance for making decisions with serious implications (Figure 2.191) (Beck *et al*, 1989).

Table 2.46 Possible interference with the masses m/z 319.8965 and 321.8936 of 2,3,7,8-TCDD and the minimum analyser resolution required for separation.

Compound	Formula	$m/z^{a)}$	Resolving power needed for separation
Tetrachlorobenzyltoluene	$C_{12}H_8Cl_4$	319.9508	5 900
Nonachlorobiphenyl	$C_{12}HCl_9$	321.8491	7 300
Pentachlorobiphenylene	$C_{12}H_3Cl_5$	321.8677	12 500
Heptachlorobiphenyl	$C_{12}H_3Cl_7$	321.8678	13 000
Hydroxytetrachlorodibenzofuran	$C_{12}H_4O_2Cl_4$	321.8936	Cannot be resolved
DDE	$C_{14}H_9Cl_5$	321.9292	9 100
DDT	$C_{14}H_9Cl_5$	321.9292	9 100
Tetrachloromethoxybiphenyl	$C_{13}H_8OCl_4$	321.9299	8 900

a) of the interfering ion.
DDE- Dichloro-diphenyl-dichloroethylene.

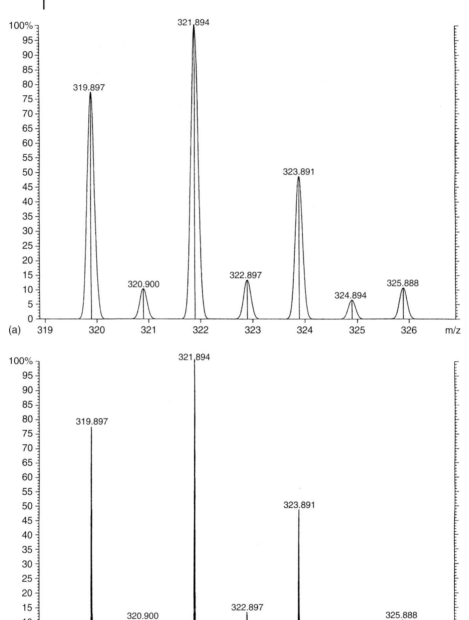

Figure 2.191 The isotope pattern for 2,3,7,8-TCDD using a double focussing magnetic sector instrument with peak widths for resolutions of 1000 (a) and 10 000 (b) (Fürst and Berns-mann, 2010).

High mass resolution is therefore also required to verify positive screening results in dioxin analysis (see also section 4.29). A comparison of the two spectrometric methods of the lower and higher resolution SIM techniques is shown in Figure 2.192 for the mass traces in the detection of TCDF (tetrachlorodibenzofuran) traces.

The resolution comparison with respect to different analyser technologies used in dioxin analysis is shown in Figures 2.193–2.195 with simulated ideal mass spectra of the TCDD isotope pattern in the molecular peak region. In practice, for real sample measurements a significant background contribution has to be kept in mind. The characteristic of the quadrupole analyser for unit mass resolution (Figure 2.193) shows the mass peaks of the isotope pattern with the base peak width of 1 Da. The enhanced resolution quadrupole technology is able to reduce the peak width down to 0.4 Da with a remarkable good mass separation (Figure 2.193).

Advances in GC-MS/MS triple quadrupole technology significantly improved the analytical performance with respect to higher sensitivity and selectivity, meeting the requirements for the control of the maximum levels for dioxins (PCDD/Fs and dl-PCBs) in food and feed (Kotz et al., 2012; Cojocariu et al., 2014). GC-HRMS and GC-MS/MS are both accepted by the EU Commission Regulation No. 589 (2014) as confirmatory methods for dioxin analysis in food and feed. According to this new EU regulation the following three specific performance criteria must be fulfilled for dioxin analysis with GC-MS/MS, in addition to previous European Commission Regulation No. 1883 (2006) and No. 252 (2012):

- The mass resolution for each quadrupole is required to be set equal to or better than unit mass resolution.
- Two specific precursor ions, each with one specific corresponding product ion for all labeled and unlabeled analytes must be acquired.

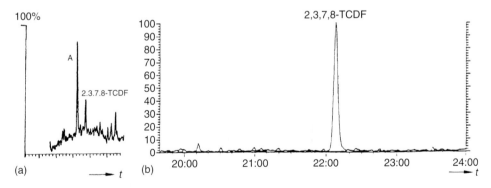

(a)

(b)

Figure 2.192 Comparison of 2,3,7,8-TCDF traces in analyses using a low resolution quadrupole GC-MS system (a = m/z 319.9) and a high resolution GC-MS system (b = m/z 319.8965, area 1.8309 E5). Both chromatograms were run on an identical human milk sample. The component marked with A in the quadrupole chromatogram is an interfering component (Fürst and Bernsmann, 2010).

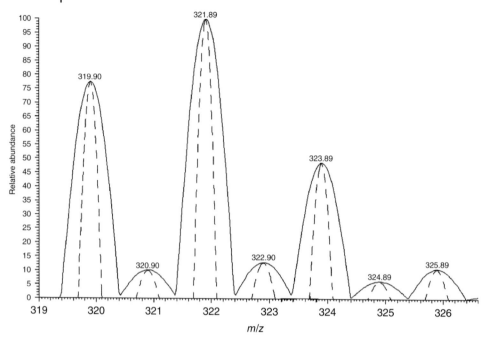

Figure 2.193 Quadrupole mass resolution at 1 and 0.4 Da peak width (dashed line) in H-SRM mode (TCDD isotope pattern).

- A maximum tolerance of ± 15% is required for the relative ion intensities of each SRM transition of an analyte in comparison to the calculated or measured average values from calibration standards under identical MS/MS conditions, in particular the collision energy and collision gas pressure.

TOF analysers show very tall peaks but typically with a broad peak base. For that reason, the TOF resolving power is measured at the half peak height. At the very good TOF resolution of 7000 FWHM (frequently named hrTOF), the peak base still is wide over almost one full mass window (Figure 2.194). The high resolution magnetic sector and Orbitrap instruments provide with distinct difference the required resolving power of 10 000 for ultimate selectivity (Figure 2.195).

2.3.2.6 Time-of-Flight Analyser

The concept of time-of-flight (TOF) MS was proposed already in 1946 by William E. Stephens of the University of Pennsylvania (Borman, 1998). In a TOF analyser, ions are separated by differences in their velocities as they fly down a field free drift region towards a collector in the order of their increasing mass-to-charge ratio (see Figure 2.196). With that principle, TOF MS is probably the simplest method of mass measurement. TOF MS is fast, offers a high duty cycle with the parallel detection of ions and is theoretically unlimited in mass range. Due to the

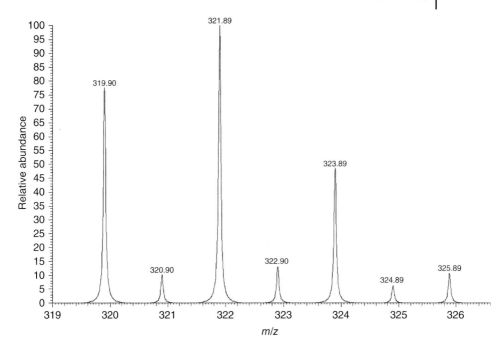

Figure 2.194 TOF mass resolution at 7000 FWHM (TCDD isotope pattern).

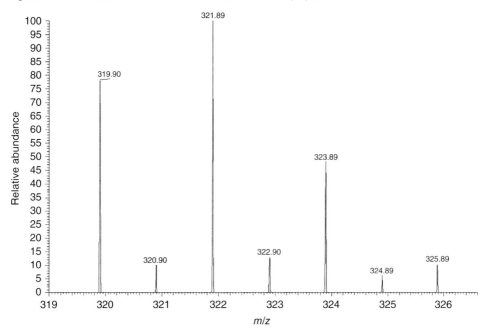

Figure 2.195 High resolution magnetic sector analyser, mass resolution at 10 000 at 10% valley (TCDD isotope pattern).

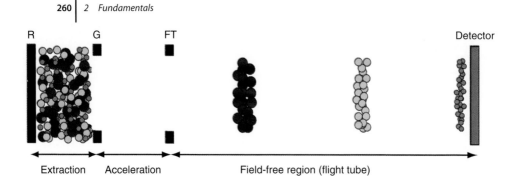

R   G   FT   Detector

Extraction region   Acceleration region   Field-free region (flight tube)

Figure 2.196 Time-of-flight operating principle R repeller, G extraction grid, FT flight tube entrance grid (McClenathan and Ray, 2004). (Reprinted with permission from the American Chemical Society.)

speed of detection and its inherent sensitivity by the parallel detection of a complete ion packet, it is specially suited for a full scan chromatographic detection, and it is capable of running fast GC because of its high sampling rates of up to more than hundred spectra per second. TOF MS is also widely used for the determination of large biomolecules (ESI; matrix assisted laser desorption ionization, MALDI), among many other low and high molecular weight applications. TOF and quadrupole-TOF analysers are inherently low resolution instruments (Webb, 2004).

TOF instruments were first designed and constructed starting in the late 1940s. Key advances were made by William C. Wiley and I.H. McLaren of the Bendix Corp., Detroit, USA, the first company to commercialize TOF mass spectrometers. According to pharmacology professor Robert J. Cotter of the Johns Hopkins University School of Medicine, Wiley and McLaren devised a time-lag focussing scheme that improved mass resolution by simultaneously correcting for the initial spatial and kinetic energy distributions of the ions. Mass resolution was also greatly improved by the 1974 invention by Boris A. Mamyrin of the Physical-Technical Institute, Leningrad, former Soviet Union, of the reflectron, which corrects for the effects of the kinetic energy distribution of the ions (Cotter, 1992, 1997; Birmingham, 2005; Guilhaus, 1995; Mamyrin, 1973, 2000, 1994).

The schematics of a TOF analyser are shown in Figure 2.197. Ions are introduced as pulsed packets either directly from the GC or LC ion source of the instrument, or with hyphenated instruments, from a previous analyser stage (e.g. quadrupole-time-of-flight hybrid, Q-TOF). Ions from the ion source are accelerated and focussed into a parallel beam that continuously enters the ion modulator region of the TOF analyser. Initially, the modulator region is fieldfree and ions continue to move in their original direction. A pulsed electric field is applied at a frequency of several kilohertz (kHz) across a modulator gap of several millimetre width, pushing ions in an orthogonal direction (to their initial movement) into an accelerating section of several kiloelectron volts. The modulator pulses serve as trigger for recording spectra at the detector.

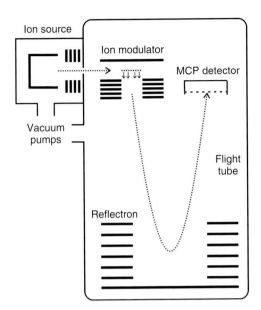

Figure 2.197 TOF mass spectrometer with GC ion source, orthogonal ion modulation into the TOF analyser and MCP detection, dotted lines indicate the ion flight path.

All ions in the pulsed ion packet receive the same initial kinetic energy, $E_{kin} = 1/2\, mv^2$. Lighter ions (low m) travel faster and reach the detector earlier. As the ions enter and move down the field free drift zone, they are separated by their mass to charge ratio in time. Commercial GC-MS and LC-MS TOF instruments are usually equipped with a reflectron. This 'ion mirror' focuses the kinetic energy distribution, originating from the small initial velocity differences due to different spatial start positions of the ions when getting pulsed from the modulator into the drift tube (The mass resolution in TOF-MS is generally limited by the initial spatial and velocity spreads). The ratio of velocity components in the two orthogonal directions of movement from source and modulator is selected such that ions are directed to the centre of the ion mirror and get focussed on a horizontal multichannel plate detector plane (MCP). The reflectron typically consists of a series of lenses with increasing potential pushing the ions back in a slight angle into the direction of the detector. Ions of higher velocity (energy) penetrate deeper into the mirror. Hence, the ion packet is getting focussed in space and in time for increased mass resolution. All ions reaching the detector are recorded explaining the inherent high sensitivity in full scan mode of TOF analysers. Duty cycles reached with orthogonal ion acceleration instruments vary in a range of 5–30% depending on the mass range and limited by the slowest ion (highest m recorded) moving across the TOF mass analyser, covering the range between quadrupole and ion trap analysers.

The TOF mass separation is characterized by the following basic equation:

$$m/z = 2e \cdot E \cdot s \, (t/d)^2$$

with

m/z	=	mass-to-charge ratio
e	=	elementary charge
E	=	extraction pulse potential
s	=	length of ion acceleration, over which E is effective
t	=	measured flight time of an ion, triggered by pulse E
d	=	length of field free drift zone.

Increasing the flight time improves the resolving power of the TOF analyser. Newest developments in commercial TOF mass analysers extend the ion flight path by the use of multiple reflectrons reaching a mass resolving power of up to 50 000. Statements on the achievable ion transmission efficiency and a potential loss of detection sensitivity by multiple reflections have not been made available yet (Xian *et al.*, 2012).

The fast data-acquisition rates make the TOF analyser the ideal mass detector for fast GC and comprehensive GC × GC (Dimandja, 2003; Hamilton *et al.*, 2004). Despite the described inherent sensitivity of TOF MS in full scan analysis, some fundamental trade-offs in terms of response and spectral quality have to be considered when setting up TOF applications. In contrast to the expectation that higher acquisition rates strongly support the deconvolution of coeluting compounds, the average ion abundance is dropping with increased scan rates and limiting the useful dynamic range. With the increase of the scan rate to 50 scans/s the ion abundance significantly drops to about 10% of a 5 scans/s intensity (see Figure 2.196). Minor components may be left unrecognized. Hence, typical reported acquisition rates with current GC-TOFMS instrumentation are 20−50 Hz, with up to 100 Hz in fast GC or GC × GC applications.

Also, the spectral quality has to be taken in account when setting up GC-TOF-MS methods for deconvolution experiments. Spectral skewing, as known from slow scanning quadrupole or sector mass spectrometers due to the increasing or decreasing substance intensity in the transient GC peak, does not occur in the ion package detection of TOF instruments. But, TOF spectral quality is limited by high data-acquisition rates. There is still the general valid rule that increased measurement times support the spectrum dynamic range and hence its quality. The effect can be demonstrated when comparing the acquired spectra at increasing scan rate against the reference spectra of the NIST library. The fit values of similarity match drop significantly with scan rates above 25, respectively, 50 scan/s, see Table 2.47 for an experiment on two common pesticides (Figure 2.198).

Table 2.47 Average similarity match values depending on the data-acquisition rate of TOF mass spectra versus NIST library for disulfoton and diazinon.

Acquisition rate (spectra/s)	Disulfoton (fit value)	Diazinon (fit value)
5	884	852
10	876	851
25	837	827[a]
50	820	623[a]
100	735	583
175	697[b]	626[b]
250	660[b]	586[b]

a) Based on four replicate spectra.
b) Based on three replicate spectra.
The average values of five replicate spectra from independent chromatographic runs are listed, except where low match quality did not provide identification.

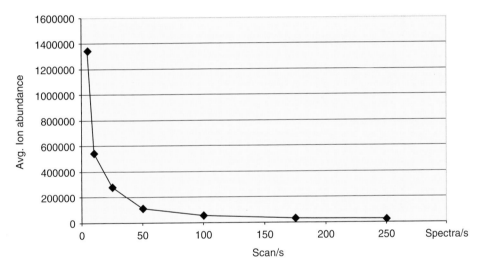

Figure 2.198 Ion abundance dependence from the scan rate in TOF MS (counts vs scan rate in spectra/s) (Meruva et al., 2000).

2.3.3
Isotope Ratio Monitoring GC-MS

The measurement of isotope ratios was the first mass spectrometry application. In 1907, Thomson for the first time showed the parabola mass spectrum of a Neon sample with his newly developed mass spectrometer. Later, in 1919, Aston concluded that the observed lines reveal the isotopes 20 and 22 of Neon. He later also discovered the isotope 21 of Ne with only 0.3 at.% abundance. The term *isotope* was coined independently by Frederick Soddy. He observed substances with identical chemical behaviour but different atomic mass in the decay of natural radioactive elements and received 1921 the Nobel Prize in Chemistry for his investigations into the origin and nature of isotopes. The term *Isotope* is derived from the Greek 'isos' for equal and 'topos' for the place in the table of elements. Consequently, the development in MS in the first half of last century was dominated by elemental analysis with the determination of the elemental isotope ratios facilitated by further improved mass spectrometer systems with higher resolving power as introduced by Alfred Nier. He carried out the first measurements on $^{13}C/^{12}C$ abundance ratios in natural samples. Already in 1938, Nier studied bacterial metabolism by using ^{13}C as the tracer.

The following investigations by Alfred Nier formed the foundation for today's high precision IRMS. In contrast to mass spectrometric organic structure elucidation and target compound quantitation, IRMS provides a different analytical dimension of precision data. Isotope ratio MS delivers information on the physiochemical history, the origin or authenticity of a sample and is essential today in multiple areas of research and control such as food, life sciences, forensics, material quality control, geology or climate research, just to name the most important applications today (Ehleringer and Cerling, 2002; Fry, 2006; Sharp, 2007).

Isotope ratio monitoring techniques using continuous flow sample introduction via GC had been introduced in the 1970s and developed into a mature analytical technology very quickly. Already in 1976, the first approach hyphenating a capillary GC with a magnetic sector MS for the systematic measurement of isotope ratios was published (Sano *et al.*, 1976). The continuous flow determination of individual compound $^{13}C/^{12}C$ ratios was introduced by Matthews and Hayes in 1978 and the term *isotope ratio monitoring GC-MS* was coined, today commonly abbreviated *irm-GC-MS* (Matthews and Hayes, 1978). The full range of $^{15}N/^{14}N$, $^{18}O/^{16}O$ and most importantly H/D determinations lasted until its publication in 1998 (Brand *et al.*, 1994; Heuer *et al.*, 1998; Hilkert *et al.*, 1999). Today irm-GC-MS is the established analytical methodology for delivering the precise ratios of the stable isotopes of the elements H, N, C and O being the major constituents of organic matter. Sulfur is not amenable to irm-GC-MS analysis due to its only low abundance in organic molecules. A compelling example was presented by Ehleringer on the geographical origin of cocaine from South America applied to determine the distribution of illicit drugs (Ehleringer *et al.*, 2000; Bradley, 2002). Organic compounds containing these elements are quantitatively converted into simple gases for mass spectrometric analysis, for example, H_2, N_2, CO, CO_2, O_2 and SO_2. For that conversion to simple gases, integrating and providing the full isotope information of a substance (in contrast to a fragmented organic mass spectrum) also the term *gas isotope ratio mass spectrometry' GIRMS* is used frequently, also including irm-GC-MS applications.

2.3.3.1 The Principles of Isotope Ratio Monitoring

Isotope ratios, although tabulated with average values for all elements, are not constant. All phase transition processes, transport mechanisms and enzymatic or chemical reactions are dependent on the physical properties of the reaction partners, that is, most importantly the mass of the molecule involved in the process or chemical reaction, for their kinetic properties.

The high precision quantitative data on isotope ratios obtained are not absolute quantitation data. Natural isotope ratios exhibit only small but meaningful variations that are measured with highest precision relative to a known standard. In most of the applications in stable isotope analysis, the differences in isotopic ratios between samples are of much more interest and significance than the absolute amount in a given sample. The isotope ratio is independent of the amount of material measured (and at the low end only determined in precision by the available ion statistics). The relative ratio measurement can be accomplished in the required and even higher precision, which is at least one order of magnitude higher than the determination of absolute values.

2.3.3.2 Notations in irm-GC-MS

The small abundance differences of stable isotopes are best represented by the delta notation (Eq. 2.29) in which the stable isotope abundance is expressed relative to an isotope standard, measured in the same run for reference:

$$\delta = \left(\frac{R_{\text{sample}}}{R_{\text{std}}} - 1 \right) \times 1000 \, [\text{‰}] \tag{2.29}$$

with R the molar ratio of the heavy to the common (light) isotope of an element, for example, $R = {}^{13}C/{}^{12}C$ or D/H or ${}^{18}O/{}^{16}O$. A graphical representation of the meaning of the δ-value is given in Figure 2.199 for the variation of the ${}^{13}C$ isotope proportion of carbon. δ-values can be positive or negative indicating a higher or lower abundance of the major isotope in the sample than in the reference.

Other systems in use for expressing isotope ratios are ppm (part per million, relative value), at.% (atom percent, absolute value) and as the most common terminology in biomedical tracer studies APE (atom percent excess, relative value). The choice is dependent primarily on the specific field of application, for example, with high degrees of enrichments being very much different from natural abundances. Also, typical in this field are historical traditions in the use of different notation systems. Values of at.% and δ-notation can be converted as follows:

$$\text{at.\%} = \frac{R_{\text{st}} \cdot (\delta/1000 + 1)}{1 + R_{\text{st}} \cdot (\delta/1000 + 1)} \cdot 100$$

with R_{st} absolute standard ratio.

2.3.3.3 Isotopic Fractionation

Isotope effects, as they can be observed in phase transition or dissociation reactions, are usually the result of incomplete processes in diffusive or equilibrium

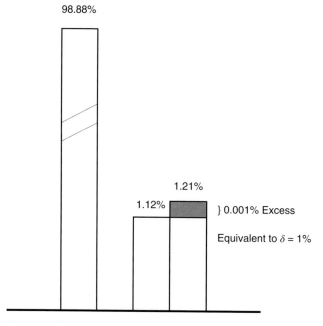

Figure 2.199 Graphical representation of the δ-notation for a ^{13}C variation of 1‰.

fractionation. This effect is caused by different translational velocities of the lighter and heavier molecule through a medium or across a phase boundary. As the kinetic energy at a given temperature is the same for all gas molecules, the kinetic energy equation

$$E_{kin} = \frac{1}{2}m \cdot v^2$$

applies for both molecules $^{12}CO_2$ and $^{13}CO_2$ with their respective masses 44 and 45 Da:

$$\frac{1}{2}(44 \cdot v_a^2) = \frac{1}{2}(45 \cdot v_b^2)$$

resulting in

$$\frac{v_a}{v_b} = \sqrt{\left(\frac{45}{44}\right)} = 1.0113$$

The velocity ratio of both CO_2 species explains that the average velocity of $^{12}CO_2$ is 1.13% higher than that of the heavier molecule.

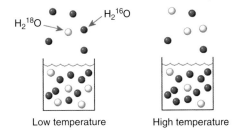

Low temperature   High temperature

Figure 2.200 Alteration of the isotope ratio during the evaporation of water resulting in a different $^{18}O/^{16}O$ ratio at low and high temperatures in vapour as well as in the liquid.

Example Water Phase Transitions

Boiling water will lose primarily the light $^1H_2{}^{16}O$ molecules as can be seen in Figure 2.200. Heavier water molecules will consequently be concentrated in the liquid; the vapour will become depleted. The reversed process is observed during condensation, for example, during the formation of raindrops from humid air. As the extent of this process is temperature dependent, isotopic 'thermometers' are formed and ultimately isotopic 'signatures' of materials and processes are created.

Equilibrium isotope effects usually are associated with phase transition processes such as evaporation, diffusion or dissociation reactions. When incomplete phase transition processes occur during sample preparation, they lead to severe alteration of the initial isotope ratio. In fact, and that is of highest importance for all sample preparation steps in IRMS, only complete conversion reactions are acceptable to maintain the integrity of the original isotope ratio of the sample.

Isotope effects are also observed in chemical and biochemical reactions (kinetic fractionation). Of particular significance is the isotope effect, which occurs during enzymatic reactions with a general depletion in the heavier isotope being key to uncover metabolic processes. Chemical bonds to the heavier isotope are stronger, more stable and need higher dissociation energies in chemical reactions due to the different vibrational energy levels involved. Hence, the rate of enzymatic reaction is faster with the light isotope leading to differences in the abundance between substrate and product, unless the substrate is fully consumed, which is not the case in cellular steady state equilibria. Kinetic fractionation effects also have to be considered when employing compound derivatization steps (e.g., silylation, methylation) during sample preparation for GC application (Meyer-Augenstein, 1997).

The natural variations in isotopic abundances can be large, depending on the relative elemental mass differences: hydrogen (100%) > oxygen (12.5%) > carbon (8.3%) > nitrogen (7.1%), see also Table 2.48 (Rossmann, 2001; Rosman and Taylor, 1998). An overview of the isotopic variations found in natural compounds is given in Figures 2.201–2.204.

Table 2.48 Natural abundances of light stable isotopes relevant to stable isotope ratio mass spectrometry.

Element	Isotope	Atomic weight	Relative abundance (%)	Elemental relative mass difference	Molecular relative mass difference (%)	Terrestrial range ‰	Terrestrial range ppm	Technical precision ‰	Technical precision ppm
Hydrogen	^1H	1.0078	99.9840	D/H	$^1HD/^1H^1H$	—	—	—	—
Deuterium	^2H (D)	2.0141	0.0156	(2/1) 100%	(3/2) 50%	700	109	1	0.16
Boron	^{10}B	10.0129	19.7	$^{11}B/^{10}B$	—	—	—	—	—
	^{11}B	11.0093	8.3	(11/10) 10%	—	60	—	—	—
Carbon	^{12}C	12.0000	98.892	$^{13}C/^{12}C$	$^{13}C^{16}O^{16}O/^{12}C^{16}O^{16}O$	—	—	—	—
	^{13}C	13.0034	1.108	(13/12) 8.3%	(45/44) 2.3%S	100	1123	0.05	0.56
Nitrogen	^{14}N	14.0031	99.635	$^{15}N/^{14}N$	$^{15}N^{14}N/^{14}N^{14}N$	—	—	—	—
	^{15}N	15.0001	0.365	(15/14) 7.1%	(29/28) 3.6%	50	181	0.10	0.72
Oxygen	^{16}O	15.9949	100	$^{18}O/^{16}O$	$^{12}C^{16}O^{18}O/^{12}C^{16}O^{16}O$	—	—	—	—
	^{17}O	16.9991	0.037	(18/16) 12.5%	(46/44) 4.6%	100	—	—	—
	^{18}O	17.9992	0.204						
Silicon	^{28}Si	27.9769	92.21	$^{29}Si/^{28}Si$	$^{29}Si^{19}F_3/^{28}Si^{19}F_3$	6	200	0.10	0.20
	^{29}Si	28.9765	4.7	(29/28) 3.6%	(86/85) 1.2%				
	^{30}Si	29.9738	3.09						
Sulfur	^{32}S	31.9721	95.02	$^{34}S/^{32}S$	$^{34}S^{16}O^{16}O/^{32}S^{16}O^{16}O$	—	—	—	—
	^{33}S	32.9715	0.76						
	^{34}S	33.9769	4.22	(34/32) 6.3%	(66/64) 3.1%	100	4580	0.20	—
	^{36}S	35.9671	0.014						
Chlorine	^{35}Cl	34.9689	75.77	$^{37}Cl/^{35}Cl$	$^{12}CH_3^{37}Cl/^{12}CH_3^{35}Cl$	100	—	0.20	9.16
	^{37}Cl	36.9659	24.23	(37/35) 5.7%	(52/50) 4.0%	10	—	0.10	—

From Gilles St. Jean, Basic Principles in Stable Geochemistry, IRMS Short Course, 9th Canadian CF-IRMS Workshop 2002.

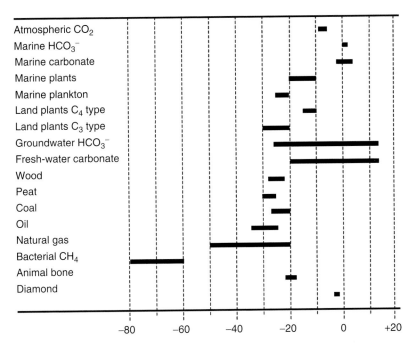

Atmospheric CO$_2$
Marine HCO$_3^-$
Marine carbonate
Marine plants
Marine plankton
Land plants C$_4$ type
Land plants C$_3$ type
Groundwater HCO$_3^-$
Fresh-water carbonate
Wood
Peat
Coal
Oil
Natural gas
Bacterial CH$_4$
Animal bone
Diamond

−80   −60   −40   −20   0   +20

Figure 2.201 δ^{13}C variations in natural compounds (δ^{13}C VPDB, ‰ scale). (de Vries, 2000, courtesy IAEA).

2.3.3.4   irm-GC-MS Technology

irm-GC-MS is applied to obtain compound specific data after a GC separation of mixtures in contrast to bulk analytical data from an EA system. The bulk analysis delivers the average isotope ratio within a certain volume of sample material. The entire sample is converted into simple gases using conventional EAs or high temperature conversion elemental analysers (TC/EAs). In contrast to a bulk analysis, irm-GC-MS delivers specific isotope ratio data in the low picomolar range of individual compounds after a conventional capillary GC separation (see Figure 2.205).

The most important step in irm-GC-MS is the conversion of the eluting compounds into simple measurement gases such as CO_2, N_2, CO and H_2. In irm-GC-MS, as a continuous flow application, this conversion is achieved online within the helium carrier gas stream while preserving the chromatographic resolution of the sample. The products of the conversion are then fed by the carrier gas stream into the isotope ratio MS.

It is important to note that during chromatographic separation on regular, fused silica capillary columns, a separation of the different isotopically substituted species of a compound already takes place. Due to the lower molar volume of the heavier components and the resulting differences in mobile/stationary phase interactions, the heavier components elute slightly earlier (Matucha, 1991). This

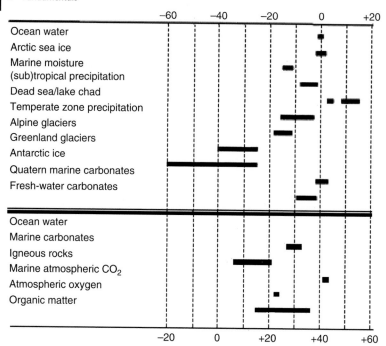

Figure 2.202 $\delta^{18}O$ variations in natural compounds (top: $\delta^{18}O$ VSMOW ‰ scale for waters, $\delta^{18}O$ VPDB ‰ scale for carbonates, bottom: $\delta^{18}O$ VSMOW ‰ scale). (de Vries, 2000, courtesy IAEA.)

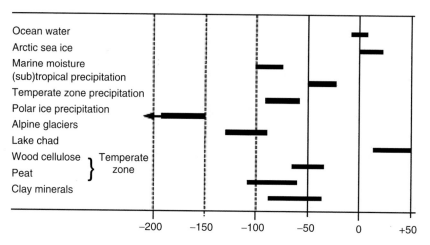

Figure 2.203 $\delta^{2}H$ variations in natural compounds ($\delta^{2}H$ VSMOW ‰ scale). (de Vries, 2000, courtesy IAEA.)

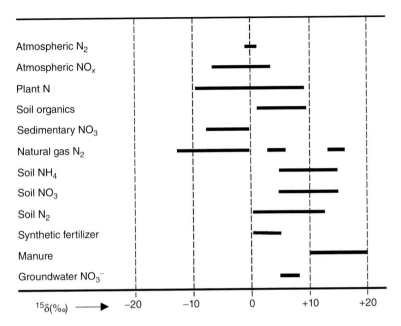

Figure 2.204 δ^{15}N variations in natural compounds (δ^{15}N Air ‰ scale). (de Vries, 2000, courtesy IAEA).

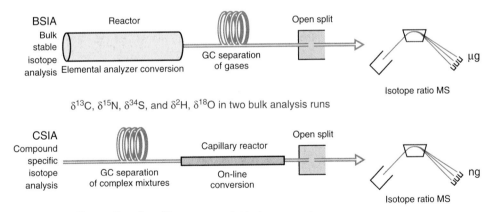

Figure 2.205 Basic scheme of bulk sample (BSIA) vs. compound specific (CSIA) isotope analysis. (Courtesy Thermo Fisher Scientific) BSIA: IRMS is coupled with an EA: The complete sample is first converted to simple gases followed by their chromatographic separation. CSIA: IRMS is coupled with a GC: The sample components are first separated by capillary chromatography and then individually converted to the measurement gases.

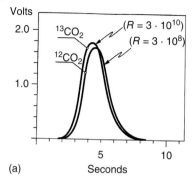 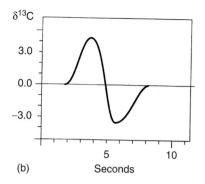

Figure 2.206 Chromatographic elution profiles of CO_2 with natural ^{13}C abundance resulting in the typical S-shaped curvature when displaying the calculated δ-values, (a) note the intensity difference of the isotopes with the choice of amplification resistors R in a difference of two orders of magnitude, (b) shows the typical ratio chromatogram $^{13}C/^{12}C$ (Rautenschlein *et al.*, 1990).

separation effect can also be observed in 'organic' GC-MS analyses when having a closer look at the peak top retention times of the ^{13}C carbon isotope mass trace. irm-GC-MS makes special use of the chromatographic isotope separation effect by displaying the typical S-shaped ratio traces during analysis indicating the substances of interest (see Figures 2.206 and 2.207).

2.3.3.5 The Open Split Interface

For high-precision isotope ratio determination, the IRMS ion source pressure must be kept absolutely constant. For this reason, it is mandatory in all continuous flow applications coupled to an isotope ratio MS, to keep the open split interface at atmospheric pressure. The open-split coupling eliminates the isotopic fractionation due to an extended impact of the high vacuum of the mass spectrometer into the chromatography and conversion reactors. The basic principle of an open split is to pick up a part of the helium/sample stream by a transfer capillary from a pressureless environment, that is, an open tube or a wide capillary, and transfer it into the IRMS (see Figure 2.208). The He stream added by a separate capillary ensures protection from ambient air. Retracting the transfer capillary into a zone of pure helium allows cutting off parts of the chromatogram. In all modes, a constant flow of helium into the isotope ratio MS, and consequently constant ion source conditions are maintained. The valve-free open split is absolutely inert and does not create any pressure waves during switching.

In addition, the open-split interface also offers an automatic peak dilution capability. Due to the transfer of the analyte gases in a helium stream more than one interface can be coupled in parallel for alternate use to an isotope ratio MS via separate needle valves.

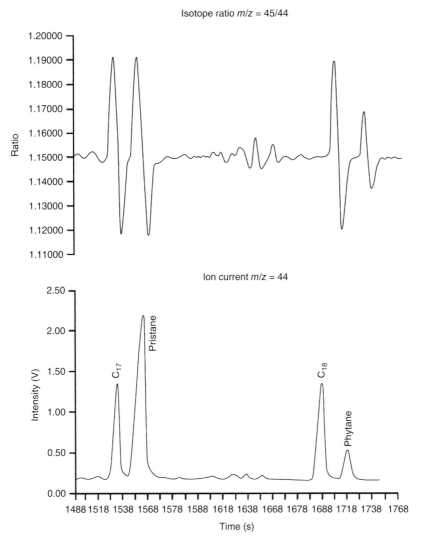

Figure 2.207 irm-GC-MS analysis of the crude oil steroid biomarkers pristane and phytane with the $\delta^{13}C$ ratio profile on top, bottom $^{12}CO_2$ intensity profile.

2.3.3.6 Compound Specific Isotope Analysis

irm-GC-MS is amenable to all GC-volatile organic compounds down to the low picomole range. A capillary design of oxidation and high temperature conversion reactors is required to guarantee the integrity of GC resolution. Table 2.49 gives an overview of the techniques available today with the achievable precision values by state of the art IRMS technology. Compound specific isotope analysis (CSIA) is one of the newest fields of isotope analysis, but today a

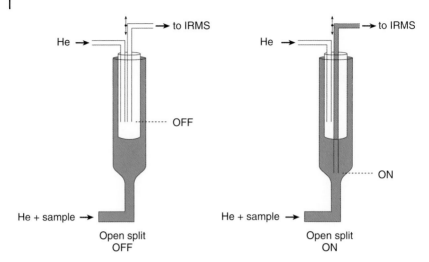

Figure 2.208 Open split interface to IRMS, effected by moving the transfer capillary to IRMS from the column inlet to the He sample flow region. (Thermo Fisher Scientific.)

Table 2.49 Analytical methods used in compound specific isotope analysis.

Element ratio	Measured species	Method	Temperature	Analytical precision[a]
$\delta^{13}C$	CO_2	Combustion	up to 1000 °C	^{13}C 0.06‰, 0.02‰/nA
$\delta^{15}N$	N_2	Combustion	up to 1000 °C	^{15}N 0.06‰, 0.02‰/nA
$\delta^{18}O$	CO	Pyrolysis	1450 °C	^{18}O 0.15‰, 0.04‰/nA
δ^2H	H_2	Pyrolysis	1280 °C	^2H 0.50‰, 0.20‰/nA

a) Ten pulses of reference gas (amplitude 3 V, for H_2 5 V) δ notation.

mature and steadily expanding application. The first GC combustion system for $\delta^{13}C$ determination was commercially introduced in 1988 by Finnigan MAT Bremen, Germany. This technique combines the resolution of capillary GC with the high precision of IRMS. In 1992, the capabilities for analysis of $\delta^{15}N$ and in 1996 for $\delta^{18}O$ were added. Quantitative pyrolysis by high temperature conversion (GC-TC IRMS) for δD analyses was introduced in 1998 introducing an energy discrimination filter in front of the HD collector at m/z 3 for the suppression of $^4He^+$ ions from the carrier gas interfering with the HD^+ signal (Hilkert *et al.*, 1999).

2.3.3.7 On-Line Combustion for $\delta^{13}C$ and $\delta^{15}N$ Determination

All compounds eluting from a GC column are oxidized in a capillary reactor to form CO_2, N_2, and H_2O as a by-product, at 940–1000°C. NO_x produced

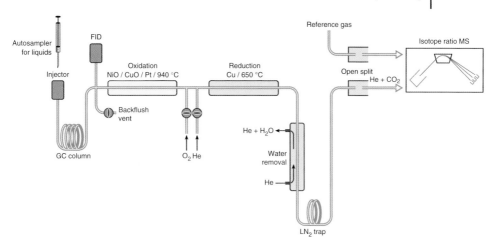

Figure 2.209 On-line oxidation of compounds eluting from the GC for the production of CO_2 and N_2 for IRMS measurement. (Thermo Fisher Scientific.)

in the oxidation reactor is reduced to N_2 in a subsequent capillary reduction reactor. The H_2O formed in the oxidation process is removed by an online Nafion dryer, a maintenance-free water removal system. For the analysis of $\delta^{15}N$, all CO_2 is retained in a liquid nitrogen cryo trap in order to avoid interferences from CO on the identical masses before transfer into the IRMS through the movable capillary open split. CO is generated in small amounts as a side reaction during the ionization process from CO_2 in the ion source of the MS. A detailed schematic of the online combustion setup is given in Figure 2.209.

2.3.3.8 The Oxidation Reactor

The quantitative oxidation of all organic compounds eluting from the GC column, including the refractory methane, is performed at temperatures up to 1000 °C. The reactor consists of a capillary ceramic tube loaded with twisted Ni, Cu and Pt wires. The resulting internal volume compares to a capillary column and secures the integrity of the chromatographic separation. The reactor can be charged and recharged automatically with O_2 added to a backflush flow (BKF) every 2–3 days, depending on the operating conditions. The built-in BKF system reverses the flow through the oxidation reactor towards an exit directly after the GC column. The BKF is activated during the analysis to cut off an eluting solvent peak in front of the oxidation furnace by flow switching. All valves have to be kept outside of the analytical flow path to maintain optimum GC performance (Figure 2.210).

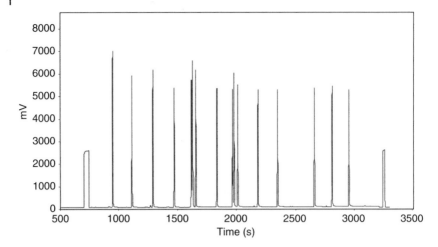

Figure 2.210 irm-GC-MS chromatogram of a fatty acid methyl ester (FAME) sample after on-line combustion at 940 °C. The rectangular peaks in the beginning and at the end of the chromatogram are the CO_2 reference gas injection peaks. (Courtesy Thermo Fisher Scientific.)

2.3.3.9 The Reduction Reactor

The reduction reactor typically comprises copper material and is operated at 650°C to remove any O_2 bleed from the oxidation reactor and to convert any produced NO_x into N_2. It is of the same capillary design as the oxidation reactor.

2.3.3.10 Water Removal

Water produced during the oxidation reaction is removed through a 300 μm inner diameter Nafion capillary, which is dried by a counter current of He on the outside (Figure 2.211). The water removal adds no dead volume and is maintenance-free.

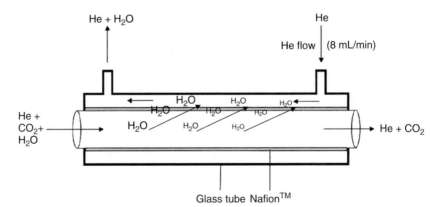

Figure 2.211 Principle of on-line removal of water from a He stream using a Nafion membrane. (Courtesy Thermo Fisher Scientific.)

On-Line Water Removal from a He Stream Using Nafion

Water is removed from a He sample stream by a gas-tight but hygroscopic Nafion tubing. The sample flow containing He, CO_2 and H_2O passes through the Nafion tubing which is mounted co-axially inside a glass tube. This glass tube, and therefore the outer surface of the Nafion tube, is constantly kept dry by a He flow of $8-10$ mL/min. Due to the water gradient through the Nafion membrane wall, any water in the sample flow will move through the membrane. A dry gas comprising of only He and CO_2 resulting from the oxidation is fed to the ion source of the mass spectrometer.

Structure of Nafion:

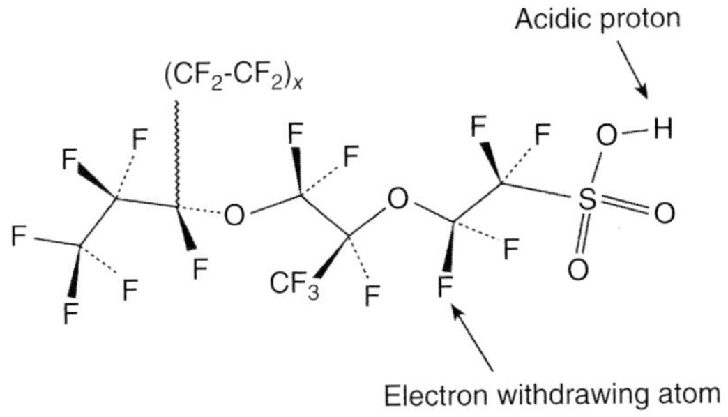

Properties:
Nafion™ is the combination of a stable Teflon™ backbone with acidic sulfonic groups. It is highly conductive to cations, making it ideal for many membrane applications. The Teflon backbone interlaced with ionic sulfonate groups gives Nafion™ a high operating temperature, for example, up to 190°C. It is selectively and highly permeable to water. The degree of hydration of the Nafion membrane directly affects its ion conductivity and overall morphology. See also: *http://de.wikipedia.org/wiki/Nafion*

2.3.3.11 The Liquid Nitrogen Trap

For the analysis of $\delta^{15}N$, all the CO_2 must be removed quantitatively to avoid an interference of CO^+ generated from CO_2 in the ion source with the N_2^+ analyte. This is achieved by immersing the deactivated fused silica capillary between the water removal and the open split into a liquid nitrogen cryo trap. The trapped CO_2 is easily released after the measurement series with no risk of CO_2 contamination of the ion source by using the movable open split.

2.3.3.12 On-Line High Temperature Conversion for δ^2H and $\delta^{18}O$ Determination

A quantitative pyrolysis by high temperature conversion of organic matter is applied for the conversion of organic oxygen and hydrogen to form the measurement gases CO and H$_2$ for the determination of δ^{18}O or δD (see Figure 2.212). This process requires an inert and reductive environment to prevent any O or H containing material from reacting or exchanging with the analyte.

For the determination of δ^{18}O, the analyte must not contact the ceramic tube that is used to protect against air. For the conversion to CO, the pyrolysis takes place in an inert platinum inlay of the reactor. Due to the catalytic properties of the platinum, the reaction can be performed at 1280°C. For the determination of δD from organic compounds, the reaction is performed in an empty ceramic tube at 1450°C. Such high temperatures are required to ensure a quantitative conversion.

Typically an online high temperature reactor is mounted in parallel to a combustion reactor at the GC oven. The complete setup for online high temperature conversion is given in Figure 2.213. A water removal step has no effect here on the dry analyte gas (Figure 2.214).

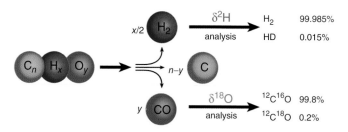

Figure 2.212 Principle of high temperature conversion (Courtesy Thermo Fisher Scientific.)

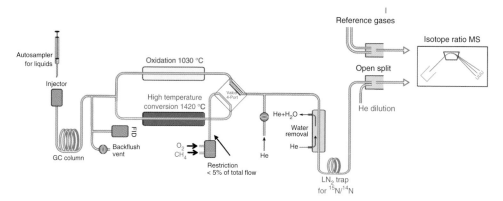

Figure 2.213 On-line high temperature conversion of compounds eluting from the GC for the production of CO and H$_2$ for IRMS measurement. (Thermo Fisher Scientific.)

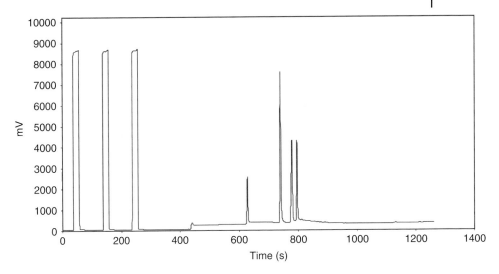

Figure 2.214 irm-GC-TC/MS chromatogram of flavour components using high temperature conversion with three CO reference gas pulses at start. (Courtesy Thermo Fisher Scientific.)

2.3.3.13 Mass Spectrometer for Isotope Ratio Analysis

Mass spectrometers employed for isotope ratio measurements are dedicated non-scanning, static magnetic sector mass spectrometer systems. The ion source is particularly optimized by a 'closed source' design for the ionization of gaseous compounds at very high ion production efficiency of $500-1000$ molecules/ion at high response linearity (Brand, 2004). After extraction from the source region, the ions are typically accelerated by $2.5-10\,kV$ to form an ion beam which enters the magnetic sector analyser through the entrance slit (see Figure 2.215).

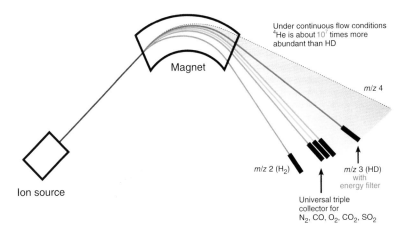

Figure 2.215 Isotope ratio mass spectrometer with ion source, magnet and the array of Faraday cups for simultaneous isotope detection. (Courtesy Thermo Fisher Scientific.)

The ion currents of the isotopes are measured simultaneously at the individual m/z values by discrete Faraday cups mounted behind a grounded slit (see Figure 2.216). An array of specially designed deep Faraday cups for quantitative measurement is precisely positioned along the optical focus plane representing the typical isotope mass cluster to be determined.

Only the simultaneous measurement of the isotope ion currents with dedicated Faraday cups for each isotope using individual amplifier electronics, cancels out ion beam fluctuations due to temperature drifts or electron beam variations and provides the required precision (see Figure 2.217, also refer to Table 2.48). Current instrumentation allows for irm-GC-MS analyses of organic compounds down to the low picomole range.

Isobaric interferences from other isotope species at the target masses require the measurement of more masses than just the targeted isotope ratio for a necessary correction. For the measurement of $\delta^{13}C$, three collectors at m/z 44, 45 and

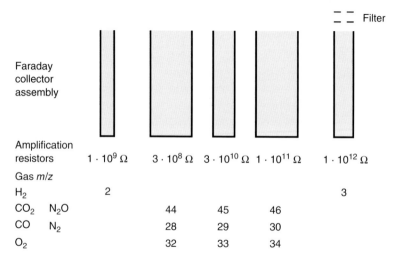

Figure 2.216 Typical Faraday cup arrangement for measurement of the isotope ratios of the most common gas species (below of the cup symbols the values of the amplification resistors are given; the HD cup is shown to be equipped with a kinetic energy filter to prevent from excess $^4He^+$).

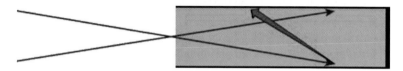

Figure 2.217 Cross-section of a Faraday cup for isotope ratio measurement, the large depth of the graphite cup prevents from losses caused by scattering. Arrows represent the ion beam with a focus point at the exit slit in front of the Faraday cup. (Courtesy Thermo Fisher Scientific.)

Table 2.50 Isobaric interferences when measuring $\delta^{13}C$ from CO_2.

m/z	Ion composition
44	$^{12}C^{16}O^{16}O$
45	$^{13}C^{16}O^{16}O$, $^{12}C^{16}O^{17}O$
46	$^{12}C^{16}O^{18}O$, $^{12}C^{17}O^{17}O$, $^{13}C^{16}O^{17}O$

46 are necessary (see Table 2.49). Algorithms for a fully automated correction of isobaric ion contributions are implemented in modern isotope ratio MS data systems, for example, for the isobaric interferences of ^{17}O and ^{13}C on m/z 46 of CO_2, see Table 2.50 (Craig, 1957). Other possible interferences such as CO on N_2, N_2O on CO_2 can be taken care of by the interface technology.

The Identical Treatment Principle (IT)

Standardization in isotope ratio monitoring measurements should be done exclusively using the principle of 'Identical Treatment of reference and sample material', the 'IT Principle'. Mostly, isotopic referencing is made with a co-injected peak of standard gas.

2.3.3.14 Injection of Reference Gases

The measurement of isotope ratios requires that sample gases be measured relative to a reference gas of a known isotope ratio. For the purpose of sample-standard referencing in irm-GC-MS, cylinders of calibrated reference gases, the laboratory standard of H_2, CO_2, N_2 or CO are used for an extended period of time. This referencing procedure turned out to be the most economic and precise compared to the addition of an internal standard to the sample, also providing necessary quality assurance purposes. An inert, fused silica capillary supplies the reference gas in the microlitre per minute range into a miniaturized mixing chamber, the reference gas injection port (see Figure 2.218 a,b). This capillary is lowered under software control into the mixing chamber, for example, 20 s, creating a He reference gas mixture which is fed into the IRMS source via a second independent gas line. This generates a rectangular, flat top gas peak without changing any pressures or gas flows in the ion source. Reference gases used are pure nitrogen (N_2) carbon dioxide (CO_2), hydrogen (H_2) and carbon monoxide (CO).

2.3.3.15 Isotope Reference Materials

In IRMS, the measurement of isotope ratios requires the samples to be measured relative to a reference of a known isotope ratio. This is the only means to achieve the required precision level of <1.5 ppm, for example, the $^{13}C/^{12}C$ isotope ratio,

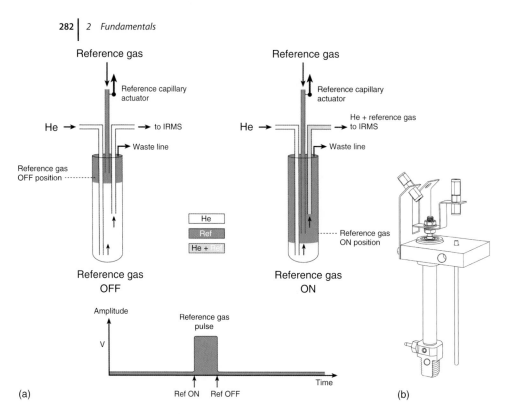

Figure 2.218 (a) Reference gas injection port in irm-GC-MS implemented as an open coupling between GC column and IRMS ion source. (b) Reference gas injection port design: left side pneumatic drive, right side valve tube for fused silica columns, top column connectors for fixed (left) and movable (right) columns. (Courtesy Thermo Fisher Scientific.)

which is 0.15‰ in the commonly used δ-notation. The employed isotopic reference scales (see Figures 2.201–2.204) are arbitrarily defined by the community relative to the isotope ratio of a selected primary reference material for a given element.

Primary reference materials are available through the International Atomic Energy Agency (IAEA) in Vienna, Austria. The most commonly used and recognized international reference materials for stable isotope ratio analysis of natural abundances are given in Table 2.51. The abbreviations used with a preceding 'V' as in 'VSMOW' refer to the reference materials prepared by the IAEA in Vienna compliant with the regular consultants meetings (for a detailed discussion see Groening, 2004).

VSMOW

Oceans contain almost 97% of the water on Earth and have a uniform isotope distribution. Oceans are the major sink in the hydrological cycle. The Standard Mean Ocean Water (SMOW) standard was initially proposed by Craig 1961 as a

Table 2.51 IRMS primary reference materials for irm-GC-MS. Absolute isotope ratios R in 1 σ standard precision.

H	$R = 0.00015576 \pm 0.000005$	VSMOW	Standard mean ocean water
C	$R = 0.011224 \pm 0.000028$	VPDB	Pee Dee Belemnite[a)]
N	$R = 0.0036765$	Air	Atmospheric air
	$R = 0.003663 \pm 0.000005$	NSVEC	N standard by Jung and Svec
O	$R = 0.0020052 \pm 0.000045$	VSMOW	Standard mean ocean water[a)]

a) As a general rule $\delta^{18}O$ data of carbonates and CO_2 gas are reported against VPDB whereas $\delta^{18}O$ data of all other materials should be reported versus VSMOW.

concept for the origin of the scale and was calculated from an average of samples taken from different oceans. The reference water VSMOW was prepared by the IAEA in 1968 with the same $\delta^{18}O$ and a $-0.2‰$ lighter δD as defined by SMOW.

VPDB

Pee Dee Belemnite is $CaCO_3$ of the rostrum of a bellemnite (Belemnita americana) from the Pee Dee Formation of South Carolina, USA. The PDB reference material has been exhausted for a long time. The new VPDB reference material was anchored by the IAEA with a fixed δ value so that it corresponds nominally to the previous PDB scale.

Atmospheric Air

Nitrogen in atmospheric air has a very homogeneous isotope composition. The atmosphere is the main nitrogen sink and the largest terrestrial nitrogen reservoir. NSVEC is the reference material initially prepared by Jung and Svec, Iowa State University, USA.

The available IAEA reference materials are intended to calibrate local laboratory standards and not for continuous quality control purposes (Groening, 2004). Every laboratory should prepare for routine work, longlasting laboratory specific standards with similar characteristics to the used references and the working standards, which are calibrated against the primary reference materials.

For comparison to results of other laboratories, the raw data of the mass spectrometer are converted into VSMOW or VPDB (Boato, 1960). Due to the relative measurement to a standard ratio, the formula contains a product term besides the sum of the δ-values:

$$\delta_3 = \delta_1 + \delta_2 + 10^{-3} \cdot \delta_1 \cdot \delta_2$$

$\delta_1 = \delta$ (sample/reference)	Measured in the lab
$\delta_2 = \delta$ (reference/standard)	Known lab calibration
$\delta_3 = \delta$ (sample/standard)	Unknown

2.3.4

Acquisition Techniques in GC-MS

In data acquisition by the mass spectrometer, there is a significant difference between detection of the complete spectrum (full scan) and the recording of individual masses (SIM/MID (selected/multiple ion monitoring) or SRM/MRM (selected/multiple reaction monitoring)). Particularly with continually operating spectrometers (ion beam instruments: magnetic sector MS, quadrupole MS) there are large differences between these two recording techniques with respect to selectivity, sensitivity and information content. For spectrometers with storage features (ion storage: ion trap MS, Orbitrap/ion cyclotron resonance (ICR)-MS), these differences are less strongly pronounced. Besides one-stage types of analyser (GC-MS), multistage mass spectrometers (GC-MS/MS) are playing an increasingly important role in residue target compound analysis and structure determination. With the MS/MS technique (tandem/multi-dimensional MS), which is available in both beam instruments and ion storage mass spectrometers, much more analytical information and a high structure-related selectivity for target compound quantitation can be obtained.

2.3.4.1 Detection of the Complete Mass Spectrum (Full Scan)

The continuous recording of mass spectra (full scan) and the simultaneous monitoring of the retention time allow the identification of analytes by comparison with libraries of mass spectra. With beam instruments it should be noted that the sensitivity required for recording the spectrum depends on the efficiency of the ion source, the transmission through the analyser and, most particularly, on the dwell time of the ions. The dwell time per mass is given by the width of the mass scan (e.g., 50–550 Da) and the scan rate of the chromatogram (e.g., 500 ms). From this, a scan rate of 1000 Da/s is calculated. Effective scan rates of modern quadrupole instruments exceed 10 000 Da/s with maximum scan rates up to 20 000 Da/s. Each mass from the selected mass range is measured only once during a scan over a short period (here 1 ms/s, Figure 2.219). All other ions from the substance formed in parallel in the ion source are not detected during the mass scan (quadrupole as mass filter).

$$\text{dwell time} = \frac{m_2 - m_1}{\text{scan rate}} \, [\text{ms}]$$

with m_1 start mass of scan

m_2 end mass of scan

Typical sensitivities for most compounds with benchtop quadrupole systems are in the mid to low pg region. Prolonging the scan time can increase the sensitivity of these systems for full scan operation (Figure 2.219). However, there is an upper limit for a useful scan time determined by the minimum required number of data points across a chromatographic peak. In practice, for coupling with capillary GC, scan rates of 0.2–1 s are used. For quantitative determinations, it should be ensured that the scan rate across chromatographic peak is adequate in order to determine the area and height correctly

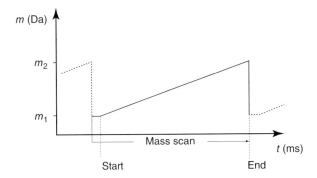

Figure 2.219 Scan function of the quadrupole analyser: each mass between the start of the scan (m_1) and the end (m_2) is only registered once during the scan.

(see also section 3.3.1). The SIM/MID mode is usually chosen to increase the sensitivity and scan rates of quadrupole systems for this reason (see also Section 2.3.4.2).

In ion storage mass spectrometers, all the ions produced on ionization of a substance are stored in parallel and detected. The mode of function is opposite to the filter character of beam instruments and particularly strong when integrating weak ion beams (Table 2.52). All the ions formed are collected in a first step in the ion trap. At the end of the storage phase all ions leave the trap, sorted according to mass during the scan, and are directed to the multiplier. This process can take place very rapidly on the ms time scale. (Figure 2.220). The scan rates of ion trap mass spectrometers are higher than 11 000 Da/s. Typical sensitivities for full spectra in ion trap mass spectrometers are in the low picograms range.

Table 2.52 Duty cycle for ion trap and beam instruments.

Scan range (Da)	Dwell time per mass Ion trap (s/Da)	Quadrupole (s/Da)	Sensitivity ratio Ion trap/Quadrupole
1	0.83	1.0	0.8
3	0.82	0.33	2.5
10	0.79	0.10	8
30	0.71	0.033	21
100	0.52	0.010	52
300	0.30	0.0033	91

The longer dwell time per mass leads to the highest sensitivity in the recording of complete mass spectra with ion trap instruments. Compared with beam instruments an increase in the duty cycle is achieved, depending on the mass range, of up to a factor of 100 and higher.

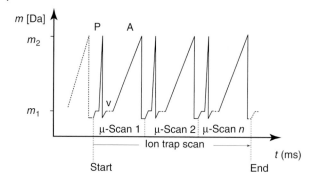

Figure 2.220 Scan function of the ion trap analyser: within an ion trap scan, several μ-scans (three μ-scans shown here) are carried out and their spectra added before storage to disk. P = pre-scan, A = analytical scan, v = variable ionization time (AGC, automatic gain control) (with ion trap instruments with external ionization v stands for the length of the storage phase, ion injection time; for internal ionization ion trap v stands for the ionization time).

**Dwell Times *t* per Ion
in Full Scan Acquisition with Ion Trap and Quadrupole MS**

Mass range 50–550 Da (500 masses wide), scan rate 500 ms

Ion trap MS	Quadrupole MS
Ion storage during ionization. Ionization time can vary, typically up to t = 25 ms at simultaneous storage of all ions formed	$t = \frac{500\,\text{ms}}{500\,\text{Da}} = 1\,\text{ms/Da}$
Detection of all stored ions	In a scan, each type of ion is measured for only 1 ms. Only a minute quantity of the ions formed are detected.

2.3.4.2 Recording Individual Masses (SIM/MID)

In the use of conventional mass spectrometers (beam instruments), the detection limit in the full scan mode is insufficient for trace analysis because the analyser only has a very short dwell time per ion available during the scan. Additional sensitivity is achieved by sharing the same dwell time between a few selected ions only by means of individual mass recording (SIM, MID) (Table 2.53 and Figure 2.221).

At the same time, a higher scan rate can be chosen so that chromatographic peaks can be plotted more precisely. The SIM technique is used exclusively for quantifying data on known target compounds, especially in trace analysis.

The mode of operation of a GC-MS system as a mass-selective detector requires the selection of certain ions (fragments, molecular ions), so that the desired analytes can be detected selectively (targeted approach). Other

Table 2.53 Dwell times per ion and relative sensitivity in SIM analysis (for beam instruments) at constant scan rates.

Number of SIM ions	Total scan time[a] [ms]	Total voltage setting time[b] [ms]	Effective dwell time per ion[c] [ms]	Relative sensitivity [%]
1	500	2	498	100
2	500	4	248	50
3	500	6	165	33
4	500	8	123	25
10	500	20	48	10
20	500	40	23	5
30	500	60	15	3
40	500	80	11	3
50	500	100	8	2
For comparison full scan				
500	500	2	1	1

a) The total scan time is determined by the necessary scan rate of the chromatogram and is held constant.
b) Total voltage setting times are necessary in order to adjust the mass filter for the subsequent SIM mass acquisition. The actual times necessary can vary slightly depending on the type of instrument.
c) Duty cycle/ion.
d) The relative sensitivity is directly proportional to the effective dwell time per ion.

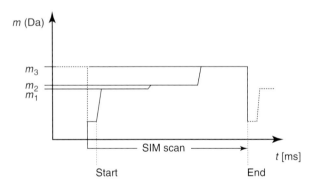

Figure 2.221 SIM scan with a quadrupole analyser: the total scan time is divided here into the three individual masses m_1, m_2 and m_3 with correspondingly long dwell times.

compounds contained in the sample besides those chosen for analysis remain undetected. Thus, the matrix present in large quantities in trace analysis is masked out, as are analytes whose appearance is not expected or planned. In the choice of masses required for detection, it is assumed that for three selective signals in the fragmentation pattern per substance a secure basis for a yes/no decision can be found in spite of small variations in the retention times (SIM, MID).

Identification of substances by comparison with spectral libraries is no longer possible. The relative intensities (ion ratios) of the selected ions serve as quality criteria (qualifiers) (1 ion ⇒ no criterion, 2 ions ⇒ 1 criterion and 3 ions ⇒ 3 criteria!). This process for detecting compounds can be affected by errors through shifts in retention times and compromised peak area determination caused by the matrix. In residue analysis, it is known that with SIM analysis false positive findings occur in about 10% of the samples. Positive SIM data are confirmed in the same way as positive results from classical GC detectors by running a complete mass spectrum of the analytes suspected. Confirmation of positive results, and statistically of negative results as well, is required by international directives either by full scan, MS/MS, ion ratios or HRMS.

Gain in Sensitivity Using SIM/MID

A typical SIM data acquisition of five selected masses at a scan rate of 0.5 s is given as a typical example:

	Ion trap MS	Quadrupole MS
Dwell time per ion	Identical ionization procedure to that of full scan, however selective and parallel or sequential storage of the selected SIM ions.	At a scan time of 500 ms, the effective dwell time per SIM mass is divided up as $t = 500\,\text{ms}/5\,\text{masses} = 100\,\text{ms/mass}$
Function	The ion trap is filled exclusively with the ions of the selected masses. If the capacity of the ion trap is not used up completely, the storage phase ends after a given time (ms).	To measure a SIM mass the quadrupole spends 100 times longer on one mass compared to full scan, and thus permits a dwell time which is 100 times longer for the selected ions to be achieved.
Sensitivity	The gain in sensitivity is most marked with matrix-containing samples, as the length of the storage phase still mainly depends on the appearance of the selected SIM ions in the sample and is not shortened by a high concentration of different mass matrix ions.	Theoretically, the sensitivity increases by a factor of 100. In practice, for real samples, a factor of up to 30–50 compared with full scan is achieved.

Consequences for trace analysis	Ion trap systems already give very high sensitivity in the full scan mode. Samples with high concentrations of matrix and detection limits below the pg level require the SIM technique (MS/MS is recommended).	Quadrupole systems require the SIM mode to achieve adequate sensitivity.
Confirmation	For 3 SIM masses by 3 intensity criteria (qualifiers), with MS/MS by means of the product ion spectra.	For 3 SIM masses by 3 ion ratio criteria (qualifiers), check of positive results after further concentration by the full scan technique or external confirmation with an ion trap instrument (see Table 2.53).

SIM Setup

1) Choice of column and oven temperature program optimization for optimal GC separation, paying particular attention to analytes with similar fragmentation patterns.
2) Full scan analysis of an average substance concentration to determine the selective ions (SIM masses, two to three ions/component); chose more potential SIM ions as finally required as special matrix conditions are to be taken into account. Chose SIM masses with highest S/N ratio.
3) Determination of the retention times of the individual components.
4) Establishment of the data acquisition interval (time window) for the individual SIM descriptors, or use the retention timed SIM technique with a symmetrical acquisition window centered to the compound retention time.
5) Test analysis of a low standard (or better, a matrix spike) for optimization (SIM masses, separation conditions).

Planning an analysis in the SIM/MID mode first requires a standard run in the full scan mode to determine both the retention times and the mass signals necessary for the SIM selection (Tables 2.54 and 2.55). As the gain in sensitivity achieved in individual mass recording with beam instruments is possible only on detection of a few ions, for the analysis of several compounds the group of masses detected must be adjusted. The more components there are to be detected, the more frequently and precisely must the descriptors be adjusted. Multi-component analyses, such as the MAGIC-60 analysis with P&T (volatile halogenated hydrocarbons see Section 4.5), cannot be dealt with in the segmented SIM mode, but are significantly facilitated using the retention timed SIM mode (t-SIM).

Table 2.54 Characteristic ions (*m/z* values) for selected polyaromatic hydrocarbons (PAHs) and their alkyl derivatives (in the elution sequence for methylsilicone phases).

Component name	m/z
Benzene-d$_6$	92, 94
Benzene	77, 78
Toluene-d$_8$	98, 100
Toluene	91, 92
Ethylbenzene	91, 106
Dimethylbenzene	91, 106
Methylethylbenzene	105, 120
Trimethylbenzene	105, 120
Diethylbenzene	105, 119, 134
Naphthalene-d$_8$	136
Naphthalene	128
Methylnaphthalene	141, 142
Azulene	128
Acenaphthene	154
Biphenyl	154
Dimethylnaphthalene	141, 155, 156
Acenaphthene-d$_{10}$	162, 164
Acenaphthene	152
Dibenzofuran	139, 168
Dibenzodioxin	184
Fluorene	165, 166
Dihydroanthracene	178, 179, 180
Phenanthrene-d$_{10}$	188
Phenanthrene	178
Anthracene	178
Methylphenanthrene	191, 192
Methylanthracene	191, 192
Phenylnaphthalene	204
Dimethylphenanthrene	191, 206
Fluoranthene	202
Pyrene	202
Methylfluoranthene	215, 216
Benzofluorene	215, 216
Phenylanthracene	252, 253, 254
Benzanthracene	228
Chrysene-d$_{12}$	240
Chrysene	228
Methylchrysene	242
Dimethylbenz[*a*]anthracene	239, 241, 256
Benzo[*b*]fluoranthene	252
Benzo[*j*]fluoranthene	252
Benzo[*k*]fluoranthene	252
Benzo[*e*]pyrene	252

Table 2.54 *(continued)*

Component name	m/z
Benzo[*a*]pyrene	252
Perylene-d$_{12}$	264
Perylene	252
Methylcholanthrene	268
Diphenylanthracene	330
Indeno[1,2,3-*cd*]pyrene	276
Dibenzanthracene	278
Benzo[*b*]chrysene	278
Benzo[*g,h,i*]perylene	276
Anthanthrene	276
Dibenzo[*a,l*]pyrene	302
Coronene	300
Dibenzo[*a,i*]pyrene	302
Dibenzo[*a,h*]pyrene	302
Rubicene	326
Hexaphene	328
Benzo[*a*]coronene	350

The use of SIM analysis with ion trap mass spectrometers has also been developed. Through special control of the analyser (waveform ion isolation) during the ionization phase only the preselected ions of analytical interest are stored (SIS, selective ion storage). This technique allows the detection of selected ions in ion storage mass spectrometers in spite of the presence of complex matrices or the co-elution of another component in high concentration. As the storage capacity of the ion trap analyser is only used for a few ions instead of for a full spectrum, very low detection limits are possible (<1 pg/component) and the usable dynamic range of

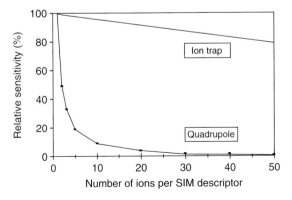

Figure 2.222 Comparison of the SIS (ion trap analyser) and the SIM (quadrupole analyser) techniques based on the effective dwell time per ion (relative sensitivity).

the analyser is extended considerably. Unlike conventional SIM operations with beam instruments, the detection sensitivity only alters slightly with the number of selected ions using the ion trap SIM technique (Figure 2.222). For the SIM technique, the sensitivity depends almost exclusively on the ionization time. The SIM technique with ion trap instruments is regarded as a necessary prerequisite for carrying out MS/MS detection.

Example of Selected Ion Monitoring

For PCB analysis taking into account the PCB replacement product Ugilec T. The analysis strategy shown here has the aim of determining a PCB pattern as completely as possible in the relevant degrees of chlorination and to test in parallel for the possible presence of Ugilec T (tetrachlorobenzyl-toluenes/trichlorobenzene). Three individual masses per time window for every two degrees of chlorination are planned for the selected SIM descriptors (scan width ±0.25 Da based on the centroid determined in the full scan mode) to determine the overlapping retention ranges of the individual degrees of chlorination (Figure 2.223). A staggered mass data acquisition is thus obtained (Figure 2.224).

No.	Substance	SIM masses (*m/z*)	Staggered time window (min:s)
1	Trichlorobenzenes	180/182/184	Start – 8:44
2	Cl_3-PCBs	256/258/260	8:44 – 14:30
3	Cl_4-PCBs	290/292/294	8:44 – 15:52
4	Cl_5-PCBs	324/326/328	14:30 – 18:34
5	Ugilec	318/320/322	15:52 – 20:15
6	Cl_6-PCBs	358/360/362	15:52 – 21:36
7	Cl_7-PCBs	392/394/396	18:34 – 23:38
8	Cl_8-PCBs	428/430/432	20:15 – 30:00
9	Cl_{10}-PCB	496/498/500	23:58 – 30:00

The retention times refer to a capillary column J&W DB5 30 m × 0.25 mm × 0.25 μm. Program: 60°C – 2 min, 20°C/min to 180°C, 5°C/min to 290°C – 5 min. Splitless injection at 280°C, column pressure 1 bar He, constant pressure. GC-MS system Finnigan INCOS 50. Sample: Aroclor 1240/1260 + Ugilec T, total concentration of the mixture about 10 ng/μL (Wagner-Redeker *et al.*, 1985).

As the SIM/MID analysis is targeted to the detection of certain ions in small mass ranges and the comparability of a complete spectrum for a library search is not required, a special SIM tuning of the mass analyzer with emphasis on the improved ion transmission in the upper mass range is recommended.

Table 2.55 Recommended SIM ions and relative intensities for major pesticides and some of their derivatives (Thier and Kirchhoff, 1992).

Compound	Molar mass	Main fragment *m/z* (intensities)					
		1	2	3	4	5	6
Acephate	183	43 (100)	44(88)	136 (80)	94 (58)	47 (56)	95 (32)
Alaclor	269	45 (100)	188 (23)	160 (18)	77 (7)	146 (6)	224 (6)
Aldicarb	190	41 (100)	86 (89)	58 (85)	85 (61)	87 (50)	44 (50)
Aldrin	362	66 (100)	91 (50)	79 (47)	263 (42)	65 (35)	101 (34)
Allethrin	302	123 (100)	79 (40)	43 (32)	81 (31)	91 (29)	136 (27)
Atrazine	215	43 (100)	58 (84)	44 (75)	200 (69)	68 (43)	215 (40)
Azinphos-methyl	317	77 (100)	160 (77)	132 (67)	44 (30)	105 (29)	104 (27)
Barban	257	51 (100)	153 (76)	87 (66)	222 (44)	52 (43)	63 (43)
Benzazolin methyl ester	257	170 (100)	134 (75)	198 (74)	257 (73)	172 (40)	200 (31)
Bendiocarb	223	151 (100)	126 (58)	166 (48)	51 (19)	58 (18)	43 (17)
Bromacil	260	205 (100)	207 (75)	42 (25)	70 (16)	206 (16)	162 (12)
Bromacil *N*-methyl derivative	274	219 (100)	221 (68)	41 (45)	188 (41)	190 (40)	56 (37)
Bromophos	364	331 (100)	125 (91)	329 (80)	79 (57)	109 (53)	93 (45)
Bromophos-ethyl	392	97 (100)	65 (35)	303 (32)	125 (28)	359 (27)	109 (27)
Bromoxynil methyl ether	289	289 (100)	88 (77)	276 (67)	289 (55)	293 (53)	248 (50)
Captafol	347	79 (100)	80 (42)	77 (28)	78 (19)	151 (17)	51 (13)
Captan	299	79 (100)	80 (61)	77 (56)	44 (44)	78 (37)	149 (34)
Carbaryl	201	144 (100)	115 (82)	116 (48)	57(31)	58 (20)	63 (20)
Carbendazim	191	159 (100)	191 (57)	103 (38)	104 (37)	52 (32)	51 (29)
Carbetamid	236	119 (100)	72 (54)	91 (44)	45 (38)	64 (37)	74 (29)
Carbofuran	221	164 (100)	149 (70)	41 (27)	58 (25)	131 (25)	122 (25)
Chlorbromuron	292	61 (100)	46 (24)	62 (11)	63 (10)	60 (9)	124 (8)
Chlorbufam	223	53 (100)	127 (20)	51 (13)	164 (13)	223 (13)	70 (10)
cis-Chlordane	406	373 (100)	375 (84)	377 (46)	371 (39)	44 (36)	109 (36)
trans-Chlordane	406	373 (100)	375 (93)	377 (53)	371 (47)	272 (36)	237 (30)
Chlorfenprop-methyl	232	125 (100)	165 (64)	75 (46)	196 (43)	51 (43)	101 (37)
Chlorfenvinphos	35	81 (100)	267 (73)	109 (55)	269 (47)	323 (26)	91 (23)
Chloridazon	221	77 (100)	221 (60)	88 (37)	220 (35)	51 (26)	105 (24)
Chloroneb	206	191 (100)	193 (61)	206 (60)	53 (57)	208 (39)	141 (35)
Chlorotoluron	212	72 (100)	44 (29)	167 (28)	132 (25)	45 (20)	77 (11)
3-Chloro-4-methylaniline (GC degradation product of Chlorotoluron)	141	141 (100)	140 (37)	106 (68)	142 (36)	143 (28)	77 (25)
Chloroxuron	290	72 (100)	245 (37)	44 (31)	75 (21)	45 (19)	63 (16)
Chloropropham	213	43 (100)	127 (49)	41 (35)	45 (20)	44 (18)	129 (16)
Chlorpyrifos	349	97 (100)	195 (59)	199 (53)	65 (27)	47 (23)	314 (21)
Chlorthal-dimethyl	330	301 (100)	299 (81)	303 (47)	332 (29)	142 (26)	221 (24)
Chlorthiamid	205	170 (100)	60 (61)	171 (50)	172 (49)	205 (35)	173 (29)
Cinerin I	316	123 (100)	43 (35)	93 (33)	121 (27)	81 (27)	150 (27)

Table 2.55 *(Continued)*

Compound	Molar mass	Main fragment *m/z* (intensities)					
		1	2	3	4	5	6
Cinerin II	360	107 (100)	93 (57)	121 (53)	91 (50)	149 (35)	105 (33)
Cyanazine	240	44 (100)	43 (60)	68 (60)	212 (48)	41 (47)	42 (34)
Cypermethrin	415	163 (100)	181 (79)	165 (68)	91 (41)	77 (33)	51 (29)
2,4-DB methyl ester	262	101 (100)	59 (95)	41 (39)	162 (36)	69 (28)	63 (25)
Dalapon	142	43 (100)	61 (81)	62 (67)	97 (59)	45 (59)	44 (47)
Dazomet	162	162 (100)	42 (87)	89 (79)	44 (73)	76 (59)	43 (53)
Demetron-*S*-methyl	230	88 (100)	60 (50)	109 (24)	142 (17)	79 (14)	47 (11)
Desmetryn	213	213 (100)	57 (67)	58 (66)	198 (58)	82 (44)	171 (39)
Dialifos	393	208 (100)	210 (31)	76 (20)	173 (17)	209 (12)	357 (10)
Di-allate	269	43 (100)	86 (62)	41 (38)	44 (25)	42 (24)	70 (19)
Diazinon	304	137 (100)	179 (74)	152 (65)	93 (47)	153 (42)	199 (39)
Dicamba methyl ester	234	203 (100)	205 (60)	234 (27)	188 (26)	97 (21)	201 (20)
Dichlobenil	171	171 (100)	173 (62)	100 (31)	136 (24)	75 (24)	50 (19)
Dichlofenthion	314	97 (100)	279 (92)	223 (90)	109 (67)	162 (53)	251 (46)
Dichlofluanid	332	123 (100)	92 (33)	224 (29)	167 (27)	63 (23)	77 (22)
2,4-D isooctyl ester	332	43 (100)	57 (98)	41 (76)	55 (54)	71 (41)	69 (27)
2,4-D methyl ester	234	199 (100)	45 (97)	175 (94)	145 (70)	111 (69)	109 (68)
Dichlorprop isooctyl ester	346	43 (100)	57 (83)	41 (61)	71 (48)	55 (47)	162 (41)
Dichlorprop methyl ester	248	162 (10)	164 (80)	59 (62)	189 (56)	63 (39)	191 (35)
Dichlorvos	220	109 (100)	185 (18)	79 (17)	187 (6)	145 (6)	47 (5)
Dicofol	368	139 (100)	111 (39)	141 (33)	75 (18)	83 (17)	251 (16)
o,p′-DDT	352	235 (100)	237 (59)	165 (33)	236 (16)	199 (12)	75 (12)
p,p′-DDT	352	235 (100)	237 (58)	165 (37)	236 (16)	75 (12)	239 (11)
Dieldrin	378	79 (100)	82 (32)	81 (30)	263 (17)	77 (17)	108 (14)
Dimethirimol methyl ether	223	180 (100)	223 (23)	181 (10)	224 (3)	42 (2)	109 (2)
Dimethoate	229	87 (100)	93 (76)	125 (56)	58 (40)	47 (39)	63 (33)
DNOC methyl ether	212	182 (100)	165 (74)	89 (69)	90 (57)	212 (48)	51 (47)
Dinoterb methyl ether	254	239 (100)	209 (41)	43 (36)	91 (35)	77 (33)	254 (33)
Dioxacarb	223	121 (100)	122 (62)	166 (46)	165 (42)	73 (35)	45 (31)
Diphenamid	239	72 (100)	167 (86)	165 (42)	239 (21)	152 (17)	168 (14)
Disulfoton	274	88 (100)	89 (43)	61 (40)	60 (39)	97 (36)	65 (23)
Diuron	232	72 (100)	44 (34)	73 (25)	42 (20)	232 (19)	187 (13)
Dodine	227	43 (100)	73 (80)	59 (52)	55 (47)	7 (46)	100 (46)
Endosulfan	404	195 (100)	36 (95)	237 (91)	41 (89)	24 (79)	75 (78)
Endrin	378	67 (100)	81 (67)	263 (59)	36 (58)	79 (47)	82 (41)
Ethiofencarb	225	107 (100)	69 (48)	77 (29)	41 (26)	81 (21)	45 (17)
Ethirimol	209	166 (100)	209 (17)	167 (14)	96 (12)	194 (4)	55 (2)
Ethirimol methyl ether	223	180 (100)	223 (23)	85 (14)	181 (12)	55 (10)	96 (9)
Etrimfos	292	125 (100)	292 (91)	181 (90)	47 (84)	153 (84)	56 (73)
Fenarimol	330	139 (100)	107 (95)	111 (40)	219 (39)	141 (33)	251 (31)

Table 2.55 *(Continued)*

Compound	Molar mass	Main fragment *m/z* (intensities)					
		1	2	3	4	5	6
Fenitrothion	277	125 (100)	109 (92)	79 (62)	47 (57)	63 (44)	93 (40)
Fenoprop isooctyl ester	380	57 (100)	43 (94)	41 (85)	196 (63)	71 (60)	198 (59)
Fenoprop methyl ester	282	196 (100)	198 (89)	59 (82)	55 (36)	87 (34)	223 (31)
Fenuron	164	72 (100)	164 (27)	119 (24)	91 (22)	42 (14)	44 (11)
Flamprop-isopropyl	363	105 (100)	77 (44)	276 (21)	106 (18)	278 (7)	51 (5)
Flamprop-methyl	335	105 (100)	77 (46)	276 (20)	106 (14)	230 (12)	44 (11)
Formothion	257	93 (100)	125 (89)	126 (68)	42 (49)	47 (48)	87 (40)
Heptachlor	370	100 (100)	272 (81)	274 (42)	237 (33)	102 (33)	
Iodofenphos	412	125 (100)	377 (78)	47 (64)	79 (59)	93 (54)	109 (49)
Loxynil isooctyl ether	483	127 (100)	57 (96)	41 (34)	43 (33)	55 (26)	37 (16)
Loxynil methyl ether	385	385 (100)	243 (56)	370 (41)	127 (13)	386 (10)	88 (9)
Isoproturon	206	146 (100)	72 (54)	44 (35)	128 (29)	45 (28)	161 (25)
Jasmolin I	330	123 (100)	43 (52)	55 (34)	93 (25)	91 (24)	81 (23)
Jasmolin II	374	107 (100)	91 (69)	135 (69)	93 (67)	55 (66)	121 (58)
Lenacil	234	153 (100)	154 (20)	110 (15)	109 (15)	152 (13)	136 (10)
Lenacil *N*-methyl derivative	248	167 (100)	166 (45)	168 (12)	165 (12)	124 (9)	123 (6)
Lindane	288	181 (100)	183 (97)	109 (89)	219 (86)	111 (75)	217 (68)
Linuron	248	61 (100)	187 (43)	189 (29)	124 (28)	46 (28)	44 (23)
MCPB isooctyl ester	340	87 (100)	57 (81)	43 (62)	71 (45)	41 (42)	69 (29)
MCPB methyl ester	242	101 (100)	59 (70)	77 (40)	107 (25)	41 (22)	142 (20)
Malathion	330	125 (100)	93 (96)	127 (75)	173 (55)	158 (37)	99 (35)
Mecoprop isooctyl ester	326	43 (100)	57 (94)	169 (77)	41 (70)	142 (69)	55 (52)
Mecoprop methyl ester	228	169 (100)	143 (79)	59 (58)	141 (57)	228 (54)	107 (50)
Metamitron	202	104 (100)	202 (66)	42 (42)	174 (35)	77 (24)	103 (19)
Methabenzthiazuron	221	164 (100)	136 (73)	135 (69)	163 (42)	69 (30)	58 (25)
Methazole	260	44 (100)	161 (44)	124 (36)	187 (31)	159 (24)	163 (23)
Methidathion	302	85 (100)	145 (90)	93 (32)	125 (22)	47 (21)	58 (20)
Methiocarb	225	168 (100)	153 (84)	45 (40)	109 (37)	91 (31)	58 (21)
Methomyl	162	44 (100)	58 (81)	105 (69)	45 (59)	42 (55)	47 (52)
Metobromuron	258	61 (100)	46 (43)	60 (15)	91 (13)	258 (13)	170 (12)
Metoxuron	228	72 (100)	44 (27)	183 (23)	228 (22)	45 (21)	73 (15)
Metribuzin	214	198 (100)	41 (78)	57 (54)	43 (39)	47 (38)	74 (36)
Mevinphos	224	127 (100)	192 (30)	109 (27)	67 (20)	43 (8)	193 (7)
Monocrotophos	223	127 (100)	67 (25)	97 (23)	109 (14)	58 (14)	192 (13)
Monolinuron	214	61 (100)	126 (63)	153 (42)	214 (34)	46 (29)	125 (25)
Napropamide	271	72 (100)	100 (81)	128 (62)	44 (55)	115 (41)	127 (36)
Nicotine	162	84 (100)	133 (21)	42 (18)	162 (17)	161 (15)	105 (9)
Nitrofen	283	283 (100)	285 (67)	202 (55)	50 (55)	139 (37)	63 (37)
Nuarimol	314	107 (100)	235 (91)	203 (85)	139 (60)	123 (46)	95 (35)
Omethoat	213	110 (100)	156 (83)	79 (39)	109 (32)	58 (30)	47 (21)
Oxadiazon	344	43 (100)	175 (92)	57 (84)	177 (60)	42 (35)	258 (22)

(continued overleaf)

Table 2.55 (Continued)

Compound	Molar mass	Main fragment m/z (intensities)					
		1	2	3	4	5	6
Parathion	291	97 (100)	109 (90)	291 (57)	139 (47)	125 (41)	137 (39)
Parathion-methyl	263	109 (100)	125 (80)	263 (56)	79 (26)	63 (18)	93 (18)
Pendimethalin	281	252 (100)	43 (53)	57 (43)	41 (41)	281 (37)	253 (34)
Permethrin	390	183 (100)	163 (100)	165 (25)	44 (15)	184 (15)	91 (13)
Phenmedipham	300	133 (100)	104 (52)	132 (34)	91 (34)	165 (31)	44 (27)
Phosalone	367	182 (100)	121 (48)	97 (36)	184 (32)	154 (24)	111 (24)
Pirimicarb	238	72 (100)	166 (85)	42 (63)	44 (44)	43 (24)	238 (23)
Pirimiphos-ethyl	333	168 (100)	318 (94)	152 (88)	304 (79)	180 (73)	42 (71)
Pirimiphos-methyl	305	290 (100)	276 (93)	125 (69)	305 (53)	233 (44)	42 (41)
Propachlor	211	120 (100)	77 (66)	93 (36)	43 (35)	51 (30)	41 (27)
Propanil	217	161 (100)	163 (70)	57 (64)	217 (16)	165 (11)	219 (9)
Propham	179	43 (100)	93 (88)	41 (42)	120 (24)	65 (24)	137 (23)
Propoxur	209	110 (100)	152 (47)	43 (28)	58 (27)	41 (21)	111 (20)
Pyrethrin I	328	123 (100)	43 (62)	91 (58)	81 (47)	105 (45)	55 (43)
Pyrethrin II	372	91 (100)	133 (70)	161 (55)	117 (48)	107 (47)	160 (43)
Quintozene	293	142 (100)	237 (96)	44 (75)	214 (67)	107 (62)	212 (61)
Resmethrin	338	123 (100)	171 (67)	128 (52)	143 (49)	81 (38)	91 (28)
Simazine	201	201 (100)	44 (96)	186 (72)	68 (63)	173 (57)	96 (40)
Tecnazene	259	203 (100)	201 (69)	108 (69)	215 (60)	44 (57)	213 (51)
Terbacil	216	160 (100)	161 (99)	117 (69)	42 (45)	41 (41)	162 (37)
Terbacil N-methyl derivative	230	56 (100)	174 (79)	175 (31)	57 (24)	176 (23)	41 (20)
Tetrachlorvinphos	364	109 (100)	329 (48)	331 (42)	79 (20)	333 (14)	93 (9)
Tetrasul	322	252 (100)	254 (67)	324 (51)	108 (49)	75 (40)	322 (40)
Thiabendazole	201	201 (100)	174 (72)	63 (12)	202 (11)	64 (11)	65 (9)
Thiofanox	218	57 (100)	42 (75)	68 (39)	61 (38)	55 (34)	47 (33)
Thiometon	246	88 (100)	60 (63)	125 (56)	61 (52)	47 (49)	93 (47)
Thiophanat-methyl	342	44 (100)	73 (97)	159 (89)	191 (80)	86 (72)	150 (71)
Thiram	240	88 (100)	42 (25)	44 (20)	208 (18)	73 (15)	45 (10)
Tri-allate	303	43 (100)	86 (73)	41 (43)	42 (31)	70 (23)	44 (21)
Trichlorfon	256	109 (100)	79 (34)	47 (26)	44 (20)	185 (17)	80 (8)
Tridemorph	297	128 (100)	43 (26)	42 (18)	44 (13)	129 (11)	55 (5)
Trietazine	229	200 (100)	43 (81)	186 (52)	229 (52)	214 (50)	42 (48)
Trifluralin	335	43 (100)	264 (33)	306 (32)	57 (7)	42 (6)	290 (5)
Vamidothion	287	87 (100)	58 (47)	44 (40)	61 (29)	59 (26)	60 (25)
Vinclozolin	285	54 (100)	53 (93)	43 (82)	124 (65)	212 (63)	187 (61)

The data refer to EI ionization at 70 eV. The relative intensities can depend in individual cases an the type of mass spectrometer or mass-selective detector used. For confirmation, mass spectra should be consulted which were run under identical instrumental conditions.

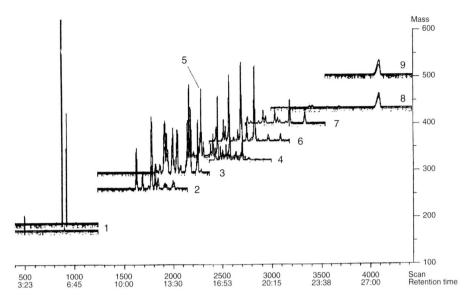

Figure 2.223 MID chart of a PCB/Ugilec analysis. In each case, two degrees of chlorination of the PCBs and Ugilec T were detected in parallel, each with three masses. The overlapping MID descriptors were switched in such a way that each degree of chlorination was detected in two consecutive time windows (see text).

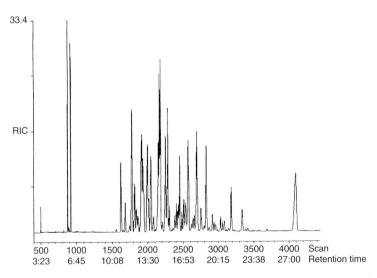

Figure 2.224 Chromatogram (RIC, reconstructed ion chromatogram) compiled after termination of the PCB/Ugilec T analysis in the MID mode from Figure 2.223.

Standard tuning aims to produce a balanced spectrum which corresponds to the data in a reference list for the reference substance FC43 (perfluorotributylamine, PFTBA), in order to guarantee good comparability of the spectra run with those in the library. Certain EPA methods require the source tuning using BFB (4-bromofluorobenzene) or DFTPP (decafluorotriphenylphosphine) for compliance with predefined mass intensities as set in the operating procedure. In the optimization of the ion source, special attention should be paid to the masses (or the mass range) involved in SIM data acquisition. The source and lens potentials should then be selected manually so that a nearby fragment of the reference compound or an ion produced by column bleed (GC temperature about 200°C) can be detected with the highest intensity but good resolution. In this way, a significant additional increase in sensitivity can be achieved with quadrupole analysers for the SIM mode.

Figure 2.225 shows the chromatogram of a PCB standard as the result of a typical SIM routine analysis. In this case, two masses are chosen as SIM masses for each PCB chlorination degree. The switching points of the individual descriptors are visible as steps in the base line. The different base line heights arise as a result of the different contributions of the chemical noise to these signals. To control the evaluation, the substance signals can be represented as peaks in the expected retention time windows (Figure 2.226). A deviation from the calibrated retention time (Figure 2.226, right segment with the masses m/z 499.8 and 497.8) leads to a shift of the peak from the middle to the edge of the window and should be a reason for further checking. If qualifiers are present (e.g., isotope patterns), these

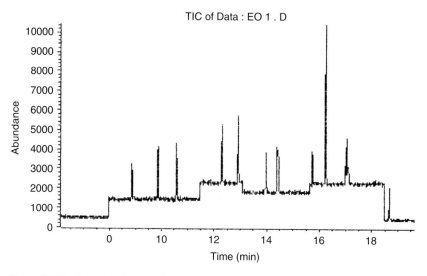

Figure 2.225 Example of a typical PCB analysis in the SIM mode. The steps in the base line show the switching points of the SIM descriptors.

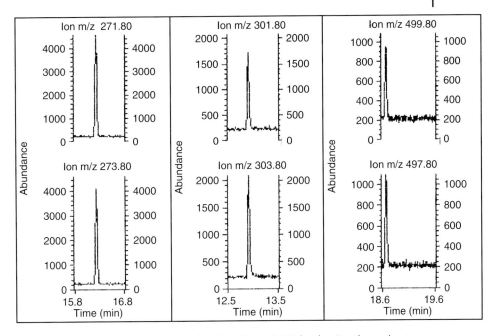

Figure 2.226 Evaluation of the PCB analysis from Figure 2.179 by showing the peaks at specific mass traces (see text).

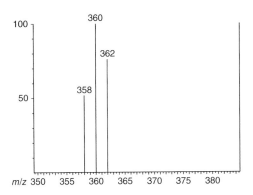

Figure 2.227 Evaluation of isotope patterns from an MID analysis of hexachlorobiphenyl by comparison of relative intensities (shown as a bar-graph spectrum).

should be checked using the relative intensities as a line spectrum, if possible (Figure 2.227), or as superimposed mass traces (Figure 2.228). In the region of the detection limit, the noise width should be taken into account in the test of agreement.

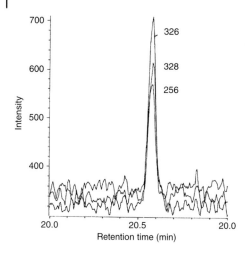

Figure 2.228 Test of PCB isotope patterns (PCB 101, pentachlorobiphenyl) in the range of the detection limit (10 pg, S/N about 4 : 1) after SIM analysis (shown as mass traces).

2.3.4.3 High Resolution Accurate Mass MID Data Acquisition

Target compound analysis using high mass resolution, for example, for PCDD/PCDFs, pesticides, POPs or pharmaceutical residues are typically performed by monitoring the compound-specific accurate mass ions at the expected retention time for each analyte. High resolution GC-MS target compound applications benefit from a unique technical feature referred to as *the lock-mass technique* for performing MID analyses. The lock mass technique provides ease of use, combined with a maximum quantitative precision and certainty in analyte confirmation.

The basic equation for sector mass spectrometers

$$m/z = c \cdot \frac{B^2}{V}$$

with

c	=	the instrument constant
B	=	magnetic field strength
V	=	acceleration voltage

shows that mass calibrations are feasible either at constant acceleration voltage with a constant calibrated magnetic field or vice versa.

For MID data acquisition, the fixed magnet setting with a variable acceleration voltage is typically used. The mass calibration for MID data acquisition follows a special procedure during the data acquisition in the form of a scan inherent mass calibration. This internal mass calibration is performed during MID analysis in every scan before acquisition of the target compound intensities. Most recent developments use two reference masses below and above the target masses referenced as the 'lock-plus-cali mass technique' or in short 'lock mode'. It provides optimum mass accuracy for peak detection even in difficult chromatographic situations. The scan-to-scan mass calibration provides the best confidence for the acquired analytical data and basically is the accepted feature for the requirement of HRGC-HRMS as a confirmation method (EPA Method 1613 b).

The internal mass calibration process is performed by the instrument control in the background without being noticed by the operator. In particular, it provides superior stability especially for high sample throughput with extended runtimes. A reference compound is leaked continuously from the reference inlet system during the GC run into the ion source. Typically perfluoro-tributylamine (FC43) is used as reference compound in HRGC-HRMS for dioxin analysis. Other reference compounds may be used to suit individual experimental conditions.

The exact ion masses of the reference compound are used in the MID acquisition windows for internal calibration. For the lock-and-cali-mass technique, two ions of the reference substance are individually selected for each MID window; one mass which is close, but below the analyte target mass, and the second, which is slightly above the analyte or internal standard target masses. Although both reference masses are used for the internal calibration, it became common practice to name the lower reference mass the 'lock mass' and the upper reference mass the 'calibration mass'.

During the MID scan, the mass spectrometer is parking (locking) the magnetic field strength at the start of each MID window and then performing the mass calibration using the lock and calibration masses followed by the acquisition of the target and internal standard mass intensities.

Lock-Mass Technique

At the start of each MID retention time window, the magnet is automatically set to one mass (Da) below the lowest mass found in the MID descriptor. The magnet is parked, or 'locked' and remains with this setting throughout the entire MID window. All analyser jumps to the calibration and target compound masses are done by fast electrical jumps of the acceleration voltage.

Internal Mass Calibration

The lock mass (L) is scanned in a small mass window starting below the mass peak by slowly decreasing the ion source acceleration voltage (see Figure 2.229, ①). The mass resolution of the lock mass peak is calculated and written to the data file. Using the lock mass setting, a second reference mass is used for building the MID mass calibration. The calibration mass is checked by an electrical jump. A fine adjustment of the electrical calibration is made based on this measurement (see Figure 2.229, ②). The electrical 'jump' (see Figure 2.229, ③) is very fast and takes only a few milliseconds. The dwell times for the sufficiently intense reference ions are short.

The resulting electrical calibration is used for subsequent MID data acquisition.

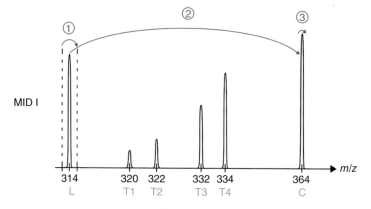

Figure 2.229 Mass detection scheme in HRMS MID calibration. The arrows show the sequence of measurement in the mass calibration steps (the magnet is 'locked' in this example at *m/z* 313).

①		Magnet locking and 'lock mass' sweep, mass calibration and resolution determination.
②		Electrical jump to calibration mass.
③		Calibration mass sweep and mass calibration.
L	from FC43	lock mass (L), *m/z* 313.983364.
C	from FC43	calibration mass (C), *m/z* 363.980170.
T1, T2	from native TCDD	analyte target masses *m/z* 319.895992, 321.893042.
T3, T4	from ^{13}C-TCDD	internal standard masses *m/z* 331.936250, 333.933300.

Data Acquisition

With the updated and exact mass calibration settings, the analyser sets the acceleration voltage to the masses of the target ions. The intensity of each ion is measured based on a preset dwell time (see Figure 2.230, ④). The dwell times to measure the analyte target ion intensities are significantly longer than the lock or calibration mass ions dwell times. This is done to achieve the optimum detection sensitivity for each analyte ion. The exact positioning on the top of the target ion mass peak allows for higher dwell times, significantly increased sensitivity and higher S/N values compared to sweep-scan techniques still used in older technology HRMS systems. It is important to note that the lock-plus-cali mass technique extends the dynamic range significantly into the lower concentration range.

Advantages of the Lock-Plus-Cali Mass Technique

The lock-plus-cali mass calibration technique provides extremely stable conditions for data acquisitions of long sequences even over days, for example, over the weekend and usually includes the performance documentation for quality control documented in the data file.

The electrical jumps of the acceleration voltage are very fast, and provide an excellent instrument duty cycle. Any outside influences from incidental background ions, long-term drift or minute electronics fluctuations are taken care of and do not influence the result.

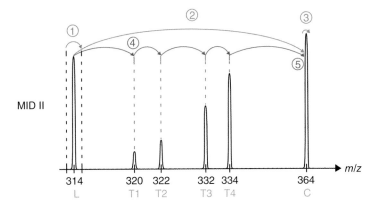

Figure 2.230 Mass detection scheme in HRMS MID data acquisition. The arrows show the sequence of measurement in the target compound and internal standard data acquisition (legend for the mass scale see Figure 2.229).
④ Consecutive electrical jumps to target and internal standard masses.
⑤ Electrical jump to calibration mass, mass calibration.

Both, the lock and cali masses are monitored in parallel during the run providing an excellent confirmation of system stability for data certainty. Together with the constant resolution monitoring, this technique provides the required traceability in MID data analysis.

Other acquisition techniques have been formerly used employing just one lock mass position. This technique requires a separate pre-run electrical mass calibration and does not allow a scan internal correction of the mass position which may arise due to long-term drifts of the analyser during data acquisition.

As a consequence, with the one mass lock techniques, the mass jumps are less precise with increasing run times. Deviations from the peak top position when acquiring data at the peak slope result in less sensitivity, less reproducibility and poor isotope ratio confirmation.

The lock-plus-cali mass technique has proven to be superior in achieving lower LOQs (limit of quantitations) and higher S/N values in high resolution MID. Figure 2.231 shows the typical chromatogram display of dioxin analysis with the TCDD target masses as well as the ^{13}C internal standard masses. In addition, the continuously monitored FC43 lock and cali masses are displayed as constant mass traces. Both traces are of valuable diagnostic use and confirm the correct measurement of the target compounds.

Setup of the MID Descriptor

The MID descriptor for the data acquisition contains all the information required by the HRMS mass analyser. Included in each descriptor is the retention time information for switching between different target ions, the exact mass calibration, the target masses to be acquired and the corresponding dwell times.

A sample chromatogram usually facilitates the setting of the retention time windows, as shown Figure 2.232. The sample chromatogram is used to optimize the

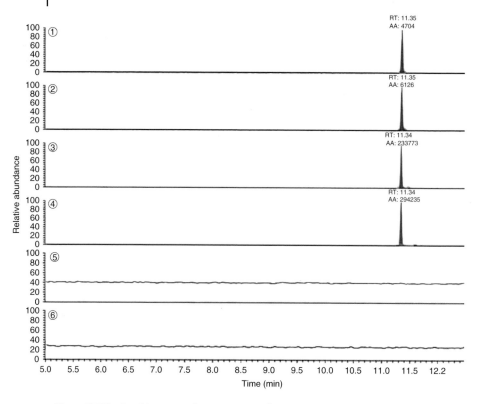

Figure 2.231 Resulting mass chromatograms of a TCDD standard solution at 100 fg/μL (DB-5MS 60 m × 0.25 mm × 0.1 μm).
① Ratio mass of 2,3,7,8-TCDD (native) *m/z* 319.8960;
② Quan mass of 2,3,7,8-tcdd (native) *m/z* 321.8930;
③ Ratio mass of 2,3,7,8-$^{13}C_{12}$-TCDD (ISTD) *m/z* 331.9362;
④ Quan mass of 2,3,7,8-$^{13}C_{12}$- TCDD (ISTD) *m/z* 333.9333;
⑤ Lock mass of FC43 *m/z* 313.983364;
⑥ Cali mass of FC43 *m/z* 363.980170.

GC component separation and set the MID windows before data analysis (often called "window finder"), and usually consists of a medium concentration standard mix.

MID Cycle Time

In order to provide a representative and reproducible GC peak integration, the total MID cycle time on the chromatographic time scale should allow for the acquisition of 8–10 data points over a chromatographic peak (Figure 2.232). The cycle time has a direct influence on the available measurement time for each ion (dwell time). If the MID cycle time is too short, the sensitivity of the instrument is compromised; too high values lead to a poor GC peak definition.

Note in Figure 2.232 the different relative time setting for the lock and calibration masses (3 ms each), the labelled internal standards (52 ms each) and the native

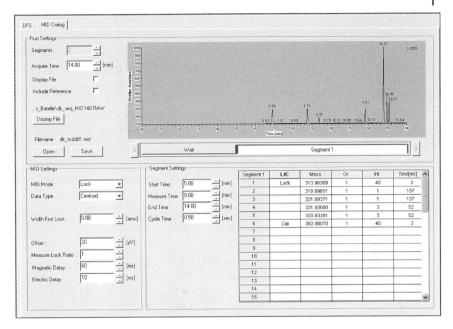

Figure 2.232 MID editor with sample chromatogram (top), target mass list and duty cycles for the highlighted 'Segment 1' retention time window (bottom right).

compounds (157 ms each). This shifts most of the available dwell time in favor for the native compounds of the sample. This differential acquisition time setup is possible due to the high intensity of the lock and cali masses and higher concentration of the labeled standards. It guarantees optimum sensitivity and quantitative precision at lowest analyte levels. A similar differential acquisition time setup is available with triple quadrupole GC-MS/MS instruments.

2.3.4.4 MS/MS – Tandem Mass Spectrometry

Can atomic particles be stored in a cage without material walls?

This question is already quite old. The physicist Lichtenberg from Göttingen wrote in his notebook at the end of the 18th century: "I think it is a sad situation that in the whole area of chemistry we cannot freely suspend the individual components of matter." This situation lasted until 1953. At that time we succeeded, in Bonn, in freely suspending electrically charged atoms, i. e. ions, and electrons using high frequency electric fields, so-called multipole fields. We called such an arrangement an ion cage.

Prof. Wolfgang Paul (in a lecture at the Cologne Lindenthal Institute in 1991).

From: Wolfgang Paul, A Cage for Atomic Particles – a basis for precision measurements in navigation, geophysics and chemistry, Frankfurter Allgemeine, Zeitung, Wednesday 15th December 1993 (291) N4.

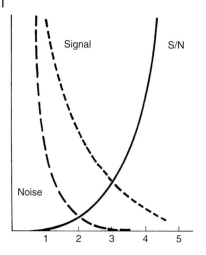

Figure 2.233 Relationship between signal, noise and the number of analytical steps (after Cooks and Busch, 1982).

As part of the further development of instrumental techniques in mass spectrometry MS/MS analysis has become the method of choice for trace analysis in complex matrices. Most of the current applications involve the determination of substances in the parts per billion and parts per trillion ranges in samples of urine, blood and animal or plant tissues, food or in many environmental analyses. In addition to trace level quantitation, the determination of molecular structures is an important area of application for GC-MS/MS analysis (McLafferty, 1983).

As an analytical background to the use of the GC-MS/MS technique in residue analysis, it should be noted that the S/N ratio increases with the number of analytical steps (Figure 2.233). Clean-up steps lower the potential signal intensity. The sequence of wet chemical or instrumental sample preparation steps can easily lead to the situation whereby, as a consequence of processing losses, a substance can no longer be detected. From this consideration, the first separation step (MS1) in a GC-MS/MS system can be regarded as a mass-specific clean-up in the analysis of extracts with large quantities of matrix. After the subsequent induced fragmentation of selected ions, an analyte is identified using the characteristic mass spectrum of the product ions, or it is quantified using the structure selective fragment ions for target compound determination in difficult matrices.

Soft ionization techniques instead of EI are the preferred ionization processes for MS/MS analysis. Although fragmentation in the EI mode of GC-MS is desirable for substance identification, frequently, only low selectivity and sensitivity are achieved in complex matrices. Soft ionization techniques, such as CI, concentrate the ion flux to a few intense ions which can form a good starting point for MS/MS

When is MS/MS Used?

In Quantitation:

- The sample matrix contributes significantly to the chemical background noise in SIM.
- Low S/N values, difficult peak integration and low-precision data in SIM.
- Quantitation with the highest sensitivity in difficult matrices is necessary.
- The SIM analysis requires additional confirmation.
- Co-elution with isobaric impurities occurs.
- Dirty matrix samples of multi methods such as QuEChERS.

In Qualitative Analysis:

- Unknown compounds need to be identified.
- The structure of the compound needs to be determined.
- Product ion spectra are used for structure elucidation.

MS/MS Scan Techniques

Scan mode in MS$_1$	MS$_2$	Result	Application
Single ion	Scan	Product ion spectrum (MS/MS spectrum)	Identification and confirmation of compounds, structure determination, create libraries of product ion spectra for comparison.
Single ion	Single ion	Individual intensities of product ions (SRM/MRM)	Structure selective, highly selective and highly sensitive target compound quantitation with complex matrices, for example, pesticides.
Scan	Single ion	Precursor masses of certain fragments (precursor ion scan)	Specific analysis of compounds (classes of substance) with common structural features, screening, for example, crude oil biomarker.
Scan	Scan-NL	Precursor ions, which undergo loss of neutral particles with NL Da (neutral loss scan)	Specific analysis of compounds (classes of substance) with common functional groups/structural features, for example, loss of COCl from PCDD/PCDF, or CO_2 from acids.

Table 2.56 Information content of mass spectroscopic techniques.

Technique	P	Factor
MS[a]	$1.2 \cdot 10^4$	0.002
Packed GC-MS[b]	$7.8 \cdot 10^5$	0.12
Capillary GC-MS[c]	$6.6 \cdot 10^6$	1
MS/MS	$1.2 \cdot 10^7$	2
Packed GC-MS/MS	$7.8 \cdot 10^8$	118
Capillary GC-MS/MS	$6.6 \cdot 10^9$	1000

a) MS: 1000 Da, unit mass resolution, maximum intensity 2^{12}.
b) Packed GC: $2 \cdot 10^3$ theoretical plates, 30 min separating time.
c) Capillary GC: $1 \cdot 10^5$ theoretical plates, 60 min separating time.
(Yost, 1983; Kaiser, 1978).

analysis. For this reason, the HPLC coupling techniques atmospheric pressure chemical ionization (APCI) and electrospray (ESI) are also in the forefront of the development and extension of MS/MS analysis. The use of the GC-MS/MS technique with PCI and NCI ionization can provide specific advantages for trace analysis. In case the EI ionisation is indispensable for multi-methods like pesticides an optimized analyzer tuning in favor of the high masses is performed (high mass tune).

The information content of the GC-MS/MS technique was already evaluated in 1983 by Richard A. Yost (University of Florida, co-developer of the MS/MS technique). On the basis of the theoretical task of detecting one of the five million substances catalogued at that time by the Chemical Abstracts Service, a minimum information content of 23 bits ($\log_2(5 \times 10^6)$) was required for the result of the chosen analysis procedure, and the MS procedures available were evaluated accordingly. The calculation showed that capillary GC-MS/MS can give 1000 times more information than the traditional GC-MS methods (Table 2.56)!

For the instrumental technique required for tandem MS, as in the consideration of the resolving power, the main differences lie between the performances of the magnetic sector and quadrupole ion trap instruments (see Section 2.3.2). For coupling with GC and HPLC, triple-quadrupole instruments have been used since the 1980s and are the dominating technology today. Ion trap MS/MS instruments have been used in research since the middle of the 1980s and are now being used in routine residue analysis as well. Tandem magnetic sector instruments are mainly used in research and development for MS (Busch *et al.*, 1988). Much higher energies (kiloelectron volt range) can be used to induce fragmentation. For the selection of precursor ions and/or the detection of the product ion spectrum, high-resolution capabilities can be used.

MS/MS Tandem Mass Spectrometry

- Ionization of the sample (EI, CI and other methods).
- Selection of a precursor ion.
- Collision-induced dissociation (CID) to product ions.
- Optimization of the collision energy.
- Mass analysis of product ions (product ion scan).
- Detection as a complete product ion spectrum (full scan), or preselected individual masses (SRM/MRM).

Ion trap MS offers a new extension to the instrumentation used in tandem MS. The methods for carrying out MS/MS analyses differ significantly from those involving triple-quadrupole instruments and reflect the mode of operation as storage mass spectrometers rather than beam instruments (March and Hughes, 1989; Plomley *et al.*, 1994).

Tandem MS consists of several consecutive processes. The GC peak with all the components (analytes, co-eluents, matrix, column bleed, etc.) reaching the ion source, all substances get ionized. From the resulting mixture of ions, the precursor ion with a particular m/z value is selected in the first step (MS_1). This ion can in principle be formed from different molecules (structures) and, even with different empirical formulae, they can be of the same nominal m/z value but different structure. The MS_1 step is identical to SIM analysis in single quadrupole GC-MS, which, for the reasons mentioned above, cannot rule out false positive signals. Fragmentation of the precursor ions to product ions occurs in a collision cell behind MS_1 through collisions of the selected ions with neutral gas molecules (CID or CAD, collision induced/activated decomposition). In this collision process, the kinetic energy of the precursor ions (ion trap: typically < 6 eV, quadrupole: up to 100 eV, magnetic sector: >1 keV) is converted into internal energy, which leads to a structure-specific fragmentation by cleavage or rearrangement of bonds, and the loss of neutral particles. A mixture of product ions with lower m/z values is formed. Following this fragmentation step, a second mass spectrometric separation (MS_2) is necessary for the mass analysis of the product ions. The spectrum of the product ions is finally detected as the product ion or MS/MS spectrum of this compound resp. precursor ion. In case of a targeted SRM/MRM analysis MS_2 selects the structure specific product ions for selective detection.

With regard to existing analysis procedures, for target compound quantitation, high speed and flexibility in the choice of precursor ions is necessary. For the analysis with internal standards (e.g., deuterated standards) and multi-component methods, the fast switch between multiple precursor ions is necessary to achieve a sufficiently high data rate for a reliable peak definition of all coeluting compounds, in particular for SRM quantitations.

Besides recording a spectrum for the product ions (product ion scan), or of an individual mass (SRM), two additional MS/MS scan techniques give valuable analysis data. By linking the scans in MS_2 and MS_1, very specific and targeted analysis routes are possible. In the precursor ion scan, the first mass analyser (MS_1) is scanned over a preselected mass range. All the ions in this mass range reach the collision chamber and form product ions (CID). The second mass analyser (MS_2) is held constant for a specific fragment mass. Only emerging ions of MS_1 which form the selected fragment are recorded. This recording technique allows the identification of substances of related structure, which lead to common fragments in the mass spectrometer (e.g., biomarkers in crude oil characterization, drug metabolites) (Noble, 1995).

In neutral loss scan, all precursor ions, which lose a particular neutral particle (that otherwise cannot be detected in MS), are detected. Both mass analysers scan, but with a constant selected mass difference, which corresponds to the mass of the neutral particle lost. This analysis technique is particularly meaningful if molecules contain the same functional groups (e.g., metabolites as acids, glucuronides or sulfates). In this way, it is possible to identify the starting ions which are characterized by the loss of a common structural element. Both MS/MS scan techniques can be used for substance-class-specific detection in triple quadrupole systems. Ion trap systems allow the mapping of these processes by linking the scans between separate stages of MS in time.

Mode of Operation of Tandem Mass Spectrometers

In tandem MS using quadrupole or magnetic sector instruments, the various steps take place in different locations in the beam path of the instrument (Yost, 1983). The term *tandem-in-space* (R. Yost) has been coined to show how they differ from ion storage mass spectrometers (Figure 2.234). In the ion trap analyser, a typical

Tandem-in-space

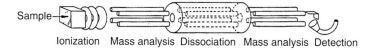

Sample → Ionization   Mass analysis   Dissociation   Mass analysis   Detection

Tandem-in-time

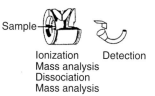

Sample → Ionization   Detection
Mass analysis
Dissociation
Mass analysis

Figure 2.234 GC-MS/MS techniques. Tandem-in-space: triple-quadrupole technique and tandem-in-time: ion trap technique.

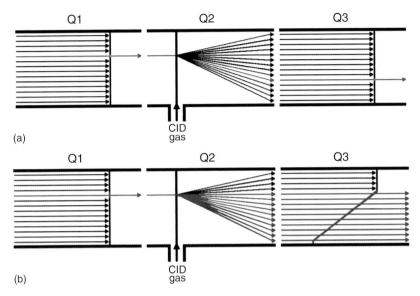

Figure 2.235 Modes of operation of a triple quadrupole mass spectrometer. Q_1 = first mass separating quad, mass selection of the precursor ion. Q_2 = collision cell, CID process. Q_3 = second mass separating quad, SRM detection (a) or product ion scan (b).

storage mass spectrometer, these processes take place in the same location, but consecutively. Richard Yost has described this as *tandem-in-time*.

In a triple-quadrupole mass spectrometer (or other beam instruments, Figure 2.235), an ion beam is passed continuously through the analyser from the ion source. The selection of precursor ions and CID takes place in dedicated devices of the analyser (Q_1, Q_2) independent of time. The only time-dependent process is the mass scan in the second mass analyser (Q_3, MS_2) for the recording of the product ion spectrum (Figure 2.236). As collision cell an enclosed quadrupole device (Q_2) typically consisting of square rods is used. Hexapole or octapole rod systems have been employed also in the past. The collision cell is operated in an RF only made without mass seperating but ion focussing capabilities. (The mass filters MS_1 (Q_1) and MS_2 (Q_3) are operated at a constant RF/DC ratio!). Helium, nitrogen, argon or xenon at pressures of about $10^{-3} - 10^{-4}$ Torr are used as collision gases (Johnson and Yost, 1985). Heavier collision gases increase the yield of product ions. The collision energies are typically in the range of 5 to 50 eV, and are controlled by a preceding lens stack accelerating the precursor ions to the set energy level (see Figure 2.236).

In instruments with ion trap analysers, the selection of the precursor ions, the CID and the analysis of the product ions occur in the same place, but with time control via a sequence of frequency and voltage values at the end caps and ring electrode of the analyser (scan function see Figure 2.237) (Soni and Cooks, 1994). The systems with internal or external ionization differ in complexity. In the case of internal ionization, the sample spectrum is first produced in the ion trap analyser

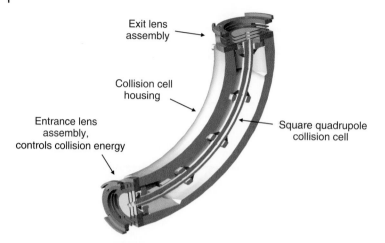

Figure 2.236 Curved collision cell of a triple quadrupole system (90°) for reduction of non-specific noise by eliminating neutral particles and photons from the ion flight path. (TSQ Quantum series, courtesy Thermo Fisher Scientific.)

and stored. The precursor ion m_p is then selected by ejecting ions above and below m_p from the trap by applying a multi-frequency signal at the end caps (waveform). Ion trap systems with external ionization employ waveforms during injection of the ions into the analyser from the external source for isolation of the precursor ion in MS/MS mode or for isolation of the desired mass scan range in other scan modes. In this way, a longer storage phase and a more rapid scan rate can be used for GC-MS in the MS/MS and SIM modes.

CID is initiated by an additional AC voltage at the end cap electrodes of the ion trap analyser. If the frequency of the AC voltage corresponds to the secular frequency of the selected ions, there is an uptake of kinetic energy by resonance. Fragmentation is effected by multiple collisions with present helium buffer gas building up the required energy level in the precursor ion. The collision energy is controlled by the level of the applied AC voltage. In the collisions, the kinetic energy is converted into internal energy and used up in bond cleavage. The product ions formed are stored in this phase at low RF values of the ring electrode and are detected at the end of the CID phase by a regular mass scan. The time required for these processes is in the lower ms range and, at a high scan rate for the GC, allows the separate monitoring of the deuterated internal standard at the same time. For the efficiency of the tandem-in-time MS/MS process the reliable choice of the frequencies for precursor ion selection and excitation is of particular importance. Instruments with external ion sources permit a calibration of the frequency scale which is generally carried out during the automatic tuning. In this way, the MS/MS operation is analogous to that of the SIM mode in that only the masses of the desired precursor ions m_p need to be known. In the case of instruments with internal ionization, where the GC carrier can cause pressure fluctuations within the analyser, the empirical determination of excitation frequencies and the broad band excitation of a mass window is necessary.

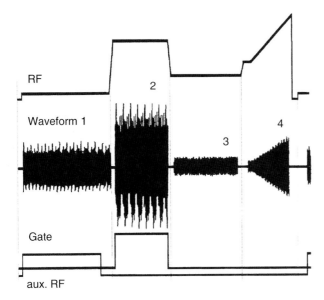

Figure 2.237 MS/MS scan function of the ion trap analyser with external ionization.
1. A gate switches the ion beam for transfer to the analyser, selection of the precursor ion m_p by a special frequency spectrum (ionization waveform), variable ion injection time up to maximum use of the storage capacity.
2. One or more m/z values are then isolated using a synthesized frequency spectrum (isolation waveform). This phase corresponds to the SIM mode.
3. Collision-induced dissociation (CID) by selective excitation to the secular frequency (activation waveform) of the precursor ion, storage of the product ions formed at a low RF value without exciting them further.
4. Product ion scan at a scan rate of about 5500 Da/s (resonance ejection waveform), detection of the ions by the multiplier.

The efficiency of the CID process is of critical importance for the use of GC-MS/MS in residue analysis. In beam instruments, optimization of the collision energy (via the entry lens potential of Q2) and the collision gas pressure is necessary. The optimization is limited by scattering effects at high chamber pressures and subsequent fragmentation of product ions. Modern collision cells are manufactured from square quadrupole rods which provide superior ion transmission characteristics even at extended length compared to former quadrupole or hexapole cells eliminating the formerly observed loss of ions due to the collision processes (see Figure 2.235). Typical collision energies for SRM quantitations are in the range of 5 to 50 eV. Cross-talk from different precursors to the same product ion is effectively eliminated by an active cleaning step between the scans (adds to the interscan time). Encapsulated collision cells using square quadrupoles are typically operated at up to 4 mbar Ar or N_2 pressure for highly sensitive target compound residue analysis.

With the ion trap analyser the energy absorbed by the ions depends on the duration of the resonance conditions for the absorption of kinetic energy and on the voltage level. Typical values for the induction phase are 3–15 ms and 500–1000 mV. With the ion trap technique there is higher efficiency in the fragmentation and transmission to product ions (Johnson and Yost, 1990). In addition, the ion trap technique has clear advantages because of the sensitivity resulting from the storage technique, which allows recording of complete product ion spectra even below the picogram range (Figure 2.238).

Latest developments in ion trap technology include the automated determination of collision energies (ACEs). This technique facilitates the set up MS/MS

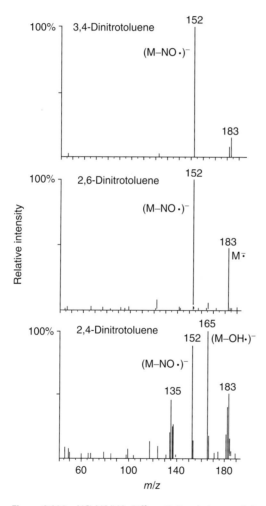

Figure 2.238 NCI-MS/MS: Differentiation between dinitrotoluene isomers by comparison of the product ion spectra (ion trap analyser, reagent gas water and detection of negative ions) (Brodbelt and Cooks, 1988).

methods by replacing the tedious manual optimization of the collision energy for every compound of a large set of targets. The program uses an empirical calibration scheme to calculate an optimum collision energy comprising all of the known instrument parameters as precursor mass and damping gas flow. Three collision energy values can be provided to check results. The middle one typically is the optimum calculated from the instrument parameters. The found levels cover the range of optimum collision energies for compounds of varying ion chemistry. In practice, a very good correlation with manual optimization, for example, for pesticides is achieved.

Low mass MS/MS fragments, that are cut from the spectrum by the former regular ion trap operation are stored and scanned for MS/MS spectra using the pulsed q-value dissociation technique (PQD). Normal CID in ion traps is done at a particular excitation q value defining the effective collision energy. Higher q values provide higher-collision energy values by getting the precursor ions moving faster but cut the low mass end from the spectrum. PQD works by first exciting the precursor ion at high q value for a short period of time (50–100 µs) using a high-collision energy as determined by ACE. During the induced fragmentation, the q value is lowered significantly to store the full range of product ions. The result is that also low m/z fragments ions are being trapped. Therefore, PQD is primarily useful as a qualitative tool for structure elucidation, but also finds utility in quantitative applications where mainly and intense low m/z fragment ions are observed as known for many N containing compounds, for example, amines. PQD is a patented technique (US 6 949 743 and 7 102 129) that was first used on the ion traps from Thermo Fisher Scientific, San Jose, CA, USA.

Today comparable results are available for both the triple-quadrupole and ion trap analysers. It is typical for the product ion spectra obtained with ion trap mass spectrometers that the intensities of the precursor ions are significantly reduced due to efficient CID. The product ion spectra show only a few well-defined but intense product ions derived directly from the precursor. Further fragmentation of product ions usually does not occur as the m/z of the precursor ion is excited exclusively. Triple quadrupole instruments show less efficiency in the CID process resulting in a higher precursor ion signal. Primary product ions, fragment consecutively in the collision cell and produce additional signals in the product ion spectrum. The efficiency of ion trap instruments is also reported to be higher than that of magnetic sector instruments. The mass range and the scan techniques of precursor scan and neutral loss scan make triple-quadrupole instruments suitable for applications outside the range of pure GC-MS use.

When evaluating MS/MS spectra, it should be noted that no isotope intensities appear in the product ion spectrum (independent of the type of analyser used). During selection of the precursor ion for the CID process, naturally occurring isotope mixtures are separated and isolated. The formation of product ions is usually achieved by the loss of common neutral species. The interpretation of these spectra is generally straightforward and less complex compared with EI spectra. When comparing product ion spectra of different instruments the acquisition parameter used must be taken into account. In particular with beam instruments, the recording parameters can be reflected in the relative intensities of the spectra. On the other hand, rearrangements and isomerizations are possible, which can lead to the same product ion spectra.

Structure Selective Detection Using MS/MS Transitions

Triple quadrupole as well as ion trap mass spectrometers besides structure elucidation find their major application in quantitation of target compounds in difficult matrices. The increased selectivity when observing specific transitions from a precursor ion to a structural related product ion provides highly confident analyses with excellent LOQs even in matrix samples. In the SRM mode, an intense precursor ion from the spectrum of the target compound is selected in the first quadrupole Q_1, fragmented in the collision cell and monitored on a selective product ion for quantitation, see the principle given graphically in Figure 2.236. The high selectivity of the SRM method is controlled by the mass resolution of the first quadrupole Q_1. While round rod quadrupoles are working at unit mass resolution, hyperbolic quadrupoles typically are set to 0.7 Da peak width at FWHM as standard and are operated in the highly resolved H-SRM mode for maximum selectivity even at narrow peak widths of 0.4 Da FWHM, also refer to Section 2.3.2.

Analogue to the SIM analysis mode used with single quadrupole instrumentation the SRM mode omits full spectral information for substance confirmation. A structure selective characteristic of the assay is given by the mass difference of the transition monitored. Typically, two independent transitions together with the chromatographic retention time provide the positive confirmation of the occurrence of a particular compound (Council Directive 96/23/EC). State-of-the-art triple quadrupole instrumentation provide transition times as low as 0.5 – 2 ms offering the potential for screening of a large number of compounds in a given chromatographic window, for example, for multi-component pesticide analysis (see Section 4.9).

Setting up MRM Methods for Quantitation

The MS/MS operation for target compound quantitation can be a very time-consuming process if a large number of compounds is involved. Each compound needs to be treated individually to define the optimum detection conditions. The MS/MS analyser runs in the SRM mode and is capable of screening a large number of transitions in one run. For each compound, one or more specific precursor ions are selected in the first quadrupole (Q1) and fragmented in the collision cell (Q2) using the optimum collision energy. Finally, from the dissociation products of the CID process the structure specific product ion is detected with the third quadrupole (Q3). The SRM MS/MS process is graphically displayed in Figure 2.239. By switching the transitions controlled by retention time an MRM method is created. Using the timed-SRM mode, each of the programmed SRM transitions is activated only in a short time window around the compound retention time. This way, complex MRM methods, for example, for pesticides analysis screening of many hundreds of target compounds can be established.

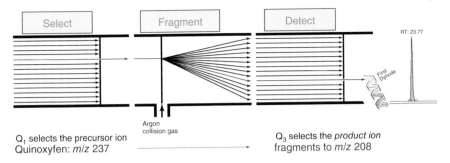

Figure 2.239 SRM process for MS/MS target compound detection, shown quinoxyfen with precursor *m/z* 237 fragmented into the detected product ion *m/z* 208.

The setup of MRM methods requires the following steps for each compound, typically analysing one or more standard mixes:

1) Get the compound retention time (first run).
2) Define the number of SRM transitions per compound
    – used for quantitation
    – and ion ratio confirmation.
3) Get the full scan spectrum from first run.
    – Note the most intense ions as precursor ions.
    – one precursor for each planned SRM transition.
4) Get the product ion spectra from each precursor ion (second run).
    – Decide on argon or nitrogen as collision gas.
    – Use a medium collision energy.
    – Note the most intense product ion for quantitation.
    – Note the other intense product ions for ion ratio confirmation.
5) Optimize the collision energies for the selected SRMs (third and more runs).
    – Start at 5 V and ramp up to 50 V in steps of 5 V (one injection per voltage setting is required if the instrument cannot change the energy during one scan), see Figure 2.240.
    – Use the collision energy of the highest product ion intensity for the method.

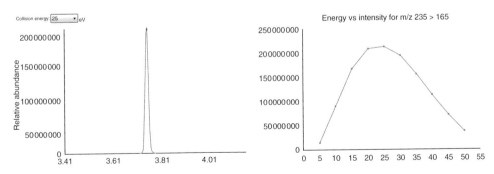

Figure 2.240 AutoSRM optimization of the collision energy between 5 and 50 V with optimum 25 V.

This very laborious procedure is necessary only once for the initial setup, or if new compounds need to be added. In many cases, already proven compound data bases are available by several GC-MS/MS vendors easing up the task, for example, for pesticides with many hundreds of the most analysed compounds. Another approach is the AutoSRM procedure launched by Thermo Fisher Scientific, which organizes the above steps automatically and comes up with the ready-to-use acquisition method. Optionally, the acquired spectra can be added to libraries for additional use.

Data Dependant Data Acquisition

The advanced electronic capabilities of modern triple quadrupole instruments offer additional features for delivering compound-specific information even during SRM quantitation analyses. Depending on the quality of a scan currently being acquired, the mode of the data acquisition for a following scan (data point) can be switched to acquire a full product ion spectrum. The level of product ion intensity that triggers the switch of acquisition mode is set method and compound specific. As a result, one data point in the substance peak is used to generate and acquire the compound product ion spectrum (see Figure 2.241). This spectrum provides the full structural information about the detected substance and is available for library search and compound confirmation. Using the data dependent acquisition mode, the final data file contains both quantitative as well as qualitative information.

For the generation of the MS/MS product ion spectrum, the applied collision energy has substantial impact on the information content of the spectrum, represented by the occurrence and intensities of fragment and precursor ions. The maximum information on the compound structure is achieved by variation of the collision energy during the product ion scan. High-collision energies lead to the generation of low mass fragments, while lower-collision energies provide the favoured main fragments and still some visible intensity of the precursor ion. The

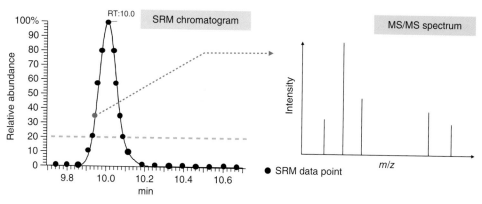

Figure 2.241 Data dependent data acquisition scheme. The first SRM scan intensity above a user defined threshold, (a) dotted line, is acquired as MS/MS product ion spectrum (b).

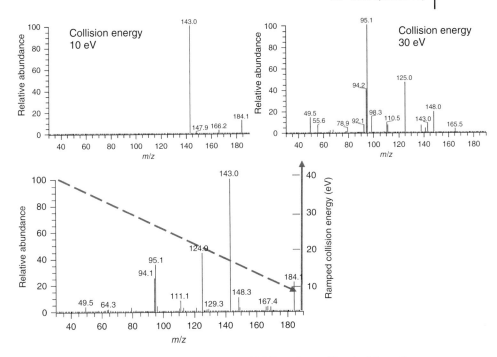

Figure 2.242 Decreasing collision energy ramp in MS/MS mode provides the information rich product ion spectrum.

instrumental approach to this solution is the variation of the collision energy during the scan from a high-energy to a low-energy level by ramping down the ion acceleration voltage using the lens stack located at the entrance of the collision cell (see Figure 2.235). The resulting product ion spectrum consequently delivers the full available information about the structural characteristics of an unknown or the compound to be confirmed (see Figure 2.242).

2.3.5
Mass Calibration

To operate a GC-MS system, calibration of the mass scale is necessary. The calibration converts the voltage or time values controlling the analyser into m/z values. For the calibration of the mass scale, a mass spectrum of a known chemical compound is used, where both the fragments (m/z values) and their intensities are known and stored in the data system in the form of a reference table.

With modern GC-MS systems, performing an up-to-date mass calibration is generally the final process in a tuning or autotune program. This is preceded by a series of necessary adjustments and optimizations of the ion source, beam focussing, and mass resolution settings, which affect the position of the signals on the mass scale. Tuning the lens potential particularly affects the transmission in individual mass areas. In particular, with beam instruments focussing must be adapted to the intensities of the reference substances in order to obtain the intensity pattern of the reference spectrum. The *m/z* values contained in a stored reference table are identified by the calibration program in the spectrum of the reference compound measured. The relevant centroid of the reference peak is calculated and correlated with the operation of the analyser. Using the stored reference table, a precise calibration function for the whole mass range of the instrument can be calculated. The actual state of the mass spectrometer at the end of the tuning procedure is thus taken into account. The data are plotted graphically and are available to the user for assessment and documentation. For quadrupole and ion trap instruments, the calibration graph is linear (Figure 2.243), whereas with magnetic sector instruments the graph is exponential (Figure 2.244).

Perfluorinated compounds are usually used as calibration standards (Table 2.57). Despit of their high molecular weights, the volatility of these compounds is sufficient to allow a controllable leak current into the ion source. In addition, fluorine has a negative mass defect ($^{19}F = 18.9984022$), so that the fragments of these standards are below the corresponding nominal mass and can easily be separated from a possible background of hydrocarbons with positive mass defects. The requirements of the reference substance are determined by the type of analyser. Quadrupole and ion trap instruments are calibrated with FC43 (PFTBA; Tables 2.58 and 2.59 and Figure 2.245) independent of their available mass range.

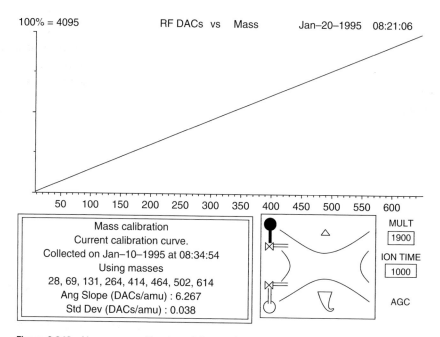

Figure 2.243 Linear mass calibration of the quadrupol ion trap analyser (voltage of ring electrode against *m/z* value).

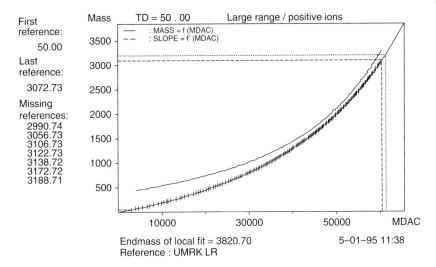

First reference:
50.00

Last reference:
3072.73

Missing references:
2990.74
3056.73
3106.73
3122.73
3138.72
3172.72
3188.71

Mass

TD = 50.00   Large range / positive ions

— : MASS = f (MDAC)
-- : SLOPE = f' (MDAC)

Endmass of local fit = 3820.70
Reference : UMRK LR

5–01–95 11:38

MDAC

Figure 2.244 Exponential mass calibration of the magnetic sector analyser.

Table 2.57 Calibration substances and their areas of use.

| Name | Formula | M | m/z max. | Instrument used | |
				Magnetic sector m/z	Quadrupole/Ion trap m/z
FC43[a]	$C_{12}F_{27}N$	671	614	< 620	> 1000
FC5311	$C_{14}F_{24}$	624	624	< 620	> 1000
PFK[b]	—	—	1017	< 1000	> 1000
(C_7)[c]	$C_{24}F_{45}N_3$	1185	1185	800–1200	—
(C_9)[d]	$C_{30}F_{57}N_3$	1485	1485	800–1500	—
Fomblin[e]	$(OCF(CF_3)CF_2)_x-(OCF_2)_y$	—	—	< 2500	—
Ultramark 1621[f]	$C_{40}H_{19}O_6N_3P_3F_{68}$	2120	2120	< 2120	—
CsI[g]	—	—	15 981	high mass range	—

a) Perfluorotributylamine.
b) Perfluorokerosene.
c) Perfluorotriheptylazine.
d) Perfluorotrinonyltriazine.
e) Poly(perfluoropropylene oxide) (also used as diffusion pump oil).
f) Fluorinated phosphazine.
g) Cesium iodide.

For the compliance of a series of EPA methods and other regulations, the targeted source tuning according to the specific manufacturers guidelines is required, see Table 2.60 (Eichelberger, Harris and Budde, 1975). This, for instance, refers to *EPA methods* 501.3, 524.2, 8260B, CLP-SOW for the determination of volatiles using BFB and the EPA methods 625, 1625, 8250, 8270 on base/neutrals/acids or

Table 2.58 Reference table FC43/PFTBA (EI, intensities >1%, quadrupole instrument).

Exact mass (Da)	Intensity (%)	Formula
68.9947	100.0	CF_3^+
92.9947	2.0	$C_3F_3^+$
99.9931	19.0	$C_2F_4^+$
113.9961	11.0	$C_2NF_4^+$
118.9915	16.0	$C_2F_5^+$
130.9915	72.0	$C_3F_5^+$
149.9899	4.0	$C_3F_6^+$
168.9883	7.0	$C_3F_7^+$
175.9929	3.0	$C_4NF_6^+$
180.9883	3.0	$C_4F_7^+$
213.9898	2.0	$C_4NF_8^+$
218.9851	78.0	$C_4F_9^+$
225.9898	2.0	$C_5NF_8^+$
263.9866	27.0	$C_5NF_{10}^+$
313.9834	3.0	$C_6NF_{12}^+$
351.9802	4.0	$C_6NF_{14}^+$
363.9802	1.0	$C_7NF_{14}^+$
375.9802	1.0	$C_8NF_{14}^+$
401.9770	2.0	$C_7NF_{16}^+$
413.9770	9.0	$C_8NF_{16}^+$
425.9770	1.0	$C_9NF_{16}^+$
463.9738	3.0	$C_9NF_{18}^+$
501.9706	8.0	$C_9NF_{20}^+$
613.9642	2.0	$C_{12}NF_{24}^+$

semi-volatiles referring to DFTPP as tuning compounds providing consistent and instrument independent ion ratio profiles for quantitation.

The precision of the mass scan and the linearity of the calibration allow the line to be extrapolated beyond the highest fragment which can be determined, which is m/z 614 for FC43. For magnetic sector instruments, the use of PFK (perfluorokerosene; Table 2.61, Figure 2.246) has proved successful besides FC43. It is particularly suitable for magnetic field calibration as it gives signals at regular intervals up to over m/z 1000. Also, perfluorophenanthrene (FC5311) is frequently used as an alternative calibration compound to FC43, which covers about the same mass range (see Table 2.62). Higher mass ranges can be calibrated using perfluorinated alkyltriazines, Ultramark or cesium iodide. Reference tables frequently also take account of the masses from the lower mass range, such as He, N_2, O_2, Ar or CO_2, which always form part of the background spectrum.

The calculated reference masses in the given tables are based on the following values for isotopic masses: ^1H 1.0078250321 Da, ^4He 4.0026032497 Da, ^{12}C 12.0000000000 Da, ^{14}N 14.0030740052 Da, ^{16}O 15.9949146221 Da, ^{19}F 18.9984032000 Da and ^{40}Ar 39.9623831230 Da. The mass of the electron

Table 2.59 Reference table FC43/PFTBA (EI, intensities >0.1%, high resolution magnetic sector instrument).

Exact mass (Da)	Intensity (%)	Formula
4.00206	—	He^+
14.01510	—	CH_2^+
18.01002	—	H_2O^+
28.00560	—	N_2^+
30.99786	—	CF
31.98928	—	O_2^+
39.96184	—	Ar^+
43.98928	—	CO_2^+
49.99626	3.0	CF_2^+
68.99466	50.7	CF_3^+
75.99933	1.0	$C_2NF_2^+$
80.99466	1.2	$C_2F_3^+$
92.99466	1.0	$C_3F_3^+$
99.99306	3.9	$C_2F_4^+$
113.99614	3.5	$C_2NF_4^+$
118.99147	5.9	$C_2F_5^+$
130.99147	49.8	$C_3F_5^+$
149.98987	2.1	$C_3F_6^+$
168.98827	5.1	$C_3F_7^+$
175.99295	1.3	$C_4NF_6^+$
180.98827	1.3	$C_4F_7^+$
199.98668	0.2	$C_4F_8^+$
213.98975	2.5	$C_4NF_8^+$
218.98508	100.0	$C_4F_9^+$
225.98975	1.4	$C_5NF_8^+$
230.98508	1.1	$C_5F_9^+$
242.98508	0.0	$C_6F_9^+$
263.98656	37.6	$C_5NF_{10}^+$
275.98656	0.2	$C_6NF_{10}^+$
280.98189	0.1	$C_6F_{11}^+$
294.98496	0.2	$C_6NF_{11}^+$
313.98336	2.7	$C_6NF_{12}^+$
325.98336	0.3	$C_7NF_{12}^+$
344.98177	0.0	$C_7NF_{13}^+$
351.98017	1.7	$C_6NF_{14}^+$
363.98017	1.0	$C_7NF_{14}^+$
375.98017	1.2	$C_8NF_{14}^+$
401.97698	2.0	$C_7NF_{16}^+$
413.97698	10.8	$C_8NF_{16}^+$
425.97698	4.0	$C_9NF_{16}^+$
451.97378	0.6	$C_8NF_{18}^+$
463.97378	7.3	$C_9NF_{18}^+$
475.97378	0.0	$C_{10}NF_{18}^+$
501.97059	22.7	$C_9NF_{20}^+$
513.97059	0.1	$C_{10}NF_{20}^+$

(*continued overleaf*)

Table 2.59 *(Continued)*

Exact mass (Da)	Intensity (%)	Formula
525.97059	0.1	$C_{11}NF_{20}^{+}$
551.96740	0.0	$C_{10}NF_{22}^{+}$
563.96740	0.1	$C_{11}NF_{22}^{+}$
575.96740	1.3	$C_{12}NF_{22}^{+}$
613.96420	3.6	$C_{12}NF_{24}^{+}$

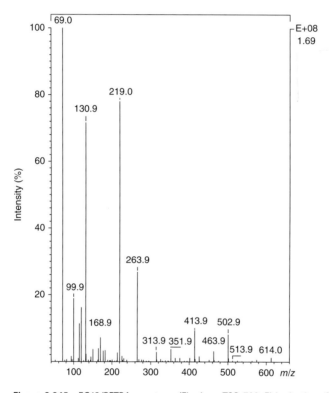

Figure 2.245   FC43/PFTBA spectrum (Finnigan TSQ 700, EI ionization, Q3).

0.00054857991 Da was taken into account for the calculation of the ionic masses (Audi and Wapstra, 1995; Mohr and Taylor, 1999).

To record high-resolution data for manual work, peak matching is employed (Webb, 2004a). At a given magnetic field strength, one or two reference peaks and an ion of the substance being analysed are alternately shown on a screen. By changing the acceleration voltage (and the electric field coupled to it), the peaks are superimposed. From the known mass and voltage difference, the exact mass of the substance peak is determined. This process is an area of the solid probe technique, as here the substance signal can be held constant over a longer period.

Table 2.60 Ion abundance criteria for the US EPA BFB and DFTPP target tuning.

Compound	m/z	Ion abundance criteria (relative abundance)	
BFB	50	15–40%	of mass 95
	75	30–60%[a]	of mass 95
	95	100%	Base peak
	96	5–9%	of mass 95
	173	<2%	of mass 174
	174	>50%	of mass 95
	175	5–9%	of mass 174
	176	>95% but <101%	of mass 174
	177	5–9%	of mass 176
DFTPP	51	30–60%	of mass 198
	68	<2%	of mass 69
	70	<2%	of mass 69
	127	40–60%	of mass 198
	197	<1%	of mass 198
	198	100%	Base peak
	199	5–9%	of mass 198
	275	10–30%	of mass 198
	365	>1%	of mass 198
	441	present	of mass 443
	442	>40%	of mass 198
	443	17–23%	of mass 442

a) Thirty to eighty percent of mass 95 for EPA 524.2.

For GC-HRMS (high mass resolution) systems, an internal mass calibration (scan-to-scan) for accurate mass determinations by control of the data system is employed. At a given resolution (e.g., 10 000), a known reference is used which is continuously leaked into the ion source during analysis. The analyser is positioned on the exact mass of the substance ion to be analysed relative to the measured centroid of the known reference. At the beginning of the next scan, the exact position of the centroid of the reference mass is determined again and is used as a new basis for the next scan (see Chapter 2.3.4.3 Lock-Plus-Cali Mass Technique).

The usability of the calibration depends on the type of instrument and can last for a period of up to several days or weeks. All tuning parameters, in particular the adjustment of the ion source, affect the calibration as described above. In particular the analyzer scan speed has a strong impact on the mass calibration with many instruments. Special attention should also be paid to a constant temperature of the ion source. Regular mass calibration using analysis conditions is recommended to comply with the lab internal QA/QC (quality assurance/quality control) procedures.

The carrier gas flow setting of the GC also can show effect on the position of the mass calibration. Ion sources with small volumes and also ion trap instruments with internal ionization show a significant drift of several tenths of a mass unit if the carrier gas flow rate is significantly changed by a temperature program.

Table 2.61 Reference table for PFK (perfluorokerosene), EI ionization high-resolution magnetic sector instrument.

Exact mass (Da)	Intensity (%)	Formula
4.002055	—	He^+
14.015101	—	CH_2^+
18.010016	—	H_2O^+
28.005599	—	N_2^+
30.007855	3.80	CF^+
31.989281	—	O_2^+
39.961835	—	Ar^+
51.004083	6.70	CHF_2^+
68.994661	100.00	CF_3^+
80.994661	0.50	$C_2F_3^+$
92.994661	3.30	$C_3F_3^+$
99.993064	5.60	$C_2F_4^+$
113.000889	0.02	$C_3HF_4^+$
118.991467	26.40	$C_2F_5^+$
130.991467	24.00	$C_3F_5^+$
142.991467	1.90	$C_4F_5^+$
154.991467	1.40	$C_5F_5^+$
168.988274	17.00	$C_3F_7^+$
180.988274	8.75	$C_4F_7^+$
192.988274	8.30	$C_5F_7^+$
204.988274	1.50	$C_6F_7^+$
218.985080	8.60	$C_4F_9^+$
230.985080	8.80	$C_5F_9^+$
242.985080	3.80	$C_6F_9^+$
254.985080	1.20	$C_7F_9^+$
268.981887	4.00	$C_5F_{11}^+$
280.981887	6.00	$C_6F_{11}^+$
292.981887	2.70	$C_7F_{11}^+$
304.981887	1.00	$C_8F_{11}^+$
318.978693	2.00	$C_6F_{13}^+$
330.978693	3.70	$C_7F_{13}^+$
342.978693	1.80	$C_8F_{13}^+$
354.978693	0.90	$C_9F_{13}^+$
368.975499	0.80	$C_7F_{15}^+$
380.975499	2.30	$C_8F_{15}^+$
392.975499	1.10	$C_9F_{15}^+$
404.975499	1.00	$C_{10}F_{15}^+$
416.975499	0.55	$C_{11}F_{15}^+$
430.972306	1.85	$C_9F_{17}^+$
442.972306	1.20	$C_{10}F_{17}^+$
454.972306	0.80	$C_{11}F_{17}^+$
466.972306	0.50	$C_{12}F_{17}^+$
480.969112	1.40	$C_{10}F_{19}^+$
492.969112	1.10	$C_{11}F_{19}^+$

Table 2.61 (Continued)

Exact mass (Da)	Intensity (%)	Formula
504.969112	0.65	$C_{12}F_{19}^{+}$
516.969112	0.50	$C_{13}F_{19}^{+}$
530.965919	0.70	$C_{11}F_{21}^{+}$
542.965919	0.60	$C_{12}F_{21}^{+}$
554.965919	0.50	$C_{13}F_{21}^{+}$
566.965919	0.60	$C_{14}F_{21}^{+}$
580.962725	0.70	$C_{12}F_{23}^{+}$
592.962725	0.65	$C_{13}F_{23}^{+}$
604.962725	0.60	$C_{14}F_{23}^{+}$
616.962725	0.50	$C_{15}F_{23}^{+}$
630.959531	0.50	$C_{13}F_{25}^{+}$
642.959531	0.50	$C_{14}F_{25}^{+}$
654.959531	0.55	$C_{15}F_{25}^{+}$
666.959531	0.50	$C_{16}F_{25}^{+}$
680.956338	0.20	$C_{14}F_{27}^{+}$
692.956338	0.25	$C_{15}F_{27}^{+}$
704.956338	0.40	$C_{16}F_{27}^{+}$
716.956338	0.25	$C_{17}F_{27}^{+}$
730.953144	0.20	$C_{15}F_{29}^{+}$
742.953144	0.25	$C_{16}F_{29}^{+}$
754.953144	0.50	$C_{17}F_{29}^{+}$
766.953144	0.20	$C_{18}F_{29}^{+}$
780.949951	0.25	$C_{16}F_{31}^{+}$
792.949951	0.30	$C_{17}F_{31}^{+}$
804.949951	0.15	$C_{18}F_{31}^{+}$
816.949951	0.05	$C_{19}F_{31}^{+}$
830.946757	0.10	$C_{17}F_{33}^{+}$
842.946757	0.10	$C_{18}F_{33}^{+}$
854.946757	0.10	$C_{19}F_{33}^{+}$
866.946757	0.05	$C_{20}F_{33}^{+}$
880.943563	0.10	$C_{18}F_{35}^{+}$
892.943563	0.10	$C_{19}F_{35}^{+}$
904.943563	0.05	$C_{20}F_{35}^{+}$
916.943563	0.05	$C_{21}F_{35}^{+}$
930.940370	0.05	$C_{19}F_{37}^{+}$
942.940370	0.05	$C_{20}F_{37}^{+}$
954.940370	0.05	$C_{21}F_{37}^{+}$
966.940370	0.05	$C_{22}F_{37}^{+}$
980.937176	0.05	$C_{20}F_{39}^{+}$
992.937176	0.05	$C_{21}F_{39}^{+}$
1004.937176	0.05	$C_{22}F_{39}^{+}$
1016.937176	0.05	$C_{23}F_{39}^{+}$

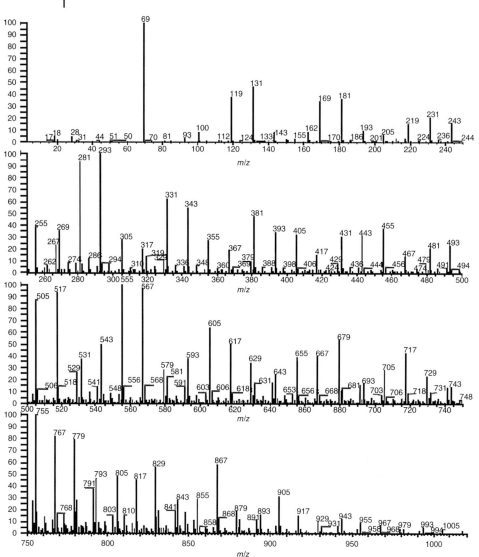

Figure 2.246 PFK spectrum (Finnigan MAT 95, EI ionization).

Calibration at an average elution temperature, the use of an open split or equipping the gas chromatograph with electronic pressure programming (EPC) for analysis at constant flow are imperative in this case. Severe contamination of the ion source or the ion optics also affects the calibration. However, reduced transmission of such an instrument should force cleaning to be carried out in good time.

For analyses with differing scan rates using magnetic instruments, calibrations are carried out at the different rates which are required for the subsequent measurements. For scan rates which differ significantly, a mass drift can otherwise occur between calibration and measurement. Calibrations of

Table 2.62   EI positive ion spectra for perfluorophenanthrene (FC5311).

m/z	Relative abundance (%)
55	1.8
56	1.1
57	3.0
69	100.0
70	1.2
93	6.0
94	1.3
100	11.1
112	1.5
119	16.8
124	1.5
131	47.1
143	5.3
155	3.1
162	5.6
169	7.7
181	11.2
193	6.7
205	2.8
217	2.1
219	2.7
231	4.1
243	7.7
255	2.1
267	2.6
286	2.1
293	7.7
305	1.1
317	2.3
331	1.1
343	1.2
367	1.9
405	4.2
455	18.3
505	4.4
517	1.2
555	4.6
605	1.6
624	1.0

Table 2.63 Reference table for PFK (perfluorokerosene) in NCI mode (CI gas ammonia).

Exact mass (Da)	Intensity (%)	Formula
168.9888	2	$C_3F_7^-$
211.9872	5	$C_5F_8^-$
218.9856	8	$C_4F_9^-$
230.9856	26	$C_5F_9^-$
249.9840	3	$C_5F_{10}^-$
261.9840	40	$C_6F_{10}^-$
280.9824	81	$C_6F_{11}^-$
292.9824	12	$C_7F_{11}^-$
311.9808	51	$C_7F_{12}^-$
330.9792	100	CF_{13}^-
342.9792	32	$C_8F_{13}^-$
361.9770	35	$C_8F_{14}^-$
380.9760	63	$C_8F_{15}^-$
392.9760	48	$C_9F_{15}^-$
411.9744	16	$C_9F_{16}^-$
423.9744	15	$C_{10}F_{16}^-$
430.9728	39	$C_9F_{17}^-$
442.9728	36	$C_{10}F_{17}^-$
454.9728	16	$C_{11}F_{17}^-$
473.9712	11	$C_{11}F_{18}^-$
480.9696	19	$C_{10}F_{19}^-$
492.9696	23	$C_{11}F_{19}^-$
504.9696	18	$C_{12}F_{19}^-$
516.9696	7	$C_{13}F_{19}^-$
523.9680	7	$C_{12}F_{20}^-$
530.9664	11	$C_{11}F_{21}^-$
535.9680	8	$C_{13}F_{20}^-$
542.9664	13	$C_{12}F_{21}^-$
554.9664	21	$C_{13}F_{21}^-$
573.9648	17	$C_{13}F_{22}^-$
585.9648	20	$C_{14}F_{22}^-$
604.9633	31	$C_4F_{23}^-$
611.9617	13	$C_{13}F_{24}^-$
623.9617	42	$C_{14}F_{24}^-$
635.9617	32	$C_{15}F_{24}^-$
654.9601	21	$C_{15}F_{25}^-$
661.9585	13	$C_{14}F_{26}^-$
673.9585	37	$C_{15}F_{26}^-$
685.9585	21	$C_{16}F_{26}^-$
699.9553	7	$C_{14}F_{28}^-$
704.9569	6	$C_{16}F_{27}^-$
711.9553	8	$C_{15}F_{28}^-$
723.9553	12	$C_{16}F_{28}^-$

Table 2.63 (*Continued*)

Exact mass (Da)	Intensity (%)	Formula
735.9553	6	$C_{17}F_{28}^{-}$
750.9601	4	$C_{23}F_{25}^{-}$
761.9521	2	$C_{16}F_{30}^{-}$
773.9521	3	$C_{17}F_{30}^{-}$
787.9489	3	$C_{15}F_{32}^{-}$
799.9489	1	$C_{6}F_{32}^{-}$

ion trap instruments are practically independent of the scan rate. Fast scanning quadrupole instruments consider the different ion flight times of low and high mass ions in the calibration algorithms.

Depending on the type of instrument, a new mass calibration is required if the ionization process is changed. While for ion trap instruments switching from EI to CI ionization is possible without alterations to the analyser, with other types, switching of the ion source or changing the ion volume are required. For an optimized CI reaction, a lower source temperature is frequently used compared with the EI mode. After these changes have been made, a new mass calibration is necessary. This involves running a CI spectrum of the reference substance and applying the CI reference table (Table 2.63). The perfluorinated reference substances can be used for both PCI and NCI (see Figures 2.247 and 2.248). The intensities given

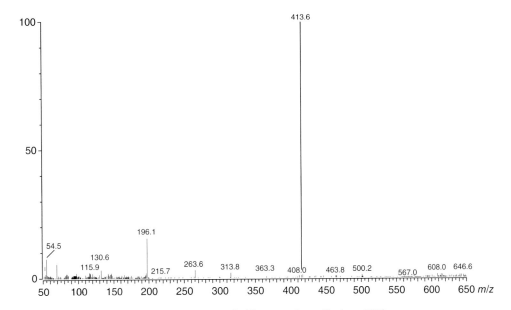

Figure 2.247 FC43/PFTBA spectrum in PCI mode (CI gas methane, Finnigan GCQ).

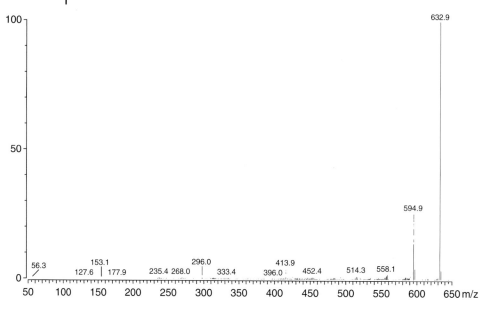

Figure 2.248 FC43/PFTBA spectrum in NCI mode (CI gas methane, Finnigan GCQ).

in the CI reference tables can also be used to optimize the flow adjustment of the reagent gas.

For reasons of quality control, mass calibrations should be carried out regularly and should be documented with a print out (see Figures 2.243 and 2.244).

2.3.6
Vacuum Systems

The carrier gas for GC-MS is either helium or hydrogen. It is well known that the use of hydrogen significantly improves the performance of the GC, lowers the elution temperatures of compounds and permits shorter analysis times because of higher flow rates. The more favourable van Deemter curve for hydrogen (see Figure 2.84) accounts for these improvements in analytical performance. As far as MS is concerned, when hydrogen is used the mass spectrometer requires a higher vacuum capacity and thus a more powerful pumping system. The type of analyser and the pumping system determine the advantages and disadvantages.

For the turbomolecular pumps mostly used at present the given rating is defined for pumping nitrogen atmosphere. The important performance data of turbomolecular pumps is the compression ratio. The compression ratio describes the ratio of the outlet pressure (forepump) of a particular gas to the inlet pressure (MS). The turbomolecular pump gives a completely background-free high vacuum and exhibits excellent start-up properties, which is important for benchtop instruments or those for mobile use. The use of helium lowers the performance of the

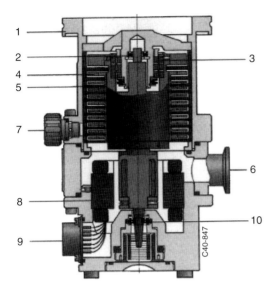

1. High vacuum connection (MS)

2. Emergency bearing

3. Permanent magnet bearing

4. Rotor

5. Stator

6. Forevacuum connection

7. Flooding connection

8. Motor

9. Electrical connection (control instrument)

10. High precision ball bearings with
    ceramic spheres (Pfeiffer Vacuum, 2003)

Figure 2.249 Construction of a turbo-molecular pump with ceramic ball bearings and permanent magnet bearings for use in mass spectrometry (Pfeiffer Vacuum, 2003).

pump. The compression ratio (e.g., for the Pfeiffer Balzer type TPH 062) decreases from 10^8 for nitrogen to 7×10^3 for helium, further to 6×10^2 for hydrogen. The reason for the much lower performance with hydrogen is its low molecular weight and high diffusion rate of hydrogen (Freeman, 1985).

A turbomolecular pump essentially consists of the rotor and a stator (Figure 2.249). Rotating and stationary discs are arranged alternately. All the discs have diagonal channels, whereby the channels on the rotor disc are arranged so that they mirror the positions of the channels of the stator discs. Each channel of the disc forms an elementary molecular pump. All the channels on the disc are arranged in parallel. A rotor disc together with a stator disc forms a single pump stage, which produces a certain compression. The pumping process is such that a gas molecule which meets the rotor acquires a velocity component in the direction of rotation of the rotor in addition to its existing velocity. The final velocity and the direction in which the molecule continues to move are determined from the vector sum of the two velocity components (Figure 2.250).

The thermal motion of a molecule, which is initially undirected, is converted into directed motion when the molecule enters the pump. An individual pumping step produces a compression of only about 30. Several consecutive pumping steps, which reinforce each other's action, lead to high compression rates.

The reduction in the performance of the pump when using hydrogen leads to a measurable increase in pressure in the analyser. Through collisions of substance ions with gas particles on their path through the analyser of the mass spectrometer, the transmission and thus the sensitivity of the instrument is reduced in the case

Particles

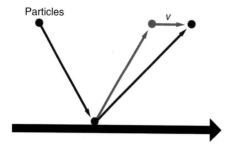

Figure 2.250 Principle of the molecular pump (Pfeiffer Vacuum, 2003).

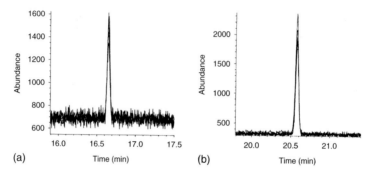

Figure 2.251 Effect of the carrier gas on the signal/noise ratio in the quadrupole GC-MS (Schulz, 1987). PCB 101, 50 pg, SIM *m/z* 256, 326, 328. (a) Carrier gas hydrogen, S/N 4 : 1. (b) Carrier gas helium, S/N 19 : 1.

of beam instruments (Figure 2.251). The mean free path L of an ion is calculated according to:

$$L = p^{-1} \cdot 5 \cdot 10^{-3} \, (\text{cm}) \tag{2.30}$$

The effect can be compensated for by using higher performance pumps with higher compression ratio or additional pumps (differentially pumped systems for source and analyser). Ion trap instruments do not exhibit this behaviour because of their ion storage principle.

When hydrogen is used as the carrier gas in GC the use of oil diffusion pumps is an advantage. The pump capacity of diffusion pumps is largely independent of the molecular weight and is therefore very suitable for hydrogen and helium. Oil diffusion pumps also have long service lives and are economical as no moving parts are involved. The pump operates using a propellant which is evaporated on a heating plate (Figure 2.252). The propellant vapour is forced downwards via a baffle and back into a fluid reservoir. Gas molecules diffuse into the propellant stream and are conveyed deeper into the pump. They are finally evacuated by

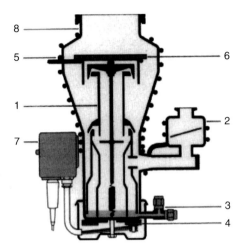

1. Jet system made of pressed aluminium components

2. Forevacuum baffle

3. Fuel top-up/measurement

4. Heating

5. Water cooling

6. Optically sealed baffle

7. Thermal protection switch/connection

8. Stainless steel pump casing

Figure 2.252 Construction of an oil diffusion pump (Pfeiffer Vacuum, 2003).

a forepump. Perfluoropolyethers (e.g., Fomblin) or polyphenyl ethers (e.g., Santovac S) are now used exclusively as diffusion pump fluids in MS. However, the favourable operation of the pump results in disadvantages for the operation of the mass spectrometer. Because a heating plate is used, the diffusion pump starts sluggishly and can only be vented again after cooling. While older models require water cooling, modern diffusion pumps for GC-MS instruments are air-cooled by a ventilator so that heating and cool-down times of 30 min and longer are unavoidable. The final vacuum of diffusion pumps is limited by the vapor pressure of the oil used to between 10^{-8} and 10^{-9} Torr. Oil vapour can easily lead to a permanent background in the mass spectrometer, which makes the use of a liquid nitrogen cooled baffle necessary, depending on the construction. The detection of negative ions in particular can be affected by the use of fluorinated polymers.

Both turbomolecular pumps and oil diffusion pumps require a mechanical forepump, as the compression is not sufficient to work against atmospheric pressure. Rotating vane pumps are generally used as forepumps (Figure 2.253). Mineral oils are used as the operating fluid. Oil vapours from the rotary vane pump passing into the vacuum tubing to the turbo pump are visible in the mass spectrometer as a hydrocarbon background, in particular, on frequent venting of the system. Special devices for separation or removal are necessary. As an alternative for the production of a hydrocarbon-free forevacuum, spiromolecular pumps can be used. These pumps can be run 'dry' without the use of oil. Their function involves centrifugal acceleration of the gas molecules through several pumping steps like the way it is in a turbomolecular pump. For the start-up, an integrated membrane pump is used. In particular, for mobile use of GC-MS systems and in

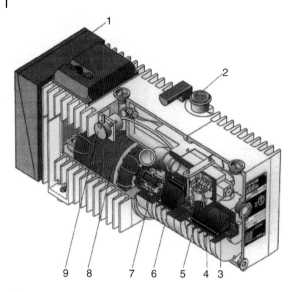

Figure 2.253 Cross-section of a two-stage rotary vane vacuum pump. 1. Start-up control, 2. exhaust, 3. aeration valve, 4. pump stage 1, 5. high vacuum safety valve, 6. pump stage 2, 7. motor coupling, 8. motor and 9 gas ballast valve. (Pfeiffer Vacuum, 2003).

other cases where systems are frequently disconnected from the mains electricity, spiromolecular pumps have proved useful.

References

Section 2.1 Sample Preparation
Section 2.1.1 Solid Phase Extraction
Section 2.1.2 Solid Phase Microextraction

Arthur, C.L. and Pawliszyn, J. (1990) Solid phase microextraction with thermal desorption using fused silica optical fibers. *Anal. Chem.*, **62**, 2145–2148.

Arthur, C.L., Potter, D.W., Buchholz, K.D., Motlagh, S., and Pawliszyn, J. (1992) Solid-phase microextraction for the direkt analysis of water: Theory and practice. *LC-GC*, **10**, 656–661.

Berlardi, R. and Pawliszyn, J. (1989) The application of chemically modified fused silica fibers in the extraction of organics from water matrix samples and their rapid transfer to capillary columns. *Water Pollut. Res. J. Can.*, **24**, 179.

Bicchi, C. *et al.* (2009) Stir-bar sorptive extraction and headspace sorptive extraction: an overview. *LCGC North Am.*, **27** (5), 376–390.

Boyd-Boland, A., Magdic, S., and Pawliszyn, J. (1996) Simultaneous determination of 60 pecticides in water using solid phase microextraction and gas chromatography-mass spectrometry. *Analyst*, **121**, 929–937.

Boyd-Boland, A. and Pawliszyn, J. (1995) Solid phase microextraction of nitrogen containing herbicides. *J. Chromatogr.*, **704**, 163–172.

Bundt, J., Herbel, W., Steinhart, H., Franke, S., and Francke, W. (1991) Structure type separation of diesel fuels by solid phase extraction and identification of the two- and three-ring aromatics by capillary GC-mass spectrometry. *J. High Resolut. Chromatogr.*, **14**, 91–98.

Deußing, G. (2005) Efficient multi-residue analysis of EDCs in drinking water. *GIT Lab. J.*, **5**, 17–19.

Eisert, R. and Pawliszyn, J. (1997) New trends in solid phase microextraction. *Crit. Rev. Anal. Chem.*, **27**, 103–135.

Farajzadeh, M.A. and Hatami, M. (2003) A new selective SPME fiber for some n-alkanes and its use for headspace sampling of aqueous samples. *J. Sep. Sci.*, **26**, 802–808.

Gorecki, T., Martos, P., and Pawliszyn, J. (1998) Strategies for the analysis of polar solvents in liquid matrices. *Anal. Chem.*, **70**, 19–27.

EPA Method 8275a (1996) Semivolatile Compounds (PAHs and PCBs) in Soils/Sludges and Solid Wastes Using Thermal Extraction/Gas Chromatography/Mass Spectrometry (TE/GC-MS), US EPA 1996, 1–23.

Gorecki, T. and Pawliszyn, J. (1997) The effect of sample volume on quantitative analysis by SPME. Part I: Theoretical considerations. *Analyst*, **122**, 1079–1086.

Hein, H. and Kunze, W. (1994) *Umweltanalytik mit Spektrometrie und Chromatographie*, Wiley-VCH Verlag GmbH, Weinheim.

Huang, G., Li, H.-F., Zhang, B.-T., Ma, Y., and Lin, J.-M. (2012) A vortex solvent bar microextraction combined with gas chromatography–mass spectrometry for the determination of phthalate esters in various sample matrices. *Talanta*, **100**, 64–70.

Hoffmann, A., Bremer, R., Sandra, p., and David, F. (2000) A Novel Extraction Technique for Aqueous Samples. *Gerstel Solutions worldwide*, **49**, 4–7.

Ligor, M. and Buszewski, B. (2008) The comparison of solid phase microextraction – GC and static headspace – GC for determination of solvent residues in vegetable oils. *J. Sep. Sci.*, **31**, 364–371.

Lord, H.L. and Pawliszyn, J. (1998) Recent advances in solid phase microextraction. *LCGC Int.*, **12**, 776–785.

Luo, Y., Pan, L., and Pawliszyn, J. (1998) Determination of five benzodiazepines in aqueous solution and biological fluids using SPME

with carbowax/DVB fibre coating. *J. Microcolumn. Sep.*, **10**, 193–201.

MacGillivray, B., Pawliszyn, J., Fowlei, P., and Sagara, C. (1994) Headspace solid-phase microextraction versus purge and trap for the determination of substituted benzene compounds in water. *J. Chromatogr. Sci.*, **32**, 317–322.

Magdic, S., Boyd-Boland, A., Jinno, K., and Pawliszyn, J. (1996) Analysis of organophosphorus insecticides in environmental samples by solid phase micro-extraction. *J. Chromatogr. A*, **736**, 219–228.

Martos, P., Saraullo, A., and Pawliszyn, J. (1997) Estimation of air/coating distribution coefficients for solid phase microextraction using retention indexes from linear temperature-programmed capillary gas chromatography. Application to the sampling and analysis of total petroleum hydrocarbons in air. *Anal. Chem.*, **69**, 402–408.

Müenster, H., Taylor, L. (2009) Mass Resolution and Resolving Power. *Thermo Fisher Scientific*, Bremen 2009, Technical Note.

Nielsson, T., Pelusio, F. *et al.* (1994) A critical examination of solid phase micro-extraction for water analysis. in *16th International Symposium on Capillary Chromatography Riva del Garda* (ed. P. Sandra), Huethig, pp. 1148–1158.

Noij, T.H.M. and van der Kooi, M.M.E. (1995) *Automated analysis of polar pesticides in water by on-line solid phase extraction and gas chromatography using the co-solvent effect*, J. High Res. Chrom., **18** (9), 535–539.

Pan, L., Chong, J.M., and Pawliszyn, J. (1997) Determination of amines in air and water using derivatization combined with SPME. *J. Chromatogr.*, **773**, 249–260.

Pawliszyn, J. (1997) *Solid Phase Microextraction: Theory and Practice*, John Wiley & Sons, Inc., New York.

Pawliszyn, J., Yang, M.J., and Orton, M.L., (1997) Quantitative determination of caffeine in beverages using a combined SPME-GC-MS method. *J. Chem. Ed.*, **74** (9), 1130–1132.

Poerschmann, J., Zhang, Z., Kopinke, F.-D., and Pawliszyn, J. (1997) Solid phase microextraction for determining the distribution of chemicals in aqueous matrices. *Anal. Chem.*, **69**, 597–600.

Pawliszyn, J. (ed.) (2012) Handbook of Solid Phase Microextraction, Elsevier, London, Waltham.

Potter, D. and Pawliszyn, J. (1992) Detection of substituted benzenes in water at the pg/ml level using solid-phase microextraction and gas chromatography-ion trap mass spectrometry. *J. Chromatogr.*, **625**, 247–255.

Risticevic, S. *et al.* (2009) Recent developments in solid-phase microextraction. *Anal. Bioanal. Chem.*, **393**, 781–795.

Sandra, P. *et al.* (2000) A Novel Extraction Technique for Aqueous Samples: Stir Bar Sorptive Extraction. Application Note 1/2000, Gerstel GmbH & Co. KG.

Saraullo, A., Martos, P., and Pawliszyn, J. (1997) Water analysis by SPME based on physical chemical properties of the coating. *Anal. Chem.*, **69**, 1992–1998.

Serodio, P. and Nogueira, J.M.F. (2004) Multi-residue screening of endocrine disrupting chemicals in water samples by stir bar sorptive extraction gas chromatography mass spectrometry detection. *Anal. Chim. Acta*, **517**, 21–32.

Shutao, W., Wang, Y., Hong, Y., and Jie, Y. (2006) Preparation of a carbon-coated SPME fiber and application to the analysis of BTEX and halocarbons in water. *Chromatographia*, **63**, 365–371.

Souza Silva, E.A. and Pawliszyn, J. (2012) Optimization of fiber coating structure enables direct immersion solid phase microextraction and high-throughput determination of complex samples. *Anal. Chem.*, **84** (16), 6933–6938.

Steffen, A. and Pawliszyn, J. (1996) The analysis of flavour volatiles using headspace solid phase microextraction. *J. Agric. Food Chem.*, **44**, 2187–2193.

Vatani, H. and Yazdi, A.S. (2014) Ionic-liquid-mediated poly(dimethylsiloxane)-grafted carbon nanotube fiber prepared by the sol–gel technique for the head space solid-phase microextraction of methyl tert-butyl ether using GC. *J. Sep. Sci.*, **37** (1-2), 127–134.

Zhang, Z. and Pawliszyn, J. (1993) Headspace solid phase microextraction. *Anal. Chem.*, **65**, 1843–1852.

Zhang, Z., Yang, M.J., and Pawliszyn, J. (1994) Solid phase microextraction – a new solvent-free alternative for sample preparation. *Anal. Chem.*, **66**, 844A–853A.

Zief, M. and Kiser, R. (1997) Solid Phase Extraction for Sample Preparation, A Technical Guide to Theory, Method Development and Use, J.T. Baker Technical Library, Phillipsburg, NJ, USA.

Section 2.1.3 Pressurized Fluid Extraction

Accelerated Solvent Extraction (ASE) (2012) Techniques for In-Line Selective Removal of Interferences. Technical Note 210, Thermo Fisher Scientific.

Curren, M.S.S. and King, J.W. (2002) New sample preparation technique for the determination of avoparcin in pressurized hot water extracts from kidney samples. *J. Chromatogr. A*, **954**, 41–49.

Dionex (2006) Dionex Product Information, *www.dionex.com* (accessed 9 May 2014).

Haglund, P. and Spinnel, E. (2010) A modular approach to pressurized liquid extraction with in-cell clean-up. *LCGC Eur.*, **23**, 292–301.

Hubert, A., Popp, P., Wenzel, K.-D., Engewald, W., and Schüürmann, G. (2001) Accelerated solvent extraction– more efficient extraktion of POPs and PAHs from real contaminated plant and soil samples. *Rev. Anal. Chem.*, **20**, 101–144.

Li, M.K., Landriault, M., Fingas, M., and Llompart, M. (2003) Accelerated Solvent Extraction (ASE) of environmental organic compounds in soils using a modified supercritical fluid extractor. *J. Hazard. Mater.*, **102**, 93–104.

Oleszek-Kudlak, S., Shibata, E., Nakamura, T., Lia, X.W., Yua, Y.M., and Donga, X.D. (2007) Review of the sampling and pretreatment methods for dioxins determination in solids, liquids and gases. *J. Chin. Chem. Soc.*, **54**, 245–262.

Ramos, J.J., Dietz, C., Gonzalez, M.J., and Ramosa, L. (2007) Miniaturised selective pressurised liquid extraction of polychlorinated biphenyls from foodstuffs. *J. Chromatogr. A*, **1152**, 254–261.

Richter, B.E., Jones, B.A., Ezzell, J.L., Porter, N.L., Avdalovic, N., and Pohl, C. (1996)

Accelerated solvent extraction: a technique for sample preparation. *Anal. Chem.*, **68**, 1033–1039.

Sanchez-Prado, L., Lamas, J.P., Lores, M., Garcia-Jares, C., and Llompart, M. (2010) Simultaneous in-cell derivatization pressurized liquid extraction for the determination of multiclass preservatives in leave-on cosmetics. *Anal. Chem.*, **82**, 9384.

Spinner, E. (2008) PLE with integrated cleanup followed by alternative detection steps for cost-effective analysis of dioxins and dioxin-like compounds. PhD thesis. Umeå University, Sweden.

Ullah, S.M.R., Srinivasan, K. and Pohl, C. (2012) Advances in Sample Preparation for Accelerated Solvent Extraction. Technical Report PN70071, Thermo Fisher Scientific, Sunnyvale, CA.

US EPA (1998) Method 3545A, January 1998, *http://www.epa.gov/osw/hazard/testmethods/sw846/pdfs/3545a.pdf* (accessed 9 May 2014).

Section 2.1.4 Online Liquid Chromatography

Che, J., Yu, C., Liang, L., and Hübschmann, H.-J., (2014) Automated Online GPC/GC-MS for the Determination of 181 Pesticides in Vegetables. *Application Note 10422*, Thermo Fisher Scientific, Shanghai.

De Paoli, M. *et al.* (1992) Determination of organophosphorus pesticides in fruit by on-line automated LC-GC. *J. Chromatogr.*, **626**, 145–150.

Kerkdijk, H., Mol, H.G.J., and van der Nagel, B. (2007) Volume overload cleanup: An approach for on-line SPE-GC, GPC-GC, and GPC-SPE-GC. *Anal. Chem.*, **79**, 7975–7983.

Majors, R.E. (1980) Multidimensional HPLC. *J. Chromatogr. Sci.*, **18**, 571–577.

Munari, F. and Grob, K. (1990) Coupling HPLC to GC: Why? How? With what instrumentation? *J. Chromatogr. Sci.*, **28**, 61–66.

Jongenotter, G.A., Kerkhoff, M.A.T., Van der Knaap, H.C.M., and Vandeginste, B.G.M. (1999) Automated On-line GPC-GC-FID involving co-solvent trapping and

the on-column interface for the determination of organophosphorus pesticides in olive oils. *J. High. Resolut. Chromatogr.*, **22**, 17–23.

Trestianu, S., Munari, F., and Grob, K. (1996) *Riva 85 – Riva 96, Ten Years Experience on Creating Instrumental Solutions for Large Volume Sample Injections into Capillary GC Columns*, CE Instruments Publication, Rodano.

Section 2.1.5 Headspace Techniques

Ashley, D.L., Bonin, M.A. *et al.* (1992) Determining volatile organic compounds in human blood from a large sample population by using purge and trap gas chromatography/mass spectrometry. *Anal. Chem.*, **64**, 1021–1029.

Ettre, L.S. and Kolb, B. (1991) Headspace gas chromatography: The influence of sample volume on analytical results. *Chromatographia*, **32** (1/2), 5–12.

Hachenberg, H. (1988) *Die Headspace Gaschromatographie als Analysen- und Meßmethode*, DANI- Analysentechnik, Mainz.

Hachenberg, H. and Schmidt, A.P. (1977) *Gaschromatographic Headspace Analysis*, John Wiley & Sons, Ltd, Chichester.

Johnson, E. and Madden, A. (1990) Efficient Water Removal for GC-MS Analysis of Volatile Organic Compounds with Tekmar's Moisture Control Module. Finnigan MAT Technical Report No. 616, Tekmar Company.

Kolb, B. (1980) *Applied Headspace Gas Chromatography*, John Wiley & Sons, Ltd, Chichester.

Kolb, B. (1992) *Headspace Gas Chromatography a Brief Introduction*, Perkin Elmer GC Applications Laboratory.

Kolb, B. (1993) Die Bestimmung des Wassergehaltes in Lebensmitteln und Pharmaka mittels der gaschromatographischen Headspace-Technik. *Lebensmittel Biotechnol.*, **1**, 17–20.

Kolb, B. and Ettre, L.S. (1991) Theory and practice of multiple headspace extraction. *Chromatographia*, **32** (11-12), 505–513.

Kolb, B. and Ettre, L.S. (2006) *Static Headspace – Gas Chromatography: Theory and Practice*, 2nd edn, John Wiley & Sons, Inc., New York.

Krebs, G., Schneider, E., and Schumann, A. (1991) Head Space GC – Analytik flüchtiger aromatischer und halogenierter Kohlenwasserstoffe aus Bodenluft. *GIT Fachz. Lab.*, **1**, 19–22.

Lin, D.P., Falkenberg, C., Payne, D.A. *et al.* (1993) Kinetics of purging for the priority volatile organic compounds in water. *Anal. Chem.*, **65**, 999–1002.

Madden, A.T. and Lehan, H.J. (1991) The effects of condensate traps on polar compounds in purge and trap analysis. Pittsburgh Conference.

Maggio, A., Milana, M.R. *et al.* (1991) Multiple headspace extraction capillary gas chromatography (MHE-CGC) for the quantitative determination of volatiles in contaminated soils. in *13th International Symposium on Capillary Chromatography, Riva del Garda, May 1991* (ed. P. Sandra), Huethig Verlag, pp. 394–405.

Matz, G. and Kesners, P. (1993) Spray and trap method for water analysis by thermal desorption gas chromatography/mass spectrometry in field applications. *Anal. Chem.*, **65**, 2366–2371.

Petersen, M.A. (2008) Quantification of volatiles in cheese using Multiple Headspace Extraction (MHE). Presentation 2008, University of Kopenhagen, Department of Food Science, Quality and Technology.

Pigozzo, F., Munari, F., and Trestianu, S. (1991) Sample transfer from head space into capillary columns. in *13th International Symposium on Capillary Chromatography, Riva del Garda, May 1991* (ed. P. Sandra), Huethig Verlag, pp. 409–416.

Tekmar, Fundamentals of Purge and Trap, TekData, Tekmar Technical Documentation B 121 988, without year.

Vulpius, T. and Baltensperger, B. (2009) Multiple Headspace Extraction (MHE) With Syringe Sampling, MSC ApS and CTC Scientific Poster.

Westendorf, R.G. (1985) A quantitation method for dynamic headspace analysis using multiple runs. *J. Chromatogr. Sci.*, **23**, 521–524.

Westendorf, R.G. (1987) Design and Performance of a Microprozessor-Based Purge and Trap Concentrator, American Laboratory 10.

Westendorf, R.G. (1989a) Automatic Sampler Concepts for Purge and Trap Gas Chromatography, American Laboratory 2.

Westendorf, R.G. (1989b) Performance of a third generation cryofocussing trap for purge and trap gas chromatography. Pittsburgh Conference.

Willemsen, H.G., Gerke, Th. and Krabbe, M.L. (1993) Die Analytik von LHKW und BTX im Rahmen eines DK-ARW(AWBR)-IKSR-Projektes, 17. Aachener Werkstattgespräch am 28. und 29. September 1993, Zentrum für Aus- und Weiterbildung in der Wasser- und Abfallwirtschaft Nordrhein-Westfalen GmbH (ZAWA), Essen 1993.

Wylie, P.L. (1988) Comparing headspace with purge and trap for analysis of volatile priority pollutants. *Res. Technol.*, **8**, 65–72.

Wylie, P.L. (1987) Comparison of headspace with purge and trap techniques for the analysis of volatile priority pollutants. in *8th International Symposium on Capillary Chromatography Riva del Garda, May 19th–21st 1987* (ed. P. Sandra), Huethig, pp. 482–499.

Zhang, B.T., Zheng, X., Li, H.-F., and Lin, J.-M. (2013) Application of carbon-based nanomaterials in sample preparation: A review. *Anal. Chim. Acta*, **783**, 1–17.

Zhu, J.Y., Chai, X.-S., Zhu, (2005) Some recent developments in headspace gas chromatography. *Curr. Anal. Chem.*, **1**, 79–83.

Section 2.1.6 Adsorptive Enrichment and Thermal Desorption

Betz, W.R., Hazard, S.A. and Yearick, E.M. (1989) Characterization and Utilization of Carbon-Based Adsorbents for Adsorption and Thermal Desorption of Volatile, Semivolatile and Non-Volatile Organic Contaminants in Air, Water and Soil Sample Matrices, International Labmate XV, p. 1.

Brown, R.H. and Purnell, C.J. (1979) Collection and analysis of trace vapour pollutants in ambient atmospheres. *J. Chromatogr.*, **178**, 79–90.

Brown, J. and Shirey, B. (2002) A Tool for Selecting an Adsorbent for Thermal Desorption Applications. Technical Report, T402025 EQF SUPELCO.

Brown, V.M., Crump, D.R., Plant, N.T. and Pengelly, I. (2014) Evaluation of the stability of a mixture of volatile organic compounds on sorbents for the determination of emissions from indoor materials and products using thermal desorption/gas chromatography/mass spectrometry. *J. Chrom. A*, **1350**, 1–9.

Figge, K., Rubel, W., and Wieck, A. (1987) Adsorptionsmittel zur Anreicherung von organischen Luftinhaltsstoffen. *Fres. Z. Anal. Chem.*, **327**, 261–278.

Health and Safety Laboratory (HSL) (1990) Generation of test atmospheres of organic vapours by the syringe injection technique. ISBN 0-11-885647-2.

Knobloch, T. and Engewald, W. (1994) Sampling and gas chromatographic analysis of Volatile Organic Compounds (VOCs) in hot and extremely humid emissions, in *16th International Symposium on Capillary Chromatography Riva del Garda* (ed. P. Sandra), Huethig, pp. 472–484.

Manura, J.J. (1995) Adsorbent Resins – Calculation and Use of Breakthrough Volume Data, Scientific Instrument Services, Ringoes, NJ, Application Note, *http://www.sisweb.com/index/referenc/resin10.htm* (accessed 9 May 2014).

Markes (2003) Sampling device. US Patent 6,564,656.

Markes (2012a) Advice on Sorbent Selection, Tube Conditioning, Tube Storage and Air Sampling. Technical Report TDTS 5, Markes International.

Markes (2012b) Diffusion-Locking Technology. Technical Report TDTS 61, Markes International.

Sigma-Aldrich (2014) Thermal Desorption Tube Selection Guide, *http://www.sigmaaldrich.com/analytical-chromatography/air-monitoring/learning-center/td-tube-selector.html* (accessed 9 May 2014).

Supelco (1986) Carbotrap – An Excellent Adsorbent for Sampling Many Airborne Contaminants. GC Bulletin 846C, Supelco Company Publication.

Supelco (1988) Efficiently Monitor Toxic Compounds by Thermal Desorption,

GC Bulletin 849C, Supelco Company Publication.

Tschickard, M. (1993) Bericht über eine Prüfgasapparatur zur Herstellung von Kalibriergasen nach dem Verfahren der kontinuierlichen Injektion, Gesch.Zchn: 35 – 820 Tsch, Mainz, Landesamt für Umweltschutz und Gewerbeaufsicht, 25 May 1993.

Supelco (1992) New Adsorbent Trap for Monitoring Volatile Organic Compounds in Wastewater, Environmental Notes, Supelco Company Publication, Bellafonte.

Kommission Reinhaltung der Luft im VDI und DIN - Normenausschuss KRdL (1981) Measurement of Gases; Calibration Gas Mixtures; Preparation by Continuous Injection Method, Directive 13.040.01., *www.vdi.de* (accessed 14 May 2014).

Woolfenden, E.A., and Cole., A. (2003) Sampling Device, Patent, US 6564656 B1, May 2003.

Section 2.1.7 Pyrolysis

Brodda, B.-G., Dix, S., and Fachinger, J. (1993) Investigation of the pyrolytic degradation of ion exchange resins by means of foil pulse pyrolysis coupled with gas chromatography/mass spectrometry. *Sep. Sci. Technol.*, **28**, 653–673.

Ericsson, I. (1978) Sequential pyrolysis gas chromatographic study of the decomposition kinetics of cis-1,4-polybutadiene. *J. Chromatogr. Sci.*, **16**, 340–344.

Ericsson, I. (1980) Determination of the temperature time profile of filament pyrolyzers. *J. Anal. Appl. Pyrolysis*, **2**, 187–194.

Ericsson, I. (1985) Influence of pyrolysis parameters on results in pyrolysis-gas chromatography. *J. Anal. Appl. Pyrolysis*, **8**, 73–86.

Ericsson, I. (1990) Trace determination of high molecular weight polyvinylpyrrolidone by pyrolysis-gas-chromatography. *J. Anal. Appl. Pyrolysis*, **17**, 251–260.

Ericsson, I. and Lattimer, R.P. (1989) Pyrolysis nomenclature. *J. Anal. Appl. Pyrolysis*, **14**, 219–221.

Fischer, W.G. and Kusch, P. (1993) An automated curie-point pyrolysis-high resolution gas chromatography system. *LCGC*, **6**, 760–763.

Galletti, G.C. and Reeves, J.B. (1991) Pyrolysis-gas chromatography/mass spectrometry of lignocellulosis in forages and by-products. *J. Anal. Appl. Pyrolysis*, **19**, 203–212.

Hancox, R.N., Lamb, G.D., and Lehrle, R.S. (1991) Sample size dependence in pyrolysis: an embarrassment or an utility? *J. Anal. Appl. Pyrolysis*, **19**, 333–347.

Hardell, H.L. (1992) Characterization of spots and specks in paper using PY-GC-MS including SPM. Poster at the 10th International Conference on Fundamental Aspects Proceeding of the Applied Pyrolysis, Hamburg, Germany pp. 28.9–2.10.

Irwin, W.J. (1982) *Analytical Pyrolysis: A Comprehensive Guide*, Marcel Dekker, New York.

Klusmeier, W., Vögler, P., Ohrbach, K.H., Weber, H., and Kettrup, A. (1988) Thermal decomposition of pentachlorobenzene, hexachlorobenzene and octachlorostyrene in air. *J. Anal. Appl. Pyrolysis*, **14**, 25–36.

Matney, M.L., Limero, T.F., and James, J.T. (1994) Pyrolysis-gas chromatography/mass spectrometry analyses of biological particulates collected during recent space shuttle missions. *Anal. Chem.*, **66**, 2820–2828.

Oguri, N. and Kirn, P. (2005) Design and Applications of a Curie Point Pirolyzer. Technical Report, Japan Analytical Industry Co., Ltd., Musashi, Mizuho, Nishitama Tokyo.

Roussis, S.G. and Fedora, J.W. (1998) Use of a thermal extraction unit for furnace-type pyrolysis: Suitability for the analysis of polymers by pyrolysis/GC-MS. *Rapid Commun. Mass Spectrom.*, **10**, 82–90.

Schulten, H.-R., Fischer, W., Wallstab, H.-J. (1987) New automatic sampler for Curie-Point pyrolysis its combination with gas chromatography. *HRC CC*, **10**, 467.

Simon, W. and Giacobbo, H. (1965) *Chem. Ing. Technol.*, **37**, 709.

Snelling, R.D., King, D.B. and Worden, R. (1994) An automated pyrolysis system for the analysis of polymers. Poster, 10P, Proceeding of the Pittsburgh Conference, Chicago, IL, 1994.

Uden, P.C. (1983) Nomenclature and terminology for analytical pyrolysis (IUPAC Recommendations 1993). *Pure Appl. Chem.*, **65**, 2405–2409.

US Environmental Protection Agency (1996) *Semivolatile Organic Compounds (PAHs and PCBs) in Soil/Sludges and Solid Wastes Using Thermal Extraction/Gas Chromatography Mass Spectrometry (TE/GC-MS), Rev.* EPA Method 8275A. US Environmental Protection Agency, 1 December 1996.

Wampler, T.P. (ed.) (1995) *Applied Pyrolysis Handbook*, Marcel Dekker, New York.

Zaikin, V.G., Mardanov, R.G. *et al.* (1990) Pyrolysis-gas chromatographic/mass spectrometric behaviour of polyvinylcyclohexane and vinylcyclohexane-styrene copolymers. *J. Anal. Appl. Pyrolysis*, **17**, 291.

Section 2.1.8 Thermal Extraction (Outgassing)

Japan Analytical Industry Co. Ltd., List of Outgas Collectors, *http://www.jai.co.jp/english/products/outgas/list.html* (accessed October 2013).

Markes (2009a) Note 69: The Micro-Chamber/Thermal Extractor – the Innovative, Rapid and Cost Effective Approach for Testing VOC Emissions from Materials, April 2009.

Markes (2009b) Direct Desorption of Car Trim Materials for Volatile Organic Compounds (VOC) and Semi-volatile Organic Compounds (SVOC) Analysis in Accordance with Method VDA 278. Application Note TDTS 59.

Markes (2011) Introducing the Micro-Chamber/Thermal Extractor™ (μ-CTE™) for Rapid Screening of Chemicals Released (emitted) by Products and Materials. Application Note 67.

Markes (2012) Rapid Aroma Profiling of Cheese Using a Micro-Chamber/Thermal Extractor with TD–GC-MS Analysis. Application Note TDTS 101.

Oguri, N. (2005) Outgas Anaiysis for Hard Disk Industries. Technical Report, Japan Analytical Industry Co., Ltd.

Roussis, S.G. and Fedora, J.W. (1998) Use of a thermal extraction unit for furnace-type pyrolysis: Suitability for the analysis of polymers by pyrolysis/GC-MS. *Rapid Commun. Mass Spectrom.*, **10**, 82–90.

Section 2.1.9 QuEChERS Sample Preparation

Anastassiades, M. (2011) QuEChERS, *http://quechers.cvua-stuttgart.de/* (accessed April 2012).

Anastassiades, M. Lehotay, S.J. and Stajnbaher, D. (2002) Quick, Easy, Cheap, Effective, Rugged and Safe (QuEChERS) approach for the determination of pesticide residues. 4th European Pesticide Residue Workshop in Rome, Italy May 28–31, 2002.

Anastassiades, M., Lehotay, S.J., Stajnbaher, D., and Schenck, F.J. (2003) Fast and easy multiresidue method employing acetonitrile extraction/partitioning and "dispersive solid-phase extraction" for the determination of pesticide residues in produce. *J. AOAC Int.*, **86**, 412.

Anastassiades, M., Scherbaum, E., Tasdelen, B., and Štajnbaher, D., (2007) Recent developments in QuEChERS methodology for pesticide multiresidue analysis, in: *Pesticide Chemistry, Crop Protection, Public Health, Environmental Safety* (eds Hideo Ohkawa, Hisashi Miyagawa, and Philip W. Lee) Wiley-VCH Verlag GmbH & Co. KGaA.

Brondi, S.H.G., de Macedo, A.N., Vicente, G.H.L., and Nogueira, A.R.A. (2011) Evaluation of the QuEChERS method and gas chromatography–mass spectrometry for the analysis pesticide residues in water and sediment. *Bull. Environ. Contam. Toxicol.*, **86**, 18–22.

Cunha, S.C., Lehotay, S.J., Mastovska, K., Fernandes, J.O., Beatriz, M., and Oliveira, P.P. (2007) Evaluation of the QuEChERS sample preparation approach for the analysis of pesticide residues in olives. *J. Sep. Sci.*, **30**, 620–632.

European Standard EN15662. (2008) *Foods of Plant Origin – Determination of Pesticide Residues Using GC-MS and/or LC-MS/MS Following Acetonitrile Extraction/Partitioning and Clean-Up by Dispersive SPE – QuEChERS Method. European Standard.*

Hetmanski, M.T., Fussel, R., Godula, M., and Hübschmann, H.-J. (2010) *Rapid Analysis of Pesticides in Difficult Matrices Using GC-MS/MS*, Application Note 51880, Thermo Fisher Scientific.

Khan, A. (2012) *QuEChERS Dispersive Solid Phase Extraction for the GC-MS Analysis of Pesticides in Cucumber*, Application Note 20549, Thermo Fisher Scientific, Runcorn.

Lehotay, S. (2007) Determination of pesticide residues in foods by acetonitrile extraction and partitioning with magnesium sulfate: Collaborative study. *JAOAC Int*, **90** (2), 485–520.

Lehotay, S.J. (2013) Revisiting the advantages of the QuEChERS approach to sample preparation, online presentation, webinar Separation Science, *http://www.sepscience.com* (accessed 14 May 2014).

Lehotay, S.J., de Kok, A., Hiemstra, M., and van Bodegraven, P. (2005) Validation of a fast and easy method for the determination of residues from 229 pesticides in fruits and vegetables using gas and liquid chromatography and mass spectrometric detection. *JAOAC Int*, **88** (2), 595–614.

Luetjohann, J., Zhang, L., Bammann, S., Kuballa, J. and Jantzen, E. (2009) Dioxins go Quechers – A new approach in the analysis of well-known contaminants. Poster RAFA Prague, Czech Republic, 2009.

Mol, H.G.J., Rooseboom, A., van Dam, R., Roding, M., Arondeus, K., and Sunarto, S. (2007) Modification and revalidation of the ethyl acetate-based multiresidue method for pesticides in produce. *Anal. Bioanal. Chem.*, **89** (6), 1715–1754.

Pinto, C.G., Laespada, M.E.F., Martin, S.H., Casas, F., Pavon, J.L.P., and Cordero, B.M. (2010) Simplified QuERhERS approach for the extraction of chlorinated compounds from soil samples. *Talanta*, **81** (2), 385–391.

Settle, V., Foster, F., Roberts, P., Stone, P., Stevens, J., Wong, J., and Zhang, K. (2010) *Automated QuEChERS Extraction for the Confirmation of Pesticide Residues in Foods using LC-MS/MS*, Application

Note 4/2010, Gerstel GmbH & Co. KG, Mülheim.

Stahnke, H., Kittlaus, S., Kempe, G., and Alder, L. (2012) Reduction of matrix effects in liquid chromatography-electrospray ionization-mass spectrometry by dilution of the sample extracts: How much dilution is needed? *Anal. Chem.*, **84**, 1474–1482.

Teledyne Tekmar (2013) Pesticide Analysis Using the AutoMate-Q40: An Automated Solution to QuEChERS Extractions, Application Note April 2013, Teledyne Tekmar, Mason, OH.

Usui, K., Hayashizaki, Y., Hashiyada, M., and Funayama, M. (2012) Rapid drug extraction from human whole blood using a modified QuEChERS extraction method. *Leg. Med.*, **14** (6), 286–296.

UTC (2011) QuEChERS Pesticide Residue Analysis Manual, UCT LLC, Bristol, PA.

Thermo Scientific (2010) *Pesticide Analysis Reference Manual*, 2nd edn, Thermo Fisher Scientific, Austin.

Section 2.2 Gas Chromatography

Desty D.H. (1991) A personal account of 40 years of development in chromatography. *LC-GC Int.*, **4** (5), 32.

Section 2.2.3 Injection Port Septa

De Zeeuw, J. (2013) Ghost peaks in gas chromatography part 2: The injection port. *Sep. Sci. Asia Pacific*, **5**, 2–7.

Grossman, S. How Hot Is Your Septum? Restek Technical Resource, www.restek.com (accessed 12 2013).

Hinshaw, J. (2008) Do's and don'ts. *LCGC Eur.*, **5**, 332.

Merlin Microseal Users Manual (2012) *Merlin Instrument Company*, Half Moon Bay, CA, http://www.merlinic.com/products/merlin-microseal (accessed 9 Decemeber 2014).

Warden, J. and Pereira, L. (2007) *An Evaluation of the Performance of GC Septa in Different Commonly Used Procedures, Poster*, Thermo Fisher Scientific, Runcorn.

Westland, J., Organtini, K., Dorman, F. (2012) Evaluation of lifetime and analytical performance of gas chromatographic inlet septa for analysis of reactive semivolatile organic compounds. *J. Chromatogr. A.*, **1239**, 72–77.

Section 2.2.4 Injection Port Liner

Thermo Fisher Scientific (2008) *Instructions AquaSil and SurfaSil Siliconizing Fluids*.

Rood, D. (2007) *The Troubleshooting and Maintenace Guide for Gas Chromatographers*, Wiley-VCH Verlag GmbH & Co. KGaA, Weinheim.

Klee, M. (2013) Liner Deactivation. *Sep. Sci. GC Solutions*, **39**, 1–3.

Section 2.2.5 Vaporizing Sample Injection Techniques

Anastassiades, M. *et al.* (2003) Fast and easy multiresidue method employing acetonitrile extraction/partitioning and 'Dispersive Solid-Phase Extraction' for the Determination of Pesticide Residues in Produce, *J. AOAC Int.*, **86** (2), 412–431.

Hetmanski, M. T. (2009) High sensitivity multi-residue pesticide analyses in foods using the TSQ Quantum GC-MS/MS FERA, York. Presented at the RAFA Conference, November 2009, Prague.

Hildmann, F., Kempe, G., and Speer, K. (2013) Application of the precolumn back-flush technology in pesticide residue analysis: a practical view, *J. Sep. Sci.*, **36**, 2128–2135.

Munari, F., Magni, P., and Facchetti, R. (2000) Improvements in PTV/PTV large volume injection and MS detection using a reverse flow injection device, ThermoQuest Italia S.p.A. ISCC 2000 Conference, Riva del Garda, Italy, poster.

Section 2.2.6 Temperature Programmable Injection Systems

Bergna, M., Banfi, S., and Cobelli, L. (1989) The Use of a Temperature Vaporizer as

Preconcentrator Device in the Introduction of Large Amount of Sample. in *12th International Symposium on Capillary Chromatography Riva del Garda* (eds P. Sandra and G. Redant), Huethig, 300–309.

Blanch, G.P., Ibanez, E., Herraiz, M., and Reglero, G. (1994) Use of a programmed temperature vaporizer for off-line SFE/GC analysis in food composition studies. *Anal. Chem.*, **66**, 888–892.

Cavagna, B., Pelagatti, S., Cadoppi, A. and Cavagnino, D. (1994) GC-ECD determination of decabromodiphenylether using direct on-column injection into capillary columns. Poster Thermo Fisher Scientific Milan, PittCon 2008, New Orleans.

Donike, M. (1973) Die temperaturprogrammierte Analyse von Fettsäuremethylsilylestern: Ein kritischer Qualitätstest für gas-chromatographische Trennsäulen. *Chromatographia*, **6**, 190.

Färber, H., Peldszus, S., and Schöler, H.F. (1991) Gaschromatographische Bestimmung von aziden Pestiziden in Wasser nach Methylierung mit Trimethylsulfoniumhydroxid. *Vom Wasser*, **76**, 13–20.

Färber, H. and Schöler, H.F. (1991) Gaschromatographische Bestimmung von Harnstoffherbiziden in Wasser nach Methylierung mit Trimethylaniliniumhydroxid oder Trimethylsulfoniumhydroxid. *Vom Wasser*, **77**, 249–262.

Färber, H. and Schöler, H.F. (1992) Gaschromatographische Bestimmung von OH- und NH-aziden Pestiziden nach Methylierung mit Trimethylsulfonium hydroxid im "Programmed Temperature Vaporizer" (PTV). *Lebensmittelchemie.*, **46**, 93–100.

Grob, K. (1983) Guidelines on how to carry out on-column injections. *HRC CC*, **6**, 581–582.

Grob, K. (1986) *Classical Split and Splitless Injection in Capillary Gas Chromatography With Some Remarks on PTV Injection*, Huethig, Heidelberg.

Grob, K. (1994) Injection techniques in capillary GC. *Anal. Chem.*, **66**, 1009A–1019A.

Grob, K. (1995) *Einspritztechniken in der Kapillar-Gaschromatographie*, Huethig, Heidelberg.

Grob, K. (2001) Split and splitless injection for quantitative gas chromatography, in *Concepts, Processes, Practical Guidelines,*

Sources of Error, 4th edn, Wiley-VCH Verlag GmbH, Weinheim.

Grob, K. and Grob, K. Jr., (1978a) Splitless Injection and the Solvent Effect. *HRC & CC*, **1**, 57–64.

Grob, K. and Grob, K. Jr., (1978b) On column injection on to glass capillary columns. I. *J. Chromatogr.*, **151**, 311.

Hinshaw, J.W. (1992) Splitless injection: Corrections and further information. *LCGC Int.*, **5**, 20–22.

Hinshaw, J. (2008) Do's and Don'ts, in *LCGC Europe May*, 332.

Hoh, E. and Mastovska, K. (2008) Large volume injection techniques in capillary gas chromatography. *J. Chromatogr. A*, **1186**, 2–15.

Karasek, F.W. and Clement, R.E. (1988) *Basic Gas Chromatography-Mass Spectrometry*, Elsevier, Amsterdam.

Klemp, M.A., Akard, M.L., and Sacks, R.D. (1993) Cryofocussing inlet with reversed flow sample collection for gas chromatography. *Anal. Chem.*, **65**, 2516–2521.

Magni, P. and Porzano, T. (2003) Concurrent solvent recondensation large sample volume splitless injection. *J. Sep. Sci.*, **26**, 1491–1498.

Mol, H.G.J., Janssen, H.G.M., and Crainers, C.A. (1993) Use of a Temperature Programmed Injector with a Packed Liner for Direct Water Analysis and On-Line Reversed Phase LC-GC. *J. High Res. Chrom.*, **16**, 459–463.

Mol, H.G.J., Janssen, H.G.M., Cramers, C.A., and Brinkman, K.A.T. (1994) Large Volume Sample Introduction Using Temperature Programmable Injectors-Implication of Line Diameter. in *16th International Symposium on Capillary Chromatography Riva del Garda* (ed P. Sandra), Huethig, pp. 1124–1136.

Müller, H.-M. and Stan, H.-J. (1987) Thermal Degradation Observed with Varying Injection Techniques: Quantitative Estimation by the Use of Thermolabile Carbamate Pesticides. in *8th International Symposium on Capillary Chromatography Riva del Garda* (ed P. Sandra), Huethig, pp. 588–596.

Müller, H.-M. and Stan, H.-J. (1989) Pesticide Residue Analysis in Food with CGC. Study of Long-Time Stability by the Use of Different Injection Techniques. in *12th International Symposium on Capillary Chromatography Riva del Garda* (eds P. Sandra and G. Redant), Huethig, pp. 582–587.

Müller, S., Efer, J., Wennrich, L., Engewald, W., and Levsen, K. (1993) Gaschromatographische Spurenanalytik von Methamidophos und Buminafos im Trinkwasser – Einflußgrößen bei der PTV-Dosierung großer Probenvolumina. *Vom Wasser*, **81**, 135–150.

Pretorius, V. and Bertsch, W. (1983) *HRC & CC*, **6**, 64.

Saravalle, C.A., Munari, F., and Trestianu, S. (1987) Multi-purpose cold injector for high resolution gas chromatography. *HRC & CC*, **10**, 288–296.

Schomburg, G. (1981) Capillary chromatography, in *4th International Symposium on Hindelang*, Vol. 371 und A921 (ed. R. Kaiser), Institut für Chromatographie.

Schomburg, G. (1983) Probenaufgabe in der Kapillargaschromatographie. *LaBo*, **7**, 2–6.

Schomburg, G. (1987) *Gaschromatographie*, Wiley-VCH Verlag GmbH, Weinheim.

Schomburg, G., Behlau, H., Dielmann, R., Weeke, F., and Husmann, H. (1977) Sampling techniques in capillary gas chromatography. *J. Chromatogr.*, **142**, 87.

Stan, H.-J. and Müller, H.-M. (1987) Cold-Splitless (PTV) and On-Column Injektion Technique Using Capillar Gas Chromatography for the Analysis of Organophosphorus Pesticides. in *8th International Symposium on Capillary Chromatography Riva del Garda* (ed P. Sandra), Huethig, pp. 406–415.

Staniewski, J. and Rijks, J.A. (1991) Potentials and Limitations of the Liner Design for Cold Temperature Programmed Large Volume Injection in Capillary GC and for LC-GC Interfacing. in *14th International Symposium on Capillary Chromatography Riva del Garda* (ed P. Sandra), Huethig, pp. 1334–1347.

Tipler, A. and Johnson, G.L. (1989) Optimization of Conditions for High Temperature Capillary Gas Chromatography Using a Split-Mode Programmable Temperature Vaporizing Injection System. in

12th International Symposium on Capillary Chromatography Riva del Garda (eds P. Sandra and G. Redant), Huethig, pp. 986–1000.

Poy, F. (1981) Practical demonstation. 4th International Symposium on Capillar Chromatography, Hindelang, Germany.

Poy, F. (1982) A new temperature programmed injection technique for capillary GC: Split mode with cold introduction and temperature programmed vaporization. *Chromatographia*, **16**, 345.

Poy, F. and Cobelli, L. (1985) Automatic Headspace and Programmed Temperature Vaporizer (PTV) Operated in Cryo-Enrichment Mode in High Resolution Gas Chromatography. *J. Chrom. Sci.*, **23**(3), 114–119.

Poy, F., Visani, S., and Terrosi, F. (1981) A universal sample injection system for capillary column GC using a programmed temperature vaporizer (PTV), *J. Chromatogr.*, **217**, 81.

Poy, F., Visani, S., and Terrosi, F. (1982) *HRC & CC*, **4**, 355.

Vogt, W., Jacob, K., and Obwexer, H.W. (1979a) Capillary Gas Chromatograhic Injection System for Large Sample Volumes, *J. Chromatogr.*, **174**, 437.

Vogt, W., Jacob, K., Ohnesorge, A.B., and Obwexer, H.W. (1979b) *J. Chromatogr.*, **186**, 197.

Section 2.2.7 Capillary Column Choice and Separation Optimization

Armstrong, D.W., He, L., and Liu, Y.S. (1999) Examination of ionic liquids and their interaction with molecules, when used as stationary phases in gas chromatography. *Anal. Chem.*, **71**, 3873–3876.

Armstrong, D.W., Payagala, T., and Sidisky, L.M. (2009) *LCGC North Am.*, **27**, 596, 598, 600.

Berthod, A. and Carda-Broch, S. (2003) Uses of Ionic Liquids in Analytical Chemistry, source not available.

Khan, A.I. (2013) *Optimizing GC Parameters for Faster Separations with Conventional Instrumentation, Technical Note 20743*, Thermo Fisher Scientific, Runcorn, Cheshire, UK.

Knapp, D.R. (1979) *Handbook of Analytical Derivatization Reactions*, John Wiley & Sons, Inc., New York.

Pierce, A.E. (1982) *Silylation of Organic Compounds*, Pierce Chemical Company, Rockford, Ill.

Restek Corporation (1993) A Capillary Chromatography Seminar. Technical Report, Restek, Bellafonte, PA.

Stenerson, K.K., Halpenny, M.R., Sidisky, L.M., and Buchanan, M.D. (2013) GC analysis of omega-3-fatty acids in fish oil capsules and farm raised salmon. *Supelco Rep.*, **31** (2), 14–15.

Stenerson, K.K., Halpenny, M.R., Sidisky, L.M., and Buchanan, M.D. (2014) Ionic liquid GC column option for the analysis of omega 3 and omega 6 fatty acids. *Supelco Rep.*, **31** (1), 20–22.

Supelco (2013) Supelco Ionic Liquid GC Columns, Introduction to the Technology, Supelco Presentation 2012, updated January 2013, *http://www.sigmaaldrich.com/analytical-chromatography/gas-chromatography.html* (accessed December 2013).

Vidal, L., Riekkola, M.-L., and Canals, A. (2012) Ionic liquid-modified materials for solid-phase extraction and separation: A review. *Anal. Chim. Acta*, **715**, 19–41.

Whitmarsh, S. (2012) Ionic liquid stationary phases: Application in gas chromatographic analysis of polar components of fuels and lubricants. *Chrom. Today*, Feb/Mar, 12–15.

van Ysacker, P.G., Janssen, H.G., Snijders, H.M.J., van Cruchten, H.J.M., Leclercq, P.A., and Cramers, C.A. (1993) High-speed-narrow-bore capillary gas chromatography with ion-trap mass spectrometric detection, in *15th International Symposium on Capillary Chromatography Riva del Garda 1993* (ed. P. Sandra), Huethig.

Section 2.2.8 Chromatography Parameters

Bock, R. (1974) *Methoden der analytischen Chemie, Bd. 1. Trennungsmethoden*, Weinheim, Verlag Chemie.

Martin, A.J.P. and Synge, R.L.M. (1941) *J. Biol. Chem.*, **35**, 1358.

Snyder, L.R., Kirkland, J.J. (1979) *Introduction to Modern Liquid Chromatography*, 2nd edn John Wiley & Sons, Inc., New York (3rd edn, 2010 with J.J. Dolan, ed.).

Section 2.2.9 Fast Gas Chromatography Solutions

Donato, P., Tranchida, P.Q., Dugo, P., Dugo, G., and Mondello, L. (2007) Review: Rapid analysis of food products by means of high speed gas chromatography. *J. Sep. Sci.*, **30**, 508–526.

EPA (1997) EPA Method 610 - Polynuclear Aromatic Hydrocarbons, *www.epa.gov/epahome/index/* (accessed 12 May 2014).

Facchetti, R. and Cadoppi, A. (2005) Ultra Fast Chromatography: A Viable Solution for the Separation of Essential Oil Samples. The Column, November 2005, 8–11, *www.thecolumn.eu.com* (accessed 14 May 2014).

Facchetti, R., Galli, S., and Magni, P. (2002) Optimization of analytical conditions to maximize separation power in ultra-fast GC, in *Proceedings of 25th International Symposium of Capillary Chromatography, KNL05, Riva del Garda, Italy, May 13–17, 2002* (ed. P. Sandra).

Magni, P., Facchetti, R., Cavagnino, D., and Trestianu, S. (2002) in *Proceedings of 25th International Symposium of Capillary Chromatography, KNL05, Riva del Garda, Italy, May 13 – 17, 2002* (ed. P. Sandra).

Warden, J., Magni, P., Wells, A. and Pereira, L. (2004) Increasing throughput in GC environmental methods. International Chromatography Symposium, Riva del Garda 2004.

Section 2.2.10 Multi-dimensional Gas Chromatography

Beens, J., Adahchour, M., Vreuls, R.J.J., van Altena, K., and Brinkman, U.A.T. (2001) Simple, non-moving modulation interface for comprehensive two-dimensional gas chromatography. *J. Chromatogr. A*, **919** (1), 127–132.

Beens, J., Blomberg, J., and Schoenmakers, P.J. (2000) Proper tuning of comprehensive two-dimensional gas chromatography (GCxGC) to optimize the separation of complex oil fractions. *J. High Resol. Chromatogr.*, **23** (3), 182–188.

Beens, H., Boelwns, R., Tijssen, R., and Blomberg, J. (1998) Simple, non-moving modulation interface for comprehensive two-dimensional gas chromatography. *J. High Res. Chromatogr.*, **21**, 47.

Bertsch, W.J. (1999) Two-dimensional gas chromatography. Concepts, instrumentation, and applications – Part 1. *J. High Res. Chromatogr.*, **22**, 647–665.

Bertsch, W.J. (2000) Two-dimensional gas chromatography. Concepts, instrumentation, and applications – Part 2: Comprehensive two-dimensional gas chromatography. *J. High Res. Chromatogr.*, **23**, 167–181.

Brechbuehler (2007) 2DGC: Multidimensional Gas Chromatography on a Single GC or Dual GC with or without MS Using a Moving Capillary Stream Switching System, *www.brechbuehler.ch/usa/mcss.htm* (accessed 12 May 2014).

Cavagnino, D., Bedini, F., Zilioli, G. and Trestianu, S. (2003) Improving sensitivity and separation power by using LVSL-GCxGC-FID technique for pollutants detection at low ppb level. Poster at the Gulf Coast Conference, Comprehensive Two-Dimensional GC Symposium, 2003.

De Alencastro, L.F., Grandjean, D., and Tarradellas, J. (2003) Application of multidimensional (heart-cut) gas chromatography to the analysis of complex mixtures of organic pollutants in environmental samples. *Chimia*, **57**, 499–504.

Deans, D.R. (1968) A new technique for heart cutting in gas chromatography. *Chromatographia*, **1**, 18–22.

Dimandja, J.M.D. (2004) GCxGC. *Anal. Chem.*, **76** (9), 167A–174A.

Focant, J., Sjödin, A., Turner, W., and Patterson, D. (2004) Measurement of selected polybrominated diphenyl ethers, polybrominated and polychlorinated biphenyls, and organochlorine pesticides in human serum and milk using comprehensive two-dimensional gas chromatography isotope dilution time-of-flight mass spectrometry. *Anal. Chem.*, **76** (21), 6313–6320.

Guth, H. (1996) Use of the Moving Capillary Switching System (MCSS) in combination with stable Isotope Dilution Analysis (IDA) for the quantification of a trace component in wine. Poster at the 18th International Symposium on Capillary Chromatography, Riva del Garda, Italy, May 20–24, 1996.

Hamilton, J., Webb, P., Lewis, A., Hopkins, J., Smith, S., and Davy, P. (2004) Partially oxidised organic components in urban aerosol using GCxGC-TOF/MS. *Atmos. Chem. Phys. Discuss.*, **4**, 1393–1423.

Hoh, E., Lehotay, S.J., Mastovska, K., and Huwe, J.K. (2007) Evaluation of automated direct sample introduction with comprehensive two-dimensional gas chromatography/time-of-flight mass spectrometry for the screening analysis of dioxins in fish oil. *J. Chromatogr. A*, **1201**, 69–77.

Hollingsworth, B.V., Reichenbach, S.E., Tao, Q., and Visvanathan, A. (2006) Comparative visualization for comprehensive two-dimensional gas chromatography. *J. Chromatogr. A*, **1105**, 51–58.

Horii, Y., Petrick, G., Katase, T., Gamo, T., and Yamashita, N. (2004) Congener-specific carbon isotope analysis of technical PCN and PCB preparations using 2DGC IRMS. *Organohalogen Comp.*, **66**, 341–348.

Kellner, R., Mermet, J.-M., Otto, M., Valcarcel, M., and Widmer, H.M. (2004) *Analytical Chemistry*, 2nd edn, Wiley-VCH Verlag GmbH, Weinheim.

Khan, A. (2013) Optimizing GC Parameters for Faster Separations with Conventional Instrumentation. Technical Note TN20743, 02-2013, Thermo Fisher Scientific.

Leco Corp (2005) The Use of Resample, a New chromaTOF Feature to Improve Data Processing for GCxGC-TOFMS. Separation Science Application Note 02/05.

Liu, Z. and Phillips, J.B. (1991) *J. Chrom. Sci.*, **29**, 227.

Marriott, P.J., Morrison, P.D., Shellie, R.A., Dunn, M.S., Sari, E., and Ryan, D. (2003) Multidimensional and comprehensive two-dimensional gas chromatography. *LCGC Eur.*, **12**, 2–10.

McNamara, K., Leardib, R., and Hoffmann, A. (2003) Developments in 2D GC with heartcutting. *LCGC Eur.*, **12**, 14–22.

Mondello, L., Tranchida, P.Q., Dugo, P., and Dugo, G. (2008) Comprehensive two-dimensional gas chromatography-mass spectrometry: A review. *Mass Spectrom. Rev.*, **27**, 101–124.

Patterson, D.G., Welch, S.M., Focant J.F. and Turner, W.E. (2006) The use of various gas chromatography and mass spectrometry techniques for human biomonitoring studies. Presented at the Dioxin 2006 Conference, FCC-2602 – 409677, Oslo, Norway, 2006.

Reichenbach, S.E., Ni, M., Kottapalli, V., and Visvanathan, A. (2004) Information technologies for comprehensive two-dimensional gas chromatography. *Chemom. Intell. Lab. Syst.*, **71**, 107–120.

Sulzbach, H. (1991) Controlling chromatographic gas streams by adjusting relative positioning off supply columns in connector leading to detector, Carlo Erba Strumentazione GmbH, Germany. Patent DE 4017909.

Welthagen, W., Schnelle-Kreis, J., and Zimmerman, R. (2003) Search criteria and rules for comprehensive two-dimensional gas chromatography-time-of-flight mass spectrometry analysis of airborne particulates. *J. Chromatogr. A*, **1019**, 233–249.

Section 2.2.11 Classical Detectors for GC-MS Systems

Bretschneider, W. and Werkhoff, P. (1988a) Progress in all-glass stream splitting systems in capillary gas chromatography part I. *HRC & CC*, **11**, 543–546.

Bretschneider, W. and Werkhoff, P. (1988b) Progress in all-glass stream splitting systems in capillary gas chromatography part II. *HRC CC*, **11**, 589–592.

Delahunty, C.M., Eyres, G., and Dufour, J.-P., (2006) Gas chromatography-olfactometry. *J. Sep. Sci.*, **29** (14), 2107–2125.

Ewender, J. and Piringer, O. (1991) Gaschromiatographische Analyse flüchtiger aliphatischer Amine unter Verwendung eines Amin-spezifischen Elektrolytleitfähigkeitsdetektors. *Dt. Lebensm. Rundsch.*, **87**, 5–7.

Hill, H.H. and McMinn, D.G. (1992) *Detctors for Capillary Gas Chromatography*, John Wiley & Sons Inc., New York.

Piringer, O. and Wolff, E. (1984) New electrolytic conductivity detector for capillary gas chromatography – Analysis of chlorinated hydrocarbons. *J. Chromatogr.*, **284**, 373–380.

Schneider, W., Frohne, J.C., and Brudderreck, H. (1982) Selektive gaschromatographische Messung sauerstoffhaltiger Verbindungen mittels Flammenionisationsdetektor. *J. Chromatogr.*, **245**, 71.

Wentworth, W.E. and Chen, E.C.M. (1981) in *Electron Capture Theory and Practice in Chromatography* (eds A. Zlatkis and C.F. Poole), Elsevier, New York, 27.

Section 2.3 Mass Spectrometry
Section 2.3.1 Ionization Processes

Borman, S. (1998) A Brief History in Mass Spectrometry, May 26, 1998, *http://masspec.scripps.edu/information/history/index.html* (accessed 14 May 2014).

Brubaker, W.M. (1968) An improved quadrupole mass spectrometer analyser. *J. Adv. Mass Spectrom.*, **4**, 293–299.

Cotter, R.J. (1997) *Time-of-Flight Mass Spectrometry*, American Chemical Society, Washington, DC.

Cotter, R.J. (1992) *Anal. Chem.*, **64**, 1027A.

Dawson, P.H. and Whetten, N.R. (1968) Mass spectrometry using radio-frequency quadrupole fields, N. R. *J. Vac. Sci. Technol.*, **5**, 1.

Dimandja, J.M. (2003) A new tool for the optimized analysis of complex volatile mixtures: Comprehensive two-dimensional gas chromatography/time-of-flight mass spectrometry. *Am. Lab.*, **2**, 42–53.

Eljarrat, E. and Barcelo, D. (2002) Congener-specific determination of dioxins and related compounds by gas chromatography coupled to LRMS, HRMS, MS/MS and TOFMS. *J. Mass Spectrom.*, **37** (11), 1105–1117.

Fürst, personal communication (1994).

Freeman, J.A. (1985) How to Select High Vacuum Pumps, Microelectronic Manufacturing and Testing, October, 1985 (Balzers reprint 1985).

GC Image 2015 — GC Image GCxGC Edition Users' Guide., *www.gcimage.com* (accessed 27 May 2014).

Fridgerio, A. (1974) *Essential Aspects of Mass Spectrometry*, John Wiley & Sons Inc.

Guilhaus, M. (1995) Review of TOF-MS. *J. Mass Spectrom.*, **30**, 1519.

Jemal, M. and Quyang, Z. (2003) *Rapid Commun. Mass Spectrom.*, **17**, 24–38.

Kaklamanos, G., Vincent, U., and von Holst, C. (2013) Multi-residue method for the detection of veterinary drugs in distillers grains by liquid chromatography-Orbitrap high resolution mass spectrometry. *J. Chromatogr. A*, **1322**, 38–48.

Liu, Y.-M., Akervik, K. and Maljers, L. (2006) Optimized high resolution SRM quantitative analysis using a calibration correction method on a triple quadrupole system. ASMS 2006 Poster Presentation, TP08, #115.

Makarov, A. (2000) Electrostatic axially harmonic orbital trapping: A high-performance technique of mass analysis. *Anal. Chem.*, **72** (6), 1156–1162.

Malavia, J., Santos, F.J., and Galceran, M.T. (2008) Comparison of gas chromatography-ion-trap tandem mass spectrometry systems for the determination of polychlorinated dibenzo-p-dioxins, dibenzofurans and dioxin-like polychlorinated biphenyls. *J. Chromatogr. A*, **1186**, 302–311.

Mamyrin, B.A. *et al.* (1973) *Sov. Phys. – JETP*, **37**, 45.

Mamyrin, B.A. *et al.* (1994) *Int. J. Mass Spectrom. Ion Processes*, **131**, 1.

Mamyrin, B.A. (2000) *Int. J. Mass Spectrom.*, **206**, 251–266.

McClenathan, D. and Ray, S.J. (2004) Plasma source TOFMS. *Anal. Chem.*, **76** (9), 159A–166A.

Meruva, N.K., Sellers, K.W., Brewer, W.E., Goode, S.R. and Morgan, S.L. (2000) Comparisons of chromatographic performance and data quality using fast gas chromatography. Paper 1397, Pittcon 2000, New Orleans, LA, 17 March 2000.

Miller, P.E. and Denton, M.B. (1986) The quadrupol mass filter: Basic operating concepts. *J. Chem. Educ.*, **63**, 617–622.

Paul, W., Reinhard, H.P., and von Zahn, U. (1958) Das elektrische Massenfilter als Massenspektrometer und Isotopentrenner. *Z. Phys.*, **152**, 143–182.

Paul, W. and Steinwedel, H. (1953) Ein neues Massenspektrometer ohne Magnetfeld. *Z. Naturforsch.*, **8 a**, 448–450.

Paul, W. and Steinwedel, H. (1956) Apparat zur Trennung von geladenen Teilchen mit unterschiedlicher spezifischer Ladung. Deutsches Patent 944 900 (U.S. Patent 2 939 952 v. 7. 6. 1960).

Peterson, A.C., McAlister, G.C., Quarmby, S.T., Griep-Raming, J., and Coon, J.J. (2010) Development and characterization of a GC-enabled QLT-orbitrap for high-resolution and high-mass accuracy GC-MS. *Anal. Chem.*, **82** (20), 8618–8628.

Peterson, A., Quarmby, S.T., McAllister, G.C. and Coon, J.J. (2009) Implementation of an EI/CI Interface on a hybrid Orbitrap system for ultra high resolution GC-MS. ASMS Presentation 2009.

Pfeiffer Vaccum (2003) *Working with Turbopumps. Technical Report PT 0053 PE*.

Schulz, J. (1987) *Nachweis und Quantifizieren von PCB mit dem Massenselektiven Detektor.*, *LaborPraxis*, **6**, 648–667.

Webb, K. (2004a) Methodology for Accurate Mass Measurement of Small Molecules, VIMMS/2004/01, LGC Ltd., Teddington, November 2004.

Webb, K. (2004b) Resolving Power and Resolution in Mass Spectrometry, Best Practice Guide, VIMMMS/2004/01, LGC Ltd., Teddington, November 2004, *www.lgc.co.uk* (accessed 14 May 2014).

Wiley, W.C. and MacLaren, I.H. (1955) *Rev. Sci. Instrum.*, **26**, 1150.

Xian, F., Hendrickson, C.L., and Marshall, A.G. (2012) High resolution mass spectrometry. *Anal. Chem.*, **84**, 708–719.

Section 2.3.2 Resolution Power

Ardenne, M., Steinfelder, K., and Tümmler, R. (1971) *Elektronenanlagerungsmassenspektrometrie organischer Substanzen*, Springer-Verlag, Berlin.

Aue, D.H. and Bowers, M.T. (1979) Stability of positive ions from equilibrium gas-phase basicity measurements, in *Gas Phase Ion Chemistry*, Vol. 2 (ed. M.T. Bowers), Academic Press, New York.

Bartmess, J. and McIver, R. (1979) The gas phase acidity scale, in *Gas Phase Ion Chemistry*, Vol. 2 (ed. M.T. Bowers), Academic Press, New York.

Beck, H., Eckart, K., Mathar, W., and Wittkowski, R. (1988) PCDD and PCDF body burden from food intake in the Federal Republic of Germany. *Chemosphere*, **18**, 417–424.

Bowadt, F.E. *et al* (1993) Combined positive and negative ion chemical ionisation for the analysis of PCB'S, in *15th International Symposium on Capillary Chromatography, Riva del Garda, May 1993* (ed. P. Sandra), Huethig.

Budzikiewicz, H. (1981) Massenspektrometrie negativer Ionen. *Angew. Chem.*, **93**, 635–649.

Budzikiewicz, H. (2005) *Massenspektrometrie*, 5 Ed., Wiley-VCH Verlag, Weinheim, Germany.

Buser, H.-R. and Müller, M. (1994) Isomer- and enantiomer-selective analyses of toxaphene components using chiral high-resolution gas chromatography and detection by mass spectrometry/mass spectrometry. *Environ. Sci. Technol.*, **28**, 119–128.

Chernetsova, E.S., Revelsky, A.I., Revelsky, I.A., Mikhasenko, I.A., and Sobolevsky, T.G. (2002) Determination of polychlorinated dibenzo-p-dioxins, dibenzofurans, and biphenyls by gas chromatography/mass spectrometry in the negative chemical ionization mode with different reagent gases. *Mass Spectrom. Rev.*, **21**, 373–387.

Class, T.J. (1991) Determination of pyrethroids and their degradation products in indoor air and on surfaces by HRGC-ECD and HRGC-MS (NCI), in *12th International Symposium on Capillary Chromatography, Riva del Garda, May 1991* (ed. P. Sandra), Huethig.

Cojocariu, C., Abalos, M., Abad Holgado, E., and Silcock, P. (2014) Meeting the European Commission Performance Criteria for the Use of Triple Quadrupole GC-MS/MS as a Confirmatory Method for PCDD/Fs in Food and Feed Samples, *Application Note 10380*, Thermo Fisher Scientific, Runcorn.

Crow, F.W., Bjorseth, A. *et al.* (1981) Determination of polyhalogenated hydrocarbons

by glass capillary gas chromatography-negative ion chemical ionization mass spectrometry. *Anal. Chem.*, **53**, 619.

DePuy, C.H., Grabowski, J.J., and Bierbaum, V.M. (1982) Chemical reactions of anions in the gas phase. *Science*, **218**, 955–960.

Dorey, R.C., Williams, K., Rhodes, C.L., Fossler, C.L., Heinze, T.M. and Freeman, J.P. (1994) High kinetic energy chemical ionization in the quadrupole ion trap: Methylamine CIMS of amines. Presented at the 42nd ASMS Conference on Mass Spectrometry and Allied Topics, Chicago, IL, June 1 – 6, 1994.

Dougherty, R.C. (1981) Negative chemical ionization mass spectrometry. *Anal. Chem.*, **53**, 625A–636A.

European Commission (2006) Commission regulation No. 1883. *Off. J. Eur. Union*, **L364**, 32–43.

European Commission (2012) Commission regulation No. 252. *Off. J. Eur. Union*, **L84**, 1–22.

European Commission (2014) Commission regulation No. 589. *Off. J. Eur. Union*, **L164**, 18–40.

Fürst, P., and Bernsmann, T. (2010) High resolution LC-MS beta-agonist screening using turboflow/exactive. Presentation at the 5th Thermo Scientific High Resolution GC-MS Meeting on POPs, Barcelona, Spain, 29-30 April, 2010.

Grimme, S. (2013) Towards first principles calculation of electron impact mass spectra of molecules. *Angew. Chem. Int. Ed.*, **52**, 6306–6312.

Gross, J.H. (2004) *Mass Spectrometry*, Springer, Heidelberg.

Hainzl, D., Burhenne, J., and Parlar, H. (1994) Isolierung von Einzelsubstanzen für die Toxaphenanalytik. *GIT Fachz. Lab.*, **4**, 285–294.

Harrison, A.G. (1992) *Chemical Ionization Mass Spectrometry*, CRC Press, Boca Raton, FL.

Horning, E.C., Caroll, D.I., Dzidic, I., and Stillwell, R.N. (1981) Negative ion atmospheric pressure ionization mass spectrometry and the electron capture detector, in *Electron Capture*, Journal of Chromatography Library, Vol. 20 (eds A. Zlatkis and C.F. Poole), Elsevier, Amsterdam.

Howe, I., Williams, D.H., and Bowen, R. (1981) *Mass Spectrometry*, 2nd edn, McGraw Hill, New York.

Hunt, D.F., Stafford, G.C., Crow, F., and Russel, J. (1976) Pulsed positive negative ion chemical ionization mass spectrometry. *Anal. Chem.*, **48**, 2098–2105.

Keller, P.R., Harvey, G.J. and Foltz, D.J. (1989) GC-MS Analysis of Fragrances Using Chemical Ionization on the Ion Trap Detector: An Easy-to-Use Method for Molecular Weight Information and Low Level Detection, Finnigan MAT Application Report No. 220.

Kotz, A., Malisch, R., Focant, J., Eppe, G., Cederberg, T.L., Rantakokko, P., Fürst, P., Bernsmann, T., Leondiadis, L., Lovasz, C., Scortichini, G., Diletti, G., di Domenico, A., Ingelido, A.M., Traag, W., Smith, F., Fernandes, A. (2012) *Organohalogen Compd.*, **74**, 156–159.

McLafferty, F.W. and Michnowicz, J.A. (1992) *Chem. Technol.*, **22**, 182.

McLafferty, F.W. and Turecek, F. (1993) *Interpretation of Mass Spectra*, 4th edn, University Science Books, Mill Valley, CA.

Munson, M.S.B. and Field, F.H. (1966) *J. Am. Chem. Soc.*, **b88**, 4337.

Smit, A.L.C. and Field, F.H. (1977) Gaseous anion chemistry. Formation and reaction of OH^-; Reactions of anions with N_2O; OH^- negative chemical ionization. *J. Am. Chem. Soc.*, **99**, 6471–6483.

Spiteller, M. and Spiteller, G. (1973) *Massenspektrensammlung von Lösungsmitteln, Verunreinigungen. Säulenbelegmaterialien und einfachen aliphatischen Verbindungen*, Springer, Wien.

Stan, H.-J. and Kellner, G. (1981) Analysis of organophosphoric pesticide residues in food by GC-MS using positive and negative chemical ionisation, in *Recent Developments in Food Analysis* (eds W. Baltes, P.B. Czedik-Eysenberg, and W. Pfannhauser), Verlag Chemie, Weinheim.

Stout, S.J. and Steller, W.A. (1984) Application of gas chromatography negative ion chemical ionization mass spectrometry in confirmatory procedures for pesticide residues. *Biomed. Mass Spectrom.*, **11**, 207–210.

Theobald, F. and Hübschmann, H.-J. (2007) *High Sensitive MID Detection Method for Toxaphenes by High Resolution GC-MS*, Application Note AN30128, Thermo Fisher Scientific, Bremen.

Yost, R.A. (1988) Analytical applications of ion trap mass spectrometry. *Spectra*, **11** (2).

Section 2.3.3 Isotope Ratio Monitoring GC-MS

Bradley, D. (2002) Tracking cocaine to its roots. *Today's Chem. Work*, **5**, 15–16.

Boato, G. (1960) Isotope fractionation processes in nature, Summer Course on Nuclear Geology Pisa, Italy.

Brand, A. (2004) in *Handbook of Stable Isotope Analytical Techniques*, Vol. 1 (ed. P.A. de Groot), Elsevier, Amsterdam, pp. 835–856.

Brand, W.A., Tegtmeyer, A.R., and Hilkert, A. (1994) *Org. Geochem.*, **21**, 585.

Craig, H. (1957) Isotopic standards for carbon and oxygen and correction factors for mass-spectrometric analysis of carbon dioxide. *Geochim. Cosmochim. Acta*, **12**, 133–149.

Ehleringer, J.R., Casale, J.F., Lott, M.J., and Ford, V.L. (2000) Tracing the geographical origin of cocaine. *Nature*, **408**, 311–312.

Ehleringer, J.R. and Cerling, T.E. (2002) in *The Earth System: Biological and Ecological Dimensions of Global Environmental Change*, Encyclopaedia of Global Environmental Change, Vol. 2 (eds H.A. Mooney and J.G. Canadell), John Wiley & Sons, Ltd, Chichester, 544–550.

Fry, B. (2006) *Stable Isotope Ecology*, Springer, New York.

Groening, M. (2004) in *Handbook of Stable Isotope Analytical Techniques*, Vol. 1 (ed. P.A. de Groot), Elsevier, Amsterdam, 875–906.

Heuer, K., Brand, W.A., Hilkert, A.W., Juchelka, D., Mosandl, A., and Podebred, F. (1998) *Z. Lebensm. Unters. Forsch.*, **206**, 230.

Hilkert, A., Douthitt, C.B., Schlüter, H.J., and Brand, W.A. (1999) Isotope ratio monitoring gas chromatography/mass spectrometry of D/H by high temperature conversion isotope ratio mass spectrometry. *Rapid Commun. Mass Spectrom.*, **13**, 1226–1230.

Matthews, D.E. and Hayes, J.M. (1978) *Anal. Chem.*, **50**, 1465.

Matucha, M. (1991) *J. Chromatogr.*, **588**, 251–258.

Meyer-Augenstein, W. (1997) The chromatographic side of isotope ratio mass spectrometry: Pitfalls and answers. *LCGC Int.*, **10**, 17–25.

Rautenschlein, M., Habfast, K., and Brand, W. (1990) in *Stable Isotopes in Paediatric, Nutritional and Metaboloc Research* (eds T.E. Chapman, R. Berger, D.J. Reijngoud, and A. Okken), Intercept Ltd, Andover, 133–148.

Rosman, K.J.R. and Taylor, P.D.P. (1998) Isotopic compositions of the elements 1997. *Pure Appl. Chem.*, **70** (1), 217–235.

Rossmann, A. (2001) Determination of stable isotope ratios in food analysis. *Food Rev. Int.*, **17** (3), 347–381.

Sano, M. *et al.* (1976) A new technique for the detection of metabolites labelled by the isotope 13C using mass fragmentography. *Biomed. Mass Spectrom.*, **3**, 1–3.

Sharp, Z. (2007) *Principles of stable isotope geochemistry*, Pearson Prentice Hall, Upper Saddle River, NJ.

St. Jean, G. (2002) Basic principles of stable isotope geochemistry. Short Course Manuscript of the 9th Canadian CF-IRMS Workshop, August 25, 2002.

de Vries, J.J. (2000) Natural abundance of the stable isotopes of C, O and H, in: Mook, W.G. Ed., *Environmental Isotopes in the Hydrological Cycle*, Vol. 1: Introduction – Theory, Methods, Review, IAEA, Vienna. *www.iaea.org* (accessed 13 May 2014).

Section 2.3.4 Acquisition Techniques in GC-MS

Brodbelt, J.S. and Cooks, R.G. (1988) Ion trap tandem mass spectrometry. *Spectra*, **11** (2), 30–40.

Busch, K.L., Glish, G.L., and McLuckey, S.A. (1988) *Mass Spectrometry/Mass Spectrometry: Techniques and Applications of Tandem Mass Spectrometry*, VCH Publishers, New York.

Cooks, R.G. and Busch, K.L. (1982) Counting molecules by desorption ionization and mass spectrometry/mass spectrometry. *J. Chem. Educ.* **59** (11), 926–933.

Thier, H.P. and Kirchhoff, J. (eds) (1992) *DFG Deutsche Forschungsgemeinschaft, Manual of Pesticide Residue Analysis*, Vol. II, Wiley-VCH Verlag GmbH, Weinheim, pp. 25–28.

Johnson, J.V. and Yost, R.A. (1985) *Anal. Chem.*, **57**, 758A.

Johnson, J.V., Yost, R.A., Kelley, P.E., and Bradford, D.C. (1990) Tandem-in-space and tandem-in-time mass spectrometry: Triple quadrupoles and quadrupole ion traps. *Anal. Chem.*, **62**, 2162–2172.

Kaiser, H. (1978) Foundations for the critical discussion of analytical methods. *Spectrochim. Acta Part B*, **33 b**, 551.

March, R.E. and Hughes, R.J. (1989) *Quadrupole Storage Mass Spectrometry*, John Wiley & Sons, Inc., New York.

McLafferty, F.E. (ed.) (1983) *Tandem Mass Spectrometry*, John Wiley & Sons, Inc., New York.

Noble, D. (1995) MS/MS flexes ist muscles. *Anal. Chem.*, **67**, 265A–269A.

Plomley, J.B., Koester, C.J., and March, R.E. (1994) Determination of N-nitroso-dimethylamine in complex environmental matrices by quadrupole ion storage tandem mass spectrometry enhanced by unidirectional ion ejection. *Anal. Chem.*, **66**, 4437–4443.

Slayback, J.R.B. and Taylor, P.A. (1983) Analysis of 2.3.7,8- TCDD and 2,3,7,8-TCDF in environmental matrices using GC-MS/MS techniques. *Spectra*, **9** (4), 18–24.

Soni, M.H. and Cooks, R.G. (1994) Selective injection and isolation of ions in quadrupole ion trap mass spectrometry using notched waveforms created using the inverse Fourier transform. *Anal. Chem.*, **66**, 2488–2496.

Wagner-Redeker, W., Schubert, R. and Hübschmann, H.-J. (1985) Analytik von Pestiziden und polychlorierten Biphenylen mit dem Finnigan 5100 GC-MS-System. Finnigan MAT Application Report No. 58, 1985.

Perfluorotributylamine (PFTBA, FC43) Reference Table, 0000 Thermo Fisher Scientific Data Sheet PS30040_E.

Thermo Electron Corp. (2004) Polychlorinated Dibenzodioxins and -furans, Data Sheet PS30042_E.

Yost, R.A. (1983) MS/MS: Tandem mass spectrometry. *Spectra*, **9** (4), 3–6.

Section 2.3.5 Mass Calibration

Audi, G. and Wapstra, A.H. (1995) The 1995 update to the atomic mass evaluation. *Nucl. Phys. A*, **595**, 409–480.

Eichelberger, J.W., Harris, L.E., and Budde, W.L. (1975) DFTPP tuning. *Anal. Chem.*, **47**, 995.

Mohr, P.J. and Taylor, B.N. (1999) CODATA recommended values of the fundamental physical constants: 1998. *J. Phys. Chem. Ref. Data*, **28** (6), 1713–1852.

3
Evaluation of GC-MS Analyses

3.1
Display of Chromatograms

Chromatograms obtained by gas chromatography (GC)/mass spectrometry (MS) are plots of the signal intensity against the retention time, as with classical GC detectors. Nevertheless, there are considerable differences between the two types of chromatogram arising from the fact that data from GC-MS analyses are in three dimensions. Figure 3.1 shows a section of the chromatogram of the total ion current (TIC) in the analysis of volatile halogenated hydrocarbons. The retention time axis also shows the number of continually registered mass spectra (scan no.). The mass axis is drawn above the time axis at an angle. The elution of each individual substance can be detected by evaluating the mass spectra using a 'maximizing masses peak finder' program and can be shown by a marker. Each substance-specific ion shows a local maximum at these positions, which can be determined by the peak finder. The mass spectra of all the analytes detected are shown in a three-dimensional representation for the purposes of screening. For further evaluation, the spectra can be examined individually.

3.1.1
Total Ion Current Chromatograms

The intensity axis in GC-MS analysis is shown as a total ion chromatogram (TIC) or as a calculated ion chromatogram (reconstructed ion chromatogram, RIC). The intensity scale may be given in absolute values, but a percentage scale is more frequently used. Both terms describe the mode of representation characteristic of the recording technique. At constant scan rates the mass spectrometer plots spectra over the preselected mass range and thus gives a three-dimensional data field arising from the retention time, mass scale and intensity. A signal parameter equivalent to FID (flame ionization detector) detection is not directly available. (Magnetic sector mass spectrometers initially were equipped with a TIC detector directly at the ion source until the end of the 1970s!) A total signal intensity comparable to the FID signal at a particular point in the scan can, however, be calculated from the sum of the intensities of all the ions at this point. All the ion intensities of a mass spectrum are added together by the data system and stored as a total

Handbook of GC-MS: Fundamentals and Applications, Third Edition. Hans-Joachim Hübschmann.
© 2015 Wiley-VCH Verlag GmbH & Co. KGaA. Published 2015 by Wiley-VCH Verlag GmbH & Co. KGaA.

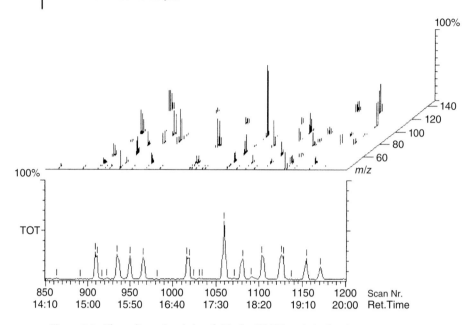

Figure 3.1 Three-dimensional data field of a GC-MS analysis showing retention time, intensity and mass axis.

intensity value (TIC) together with the spectrum. The TIC chromatogram thus constructed is therefore dependent on the scan range used for data acquisition. When making comparisons it is essential to take the data acquisition conditions into consideration (Barwick *et al.*, 2006).

SIM (selected ion monitoring)/MID (multiple ion detection) or SRM (selected reaction monitoring)/MRM (multiple reaction monitoring) analyses give a chromatogram in the same way but no mass spectrum is retrievable. The TIC in this case is composed of the intensities of the selected ions. Analyses, where switching from individual masses to fixed retention times is planned, often show clear jumps in the base line (see Figure 2.162).

The appearance of a GC-MS chromatogram (TIC/RIC) showing the peak intensities is therefore strongly dependent on the mass range chosen for display. The repeated GC-MS analysis of one particular sample employing mass scans of different widths leads to peaks of different heights above the base line of the TIC. The starting mass of the scan has a significant effect here. The result is a more or less strong recording of an unspecific background which manifests itself in a higher or lower base line of the TIC chromatogram. Peaks of the same concentration are therefore shown with different signal/noise (S/N) ratios in the TIC at different scan ranges. In spite of differing representation of the substance peaks, the detection limit of the GC-MS system naturally does not change. Particularly in trace analysis the concentration of the analytes is usually of the same order of magnitude or even below that of the chemical noise (matrix) in spite of good sample processing so that the TIC cannot represent the elution of these analytes. Only the use of selective information from the mass chromatogram (see Section 3.1.2) brings the substance peak sought on to the screen for further evaluation.

In the case of data acquisition using selected individual masses (SIM/MID or SRM/MRM), only the intensity changes of the masses selected by the data acquisition parameter are shown. Already during data acquisition only those signals (ion intensities) which correspond to the acquisition mass range selected by the user are recorded from the TIC. The greater part of the ion current generated by the ion source is therefore not detected using the SIM/MID or SRM/MRM techniques (see Section 2.3.3). Only substances which give signals of the selected masses as fragment or molecular ions are shown as peaks. A mass spectrum for the purpose of checking identity is therefore not available. For confirmation, a mass spectrum should be acquired using an alternating full scan/SIM mode or separately in a subsequent analysis. The retention time and the relative intensities of two or three specific ion intensities are used as qualifying features. In trace analysis unambiguous detection is not possible just using one ion signal only. Positive results of an SIM/MID analysis basically require additional confirmation by a mass spectrum. SRM/MRM analyses offer the recording of an MS/MS product ion spectrum or require the monitoring of multiple transitions as additional qualifiers.

3.1.2
Mass Chromatograms

The three-dimensional data field of GC-MS analyses in the full-scan mode does not only allow the determination of the total ion intensity at a point in the scan. To show individual analytes selectively, the intensities of selected ions (masses) from the TIC are shown and plotted as an intensity/time trace (EIC or XIC extracted ion chromatogram, also mass chromatogram). A meaningful assessment of S/N ratios of certain substance peaks can only be carried out using mass chromatograms of substance-specific ions (fragment/molecular ions).

The evaluation of these mass chromatograms allows the exact determination of the detection limit using the S/N ratio of the substance-specific ion produced by a compound. With the SIM/MID mode, this ion would be detected exclusively, but a complete mass spectrum for confirmation would not be available. In the case of complex chromatograms of real samples, mass chromatograms offer the key to the isolation of co-eluting components so that they can be integrated perfectly and quantified.

In the analysis of lemons for pesticide residues, a co-elution situation was discovered by data acquisition in the full scan mode of the ion trap detector and was evaluated using mass chromatograms.

The routine testing with GC-MS in full scan mode gives a trace which differs from that using an element-specific NPD (nitrogen-phosphorous detector) detector (Figure 3.2). A large number of different peaks appear in the retention region of Quinalphos (Figure 3.3) while the NPD chromatogram shows a very clean chromatogram. The Quinalphos peak has a shoulder on the left side and is closely

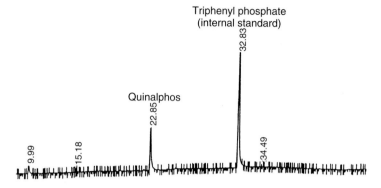

Figure 3.2 Analysis of a lemon extract using the NPD as detector. The chromatogram shows the elution of a pesticide as well as the internal standard.

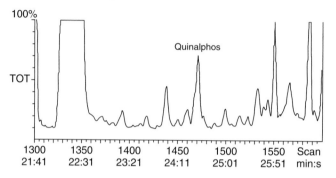

Figure 3.3 Confirmation of the identity of a lemon extract by GC-MS. The total ion current clearly shows the questionable peak with a shoulder. A co-eluting second substance, Chlorfenvinphos, gives rise to the shoulder.

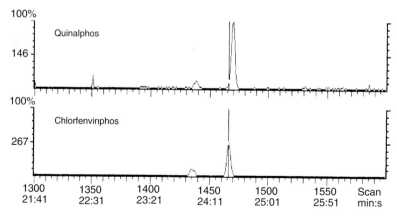

Figure 3.4 Mass chromatograms for the specific masses show the co-elution of Quinalphos (*m/z* 146) and Chlorfenvinphos (*m/z* 267). The retention time range is identical with that in Figure 3.3. The selective plot of the mass signals (EIC) confirms the coelution of both substances.

followed by another less intense component. In the mass chromatogram of the characteristic individual masses (EIC of fragment ions), it can be deduced from the TIC that another eluting substance is present (Figure 3.4). Unlike NPD detection, with GC-MS analysis, it becomes clear after evaluating the mass chromatograms and mass spectra that the co-eluting substance is Chlorfenvinphos.

In routine analysis this evaluation is carried out for target compound analysis by the data system. If the information on the retention time of an analyte, the mass spectrum, the selective quantifying mass and a valid calibration are supplied, a chromatogram can be evaluated in a very short time for a large number of components (Figure 3.5, see also Section 3.3).

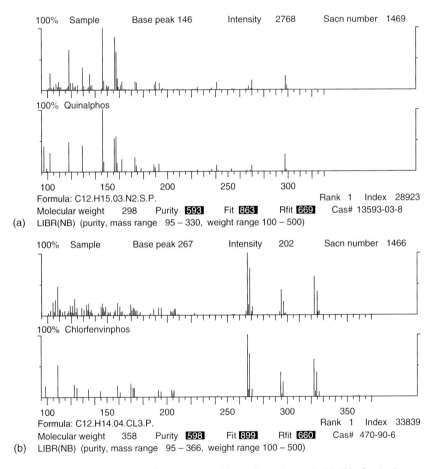

(a) Formula: C12.H15.03.N2.S.P.
Rank 1   Index 28923
Molecular weight   298   Purity **593**   Fit **863**   Rfit **669**   Cas# 13593-03-8
LIBR(NB) (purity, mass range   95 – 330, weight range 100 – 500)

(b) Formula: C12.H14.04.CL3.P.
Rank 1   Index 33839
Molecular weight   358   Purity **598**   Fit **899**   Rfit **660**   Cas# 470-90-6
LIBR(NB) (purity, mass range   95 – 366, weight range 100 – 500)

Figure 3.5 The phosphoric acid esters are successfully identified by a library comparison after extraction of the spectra by a background subtraction. (a) Quinalphos is confirmed by comparison with the NBS library (FIT value 863). (b) Chlorfenvinphos is confirmed by a comparison with the NBS library (FIT value 899). (former NBS is the predecessor of the current NIST mass spectral library).

3.2
Substance Identification

3.2.1
Extraction of Mass Spectra

One of the great strengths of mass spectrometry is the provision of direct information about an eluting component. The careful extraction of the substance-specific
signals from the chromatogram is critical for reliable identity determination. For
identification or confirmation of individual GC peaks, recording mass spectra
which are as complete as possible is an important basic prerequisite.

By plotting mass chromatograms co-elution situations can be discovered, as
shown in Section 3.1.2. The mass chromatograms of selected ions give important
information via their maximizing behaviour. Only when maxima of different
ions (m/z values) showup at exactly the same retention time can it be assumed
that the fragments observed originate from a single substance, that is, from
the same chemical structure. The only exception is the ideal simultaneous
co-elution of compounds. If peak maxima with different retention times are
shown by various ions, it must be assumed that there are co-eluting components
(see Figures 3.6 and 3.7).

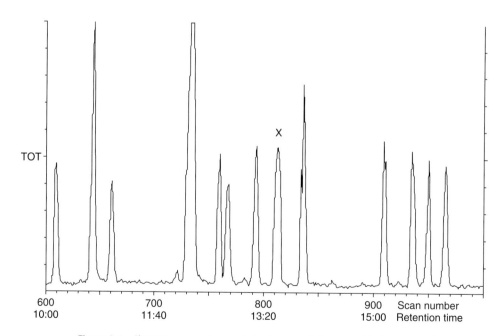

Figure 3.6 Chromatogram of an analysis of volatile halogenated hydrocarbons by purge
and trap GC-MS. The component marked with X has a larger half width than the neighbouring peaks.

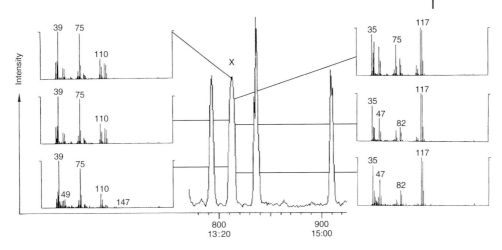

Figure 3.7 Continuous plotting of the mass spectra in the peak marked X in Figure 3.6.

3.2.1.1 Manual Spectrum Subtraction

By subtraction of the background or the co-elution spectra before or behind a questionable GC peak, the mass spectrum of the substance sought is extracted from the chromatogram as free as possible from other signals. All substances co-eluting with an unknown substance including the matrix components and column bleed are described in this context as chemical background. The differentiation between the substance signals and the background and its elimination from the substance spectrum is of particular importance for successful spectroscopic comparison in a library search. In the example of the GC-MS analysis of lemons for pesticides described earlier, this procedure is used to determine the identity of the active substances.

The possibilities for subtraction of mass spectra are shown in the following real example of the analysis of volatile halogenated hydrocarbons by purge and trap-GC-MS. Figure 3.6 shows part of a TIC chromatogram. The peak marked with X shows a larger width than that of the neighbouring components. On closer inspection of the individual spectra in the peak, it can be seen that in the rising slope of the peak ions with m/z 39, 75, 110 and 112 dominate. As the elution of the peak continues, other ion signals appear. The ions with m/z 35, 37, 82, 84, 117, 119 and 121 appear in increased strength, while the previously dominant signals decrease. Figure 3.7 shows this situation using the continuing presentation of individual mass spectra in this GC peak.

From the individual mass spectra (Figure 3.7), it can be recognized that some signals obviously belong together. Figure 3.8 shows the mass chromatograms of the ions m/z 110/112 and m/z 117/119 (as a sum in each case) above the TIC from the detected mass range of m/z 33–260. The mass chromatograms show an intense GC peak at the questionable retention time in each case. The peak maxima are not superimposed and are slightly shifted towards each other. This is an important indication of the co-elution of two components (Figure 3.9).

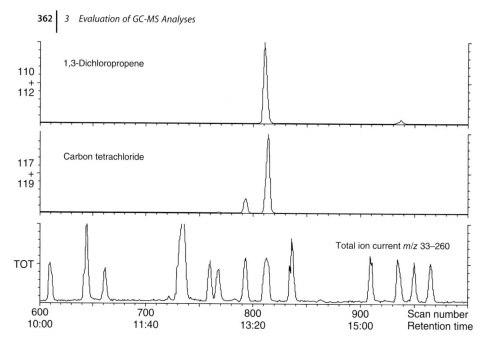

Figure 3.8 Mass chromatograms for *m/z* 110/112 and *m/z* 117/119 shown above a total ion current chromatogram.

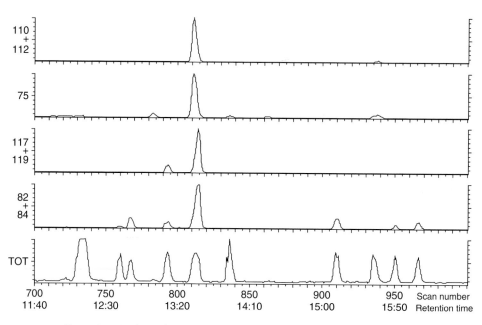

Figure 3.9 Analysis of a co-elution situation by inclusion of additional fragment ions.

If other ions are included in this first mass analysis, it can be concluded that the fragments belong together from their common maximizing masses behaviour (Biller and Biemann, 1974).

After the individual mass signals have been assigned to the two components, the extraction of the spectrum for each compound can be performed. Figure 3.10 shows the division of the peak into the front peak slope A and the back peak slope B. With the background subtraction function contained in all data systems, the spectra in the shaded areas A and B are added and subtracted from one another.

The subtraction of the areas A and B gives the clean spectra of the co-eluting analytes. In Figure 3.11 the subtraction A − B shows the spectrum of the first component, which is shown to be 1,3-dichloropropene (Figure 3.12) from a library comparison. The reverse procedure, that is, the subtraction B − A, gives the identity of the second component (Figures 3.13 and 3.14).

A further frequent use of spectrum subtraction allows the removal of background signals caused by the matrix or column bleed. Figure 3.15 shows the elution of a minor component from the analysis of volatile halogenated hydrocarbons, which elutes in the region where column bleed begins. For background subtraction, the spectra in the peak and from the region of increasing column bleed are subtracted from one another.

The result of the background subtraction is shown in Figure 3.16. While the spectrum clearly shows column bleed from the substance peak with m/z 73, 207 and a weak CO_2 background at m/z 44, the resulting substance spectrum is free from the signals of the interfering chemical background after the subtraction. This

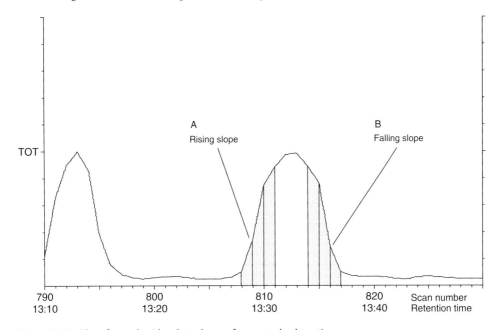

Figure 3.10 Plot of a peak with selected areas for spectral subtraction.

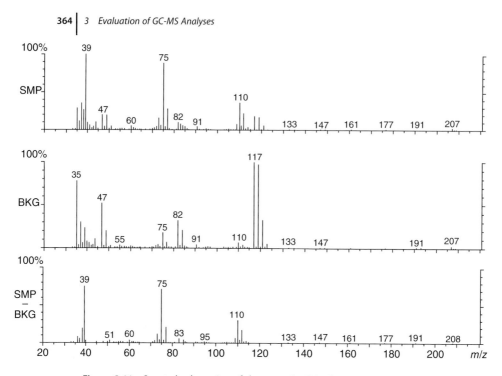

Figure 3.11 Spectral subtraction of the areas A − B in Figure 3.10:
SMP   = spectra of the rising peak slope (A), sample
BKG   = spectra of the falling peak slope (B), background
SMP-BKG = resulting spectrum of the component eluting first.

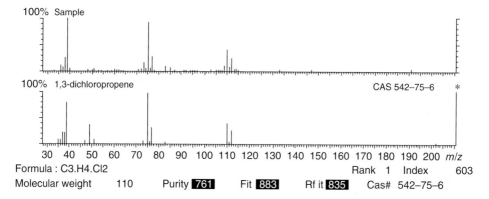

Formula : C3.H4.Cl2   Rank 1   Index   603
Molecular weight   110   Purity 761   Fit 883   Rf it 835   Cas# 542–75–6

Figure 3.12 Identification of the first component by library comparison.

clean spectrum can then be used for a library search in which it can be confirmed
as 1,2-dibromo-3-chloropropane.

In the subtraction of mass spectra, it should generally be noted that in certain
cases primary substance signals can also be reduced. In these cases, it is necessary
to choose another background area. If changes in the substance spectrum cannot

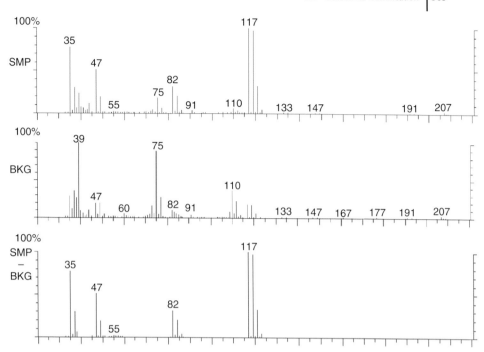

Figure 3.13 Spectral subtraction of the areas B – A in Figure 3.10:
SMP = spectra of the falling peak slope (B), sample
BKG = spectra of the rising peak slope (A), background
SMP-BKG = resulting spectrum of the component eluting second.

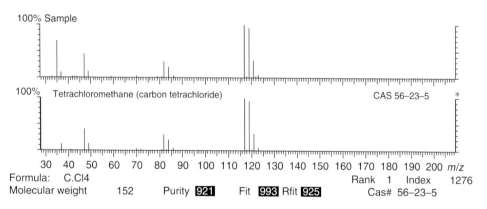

Figure 3.14 Identification of the second component by library comparison.

be prevented in this way, the library search should be carried out with a small proportion of chemical noise. In the library search programs of individual manu-facturers, there is also the possibility of editing the spectrum before the start of the search. In critical cases, this option should also be followed to remove known interfering signals resulting from the chemical noise from the substance spectrum.

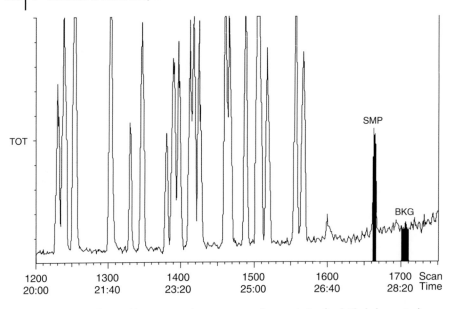

Figure 3.15 The total ion current chromatogram of an analysis of volatile halogenated hydrocarbons shows the elution of a minor component at the beginning of column bleed (the areas of background subtraction are shown in black).

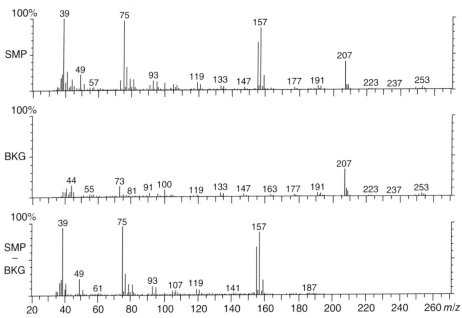

Figure 3.16 Result of background subtraction from Figure 3.15:
SMP   = spectra from the substance peak, sample
BKG   = spectra from the background (column bleed)
SMP – BKG = resulting substance spectrum.

3.2.1.2 **Deconvolution of Mass Spectra**

The advancements in full-scan sensitivity, especially in ion trap, Orbitrap and time-of-flight (TOF) MS instrumentation as well as the increased application of fast and two-dimensional GC methods is creating a strong demand for post-acquisition deconvolution methods (Shao and Wang, 2004, Dimandja, 2004, Figure 3.17). The extraction of pure spectra from compounds co-eluting with other analytes or interferences with conventional background subtraction methods of GC-MS data systems is of limited use and cannot recognize the transient dependence of ion intensities of multiple compounds eluting close together.

An automated mass spectra deconvolution and identification system (AMDIS) was developed at the National Institute of Standards and Technology (NIST) with the support of the Special Weapons Agency of the Department of Defense, for the critical task of verifying the Chemical Weapons Convention ratified by the United States Senate in 1997. In order to meet the rigorous requirements for this purpose, AMDIS was tested against more than 30 000 GC-MS data files accumulated by the EPA Contract Laboratory Program without a single false positive for the target set of known chemical warfare agents. While this level of reliability may not be required for all laboratories, this shows the degree to which the algorithms have been tested. After 2 years of development and extensive testing, it has been made available to the general analytical chemistry community for download on the Internet.

The AMDIS program analyses the individual ion signals and extracts and identifies the spectrum of each component in a mixture analysed by GC-MS. The software comprises an integrated set of procedures for first extracting the pure component spectra from the chromatogram and then to identify the compound by a reference library (Figure 3.18–3.23).

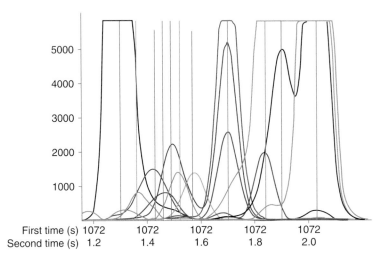

| First time (s) | 1072 | 1072 | 1072 | 1072 | 1072 |
| Second time (s) | 1.2 | 1.4 | 1.6 | 1.8 | 2.0 |

Figure 3.17 Deconvolution of overlapping peaks in GC × GC/TOF-MS. Every vertical line indicates the peak of an identified component. (Dimandja, 2004, reprinted with permission from Analytical Chemistry, Copyright 2004 by American Chemical Society.)

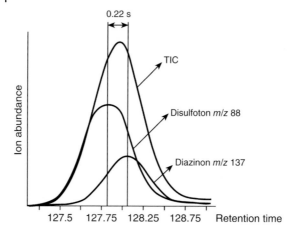

Figure 3.18 Deconvolution example (Meruva, 2002).

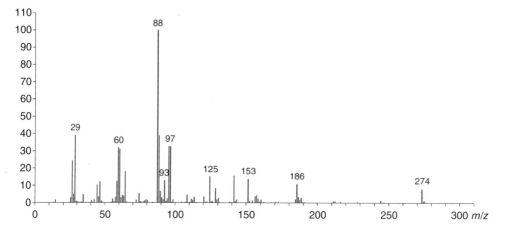

Figure 3.19 Disulfoton, spectrum pure compound, NIST#: 52689.

The overall process involves four sequential steps in spectrum purification and identification:

1) Noise analysis by a complete analysis of noise signals with the use of this information for component perception. A correction for baseline drift is done for each component in case the chromatogram does not have a flat baseline.
2) Component perception identifies the location of each of the eluted components on the retention time scale by investigating the elution peak profile.
3) True spectral 'deconvolution' of the data. Even if there is no available constant background for subtraction, AMDIS extracts clean spectra. The extraction of closely co-eluting components is possible even for analytes that peak within a single scan of each other in a wide range of each component's concentration.
4) Library search for compound identification to match each deconvoluted spectrum to a reference library spectrum.

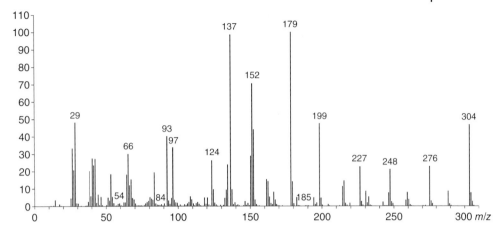

Figure 3.20 Diazinone, spectrum pure compound, NIST#: 147231.

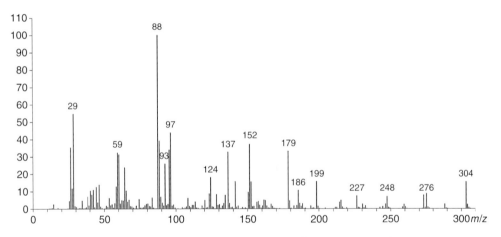

Figure 3.21 Co-elution spectrum 3 : 1 (at peak maximum of disulfoton in Figure 3.18).

Unlike a traditional identification algorithm, AMDIS includes uncertainties in the deconvolution, purity and retention times in the match factor. The final match factor is a measure of both the quality of the match and of the confidence in the identification.

AMDIS can operate as a 'black box' chemical identifier, displaying all identifications that meet a user-selectable degree of confidence. Identification can be aided by internal standards and retention times. Also employed can be retention index (RI) data when identifying target compounds and internal and external standards are maintained in separate libraries. AMDIS reads GC-MS raw data files in the formats of the leading GC-MS manufacturer or is already integrated in the instrument data systems.

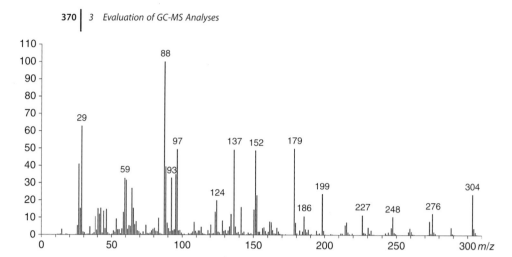

Figure 3.22 Co-elution spectrum 2:1 (at peak maximum of diazinone in Figure 3.18).

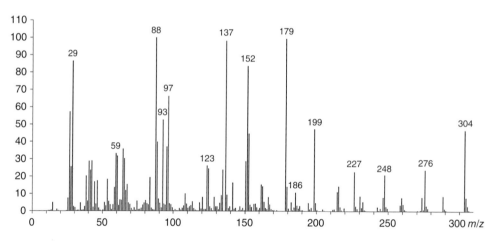

Figure 3.23 Co-elution spectrum 1:1 (at right peakside in Figure 3.18).

With its unique deconvolution algorithms, AMDIS has proven its capabilities for the efficient removal of overlapping interferences in many GC-MS applications (Mallard and Reed, 1997). The deconvolution process is independent from the type of analyser and scan rate used to resolve overlapping peaks for substance identification as well as multi component residue analysis (Dimandja, 2004; Mallard *et al.*, 2005; Zhang, Wu and Li, 2006). Without time-consuming manual data evaluation, AMDIS provides sensitive compound information even with complex background present (Halket *et al.*, 1999).

AMDIS

Has been designed to reconstruct 'pure component' spectra from complex GC-MS chromatograms even when components are present at trace levels. For this purpose, observed chromatographic behaviour is used along with a range of noise-reduction methods. AMDIS works as well with specialized libraries (environmental, flavour and fragrance and drugs and toxins) which can be linked into the search with the NIST Library. AMDIS has a range of other features, including the ability to search the entire NIST Library with any of the spectra extracted from the original data file. It can also employ RI information when identifying target compounds and can make use of internal and external standards maintained in separate libraries. A history list of selected performance standards is also maintained.

As of version 2.72, AMDIS reads data files in the following formats:

Agilent Files	Micromass Files
Agilent MS Engine Files	mzXML/mzData Files
Bruker Files	NetCDF Files
Finnigan GCQ Files	PerkinElmer Files
Finnigan INCOS Files	Shimadzu MS Files
Finnigan ITDS Files	Schrader/GCMate Files
INFICON Files	Varian MS Files
JEOL/Schrader Files	Varian SMS Files
Kratos Mach3 Files	Varian XMS Files
MassLynx NT Files	Xcalibur Raw Files

3.2.2
The Retention Index

If the chromatographic conditions are kept constant, the retention times of the compounds remain the same. All identification concepts using classical detectors function on this basis. The retention times of compounds, however, can change through ageing of the column and more particularly through differing matrix effects between samples.

The measurement of the retention times relative to a co-injected standard can help to overcome these difficulties. Fixed retention indices (RI) are assigned to these standards. An analyte is included in a RI system with the RI values of the standards eluting before and after it. It is assumed small variances in the retention times affect both the analyte and the standards so that the RI values calculated remain constant (Deutsche Forschungsgemeinschaft, 1982).

The first RI system to become widely used was developed by Kovats (Zenkevich, 1993). In this system, a series of *n*-alkanes is used as the standard. Each *n*-alkane is assigned the value of the number of carbon atoms multiplied by 100 as the retention index (pentane 500, hexane 600, heptane 700, etc.). For isothermal operations, the RI values for other substances are calculated as follows:

$$\text{Kovats index RI} = 100 \cdot c + 100 \frac{\log (t'_R)_x - \log (t'_R)_c}{\log (t'_R)_{c+1} - \log (t'_R)_c} \tag{3.1}$$

The t'_R values give the retention times of the standards and the substance corrected for the dead time t_0 ($t'_R = t_R - t_0$). As the dead time is constant in the cases considered, uncorrected retention times are mostly used. The determination of the Kovats indices (Figure 3.24) can be carried out very precisely and on comparison is reproducible within ± 10 RI units between various different laboratories. In libraries of mass spectra, the RIs are also given (see NIST, the terpene library by Adams, the toxicology library by Pfleger/Maurer/Weber).

On working with linear temperature programs (Bemgard *et al.*, 1994; van den Dool and Kratz, 1963) a simplification is used which was introduced by van den Dool and Kratz, whereby direct retention times are used instead of the logarithmic terms used by Kovats (van den Dool and Kratz, 1963; Bianchi *et al.* 2007):

$$\text{Modified Kovats index RI} = 100 \cdot c + 100 \frac{(t'_R)_x - (t'_R)_c}{(t'_R)_{c+1} - (t'_R)_c} \tag{3.2}$$

The weakness of RI systems using alkanes lies in the fact that not all analytes are affected by variances in the measuring system to the same extent, or selective

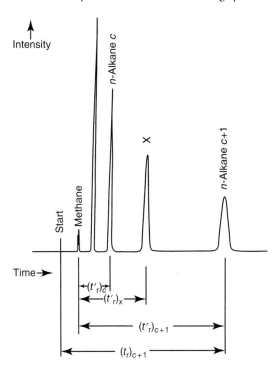

Figure 3.24 Determination of the Kovats index for a substance X by interpolation between two *n*-alkanes (Schomburg, 1987).

detectors are increasingly used. For these special purposes, homologous series of substances which are as closely related as possible have been developed (Hall *et al.*, 1986; Lipinski and Stan, 1989; Kostiainen and Nokelainen, 1990). For use in trace analysis in environmental chemistry and particularly for the analysis of pesticides and chemical weapons, the homologous M-series (Figure 3.25) of *n*-alkylbis(trifluoromethyl)phosphine sulfides was developed (Manninen *et al.*, 1987).

The molecule of the M-series contains active groups which also respond to the selective detectors ECD (electron capture detector), NPD (nitrogen/phosphorous detector), FPD (flamephotometric detector) and PID (photo ionization detector) and naturally also give good responses in FID and MS (detection limits: ECD about 1 pg, FID about 300 pg, Figure 3.26). In the mass spectrometer, all components of the M-series show intense characteristic ions at M − 69 and M − 101 and a typical fragment at m/z 147 (Figure 3.27). The M-series can be used with positive and negative chemical ionization (CI).

The use of RIs in spite of, or perhaps because of, the wide use of GC-MS systems is now becoming more important again as a result of the outstanding stability of

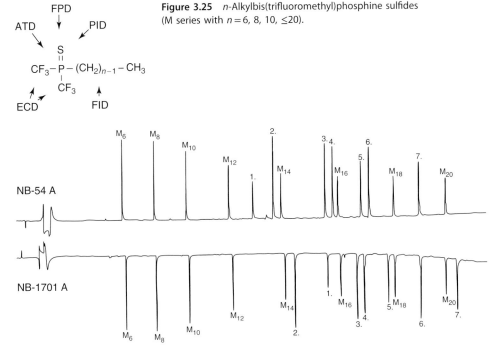

Figure 3.25 *n*-Alkylbis(trifluoromethyl)phosphine sulfides (M series with $n = 6, 8, 10, \leq 20$).

Figure 3.26 Chromatograms of the M series and of pesticides (phosphoric acid esters) on columns of different polarities (HNU/Nordion). Carrier gas He, detector NPD, program: 50 °C (2 min), 150 °C (20 °C/min), 270 °C (6 °C/min). Components: M series M_6, M_8 to M_{20}, 1. dimethoate, 2. diazinon, 3. fenthion, 4. trichloronate, 5. bromophos-ethyl, 6. ditalimphos and 7. carbophenothion.

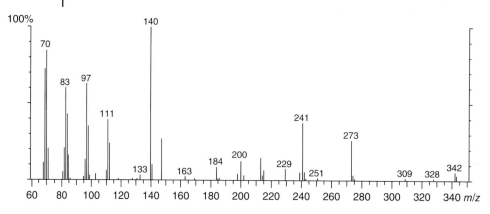

Figure 3.27 M series: El Mass Spectrum of the component M_{10} (HNU).

fused silica capillaries and the good reproducibility of gas chromatographs now available. The broadening of chromatography data systems with optional evaluation routines is just beginning. These are especially dedicated to the processing of RIs, for example, for two-column systems.

If the RI of a compound is not known, it can be estimated from empirical considerations of the elements and partial structures present in the molecule (Katritzky *et al.*, 1994; Weber and Weber, 1992) (Tables 3.1 and 3.2). A first approximation can already be made using the empirical formula of an analyte. This is particularly valuable for assessing suggestions from the GC-MS library search because, besides good correspondence to a spectrum, plausibility with regard to the retention behaviour can be tested (see also Section 4.15). According to Weber and Weber (1992) the values determined give a correct estimation within 10% (Table 3.3) using also structural characteristics (Table 3.2).

Table 3.1 Contributions for the determination of the retention index from the empirical formula (Weber and Weber, 1992).

Element	Index contribution
H, F	0
C, N, O	100
Si in $Si(CH_3)_3$	0
P, S, Cl	200
Br	300
J	400

Table 3.2 Retention behaviour of structural isomers (Weber and Weber, 1992).

Alkyl branches	Tertiary < secondary < *n*-alkyl
Disubstituted aromatics	Ortho < meta, para

Table 3.3 Comparison of calculated retention indices with empirically determined values.

Substance	Calculated RI	Determined RI
Atrazine	1500	1716
Parathion-ethyl	2000	1970
Triadimefon	2000	1979

Examples of retention index calculations

Number	Element	Contribution	Total
Parathion-ethyl		$C_{10}H_{14}NO_5PS$	
10	C	100	1000
14	H	—	—
1	N	100	100
5	O	100	500
1	P	200	200
1	S	200	200
		Sum	2000
Atrazine		$C_8H_{14}ClN_5$	
8	C	100	800
14	H	—	—
1	Cl	200	200
5	N	100	500
		Sum	1500
Triadimefon		$C_{14}H_{14}ClN_3O_2$	
14	C	100	1400
14	H	—	—
1	Cl	200	200
3	N	100	300
2	O	100	200
1	*tert*-C	−100	−100
		Sum	2000

Polar groups with hydrogen bonding increase the boiling point of a compound and are thus responsible for stronger retention. For the second and every additional polar group, the RI increases by 150 units. Branches in the molecule increase the volatility. For each quaternary carbon atom present in a *t*-butyl group, the RI is reduced by 100 units. Values can be estimated with higher precision from RIs for known structures by calculating structure elements according to Tables 3.1 and 3.2.

3.2.3
Libraries of Mass Spectra

In electron ionization (70 eV) a large number of fragmentation reactions take place with organic compounds. These are independent of the manufacturer's design of the ion source. The focusing of the ion source has a greater effect on the characteristics of a mass spectrum, which leads to a particularly wide adjustment range, especially in the case of quadrupole analysers. The relative intensities of the higher and lower mass ranges can easily be reversed. In the early days of use of quadrupole instruments, this possibility was highly criticized by those using the established magnetic sector instruments. The problem is resolved in so far as both the manual and the automatic tuning of the instruments are aimed at giving the intensities of a reference compound (in contrast SIM tuning aims to give high sensitivity within a specific mass range, see Section 2.3.4.2). Perfluorotributylamine (PFTBA, FC43) is used as the reference substance in all GC-MS systems. Other influences on the mass spectrum in GC-MS systems are caused by the changing substance concentration during the mass scan (beam instruments). On running spectra over a large mass range (e.g. in the case of methyl stearate with a scan of 50–350 u in 1 s), sharp GC peaks lead to a mismatch of intensities (skewing) between the front and back slopes of the peak. The skew of intensities is thus the opposite of the true situation. This effect can only be counteracted by the use of fast scan rates, which, however, result in lower sensitivity. In practice, standardized spectra must be used for these systems in order to calculate the compensation (for background subtraction see Section 3.2.1). Ion trap mass spectrometers do not show this skewing of intensities because there is parallel storage of all the ions formed. A mass spectrum should therefore not be regarded as naturally constant, but the result of an extremely complex process.

In practice, the variations observed affect the relative intensities of particular groups of ions in the mass spectrum. The fragmentation processes itself are not affected (the same fragments are found with all GC-MS instruments) nor are the isotope ratios which result from natural distribution. Only adherence to a parameter window which is as narrow as possible (so-called standard conditions) during data acquisition creates the desired independence from the external influences described.

The comparability of the mass spectra produced is thus ensured for building up libraries of mass spectra. All commercially available libraries were run under the standard conditions mentioned and allow the comparison of the fragmentation pattern of an unknown substance with those available from the library. For the large universal libraries, it should be assumed that most of the spectra were initially not run with GC-MS systems, and that still today many reference spectra are run using a solid sample inlet or similar inlet techniques. For example, a former reference spectrum of Aroclor 1260 (a mixture of PCBs with a 60% degree of chlorination) can only be explained in this way. Information on the inlet system used is rarely found in library entries.

EI spectra are particularly informative because of their fragmentation patterns. All search processes through libraries of spectra are mainly based on EI spectra. With the introduction of the reproducible chemical ionization into ion trap mass spectrometers, the first commercial CI library with over 300 pesticides was produced by Finnigan Corp. in 1992. The introduction of substructure libraries (MS/MS product ion spectra) and accurate mass spectra libraries is currently ongoing (NIST). The commercially available libraries are divided into general extensive collections and special task-related collections with a narrow range of applications.

3.2.3.1 Universal Mass Spectral Libraries

NIST/EPA/NIH Mass Spectral Library

The NIST/EPA/NIH Mass Spectral Library is probably the most popular and most widely distributed library for GC-MS instruments. The 2011 edition has been largely expanded by the number of EI (electron-ionization) mass spectra with the addition of Kovats RIs. MS/MS mass spectra are increasingly included. Extensive spectra evaluation and quality control has been involved in the new edition of the NIST database. Each spectrum was critically examined by experienced mass spectrometrists, and each chemical structure has been examined for correctness and consistency, using both human and computer methods (Ausloos *et al.*, 1999). Spectra of stereoisomers have been intercompared, chemical names have been examined by experts and IUPAC names provided. CAS registry numbers have been verified.

The NIST library is available with a new version of the NIST Mass Spectral (MS) Program (v.2.0g) and the enhanced versions of MS Interpreter and AMDIS, the mass spectral interpretation tools with thermodynamics-based interpretation of fragmentation and chemical substructure analysis. The binary format has not changed from the 2002 version, although several new files have been added that associate equivalent compounds and link individual compounds to the RI library. Raw data files are provided in both an SDFile format (structure and data together) as well as in earlier formats. The SDFile format holds the chemical structure as a MOLFile and the data in a simple ASCII format. The NIST MS Search Program is also part of many commercial instrumental GC-MS software suites.

The 2014 edition of the NIST database now contains 276,248 spectra of 242,466 compounds (mainlib), an increase of 13% from the 2011 version. The library includes also 33,782 replicate spectra (replib) of differing spectral quality.

Other major enhancements have been made to the prior version including many replacements with higher quality spectra, a thorough review of chemical names and merging of the previous salts library into the main library.

The NIST database of MS/MS spectra has undergone an even greater enhancement. The new collection contains 234,284 Spectra of 45,298 ions (nist_msms). Spectra have been acquired on ion trap, QToF and triple quadrupole instruments. Spectra for the latter instrument classes have been acquired over a wide range of collision energies to ensure matching regardless of instrument settings. Also, when available, high mass accuracy spectra are stored. New spectra include metabolites, peptides, contaminants, lipids and more.

The contents of the NIST 14 Mass Spectral and RI Libraries are specified as follows:

Library description	Library file name	Library contents
NIST/EPA/NIH mass spectral library (EI)		
Main EI MS library	mainlib	276,248 spectra
Replicate EI MS spectra	replib	33,782 spectra
Tandem (MS/MS) library		
MS/MS Library 2012	nist_msms and nist_msms2	234,284 Spectra of 45,298 ions
GC method/retention index library		
Retention index library	nist_ri	346 757 Kovats RI values for 82,868 compounds

A breakdown of the NIST library entries by molecular weight is shown in Figure 3.28 with the highest frequency of entries in the molecular weight range 200–350 Da (SIS, 2013).

The increase in the number of spectra in the NIST library was accomplished by the addition of complete, high-quality spectra either measured specifically for the library or taken from major practical collections, including:

- Chemical Concepts – including Prof. Henneberg's industrial chemicals collection (Max-Planck-Institute for Coal Research, Muelheim, Germany), see details given later
- Georgia and Virginia Crime Laboratories
- TNO Flavours and Fragrances
- AAFS Toxicology Section, Drug Library
- Association of Official Racing Chemists
- St. Louis University Urinary Acids
- VERIFIN and CBDCOM Chemical Weapons

The new addition of Kovats RI values contains 346 757 Kovats RI values for 70 835 compounds on non-polar columns, 12 452 of which are compounds represented in the EI library. Full annotation is provided, including literature source

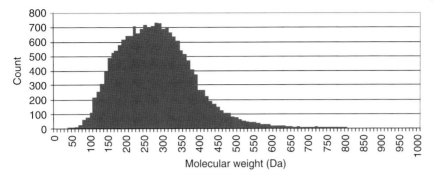

Figure 3.28 Molecular weight distribution in the NIST library.

and measurement conditions. These are provided in a format accessible by the NIST Search Program and separately as an ASCII SDFile.

The new addition of MS/MS Spectra provides 121 586 spectra of 15 180 precursor ions (positive and negative ions). A range of instruments is represented, including ion trap and triple quadrupole mass spectrometers. Spectra have been provided by contributors, measured at NIST and extracted from the literature. It also documents spectrum variations between instrument classes at different conditions. It was found that at sufficiently high signal-to-noise measurement conditions, modern instruments are capable of providing very reproducible, library searchable spectra. While collision energy can be an important variable, spectra varies in an understandable way depending on compound and instrument class and conditions. This library is provided in formats equivalent to the EI library but with new fields added to describe the instrument and analysis conditions. A small number of MS1 spectra are also included for reference purposes. These generally contain the precursor ions used for MS/MS.

Wiley Registry of Mass Spectral Data

The Wiley Registry™ of Mass Spectral Data has been published in its 10th Edition (John Wiley & Sons, 2013, Editor: Fred W. McLafferty) (McLafferty and Stauffer, 1989). It is the largest and most comprehensive mass spectral library ever made commercially available in the most common mass spectrometry software formats and compatible with most manufacturer data systems. Applications include pathology, toxicology, forensics, quality assurance, border control, research and development, food safety and environmental sciences. The 10th edition of the Wiley Registry contains:

- Over 719 000 mass spectra
- Over 638 000 compounds
- With over 684 000 searchable chemical structures.

Most spectra are accompanied by the structure and trivial names, molecular formula, molecular weight, nominal mass and base peak. Included are:

- Chemical warfare precursors
- Combinatorial library compounds
- High molecular diversity for fragmentation analysis
- New high-resolution organics
- Structure and substructure searchable
- Spectra with RIs.

Also available is the combination of the large Wiley Registry with the current NIST database. The Wiley Registry 10th Edition/NIST 2012 (W10/N12, ISBN: 978-1-118-61611-6 of June 2013) provides today the most extensive mass spectral library with:

- More than 870 000 mass spectra, chemical formulas, and the exact masses of more than 73 600 unique compounds
- More than 3 million chemical names and synonyms
- High-resolution spectra with most new spectra containing over 125 peaks per spectrum.

The Wiley/NIST library provides comprehensive coverage of small molecule organics, pharmaceutical drugs, illegal drugs, poisons, pesticides, steroids, natural products, organic compounds and chemical warfare agents for different applications:

- *Toxicology/Forensics/Public Health*: The library contains a wide scope of spectra covering drugs, poisons, pesticides and metabolites.
- *Industrial R&D/Quality Assurance*: The library contains a comprehensive collection of small organic compounds and their metabolites, including a combinatorial library appropriate for fragmentation analysis.
- *Research/teaching*: The library contains data for fragmentation analysis as well as comprehensive coverage of most compounds measurable by GC-MS.
- *Environmental*: The library contains most known pesticides and includes precursors being used in the production of new pesticide classes.
- *Available formats and compatibility*: The library is available in two formats: Chemstation (Agilent) and the NIST MS Search (Bruker, JEOL, LECO, Perkin Elmer, Thermo, Varian, Waters). Other formats are available on request.

3.2.3.2 Application Libraries of Mass Spectra

Mass Spectra of Geochemicals, Petrochemicals and Biomarkers
This database is focused on organic, geochemical and petrochemical applications (De Leeuw, 2004), and comprises:

- 1100 mass spectra of well-defined compounds.
- Information including mass spectra, chemical structure, chemical name, molecular formula, molecular weight (nominal mass), base peak, reference and measurement condition.
- Chemical structures elucidated, if necessary, by a variety of techniques including NMR spectroscopy and single-crystal X-ray structure analysis (Wiley).

Chemical Concepts Library of Mass Spectra

The Mass Spectra Chemical Concepts has been updated and consists of mass spectra of more than 40 000 compounds (Henneberg, 1998). It is included in the new release of the NIST 2011 library.

The main part of this mass spectra reference library comes from the Industrial Chemicals Collection of Prof. Dr. Henneberg, Max-Planck-Institute for Coal Research, Muelheim, Germany. Also universities and institutes such as ETH, Zürich, Switzerland and ISAS, Dortmund, Germany have contributed their research spectra to this collection. Before being included into the library, the data pass consistency and quality checks were performed at the Max-Planck-Institute.

Additional information included with the mass spectra are (Chemical Concepts, 2004):

- Chemical structure
- Chemical name
- Molecular formula
- Molecular weight (nominal mass)
- Base peak
- Reference
- Measurement condition.

Alexander Yarkov – Mass Spectra of Organic Compounds

The new specialized data collection contains 37 055 mass spectra of physiologically active organic compounds. The data resulted from quality control in combinatorial synthesis and cover a wide range of compound classes.

Additional information included with the mass spectra are (Yarkov, 2008):

- Chemical structure
- Chemical name
- Molecular formula
- Molecular weight (nominal mass)
- Base peak
- Reference
- Measurement condition.

Mass Spectra of Designer Drugs

This mass spectrum collection edited by Peter Rösner covers the entire range of designer drugs with a most recent update in (Roesner, 2007). It is the first database featuring systematic structures in depth. Carefully compiled by the mass spectral experts at the Regional Departments of Criminal Investigation in Kiel, Hamburg and the Federal Criminal Laboratory in Wiesbaden, Germany, this database includes 19,037 mass spectra of 15,556 chemical compounds like designer drugs and medicinal drugs. Chemical warfare agents are added due to the recent interest in homeland security. All data has been taken from both legal and underground literature, providing the most comprehensive picture of these compounds available worldwide. Even highly potential hallucinogens like the Bromo-DragonFly are covered.

Mass Spectra of Volatiles in Food

This mass spectral database is dedicated to the application areas of the food and flavour industries, and was selected and quality controlled by the mass spectral experts at the Central Institute of Nutrition and Food Research in the Netherlands. The collection includes 1620 reference mass spectra and covers the whole range of volatile compounds in food. Apart from the large number of natural, nature-identical and artificial flavours and aromas, there are – among others – food additives and solvents, pesticides and veterinary pharmaceutical compounds, which are frequently found as residues. Derivatives of non-volatile compounds such as sugars or polyhydroxyphenols are also available. The database is now available in its second edition (Central Institute of Nutrition and Food Research, 2003, ISBN 978-0-471-64825-3).

Flavours and Fragrances of Natural and Synthetic Compounds

The library of Flavours and Fragrances of Natural and Synthetic Compounds, edited by Prof. Mondello contains more than 3000 mass spectra in its second edition, linear retention index data, calculated Kovats RI and searchable chemical structures of compounds of interest for the flavours and fragrances industry as well as research applications (Wiley, ISBN: 978-1-118-14583-8).

Adams Essential Oil Components Library

The Adams Essential Oil mass spectral library is available in its fourth edition (Adams, 1989, 2007). This comprehensive collection of mass spectra and retention times of common components in plant essential oils covers 2205 compounds, each including:

> *RT*: retention time on DB-5 capillary column
> *AI*: arithmetic retention index
> *KI*: Kovat's retention index
> CAS#: chemical abstracts service number
> *MF*: molecular formula
> *FW*: formula weight
> MSD LIB#: entry number in library
> *CN*: chemical name

Also included are a list of synonyms and the source of compounds used for spectrum. If the compound occurs in nature, two additional sources for the compound (concentration at % oil, plant name, literature reference) are included. The library (including retention times) is available for the most common mass spectrometer computer systems (Allured Books, ISBN 1932633219).

Mass Spectral and GC Data of Drugs, Poisons, Pesticides, Pollutants and Their Metabolites

This specialized collection is dedicated to environmental and forensic analysis, occupational toxicology and food analysis and contains data obtained from clinical samples over the course of more than 20 years. It encompasses almost 9000 substances from simple analgesics to designer drugs, and from pesticides and pollutants to chemical warfare agents, including metabolites to allow the identification of the mother substance. Prof. Karl Pfleger is the former and Prof. Hans H. Maurer the current, head of the Clinical Toxicology Laboratory at the clinical campus of Saarland University in Homburg, Germany. Together with Armin Weber, they have developed this unique and most comprehensive toxicological database (Pfleger *et al.*, 2011).

The new release covers 800 additional spectra, including novel designer drugs and a broad range of AIDS therapeutics, for a total of 8650 clinically relevant substances, including 5200 metabolites.

- Data of nearly all the new drugs relevant to clinical and forensic toxicology, doping control, food chemistry, and so on.
- Nearly complete coverage of trimethylsilylated, perfluoroacylated, perfluoroalkylated and methylated compounds.
- Sections on sample preparation and GC-MS methods.

Mass Spectra of Physiologically Active Substances

This mass spectral data base is focused on doping control, endocrinology, and clinical toxicology. It has been complied by the Institute of Biochemistry, German Sport University Cologne and covers more than 4000 quality mass spectra of drugs, steroid hormones, including their trimethylsilyl-, O-methoxyoxime- and acetal derivatives, endocrine disruptors, and β-2-agonists compounds. The library includes chemical names, structures, molecular formula, and synonyms (Parr *et al.*, 2011).

The database is available in multiple MS datasystem formats like ACD, Agilent Chemstation, NIST, Finnigan GCQ, SSQ, TSQ, ICIS, ITS40, Magnum, INCOS, PE Turbomass, Shimadzu QP-5000, Thermo Scientific Xcalibur, Thermo Galactic SpectralID, Varian Saturn, VG Labbase, Masslab, Waters Masslynx.

Mass Spectra of Drugs, Pharmaceuticals and Metabolites

The collection edited by Rolf Küehnle contains 2200 mass spectra of drugs, pharmaceuticals and metabolites. The inclusion of the silylated derivatives as used for GC-MS analyses is of special value in this database. Additional information included are chemical structure, chemical name, molecular formula, molecular weight (nominal mass), base peak and the reference (Küehnle, R. (2006)).

Mass Spectra of Pharmaceuticals and Agrochemicals

The collection of 4563 unreduced spectra includes compounds that are subject to the drug trafficking laws as well as their precursors, by-products and metabolites. Other compound groups included are medical drugs, drugs with psychotropic effect, anabolics and pesticides. Chemical structures, synonym and systematic name, molecular weight, formula and experimental conditions complete the data record (Wiley, ISBN 978-3-527-31615-1).

Mass Spectra of Androgenes, Estrogens and other Steroids

The collection edited by Hugh L. J. Makin contains 2979 EI mass spectra of androgens and estrogens and their trimethylsilyl-, *O*-methoxyoxime- and acetal derivatives. Each spectrum is accompanied by the structure and trivial name, molecular formula, molecular weight, nominal mass and base peak. All spectra of androgens and estrogens have been obtained on the same mass spectrometer under identical conditions (Makin, 2005). The database is available in the formats for many popular MS data system.

AAFS SWGDRUG Library

The American Academy of Forensic Sciences, Toxicology Section committee was set up to coordinate the generation of reliable mass spectra of new drugs and metabolite standards, and to make these available to the profession on a timely basis. The Scientific Working Group for the analysis of Seized Drugs (SWGDRUG) has compiled a mass spectral library from a variety of sources, containing drugs and drug-related compounds. The mass spectral database and the list of entries are available for download from the internet as zip file. The library is supported by the NIST MSSEARCH program.

This library is a 'subset' of one that has been compiled over a period of many years by Dr. Graham Jones and colleagues in Edmonton, Alberta, Canada. Pure drug spectra, plus GC breakdown products and pure metabolite standards have been edited into this compilation of 1885 mass spectra. The collection will be updated on a regular basis. All spectra were run on Agilent quadrupole GC-MS instruments tuned against PFTBA (FC43). The current version of the full spectra library was last updated as the SWGDRUG MS Library Version 1.9 in July 2013 (AAFS, see *http://www.swgdrug.org/ms.htm*).

The Fiehn Library

The Fiehn Library is a mass spectral and retention index library for comprehensive metabolic profiling. The library comprises over 1000 identified metabolites that are currently screened by the Fiehn laboratory. The compound list is continually extended. The Fiehn mass spectral library is commercially available for GC-quadrupole mass spectrometers from Agilent Technologies, and for GC-TOF mass spectrometers from Leco Corporation, see: *http://fiehnlab.ucdavis.edu/Metabolite-Library-2007/*

The Golm Metabolome Database

The Golm Metabolome Database (GMD) provides public access to custom mass spectra libraries, metabolite profiling experiments and other necessary information related to the field of metabolomics. The main goal is the representation of an exchange platform for experimental research activities and bioinformatics to develop and improve metabolomics by multidisciplinary cooperation. GMD is maintained by the GMD Consortium, a joint venture of researchers from various research areas having different research interest. The consortium based on an interdisciplinary cooperation to joining expertises from these research interests. See *http://csbdb.mpimp-golm.mpg.de/csbdb/gmd/gmd.html*.

The Lipid Library
The Archives of Mass Spectra by W.W. Christie comprise ~1670 mass spectra in total (AOCS, 2014). They are made available on the web for study but without interpretation for the following compound groups:

- Methyl esters of fatty acids (FAMEs)
- Picolinyl esters
- DMOX (4,4-dimethyloxazoline) derivatives
- Pyrrolidine derivatives
- Miscellaneous fatty acid derivatives, lipids, artefacts, and so on.

All the mass spectra illustrated on these pages were obtained by electron ionization at an ionization potential of 70 eV on quadrupole mass spectrometers. The website also offers the Bibliography of Mass Spectra with lists of references mainly concerning the use of mass spectrometry for structural analysis of fatty acids.

3.2.4
Library Search Programs

In general, it is expected that the identity of an unknown compound will be found in a library search procedure. However, it is better to consider the results of a search procedure from the aspect of similarity between the reference and the unknown spectrum (Stein and Scott, 1994). Other information for confirmation of identity, such as retention time, processing procedure and other spectroscopic data should always be consulted. A review in the journal Analytical Chemistry (Warr, 1993a) began with the sentence.

> "Library searching has limitations and can be dangerous in novice hands."

Examples of critical cases are different compounds which have the same spectra (isomers), the same compounds with different spectra (measuring conditions, reactivity and decomposition) or the fact that a substance being searched for is not in the library but similar spectra are suggested. In particular, the limited scope of the libraries must be taken into account.

A good overview over the total number of known and well characterized substances can be found at the Chemical Abstract Service (CAS) statistics. The majority of the registered substances is of organic nature and hence applicable to mass spectrometric analysis. In 2008, CAS announced the registry has reached 40 million substances with the CAS Registry Number 1073662-18-6 for an azulenobenzofuran derivate. The strong growth in new registered compounds became evident with 60 million substances registered in 2011 with 1298016-92-8

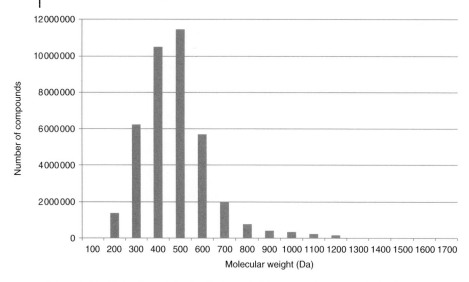

Figure 3.29 Molecular weight distribution of CAS registered compounds (Little, 2013).

for an amino-thiadiazine derivative, and 70 million substances have been registered already by end of 2012 with the CAS Registry Number 1411769-41-9 for a
pyrazolyl-piperazine compound, by September 2014 already over 90 million. The
CAS website features an always current database counter, see *www.cas.org*.

A closer look at the molecular weight distribution gives some background on the
limited compound coverage of the available GC-MS libraries, and reveals the large
potential for small molecule analysis to be run by GC-MS or LC-MS, depending
on the polarity of the compounds. Ninety-nine percent of all CAS registered substances are in the molecular weight range of up to 1000 Da as shown in Figure 3.29
(Little, 2013), a typical mass range capable for most of the modern GC-MS and
LC-MS instrumentation today. For unknown compounds detected, the search in
large spectra libraries will stay as an important step in compound identification,
but the manual skills for spectra interpretation will prevail (see McLafferty and
Turecek, 1993).

Today the library search program provided by the National Institute of
Standards and Technology (NIST) is undoubtably of widest distribution. It
was introduced in 1995 together with the NIST/U.S. Environmental Protection
Agency (EPA)/National Institutes of Health (NIH) Mass Spectral Database for
use on the Microsoft Windows™ platform. Certain algorithm components of
the formerly widely used INCOS library search program still build a basis of
the NIST search procedure (Davies 1993; Warr, 1993a/b; Stein, 1994, 1999).
Today the data systems of the leading GC-MS manufacturer provide a direct data
link and integrate by this way the NIST library search into the proprietary data
processing software. Mass spectra can also be imported from a text file. Probability based components later have significantly improved the recent versions
of the NIST search (Stein, 1994) and have increasingly replaced the independent

Probability Based Match (PBM) search program (Stauffer *et al.*, 1985a/b). The SISCOM procedure (Search for Identical and Similar Compounds) developed by Henneberg/Weimann is available on stand-alone work station (Henneberg and Weimann, 1984). It stands out on account of its excellent performance for data system supported interpretation of mass spectra. The procedures for determination of similarity between spectra are based on the classical spectrum interpretation considerations (Neudert *et al.*, 1987). The NIST and PBM procedures aim to give suggestions of possible substances to explain an unknown spectrum. Both algorithms dominate in the qualitative evaluation using magnetic sector, quadrupole and ion trap GC-MS systems. Other search procedures, such as the Biemann search, have been replaced by newer developments and broadening of the algorithms by the manufacturers of spectrometers.

3.2.4.1 The NIST Search Procedure

At the beginning of the 1970s, the INCOS company presented a search procedure which operated both on the principles of pattern recognition and with the components of classical interpretation techniques, which could reliably process data from different types of mass spectrometer (Davis, 1993). The newest development in the area of computer-supported library searches, which is the further development of the INCOS procedure, has been presented by Steven Stein (NIST) through targeted optimization of the weighting and combination with probability values (Stein *et al.*, 1994a/b, 1999). The early years of GC-MS were characterized by the rapid development of quadrupole instruments, which were ideal for coupling with gas chromatographs, because of their scan rates, which were high compared with the magnetic sector instruments of that time. The spectral libraries available had been drawn up from spectra run on magnetic sector instruments.

From the beginning the INCOS procedure was able to take into account the relatively low intensities of the higher masses in spectra run on quadrupole systems, besides the typical high mass intensity magnetic sector spectra. The INCOS search has remained virtually unchanged since the 1970s. The search was known for its high hit probability, even with mass spectra with a high proportion of matrix noise obtained in residue analysis, and its independence from the type of instrument.

After a significance weighting (square root of the product of mass and intensity), data reduction by a noise filter and a redundancy filter, the extensive reference database is searched for suitable candidates for a pattern comparison in a rapid pre-search. The NIST search uses all masses of the unknown spectrum in this pre-search (former INCOS only the eight most intense signals). The intensity ratios are not yet considered at this point, only the occurrence of the mass signal. Reference spectra, which only contain a small number or no matching masses are excluded from the list of possible candidates and are not further processed.

The main search is the critical step in the search algorithm, in which the candidates selected in the pre-search are compared with the unknown spectrum and arranged in a prioritized list of suggestions. Of critical importance for the

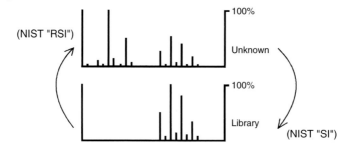

Figure 3.30 Diagram showing local normalization and the principle of match and reverse match calculation. Reverse match value high: All masses in the library spectrum are present in the unknown spectrum and the isotope pattern also fits after 'local' normalization of the intensities. Match value low: The unknown spectrum has more mass signals than the library spectrum. Only some of the masses from the unknown spectrum are present in the library spectrum.

tolerance for different types of mass spectrometer and differing conditions of data acquisition (and thus for the high hit rate) is a process known as *local normalization*.

Local normalization introduces an important component into the search procedure, which is comparable to the visual comparison of two patterns (Figure 3.30). Individual clusters of ions and isotope patterns are compared with one another in a local mass window. The central mass of such a window from the reference spectrum is compared with the intensity of this mass in the unknown spectrum in order to assess the matching of the line pattern to the left and right. In this way, the nearby region of each mass signal is examined and, for example, the matching of isotope patterns (e.g. C, Cl, Br, S, Si) and cleavage reactions are assessed.

The advantage of this procedure lies in the fact that deviating relative intensities caused by a high proportion of chemical noise, spectral skewing, or the type of data acquisition do not have any effect on the result of the search. A variance in the relative signal intensities in a mass spectrum can be caused by varying the choice of spectra from the rising or falling slopes of the peak in the case of quadrupole and magnetic sector instruments (skewing) and by changes in the tuning parameters of the ion source or its increasing contamination. Furthermore, local normalization has a positive effect on spectra with a high proportion of noise (trace analysis, chemical background). Local normalization is the major reason why spectral libraries require only one mass spectrum per substance entry.

Two values are determined for spectral comparison as a result of the main search. The reverse match value (NIST "RSI", former INCOS "FIT") value gives a measure of how well the reference spectrum is represented with its masses in the unknown spectrum (reverse search procedure, ignoring all peaks that are in the sample spectrum but not the reference spectrum). The forward looking mode of searching, whereby the presence of the unknown spectrum in the reference spectrum is examined (forward search procedure, all peaks of the sample spectrum are compared), is expressed as the match value (NIST "SI", former INCOS reverse fit "RFIT"). The combination of the two values gives information on the purity of the unknown spectrum (Figure 3.30). If the reverse match value is high (NIST "RSI") and the match value lower (NIST "SI"), it can be assumed that the spectrum measured contains considerably more mass signals than the reference spectrum used for comparison. It would be necessary to evaluate whether a co-eluate, chemical noise, the presence of a homologous substance or another reason is responsible for the appearance of the additional lines. This is typically done by displaying mass chromatograms and performing background subtractions.

All the candidates found in the pre-search are processed in the main search as described. As a result a sorted list according to match "SI" and reverse match "RSI" values is displayed. Sorting can be selected by "SI" or "RSI" values. (Table 3.4). The initial sorting according to reverse match is recommended because with this value the best estimation of the possible identity is achieved. Further sorting according to the match values gives additional solutions, which generally supplement the further steps towards identification with valuable information on partial structures or identifying a particular class of compound.

The maximum achievable match value for identical spectra is 999. As of general experience, values greater than 900 indicate an excellent match. Good matches to consider give values greater than 800. Values below 800 are considered fair, below 600 very poor. Unknown spectra with a large number of peaks tend to yield lower match factors than spectra with fewer peaks. Fewer peaks in a spectrum tend to increase the RSI match factors, as can be experienced with polyaromatic hydrocarbon (PAH) spectra. On the contrary, spectra with many ions, especially those with background or matrix interferences from highly sensitive full scan measurements of real-life samples tend to lower SI match values, compared to clean spectra, due to the additional chemical interferences (Figure 3.31). A spectrum subtraction has to be done manually in the chromatogram (Figure 3.32). NIST does not offer the function of searching difference spectra to the components of the hit list.

An additional probability value is given with the NIST hit list. The probability value ("Prob.%") for a certain compound hit is derived as a relative probability from the spectral differences between adjacent hits in the list. Several high values indicate high spectra similarities of the hits, low values indicate different spectral patterns (see also Figure 3.33i and g). The probability calculation requires

Table 3.4 Results of the NIST spectral comparison.

SI	RSI	Prob.	Assessment
High	High	High	Identification or that of an isomer very probable
Low	High	Low	Identification possible, but homologues, co-elution, or noise present
High	Low	Low	Possibly an incomplete spectrum

Sorting of the suggestions should first be carried out according to RSI and then according to SI values.

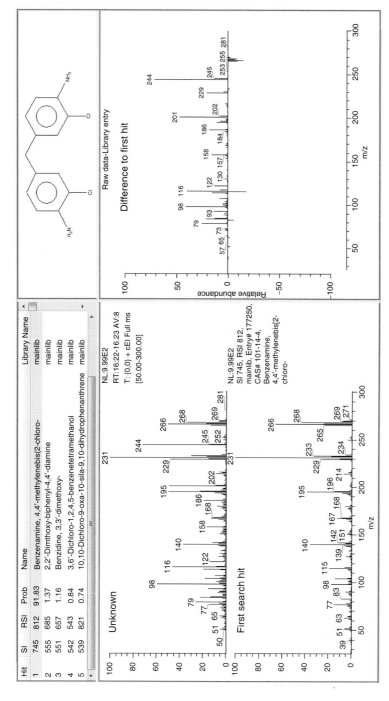

Hit	SI	RSI	Prob	Name	Library Name
1	745	812	91.83	Benzenamine, 4,4'-methylenebis(2-chloro-	mainlib
2	555	685	1.37	2,2'-Dimthoxy-biphenyl-4,4'-diamine	mainlib
3	551	657	1.16	Benzidine, 3,3'-dimethoxy-	mainlib
4	542	543	0.84	3,6'-Dichloro-1,2,4,5-benzenetetramethanol	mainlib
5	539	821	0.74	10,10-Dichloro-9-oxa-10-sila-9,10-dihydrophenanthrene	mainlib

Figure 3.31 NIST library search result of a co-elution spectrum. The hit list upper left is sorted by SI match value. The RSI match value with 812 is high, SI value 745 low due to additional masses in the unknown spectrum (upper left) which are not in the library reference (bottom left). The probability with 91.8% is very high, indicating a unique spectrum with significant difference to the next search hit. The difference spectrum (upper right) can be retrieved for search by spectrum subtraction in the chromatogram.

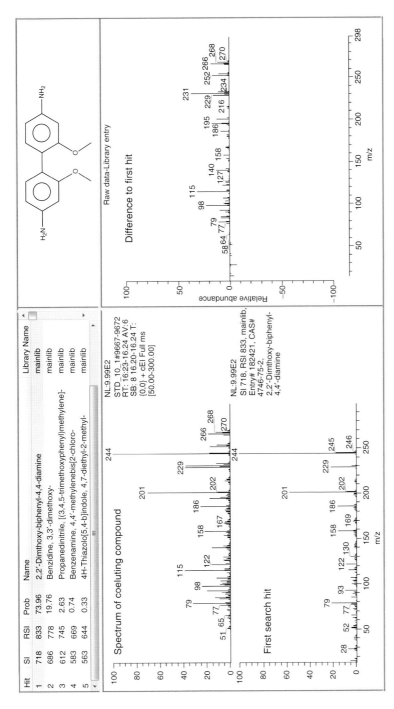

Figure 3.32 NIST library search of the difference spectrum of Figure 3.31 after careful background subtraction. The RSI match value 833 is high, SI value with 718 is low due to other fragments from background ions, the probability value of 74% is high due to the unique spectrum of the first hit. The difference spectrum shows signals of the coeluting compound of Figure 3.31.

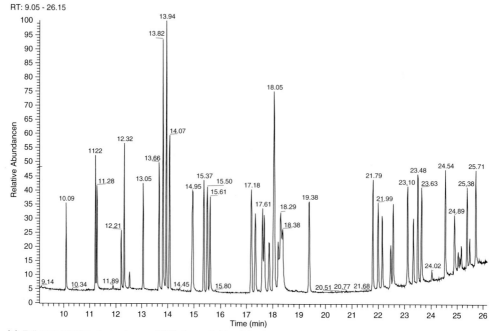

(a) Full scan total ion chromatogram (TIC) of a pesticide mixture.

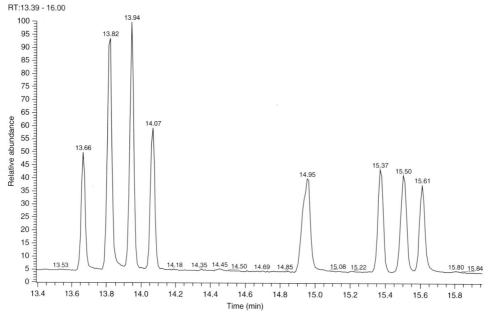

(b) Chromatogram detail, peak at RT 14.95 min shows unsymmetrical peak shape.

Figure 3.33 (a-i) illustrate the peak analysis of coeluting compounds using mass chromatograms, spectrum background subtraction and library search.

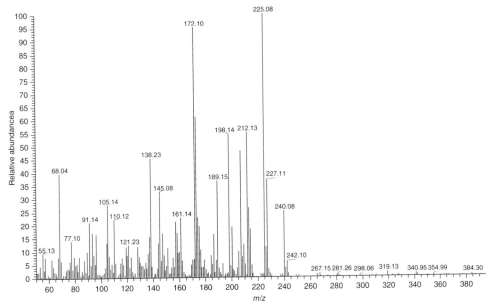

(c)   Mass spectrum of peak at RT 14.95 min.

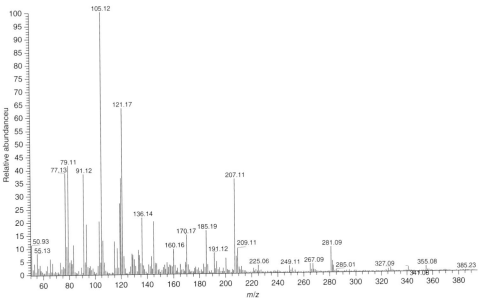

(d)   Background spectrum before peak at RT 14.95 min (m/z 77, 91, 105, 121 indicate alkyl-phenylen, m/z 207, 281 polysiloxane components of the stationary phase of the used Rtx-5ms column).

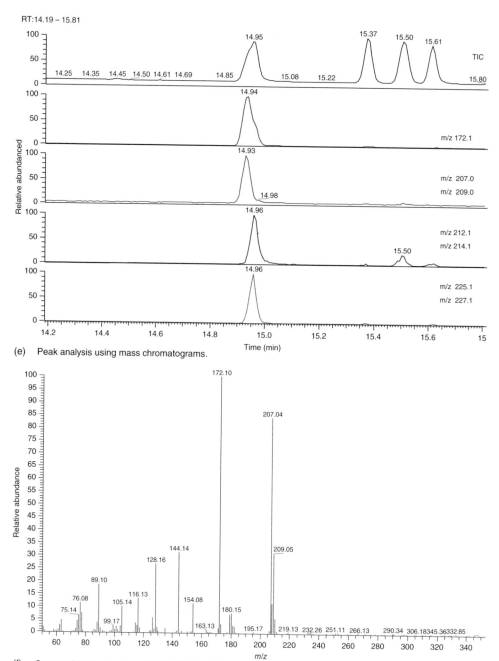

(e) Peak analysis using mass chromatograms.

(f) Corrected mass spectrum at RT 14.93 min (left peak slope subtracted by right slope).

Hit	SI	RSI	Prob	Name	Library Name
1	826	838	96.10	1,4-Naphthalenedione, 2-amino-3-chloro-	mainlib
2	649	658	1.84	N-[2-[3-Chloro-1,4-naphthoquinonyl]-a-amino-p-toluenesulf...	mainlib
3	609	636	0.42	2-Chlorobenzo[b]thiophene-3-acetonitrile	mainlib
4	607	744	0.39	2-Phenyl-4-chloromethyl-5-methyl-1,2,3-triazole	mainlib
5	601	607	0.31	5-Chlorobenzo [b] thiophene-3-acetonitrile	mainlib

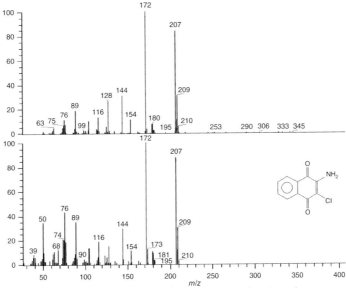

(g) Library search result (NIST), first hit Quinoclamine with spectrum comparison, top unknown (Figure 3.32f), bottom first hit.

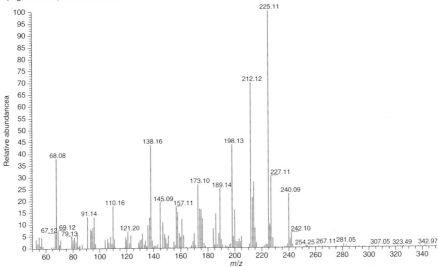

(h) Corrected mass spectrum at RT 14.96 min (right peak slope subtracted by left slope and column background Figure 3.32d).

Hit	SI	RSI	Prob	Name	Library Name
1	850	862	96.47	Cyanazine	mainlib
2	543	548	0.96	Oxazolidin-2-one, 3-[2-(5,7-dichloro-2-methyl-3-indol...	mainlib
3	521	523	0.38	2,3-Dicyano-2-(2-oxo-cyclododecyl)-succinonitrile	mainlib
4	505	585	0.21	Cyprazine	mainlib
5	499	549	0.17	Cyanamide,N-[4-chloro-6-(dimethylamion)-1,3,5-tria...	mainlib

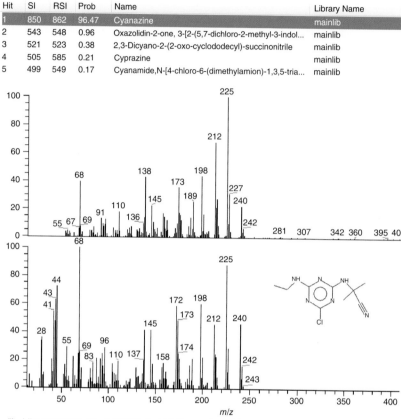

(i) Library search result (NIST), first hit Cyanazine with spectrum comparison, top unknown
(Figure 3.32h), bottom first hit. Note, m/z 172 in the unknown spectrum got lost due to
the manual substraction process.

the search of the NIST database with the set of replicate spectra. Another
separate probability factor ("InLib") is provided as a measure of the probability that the unknown compound is contained in the searched databases
(Stein, 2011).

In the 1990s, Steven Stein from NIST took the former INCOS approach and
extended it to the most common situations when unknown compounds are not
present in the library. Structurally similar compounds can appear in the NIST
library search hit list. The identity search presumes that the unknown compound
is represented in the reference database, as it is designed to find the exact match to
the unknown spectrum. The extended search mode for similarity should be used if
it becomes apparent that the spectrum of the unknown compound is not present
in the library. Stein also added probabilities to the hit list that give information
about common substructures which may be present or absent in the unknown

compound (Stein, 1994, 1995). On the basis of this advanced performance, the NIST library search is recommended as the first step in the structural elucidation of compounds not found in reference libraries (Stein, 2011).

INCOS Library Search

Principle	Pattern recognition (after Joel Karnovsky, INCOS) (Sokolow *et al.*, 1978)

Course

1. Significance weighting	$\sqrt{m\ I}$
2. Noise filter	Window $\pm\,50$ Da, ≥ 40 masses
3. Redundancy filter	Window $\pm\,7$ Da, ≥ 6 masses
4. Pre-search	8 masses + molecular weight
5. Main search	Local normalization FIT, RFIT and PUR[*] calculation (* PUR as normalized product of FIT and RFIT)
6. Sorting and display	

Advantages

+ high hit probability
+ secure identification even with spectra with high noise levels
+ search is independent of scan times and type of instrument because of local normalization
+ only one spectrum per substance necessary in the library
+ very fast
+ available significantly advanced for a variety of data systems in the form of NIST algorithm.

Limitations

− manual difference calculation necessary for co-eluates
− with spectra with many equally distributed fragments.

3.2.4.2 The PBM Search Procedure

A completely different search strategy forms the basis of the PBM algorithm (probability based match). The statistical mathematical treatment by Prof. McLafferty allows predictions to be made on the probable identity of a substance suggestion (Atwater *et al.*, 1985; Palisade Corporation, 1994). The search procedure was developed in the 1970s at Cornell University as part of the Cornell algorithm (STIRS, the self-training interpretative and retrieval system as an interpretative system). In the subsequent years, parts of the PBM procedure were adapted for PCs, also under the name PBM. These only contained the less powerful mode, 'pure search'. At the beginning of the 1990s, a PC version known as *Benchtop/PBM* (Palisade, 1996) was released, which now also provided the mode 'mixture search' for the data systems of commercial GC-MS systems. Today the Benchtop/PBM library search system is a legacy product with minimal support (SIS, 2013).

A significance weighting of the mass signals (here, a sum of the mass m and the intensity I) is also carried out first with the PBM search procedure. In addition, the frequency of individual mass signals based on their appearance in the whole spectral library is also taken into account (the reference is the latest version of the Wiley library). The pre-search of the PBM procedure is orientated towards the method of significance weighting. In the database used for PBM, the reference spectra are sorted according to the values of maximum significance. For a given maximum significance of the weighted masses of an unknown spectrum, a set of reference spectra can thus be selected. The choice of several sets of mass spectra is specified as the search depth, which, starting from $(m + I)_{max}$, can reach search depths of 3 to $(m + I)_{max} - 3$. The number of possible candidates for the main search is thus broadened.

The main search in the PBM procedure can be carried out in two ways. In the pure search mode, only the fragments of the unknown spectrum are searched for in the reference spectra and compared (forward search). This procedure requires mass spectra, which are free from overlap and matrix signals. This only is the case with simple separations at a medium concentration range.

The mixture search mode first tests whether the mass signals of the reference spectrum are present in the unknown spectrum (reverse search). Local normalization analogous to the INCOS procedure is also carried out from version 3.0 (1993) onwards. With every spectrum selected from the pre-search, a subtraction from the unknown spectrum is carried out in the course of the procedure. The result of the subtraction is, in turn, compared with the candidates remaining from the pre-search and matching criteria are met. In this way, even with high resolution GC, the possibility of detecting mixed spectra due to sample matrix or co-elution is accounted for. A successful search requires that the second (or third) component of a mixture is contained in the pre-search result (hit list). Combinations of spectra with markedly different highest significances do not appear together in the pre-search hit list because the pre-search is limited by the depth of the search. Only by using this procedure, the probability values of the hit list are improved. This constitutes a major difference between this PBM and the SISCOM search (see Section 3.2.4.3).

The sorting of results from the PBM search procedure takes place on the basis of probabilities which are determined in the course of spectrum comparison (Table 3.5). In the forefront is the aim of giving a statement about the identity of a suggestion. The assignment to class I gives information as to which degree

Table 3.5 Results of the benchtop/PBM spectral comparison.

Class I	The probability that the suggestion is identical or a stereoisomer
Class IV	Extension of the probability from class I to compounds having structural differences compared with the reference, which only have a small effect on the mass spectrum (homologues, positional isomers)
% contamination	Gives the proportion of ions which are not present in the reference

a spectrum is the same or stereoisomeric with the suggestion. Here maximum values can reach 80%, but not higher. The extension of the definition to the ring positions of positional isomers, to homologous compounds, and the change in position of individual C-atoms or of double bonds gives a probability defined as class IV. A higher value implies that a compound with structural features, which has little or no effect on the appearance of the mass spectrum, is present. In such cases, mass spectroscopy a analytical method shows itself to be insensitive to the differences in structural details between individual molecules. The user is thus given a value for assessment, which indicates that the mass spectrum being considered matches the spectrum well on account of its few specific fragments, but probably cannot be identified conclusively. The literature should be consulted for details of the calculation of the probability values in PBM. (Stauffer, 1985a/b)

The clear grading of the probabilities is typical for the PBM procedure (Figure 3.34). Sensible suggestions with high values of 70% to over 90% (class IV) are given. Then in the hit list, there is the rapid lowering of the values to below 20%, which avoids the occurrence of false positive suggestions. It is obvious that, as a result of the procedure in individual cases, recognizably correct suggestions are given with low probability values (false negative suggestions). Therefore, poorly placed suggestions must also be included in the discussion of the search results (Figure 3.35). In such cases, the framework conditions, such as measuring range, quality of the library spectrum or possibly a large quantity of noise in the extracted spectrum should be investigated.

In practice, the hit quota of the PBM search depends on the spectrum quality, which is also affected by the data acquisition parameters of the instrument, the choice of the spectrum in the peak (rising/falling slope, maximum), scan time and the type of instrument. In this context, it is useful that in PBM libraries several spectra per substance are available from different sources.

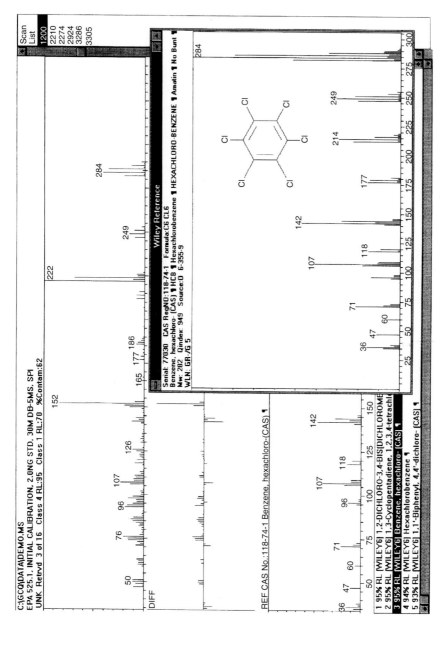

Figure 3.34 Screen display of suggestions arising from a PBM library search (above: unknown spectrum; middle: difference spectrum; below and in the reference window: library suggestion; underneath: list of suggestions with probabilities).

Unknown spectrum

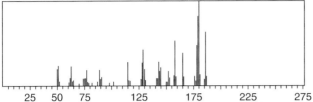

Hit :1 CAS : 5325–97–3 RL : 95 Phenanthrene, 1,2,3,4,5,6,7,8–octahydro–

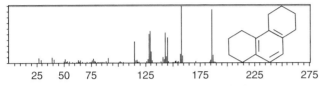

Hit : 2 CAS : 29966–04–9 RL : 94 Phenanthrene, octahydro–

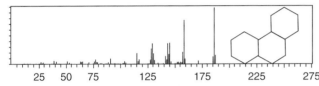

Hit : 3 CAS : 103–30–0 RL : 89 Benzene, 1,1′ – (1,2-ethenediyl)bis–, (E) –

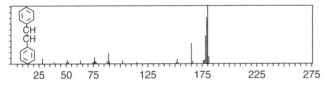

Hit : 4 CAS : 5325–97–3 RL : 76 Phenanthrene, 1,2,3,4,5,6,7,8 –octahydro–

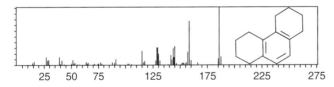

Figure 3.35 Hit list from a PBM search with probabilities % (RL). The mixture search mode gives suggestions for both co-eluting compounds.

PBM Library Search

Principle	**Statistical mathematical tests after Prof. McLafferty**

Course

Modi:	Pure search or mixture search
Both modi:	Significance weighting $(m + I)$
Both modi:	Pre-search by search depths 1 to 3
	$(m + I)_{max}$ to $(m + I)_{max} - 3$

Pure search mode	**Mixture search mode**
Forward search	Reverse search
Fragments of the unknown spectrum must be present in the reference	Fragments of the reference spectrum must be present in the unknown spectrum
	Spectrum subtraction with each reference from the pre-search Comparison of results by forward search in the remaining references
Sorting	Sorting

Advantages

+ the mixture search is more meaningful
+ information on multiple occurrence of similar spectra
+ suggestions with regard to structural formulae and trivial names
+ available as complete additional software for many MS and GC-MS systems.

Limitations

− pure search is less useful for complex GC-MS analyses
− very dependent on the quality of spectra and thus on recording parameters, choice of spectrum in the GC peak and on the type of instrument
− therefore, multiple spectra per substance in PBM libraries.

Note

There are different PBM programs among the data systems of MS manufacturers available; some are more powerful than others; the mixture search mode is often omitted for reasons of speed.

3.2.4.3 **The SISCOM Procedure**

The SISCOM procedure developed by Henneberg and Weimann (Max Planck Institute for Coal Research, Mülheim a. d. Ruhr, Germany) was targeted to structural determination in industrial MS laboratories. The procedure is commercially available in the form of MassLib in association with various libraries. Its primary goal is to give the most plausible suggestion for the structure of an unknown component and offers information to support the interpretation of the spectra from a comprehensive knowledge base, that is, the collection of spectra which have already been interpreted: the Search for Identical and Similar Compounds.

One of the great strengths of the procedure is the deconvolution of the mass spectrum of an analyte from chromatographic data using an automatic background correction. Within this process, the variation of mass intensities during a GC peak is used to determine a reliable substance spectrum (Figure 3.36). In addition to the identity search, the search procedure offers a similarity search. Through this, account is taken of the fact that mass spectra can differ from a reference spectrum depending on the conditions of measurement, the compound can be contaminated by other compounds or by the matrix or the quality of the

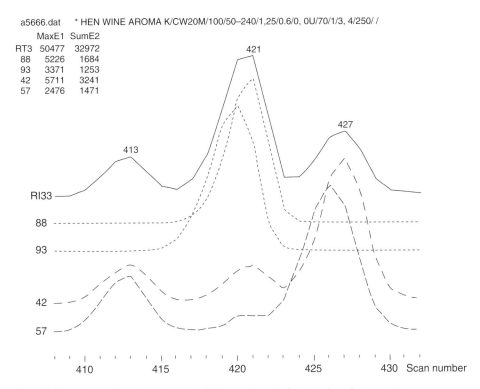

Figure 3.36 Automatic peak purity control (deconvolution) showing the relevant mass chromatograms with *m/z* values on co-elution using the SISCOM procedure (MassLib, Chemical Concepts).

reference spectrum can be in doubt. A similarity search also allows the limitation caused by the limited scope of spectral libraries to be counteracted. All the search systems mentioned earlier are limited by this handicap. If the aim of the search is only identification, the results of the similarity search can be rearranged in the direction of identity using a special algorithm. Complete spectra are used here and weightings with regard to mass and intensity are undertaken. A mixture correction procedure is also used to determine the nature of co-eluates. The retention index can also be included in the evaluation of GC-MS analyses.

An essential part of the recognition of similarities between an unknown spectrum and a reference spectrum is the simultaneous use of different methods of comparison, which evaluate different aspects of the similarities between two spectra. In the SISCOM procedure, the results of these methods are combined finally in such a way that the contribution of the part of the comparison which does not agree can be compensated for by the part which does agree. The differences between an unknown spectrum and the reference spectrum can thus be tolerated and do not hinder the selection of a relevant reference. The results of the different methods of comparison are represented by factors. The order of the hit list is based on a weighted function of these comparison factors. The calculation of a comparison value as a combination of factors with different relative importance is a multidimensional consideration. The ordering of references in a one-dimensional list of suggestions can therefore only be a compromise. The recognition of relevant reference spectra, which are arranged as subsequent hits in the similarity evaluation, is possible for an experienced user when employing the whole range of comparison factors.

The libraries used for the SISCOM search are divided into 14 sections of ion series. The assignment of a reference spectrum to one of these sections is based on the strongest peak in the ion series spectrum. Ion series spectra consist of 14 peaks. Each of these peaks (or ion series) represents the sum of all the intensities of the original spectrum and with a mass of m/z modulo $14 = n$. Thus, for example, the ion series 1 corresponds to the sum of the intensities of the masses 15, 29, 43, ... The strongest ion series in the spectrum of the unknown substance decides which section of the library is to be used for further searching. As the ion series spectra of related compounds (similar structures) are almost identical (even when the intensity ratios in the spectra differ), the ion series spectra are particularly suitable as a filter for a similarity search.

To compare the unknown spectrum (U) with the reference (R) from the library, SISCOM uses coded spectra, which are calculated from the original spectra by reduction to characteristics. To be determined as characteristic in the SISCOM sense, a mass peak must be larger than the arithmetic mean of its neighbouring homologues (in ±14 Da intervals). The characteristics used in a spectroscopic comparison are divided into three groups. One group contains the characteristics present in U and R; and the other two groups contain characteristics present only in U and only in R, respectively. Three comparison factors arise from these three groups:

NC: The number of common characteristics contained in both the unknown spectrum and the reference spectrum. High NC values indicate similar structures.

NR: The number of characteristics from the reference spectrum which are not present in the unknown spectrum. NR is a measure in relation to NC of the extent to which the reference spectrum is represented in the unknown spectrum or is part of it.

NU: The number of characteristics in the unknown spectrum which are not present in the reference spectrum. NU gives the part of the unknown spectrum which is not explained by the reference spectrum. Contamination or a mixture may be present.

These three factors NC, NR and NU do not involve any intensity parameters. 150 candidates result from the comparison factor calculation (pre-search).

Three other factors are derived from these first three which use intensities. All six factors are used together to order the 150 candidates in the pre-search for the hit list.

IR: The relative intensity in % of the characteristics of NR. IR completes or differentiates the importance of NR. A few characteristics not contained can have high peak intensities or many characteristics which are absent only consist of peaks of low intensity.

IU: The relative intensity as a % of the characteristics of NC. IU is equivalent to IR and is a measure of the purity of the spectrum, in the case where the suggestion is contained as part of a mixture.

PC: The measure of the correlation of the peak patterns of the relative intensities in the characteristics of NC. As PC is essentially calculated from the occurrence of intense peaks, high values are of only limited importance in spectra with few dominant masses. The PC only makes a limited contribution to the SI, that is, deviations for example, as a result of measuring conditions or different types of instrument, are tolerated better.

The evaluation of the hit list requires experience and sound knowledge of mass spectrometry. In the case of identical reference spectra, NR and IR are small but PC is high. For a pure spectrum, IU is small. If a mixture is present, IU is large. Components of a mixture can also appear in the suggestion list of SISCOM, provided that NR and IR are very small and PC correspondingly high. Substances with similar structures are assigned comparable factors, as are components of mixtures. Particularly high NC values indicate high similarity even when PC is negligibly small. Isomeric compounds are an example of this. In the evaluation, it should also be noted that a search result depends on the start of the mass range measured in the unknown mass spectrum and on the reference spectrum as well as on the presence of reduced reference spectra. For optimal search results, the unknown and the reference spectra should both contain the lower mass region from m/z 25 and be present in nonreduced form. Many important characteristics from this region (e.g. m/z 27, 30, 31, 35 ...) can contribute to similarity searches or identification of spectra.

SISCOM provides a powerful automatic mixture correction. The subtraction procedure is based on the assumption that specific ions exist for a substance which is removed. These ions disappear completely if the exact percentage proportion is removed. If larger proportions are removed, negative peaks are formed. The mixture correction uses an iterative procedure based on these, which is orientated towards the sum of the negative intensities.

The SISCOM search with mixture correction subtracts the spectrum of the best hits from the unknown spectrum. According to experience, this is almost always one of the components present (if a corresponding spectrum is present in the library) or an isomer with almost the same spectrum. With the spectrum thus corrected, a completely new search is carried out! In this way, a second component is then found if it was not present in the first hit list. The best hit in this second search is also subtracted from the original unknown spectrum and a third complete search is carried out so that in the case of a mixture, a purified spectrum is now also present for the first component, which leads to higher identity values. This threefold search gives high identity values from the purified spectra. For identity values of less than 80%, identification is very improbable. This makes it easier for only the relevant results to be produced in automatic evaluation processes. A further advantage of this method is that the result of the correction procedure is independent of whether the unknown spectrum is pure or the spectrum of a mixture. In the case of mixtures, the MassLib program package (Chemical Concepts, Weinheim, Germany) has a module available for determining the ratio of components in a GC peak using regression analysis (Beynon and Brenton, 1982).

3.2.5
Interpretation of Mass Spectra

There are no hard and fast rules about fully interpreting a mass spectrum. Unlike spectra obtained using other spectroscopic procedures, such as UV, IR, NMR or fluorescence, mass spectra do not show uptake or emission of energy by the compound (i.e. the intact molecule), but reflect the qualitative and quantitative analysis of the processes accompanying ionization (fragment formation, rearrangements, chemical reaction) by the direct measurement of the involved ionic species (see Section 2.3.2). The time factor and the energy required for ionization (electron beam, temperature, pressure) also play a role. With all other spectroscopic methods the intensity of interaction with certain functional groups or other structural elements is determined. In mass spectrometry, however, the appearance of certain characteristics of a structure in the mass spectrum always depends on the total structure of the compound. The failure of expected signals to appear generally does not prove in mass spectrometry that certain structural elements are not present; only positive signals count. It is also true that a mass spectrum cannot be associated with a particular chemical structure without additional information.

Nevertheless, procedures are recommended for deciphering information coded in a mass spectrum. (Certain users accuse experienced mass spectrometrists of having a criminological feel for the subject – and they are justified!) In this spectroscopic discipline, the experience of a frequently investigated class of substances is rapidly built up. New groups of substances usually require new methods of resolution. It is therefore of high importance that other parameters relating to the substance besides mass spectrometry (spectrum and accurate mass), such as UV, IR, NMR spectra, solubility, elution temperature, acidic or basic clean-up, synthesis reaction equations or those of conversion processes, should be incorporated into the interpretation of the spectrum.

The procedure shown in the following scheme has proved to be effective:

1) *Spectrum display*
   Does the mass spectrum originate from a single substance or are there signals which appear not to belong to it? Can a deconvolution or the subtraction of the background give a clearer representation? Is the spectrum possibly falsified by the subtraction? What information do the mass chromatograms give about the most significant ions?

2) *Library search*
   As all GC-MS systems are connected to very powerful computer systems, each interpretation process for EI spectra should begin with a search through available spectral libraries (see Section 3.2.3). Spectral libraries are an inestimable source of knowledge, which can give information as to whether the substance belongs to a particular class or on the appearance of clear structural features, even when identification seems improbable. Careful use of the database spares time and gives important suggestions. The different search procedures especially the similarity search modes can all help.

3) *Molecular ion*
   Which signal could be that of the molecular ion? Is an $M^{+/-}$, $(M + H)^+$ or an $(M - H)^{+/-}$ present? Which signals are considered to be noise or chemical background? Mass chromatograms can also help here. If the molecular weight is known, for example, from CI results, the library search should be carried out again limited to the molecular mass.

4) *Isotope pattern*
   Is there an obvious isotope pattern, for example, for chlorine, bromine, silicon or sulfur? The molecular ion shows all elements with stable isotopes in the compound. Is it possible to find out the maximum number of carbon atoms? This is a typical limitation with residue analysis as usually it is not possible to detect ^{13}C signal intensities with certainty. Also a noticeable absence of isotope signals, particularly with individual fragments, can be important for identifying the presence of phosphorus, fluorine, iodine, arsenic and other monoisotopic elements. Only the molecular or quasimolecular ions give complete information on all isotopes in the elemental formula.

5) *Nitrogen rule*
   Is nitrogen present? An uneven molecular mass indicates an uneven number of nitrogen atoms in the empirical formula.

6) *Fragmentation pattern*
   What information does the fragmentation pattern give? Are there pairs of fragments, the sum of which give the molecular weight? Which fragments could be formed from an α-cleavage? Here the use of a table giving details of neutral losses from molecular ions $(M - X)^+$ is advisable (see Table 3.6).

Table 3.6 Mass correlations to explain cleavage reactions $(M - X)^+$ and key fragments X^+.

m/z	Fragment X^+	$M^+ - X$	Explanations
12	C		
13	CH		
14	CH_2, N, N_2^{2+}		
15	CH_3	$M^+ - 15$	Nonspecific, CH_3 at high intensity
16	O, NH_2, O_2^{2+}, CH_4	$M^+ - 16$	Rarely CH_4, (but frequently $R^+ - CH_4$ in alkyl fragments), O from N-oxides and nitro compounds, NH_2 from anilines
17	OH, NH_3	$M^+ - 17$	Nonspecific O-indication, NH_3 from primary amines
18	H_2O, NH_4	$M^+ - 18$	Nonspecific O-indication, strong for many alcohols, some acids, ethers and lactones
19	H_3O, F	$M^+ - 19$	F-indication
20	HF, Ar^{2+}, CH_2CN^{2+}	$M^+ - 20$	F-indication
21	$C_2H_2O^{2+}$		(rarely)
22	CO_2^{2+}		
23	Na		(rarely)
24	C_2		
25	C_2H	$M^+ - 25$	Rarely with a terminal C≡CH group
26	C_2H_2, CN	$M^+ - 26$	From purely aromatic compounds, rarely from cyanides
27	C_2H_3, HCN	$M^+ - 27$	CN from cyanides, C_2H_3 from terminal vinyl groups and some ethyl esters
28	C_2H_4, N_2, CO	$M^+ - 28$	CO from aromatically bonded O, ethylene through RDA from cyclohexenes, by H-migration from alkyl groups, nonspecific from alicyclic compounds
29	C_2H_5, CHO	$M^+ - 29$	Aromatically bonded O, nonspecific with hydrocarbons
30	C_2H_6, H_2NCH_2, NO, CH_2O, BF (N-fragment)	$M^+ - 30$	CH_2O from cyclic ethers and aromatic methyl ethers, NO from nitro compounds and nitro esters
31	CH_3O, CH_2OH, CH_3NH_2, CF, (O-fragment)	$M^+ - 31$	Methyl esters, methyl ethers, alcohols
32	O_2, CH_3OH, S	$M^+ - 32$	Methyl esters, some sulfides and methyl ethers
33	CH_3OH_2, SH, CH_2F	$M^+ - 33$	SH nonspecific S-indication, $M^+ - 18 - 15$ nonspecific O-indication, strong with alcohols
34	SH_2 (S-fragment)	$M^+ - 34$	Nonspecific S-indication, strong with thiols
35	^{35}Cl, SH_3	$M^+ - 35$	Chlorides, nitrophenyl-compounds $(M^+ - 17 - 18)$
36	HCl, C_3	$M^+ - 36$	Chlorides

Table 3.6 (Continued)

m/z	Fragment X$^+$	M$^+$ − X	Explanations
37	^{37}Cl, C$_3$H		
38	H^{37}Cl, C$_3$H$_2$		
39	C$_3$H$_3$	M$^+$ − 39	Weak with aromatic hydrocarbons
40	Ar, C$_3$H$_4$	M$^+$ − 40	Rarely with CH$_2$CN
41	C$_3$H$_5$, CH$_3$CN	M$^+$ − 41	C$_3$H$_5$ from alicyclic compounds, CH$_3$CN from aromatic N-methyl and o-C-methyl heterocycles
42	CH$_2$=C=O, C$_3$H$_6$, C$_2$H$_4$N	M$^+$ − 42	Nonspecific with aliphatic and alicyclic systems, strong through RDA from cyclohexenes, by rearrangement from α-, β-cyclo-hexenones, enol and enamine acetates
43	CH$_3$CO, C$_3$H$_7$, C$_2$H$_4$N, CONH	M$^+$ − 43	Acetyl, propyl, aromatic methyl ethers (M$^+$ − 15 − 28), nonspecific with aliphatic and alicyclic systems
44	CO$_2$, CH$_3$NHCH$_2$ (N-fragment), CH$_2$CHOH	M$^+$ − 44	CO$_2$ from acids, esters, butane from aliphatic hydrocarbons
45	C$_2$H$_5$O, HCS (Sulfides)	M$^+$ − 45	Ethyl esters, ethyl ethers, lactones, acids, CO$_2$H from some esters; CH$_3$NHCH$_3$ from dimethylamines
46	C$_2$H$_5$OH, NO$_2$	M$^+$ − 46	Ethyl esters, rarely acids, nitro compounds, n-alkanols (M$^+$ − 18 − 28)
47	CH$_3$S (S-fragment), C^{35}Cl, C$_2$H$_5$OH$_2$, CH(OH)$_2$		
48	CH$_3$SH, CH^{35}Cl		
49	C^{37}Cl, CH$_2$ ^{35}Cl		
50	C$_4$H$_2$, CH ^{35}Cl		
51	C$_4$H$_3$ (Aromatic fragment)		
52	C$_4$H$_4$, CH$_3$ ^{37}Cl (Aromatic fragment)		
53	C$_4$H$_5$		
54	⟋⟍, C$_2$H$_4$CN	M$^+$ − 54	Cyclohexene (RDA)
55	C$_4$H$_7$, C$_2$H$_3$CO	M$^+$ − 55	C$_4$H$_7$ from alicyclic systems and butyl esters
56	C$_4$H$_8$, C$_2$H$_4$CO	M$^+$ − 56	Nonspecific with alkanes and alicyclic systems
57	C$_4$H$_9$, C$_2$H$_5$CO, C$_3$H$_2$F	M$^+$ − 57	Nonspecific with alkanes and alicyclic systems
58	CH$_3$COHCH$_2$, C$_2$H$_5$CHNH$_2$, C$_2$H$_6$NCH$_2$	M$^+$ − 58	C$_3$H$_6$O from α-methylaldehydes and acetonides
59	C$_2$H$_6$COH, C$_2$H$_5$OCH$_2$, CO$_2$CH$_3$, CH$_3$CONH$_2$	M$^+$ − 59	Methyl esters
60	CH$_2$CO$_2$H$_2$, CH$_2$ONO	M$^+$ − 60	O-acetates (M$^+$ − AcOH), methyl esters (M$^+$ − CH$_3$OH − CO)

(continued overleaf)

Table 3.6 (*Continued*)

m/z	Fragment X$^+$	M$^+$ – X	Explanations
61	$CH_3CO_2H_2$, C_2H_4SH		
62	$HOCH_2CH_2OH$	M$^+$ – 62	Ethylene ketals
63	C_5H_3		
64	SO_2	M$^+$ – 64	SO_2 cleavage from sulfonic acids
65	C_5H_5		
67			
68	, C_4H_4O, C_3H_6CN		
69	C_5H_9, C_3H_5CO, CF_3, C_3HO_2 (1,3-dioxyaromatics)		
70	C_5H_{10},		
71	C_5H_{11}, C_4H_7CO,		
72	$C_4H_{10}N$, $C_3N_7NHCH_2$, $C_2H_5COHCH_2$		
73	$CO_2C_2H_5$, $C_3H_7OCH_2$, $CH_2CO_2CH_3$, C_4H_8OH (O-fragments)		
74	CH_2=$COHOCH_3$, CH_3CH=$COHOH$		
75	$C_2H_5CO_2H_2$, $C_2H_5SCH_2$, $CH_3OCHOCH_3$, (dimethyl acetates)		
76	C_6H_4		
77	C_6H_5		
78	C_6H_6		
79	C_6H_7, ^{79}Br		
80	C_6H_8, $H^{79}Br$, , , CH_3S_2H		
81	C_6H_9, ^{81}Br,		
82	C_6H_{10}, $H^{81}Br$		
83	C_6H_{11}, C_4H_7CO		
84	,		
85	C_6H_{13}, C_4H_9CO		
86	C_3H_7COH=CH_2		
87	$CO_2C_3H_7$, $CH_2CO_2C_2H_5$, $CH_2CH_2CO_2CH_3$, (O-fragments)		
88	CH_2=$COHOC_2H_5$, CH_3CH=$COHOCH_3$		
91	, , n-Alkyl chlorides		
92	,		

Table 3.6 (Continued)

m/z	Fragment X+	m/z	Fragment X+

93	$CH_2{}^{79}Br$, 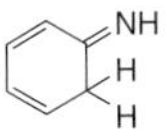 (structure)		
94	$CH_3{}^{79}Br$, 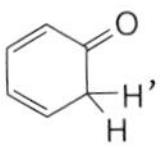 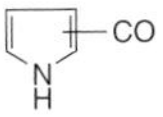 (structure), (structure) —CO	120	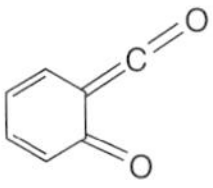 (structure)
95	$CH_2{}^{81}Br$, 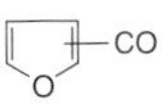 (structure) —CO	121	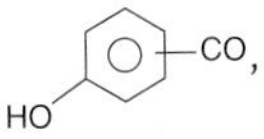 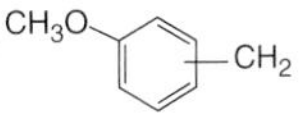 (structure) —CO, CH_3O— (structure) —CH_2
96	$C_5H_{10}CN$, $CH_3{}^{81}Br$	127	J
97	C_7H_{13}, 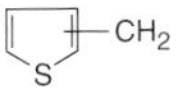 (structure) —CH_2	128	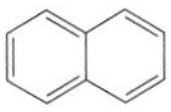 HJ, (structure)
98	(structure)	130	(structure)
99	C_7H_{15}, 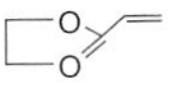 (structure) (Ethylene ketals)	131	C_3F_5
104	$C_2H_5CHONO_2$, 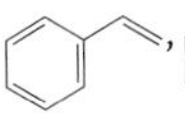 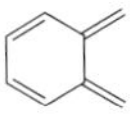 (structure), (structure)	135	(structure) (n-Alkyl bromides)
105	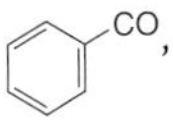 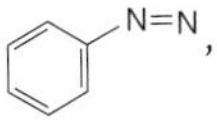 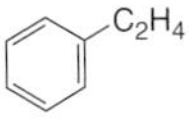 (structure) —CO, (structure) —N=N, (structure) —C_2H_4	141	 (structure) —CH_2
111	(structure) —CO	142	(structure) —CH_2
115	(structure)	149	(structure)
119	(structure) —$C(CH_3)_2$, (structure) —CO	152	(structure)

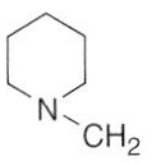

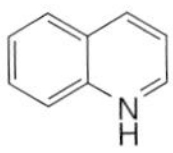

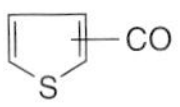

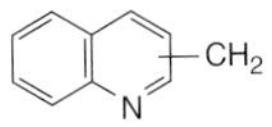

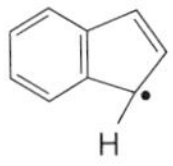

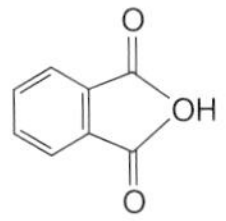

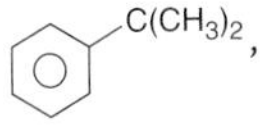

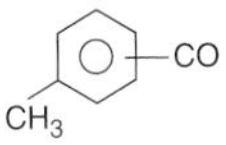

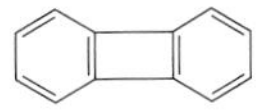

7) *Key fragments*

What can be said about characteristic fragments in the lower mass region? Is there information on aromatic building blocks or even ions formed through rearrangements (McLafferty, retro-Diels-Alder)? Also the application of MS/MS with the registration of product ion spectra from major ions is a successful route for structure elucidation. Here tables with plausible explanations of fragment ions are helpful (see Table 3.6).

8) *Structure postulate*

Bringing together information rapidly leads to a rough interpretation, which initially gives a partial structure and finally it postulates the molecular structure. Two possibilities test this postulated structure: which fragmentation pattern would the proposed structure give? Is the proposed substance available as a reference and does this correspond to the unknown spectrum?

Within this interpretation scheme, spectroscopic comparison through library searching is definitely placed at the beginning of the interpretation. Confirmation of a suggestion from a library search is achieved through an assessment using the scheme mentioned earlier. Interpretation has never finished simply with the print-out of the list of suggestions!

Steps to the Interpretation of Mass Spectra

1) Library search!
2) Only one substance?
3) Molecular ion?
4) Isotope pattern?
5) Nitrogen?
6) Fragmentation pattern?
7) Fragments? MS/MS product ion spectra?
8) Reference spectrum?

Why is ^{12}C the Official Reference Mass for Atomic Mass Units?

"Before the 1970s, two conventions were used for determining relative atomic masses. Physicists related their mass spectrometric determinations to the mass of ^{16}O (i.e. ^{16}O has a mass of exactly 16 on the amu scale), the most abundant isotope of oxygen, and chemists used the weighted mass of all three isotopes of oxygen: ^{16}O, ^{17}O and ^{18}O. At an international congress devoted to the standardization of scientific weights and measures, the redoubtable A.O. Nier proposed a solution to these disparate conventions whose negative consequences were becoming serious. He suggested that the carbon-12 isotope (^{12}C) be the reference for the atomic mass unit (amu). By definition, its mass would be exactly 12 amu, a convention that would be acceptable to the physicists. In accordance with this convention, the average mass for oxygen (the weighted sum of the three naturally occurring isotopes) becomes 15.9994 amu, a number close enough to 16 to satisfy the chemists." (Sharp, 2007)

3.2.5.1 Isotope Patterns

For organic mass spectrometry, only a few elements with noticeable isotope patterns are important, while in inorganic mass spectrometry there are many isotope patterns of metals, some of them very complex. From Table 3.7 it can be seen that the elements carbon, sulfur, chlorine, bromine and silicon consist of naturally occurring stable, nonradioactive isotopes. The elements fluorine, phosphorus and iodine are among the few monoisotopic elements in the periodic table (Rosman and Taylor, 1998).

If isotope signals appear in mass spectra, there is the possibility that these elements can be recognized by their typical pattern and also the number of them in molecular and fragment ions can be determined. With carbon, there are limitations to this procedure in trace analysis, because usually the quantity of substance available is too small for a sufficiently stable analysable signal to be obtained. In order to allow conclusions to be drawn on the maximum number of carbon atoms, a larger quantity of substance is necessary. For this, the technique of individual mass registration (SIM, MID) with longer dwell times is particularly suitable for giving good ion statistics. In the evaluation, in the case of carbon, only the maximum number of carbon atoms (isotope intensity/1.1%) can be calculated, as contributions from other elements must be taken into account.

Some elements, in particular the halogens chlorine and bromine, which are contained in many active substances, plastics and other technical products, can be recognized by the typical isotope patterns. These easily recognized patterns are shown in Figures 3.37 – 3.43. The intensities shown are scaled down to a unit ion stream of the isotope pattern. The lowering of the specific response of the compound as a function of, for example, the degree of chlorination, is shown. The simple occurrence of the elements shown as a series is used as a reference in each case. The relative intensities within an isotope pattern are given as a percentage in the caption underneath the isotope lines with the mass contribution to the molecular weight based on the most frequent occurrence in each case. What is noticeable for chlorine and bromine is the distance between the isotopes of two mass units. Provided that chlorine and bromine occur separately, the degree of chlorination or bromination can be easily determined by comparison of the variation in the intensities. For compounds which contain both chlorine and bromine the degree of substitution cannot be determined by comparison of the patterns alone. In these cases the high atomic weight of bromine is a help (for calculation of RIs from the empirical formula see Section 3.2.2). In GC-MS coupling, relatively high molecular weights (fragment ions) are detected but through the presence of bromine in a molecule at earlier retention times than can be expected by the molecular weight. In library searches, mixed isotope patterns of chlorine and bromine are reliably recognized.

With sulfur, the distance between the isotope peaks is also two mass units. However, when the proportion of sulfur in the molecule is low (e.g. in the case of

Table 3.7 Exact masses and natural isotope frequencies.

Element	Isotope	Nominal mass[a] [g/mol]	Exact mass [g/mol]	Abundance [b] [%]	Factors for calculating the isotope intensity[c]	
					M + 1	M + 2
Hydrogen	^1H	1	1.007825	99.99		
	D, ^2H	2	2.014102	0.01		
Carbon	^{12}C	12	12.000000[d]	98.9		
	^{13}C	13	13.003354	1.1	1.1	0.006
Nitrogen	^{14}N	14	14.003074	99.6	0.4	
	^{15}N	15	15.000108	0.4	0.4	
Oxygen	^{16}O	16	15.994915	99.76		
	^{17}O	17	16.999133	0.04	0.04	
	^{18}O	18	17.999160	0.20		0.20
Fluorine[e]	F	19	18.998405	100		
Silicon	^{28}Si	28	27.976927	92.2		
	^{29}Si	29	28.976491	4.7	5.1	
	^{30}Si	30	29.973761	3.1		3.4
Phosphorus	P	31	30.973763	100		
Sulfur	^{32}S	32	31.972074	95.02		
	^{33}S	33	32.971461	0.76	0.8	
	^{34}S	34	33.976865	4.22		4.4
Chlorine	^{35}Cl	35	34.968855	75.77		
	^{37}Cl	37	36.965896	24.23		32.5
Bromine	^{79}Br	79	78.918348	50.5		
	^{81}Br	81	80.916344	49.5		98.0
Iodine	I	127	126.904352	100		

a) The calculation of the nominal mass of an empirical formula is carried out using the mass numbers of the most frequently occurring isotope, for example, Lindane $^{12}C_6\,^1H_6\,^{35}Cl_6$ exact mass of M$^+$ 287.860065 g/mol.
b) The isotope abundance is a relative parameter. The abundances of an element add up to 100%.
c) In the isotope pattern of the ion the intensity of the first mass peak (M) is assumed to be 100%. The intensities of the isotope peaks (satellites) M + 1 and M + 2 are given by multiplying the factors with the number of atoms of an element in the ion.

Example	$C_{10}H_{22}$	M$^+$ 142	Intensity m/z 143: $10 \cdot 1.1 = 11\%$
	C_6Cl_6	M$^+$ 282	Intensity m/z 284: $6 \cdot 32.5 = 195\%$
	S_6	M$^+$ 192	Intensity m/z 194: $6 \cdot 4.4 = 26.4\%$

d) See box on p. 408 on ^{12}C as the reference mass.
e) Fluorine, sodium, aluminium, phosphorus, manganese, arsenic and iodine, for example, appear as monoisotopic elements in mass spectrometry.

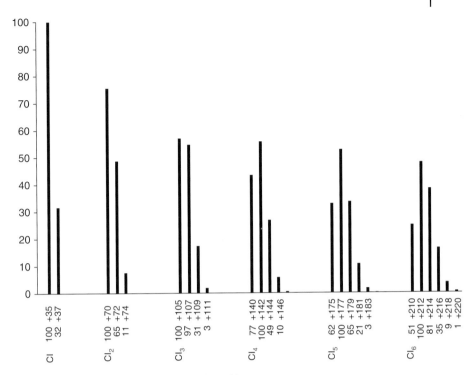

Figure 3.37 Isotope pattern of chlorine (Cl to Cl_6).

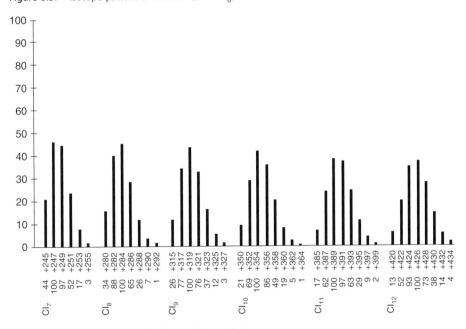

Figure 3.38 Isotope pattern of chlorine (Cl_7 to Cl_{12}).

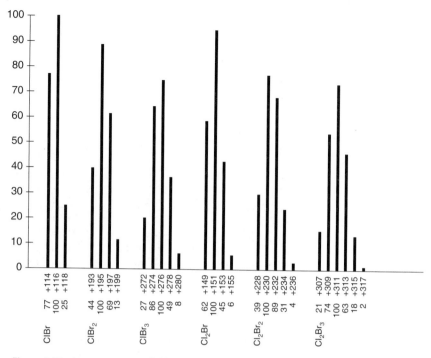

Figure 3.39 Isotope pattern of chlorine/bromine (ClBr to Cl$_2$Br$_3$).

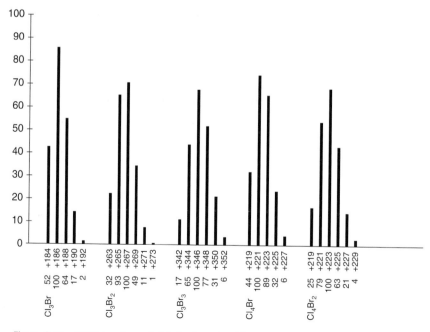

Figure 3.40 Isotope pattern of chlorine/bromine (Cl$_3$Br to Cl$_4$Br$_2$).

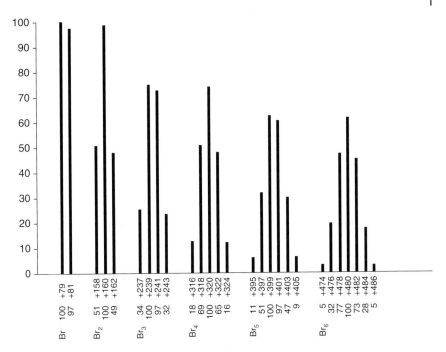

Figure 3.41   Isotope pattern of bromine (Br to Br$_6$).

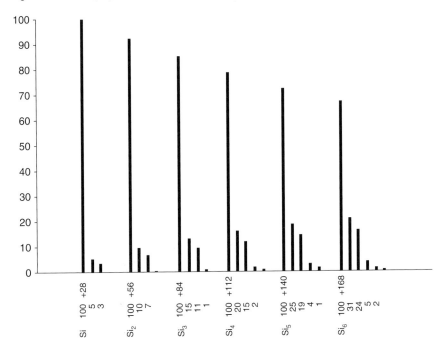

Figure 3.42   Isotope pattern of silicon (Si to Si$_6$).

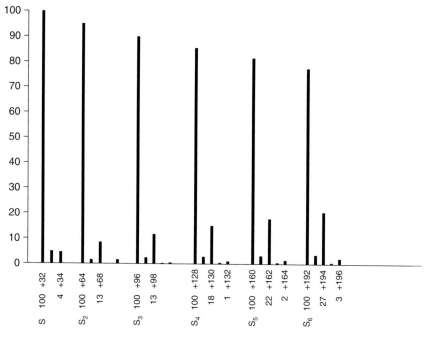

Figure 3.43 Isotope pattern of sulfur (S to S_6).

phosphoric acid esters), it is difficult to be sure of the presence of sulfur (Figure 3.43). Here a detailed investigation of the fragmentation is necessary. Higher sulfur contents (see also Figure 3.164) give clear information. The combinations of sulfur and chlorine in which chlorine is clearly dominant are also shown (Figure 3.44). Differences are extremely difficult to see with the naked eye and with computers only when measurements are obtainable with good ion statistics.

Silicon occurs very frequently in trace analysis. Silicones get into the analysis through derivatization (silylation), partly through clean-up (joint grease), but more frequently through bleeding from the septum or the column (septa of autosampler vials, silicone phases). The typical isotope pattern of all silicone masses can be recognized rapidly and excluded from further evaluation measures (Figure 3.42).

3.2.5.2 Fragmentation and Rearrangement Reactions
The starting point for a fragmentation is the molecular ion (EI) or the quasimolecular ion (CI). A large number of reactions which follow the primary ionization need to be described here. All the reactions follow the thermodynamic aim of achieving the most favourable energy balance possible. The basic mechanisms which are involved in the generation of spectra of organic compounds will be

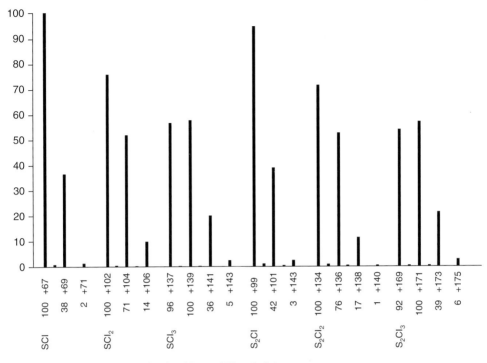

Figure 3.44 Isotope pattern of sulfur/chlorine (SCl to S_2Cl_3).

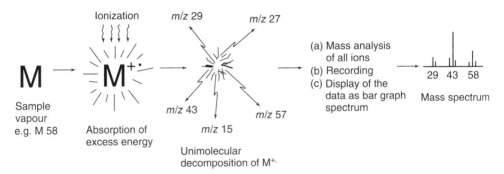

Figure 3.45 Principle of mass spectrometry: formation of molecular ions as the starting point of a fragmentation (after Frigerio).

discussed briefly here (Figure 3.45). For more depth, the references cited should be consulted (Budzikiewicz, 1998; Pretsch *et al.*, 1990; Howe *et al.*, 1981; McLafferty and Turecek, 1959).

There are two possible mechanisms for the cleavage of carbon chains following ionization, known as *α-cleavage* and *formation of carbenium ions*. The starting point after ionization is the localization of positive charge on electron-rich structures in the molecule.

α-Cleavage

α-Cleavage takes place after ionization by loss of one nonbonding electron from a heteroatom (e.g. in amines, ethers, ketones, see Tables 3.8–3.10) or on formation of an allylic or benzylic carbenium ion (from alkenes, alkylaromatics):

Table 3.8 Characteristic ions from the α-cleavage of amines.

Amines		
$R_1-\overset{+\bullet}{\underset{R_2}{N}}-CH_2-R \longrightarrow R_1-\overset{+}{\underset{R_2}{N}}=CH_2 + R^\bullet$		
m/z	R_1	R_2
30	H	H
44	CH_3	H
58	CH_3	CH_3

Table 3.9 Characteristic ions from the α-cleavage of ethers.

Ethers	
$R_1-\overset{+\bullet}{\underline{O}}-CH_2-R \longrightarrow R_1-\overset{+}{\underline{O}}=CH_2 + R^\bullet$	
m/z	R_1
31	H
45	CH_3
59	C_2H_5
73	C_3H_7
87	C_4H_9
etc.	

Table 3.10 Characteristic ions from the α-cleavage of ketones.

Ketones	
$R_1-\overset{\overset{\displaystyle O^{+\bullet}}{\|}}{C}-R \longrightarrow R_1-C\equiv\overset{+}{O}| + R^\bullet$	
m/z	R_1
29	H
43	CH_3
57	C_2H_5
71	C_3H_7
85	C_4H_9

Formation of Carbenium Ions

Fragmentation involving formation of carbenium ions takes place at a double bond in the case of aliphatic carbon chains (allylic carbenium ion) or at a branch. With alkylaromatics, side chains are cleaved giving a benzylic carbenium ion (benzyl cleavage), which dominates as the tropylium ion m/z 91 in many spectra of aromatics.

Alkenes

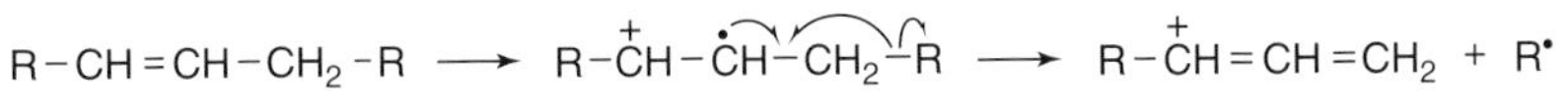

$$R-CH=CH-CH_2-R \longrightarrow R-\overset{+}{C}H-\overset{\bullet}{C}H-CH_2-R \longrightarrow R-\overset{+}{C}H=CH=CH_2 + R^{\bullet}$$

Branched carbon chains

$$R_3C\overset{+}{\cdot}R \longrightarrow R_3C^+ + R^{\bullet}$$

The formation of carbenium ions occurs preferentially at tertiary branches rather than secondary ones.

Alkylaromatics (benzyl cleavage)

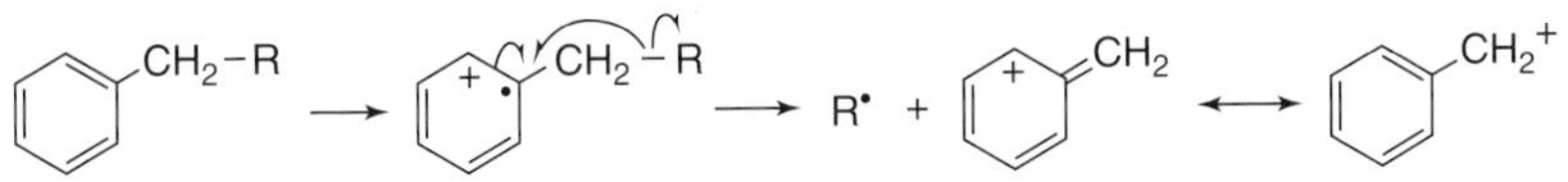

Loss of Neutral Particles

The elimination of stable neutral particles is a common fragmentation reaction. These include H_2O, CO, CO_2, NO, HCN, HCl, RCOOH and alkenes (McLafferty, 1959). These reactions can usually be recognized from the corresponding ions and mass differences in the spectra. Eliminations are particularly likely to occur when α-cleavage is impossible. MS/MS mass spectrometer providing the neutral loss scan mode allow a substance class-specific detection based on the elimination of neutrals.

Alcohols (loss of water)

$$[C_nH_{2n+1}OH]^{\ddagger} \longrightarrow [C_nH_{2n}]^{\ddagger} + H_2O$$

Carbonyl functions (loss of CO)

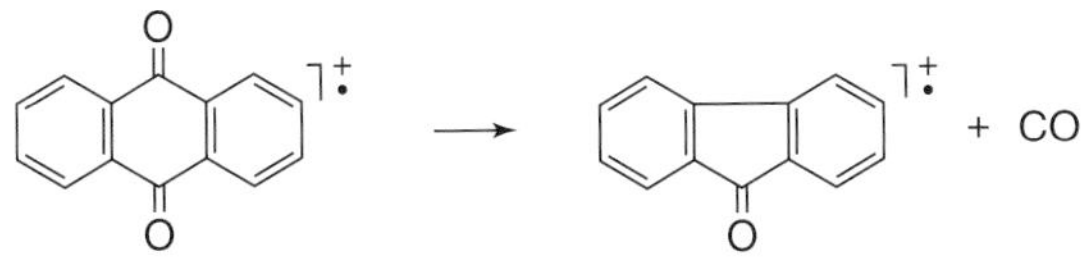

Heterocycles (loss of HCN)

$$\text{(pyridine ring)}^{+\cdot} \longrightarrow C_4H_4^{+\cdot} + HCN$$

Retro-Diels-Alder (loss of alkenes; in short RDA)

$$\text{(cyclohexene)}^{+\cdot} \longrightarrow \text{(butadiene)}^{+\cdot} + \begin{array}{c} CH_2 \\ \| \\ CH_2 \end{array}$$

The McLafferty Rearrangement

The McLafferty rearrangement involves the migration of an H-atom in a six-membered ring transition state. The following conditions must be fulfilled for the rearrangement to take place:

- The double bond C=X is C=C, C=O or C=N.
- There is a chain of three σ-bonds ending in a double bond.
- There is an H-atom in the y-position relative to the double bond which can be abstracted by the element X of the double bond.

According to convention, the McLafferty rearrangement is classified as the loss of neutral particles (alkene elimination with a positive charge remaining on the fragment formed from the double bond, see Table 3.11).

$$\begin{array}{c} \text{(McLafferty transition state)} \end{array}^{+\cdot} \longrightarrow \begin{array}{c} \text{(enol ion)} \end{array}^{+\cdot} + \begin{array}{c} D \\ \| \\ B \end{array}$$

Table 3.11 Characteristic ions formed in the McLafferty rearrangement.

m/z	R_1	Found
44	H	in aldehydes
60	OH	in organic acids
74	$O-CH_3$	in methyl esters

3.2.5.3 DMOX Derivatives for Location of Double Bond Positions

The location of double bonds in polyunsaturated fatty acids by GC-MS involves many different methodologies either by suitable derivatization or CI techniques (Christie, 1996; López and Grimalt, 2004; Jhama et al., 2005). Derivatization reactions include the reaction with DMOX (Fay and Richli, 1991; Dobson and Christie, 2002), dimethyl disulfide (DMDS) (Moss and Lambert-Fair, 1989), as well as methoxy and methoxybromo derivatives (Shantha and Kaimal, 1984) besides other known derivatives. CI allows specific reactions with C,C-double and triple bonds with suitable reagent gases, which in many cases permit a location of the sites of unsaturation in organic molecules (Budzikiewicz, 1985).

Methoxybromo derivatives of unsaturated fatty acids including conjugated acids yield simple mass spectra to locate the position of double bonds in these acids. Unlike other methods using methoxy derivatives, the methoxybromo derivatives yield fewer ions, the diagnostic peaks forming the most intense ions of the spectra with the characteristic appearance of fragments corresponding to $[CH_3(CH_2)_nCH(OMe)CH(Br)CH_22H]^+$.

The preparation of DMOX derivatives is most widely applied for routine identification of fatty acids in unknown samples. A relatively mild reaction minimizes possible isomerization reactions. The DMOX derivatives are comparable to FAMEs in volatility and hence in chromatographic resolution. When ionized under regular EI conditions, radical-induced cleavage processes give rise to mass spectra that are easy to interpret in terms of locating double bond positions in the hydrocarbon chain. The total number of carbons and the degree of unsaturation can be taken from the molecular ion information. Typical mass distances of 12 Da unveil double bonds corresponding to fragments containing n and $n − 1$ carbons (see the mass peaks marked by * with the pairs m/z 196/208, 236/248 and 276/288 in Figure 3.46). Monoenes with double bonds between C7 and C15 follow these rules and exhibit intense allylic ions. Especially with an increasing degree of unsaturation and for conjugated double bonds, DMOX spectra are more informative compared to other derivatives. With double bonds closer to the carboxyl end, the spectra show at C4, C5 and C6 characteristic odd-numbered ions at m/z 139, 153 and 167, respectively. At C3 the base peak m/z 152 and at C2 m/z 110 are prominent ions (Christie, 2001).

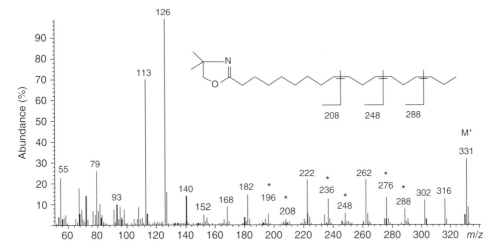

Figure 3.46 Spectrum of the DMOX derivative of 9,12,15-octadecatrieenoate (C18: 3, ω3). Masses indicated by * are the diagnostic ions with 12 Da mass distance for locating double bonds (Christie – The Lipid Library, AOCS, 2014).

3.2.6

Mass Spectroscopic Features of Selected Substance Classes

The compounds involved and detected by analytical GC-MS workflows like volatiles, PCBs, or pesticides are discussed with representative mass spectra in the following sections. These compounds do not necessarily belong to the same functional compound class with a chemical classification for instance of amines, ester, phenols, hydrocarbons and others. Many standard references cover these characteristic fragmentation patterns of specific functional groups and compound classes in detail (Howe, 1981; McLafferty and Turecek, 1993; Budzikiewicz, 1998).

3.2.6.1 Volatile Halogenated Hydrocarbons

This group of compounds does not belong to a single class of compounds (Figures 3.47–3.56). In a single analysis, more than 60 aliphatic and aromatic compounds (Magic 60) are typically determined by headspace GC-MS (static/purge and trap) more than 300 compounds were collected and identified from air samples (Ciccioli *et al.*, 1993). A noticeable feature is the frequent appearance of chlorine and bromine isotope patterns in the mass spectra. With aliphatic compounds, molecular ions do not always appear. With increasing

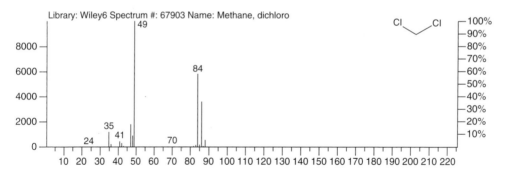

Figure 3.47 Dichloromethane (R30) CH_2Cl_2, M: 84, CAS Reg. No.: 75-09-02.

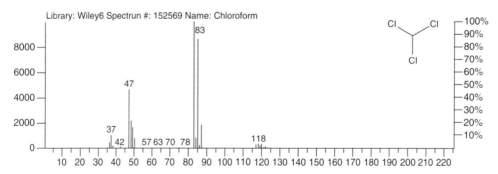

Figure 3.48 Chloroform $CHCl_3$, M: 118, CAS Reg. No.: 67-66-3.

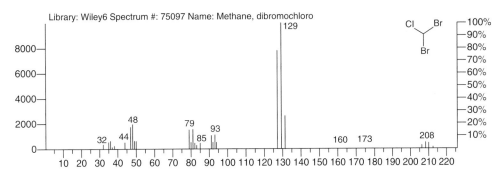

Figure 3.49 Dibromochloromethane CHBr$_2$Cl, M: 206, CAS Reg. No.: 124-48-1.

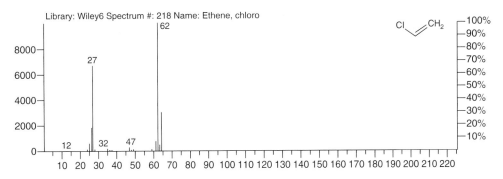

Figure 3.50 Vinyl chloride C$_2$H$_3$Cl, M: 62, CAS Reg. No.: 75-01-4.

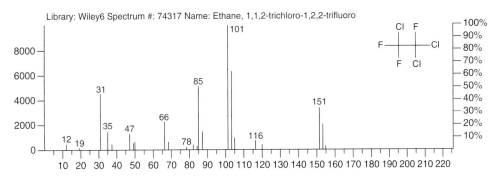

Figure 3.51 1,1,2-Trifluoro-1,2,2-trichloroethane (R113) C$_2$Cl$_3$F$_3$, M: 186, CAS Reg. No.: 76-13-1.

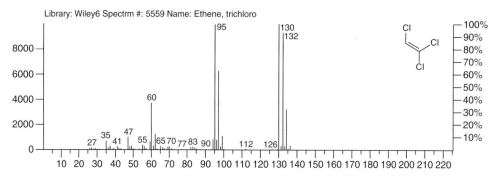

Figure 3.52 Trichloroethylene C_2HCl_3, M: 130, CAS Reg. No.: 79-01-6.

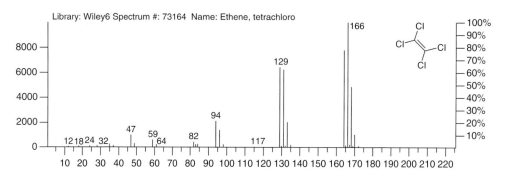

Figure 3.53 Tetrachloroethylene (Per) C_2Cl_4, M: 164, CAS Reg. No.: 127-18-4.

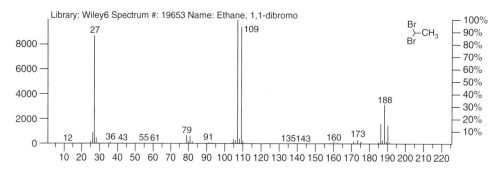

Figure 3.54 1,1-Dibromoethane $C_2H_4Br_2$, M: 186, CAS Reg. No.: 557-91-5.

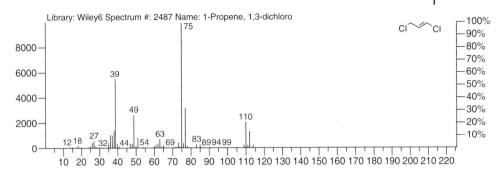

Figure 3.55 Dichloropropene $C_3H_4Cl_2$, M: 110, CAS Reg. No.: 542-75-6.

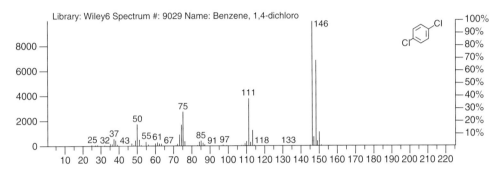

Figure 3.56 p-Dichlorobenzene $C_6H_4Cl_2$, M: 146, CAS Reg. No.: 106-46-7.

molecular size, the M^+ intensities decrease. Usually, the loss of Cl (and also Br, F) as a radical from the molecular ion occurs. Fluorine can be recognized as HF elimination from M^+ with a difference of 20 Da or as the CF fragment m/z 31. Bromine should be assumed from signals with significantly higher masses but relatively short retention times. For detection, it is a good recommendation to include the masses m/z 35/37 in the scan to guarantee ease of identification during library search.

Aromatic halogenated hydrocarbons generally show an intense molecular ion. There is successive radical cleavage of chlorine. In the lower mass range, the characteristic aromatic fragments appear with lower intensity.

3.2.6.2 Benzene/Toluene/Ethylbenzene/Xylenes (BTEX, Alkylaromatics)

Alkylaromatics form very stable molecular ions which can be detected with very high sensitivity (Figures 3.57–3.65). The tropylium ion occurs at m/z 91 as the base peak, which is, for example, responsible for the uneven base peak in the toluene spectrum (M – 92). The fragmentation of the aromatic skeleton leads to typical series of ions with m/z 38–40, 50–52, 63–67, 77–79 ('aromatic rubble'). Ethylbenzene and the xylenes cannot be differentiated from their spectra because they are isomers. In these cases, the retention times of the components are more meaningful.

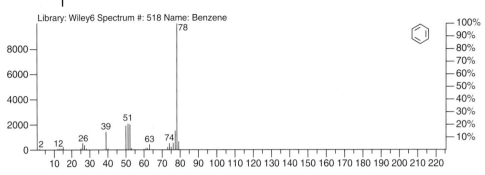

Figure 3.57 Benzene C_6H_6, M: 78, CAS Reg. No.: 71-43-2.

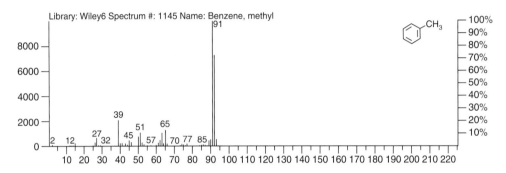

Figure 3.58 Toluene C_7H_8, M: 92, CAS Reg. No.: 108-88-3.

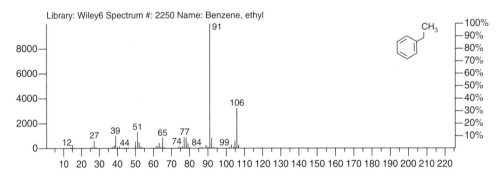

Figure 3.59 Ethylbenzene C_8H_{10}, M: 106, CAS Reg. No.: 100-41-4.

With alkyl side chains, the problem of isomerism must be taken into account. For di-methylnaphthalene, for example, there are 10 isomers! Alkylaromatics fragment through benzyl cleavage. From a propyl side chain onwards benzyl cleavage can take place with H-transfer to the aromatic ring (to C_8H_{10}, *m/z* 106) or without transfer (to C_8H_9, *m/z* 105) depending on the steric or electronic conditions.

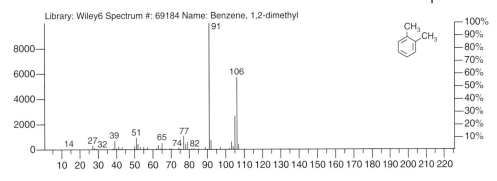

Figure 3.60 *o*-Xylene C$_8$H$_{10}$, M: 106, CAS Reg. No.: 95-47-6.

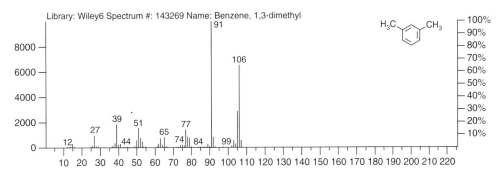

Figure 3.61 *m*-Xylene C$_8$H$_{10}$, M: 106, CAS Reg. No.: 108-38-3.

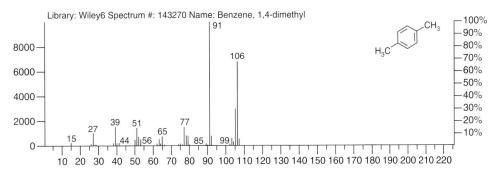

Figure 3.62 *p*-Xylene C$_8$H$_{10}$, M: 106, CAS Reg. No.: 106-42-3.

3.2.6.3 Polyaromatic Hydrocarbons (PAHs)

PAHs form very stable molecular ions (Figures 3.66–3.74). They can be recognized easily from a 'half mass' signal caused by doubly charged molecular ions as $m/2z = 1/2\, m/z$. (Be aware of heterocycles with odd numbers of nitrogen, the doubly charged ion appears on two masses in LRMS, see Figure 3.72.) Masses in

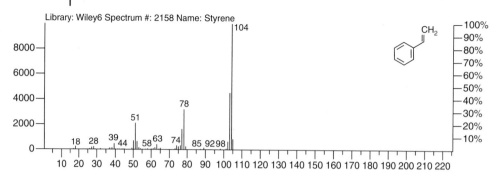

Figure 3.63 Styrene C_8H_8, M: 104, CAS Reg. No.: 100-42-5.

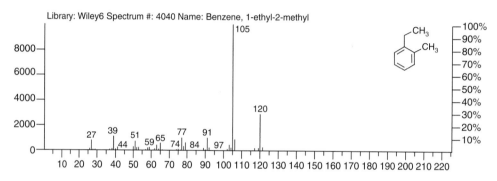

Figure 3.64 1-Ethyl-2-methylbenzene C_9H_{12}, M: 120, CAS Reg. No.: 611-14-3.

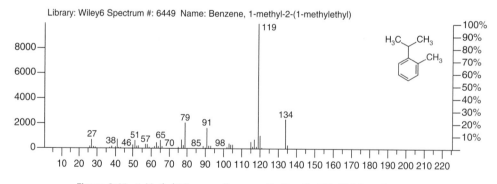

Figure 3.65 1-Methyl-2-isopropylbenzene $C_{10}H_{14}$, M: 134, CAS Reg. No.: 1074-17-5.

the range m/z 100 to m/z 320 should be scanned for PAH full scan analysis, if not SIM or SRM is used. In this range, all polycondensed aromatics from naphthalene to coronene (including all 16 EPA components) can be determined and a possible matrix background of hydrocarbons with aliphatic character can be almost completely excluded from detection.

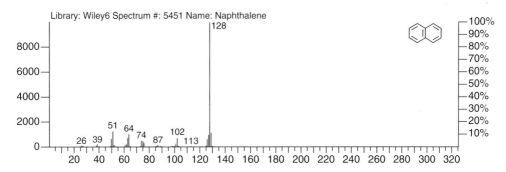

Figure 3.66   Naphthalene $C_{10}H_8$, M: 128, CAS Reg. No.: 91-20-3.

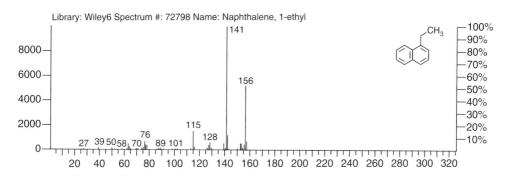

Figure 3.67   1-Ethylnaphthalene $C_{12}H_{12}$, M: 156, CAS Reg. No.: 1127-76-0.

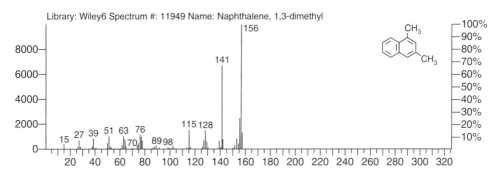

Figure 3.68   1,3-Dimethylnaphthalene $C_{12}H_{12}$, M: 156, CAS Reg. No.: 575-41-7.

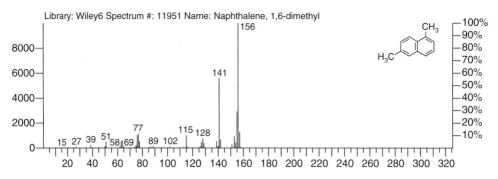

Figure 3.69 1,6-Dimethylnaphthalene $C_{12}H_{12}$, M: 156, CAS Reg. No.: 575-43-9.

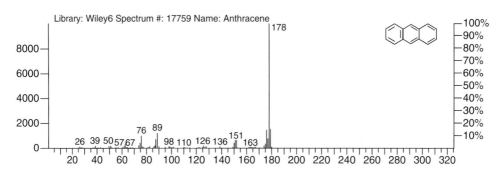

Figure 3.70 Anthracene $C_{14}H_{10}$, M: 178, CAS Reg. No.: 120-12-7.

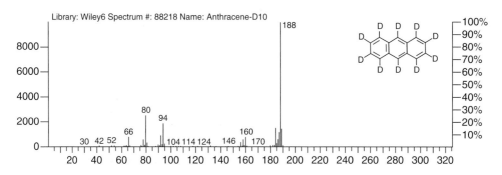

Figure 3.71 Anthracene-d_{10} $C_{14}D_{10}$, M: 188, CAS Reg. No.: 1719-06-8.

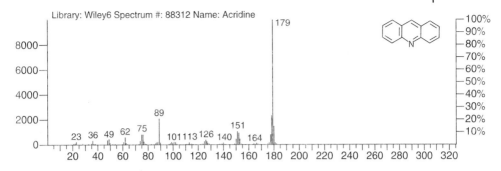

Figure 3.72 Acridine $C_{13}H_9N$, M: 179, CAS Reg. No.: 260-94-6.

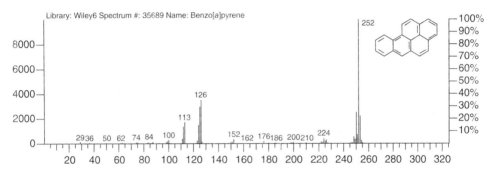

Figure 3.73 Benzo[a]pyrene $C_{20}H_{12}$, M: 252, CAS Reg. No.: 50-32-8.

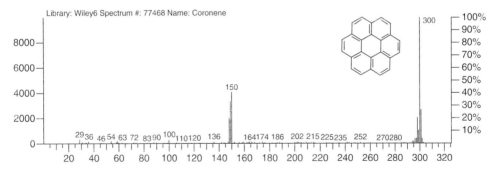

Figure 3.74 Coronene $C_{24}H_{12}$, M: 300, CAS Reg. No.: 191-07-1.

3.2.6.4 Phenols

In mass spectrometry, phenolic substances are determined by their aromatic character (Figures 3.75–3.80). Depending on the side chains, intense molecular ions and less intense fragments appear. In GC-MS, phenols are usually chromatographed as their methyl esters or acetates. Phenols especially halogenated phenols are acidic and tend to severe peak tailing. In trace analysis, chlorinated and brominated phenols are the most important and can be recognized by their clear isotope patterns. The loss of CO (M − 28) gives a less intense

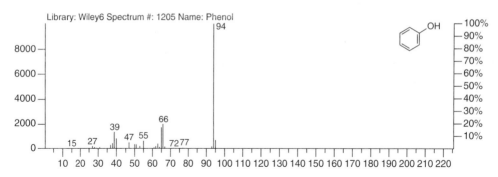

Figure 3.75 Phenol C_6H_6O, M: 94, CAS Reg. No.: 108-95-2.

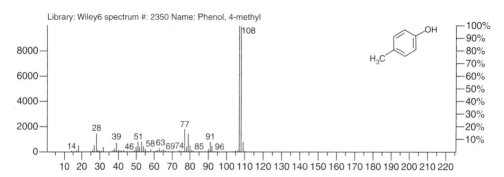

Figure 3.76 *p*-Cresol C_7H_8O, M: 108, CAS Reg. No.: 106-44-5.

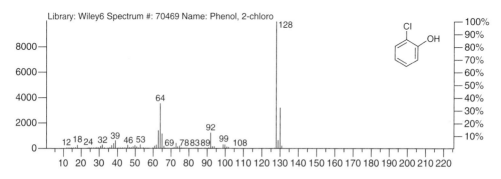

Figure 3.77 *o*-Chlorophenol C_6H_5ClO, M: 128, CAS Reg. No.: 95-57-8.

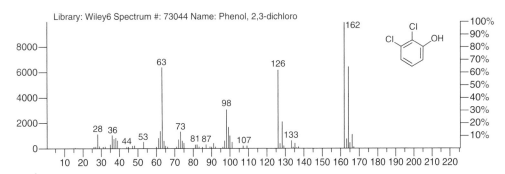

Figure 3.78 2,3-Dichlorophenol $C_6H_4Cl_2O$, M: 162, CAS Reg. No.: 576-24-9.

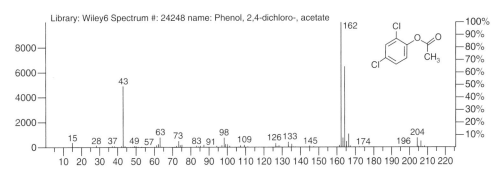

Figure 3.79 2,4-Dichlorophenyl acetate $C_8H_6Cl_2O_2$, M: 204, CAS Reg. No.: 6341-97-5.

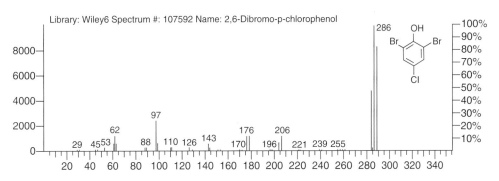

Figure 3.80 2,6-Dibromo-4-chlorophenol $C_6H_3Br_2ClO$, M: 284.

signal but is a clear indication of the presence of phenols. Halogenated phenols clearly show the loss of HCl (M − 36) and HBr (M − 80) in their spectra. With phenols isomers are also best recognized from their retention times rather than their mass spectra.

3.2.6.5 Pesticides

The common use as plant protection agents forms the basis of the classification of these compounds (Figures 3.81–3.112). In a collection of pesticide spectra there is therefore a wide variety of compound classes, which are covered to some extent by the other substance classes described here. Even when considering what appears to be a single group, such as phosphoric acid esters, it is virtually impossible to give generalizations on fragmentation. Usually, only stable compounds with aromatic character form intense molecular ions. In other cases, molecular ions are of low intensity and cannot be isolated from the matrix in trace analyses. For many pesticides (phosphoric acid esters, triazines, phenylureas, chlorinated hydrocarbons, etc.), the use of PCI is advantageous for confirming identities or allowing selective detection by NCI.

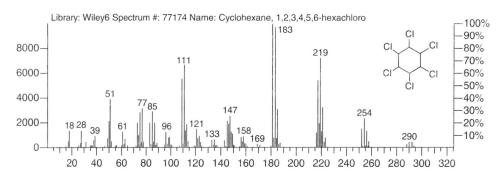

Figure 3.81 Lindane $C_6H_6Cl_6$, M: 288, CAS Reg. No.: 58-89-9 (isomers with similar spectra).

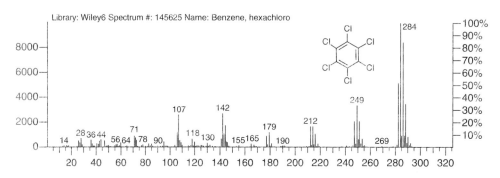

Figure 3.82 Hexachlorobenzene (HCB) C_6Cl_6, M: 282, CAS Reg. No.: 118-74-1.

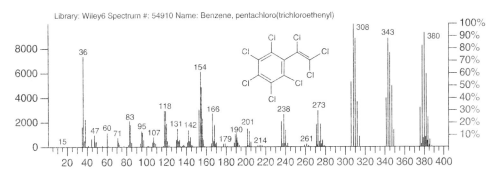

Figure 3.83 Octachlorostyrene C_8Cl_8, M: 376, CAS Reg. No.: 29082-74-4.

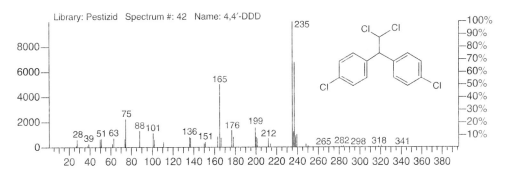

Figure 3.84 4,4'-DDD $C_{14}H_{10}Cl_4$, M: 318, CAS Reg. No.: 72-54-8.

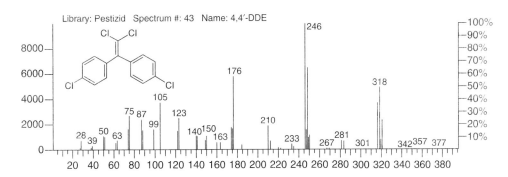

Figure 3.85 4,4'-DDE $C_{14}H_8Cl_4$, M: 316, CAS Reg. No.: 72-55-9.

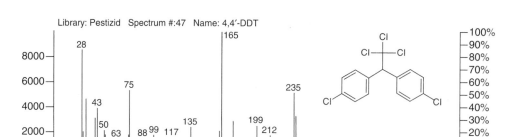

Figure 3.86 4,4'-DDT $C_{14}H_9Cl_5$, M: 352, CAS Reg. No.: 50-29-3.

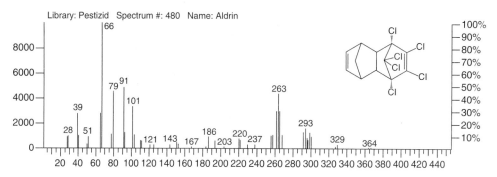

Figure 3.87 Aldrin $C_{12}H_8Cl_6$, M: 362, CAS Reg. No.: 309-00-2.

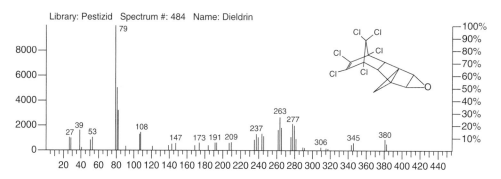

Figure 3.88 Dieldrin $C_{12}H_8Cl_6O$, M: 378, CAS Reg. No.: 60-57-1.

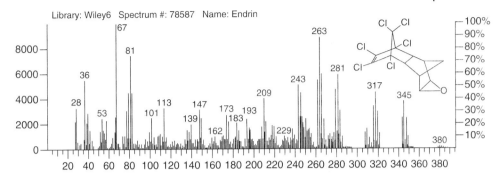

Figure 3.89 Endrin $C_{12}H_8Cl_6O$, M: 378, CAS Reg. No.: 72-20-8.

Chlorinated Hydrocarbons

In this group of organochlorine pesticides (OCPs), there is a large number of different types of compounds (Figures 3.81–3.90). In the example of Lindane and HCB (hexachlorobenzene), the difference between compounds with saturated and aromatic character can clearly be seen (molecular ion, fragmentation).

The high proportion of chlorine in these analytes appears with intense and characteristic isotope patterns. With the nonaromatic compounds (polycyclic polychlorinated alkanes made by Diels-Alder reactions, e.g. Dieldrin, Aldrin) spectra with large numbers of lines are formed through extensive fragmentation of the molecule. These compounds can be readily analysed using negative CI, whereby the fragmentation is prevented.

Triazines

Triazine herbicides are substitution products of 1,3,5-triazines and thus belong to a single series of substances (Figures 3.91–3.93). Hexazinon is also determined together with the triazine group in analysis (Figure 3.94). Without exception the EI spectra of triazines show a large number of fragment ions and usually also contain the molecular ion with varying intensity. The high degree of fragmentation is responsible for the low specific response of triazines in trace analysis. All triazine

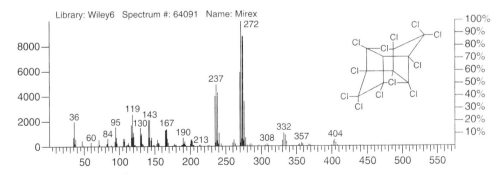

Figure 3.90 Mirex $C_{10}Cl_{12}$, M: 540, CAS Reg. No.: 2385-85-5.

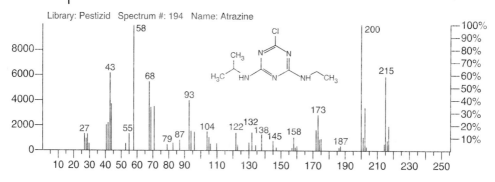

Figure 3.91 Atrazine $C_8H_{14}ClN_5$, M: 215, CAS Reg. No.: 1912-24-9.

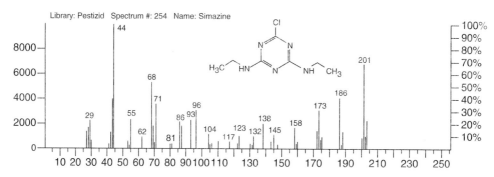

Figure 3.92 Simazine $C_7H_{12}ClN_5$, M: 210, CAS Reg. No.: 122-34-9.

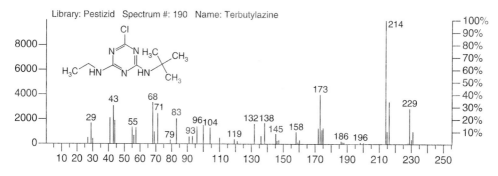

Figure 3.93 Terbutylazine $C_9H_{16}ClN_5$, M: 229, CAS Reg. No.: 5915-41-3.

analyses can be confirmed and quantified readily by positive CI (e.g. with NH_3 as the reagent gas).

Carbamates

The highly polar carbamate pesticides cannot always be analysed by GC-MS (Figures 3.95–3.97). The low thermal stability leads to decomposition even in the injector. Substances with definite aromatic character, however, form stable intense molecular ions.

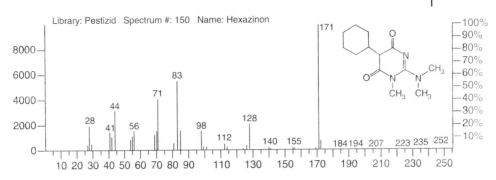

Figure 3.94 Hexazinon $C_{12}H_{20}N_4O_2$, M: 252, CAS Reg. No.: 51235-04-2.

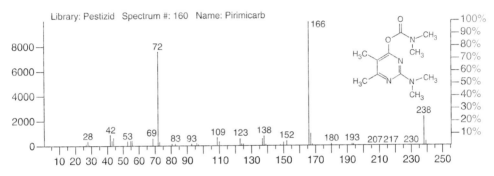

Figure 3.95 Pirimicarb $C_{11}H_{18}N_4O_2$, M: 238, CAS Reg. No.: 23103-98-2.

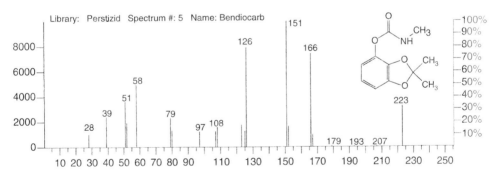

Figure 3.96 Bendiocarb $C_{11}H_{13}NO_4$, M: 223, CAS Reg. No.: 22781-23-3.

Phosphoric Acid Esters

This large group of organophosphorous pesticides (OPs) does not exhibit uniform behaviour in mass spectrometry (Figures 3.98–3.103). In trace analysis, the detection of molecular ions is usually difficult, except in the case of aromatic compounds (e.g. Parathion). The high degree of fragmentation frequently extends into the area of matrix noise. Because of this it is more difficult to detect individual compounds, as in full-scan analysis, low starting masses must be used. Positive CI is suitable for phosphoric acid esters because generally a strong CI

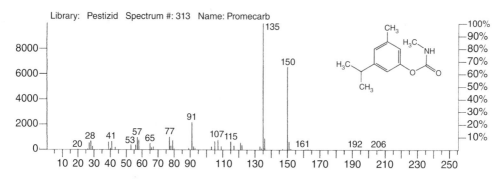

Figure 3.97 Promecarb $C_{12}H_{17}NO_2$, M: 207, CAS Reg. No.: 2631-37-0.

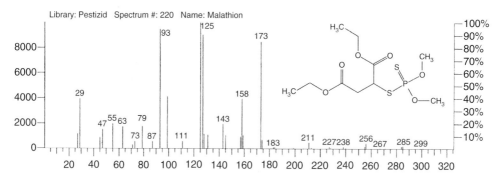

Figure 3.98 Malathion $C_{10}H_{19}O_6PS_2$, M: 330, CAS Reg. No.: 121-75-5.

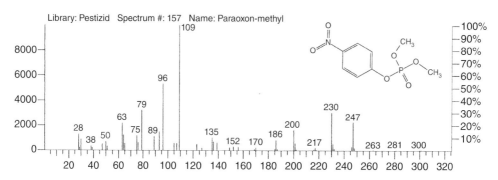

Figure 3.99 Paraoxon-methyl $C_8H_{10}NO_6P$, M: 247, CAS Reg. No.: 950-35-6.

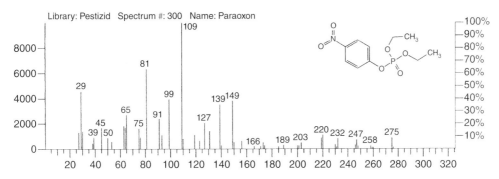

Figure 3.100 Paraoxon(-ethyl) $C_{10}H_{14}NO_6P$, M: 275, CAS Reg. No.: 311-45-5.

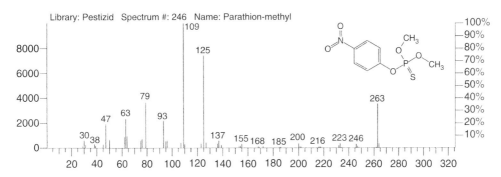

Figure 3.101 Parathion-methyl $C_8H_{10}NO_5PS$, M: 263, CAS Reg. No.: 298-00-0.

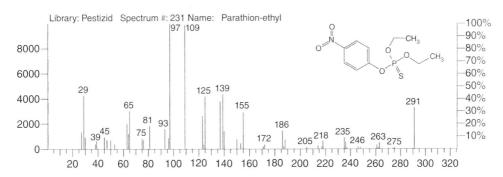

Figure 3.102 Parathion(-ethyl) $C_{10}H_{14}NO_5PS$, M: 291, CAS Reg. No.: 56-38-2.

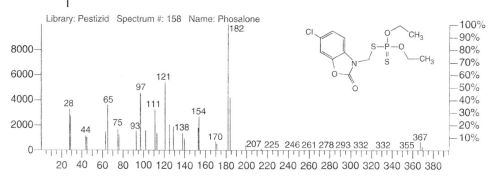

Library: Pestizid   Spectrum #: 158   Name: Phosalone

Figure 3.103   Phosalone $C_{12}H_{15}ClNO_4PS_2$, M: 367, CAS Reg. No.: 2310-17-0.

Table 3.12   Fragments typical of various groups from phosphoric acid ester (PAE) pesticides (Stan, 1977).

Group	R		*m/z* 93	*m/z* 97	*m/z* 109	*m/z* 121	*m/z* 125
Dithio-PAE	Ia	CH_3	+	–	–	–	+
	Ib	C_2H_5	+	+	–	+	+
Thiono-PAE	IIa	CH_3	+	–	+	–	+
	IIb	C_2H_5	+	+	+	+	+
Thiol-PAE	IIIa	CH_3	(+)	–	+	–	+
	IIIb	C_2H_5	–	+	+	+	–
PAE	IVa	CH_3	(+)	–	+	–	–
	IVb	C_2H_5	+/–	–	+	–	–

reaction can be expected as a result of the large number of functional groups. The presence of the halogens Cl or Br can only be determined with certainty from the (quasi)molecular ions. Phosphoric acid esters are also used as highly toxic chemical warfare agents (Tabun, Sarin and Soman, see Section 3.2.6.10).

The occurrence of fragments belonging to a particular group in the spectra of phosphoric acid esters has been intensively investigated (Table 3.12). Phosphoric acid esters are subdivided as follows with R the short methyl- or ethyl-, and Z the compound determining side chains (Stan *et al.*, 1977; Stan and Kellner, 1981):

I	Dithiophosphoric acid esters	$(RO)_2$-P(S)-S-Z
II	Thionophosphoric acid esters	$(RO)_2$-P(S)-O-Z
III	Thiophosphoric acid esters	$(RO)_2$-P(O)-S-Z
IV	Phosphoric acid esters	$(RO)_2$-P(O)-O-Z

Phenylureas

GC-MS can be used for the determination of phenylureas only after derivatization of the active substances (Färber and Schöler, 1991). HPLC or HPLC-MS are currently the most suitable analytical methods because of the thermal lability and polarity of these compounds. A spectrum with few lines is obtained in GC-MS analyses, which is dominated by the dimethylisocyanate ion *m/z* 72, which is specific to the group. The molecular ion region is of higher specificity but usually lower intensity (Figures 3.104–3.108). The phenylureas are rendered more suitable for GC by methylation of the azide hydrogen (e.g. with trimethylsulfonium hydroxide (TMSH) in the PTV (programmed temperature vaporizer) injector after Färber). The mass spectra of the methyl derivatives correspond to those of the parent substances, except that the molecular ions are 14 masses higher with the same fragmentation pattern (Figures 3.105, 3.107 and 3.109).

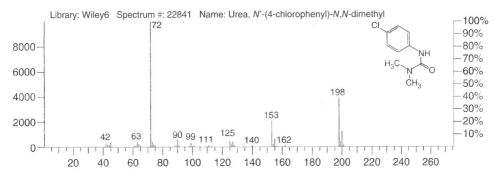

Figure 3.104 Monuron $C_9H_{11}ClN_2O$, M: 198, CAS Reg. No.: 150-68-5.

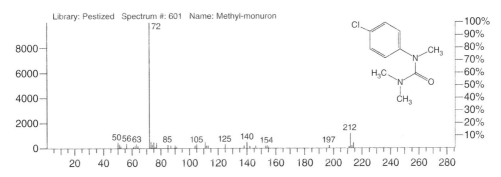

Figure 3.105 Methyl-monuron $C_{10}H_{13}ClN_2O$, M: 212.

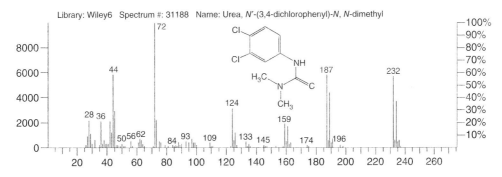

Figure 3.106 Diuron $C_9H_{10}Cl_2N_2O$, M: 232, CAS Reg. No: 330-54-1.

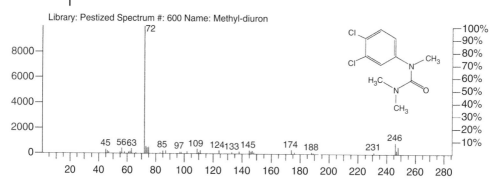

Figure 3.107 Methyl-diuron $C_{10}H_{12}Cl_2N_2O$, M: 246.

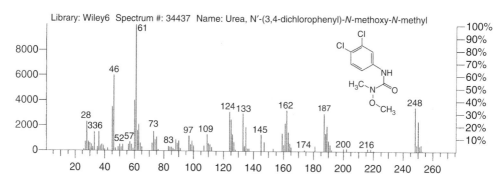

Figure 3.108 Linuron $C_9H_{10}Cl_2N_2O_2$, M: 248, CAS Reg. No.: 330-55-2.

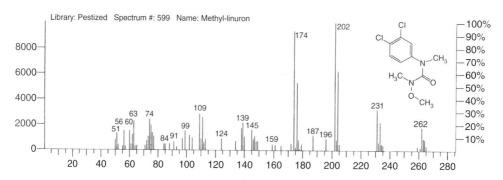

Figure 3.109 Methyl-linuron $C_{10}H_{12}Cl_2N_2O_2$, M: 262.

Phenoxyalkylcarboxylic Acids

The free acids cannot be analyzed by GC-MS in the case of trace analysis. They are determined using the methyl ester (Figures 3.110–3.112). If the aromatic character predominates, intense molecular ions occur in the upper mass range. Increasing the length of the side chains significantly reduces the intensity of the molecular

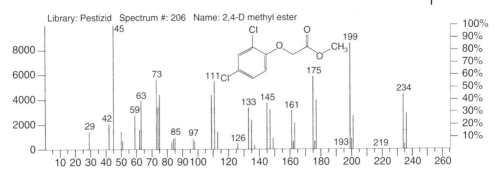

Figure 3.110 2,4-D methylester $C_9H_8Cl_2O_3$, M: 234, CAS Reg. No.: 1928-38-7.

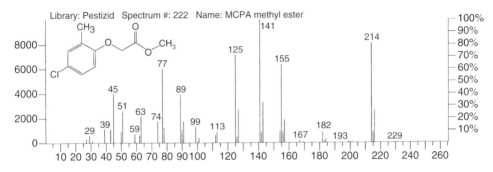

Figure 3.111 MCPA methylester $C_{10}H_{11}ClO_3$, M: 214, CAS Reg. No.: 2436-73-9.

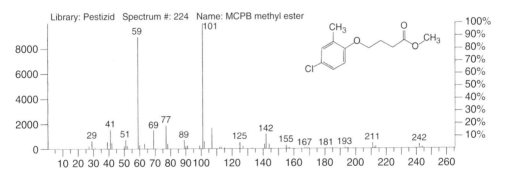

Figure 3.112 MCPB methylester $C_{12}H_{15}ClO_3$, M: 242, CAS Reg. No.: 57153-18-1.

ion and leads to signals in the lower mass range. It should be noted that the presence of Cl or Br can be determined with certainty only from the (quasi)molecular ion. With EI the molecular ion fragments thereby losing a Cl radical. Because of this, the isotope signals of the fragments cannot be evaluated conclusively (see MCPB methyl ester). A final confirmation can be achieved through CI.

3.2.6.6 Polychlorinated Biphenyls (PCBs)

The spectra of polychlorinated biphenyls (PCBs) have similar features independent of the degree of chlorination (Figures 3.113–3.122). As they are all derived from the same aromatic backbone, their molecular ions are strongly pronounced. The degree of chlorination can clearly be determined from the isotope pattern. Fragmentation involves successive loss of Cl radicals and, in the lower mass range, degradation of the basic skeleton. For data acquisition in full scan, the mass range above m/z 100 or 150 is required, so that detection of PCBs is usually possible above an accompanying matrix background. SIM or SRM analysis is applied to achieve a high selectivity. Individual isomers with a particular degree of chlorination have almost identical mass spectra. They can be differentiated on the basis of their retention times and therefore good gas chromatographic separation is a prerequisite for the determination of PCBs (see Figure 2.80). The isomers 31 and 28 (nomenclature after Ballschmitter and Zell) are used as resolution criteria (Table 3.13). The spectra of compounds with different degrees of chlorination are shown in the following figures.

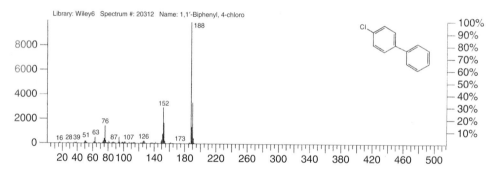

Figure 3.113 Monochlorobiphenyl $C_{12}H_9Cl$, M: 188.

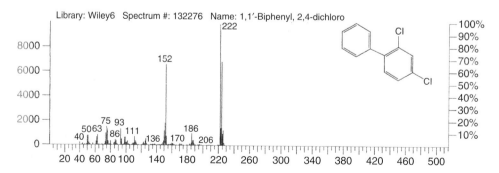

Figure 3.114 Dichlorobiphenyl $C_{12}H_8Cl_2$, M: 222.

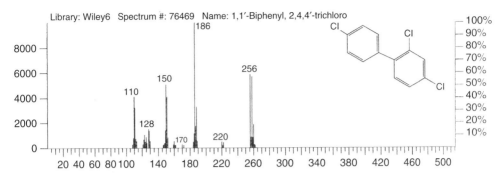

Figure 3.115 Trichlorobiphenyl (e.g. PCB 28, 31) $C_{12}H_7Cl_3$, M: 256.

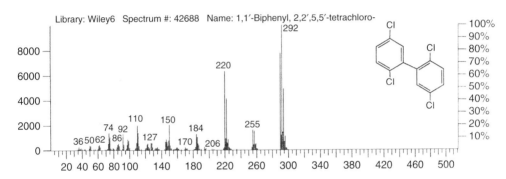

Figure 3.116 Tetrachlorobiphenyl (e.g. PCB 52) $C_{12}H_6Cl_4$, M: 290.

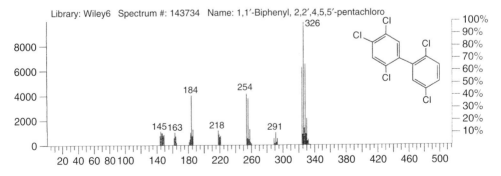

Figure 3.117 Pentachlorobiphenyl (e.g. PCB 101, 118) $C_{12}H_5Cl_5$, M: 324.

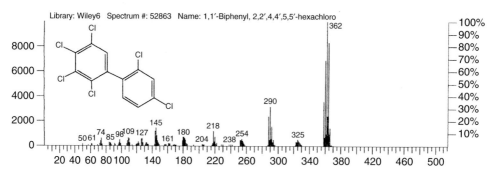

Figure 3.118 Hexachlorobiphenyl (e.g. PCB 138, 153) $C_{12}H_4Cl_6$, M: 358.

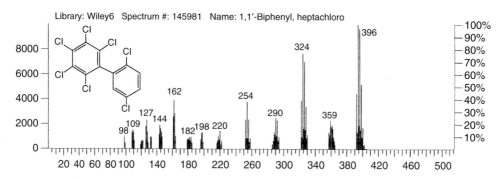

Figure 3.119 Heptachlorobiphenyl (e.g. PCB 180) $C_{12}H_3Cl_7$, M: 392.

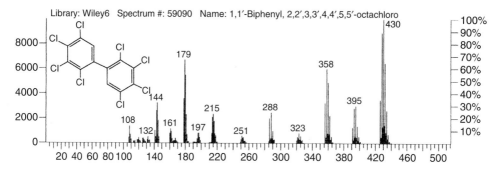

Figure 3.120 Octachlorobiphenyl $C_{12}H_2Cl_8$, M: 426.

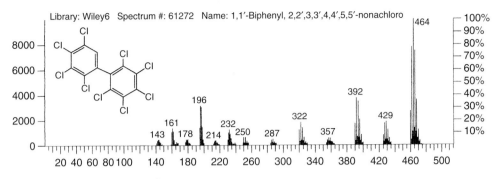

Figure 3.121 Nonachlorobiphenyl $C_{12}HCl_9$, M: 460.

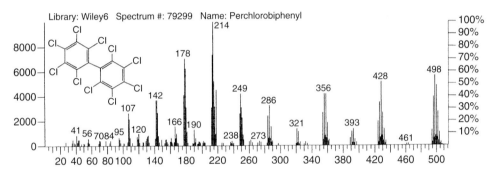

Figure 3.122 Decachlorobiphenyl (PCB 209) $C_{12}Cl_0$, M: 494.

Table 3.13 Indicator PCB congeners for quantitation (indicate the presence or absence of PCBs).

PCB No.	Structure (Cl substitution)	
28	2,4,4′	Cl_3–PCB
52	2,2′,5,5′	Cl_4–PCB
101	2,2′,4,5,5′	Cl_5–PCB
118	2,3′,4,4′,5	Cl_5–PCB
138	2,2′,3,4,4′,5	Cl_6–PCB
153	2,2′,4,4′,5,5′	Cl_6–PCB
180	2,2′,3,4,4′,5,5′	Cl_7–PCB
209	2,2′,3,3′,4,4′,5,5′,6,6′	Cl_{10}–PCB[a]

a) Used as internal standard.

3.2.6.7 Polychlorinated Dioxins/Furans (PCDDs/PCDFs)

The persistence of this class of substance in the environment parallels their mass spectroscopic stability. As they are aromatic, dioxins and furans give good molecular ion intensities with pronounced isotope patterns and only low fragmentation (Figures 3.123 and 3.124). Internal standards with 6- and 12-fold isotopic labelling (^{13}C) are used as internal standards for quantitation. High levels of labelling are necessary in order to obtain mass signals for the standard above the native Cl isotope pattern.

3.2.6.8 Drugs

Various classes of active substances are assigned to the group of drugs. The amphetamines are of particular interest with regard to mass spectrometry because their EI spectra are dominated by α-cleavage (Figures 3.125–3.127). In this case, the mass scan must be started at a correspondingly low mass to determine the ions *m/z* 44 and 58. The drugs of the morphine group (morphine, heroin, codeine and cocaine) give stable molecular ions and can be recognized with certainty on the basis of their fragmentation pattern (Figures 3.128–3.131). The selective detection and confirmation of the identity of drugs is also possible in complex matrices using CI with ammonia or isobutane as the CI gas.

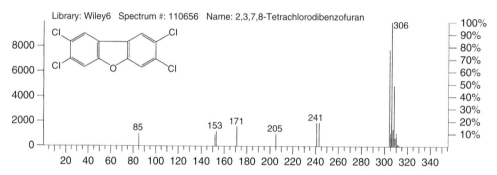

Figure 3.123 2,3,7,8-Tetrachlorodibenzofuran (2,3,7,8-TCDF) $C_{12}H_4Cl_4O$, M: 304, CAS Reg. No.: 51207-31-9.

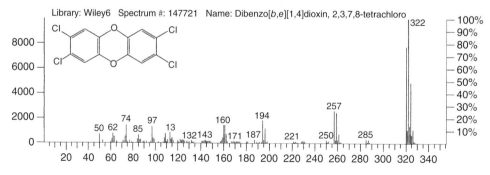

Figure 3.124 2,3,7,8-Tetrachlorodibenzodioxin (2,3,7,8-TCDD) $C_{12}H_4Cl_4O_2$, M: 320, CAS Reg. No.: 1746-01-6.

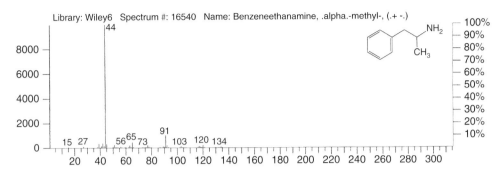

Figure 3.125 Amphetamine $C_9H_{13}N$, M: 135, CAS Reg. No.: 300-62-9.

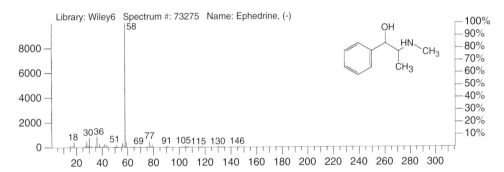

Figure 3.126 Ephedrine $C_{10}H_{15}NO$, M: 165, CAS Reg. No.: 299-42-3.

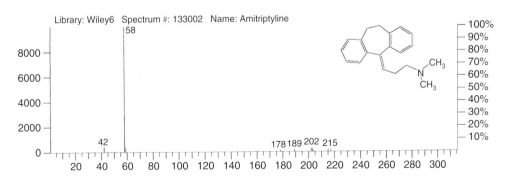

Figure 3.127 Amitriptyline $C_{20}H_{23}N$, M: 277, CAS Reg. No.: 50-48-6.

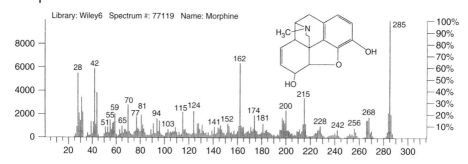

Figure 3.128 Morphine $C_{17}H_{19}NO_3$, M: 285, CAS Reg. No.: 57-27-2.

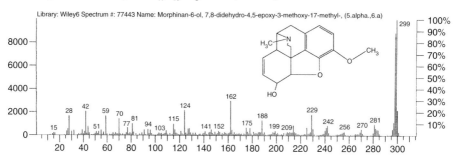

Figure 3.129 Codeine $C_{18}H_{21}NO_3$, M: 299, CAS Reg. No.: 76-57-3.

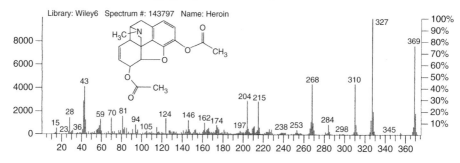

Figure 3.130 Heroin $C_{21}H_{23}NO_5$, M: 369, CAS Reg. No.: 561-27-3.

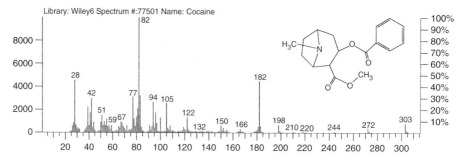

Figure 3.131 Cocaine $C_{17}H_{21}NO_4$, M: 303, CAS Reg. No.: 50-36-2.

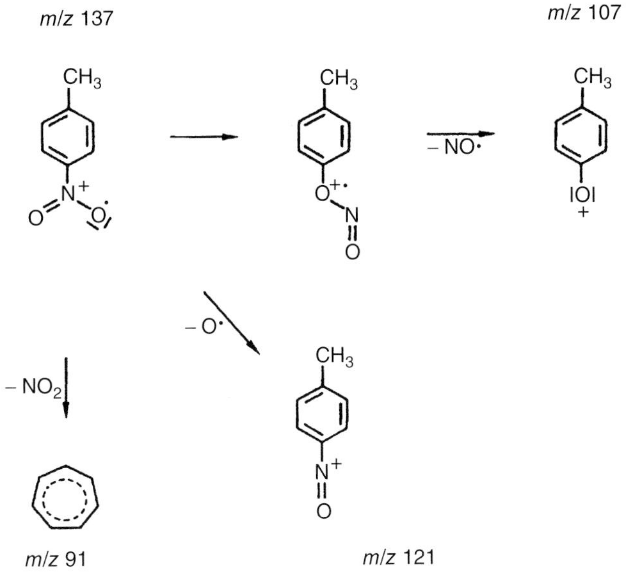

m/z 137   *m/z* 107

m/z 91   *m/z* 121

Figure 3.132   Fragmentation of 4-nitrotoluene.

3.2.6.9 Explosives

All explosives show a high proportion of oxygen in the form of nitro groups. GC-MS analysis of the nonaromatic compounds (Hexogen, Octogen, Nitropenta, etc.) is problematic because of decomposition, even in the injector. However, aromatic nitro compounds and their metabolites (aromatic amines) can be detected with high sensitivity because of their stability (Figures 3.132–3.147). With EI, molecular ions usually appear at low intensity because nitro compounds eliminate NO (M − 30). *o*-Nitrotoluenes are an exception because a stable ion is formed after loss of an OH radical (M − 17) as a result of the proximity of the two groups (ortho effect) (Figures 3.132 and 3.133). If the scan is run including the mass *m/z* 30, the mass chromatogram can indicate the general elution of the nitro compounds. In this area CI is useful for confirming and quantifying results. In particular, water has proved to be a useful CI gas in ion trap systems for residue analysis of explosives.

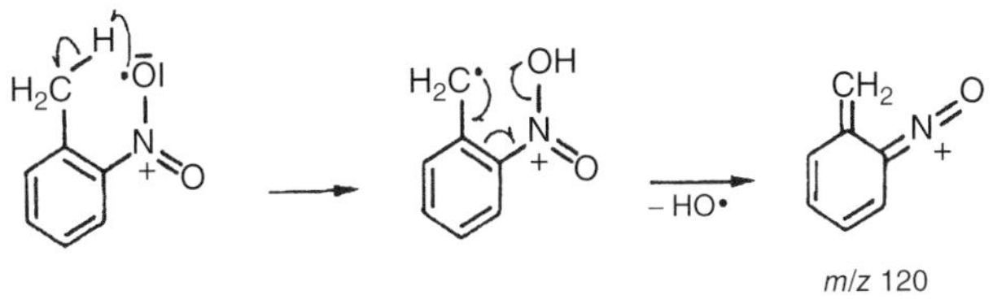

m/z 120

Figure 3.133   Fragmentation of 2-nitrotoluene (ortho effect).

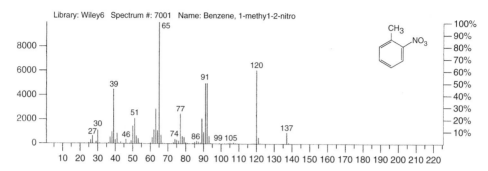

Figure 3.134 2-Nitrotoluene $C_7H_7NO_2$, M: 137, CAS Reg. No.: 88-72-2.

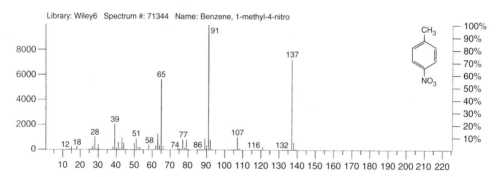

Figure 3.135 4-Nitrotoluene $C_7H_7NO_2$, M: 137, CAS Reg. No.: 99-99-0.

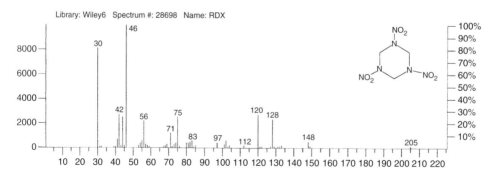

Figure 3.136 Hexogen (RDX) $C_3H_6N_6O_6$, M: 222, CAS Reg. No.: 121-82-4.

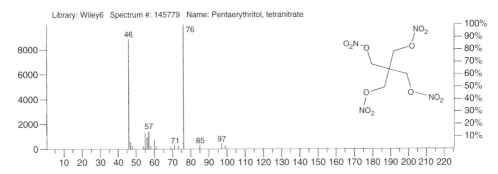

Figure 3.137 Nitropenta (PETN) $C_5H_8N_4O_{12}$, M: 316, CAS Reg. No.: 78-11-5.

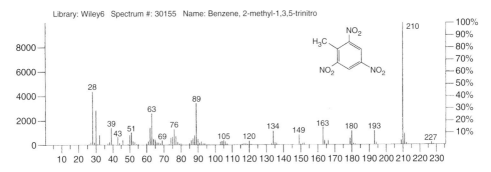

Figure 3.138 Trinitrotoluene (TNT) $C_7H_5N_3O_6$, M: 227, CAS Reg. No.: 118-96-7.

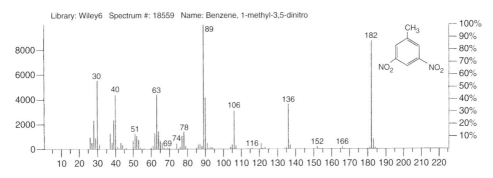

Figure 3.139 3,5-Dinitrotoluene (3,5-DNT) $C_7H_6N_2O_4$, M: 182, CAS Reg. No.: 618-85-9.

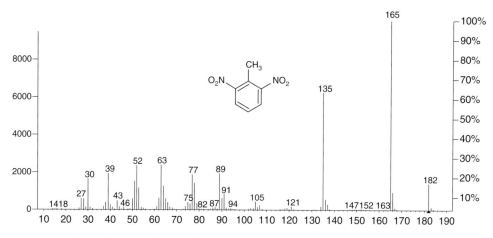

Figure 3.140 2,6-Dinitrotoluene (2,6-DNT) $C_7H_6N_2O_4$, M: 182, CAS Reg. No.: 606-20-2.

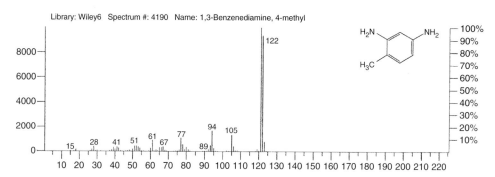

Figure 3.141 2,4-Diaminotoluene (2,4-DAT) $C_7H_{10}N_2$, M: 122, CAS Reg. No.: 95-80-7 (Note: 2,4-DAT and 2,6-DAT form a critical pair during GC separation).

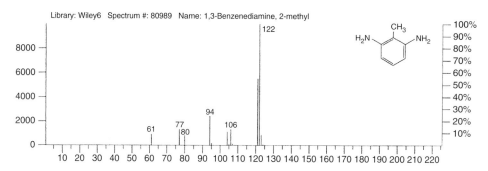

Figure 3.142 2,6-Diaminotoluene (2,6-DAT) $C_7H_{10}N_2$, M: 122, CAS Reg. No.: 823-40-5.

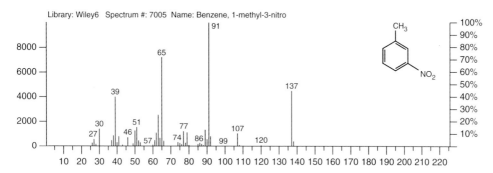

Figure 3.143 (Mono-)3-nitrotoluene (3-MNT) $C_7H_7NO_2$, M: 137, CAS Reg. No.: 99-08-1.

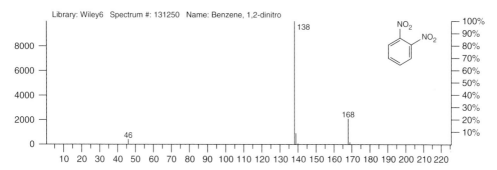

Figure 3.144 1,2-Dinitrobenzene (1,2-DNB) $C_6H_4N_2O_4$, M: 168, CAS Reg. No.: 528-29-0.

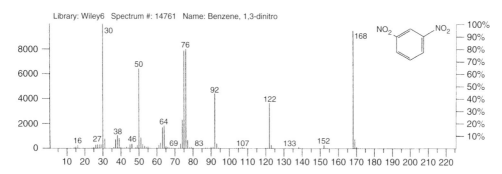

Figure 3.145 1,3-Dinitrobenzene (1,3-DNB) $C_6H_4N_2O_4$, M: 168, CAS Reg. No.: 99-65-0.

Library: Wiley6 Spectrum #: 14762 Name: Benzene, 1,4-dinitro

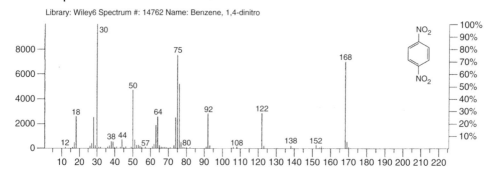

Figure 3.146 1,4-Dinitrobenzene (1,4-DNB) $C_6H_4N_2O_4$, M: 168. CAS Reg. No.: 100-25-4.

Library: Wiley6 Spectrum #: 4322 Name: Benzene, nitro

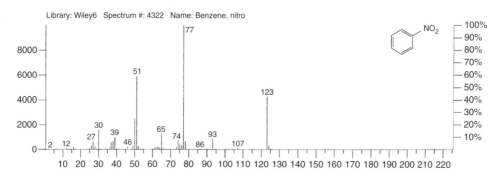

Figure 3.147 (Mono-)nitrobenzene (MNB) $C_6H_5NO_2$, M: 123, CAS Reg. No.: 98-95-3.

3.2.6.10 Chemical Warfare Agents

The identification of chemical warfare agents is important for testing disarmament measures and for checking disused military sites. Here also there is no single chemical class of substances. Volatile phosphoric acid esters and organoarsenic compounds belong to this group (Figures 3.148–3.153). CI is the method of choice for identification and confirmation of identity.

Library: Wiley6 Spectrum #: 216840 Name: Isopropoxy-methylphosphoryl fluoride

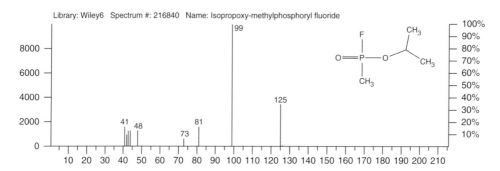

Figure 3.148 Sarin $C_4H_{10}FO_2P$, M: 140, CAS Reg. No.: 107-44-8.

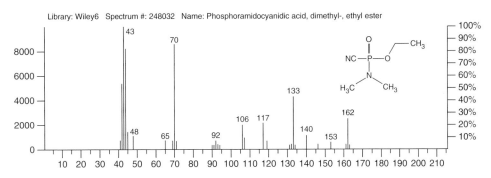

Figure 3.149 Tabun $C_5H_{11}N_2O_2P$, M: 162, CAS Reg. No.: 77-81-6.

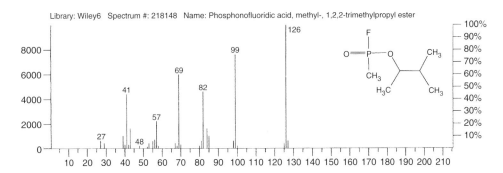

Figure 3.150 Soman $C_7H_{16}FO_2P$, M: 182, CAS Reg. No.: 96-64-0.

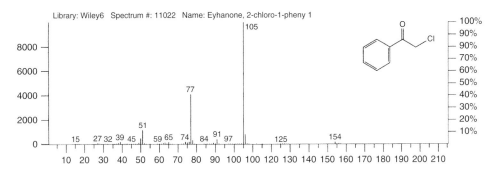

Figure 3.151 Chloroacetophenone (CN) C_2H_7ClO, M: 154, CAS Reg. No.: 532-27-4.

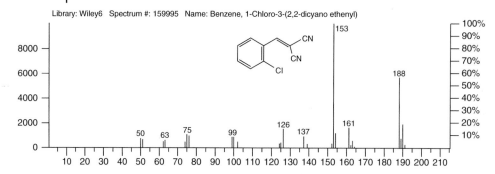

Figure 3.152 *o*-Chlorobenzylidenemalnonitrile (CS) $C_{10}H_5ClN_2$, M: 188, CAS Reg. No.: 2698-41-1.

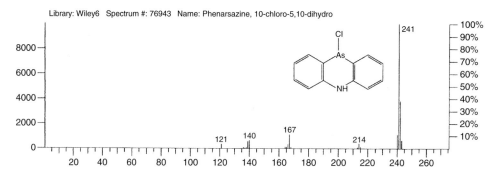

Figure 3.153 Adamsite (DM) $C_{12}H_9AsClN$, M: 277, CAS Reg. No.: 578-94-9.

3.2.6.11 Brominated Flame Retardants (BFR)

Highly brominated compounds are used as flameproofing agents (Figures 3.154 and 3.155). The polybrominated biphenyls (PBBs) and polybrominated diphenylether (PBDE), which are mostly used, have molecular weights of up to 1000 (decabromobiphenyl M 950). Brominated flame retardants have recently become of interest to analysts because burning materials containing PBB and PBDE can lead to the formation of polybrominated dibenzodioxins and furans. Decabromodiphenylether (PBDE 209) is thermolabile and tend to disintegrate in the GC injector and during separation on the column. A short column with higher flows for low elution temperatures are required for sensitive analysis. The spectra of PBB and PBDE are characterized by the symmetrical isotope pattern of the bromine and the high stability of the aromatic molecular ion. Flameproofing agents based on brominated alkyl phosphates have a greater tendency to fragment.

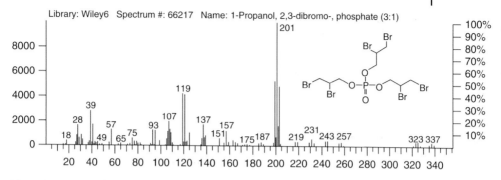

Figure 3.154   Bromkal P67–6HP (Tris) $C_9H_{15}Br_6O_4P$, M: 692, CAS Reg. No.: 126-72-7.

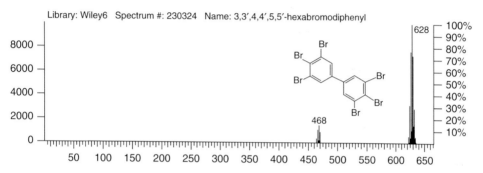

Figure 3.155   Hexabromobiphenyl (HBB) $C_{12}H_4Br_6$, M: 622.

3.3
Quantitation

> *It is important to point out that accuracy assessment is a continuous process, which should be implemented in the routine work as a part of the QA/QC set-up of the laboratory.*

Richard Boqué,
Universitat Rovira i Virgili, Tarragona,
Spain (Boqué, Marato and Vander Heyden, 2008)

Besides the identification of components of a mixture, the use of GC-MS systems to determine the concentration of target compounds according to legal requirements is a primary application in governmental, industrial as well as private control labs. The need to determine quantitatively an increasing number of components in complicated matrices in ever smaller concentrations makes the use of GC-MS systems in routine analysis appropriate for economic reasons. Gone are the days when only positive results from classical GC were confirmed by GC-MS. In many areas of trace analysis, the development of routine multicomponent methods has become possible only through the selectivity of detection and the specificity of identification of GC-MS and GC-MS/MS systems. The development of GC-MS data systems has therefore been successful in recent years, particularly in areas where the use of the mass spectrometric substance information coincides with the integration of chromatographic peaks and thus increases the certainty of quantitation. Compared with chromatography data systems for stand-alone GC and HPLC, there are therefore differences and additional features which arise from the use of the mass spectrometer as the detector.

3.3.1
Acquisition Rate

Mass spectrometers do not continuously record the substance stream arriving in the detector (as, for example, with FID, ECD, etc.). The chromatogram is comprised of a series of measurement points which are represented by mass spectra. The scan rate chosen by the user establishes the time interval between the data points. The maximum possible scan rate depends on the scan speed of the spectrometer. It is determined by the width of the mass range to be acquired and the necessity of achieving an analytical detection capacity of the instrument which is as high as possible. For routine measurements, scan rates below 1.0 s/scan are usually chosen. Compared with a sharp concentration change in the slope of a GC peak, these scan rates are only slow.

In the integration of GC-MS chromatograms, the choice of scan rate is particularly important for the determination of the peak area values. With the scan rate of 1s/scan, which is frequently used, the peak area of rapidly eluting components cannot be determined reliably as too few data points for the correct plotting of the GC peak are recorded (Figure 3.156). The incorporation of the top of the peak in the calculation of height and area cannot be carried out correctly in these cases. Under certain circumstances small peaks

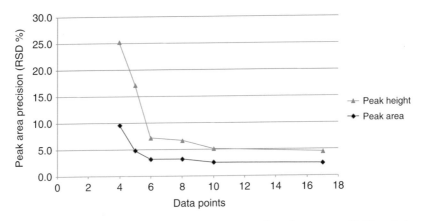

Figure 3.156 Precision of the peak area determination from the scan rate (Dallüge *et al.*, 2002).

Figure 3.157   (a) Peak area precision vs. acquisition data points in Ultra Fast GC (Dallüge *et al.*, 2002), (b) Comparison of an actual chromatogram with one reconstructed from data points at a low scan rate (Chang, 1985).

next to sharp peaks can be lost in the plot. In the case of a determination of area values in quantitative analyses, special attention should first be paid to the recording parameters before a possible cause is looked for in the detector itself.

To determine the optimal scan rate and sampling frequency, the base peak width of sharp early eluting components should be investigated. Little informa-tion is available from systematic investigations on the impact on the GC-MS sampling frequency on the precision of the peak area or peak height determi-nation, but deal with dwell time optimization for sensitivity and S/N (Kirchner *et al.*, 2005). With the advent and high speed of resistively heated Ultra Fast GC modules also the discussion on the minimum number of data points required for reliable peak area determination started (Dallüge *et al.*, 2002). It could be shown that six data points across a fast GC peak already provides acceptable area precision in the range of 3% RSD. Increasing the number of data points did not further increase the area or height precision significantly (see Figure 3.157). In practice up to 10 data points/peak are commonly used. This consideration allows the easy optimization of scan times (full scan) or dwell times (SIM, MRM) for sensitivity with different column separation conditions:

$$\text{max. dwell time [s]} = \frac{\text{base peak width [s]}}{\text{min. \# data points}}$$

3.3.2
Decision Limit

The question of when a substance can be said to be detected cannot be answered differently for quantitative GC-MS compared with all other chromatographic sys-tems. The answer lies in the determination of the S/N ratio.

In the basic adjustment of mass spectrometers, unlike classical GC detec-tors, the zero point is adjusted correctly (electrometer zero) to ensure the exact plot of isotope patterns. For this, the adjustment is chosen in such a way that minimum noise of the electronics is determined which can then be removed during data acquisition by software filters. Besides electrical noise, there is also chemical noise (matrix, column bleed, leaks, etc.), particularly in trace analysis.

The decision as to whether a substance has been detected or not is usually assessed in the signal domain. Here it is established that the decision limit is such that the smallest detectable signal from the substance can be clearly differentiated from a blank value (critical value).

In the measurement of a substance-free sample (blank sample), an average signal is obtained which corresponds to a so-called blank value. Multiple measurements confirm this blank value statistically and give its standard deviation. With GC-MS, this blank value determination is carried out in practice in the immediate vicinity to the substance peak, whereby the noise widths before and after the peak are taken into consideration. The average noise and the signal intensity can also be determined manually from the print-out; also, suitable S/N algorithms are available in data systems. A substance can be said to be detected if the substance signal exceeds a certain multiple of the noise width (standard deviation). This value is chosen arbitrarily as 2, 3 or higher (depending on the laboratory SOP or on the application) and is used as the deciding criterion in most routine evaluations of GC-MS data systems (Figure 3.158).

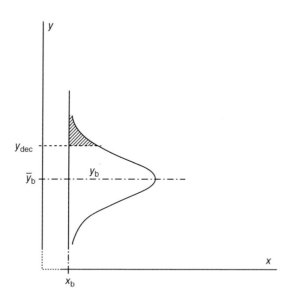

Figure 3.158 Statistical interpretation of the decision limit. y_b = random sample of blank values, y_{dec} = decision limit. In a statistically defined system the decision limit, that is, the smallest detectable signal which can be differentiated from a blank value, can only be obtained from multiple measurements of the blank value and the analysis at a given error probability (x substance domain, y signal domain, Ebel, 1987).

3.3.3
Limit of Detection

How does the decision limit relate to the limit of detection (LOD)? Usually both terms are used synonymously, which is incorrect! The LOD of a method is never given in counts or parts of a scale but is given a dimension, such as pg/μL, ng/L or ppb, in any case the LOD is expressed in the substance domain! The transfer from the signal domain of the decision limit into the substance domain of the LOD is accomplished by a valid calibration function (see Figure 3.159).

For practical reasons, the signal domain is used to calculate the LOD on screen or printouts. In a chromatogram of a low concentration standard, the width of the noise band is determined, and from the average noise the peak height is taken. This can simply be done in centimeter or counts, or a suitable software routine used. The result is an S/N ratio for a certain peak and concentration injected, often the analyte of interest. Here, the LOD can only be regarded as a qualitative parameter from just one measurement as here the uncertainty of the calibration function is not taken into account. In many cases, an S/N ratio of 3 is used as a practical criterion for the analyte LOD.

Another approach to determine the LOD is based statistically on the results of a series of measurements on low concentration level. Especially with MS/MS instruments and HRMS accurate mass acquisitions, the noise band is not continuously accessible anymore due to the high selectivity of these analyser types and a very noise-free digital signal processing. With a noise band tending to zero, the S/N calculation becomes increasingly meaningless.

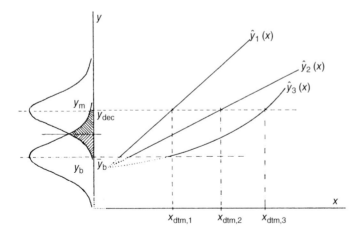

Figure 3.159 Definition of the limit of detection (LOD) from the decision limit. The limit of detection (LOD) is defined as that quantity of substance, concentration or content x_{dtm} which is given using the calibration function from the smallest detectable signal y_{dec} (decision limit). Calibration functions with different sensitivities (slopes) lead to different limits of detection at the same signal height (x substance domain, y signal domain, Ebel, 1987).

The variance of peak areas from consecutive measurements also shows all 'noise' influences for each injection, the overall precision of the measurement expressed in the relative standard deviation (%RSD) (Glaser *et al.*, 1981). The theoretical instrument detection limit (IDL) can be calculated from the peak variance of a series of measurements. In these cases, the Student t-distribution is used to calculate the 99% confidence factor t_α, taken from a Student t-distribution table for $n-1$ measurements:

$$IDL = t_\alpha, x \quad RSD$$

with

IDL instrument detection limit

t_α confidence factor for the desired level of 99% from the Student t-distribution
   for $n-1$ measurements

RSD relative standard deviation in % from n consecutive measurements.

According to the example shown in Table 3.14, the calculated IDL is given with 1.8 fg of OFN. This result is based on the estimate that this 1.8 fg substance signal can be statistically distinguished from the mean value of a series of blank measurements. With higher precision of the measurements, hence lower RSDs, the IDL value tends to be lower. Yet, it is not specified at which concentration the measurements are carried out. This number does not specify a real injection of this calculated very low concentration.

Table 3.14 Example of consecutive measurements of 10 fg OFN, peak area and height determination on m/z 272.

| Injection # | Response of OFN ion m/z 272 | | Moving RSD[a] | |
	Peak area	Peak height	Peak area	Peak height
1	2074	2629	—	—
2	1859	2250	—	—
3	1553	1953	—	—
4	1640	1948	—	—
5	1690	2024	—	—
6	1792	2221	—	—
7	1559	2028	—	—
8	1531	1869	11.0%	11.6%
9	1738	2151	7.2%	6.7%
10	1566	2041	6.0%	5.6%
11	1814	2242	6.6%	6.3%
12	1871	2579	7.7%	10.0%
		Lowest RSD	6.0%	
		IDL	1.8 fg	

a) A moving RSD from 8 runs is calculated, the lowest RSD taken for IDL calculation.

3.3.4
Limit of Quantitation

The limit of quantitation (LOQ) of a method is also given as a quantity of substance or concentration in the substance domain. This limit incorporates the calibration and thus also the uncertainty (error consideration) of the measurements (Ebel and Kamm 1983). Unlike the LOD it is guaranteed statistically and gives the lower limiting concentration which can be unambiguously determined quantitatively. It can differ significantly from the blank value (Montag, 1982; ISO 11843, 1997).

As a component can be determined only after it has been detected, the LOQ cannot be lower than the LOD. As there is a relative uncertainty in the result of about 100% at a concentration of a substance corresponding to the LOD in an analysis sample, the LOQ must be correspondingly higher than the LOD, depending on the requirement (Figure 3.160).

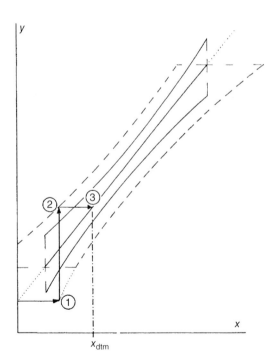

Figure 3.160 Statistical definition of the limit of quantitation. 1. Measured value at quantity of substance = 0 (signal domain). 2. Signal size which can be differentiated significantly taking the standard deviation into account. 3. Limit of quantitation x_{dtm} from the calibration function. The limit of quantitation is the lower limiting concentration x_{dtm} at a fixed statistical error probability which can definitely be quantitatively determined and be differentiated significantly from zero concentration (x substance domain, y signal domain, Ebel, 1987).

3.3.5
Sensitivity

The term *sensitivity*, which is frequently used to describe the quality of a residue analysis, or the current state of a measuring instrument, is often used incorrectly as a synonym for the lowest possible LOD or LOQ. A sensitive analysis procedure, however, exhibits a large change in signal with a small change in substance concentration. The sensitivity of a procedure thus describes the slope of a linear calibration function (see Section 3.3.6). At the same confidence interval of the measured points of a calibration function (Figure 3.161), sensitive analysis procedures give a narrower confidence interval than less sensitive ones! The LOD is independent of the sensitivity of an analytical method.

3.3.6
The Calibration Function

For an analysis procedure that has been developed, the calibration function is built from the measurement of known concentrations. Fixed response factors are used only to a limited degree in GC-MS because the intensity of the signal depends on the operating parameters of the mass spectrometer. The calibration function generally describes the dependence of the signal on the substance concentration. In the case of a linear dependence, the regression calculation gives a straight line for the calibration function, the equation for which contains the blank value a_0 and the sensitivity a:

$$f(x) = a_0 + a \cdot x \tag{3.1}$$

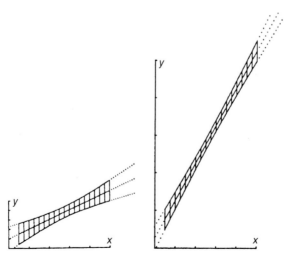

Figure 3.161 Confidence intervals for calibration curves with different sensitivity at the same distribution of the measured values but with different slopes (x substance domain, y signal domain) (Ebel, 1987).

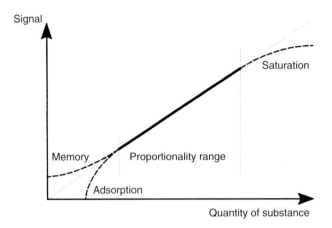

Figure 3.162 Variation of the signal intensity with the quantity of substance.

The calibration function is defined exclusively within the working range given by the experimental calibration. GC-MS systems achieve very low LODs so that, at a correspondingly dense collection of calibration points near the blank value, a nonlinear area is described. This area can be caused by unavoidable active sites (residual activities) in the system and 'swallows up' a small but constant quantity of substance. Such a calibration function tends to approach the x-axis before reaching the origin.

In the upper concentration range, the signal hardly increases at all with increasing substance concentration because of increasing saturation of the detector. The calibration function stops increasing and tends to be asymptotic (Figure 3.162).

The best fit of the calibration curve to the measured points is determined by a regression calculation. The regression coefficient gives information on the quality of the curve fit. Linear regressions are not always suitable for the best fit in GC-MS analysis; quadratic regressions frequently give better results. Particularly in the area of trace analysis, the type of fit within the calibration can change through the nonlinear effects described earlier. In such cases it is helpful to limit the regression calculation to that of a sample concentration lying close to the calibration level (local linearization, next three to four data points). If individual data systems do not allow this, a point-to-point calibration can be carried out instead, provided an acceptable density of data points.

Important aspects regarding the optimization of a calibration can be derived from the facts discussed here:

- One-point calibrations have no statistical confirmation and can therefore only be used as an orientation.
- Multipoint calibrations must cover the expected concentration range. For regulated methods, a factor of 10 below the regulated concentration (maximum residue level, MRL) is required.
- Extrapolation beyond the experimentally measured points is not justified. Calibration functions do not have to pass through the origin.

- Multiple measurements at an individual calibration level define the confidence interval which can be achieved.
- A dense distribution of calibration points near the LOQ give an improvement in the regression fit (unlike the equidistant position of the calibration level).

3.3.7
Quantitation and Standardization

In order to determine the substance concentration in an unknown analysis sample, the peak areas of the sample are calculated using the calibration function and the results are given in terms of quantity or concentration. Many data systems also take into account the sample amount as weighed out and dilution or concentration steps in order to be able to give the concentration in the original sample.

For GC-MS, the methods of external or internal standard calibration or standard addition are used as standardization procedures (Funk *et al.*, 2008).

3.3.7.1 External Standardization
External standardization corresponds to the classical calibration procedure (Figure 3.163). The substance to be determined is used to prepare standard solutions with a known concentration. Measurements are made on standard solutions of different concentrations (calibration steps, calibration levels). For calibration, the peak areas determined are plotted against the concentrations of the different calibration levels.

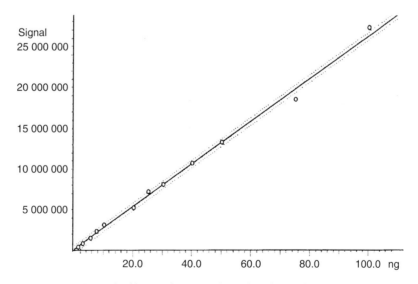

Figure 3.163 External calibration for PCB 28 by on-line thermodesorption GC-MS coupling. (Tschickard, State Office for the Environment, Water Management and Trade Supervision, Mainz, Germany).

With external standardization, standard deviations below 10% are obtained, which occur even using reproducible injection techniques involving autosamplers. The causes of this distribution are injection errors and small changes in the mass spectrometer. As absolute values from different analysis runs are used for external standardization, the calibration function and the sample measurement show all the effects, which can cause a change in response between the mass spectrometric measurements. Various factors can contribute to this, for example, slightly different ionization efficiency through geometric changes in the filament, slightly varying transmission due to contamination of the lens systems with the matrix, or a contribution from the multiplier on signal production. These factors can be compensated for by internal standardization (Guichon and Guillemin, 1988; Naes and Isakson, 1992).

3.3.7.2 Internal Standardization

The principle of the internal standard is based on the calculation of relative values, which are determined within the same analysis. One or more additional substances are introduced as a fixed reference parameter, the concentration of which is kept constant in the standard solutions and is always added to the analysis sample at the same concentration. For the calculation, the peak area values (or peak heights) of the substance being analysed relative to the peak area (or height) of the internal standard are used. In this way, potential volume errors and variations in the function of the instrument are compensated for and quantitative determinations of the highest precision are achieved (see also ISO 5725-6, 1994). Standard deviations of less than 5% can be achieved with internal standardization.

The time at which the internal standard is added during the sample preparation depends on the analysis requirement. For example, the internal standard can be added to the sample at an early stage (surrogate standard added before sample preparation) to simplify the clean-up. The addition of different standards at different stages of the clean-up allows the efficiency of individual clean-up steps to be monitored. Addition of the quantitation standard to the extract directly before the measurement serves to calculate the recoveries in the sample processing steps. This so called 'syringe standard' can be added very precisely by many autosamplers in the sandwich mode from a separate internal standard vial right before injection.

The choice of the internal standard is particularly important also in GC-MS. Basically the internal standard should behave as far as possible in the same way as the substance being analysed. Unlike classical GC detectors, the GC-MS procedure offers the unique possibility of using isotopically labelled, but nonradioactive analogues of the substances being analysed. Deuterated or ^{13}C carbon standards are frequently used for this purpose as they fulfil the requirement of comparable behaviour during clean-up and analysis to the greatest degree. The degree of labeling should always be sufficiently high to avoid interference with the natural isotope intensities of the unlabelled substance. The isotopically labeled internal standards thus chosen can thus be detected selectively through its own mass trace and integrated (Figure 3.164), see also section 'Isotope Dilution'.

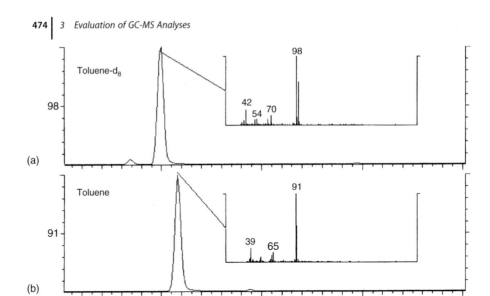

Figure 3.164 Elution of toluene-d$_8$ as the internal standard using purge and trap GC-MS (mass chromatograms). (a) Toluene-d$_8$, *m/z* 98 and (b) toluene *m/z* 91.

Requirements of the Internal Standard in GC-MS Analysis

- The internal standard chosen must be stable to clean-up and analysis and as inactive as possible.
- As far as possible the properties of the standard should be comparable to those of the analyte with regard to sample preparation and analysis; therefore isotopically labelled standards should ideally be used.
- The standard itself must not be present in the original sample.
- The retention behaviour in the GC should be adjusted so that elution occurs in the same section of the program (isothermal or heating ramp).
- The use of several standards allows them to be used as retention time standards.
- The retention behaviour of the internal standard should be adjusted to ensure that overlap with matrix peaks or other components to be determined is avoided and faultless integration is possible.
- The ionisation and fragmentation behaviour (mass spectrometric response) should be comparable.
- The choice of the quantifying mass of the standard should be in the same mass range and exclude interference by the matrix or other components.

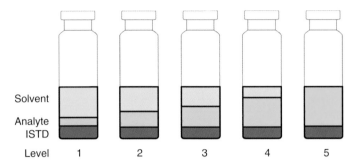

Figure 3.165 Preparation of solutions for calibration with internal standards (*c* (ISTD) = constant, *c* (analytes) = variable, total volume = constant, fill with solvent, calculate concentrations on the basis of total volumes).

In the preparation of standard solutions for calibration with internal standards, their concentration must be kept constant in all the calibration levels. Pipetting the same volumes of the internal standard followed by different volumes of a mixed analyte standard stock solution and then making up with solvent to the same fixed volume in the sample vial has proved to be a good method (Figure 3.165). The preparation of the internal standard calibration vials can be performed in routine analysis by program control of a robotic liquid autosampler as well (Zhang *et al.*, 2014). Keeping the mixed standard and the internal standard vials available on an autosampler platform simplifies many error prone manual steps.

The internal standard calibration curves plots the area ratio of the analyte relative to the area of the internal standard against the concentration in the sample (Figure 3.166). The parameter determined relative to the internal standard is thus independent of deviations in the injection volume and possible variations in the performance of the detector, as all these influences affect the analyte and the internal standard to the same extent. To calculate the analysis results, the ratio of the peak area of the analyte to that of the internal standard is determined and the concentration calculated using the calibration function (Table 3.15).

Isotope Dilution

The 'isotope dilution' quantitation method provides the highest accuracy of the internal standard quantification methods. As internal standard compounds, the most similar analogues of the native analytes labeled with stable isotopes are administered. Widely used are ^{13}C-labeled or deuterated compounds, for example, ^{13}C$_{12}$-TCDD or d$_{12}$-benzo[*a*]pyrene. The chemical and physical behaviour of the native and labeled compound is almost identical as ideally required by the concept of internal standardization.

The term *isotope dilution* was borrowed from isotope ratio mass spectrometry for quantitative organic analysis. The basic definition asserts that it is not necessary to carry out a standard curve by different standard dilutions. Instead a known quantity of a rare isotope is added as a spike to each sample. The measurement of the isotope ratio of the resulting mix compared to the known isotope ratio

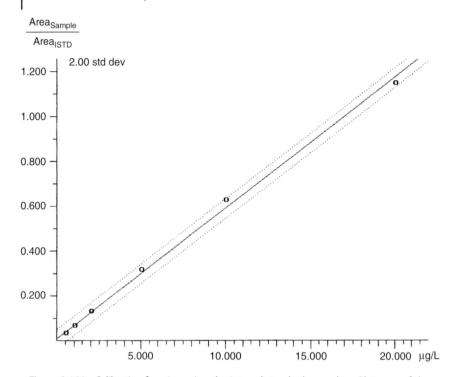

Figure 3.166 Calibration function using the internal standard procedure. Plot: area of the sample/area of the internal standard (ISTD) against concentration in the sample. The dotted lines indicate two standard deviations calibration precision.

of the spike and natural abundances allows the calculation of the concentration of the unknown in the individual sample. In organic quantitation methods, typically the average relative response of standards in different concentrations (native ag. labeled ISTD (internal standard)) is used as the basis for quantitative calibration.

A typical example for extended use of the isotope dilution method is the quantification of polychlorinated dioxins. Sixfold and twelvefold labeled dioxins and furans in the required chlorination degrees are used as recovery and surrogate standards during sample preparation and GC-MS analysis, as described in the widely applied EPA 1613 method. The nearly identical compound characteristics during clean-up, the chromatographic process and detection, guarantees most reliable quantitation results based on the sample individual recovery values in one analysis run (see details in the application Section 4.14 on dioxin analysis).

In the clean-up of biological material the carrier effect can be exploited through the addition of an internal standard. The standard added at comparatively high concentrations can cover up active sites in the matrix and thus improve the extraction of the substance being analysed. The result is a significantly improved extraction recovery from active matrices for the native compound for a reliable representation of the given content in the sample.

Table 3.15 List of volatile halogenated hydrocarbons found in a drinking water analysis using the ISTD procedure (Types: I: internal standard, S: surrogate standard, A: analyte).

Number	Substance	Type	Scan #	Retention time	Me	Calculated concentration	Unit
1	Fluorobenzene IS1	I	680	17:00	BB	5.000	g/L
2	1,2-Dichlorobenzene IS2	I	1050	26:15	BB	5.001	g/L
3	Bromofluorobenzene SS1	S	943	23:34	BB	5.000	g/L
9	Trichlorofluoromethane	A	424	10:36	BB	0.045	g/L
11	Dichloromethane	A	522	13:03	BB	0.518	g/L
17	Chloroform	A	630	15:45	BB	8.714	g/L
18	1,1,1-Trichloroethane	A	641	16:01	BB	0.071	g/L
21	Benzene	A	664	16:36	BB	0.056	g/L
22	Carbon tetrachloride	A	651	16:17	BB	0.023	g/L
24	Trichloroethylene	A	702	17:33	BV	0.060	g/L
26	Dichlorobromomethane	A	732	18:18	BB	7.415	g/L
29	Toluene	A	779	19:28	BB	0.167	g/L
32	Dibromochloromethane	A	829	20:44	BB	4.860	g/L
34	Tetrachloroethylene	A	813	20:19	BV	0.058	g/L
35	Chlorobenzene	A	868	21:42	BB	0.032	g/L
37	Ethylbenzene	A	874	21:51	BV	0.037	g/L
38	*meta*, *para*-Xylene	A	881	22:02	VB	0.054	g/L
39	Bromoform	A	923	23:05	BB	0.635	g/L
45	Bromobenzene	A	954	23:51	BB	0.034	g/L
51	1,2,4-Trimethylbenzene	A	998	24:57	BB	0.084	g/L
55	Isopropyltoluene	A	1019	25:28	BB	0.057	g/L
59	Naphthalene	A	1182	29:33	BB	0.155	g/L
61	Hexachlorobutadiene	A	1176	29:24	BV	0.126	g/L

In chromatography the Deuterium and ^{13}C labeled standards have a slightly shorter retention time than the native analytes. This retention time difference, though small, is always visible in the mass chromatogram, which is used for selective integration of the individual components (see also Figure 2.206).

3.3.7.3 The Standard Addition Procedure

Matrix effects frequently lead to varying extraction yields. The headspace and purge and trap techniques in particular are affected by this type of problem. If measures for standardizing the matrix are unsuccessful, the standard addition procedure can be used, as in, for example, atomic absorption (AAS) for the same reason. Here the calibration is analogous to the external standardization described earlier and involves addition of known quantities of the analyte to be determined (Miller, 1992). The calibration samples are prepared with constant quantities of the sample material by addition of corresponding volumes of the standard solution. One sample is left as it is, that is, no standard is added.

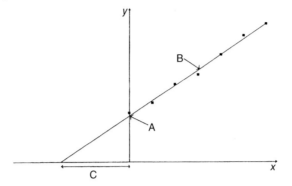

Figure 3.167 Calibration using standard addition. x axis: quantity of substance, y axis: signal intensity. (A) Intercept with the y axis: signal of the unaltered sample, quantity added = 0 (B) Calibration function built by adding increasing amounts of the analyte to the sample material under investigation (C) Content in the sample with the absolute value of the intercept with the x axis.

The analysis results are calculated by plotting the peak areas against the quantity added. The result of the unaltered sample defines the intercept with the y axis. The calibration function cuts the y axis at a height corresponding to the concentration of analyte in the sample. The concentration in the sample can be read off as the absolute value at the intercept of the extrapolated calibration function with the x axis (Figure 3.167).

In spite of the good results obtained with this procedure, in practice there is one major disadvantage. A separate calibration must be carried out for every sample. Today robotic autosampler allow the automated handling of the standard addition within a sequence controlled operation. A further criticism of the addition procedure arises from the statistical aspect. The linear extrapolation of the calibration function is only carried out assuming its validity for this area also. A possible non-linear deviation in this lower concentration range would give rise to considerable errors. To allow a better fit of the calibration function, a larger number of additions should be analysed.

3.3.8
The Accuracy of Analytical Data

In spite of precise measuring procedures and careful evaluation of the data, the accuracy of analytical data is highly dependent on the sampling procedure, transportation to the laboratory, choice of a representative laboratory sample and the sample preparation procedure (Figure 3.168)! Even powerful instrumental analysis cannot correct errors that have already occurred.

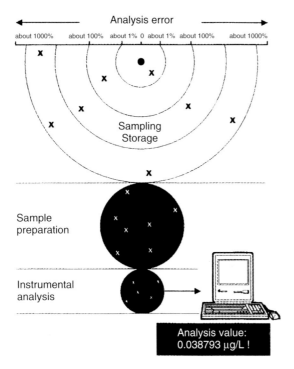

Figure 3.168 Sources of error in analysis (Hein and Kunze, 1994).

3.4
Frequently Occurring Impurities

With increasing concentration during sample preparation and increasing sensitivity of GC-MS systems, the question of what are the necessary laboratory conditions for trace analysis is becoming more important. Interfering signals from impurities are increasingly appearing (Table 3.16). These arise from 'cleaned' glass apparatus (e.g. rotary evaporators, pipettes), through contact with polymers (e.g. cartridges from solid-phase extraction, septa, solvent and sample bottles), from the solvents themselves (e.g. stabilizers) or even from the laboratory surroundings (e.g. dust, solvent vapours, etc.). There are also sources of interfering signals in the GC-MS system itself. These are often found in the surroundings of the GC injector. They range from septum bleed or decomposing septa to impurities which have become deposited in the split vent and which get into the measuring system on subsequent injections. Also the closures of autosampler vial should be considered as a common source of contamination. The capillary columns used are well known for causing sometimes high background noise by column bleed (esp. polar phases). As these typical mass signals constantly occur in trace analyses, the structures of the most frequently occurring ions are listed (Figure 3.169). Sources of

Table 3.16 Mass signals of frequently occurring contaminants.

m/z values	Possible cause
149, 167, 279	Phthalate plasticizers, various derivatives
129, 185, 259, 329	Tri-*n*-butyl acetyl citrate plasticizer
99, 155, 211, 266	Tributyl phosphate plasticizer
91, 165, 198, 261, 368	Tricresyl phosphate plasticizer
108, 183, 262	Triphenylphosphine (synthesis by-product)
51, 77, 183, 201, 277	Triphenylphosphine oxide (synthesis by-product)
41, 55, 69, 83, …	Hydrocarbon background series (forepump oil,
43, 57, 71, 85, …	greasy fingers, tap grease, etc.)
99, 113, 127, 141, …	
285, 299, 313, 327, …	
339, 353, 367, 381, …	
407, 421, 435, 449, …	
409, 423, 437, 451, …	
64, 96, 128, 160, 192, 224, 256	Sulfur (as S_8)
205, 220	Antioxidant 2,6-di-*t*-butyl-4-methylphenol (BHT, Ionol) and isomers (technical mixture)
115, 141, 168, 260, 354, 446	Poly(phenyl ether) (diffusion pump oil)
262, 296, 298	Chlorophenyl ether (impurity in diffusion pump oil)
43, 59, 73, 87, 89, 101, 103, 117, 133	Poly(ethylene glycol) (all Carbowax phases)
73, 147, 207, 221, 281, 355, 429	Silicone rubber (all silicone phases)
133, 207, 281, 355, 429	Silicone grease
233, 235	Rhenium oxide ReO^- (from the cathode in NCl)
217, 219	Rhenium oxide ReO^-
250, 252	Rhenium oxide $HReO^-$

contamination in the mass spectrometer are also known. Among these are background signals arising from pump oil (rotary vane pump, diffusion pump) and degassing products from sealing materials (e.g. Teflon) and ceramic parts after cleaning.

For trace analysis, the carrier gas used (helium, hydrogen) should be of the highest available purity from the beginning of installation of the instrument. Contamination from the gas supply tubes (e.g. cleaned up for ECD operations!) leads to ongoing interference and can be removed only at great expense. With central gas supply, plants in particular gas purification (irreversible binding of organic contaminants to getter materials) directly at the entry to the GC-MS system is recommended for trace analysis. Leaks in the vacuum system, in the GC system and in the carrier gas supply always lead to secondary effects and should therefore be carefully eliminated. Under no circumstances should plastic tubing be used for the carrier gas supply to GC-MS systems (Table 3.17).

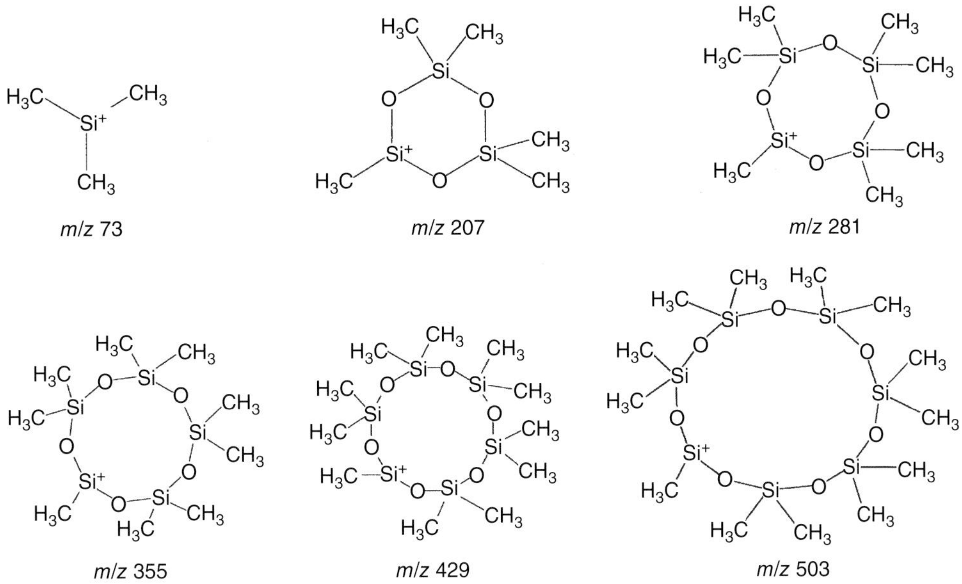

Figure 3.169 Structures of the most important signals resulting from column bleed of silicone phases (Spiteller and Spiteller, 1973).

Table 3.17 Diffusion of oxygen through various tubing materials (air products).

Line material	Contamination by O_2 (ppm)
Copper	0
Stainless steel	0
Kel-F	0.6
Neoprene	6.9
Polyethylene (PE)	11
Teflon (PTFE)	13
Rubber	40

Note: measured in argon 6.0 at a tubing diameter of 6 mm and 1 m length with a flow rate of 5L/h.

Some typical interfering components frequently occurring in residue analysis are listed together with their spectra (Figures 3.170–3.184). They are listed in order of the mass of the base peak (Figures 3.185 and 3.186). In GC-NCI/MS an intense contamination can be observed exhibiting m/z 166 with a significant isotope cluster. This compound was found to migrate from butyl rubber septa of autosampler vials into the sample and can be explained as artefacts from vulcanization facilitators (Kapp and Vetter, 2006) (Figures 3.170 and 3.171).

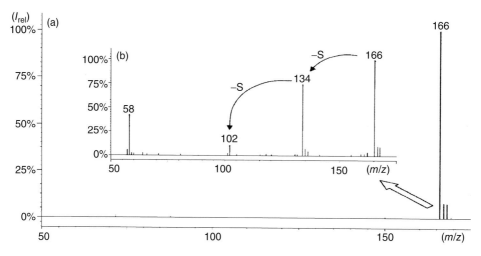

Figure 3.170 (a) GC–NCI–MS spectrum of a contaminant from buthyl rubber septa. (b) Confirmation of two S atoms by GC-MS/MS.

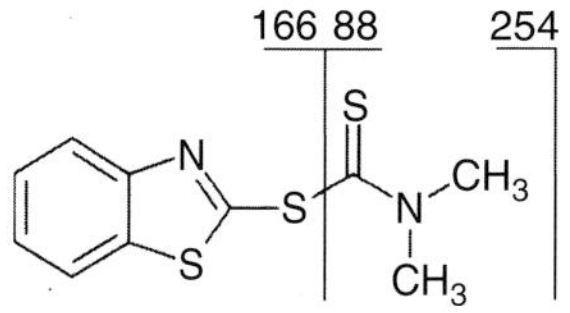

Figure 3.171 Structure of the artefact 2-benzothiazolyl-*N,N*-dimethyldithiocarbamate.

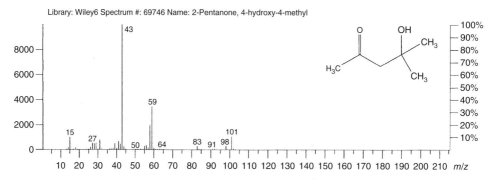

Figure 3.172 Diacetone alcohol $C_6H_{12}O_2$, M: 116, CAS: 123-42-2. Occurrence: acetone dimer, forms from the solvent under basic conditions.

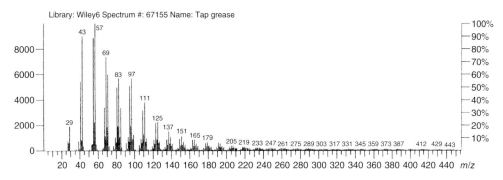

Figure 3.173 Hydrocarbon background. Occurrence: greasy fingers as a result of maintenance work on the analyser, the ion source or after changing the column, also from background of forepump oil (rotatory vane pumps), joint grease.

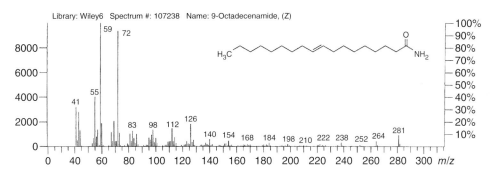

Figure 3.174 Oleic acid amide. Occurrence: lubricant for plastic sheeting, erucamide occurs just as frequently.

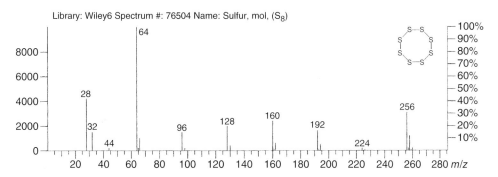

Figure 3.175 Molecular sulfur S_8, M: 256, CAS: 10544-50-0. Occurrence: from soil samples (microbiological decomposition), impurities from rubber objects, and by-product in the synthesis of thio-compounds.

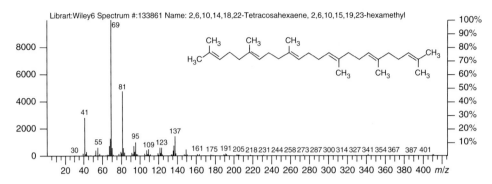

Figure 3.176 Squalene $C_{30}H_{50}$, M: 410, CAS: 7683-64-9. Occurrence: stationary phase in gas chromatography.

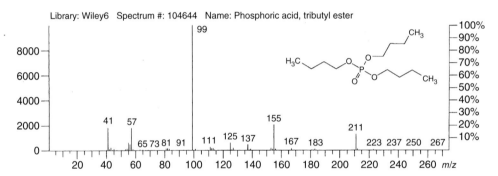

Figure 3.177 Tributyl phosphate $C_{12}H_{27}O_4P$, M: 266, CAS: 126-73-8. Occurrence: widely used plasticizer.

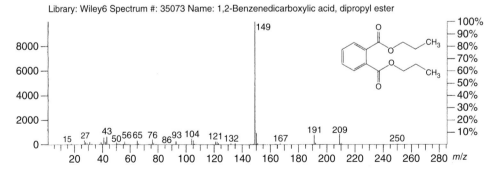

Figure 3.178 Propyl phthalate $C_{14}H_{18}O_4$, M: 250, CAS: 131-16-8. Occurrence: plasticizer, widely used in many plastics (also SPE cartridges, bottle tops, etc.), typical mass fragment of all dialkyl phthalates *m/z* 149.

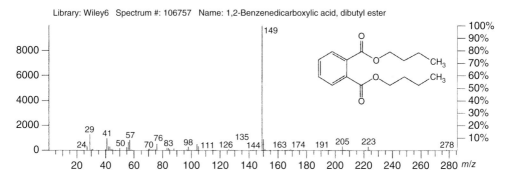

Figure 3.179 Dibutyl phthalate $C_{16}H_{22}O_4$, M: 278, CAS: 84-74-2.

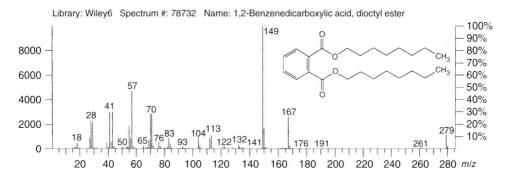

Figure 3.180 Dioctyl phthalate $C_{24}H_{38}O_4$, M: 390, CAS: 117-84-0.

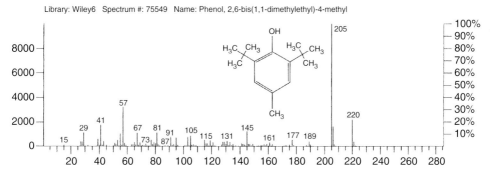

Figure 3.181 Ionol (BHT) $C_{15}H_{24}O$, M: 220, CAS: 128-37-0. Occurrence: antioxidant in plastics, stabilizer (radical scavenger) for ethers, THF, dioxan, technical mixture of isomers.

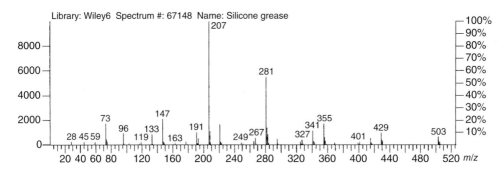

Figure 3.182 Silicones, silicone grease. Occurrence: typical column bleed from many silicone phases, from septum caps of sample bottles, injectors, and so on (see also Figure 3.165).

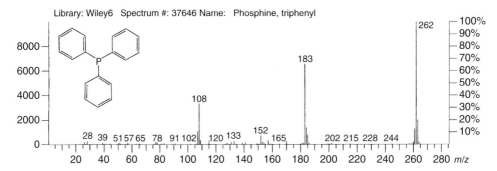

Figure 3.183 Triphenylphosphine $C_{18}H_{15}P$, M: 262, CAS: 603-35-0. Occurrence: catalyst for syntheses.

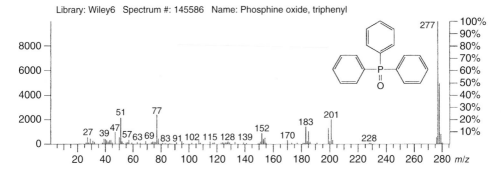

Figure 3.184 Triphenylphosphine oxide $C_{18}H_{15}OP$, M: 278, CAS: 791-28-6. Occurrence: forms from the catalyst in syntheses (e.g. the Wittig reaction).

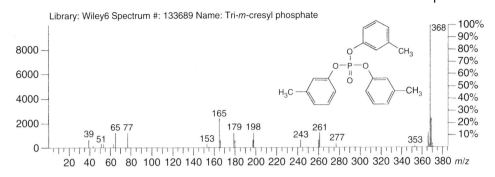

Figure 3.185 Tri-*m*-cresyl phosphate $C_{21}H_{21}O_4P$, M: 368, CAS: 563-04-2. Occurrence: plasticizer (e.g. in PVC (polyvinylchloride), nitrocellulose, etc.).

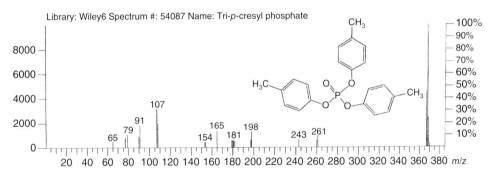

Figure 3.186 Tri-*p*-cresyl phosphate $C_{21}H_{21}O_4P$, M: 368, CAS: 78-32-0. Occurrence: plasticizer (e.g. in PVC, nitrocellulose, etc.).

References

Section 3.1.1 Total Ion Current Chromatograms

Barwick, V. *et al.* (2006) *Best Practice Guide for Generating Mass Spectra*, LGC, Teddington, December 2006, *www.lgc.co.uk* (accessed 13 May 2014).

Bemgard, A., Colmsjö, A., and Wrangskog, K. (1994) Prediction of temperature-programmed retention indexes for polynuclear aromatic hydrocarbons in gas chromatography. *Anal. Chem.*, **66**, 4288–4294.

Ciccioli, P., Brancaleoni, E., Cecinato, A., and Frattoni, M. (1993), A Method for the Selective Identification of Volatile Organic Compounds (VOC) in Air by HRGC-MS, in *15th International Symposium Capillary Chromatogrtography, Riva del Garda, May 1993* (ed. P. Sandra), Huethig, Heidelberg, pp. 1029–1042.

Deutsche Forschungsgemeinschaft (1982) *Gaschromatographische Retentionsindizes toxikologisch relevanter Verbindungen auf SE-30 oder OV-1, Mitteilung 1 der Kommission für Klinisch-toxikologische Analytik*, Verlag Chemie, Weinheim.

Hall, G.L., Whitehead, W.E., Mourer, C.R., and Shibamoto, T. (1986) A new gas chromatographic retention index for pesticides and related compounds. *HRC & CC*, **9**, 266–271.

Katritzky, A.L., Ignatchenko, E.S., Barcock, R.A., Lobanov, V.S., and Karelson, M. (1994) Prediction of gas chromatographic retention times and response factors using a general quantitative structure property relationship treatment. *Anal. Chem.*, **66**, 1799–1807.

Kostiainen, R. and Nokelainen, S. (1990) Use of M-series retention index standards in the identification of trichothecenes by electron impact mass spectrometry. *J. Chromatogr.*, **513**, 31–37.

Lipinski, J. and Stan, H.-J. (1989) Compilation of Retention Data for 270 Pesticides on Three Different Capillary Columns, in *10th International Symposium Capillary Chromatogrphy, Riva del Garda, May 1989* (ed. P. Sandra), Huethig, Heidelberg, pp. 597–611.

Manninen, A. *et al.* (1987) Gas chromatographic properties of the M-series of universal retention index standards and their application to pesticide analysis. *J. Chromatogr.*, **394**, 465–471.

Schomburg, G. (1987) *Gaschromatographie*, 2. Ed., VCH Publishers, Weinheim, pp. 54–62.

Van Den Dool, H. and Dec. Kratz, P. (1963) A generalization of the retention index system including linear temperature programmed gas-liquid partition chromatography, *J. Chrom. A*, **11**, 463–471.

Weber, E. and Weber, R. (1992) *Buch der Umweltanalytik, Methodik und Applikationen in der Kapillargaschromatographie*, Band **4**, GIT Verlag, Darmstadt, pp. 64–65.

Zenkevich, I.G. (1993) The Exhaustive Database for Gaschromatographic Retention Indizes of Low-Boiling Halogenhydrocarbons at PLOT Alumina Columns, in *15th International Symposium on Capillary Chromatography, Riva del Garda, May 1993* (ed. P. Sandra), Huethig, Heidelberg, pp. 181–186.

Section 3.2.1 Extraction of Mass Spectra

Ausloos, P., Clifton, C.L., Lias, S.G., Mikaya, A.I., Stein, S.E., Tchekhovskoi, D.V., Sparkman, O.D., Zaikin, V., and Zhu, D. (1999) The critical evaluation of a comprehensive mass spectral library. *J. Am. Soc. Mass Spectrom.*, **10**, 287–299.

Biller, J.E. and Biemann, K. (1974) Reconstructed Mass Spectra, A Novel Approach for the Utilization of Gas Chromatograph-Mass Spectrometer Data. *Anal. Lett.*, **7**, 515–528.

Dimandja, J.M.D. (2004) GCxGC. *Anal. Chem.*, **76** (9), 167A–174A.

Halket, J.M., Przyborowska, A., Stein, S.E., Mallard, W.G., Down, S., and Chalmers, R. (1999) Deconvolution gas chromatography/mass spectrometry of urinary organic acids – potential for pattern recognition and automated identification of metabolic disorders. *Rapid Commun. Mass Spectrom.*, **13**, 279–284.

Mallard, W.G. and Reed, J. (1997) AMDIS, Users Guide U.S. Department of Commerce, Technology Administration, National Institute of Standards and Technology (NIST), Standard Reference Data Program, Gaithersburg, MD, *www.chemdata.nist.gov/mass-spc/amdis/AMDIS.pdf* (accessed 13 May 2014).

Mallard, G., Stein, S. and Toropov, O. (2005) AMDIS – Automatic Mass Spectral Deconvolution and Identification Software, Agilent Technologies Product Information Deconvolution Reporting Software.

Shao, X., Wang, G., Wang, S., and Su, Q. (2004) Extraction of mass spectra and chromatographic profiles from overlapping GC-MS signals with background. *Anal. Chem.*, **76**, 5143–5148.

Stein, S.E. (1994) Estimating probabilities of correct identification from results of mass spectral library searches. *J. Am. Soc. Mass Spectrom.*, **5**, 316–323.

Stein, S.E. (1995) Chemical substructure identification by mass spectral library searching. *J. Am. Soc. Mass Spectrom.*, **6**, 644–655.

Stein, S.E. (1999) An integrated method for spectrum extraction and compound identification from gas chromatography/mass spectrometry data. *J. Am. Soc. Mass Spectrom.*, **10**, 770–781.

Stein, S.E. and Scott, D.R. (1994) Optimization and testing of mass spectral library search algorithms for compound identification. *J. Am. Soc. Mass Spectrom.*, **5**, 859–866.

Stein, S. (2011) *NIST Standard Reference Database 1A, User's Guide, U.S. Department of Commerce, National Institute of Standards and Technology.* Standard Reference Data Program, Gaithersburg, MD, USA.

Zhang, W., Wu, P., and Li, C. (2006) Study of automated mass spectral deconvolution and identification system (AMDIS) in pesticide residue analysis. *Rapid Commun. Mass Spectrom.*, **20**, 1563–1568.

Section 3.2.2 The Retention Index

Bianchi, F., Careri, M., Mangia, A., and Musci, M. (2007) Retention indices in the analysis of food volatiles in temperature-programmed gas chromatography: Database creation and evaluation. *J. Sep. Sci.*, **30**, 563–572.

Meruva, N.K., Goode, S.R., Brewer, W.E. and Morgan, S.L. (2002) *Chromatographic Performance Comparisons of Methods for Fast Gas Chromatography of complex mixtures*, manuscript in preparation. See: *http://www.chem.sc.edu/faculty/morgan/students/meruva.html.*

Van Den Dool, H. and Kratz, P.D. (1963) A generalization of the retention index system including linear temperature programmed gas – liquid partition chromatography, *J. Chromatogr.*, **11**, 463–471.

Weber, E. and Weber, R. (1992) *Buch der Umweltanalytik, Methodik und Applikationen in der Kapillargaschromatographie*, Band **4**, GIT Verlag, Darmstadt.

Section 3.2.3 Libraries of Mass Spectra

AAFS Drug Library *http://www.ualberta.ca/~gjones/mslib.html* (accessed 10 May 2014).

Adams, R.P. (1989) *Identification of Essential Oils by Ion Trap Mass Spectrometry*, Academic Press, San Diego, CA, USA.

Adams, R.P. (2007) *Identification of essential oil components by gas chromatography/mass spectrometry*, 4th ed., Allured Books, Carol Stream, IL, USA.

Central Institute of Nutrition and Food Research (2003) *Mass Spectra of Volatiles in Food (SpecData)*, 2nd edn, John Wiley & Sons, Inc., ISBN: 978–0-471–64825-3, November 2003.

Chemical Concepts (2004) *Mass Spectra Chemical Concepts (SpecInfo)*, John Wiley & Sons, Inc., ISBN: 978–0-471–66229-7.

AOCS (2014) Christie, W.W., Mass Spectrometry of Fatty Acid Derivatives, *http://lipidlibrary.aocs.org/ms/masspec.html* (accessed 10 May 2014).

De Leeuw, J.W. (2004) *Mass Spectra of Geochemicals, Petrochemicals and Biomarkers (SpecData)*, John Wiley & Sons, Ltd, ISBN: 978-0-471-64798-0, August 2004.

John Wiley & Sons Inc. (2013) *Wiley Registry of Mass Spectral Data*, 10th edn, John Wiley & Sons, Inc., ISBN: 978-0-470-52037-6.

Küehnle, R. (2006) *Mass Spectra of Drugs, Pharmaceuticals and Metabolites (SpecInfo)*, ISBN: 0-471-66230-5, John Wiley & Sons, Inc.

Makin, H.L.J. (2005) *Mass Spectra of Androgenes, Estro-gens and other Steroids*, John Wiley & Sons, Inc., ISBN: 978-0-471-74945-5.

McLafferty, F.W. and Stauffer, D.B. (1989) *The Wiley/NBS Registry of Mass Spectral Data*, John Wiley & Sons, Inc., New York.

National Institute of Science and Technology (NIST) (2014) NIST Standard Reference Database 1A, NIST/EPA/NIH Mass Spectral Library with Search Program, *http://www.nist.gov/srd/nist1a.cfm* (accessed 10 May 2014).

Palisade (1996) BenchTop/PBM : Mass Spectrometry Library search system, *http://www.sisweb.com/software/ms/benchtop.htm* (accessed 10 May 2014).

Parr, M.K., Opfermann, G., Schänzer, W. and Makin, H.L.J. (2011) *CD ROM*, ISBN 978-3-527-33080-5, Wiley-VCH, Weinheim, Germany.

Pfleger, K., Maurer, H.H., and Weber, A. (2011) *Mass Spectral and GC Data of Drugs, Poisons, Pesticides, Pollutants and Their Metabolites*, 4th edn, John Wiley & Sons, Inc., ISBN: 978-3-527-32398-2.

Roesner, P., Junge, T., Westphal, F., Fritschi, G. and Tenczer, J. (2007) *Designer Drugs*, Wiley-VCH, Weinheim, Germany (CD ROM version, Wiley-VCH Verlag GmbH & Co. KGaA, edition 2013), From: *http://www.designer-drugs.de/about.pl*.

SIS (2013) BenchTop/PBM mass spectrometry library search system, scientific instrument services, *http://www.sisweb.com/software/ms/nist.htm* (accessed 10 May 2014).

Stein, S.E. (1994) Estimating probabilities of correct identification from results of mass spectral library searches. *J. Am. Soc. Mass Spectrom.*, **5**, 316–323.

Stein, S.E. (1999) An integrated method for spectrum extraction and compound identification from gas chromatography/mass spectrometry data. *J. Am. Soc. Mass Spectrom.*, **10**, 770–781.

Stein, S.E. and Scott, D.R. (1994) Optimization and testing of mass spectral library search algorithms for compound identification. *J. Am. Soc. Mass Spectrom.*, **5**, 859–866.

Yarkov, A. (2008) *Mass Spectra of Organic Compounds (SpecInfo)*, John Wiley & Sons, Inc., ISBN: 978-0-471-66773-0.

Section 3.2.4 Library Search Programs

Atwater, B.L., Stauffer, D.B., McLafferty, F.W., and Peterson, D.W. (1985) Reliability Ranking and Scaling Improvements to the Probability Based Matching System for Unknown Mass Spectra. *Anal. Chem.*, **57**, 899–903.

Beynon, J.H. and Brenton, A.G. (1982) *Introduction to Mass Spectrometry*, University Wales Press, Swansea.

Davies, A.N. (1993) Mass spectrometric data systems. *Spectrosc. Eur.*, **5**, 34–38.

Henneberg, D. and Weimann, B. (1984) *Search for Identical and Similar Compounds in Mass Spectral Data Bases*, Spectra, pp. 11–14.

Little, J.L., Cleven, C.D., Howard, A.S., Yu, K. (2013) *Identifying 'Known Unknowns'* in Commercial Products by Mass Spectrometry, LCGC North America **31**, 114–125.

McLafferty, F.W. and Turecek, F. (1993) *Interpretation of Mass Spectra*, 4th edn, University Science Books, Mill Valley, CA.

Neudert, B., Bremser, W., and Wagner, H. (1987) Multidimensional computer evaluation of mass spectra. *Org. Mass Spectrom.*, **22**, 321–329.

Palisade Corporation (1994) Benchtop/PBM Users Guide.

Sokolow, S., Karnovsky, J. and Gustafson, P. (1978) The Finnigan Library Search Program. Finnigan MAT Applikation No. 2.

Stauffer, D.B., McLafferty, F.W., Ellis, R.D., and Peterson, D.W. (1985a) Adding forward searching capabilities to a reverse search algorithm for unknown mass spectra. *Anal. Chem.*, **57**, 771–773.

Stauffer, D.B., McLafferty, F.W., Ellis, R.D., and Peterson, D.W. (1985b) Probability-based-matching algorithm with forward searching capabilities for matching unknown mass spectra of mixtures. *Anal. Chem.*, **57**, 1056–1060.

Stein, S.E. (1994) Estimating probabilities of correct identification from results of mass spectral library search. *J. Am. Soc. Mass Spectrom.*, **5**, 316–323.

Stein, S.E. and Scott, D.R. (1994) Optimization and testing of mass spectral library search algorithms for compound identification. *J. Am. Soc. Mass Spectrom.*, **5**, 859–866.

Warr, W.A. (1993a) Computer-assisted structure elucidation-Part 1: library search and spectral data collections. *Anal. Chem.*, **65**, 1045A–1050A.

Warr, W.A. (1993b) Computer-assisted structure elucidation –Part 2: indirect database approaches and established systems. *Anal. Chem.*, **65**, 1087A–1095A.

Section 3.2.5 Interpretation of Mass Spectra

Budzikiewicz, H. (1985) Structure elucidation by ion-molecule reactions in the gas phase: the location of C, C-double and triple bonds. *Fresenius J. Anal. Chem.*, **321** (2), 150–158.

Budzikiewicz, H. (1998) *Massen-spektrometrie*, 4. Aufl, Wiley-VCH Verlag GmbH, Weinheim.

Christie, W.W. (1996) Mass spectrometry of fatty acids – Part 1. *Lipid Technol.*, **8**, 18–20, substantially re-written see: *http://lipidlibrary.aocs.org/ms/masspec.html* (accessed 10 May 2014).

Christie, W.W. (2001) A practical guide to the analysis of conjugated linoleic acid. *Inform*, **12** (2), 147–152.

Dobson, G. and Christie, W.W. (2002) Spectroscopy and spectrometry of lipids – Part 2, mass spectrometry of fatty acid derivatives. *Eur. J. Lipid Sci. Technol.*, **104**, 36–43.

Fay, L. and Richli, U. (1991) Location of double bonds in polyunsaturated fatty acids by gas chromatography-mass spectrometry after 4,4-dimethyloxazoline derivatization. *J. Chromatogr.*, **541**, 89–98.

Jhama, G.N., Attygalleb, A.B., and Meinwalda, J. (2005) Location of double bonds in diene and triene acetates by partial reduction followed by methylthiolation. *J. Chromatogr. A*, **1077** (1), 57–67.

López, J.F. and Grimalt, J.O. (2004) Phenyl- and cyclopentylimino derivatization for double bond location in unsaturated C (37)-C (40) alkenones by GC-MS. *J. Am. Soc. Mass Spectrom.*, **15** (8), 1161–1172.

McLafferty, F.W. (1959) Mass spectrometric analysis — molecular rearrangements, *Anal. Chem.*, **31**, 82–87.

McLafferty, F.W. and Turecek , F. (1993) *Interpretation of Mass Spectra*, 4th ed., Mill Valley, CA, USA: University Science Books.

Moss, C.W. and Lambert-Fair, M.A. (1989) Location of double bonds in monounsaturated fatty acids of Campylobacter cryaerophila with dimethyl disulfide derivatives and combined gas chromatography-mass spectrometry. *J. Clin. Microbiol.*, **27** (7), 1467–1470.

Rosman, K.J.R. and Taylor, P.D.P. (1998) Isotopic compositions of the elements 1997. *Pure Appl. Chem.*, **70** (1), 217–235.

Shantha, N.C. and Kaimal, T.N.B. (1984) Mass spectrometric location of double bonds in unsaturated fatty acids including conjugated acids as their methoxy-bromo derivatives. *Lipids*, **19** (12), 971–974.

Sharp, Z. (2007) *Principles of Stable Isotope Geochemistry*, Pearson Prentice Hall, Upper Saddle River, NJ.

Section 3.2.6 Mass Spectroscopic Features of Selected Substance Classes

Budzikiewicz, H. (1998) *Massenspektrometrie*, 4. Aufl, Wiley-VCH Verlag GmbH, Weinheim.

Färber, H. and Schöler, F. (1991) Gaschromatographische Bestimmung von Harnstoffherbiziden in Wasser nach Methylierung mit Trimethylanilinium-hydroxid oder Trimethylsulfonium-hydroxid. *Vom Wasser*, **77**, 249–262.

Howe, I. *et al.* (1981) *Mass Spectrometry*, McGraw-Hill, New York.

McLafferty, F.W. and Turecek, F. (1993) *Interpretation of Mass Spectra*, 4th ed., Mill Valley, CA, USA: University Science Books.

Pretsch, E. *et al.* (1990) *Tabellen zur Strukturaufklärung organischer Verbindungen mit spektrometrischen Methoden*, 3. Aufl., 1. korr. Nachdruck, Springer, Berlin.

Stan, H.-J., Abraham, B., Jung, J., Kellert, M., and Steinland, K. (1977) Nachweis von Organophosphorinsectici-den durch Gas-Chromatographie-Massenspektroskopie. *Fresenius Z. Anal. Chem.*, **287**, 271–285.

Stan, H.-J. and Kellner, G. (1981) in *Recent Development in Food Analysis*, Proceedings of the First European Conference on Food Chemistry (EURO FOOD CHEM I), Vienna, February 17 – 20, 1981 (eds W. Baltes, P.B. Czodik-Eysenberg, and W. Pfannhauser), Verlag Chemie, Weinheim, pp. 183–189.

Section 3.3 Quantitation

Boqué, R., Marato, A., and Vander Heyden, Y. (2008) Assessment of accuracy in chromatographic analysis. *LCGC Eur.*, **21**, 264–267.

Chang, C. (1985) Parallel Mass Spectrometry for High Performance GC and LC Detection, *Int. Lab.* **5**, 58–68.

Dallüge, J., Vreuls, R.J., van Iperen, D.J., van Rijn, M. and Brinkman, U.A.Th (2002) Resistively Heated Gas Chromatography Coupled to Quadrupole Mass Spectrometry, *J. Sep. Sci.* **25** (9), 608–614.

Glaser, J., Foerst, D.L., McKee, G.D., Quave, S.A. and Budde, W.L. (1981) Trace Analyses for Wastewaters, *Env. Sci. Techn.* **15** (12), 1426–1435.

ISO 5725-6:1994. (1994) *Accuracy (Trueness and Precision) of Measurement Methods and Results (Parts 1 – 4)*, International Organization for Standardisation, Geneva.

Kirchner, M., Matisova, E., Hrouzkova, S., and de Zeeuw, J. (2005) Possibilities and Limitations of Quadrupole Mass Spectrometric Detector in Fast Gas Chromatography, *J.Chrom. A* **1090**, 126–132.

Section 3.3.2 Decision Limit

Ebel, S. and Dorner, W. (1987) *Jahrbuch Chemielabor 1987*. VCH Publishers, Weinheim.

Section 3.3.4 Limit of Detection

ISO 11843. (1997) Capability of Detection, International Organization for Standardization, Geneva.

Section 3.3.6 The Calibration Function

Ebel, S. and Dorner, W. (1987) *Jahrbuch Chemielabor 1987*, VCH Publishers, Weinheim.

Ebel, S. and Kamm, U. (1983) Statistische Definition der Bestimmungsgrenze. *Fresenius Z. Anal. Chem.*, **316**, 382–385.

Guichon, G. and Guillemin, C.L. (1988) *Quantitative Gas Chromatography*, Elsevier, Amsterdam, Oxford, New York, Tokyo.

Miller, J.N. (1992) The method of standard additions. *Spectrosc. Eur.*, **4** (6), 26–27.

Montag, A. (1982) Beitrag zur Ermittlung der Nachweis-und Bestimmungsgrenze analytischer Meßverfahren. *Fresenius Z. Anal. Chem.*, **312**, 96–100.

Naes, T. and Isakson, T. (1992) The importance of outlier detection in spectroscopy. *Spectrosc. Eur.*, **4** (4), 32–33.

Section 3.3.7 Quantitation and Standardization

Funk, C. Dammann, V. and Donnevert, G. (2008) *Quality Assurance in Analytical Chemistry*, 2. ed., Wiley-VCH, Weinheim, Germany.

Zhang, X. Chongtian, Y., Liang, L. and Hübschmann, H-J. (2014) Determination of BTEX, in *Cigarette Filter Fibers by GC-MS with Automated Calibration using the TriPlus RSH Autosampler, Thermo Fisher Scientific Application Note 10399*, Shanghai, China.

Section 3.3.8 The Accuracy of Analytical Data

Hein, H. and Kunze, W. (1994) *Umweltanalytik mit Spektrometrie und Chromatographie*, VCH Publishers, Weinheim.

Section 3.4 Frequently Occurring Impurities

Kapp, T. and Vetter, W. (2006) Migration von Additiven aus Buthylgummihaltigen Verschlüssen von Probegläschen. *Lebensmittelchem.*, **60**, 152.

Spiteller, M. and Spiteller, G. (1973) *Massenspektrensammlung von Lösemitteln,Verunreinigungen, Säulenbelegmaterialien und einfachen aliphatischen Verbindungen*, Springer, Wien.

Further Reading

Barker, J. (1999) *Mass Spectrometry*, 2nd edn, John Wiley & Sons, Ltd, Chichester.

4
Applications

The applications given here have been chosen in order to describe typical areas of use of GC-MS, such as air, water, soil, foodstuffs, the environment, waste materials, drugs or pharmaceutical products. Special emphasis has been placed on current and reproducible examples which give successful templates for routine laboratories. The selection cannot be totally representative of the use of modern GC-MS, but shows the main areas into which the methodology has spread and will continue to do so. In addition, in special areas of application, such as the analysis of isotope-specific measuring procedures and the isotope dilution method for dioxin analysis are described.

Most of the applications described are compiled from the references cited and documented with various graphics and tables. The analysis conditions are described in full to allow adaptation of the methods. If any of the methods have been published, the sources are given for each section. References to other directly related literature are also given.

4.1
Air Analysis According to EPA Method TO-14

Keywords: air; SUMMA canister; thermodesorption; volatile halogenated hydrocarbons; cryofocusing; water removal; thick film column.

Introduction

The EPA (US Environmental Protection Agency) describes a process for sampling and analysis of volatile organic compounds (VOCs) in the atmosphere. It is based on the collection of air samples in passivated stainless steel canisters (SUMMA canisters). The organic components are separated by GC and determined using conventional GC detectors or by mass spectrometry (Figure 4.1). The use of mass spectrometers allows the direct positive detection of individual components (Madden, 1994).

For a SIM (selected ion monitoring) analysis the mass spectrometer is programmed in such a way that a certain number of compounds in a defined retention time range are detected. These SIM segments are switched at the programmed retention time so that a list of target compounds can be worked

Handbook of GC-MS: Fundamentals and Applications, Third Edition. Hans-Joachim Hübschmann.
© 2015 Wiley-VCH Verlag GmbH & Co. KGaA. Published 2015 by Wiley-VCH Verlag GmbH & Co. KGaA.

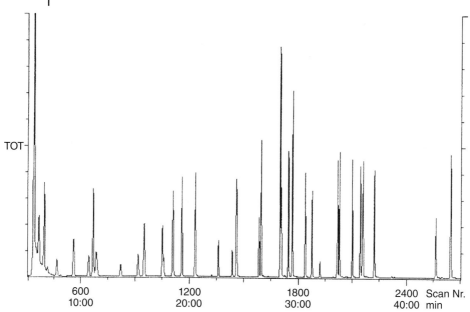

Figure 4.1 A typical chromatogram of a 10 ppbv VOC standard.

through. Alternatively a timed-SIM with a short acquisition window centered to the compound retention time can be used. In the full scan mode, the mass spectrometer works as a universal detector acquiring complete mass spectra for analyte identification and confirmation.

A cryoconcentrator with a three-way valve system is used for concentration. The sample can be applied by two routes without having to alter the screw joints on the tubing. Usually the inlet is via control of a mass flow regulator to a cryofocusing unit. The direct measurement of the sample volume provides very precise data. A Nafion drier is used to dry the air (Schnute and McMillan, 1993).

SUMMA canisters

The analysis of VOCs is frequently carried out by adsorption on to suitable materials. For this Tenax is mainly used (see Section 2.1.5.2). The limits of this adsorption method lie in the adsorption efficiency, which is dependent upon the compound, the breakthrough of the sample at higher air concentrations, the impossibility of multiple measurements on a sample and the possible formation of artefacts. Stainless steel canisters, whose inner surfaces have been passivated by the SUMMA process, do not exhibit these limitations. This passivation process involves polishing the inner surface and applying a Cr/Ni oxide layer. Containers treated in this way have been used successfully for the collection and storage of air samples. Purification and handling of these canisters and the sampling apparatus must be carried out carefully, however, because of possible contamination problems.

Another means of sample injection is the loop injection. The sample is drawn through a 5 mL sample loop directly into the cryoconcentrator. This method is suitable for highly concentrated samples, as only a small quantity of sample is required. In this case the drier is avoided.

The Nafion drier is a system for removal of water from the air sample, which uses a semi-permeable membrane (see Figure 2.211). Nonpolar compounds pass the membrane unaffected while polar ones, such as water, are held by the membrane and diffuse outwards. The outer side of the membrane is dried by a clean air stream and the water thus separated is removed from the system. The Nafion drier is recommended by the TO-14 method to prevent blockage of the cryofocusing unit by the formation of ice crystals.

In this application the GC is coupled to the mass spectrometer by an open split interface. A restrictor limits the carrier gas flow. Open coupling was chosen because the sensitivity of the ion trap GC-MS makes the concentration of large quantities of air superfluous. Open coupling dilutes the moisture, which may be contained in the sample, to an acceptable level so that cryofocusing can be used without additional drying.

Analysis Conditions

Sample material			Air
Sample concentration	Sampling Concentrator		SUMMA canister Grasby Nutech model 3550A cryoconcentrator with 354A cryofocusing unit, Nafion drier
	Autosampler		Nutech 3600 16-position sampler
GC Method	Column	Type	J&W DB-5
		Dimensions	60 m length $\times$ 0.25 mm ID $\times$ 1.0 µm film thickness
			The thick film columns DB-1 or DB-5 guarantee chromatographic separation even at start temperatures just above room temperature if cryofocusing is used
	Pre-column		2 m, 0.53 mm ID, deactivated
	Carrier gas		Helium
	Flow		Constant flow, 1 mL/min
	Sample injection		The sample is drawn from the SUMMA sampling canister through the Nafion drier and reaches the cryoconcentrator cooled to $-160\,°C$. It is then heated rapidly to transfer the sample to the cryofocusing unit of the GC (Figure 4.2)
	Cryofocusing		Liquid N_2, $-190\,°C$ sample focusing, heated to $150\,°C$ for injection
	Oven program	Start	$35\,°C$, 6 min

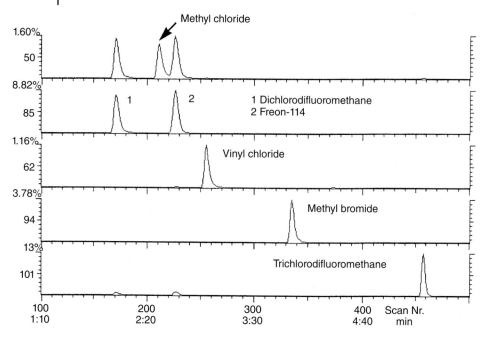

Figure 4.2 Elution of gaseous VOCs following injection using cryofocusing.

(Continued)			
		Ramp 1	8 °C/min
		Final temperature	200 °C
	Transferline	Temperature	250 °C
	Open split interface		SGE type GMCC/90, mounted in the transfer line to the MS restrictor capillary 0.05 mm ID, adjusted to 2.5% transmission
MS method	System		Finnigan MAGNUM
	Analyzer type		Ion trap MS with internal ionization
	Ionization		EI
	Electron energy		70 eV
	Ion trap	Temperature	250 °C
	Acquisition	Mode	Full scan
	Mass range		35–300 Da
	Scan rate		1 s/scan
	Resolution		Nominal mass
Calibration	External standard		Based on 1 L samples from a SUMMA canister
	Calibration range		0.1 and 20 ppbv (Figure 4.3)
	Data points		6

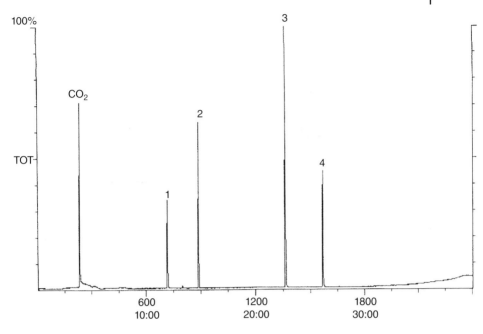

Figure 4.3   Continuous flushing of the gas lines ensures contamination-free analyses. 1. bromochloromethane, 2. 1,4-difluorobenzene, 3. chlorobenzene-d₅ as internal standard and 4. bromofluorobenzene.

Limit of Detection

For the EPA method TO-14 (Table 4.1 and Figure 4.4), a limit of detection of 0.1 ppbv is required. This requirement is achieved even at the high split rate of the interface.

Results

With this method, data acquisition in the full-scan mode allows mass spectra to be acquired for subsequent identification through library searching even at the required limit of detection of 0.1 ppbv. Figures 4.5 and 4.6 show examples of comparability of the spectra and the identification of dichlorobenzene at 20.0 and 0.1 ppbv. With the procedure described, both the compounds required according to TO-14 and other unexpected components in the critical concentration range can be identified.

Table 4.1 Limits of detection for compounds used in the EPA method TO-14.

Dichlorodifluoromethane	0.01 ppbv
Chloromethane	0.01 ppbv
Freon-114	0.02 ppbv
Vinyl chloride	0.01 ppbv
Bromomethane	0.01 ppbv
Chloroethane	0.08 ppbv
Trichlorofluoromethane	0.01 ppbv
1,1-Dichloroethene	0.02 ppbv
Methylene chloride	0.01 ppbv
3-Chloropropane	0.02 ppbv
Freon-113	0.01 ppbv
1,1-Dichloroethane	0.01 ppbv
cis-1,2-Dichloroethene	0.01 ppbv
Chloroform	0.01 ppbv
1,1,1-Trichloroethane	0.01 ppbv
1,2-Dichloroethane	0.01 ppbv
Benzene	0.01 ppbv
Tetrachloromethane	0.01 ppbv
1,2-Dichloropropane	0.02 ppbv
Trichloroethene	0.02 ppbv
cis-1,3-Dichloropropene	0.01 ppbv
trans-1,3-Dichloropropene	0.01 ppbv
Toluene	0.01 ppbv
1,1,2-Trichloroethane	0.01 ppbv
1,2-Dibromoethane	0.01 ppbv
Tetrachloroethene	0.01 ppbv
Chlorobenzene	0.01 ppbv
Ethylbenzene	0.01 ppbv
m/p-Xylene	0.01 ppbv
Styrene	0.02 ppbv
o-Xylene	0.01 ppbv
1,1,2,2-Tetrachloroethane	0.09 ppbv
4-Ethyltoluene	0.02 ppbv
1,3,5-Trimethylbenzene	0.02 ppbv
1,2,4-Trimethylbenzene	0.01 ppbv
1,3-Dichlorbenzene	0.01 ppbv
Benzyl chloride	0.08 ppbv
1,4-Dichlorbenzene	0.01 ppbv
1,2-Dichlorbenzene	0.01 ppbv
1,2,4-Trichlorobenzene	0.01 ppbv
Hexachlorobutadiene	0.02 ppbv

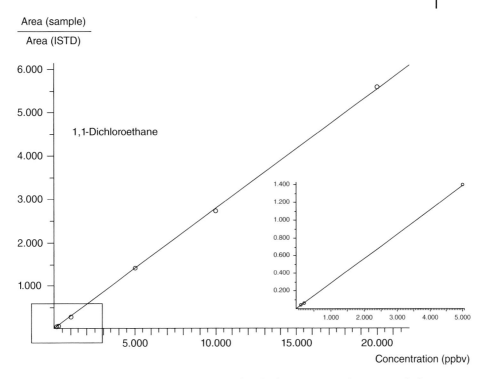

Figure 4.4 Six point calibration from 0.1 to 20 ppbv. The lower region is shown magnified. ppbv = parts per billion in volume.

4.2
BTEX in Surface Water as of EPA Method 8260

Keywords: environmental analysis; surface water; waste water; BTEX; purge and trap; BFB tuning.

Introduction

The EPA method 8260 is used to determine VOCs by purge and trap (P&T) GC-MS in a wide variety of solid and aqueous sample matrices (EPA Method 8260C, 2006). The method covers more than 100 VOC compounds and is applied for the trace determination of the BTEX components benzene, toluene, ethylbenzene and isomeric xylenes as well.

BTEX compounds can be released to surface water streams from, for example oil spills or industrial processes and get monitored on low levels. In particular, during the production of gasoline products, oil refineries generate a waste stream that contains petroleum by-products such as BTEX. These volatile, monoaromatic hydrocarbons can be toxic to receiving streams.

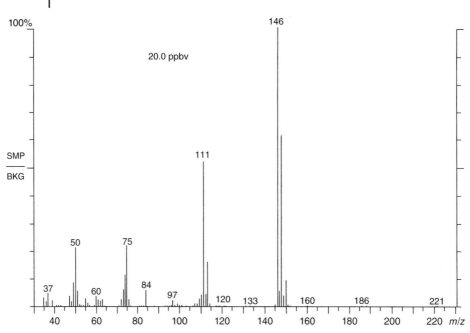

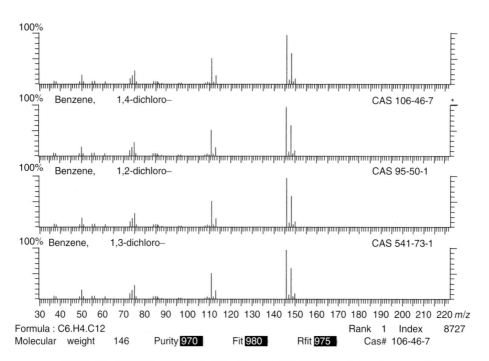

Figure 4.5 Spectrum and result of the library search (NIST) for dichlorobenzene at a concentration of 20 ppbv.

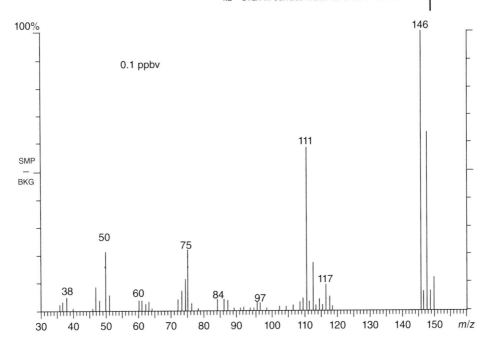

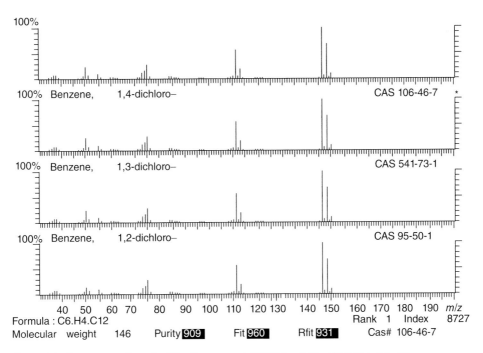

Figure 4.6 Spectrum and result of the library search (NIST) for dichlorobenzene at a concentration of 0.1 ppbv.

Sample Preparation

Before analysis, samples containing high levels of BTEX in the ppm range were diluted with organic free water. The applied P&T unit automatically spikes the internal standards and surrogates into a 5 mL sample.

For calibration, a solution of 250 ng/μL is prepared in P&T grade methanol and 1 μL is spiked into 5 mL of organic free water at a final concentration of 50 μg/L. A calibration curve is generated from 20 to 200 μg/L.

Standards are prepared in organic free water. Each standard gets transferred into a 40 mL sampler vial and loaded onto the P&T autosampler. A 5 mL aliquot is transferred automatically from the 40 mL vial into a fritted sparger for analysis.

Experimental Conditions

The GC was configured for P&T analysis by installing the transfer line of the P&T unit with a dedicated adapter to the SSL injector (Butler, 2013). The selected internal trap #10 containing three different trapping materials of increasing strength is recommended for the EPA methods 524.2, 624 and 8260 with MS detection (OI Analytical).

Analysis Conditions

Sample material			Water, surface water, waste water
Purge and Trap	System		O.I Analytical Eclipse model 4660
	Autosampler		O.I Analytical model 4551A
	Sample	Volume	5 mL
		Purge temperature	40 °C
		Purge flow	40 mL/min
		Purge time	11 min
	Trap	Type, material	#10 (Tenax, silica gel, carbon molecular sieve)
		Water management	Purge 110 °C, bake 240 °C
		Desorb preheat	180 °C
		Desorb time	30 s
	Bake	Rinse cycles	Two times
		Temperature, cycle time	210 °C, 10 min
GC Method	System		Thermo Scientific™ TRACE 1300 GC
	Column	Type	TG-VMS
		Dimensions	20 m length × 0.18 mm ID × 1.0 μm film thickness

(Continued)

		Carrier gas	Helium
		Flow	Constant flow, 1 mL/min
	SSL Injector	Injection mode	Split
		Split flow	30 mL/min
		Injection temperature	200 °C
		Injection volume	1 µL
	Oven program	Start	45 °C, 4.5 min
		Ramp 1	8 °C/min to 100 °C
		Ramp 2	25 °C/min to 230 °C
		Final temperature	230 °C, 2 min
	Transferline	Temperature	200 °C
			SGE type GMCC/90, mounted in the transfer line to the MS restrictor capillary 0.05 mm ID, adjusted to 2.5% transmission
MS Method	System		Thermo Scientific™ ISQ
	Analyzer type		Single quadrupole MS
	Ionization		EI
	Electron energy		70 eV
	Ion source	Temperature	230 °C
	Acquisition	Mode	Full scan
	Mass range		35–350 Da
	Scan rate		0.2 s (5 spectra/s)
	Resolution	Setting	Normal (0.7 Da)
Calibration	Internal standard	Compounds	Deuterium and fluorine labelled analogues
		Range	20–200 µg/L

The analyte separation was carried out using a thick film column TG-VMS of 20 m length, 0.18 mm ID and 1.0 µm film thickness, installed to a regular split/splitless inlet with a narrow ID inlet liner. A constant split vent of 30 mL/min was used and the column flow set to constant 1 mL/min. The list of compounds analyzed is given in Table 4.2 with retention times.

For compliance with the EPA 8260 method requirements on the MS tune conditions, a 1 µL injection of 25 ng/µL 4-bromofluorobenzene solution (BFB) was injected at the beginning of each analysis shift for a continuous performance documentation.

Table 4.2 List of BTEX compounds and standards analyzed with achieved precision and MDL data.

Retention time (min)	Compound	% RSD calibration	MDL (µg/L)	% RSD at MDL
4.24	Dibromofluoromethane (surrogate)	2.13	—	2.32
4.79	Benzene	6.16	0.008	3.15
4.99	1,2-Dichloroethane-d4 (surrogate)	2.65	—	2.13
5.41	Fluorobenzene (ISTD)	9.38	—	4.51
7.86	Toluene-d8 (surrogate)	1.00	—	1.30
7.94	Toluene	5.67	0.010	3.48
10.31	Chlorobenzene-d5 (ISTD)	11.06	—	4.78
10.46	Ethyl benzene	5.98	0.009	3.46
10.72	*m*- and *p*-Xylene	6.49	0.024	4.70
11.41	*o*-Xylene	7.12	0.015	5.85
12.23	BFB (surrogate)	1.18	—	1.93
13.39	1,4-Dichlorobenzene-d4 (ISTD)	10.26	—	4.12
	Average values	4.27	0.013	3.48

Results

Before sample analysis, the tuning of the MS was checked by using the autotune function and run the BFB check sample. The used ISQ MS met the tuning criteria for BFB listed in the method (see Figure 4.7).

The entire run-time of one analysis cycle on the 20 m column was 16 min as shown of the 200 µg/L BTEX standard in Figure 4.8. The analyzed surface water sample shown in Figure 4.9 contained 0.62 ppm benzene. Method detection limits (MDLs) were generated by running replicate samples at 0.1 µg/L BTEX concentration (see Tables 4.2 and 4.3).

Conclusions

The applied mass spectrometer met the quality control criteria of the EPA method 8260 and generated a typical spectrum for BFB within the method criteria limits.

The BTEX calibration curve from 20 to 200 µg/L in water showed a wide linear range and good precision with an average RSD (relative standard deviation) of 4.3%.

The average MDL for water produced from eight replicates of 0.1 µg/L spiked organic free water was 0.013 µg/L, showing very good sensitivity and precision with an average standard deviation of better than 4% RSD (Butler, 2013).

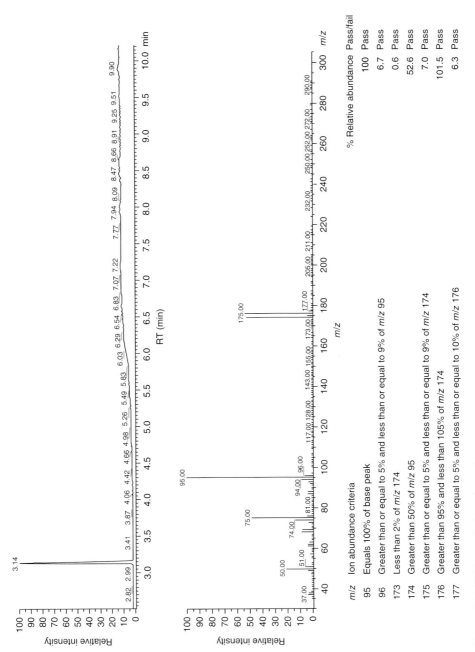

Figure 4.7   BFB tune criteria and report.

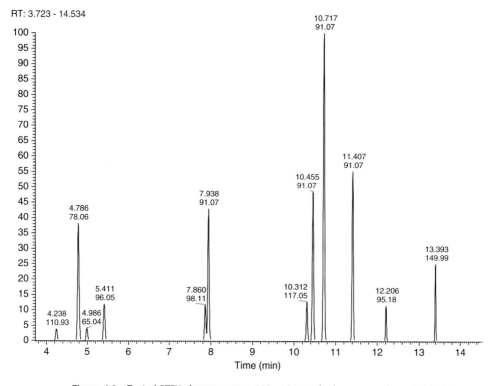

Figure 4.8 Typical BTEX chromatogram, 200 µg/L standard, compounds see Table 4.2.

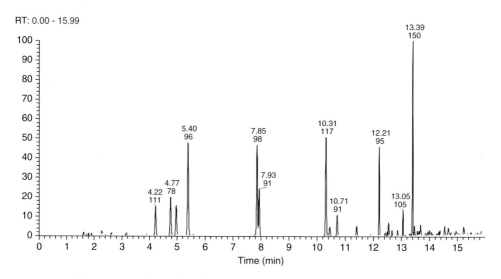

Figure 4.9 Analysis of a surface water sample.

Table 4.3 Results of the surface water sample of Figure 4.9.

Compound	Sample (ppm)
Benzene	0.62
Toluene	1.163
Ethylbenzene	0.29
o-Xylene	0.32
m- and *p*-Xylene	0.73

4.3
Simultaneous Determination of Volatile Halogenated Hydrocarbons and BTEX

Keywords: volatile halogenated hydrocarbons; BTEX; gas from landfill sites; seepage water; P&T; Tenax; internal standard.

Introduction

The analysis of seepage water (Figure 4.10) and the condensate residue from the incineration of gas from landfill sites (Figure 4.11) are two examples of the effective simultaneous control of the volatile halogenated hydrocarbon and BTEX concentrations. The P&T technique was chosen here so that as wide a spectrum

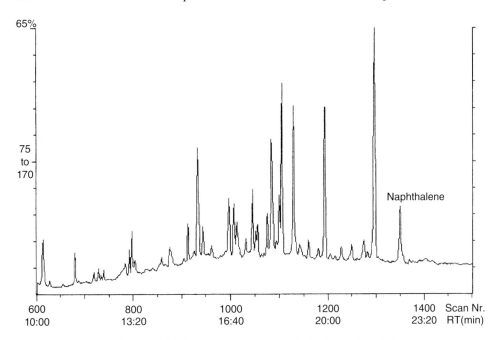

Figure 4.10 Purge and trap GC-MS chromatogram for the analysis of a condensate from a block heating power station run on gas from landfill sites.

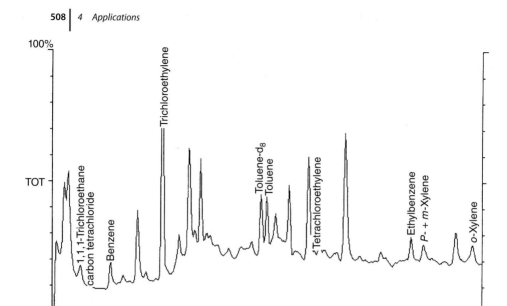

Figure 4.11 Purge and trap GC-MS chromatogram for the analysis of a seepage water sample from a landfill site (section).

of unknown components as possible could be determined together. The complex elution sequence can be worked out easily using GC-MS by means of selective mass signals. Toluene-d_8 is used as the internal standard for quantitation.

Analysis Conditions

Sample material			Air
Sample concentration		System	Tekmar purge and trap concentrator LSC 2000
		Trap	Tenax
	Sample	Volume	5 mL
		Standby temperature	30 °C
	Purge	Sample preheat time	2.50 min
		Sample temperature	40 °C
		Purge duration	10 min
		Purge flow	40 mL/min
	Desorb	Preheat	175 °C
		Desorb temperature	180 °C
		Desorb time	4 min
	Water management	MCM Desorb	5 °C
		Trap bake	10 min at 200 °C
		MCM bake	90 °C

(*Continued*)

	Instrument	BGB	off
		Mount	60 °C
		Valve	220 °C
		Transfer line	200 °C
		Connection to GC	LSC 2000, linked into the carrier gas supply of the injector
GC Method	System		Finnigan MAGNUM
	Column	Type	Rtx-624
		Dimensions	60 m length × 0.32 mm ID × 01.8 µm film thickness
	Carrier gas		Helium
	Flow		1 mL/min, constant pressure
	PTV Injector	Injection mode	split
		Base temperature	200 °C
		Transfer rate	Isothermal
	Split	Open	Until end of run
		Split flow	20 mL/min
		Purge flow	5 mL/min
	Oven program	Start	40 °C, 5 min
		Ramp 1	15 °C/min
		Final temperature	200 °C, 9.5 min
	Transferline	Temperature	250 °C
MS Method	System		Finnigan MAGNUM
	Analyzer type		Ion trap with internal ionization
	Ionization		EI
	Electron energy		70 eV
	Ion source	Temperature	280 °C
	Mass range		45–220 Da
	Scan rate		1 s/scan
	Resolution	Setting	Normal (0.7 Da)
Calibration	Internal standard		Toluene-d8 (1 µg/L)

PTV, programmed temperature vaporizer.

Results

The elution sequence of the volatile halogenated hydrocarbons compared to the BTEX aromatics is shown in Figure 4.12 in the form of mass chromatograms using TIC in a standard analysis. The internal standard, toluene-d$_8$ (1 µg/L) elutes immediately before the toluene peak (see TIC). The column used has very high thermal stability until the end of the temperature program. This is shown by a completely flat base line. Benzene elutes here between tetrachloroethane and trichloroethylene, *ortho*-xylene, as the last BTEX component, between dibromochloromethane and bromoform. In the concentration range of around 1 µg/L, volatile halogenated

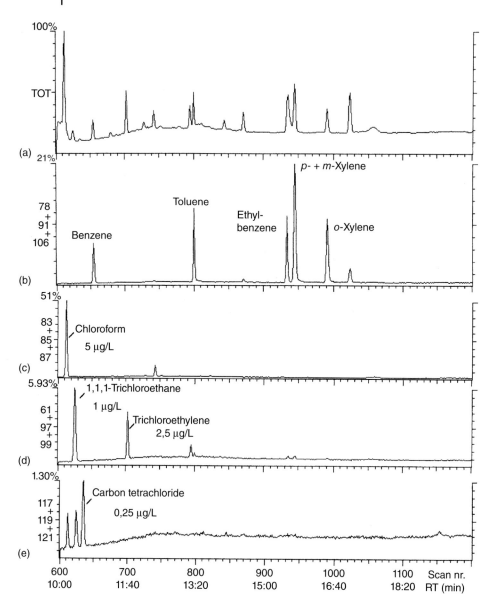

Figure 4.12 Analysis of a standard volatile halogenated hydrocarbon/BTEX mixture. (a) Total ion current (for conditions see text); (b) BTEX components (*m/z* 78, 91 and 106); (c) chloroform (*m/z* 83, 85 and 87); (d) 1,1,1-trichloroethane (*m/z* 61, 97 and 99); (e) carbon tetrachloride (*m/z* 117, 119 and 121); (f) trichloroethylene (*m/z* 95, 130 and 132); (g) bromodichloromethane (*m/z* 83, 85 and 129); (h) tetrachloroethylene (*m/z* 166); (i) dibromochloromethane (*m/z* 79, 127 and 129) and (j) bromoform (*m/z* 173).

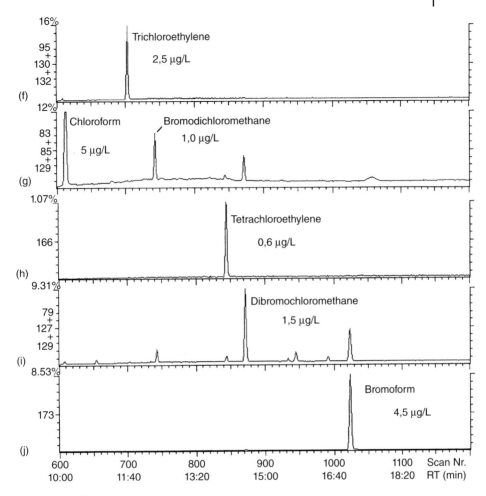

Figure 4.12   *(Continued)*

hydrocarbons and BTEX are determined together in an analysis with a good signal/noise (S/N) ratio.

4.4
Static Headspace Analysis of Volatile Priority Pollutants

Keywords: headspace; VOCs; water; soil; quantitation; internal standard.

Introduction

Many analytical techniques have been used for the quantitation of VOC in water and soil, including liquid–liquid microextraction (LLME), solid phase microextraction (SPME) and P&T. Automated static headspace analysis offers the

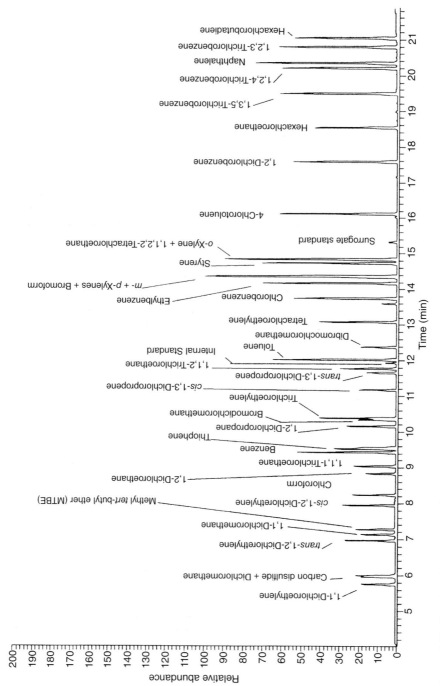

Figure 4.13 Total ion chromatogram obtained for water analysis by static headspace GC-MS. The concentration of each analyte is 20 µg/L. The column (DB1, 30 m × 0.32 mm × 1 µm) is programmed from −40 to 260 °C at 10 °C/min. 1 mL of headspace is injected in the split mode.

advantages of simplicity and robustness especially when large sample throughput is required (Rinne, 1986). A typical application of the static headspace method described here is the analysis of surface, and waste water see Figure 4.13, other techniques such as P&T GC-MS are better designed for ultra-trace analysis of drinking water. The samples were transferred to 20 mL headspace vials, together with an internal standard (e.g. toluene-d_8). 'Salting-out' was achieved by saturating the sample with sodium sulfate. Using the sample agitation feature, headspace equilibrium is reached very quickly, allowing all sampling operations to take place during the GC run time (Belouschek *et al.*, 1992).

The GC was operated in accordance with the 'whole column trapping technique' using a standard medium bore capillary column with a 1 µm stationary phase and liquid CO_2 as coolant.

The analytes were trapped in the column inlet at subambient temperature, without the need for a dedicated cold trapping device. Trapping of the VOCs helped to maintain optimum chromatographic efficiency by focusing the analytes at the column inlet. High sensitivity was achieved by injecting a relatively high volume of headspace using a low split flow. The injecting speed was controlled in order to prevent the injector from overflowing. The MS was operated in scanning electron ionisation (EI) mode, allowing acquisition of full mass spectra and thereby enabling both targeting analysis and identification of unknowns during a single analysis (Figure 4.14). At 0.5 µg/L, all compounds were identified by automated library searching, using the NIST 98 library (Figure 4.15).

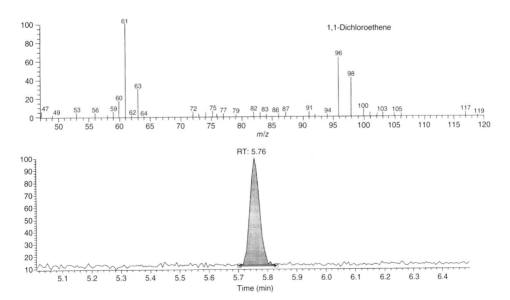

Figure 4.14 Mass spectrum and chromatogram obtained for 1,1-dichloroethylene at 0.5 µg/L.

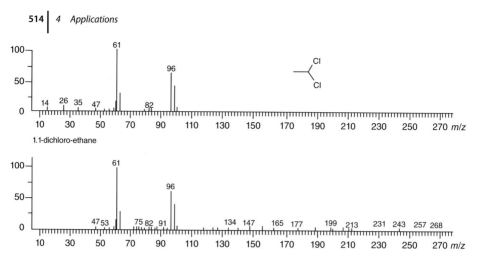

1.1-dichloro-ethane

Figure 4.15 Library search result using the NIST 98 library.

The quantitation was based on an internal standard method, using an isotopically labelled analogue of toluene. A surrogate standard was also added to the samples in order to control the long-term spectral balance stability and to check the MS tuning criteria.

Analysis Conditions

Sample material			Water, soil
Sample preparation	Headspace	System	CE Instruments HS2000
		Incubation	40 °C, 10 min
		Sample volume	1 mL
GC Method	System		Finnigan Trace MS
	Column	Type	DB-1
		Dimensions	30 m length × 0.32 mm ID × 1.0 μm film thickness
	Cryofocusing		liquid CO_2 oven cooling
	Carrier gas		Helium
	Flow		Constant flow, 1 mL/min
	SSL injector	Injection mode	Splitless
		Injection temperature	200 °C
		Injection volume	1 μL
	Split	Closed until	1 min
		Open	1 min to end of run

(Continued)

	Oven program	Start	−40 °C
		Ramp 1	10 °C/min to 260 °C
		Final temperature	260 °C
	Transferline	Temperature	260 °C
MS Method	System		Finnigan Trace MS
	Analyzer type		Single quadrupole MS
	Ionization		EI
	Electron energy		70 eV
	Ion source	Temperature	200 °C
	Acquisition	Mode	Full scan
	Mass range		45–270 Da
	Scan rate		0.4 s/scan
Calibration	External standard	Range	0.1–100 µg/L
	Data points		6

Results

The technique of static headspace GC-MS offers significant benefits for laboratories tasked with running VOC analyses. The virtual elimination of sample carry-over, thanks to the programmable temperature cleaning cycle of the syringe and needle heaters in the headspace autosampler, obviates the need for running blank samples between specimen samples.

The wide linear dynamic range of the mass spectrometer, even running in full scan mode, permits target compounds to be accurately quantified over a concentration range of at least three decades. Full-scan operation enables unknown peaks to be automatically detected and identified. Even after a period of several weeks of unattended operation, highly sensitive and reproducible results are achieved.

The precision study, based on 72 replicate injections of the low standard (0.5 µg/L), represents analyses acquired over 3 weeks of continuous operation (see Table 4.4). Such high precision is normally associated with quadrupole mass spectrometers running in the SIM mode; yet these results were obtained in full-scan mode. The R-square results show the linearity of the system. The figures were derived from multiple calibration curves spanning 3 orders of magnitude (0.1–100 µg/L) and accumulated over 3 days (Figure 4.16). The calculated limits of detection (LOD) for each target compound lie in the low parts per billion range, making this technique suitable for analyses of VOCs in water or other materials like soil or sediments (Figure 4.17).

Table 4.4 Substance list, LODs are given for the static headspace method in water.

Substance name	% RSD	*R*-square	LOD (µg/L)
1,1-Dichloromethane	0.552	0.9998	0.010
Carbon disulfide	0.945	0.9990	0.013
Dichloromethane	1.150	0.9992	0.018
trans-1,2-Dichloroethane	0.698	0.9997	0.012
MTBE	0.697	0.9993	0.013
cis-1,2-Dichloroethane	1.318	0.9963	0.022
Chloroform	0.578	0.9996	0.013
1,2-Dichloroethane	0.362	0.9989	0.011
1,1,1-Trichloroethane	0.747	0.9998	0.014
Benzene	0.568	0.9994	0.016
Thiophene	0.793	0.9992	0.012
Carbon tetrachloride	0.385	0.9996	0.008
Bromodichloromethane	0.397	0.9995	0.010
Trichloroethane	0.585	0.9994	0.008
cis-1,3-Dichloropropene	0.909	0.9933	0.019
trans-1,3-Dichloropropene	1.017	0.9883	0.026
1,1,2-Trichloroethane	0.772	0.9989	0.015
Toluene	0.409	0.9997	0.016
Dibromochloromethane	0.225	0.9994	0.006
Tetrachloroethane	0.432	0.9997	0.006
Chlorobenzene	0.792	0.9990	0.007
Ethylbenzene	0.365	0.9996	0.011
Bromoform	0.397	0.9987	0.013
m- + *p*-Xylenes	0.219	0.9999	0.011
Styrene	0.833	0.9998	0.013
o-Xylene	0.193	0.9993	0.005
1,1,2,2-Tetrachloroethane	0.663	0.9983	0.013
4-Chlorotoluene	0.178	0.9996	0.007
1,2-Dichlorobenzene	0.605	0.9996	0.006
Hexachloroethane	0.496	0.9996	0.007
1,3,5-Trichlorobenzene	2.032	0.9936	0.021
1,2,4-Trichlorobenzene	2.550	0.9912	0.026
Naphthalene	1.216	0.9950	0.016
1,2,3-Trichlorobenzene	2.825	0.9933	0.027
Hexachlorobutadiene	1.569	0.9933	0.020

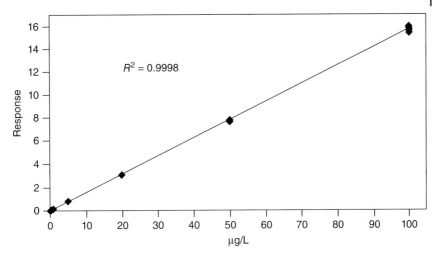

Figure 4.16   Calibration curve obtained for 1.1-dichloroethylene over the 0.1–100 µg/L concentration range. The graph is based on nine calibrations (consecutive calibrations were repeated over 3 days).

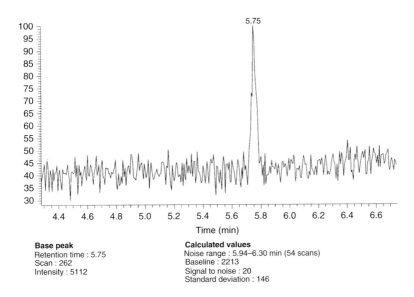

Base peak	Calculated values
Retention time : 5.75	Noise range : 5.94–6.30 min (54 scans)
Scan : 262	Baseline : 2213
Intensity : 5112	Signal to noise : 20
	Standard deviation : 146

Figure 4.17   Signal/noise calculation for 1,1-dichloroethylene at 0.1 µg/L.

The system yields very high stability and sensitivity. The % RSDs and LOD were based on analyses performed at 0.5 µg/L ($n = 72$, over 3 weeks of continuous operation). The LOD computed according to LOD $= 3\sigma$. R-square values were obtained as described in the text.

4.5

MAGIC 60 – Analysis of Volatile Organic Compounds

Keywords: purge and trap; water; soil; volatile halogenated hydrocarbons; BTEX; drinking water; EPA; internal standards

Introduction

The method described here for identification and determination of VOCs in water and soil is taken from the EPA methods 8260, 8240 and 524.2. These ~60 substances are referred to by the term *MAGIC 60*. The method also allows the determination of compounds in solid materials, such as soil samples, and is generally applicable for a broad spectrum of organic compounds, which have sufficiently high volatility and low solubility in water for them to be determined effectively by the P&T procedure. Vinyl chloride can also be determined with certainty by this method (Figures 4.18 and 4.19).

Gastight syringes (5 and 25 mL) with open/shut valves have been shown to be useful for the preparation of water samples. The syringe is carefully filled

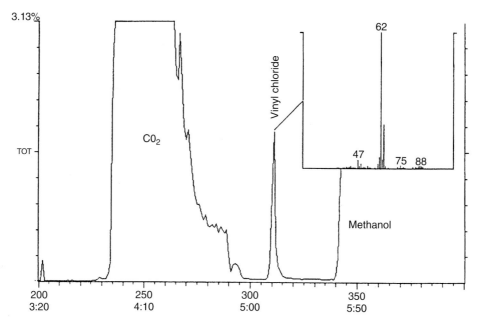

Figure 4.18 Elution of vinyl chloride (50 µg/L) between CO_2 and methanol (DB-624, for analysis parameters see text). The methanol peak can be avoided by using poly(ethylene glycol) as the solvent for the preparation of standard solutions (Figure 4.19).

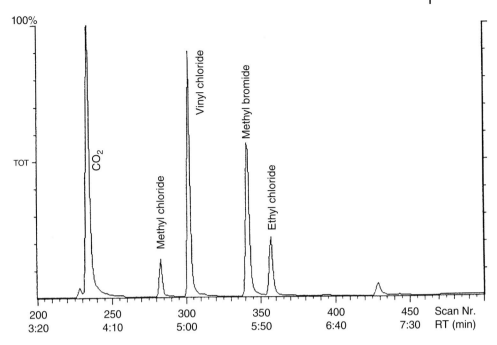

Figure 4.19 Elution of vinyl chloride (50 µg/L) from a DB-VRX column (for analysis parameters see text) with poly(ethylene glycol) as the solvent.

with the sample from the plunger, the plunger inserted and the volume taken up adjusted. The standard solutions are added through the valve with a microliter syringe. The syringe is then connected to the P&T unit via the open/shut valves and the prepared sample transferred to a needle sparger or frit sparger vessel.

The analysis of soil samples is similar to that of water, but, depending on the expected concentration of volatile halogenated hydrocarbons in the sample, a different preparation procedure is chosen. For samples where a concentration of less than 1 mg/kg is expected, 5 g of the sample are placed directly in a 5 mL needle sparger and treated with 5 mL reagent water. Solutions of the internal standards are added with a microliter syringe. At higher concentrations 4 g of the sample are weighed into a 10 mL vessel, which can be closed with a Teflon-coated septum (e.g. headspace vial), and are treated with 9.9 mL methanol and 0.1 mL of the surrogate standard solution. After the solid phase has settled, 5–100 µL of the methanol phase are taken up, depending on the concentration, in a prepared 5 mL (25 mL) syringe together with reagent water. After addition of the internal standard, the sample is injected into the P&T apparatus (needle sparger).

Analysis Conditions

Sample material			Drinking water, soil
Sample concentration	System		Tekmar purge and trap concentrator LSC 2000
	Autosampler		Tekmar ALS 2016 With 25 mL frit sparge glass vessels for water, needle sparge vessels for soil samples
	Trap		Supelco Vocarb 3000
	Sample	Volume	5 mL
		Standby temperature	Room temperature for water and soil
	Purge	Sample preheat time	2.50 min
		Sample temperature	40 °C
		Pre-purge	0 min
		Purge duration	12 min
		Purge flow	40 mL/min
		Dry purge	0 min
	Desorb	Preheat	255 °C for Vocarb 3000
		Desorb temperature	260 °C
		Desorb time	4 min
	Water management	MCM Desorb	0 °C
		Trap bake	20 min at 260 °C
		MCM bake	90 °C
	Instrument	Bake gas bypass (BGB)	On after 120 s
		Mount	110 °C
		Valve	110 °C
		Transfer line	110 °C
		Connection to GC	LSC 2000, linked into the carrier gas supply of the injector
	Cleaning cycle	Sample	Blank DI water
		Purge time	5 min
		Desorption	1 min
		Bake mode	5 min
		Valves, transfer line	200 °C
		All other parameters remain unchanged	
GC Method	System		Finnigan MAGNUM
	Analyzer type		Ion trap with internal ionisation
	Column	Type	DB-VRX, DB-624 or Rtx-624
		Dimensions	60 m length × 0.32 mm ID × 1.8 µm film thickness

(Continued)

	Carrier gas		Helium
	Flow		1 mL/min, constant pressure
	PTV Injector	Injection mode	Split
		Base temperature	200 °C
		Transfer rate	Isothermal
	Split	Open	Until end of run
		Split flow	20 mL/min
		Purge flow	5 mL/min
	Oven program	Start	40 °C, 5 min
		Ramp 1	7 °C/min to 180 °C
		Ramp 2	15 °C/min to 220 °C
		Final temp.	220 °C, 5 min
	Transferline	Temperature	250 °C
MS Method	System		Finnigan MAGNUM
	Analyzer type		Ion trap with internal ionization
	Ionization		EI
	Electron energy		70 eV
	Ion source	Temperature	250 °C
	Mass range		33–260 Da
	Scan rate		1 s
	Resolution	Setting	Normal (0.7 Da)
Calibration	Internal standards		Toluene-d8, fluorobenzene, 4-bromofluorobenzene, 1,2-dichloroethane-d4 Solvent for all standard solutions is methanol

Results

For quantitation, a calibration with eight steps from 0.1 to 40 µg/L is constructed (Figure 4.20). The P&T procedure gives very good linearity over this range. A standard analysis of the compounds listed in Table 4.5 in water is shown in Figure 4.21. The search and identification of individual analytes in the chromatograms (Figure 4.22) are carried out automatically on the basis of reference data, such as the retention time and mass spectrum, which are stored in a calibration file (Figure 4.23).

Table 4.5 MAGIC 60 substance list with details of quantitation masses, CAS numbers, limits of detection (LOD, limit of detection) and limits of quantitation (LOQ), arranged alphabetically (for the method, see text).

Compound	*m/z*	CAS No.	LOD (μg/L)	LOQ (μg/L)
Benzene	77, 78	71-43-2	0.05	0.1
Bromobenzene	77, 156, 158	108-86-1	0.1	0.2
Bromochloromethane	49, 128, 130	74-97-5	0.05	0.1
Bromodichloromethane	83, 85, 127	75-27-4	0.05	0.1
Methyl bromide	94, 96	74-96-4	n.b.	n.b.
Bromoform	173, 175, 252	95-25-2	0.05	0.1
n-Butylbenzene	91, 134	104-51-8	0.05	0.1
sec-Butylbenzene	105, 134	135-98-8	0.05	0.1
t-Butylbenzene	91, 119	98-06-6	0.05	0.1
Chlorobenzene	77, 112, 114	108-90-7	0.1	0.2
Ethyl chloride	64, 66	75-00-3	n.a.	n.a.
Methyl chloride	50, 52	74-87-3	n.a.	n.a.
Chloroform	83, 85	67-66-3	0.05	0.1
2-Chlorotoluene	91, 126	95-49-8	0.05	0.1
4-Chlorotoluene	91, 126	106-43-4	0.05	0.1
Dibromochloromethane	127, 129	124-48-1	0.05	0.1
1,2-Dibromo-3-chloropropane	75, 155, 157	96-12-8	0.25	0.5
Dibromomethane	93, 95, 174	74-95-3	0.1	0.2
1,2-Dibromoethane	107, 109	106-93-4	0.1	0.2
1,2-Dichlorobenzene	111, 146	95-50-1	0.15	0.2
1,3-Dichlorobenzene	111, 146	541-73-1	0.1	0.2
1,4-Dichlorobenzene	111, 146	106-46-7	0.1	0.2
1,1-Dichloroethane	63, 112	75-34-3	0.05	0.1
1,2-Dichloroethane	62, 98	107-06-2	0.1	0.2
1,1-Dichloroethylene	61, 63, 96	75-35-4	0.05	0.1
cis-1,2-Dichloroethylene	61, 96, 98	156-59-4	0.05	0.1
trans-1,2-Dichloroethylene	61, 96, 98	156-60-5	0.05	0.1
Dichlorodifluoromethane	85, 87	75-71-8	0.1	0.2
Dichloromethane	49, 84, 86	75-09-2	0.5	0.5
1,2-Dichloropropane	63, 112	78-87-5	0.1	0.2
1,3-Dichloropropane	76, 78	142-28-9	0.05	0.1
2,2-Dichloropropane	77, 97	590-20-7	0.05	0.1
1,1-Dichloropropylene	75, 110, 112	563-68-6	0.05	0.1
cis-1,3-Dichloropropylene	75, 110, 112	10061-01-5	0.05	0.1
trans-1,3-Dichloropropylene	75, 110, 112	10061-01-5	0.05	0.1
Ethylbenzene	91, 106	100-41-4	0.05	0.1
Hexachlorobutadiene	225, 260	87-68-3	0.15	0.2
Isopropylbenzene	105, 120	48-82-8	0.1	0.2
4-Isopropyltoluene	91, 119, 134	99-87-6	0.05	0.1
Naphthalene	128	91-20-3	0.5	0.5
Styrene	78, 104	100-42-5	0.1	0.2
1,1,1,2-Tetrachloroethane	131, 133	620-30-6	0.1	0.2
1,1,2,2-Tetrachloroethane	83, 85, 131	79-34-5	0.1	0.2

Table 4.5 *(Continued)*

Compound	m/z	CAS No.	LOD (µg/L)	LOQ (µg/L)
Tetrachloroethylene	129, 166, 168	127-18-4	0.1	0.2
Carbon tetrachloride	117, 119	56-23-5	0.05	0.1
Toluene	91, 92	108-88-3	*	*
1,2,4-Trichlorobenzene	180, 182	120-82-1	0.3	0.3
1,2,3-Trichlorobenzene	180, 182	87-61-6	0.4	0.4
1,1,1-Trichloroethane	61, 97, 99	71-55-6	0.05	0.1
1,1,2-Trichloroethane	83, 85, 97	79-020-5	0.1	0.2
Trichloroethylene	95, 130, 132	79-01-6	0.05	0.1
Trichlorofluoromethane	101, 103	75-69-4	0.05	0.1
1,2,3-Trichloropropane	75, 77	96-18-4	0.1	0.2
1,2,4-Trimethylbenzene	105, 120	95-63-6	0.1	0.2
1,3,5-Trimethylbenzene	105, 120	108-67-8	0.1	0.2
Vinyl chloride	62, 64	75-01-4	n.b.	n.b.
m-Xylene	91, 106	108-38-3	0.1	0.2
o-Xylene	91, 106	95-47-6	0.1	0.2
p-Xylene	91, 106	95-47-6	0.1	0.2
Internal standards				
4-Bromofluorobenzene	95, 174	—	—	—
1-Chloro-2-bromopropane	77, 79	—	—	—
1,2-Dichlorobenzene-d_4	115, 150, 152	—	—	—
1,2-Dichloroethane-d_4	65, 102	—	—	—
Fluorobenzene	77, 96	—	—	—
Toluene-d_8	70, 98, 100	—	—	—

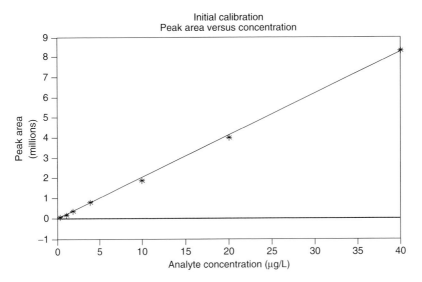

Figure 4.20 Calibration function for *cis*-1,2-dichloroethylene based on the masses *m/z* 61, 96 and 98. Correlation 0.9991 and relative standard deviation 4.96%.

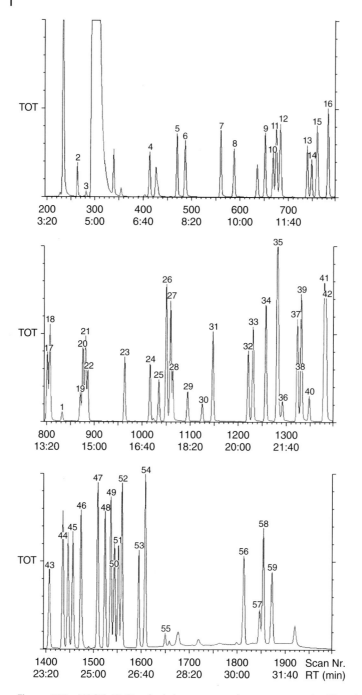

Figure 4.21 MAGIC 60: Standard chromatogram for a water sample. All analytes: 20 µg/L, internal standard: 2 µg/L.

No.	Compound name	Status	Type	RT (min)
1	Benzene	Active	I	13:52
2	Fluorobenzene	Active	A	4:24
3	Dichlorodifluoromethane	Active	A	4:42
4	Chloromethane	Active	A	6:54
5	Trichlorofluoromethane	Active	A	7:50
6	1,1–Dichloroethylene	Active	A	8:07
7	Methylene chloride	Active	A	9:20
8	*trans*–1,2–Dichloroethylene	Active	A	9:48
9	1,1–Dichloroethylene	Active	A	10:52
10	*cis*–1,2–Dichloroethylene	Active	A	11:09
11	Bromochloromethane	Active	A	11:15
12	Chloroform	Active	A	11:24
13	2,2–Dichloropropane	Active	A	12:20
14	1,2–Dichloroethane-d4	Active	A	12:29
15	1,2–Dichloroethane	Active	A	12:40
16	1,1,1–Trichloroethane	Active	A	13:02
17	1,1,–Dichloropropylene	Active	A	13:22
18	Carbon tetrachloride	Active	A	13:27
19	Dibromomethane	Active	A	14:31
20	1,2–Dichloropropane	Active	A	14:36
21	Trichloroethylene	Active	A	14:42
22	Bromodichloromethane	Active	A	14:47
23	*cis*-1,3–Dichloropropylene	Active	A	16:04
24	*trans*-1,3–Dichloropropyle	Active	A	16:56
25	1,1,2–Trichloroethane	Active	A	17:14
26	Toluene-d8	Active	A	17:30
27	Toluene	Active	A	17:38
28	1,3–Dichloropropane	Active	A	17:44
29	Dibromochloromethane	Active	A	18:15
30	1,2–Dibromoethane	Active	A	18:44
31	Tetrachloroethylene	Active	A	19:07
32	1,1,1,2–Tetrachloroethane	Active	A	20:21
33	Chlorobenzene	Active	A	20:30
34	Ethylbenzene	Active	A	20:56
35	*m,p*–Xylene	Active	A	21:21
36	Bromoform	Active	A	21:32
37	Styrene	Active	A	22:02
38	1,1,2,2–Tetrachloroethane	Active	A	22:09
39	*o*–Xylene	Active	A	22:11
40	1,2,3–Trichorpropane	Active	A	22:28
41	Isopropylbenzene	Active	A	22:58
42	4–Bromofluorobenzene	Active	A	23:02
43	Bromobenzene	Active	A	23:27
44	2–Chlorotoluene	Active	A	23:55
45	4–Chlorotoluene	Active	A	24:16
46	1,3,5–Trimethylbenzene	Active	A	24:34
47	*tert*–Butylbenzene	Active	A	25:09
48	1,2,4–Trimethylbenzene	Active	A	25:24
49	*sec* –Butylbenzene	Active	A	25:37
50	1,3–Dichlorobenzene	Active	A	25:44
51	1,4–Dichlorobenzene	Active	A	25:53
52	4–Isopropyltoluene	Active	A	26:00
53	1,2–Dichlorobenzene	Active	A	26:35
54	*n*–Butylbenzene	Active	A	26:48
55	1,2–Dibromo–3–Chloropropa	Active	A	27:29
56	1,2,4–Trichlorobenzene	Active	A	30:13
57	Naphthalene	Active	A	30:46
58	Hexachlorobutadiene	Active	A	30:53
59	1,2,3–Trichlorobenzene	Active	A	31:12

Figure 4.21 (*Continued*)

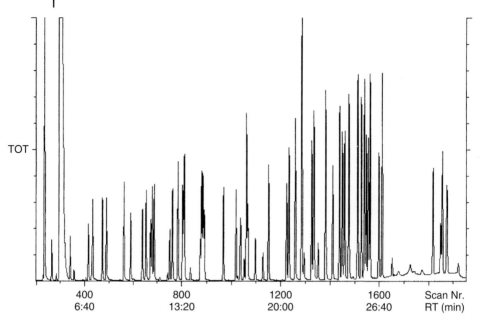

Figure 4.22 MAGIC 60: standard chromatogram of a soil sample. All analytes: 100 µg/kg, internal standard: 10 µg/kg (for analysis parameters see text).

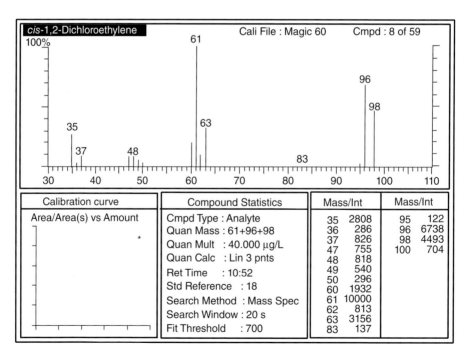

Figure 4.23 Entry for *cis*-1,2-dichloroethylene from the calibration file.

4.6
irm-GC-MS of Volatile Organic Compounds Using Purge and Trap Extraction

Keywords: compound specific isotope analysis, VOC, purge and trap, sources of contamination, degradation pathways

Introduction

The compound specific isotope analysis (CSIA) has been successfully used in the assessment of *in situ* remediation of contaminated environments, identification of pollutant degradation pathways, or the verification of contaminant sources. In these types of studies, the sensitivity of isotope ratio monitoring GC-MS (irm-GC-MS) is often a limiting factor, since concentrations of organic contaminants in groundwater are very often in the low μg/L range. Hence, in order to be able to routinely use irm-GC-MS techniques in environmental studies, efficient extraction procedures are required.

While P&T, providing the lowest MDLs for VOCs, is a routinely used extraction method for the trace level quantification, the on-line coupling with irm-GC-MS has rarely been reported (Schmidt *et al.*, 2004).

Since the P&T procedure includes various phase transition steps that may shift the isotopic signature of the analytes (evaporation, sorption and condensation), the P&T method parameters of purge time, desorption time and injection temperature have been carefully evaluated for the determination of the $\delta^{13}C$-values. The compound specific isotope ratios of 10 different VOCs ranging from the unpolar benzene to the polar MTBE as listed in Table 4.6 were determined (Zwank, 2003; Zwank and Berg, 2013).

Analysis Conditions

Sample material			Groundwater
Sample concentration		System	Tekmar purge and trap concentrator LSC 3100
		Autosampler	Tekmar AQUATek 70
			Aqueous samples were filled into 40 mL vials without headspace
			25 mL automatically transferred to the frit sparge glass vessel
		Trap	Supelco Vocarb 3000
	Sample	Volume	25 mL
		Standby temperature	Room temperature

(Continued)

	Purge	Sample temperature	Room temperature
		Pre-purge	0 min
		Purge duration	30 min
		Purge flow	40 mL/min, N_2
		Dry purge	0 min
	Desorb	Preheat	none
		Desorb temperature	250 °C
		Desorb time	1 min
GC Method	System		Thermo Scientific Trace GC Ultra
	Column	Type	Rtx-VMS
		Dimensions	60 m length × 0.32 mm ID × 1.8 μm film thickness coupled on-line to the pre-column of the GC system
	Pre-column		0.5 m length × 0.53 mm ID, deactivated
	Cryofocusing		−120 °C, liquid N_2
	Carrier gas		Helium
	Flow		180 kPa, constant pressure
	Oven program	Start	40 °C, 2 min
		Ramp 1	2 °C/min to 50 °C, 4 min
		Ramp 2	8 °C/min to 100 °C
		Ramp 3	40–210 °C
		Final temperature	210 °C, 3.5 min
	Interface	Type	Thermo Scientific Combustion interface GC-C III
		Temperature	940 °C
MS Method	System		Thermo Scientific DELTAplus XL
	Analyzer type		Magnetic sector MS
	Ionization		EI
	Scan mode		Simultaneous ion detection
	Mass range		m/z 44, 45, 46

Table 4.6 Compounds in order of GC elution, method-detection limits in water (MDL), accuracy and reproducibility of purge and trap extraction coupled to irm-GC-MS compared to elemental analyser technique (EA).

Compound	MDL (μg/L)	δ^{13}C EA-IRMS VPDB (‰)	δ^{13}C P&T-irm-GC-MS VPDB (‰)
1,1-Dichloroethylene	3.6	−29.25 + 0.14	−29.07 + 0.08
trans-1,2-Dichloroethylene	1.5	−26.42 + 0.17	−25.61 + 0.22
Methyl-*t*-butyl ether	0.63	−28.13 + 0.15	−27.75 + 0.09
cis-1,2-Dichloroethylene	1.1	−26.61 + 0.06	−25.96 + 0.07c
Chloroform	2.3	−45.30 + 0.19	−46.22 + 0.14c
Tetrachloromethane	5.0	−38.62 + 0.01	−38.37 + 0.27
Benzene	0.3	−27.88 + 0.20	−27.27 + 0.20
Trichloroethylene	1.4	−26.59 + 0.08	−26.11 + 0.20
Toluene	0.25	−27.90 + 0.24a	−27.16 + 0.35
Tetrachloroethylene	2.2	−27.32 + 0.14	−26.76 + 0.19

Aqueous solutions of the target analytes were obtained by spiking aliquots of methanolic stock solutions into tap water for method setup and parameter optimization. The isotopic signatures of all the compounds relative to Vienna Pee Dee Belemnite (VPDB) were obtained using CO_2 that was calibrated against referenced CO_2.

Results

P&T allowed MDLs ranging from 0.25 to 5.0 µg/L, depending on the analyte, which corresponds to the highest sensitivity of CSIA for volatile compounds, reported so far. These results were due both to the high sample volume (25 mL) as well as the high extraction efficiency (up to 80%) of the analytes. Thus, P&T-irm-GC-MS allows determining compound-specific stable isotope signatures of contaminant concentrations frequently found in groundwater. As Figure 4.24 and Table 4.6 show, P&T allowed highly reproducible CSIA measurements.

Expressed as the absolute amount of carbon injected on-column, the averaged MDLs for the analytes correspond to 0.4 + 0.1 nmol C on-column. As can be seen from Figure 4.25, the P&T method yields very clean chromatograms with very sharp peaks due to cryofocusing. The detection limit could even be slightly lowered.

Since the validated extraction technique shows little if any carbon isotopic fractionations due to the high extraction efficiency, it is also applicable for CSIA of

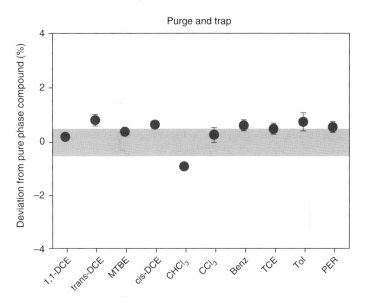

Figure 4.24   Reproducibility and accuracy of P&T-irm-GC-MS. Plotted are the differences from the pure liquid standards measured by EA-IRMS and the horizontal bars correspond to a ^{13}C-measurement within a +0.5‰ interval of the EA-IRMS measurements allowing a direct comparison between the different compounds.

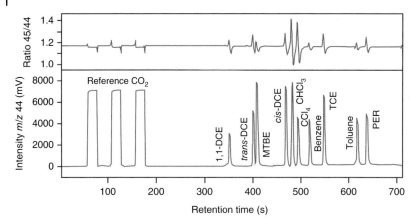

Figure 4.25 Chromatogram of an online P&T-irm-GC-MS analysis. The concentrations of the different analytes were adjusted to achieve similar signal intensities (1.7 μg/L (toluene) −32 μg/L (CCl₄)). The three first peaks correspond to the reference CO_2 gas pulses.

D/H-ratios, which require 10−20 times higher analyte concentrations than $\delta^{13}C$ analysis. P&T pre-concentration methods could also be used to lower analyte concentrations needed for $\delta^{15}N$, and is expected to work also for $\delta^{18}O$ analysis.

4.7
Geosmin and Methylisoborneol in Drinking Water

Keywords: drinking water; off odour contamination; SPME; headspace; automation: quantitation

Introduction

Most complaints about the quality of drinking water found relate to odour and taste. Geosmin (1,2,7,7-tetramethyl-2-norborneol) and 2-MIB (2-methylisoborneol) are compounds mainly produced by blue-green algae (cyanobacteria) and actinomycete bacteria that cause musty, earthy odours in water supply reservoirs. Although these compounds have not been shown to be a health concern in public water supplies, the caused odours require removal and thus concentrations of geosmin and 2-MIB are monitored routinely in areas where they occur. The odour threshold for these compounds is very low and humans can typically detect them in drinking water at 30 and 10 ng/L (ppt) for geosmin and 2-MIB, respectively.

Past analytical techniques to determine concentrations of these off odour compounds used closed-loop stripping (CLS) and P&T sample preparation that provided the required sensitivity but have been either time consuming (CLS) or required a significant technical effort (P&T) (Preti *et al.*, 1993; McMillan, 1994). SPME has shown in many applications to be a very versatile and sensitive

Table 4.7 Geosmin and MIB compound characteristics.

Compound name	Chemical structure	Molecular weight (Da)	USGS parameter code	CAS registry number
Geosmin $C_{12}H_{22}O$		182.3	62719T	23333-91-7
2-Methyisoborneol (MIB) $C_{11}H_{20}O$		168.3	62749T	2371-42-8
2-Isopropyl-3-methoxypyrazine (IPMP, surrogate compound) $C_3H_{12}N_2O$		152.2	—	25773-40-4

sampling strategy with the potential of high sensitivity and full automation (Chang *et al.*, 2007). This application describes the automated SPME technique from the sample headspace for the sensitive analysis of these compounds with high sample throughput (Table 4.7).

Analysis Conditions

Sample material		Drinking water
Sample concentration SPME	Fibre type	Supelco DVB/CAR/PDMS 50/30 cm
	Incubation	60 °C
	Agitator on time	10 s
	Agitator off time	10 s
	Incubation time	30 min
	Injection depth	20 mm
	Pre- or post-injection	No delay
	Needle speed in vial	20 mm/s
	SPME extraction time	30 min
	SPME desorption time	4 min
	Fibre conditioning time (station)	10 min
	Fibre conditioning temperature	250 °C

(Continued)

GC Method	System		Thermo Scientific Trace GC Ultra
	Autosampler		Thermo Scientific TriPlus RSH
	Column	Type	Rxi-5MS
		Dimensions	30 m length × 0.25 mm ID × 0.25 µm film thickness
	Carrier gas		Helium
	Flow		constant flow, 1.5 mL/min
	PTV Injector	Injection mode	Splitless
		Base temperature	60 °C
		Transfer rate	14.5 °C/s
		Transfer temperature	250 °C
		Transfer time	4 min
	Split	closed until	1 min
		open	1 min to end of run
		Split flow	15 mL/min
		Purge flow	5 mL/min
	Oven program	Start	60 °C, 4.0 min
		Ramp 1	20 °C/min to 250 °C
		Final temperature	250 °C, 1.5 min
	Transferline	Temperature	250 °C
MS Method	System		Thermo Scientific DSQ II
	Analyzer type		Single quadrupole MS
	Ionization		EI
	Electron energy		70 eV
	Ion source	Temperature	230 °C
			Closed EI ion volume
	Acquisition	Mode	SIM
	Mass range		Segment 1 (7 min): m/z 137, 152, 100 ms dwell time
			Segment 2 (8 min): m/z 95, 107, 100 ms dwell time
			Segment 3 (10 min): m/z 112, 149, 100 ms dwell time
Calibration	Internal standard		Isopropylmethoxypyrazine
		Range	1–1000 ppt
	Data points		7

DVB, Divinylbenzene; CAR, Carboxen and PDMS, polydimethylsiloxane.

SPME Method

All standards were dissolved in methanol as solvent. The internal standard isopropylmethoxypyrazine was spiked to each sample at 300 ppt level. The addition of 30% (w/v) salt to the sample was used for salting out the analytes. The reproducibility of the SPME headspace analysis depended on the same methanol content in each of the samples. The incubation temperatures and time were varied to maximize analyte concentrations. 60 °C for 30 min was found optimal. Each calibration level (1, 10, 50, 100, 200, 500 and 1000 ppt) was prepared in 100 mL volumetric flask containing 300 ppt of internal standard. A 10 mL aliquot was transferred to a headspace vial with 3 g of salt. Each calibration standard was run in duplicate.

Results

SPME as a solventless sample preparation technique includes extraction, concentration and sample GC injection of the analytes in a single procedure. Automation of this sample preparation process offers a valuable tool to greatly extend sample throughput. SPME has gained widespread acceptance as the technique of preference for the analysis of these odour compounds.

Excellent linearity was obtained for both Geosmin and 2-MIB with excellent linearity of the quantitative calibration over a range of 1 – 1000 ppt (see Figures 4.26 and 4.27). Seven replicates were analyzed for method-detection limits according to 40 CFR Part 136.

Average RSDs of the replicate injections were 0.6% for geosmin and 1.6% for 2-MIB.

These detection limits were run as 1 : 10 split injections, thus lower detection limits are possible with splitless injections. Excellent sensitivity and

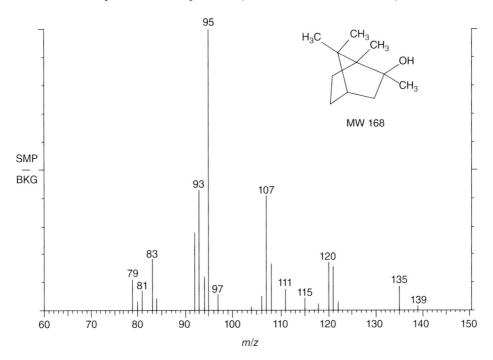

Figure 4.26 Structure and mass spectrum of 2-MIB.

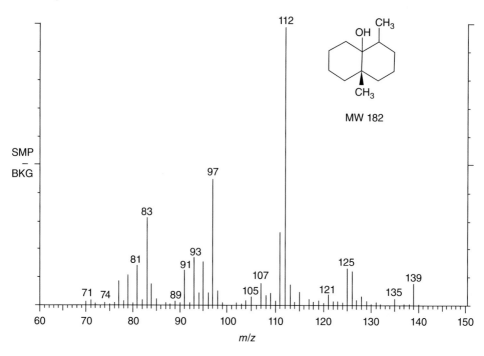

Figure 4.27 Structure and mass spectrum of Geosmin.

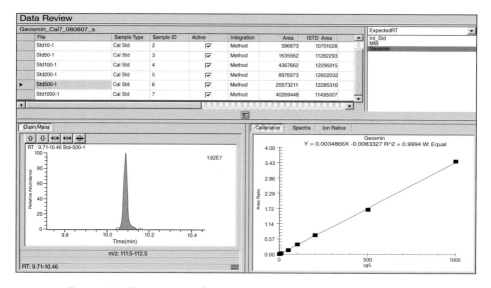

Figure 4.28 Chromatogram of a geosmin purge and trap analysis at 2.5 ppt (50 pg/20 mL).

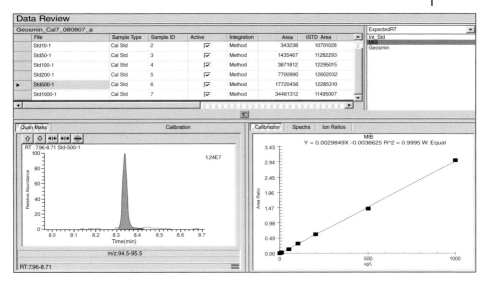

Figure 4.29 Chromatogram of a MIB purge and trap analysis at 2.5 ppt (50 pg/20 mL).

chromatographic performance was achieved across the calibration range for both 2-MIB and geosmin. MDLs of less than 0.3 ppb – less than the concentrations typically detected by humans – can be achieved by the described method (Figures 4.28 and 4.29).

4.8
Polycyclic Musks in Waste Water

Keywords: musk perfumes; surface water; waste water; solid phase extraction

Introduction

Synthetic musk perfumes are added to washing and cleaning agents as well as fabric conditioners and many bodycare products to improve their odours. Because of the ecotoxicological problems caused by their poor degradability and lipophilic properties, there are reservations about their widespread use. In the examination of water, sewage sludge, sediments and fish for nitro-musk compounds (e.g. musk xylene, 1-*t*-butyl-3,5-dimethyl-2,4,6-trinitrobenzene), three further compounds were found in the GC-MS chromatograms. These were the polycyclic nitro-musk compounds HHCB (1,3,4,6,7,8-hexahydro-4,6,6,7,8,8-hexamethylcyclo-penta-(*g*)-2-benzopyran), AHTN (7-acetyl-1,1,3,4,4,6-hexa-methyltetralin) and ADBI (4-acetyl-1,1-dimethyl-6-*tert*-butylindane) (Table 4.8). These substances are widely used in the cosmetics and perfumes industry and have already been detected in river water and various species of fish (Eschke *et al.*, 1994 a/b; Rimkus and Wolf, 1994).

Table 4.8 Polycyclic musk perfumes.

Compound name	CAS-No.	Empirical formula	M	Structure
HHCB, 1,3,4,6,7,8-Hexahydro-4,6,6,7,8,8-hexamethylcyclopenta-(*g*)-2-benzopyran	1222-05-5	$C_{18}H_{26}O$	258	
AHTN, 7-Acetyl-1,1,3,4,4,6-hexamethyltetralin	1506-02-1	$C_{18}H_{26}O$	258	
ADBI, 4-Acetyl-1,1-dimethyl-6-tert.butylindane	13171-00-1	$C_{17}H_{24}O$	244	

Analysis Conditions

Sample material			Waste and surface water
Sample preparation	SPE clean-up		2 L water are enriched using SPE with 2 g C_{18}-cartridges, elute with 10 mL acetone, evaporate eluate to dryness with nitrogen, take up the residue in n-hexane
GC Method	System		Varian 3400
	Injector		PTV cold injection system, splitless mode
	Column	Type	DB-5MS
		Dimensions	30 m length × 0.32 mm ID × 0.25 μm film thickness
			HT8
			25 m length × 0.22 mm ID × 0.25 μm film thickness
	Carrier gas		Helium

(Continued)

	Oven program	Start	60 °C
		Ramp 1	10 °C/min to 300 °C
		Final temperature	300 °C, 6 min
MS Method	System		Finnigan ITS 40
	Analyzer type		Ion trap MS with internal ionization
	Ionization		EI
	Acquisition	Mode	Full scan
	Mass range		40–400 Da
	Scan rate		1 s (1 spectrum/s)

SPE; solid phase extraction.

Results

The EI mass spectra of the compounds taken in the full-scan mode are suitable for unambiguous identification because of their typical fragmentation patterns. Intense fragment signals are seen at m/z 43, 258, 243 and 229. The mass spectrum of HHCB also contains the fragment m/z 213, unlike AHTN (Figures 4.30–4.32). HHCB and AHTN are only partially separated on a 5%-phenyl column. All three substances are separated well on the slightly polar phase of the HT8 column. Figure 4.33 shows the chromatogram of a waste water sample with the mass chromatograms of the selective ions under the TIC. In random samples, significant measurable concentrations in the ppb range were found both in

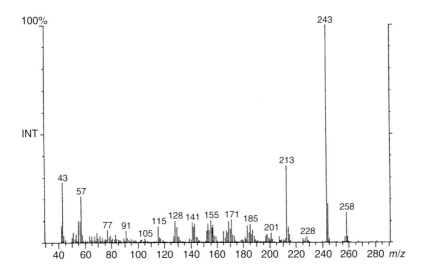

Figure 4.30 Mass spectrum of HHCB, M 258.

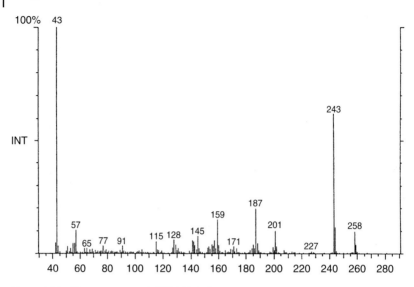

Figure 4.31 Mass spectrum of AHTN, M 258.

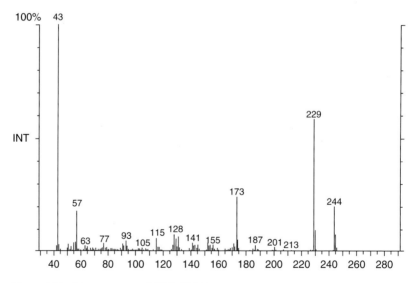

Figure 4.32 Mass spectrum of ADBI, M 244.

surface water and in the inflow and outflow of municipal sewage works. Very high concentrations were found in the muscle tissue of fish from these types of water (Figure 4.34), indicating the low degradability and bioaccumulation of these compounds (Eschke, 1994a,b).

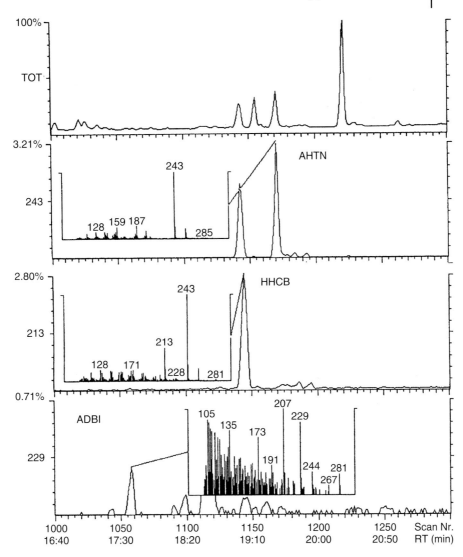

Figure 4.33 Chromatogram of a waste water sample with mass chromatograms of HHCB (*m/z* 213/243), AHTN (*m/z* 243) and ADBI (*m/z* 229).

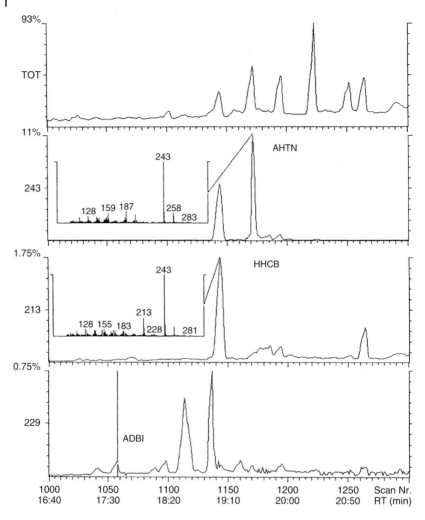

Figure 4.34 Chromatogram of a fish sample (muscle tissue, eel) with the mass chromatograms of HHCB (*m/z* 213/243), AHTN (*m/z* 243) and ADBI (*m/z* 229).

4.9

Organotin Compounds in Water

Keywords: food safety; European Union Water Framework Directive; triple quadrupole GC-MS/MS.

Introduction

The monitoring of organotin compounds is regulated by the European Union Water Framework Directive (European Commission, 2008). This includes the mono-, di-, tri-, tetrabutyl and triphenyl tin compounds. Tributyltin compounds are considered to be most hazardous. Studies have reported adverse health effects causing kidney and central nervous disorders in humans (Ostrakhovitch and Cherian, 2007; OSPAR Commission, 2011).

Organotins are used for a variety of applications. Tributyl and triphenyltin are well known as *antifouling agents* on underwater structures, triphenyltin as well as a fungicide in crop protection. Mono- and dibutyltin have uses as stabilizers in plastics and catalysts in soft foam production (European Commission, 2008; OSPAR Commission, 2011).

Organotin compounds are lipophilic and get accumulated in adipose tissue. The toxicity of these compounds at low concentrations drives the requirement for accurate and sensitive analytical methods for their detection, quantitation and research for less-toxic replacements (Takeuchi, Mizuishi and Hobo, 2000; Morabito and Quevauviller, 2002). The low concentration with ppb to ppt levels in water samples creates a particular analytical challenge for the sample preparation and a very selective and sensitive detection in matrix samples (Butler and Phillips, 2009).

Sample Preparation

The extraction of organotin compounds from water samples involves the analyte ethylation with tetraethyl borate as an in-situ derivatization step, followed by liquid extraction with pentane and subsequent extract concentration.

Experimental Conditions

Special attention is required for an inert GC inlet system to achieve a tailing free peak shape and high compound response. For this purpose the PTV injector as a temperature programmable injection system was chosen, with a special liner deactivation using the SurfaSil treatment for high inertness (Thermo Scientific, 2014).

The deteriorating impact of the sample matrix on the chromatographic performance was reduced by the use of a guard column (pre-column) connected in front of the analytical column. The pre-column allows a quick preventive maintenance without replacing the analytical column, while keeping the compound retention times constant.

The GC-MS/MS detection method described uses a timed-SRM (selected reaction monitoring) setup with three transitions for each analyte for a confirming ion ratio confirmation. The t-SRM window is typically set to 30–60 s width for all compounds. The compliance with the calibrated compound ion ratios of standards is checked by the processing software for the peaks identified in the samples (Figure 4.35 and Table 4.9).

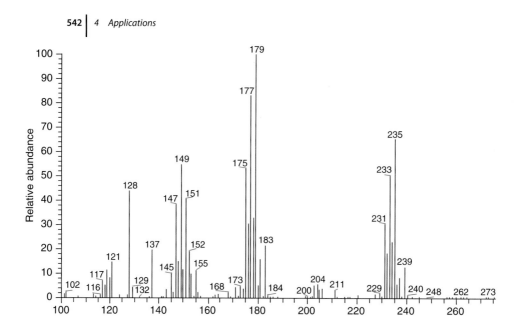

Figure 4.35 Full scan spectrum of monobutyltin (MBT) with the clearly visible Sn isotope pattern.

Sample Measurements

A single-point calibration was used at a level of 10 ng/L. The concentration is calculated on the actual levels in the water sample. The result shown in Figure 4.36 demonstrates the method capability to reach 0.05 ng/L of organotins in the water sample and below. The actual amount injected on column was 0.2 pg for each organotin compound. Calculations were performed using tripropyltin as internal standard.

Analysis Conditions

Sample material		Sea water, surface water
Sample preparation	Sample	400 mL water Adjust pH to 5 using a 1 M acetic acid/sodium acetate buffer
	Derivatization	Ethylation, add a 2% w/v sodium tetraethyl borate solution in 0.1 M NaOH
	Liquid extraction	Shake with pentane, 10 min Concentrate organic phase to 400 μL

Table 4.9 MRM acquisition method for organotin compounds.

Compound name	Precursor ion (*m/z*)	Product ion (*m/z*)	Collision energy (V)	Retention time (min)
Monobutyltin	235.08	150.98	6	7.94
Monobutyltin	233.08	176.95	6	7.94
Monobutyltin	235.08	178.95	6	7.94
Tripropyltin, ISTD	249.08	164.91	8	8.89
Tripropyltin, ISTD	245.08	160.91	8	8.89
Tripropyltin, ISTD	247.08	162.91	8	8.89
Tetrapropyltin, ISTD	249.08	164.91	8	8.89
Tetrapropyltin, ISTD	245.08	160.91	8	8.89
Tetrapropyltin, ISTD	247.08	162.91	8	8.89
Dibutyltin	261.03	205.03	8	9.42
Dibutyltin	263.03	150.98	8	9.42
Dibutyltin	263.03	207.03	8	9.42
Monoheptyltin, ISTD	178.95	150.98	8	10.55
Monoheptyltin, ISTD	275.07	176.95	8	10.55
Monoheptyltin, ISTD	277.07	178.95	8	10.55
Tributyltin	287.09	174.94	8	11.21
Tributyltin	291.08	235.08	8	11.21
Tributyltin	289.09	176.95	8	11.21
Monooctyltin	287.09	174.94	8	11.21
Monooctyltin	289.09	176.95	8	11.21
Monooctyltin	291.08	178.95	8	11.21
Tetrabutyltin	287.09	174.94	8	11.21
Tetrabutyltin	289.09	176.95	8	11.21
Tetrabutyltin	291.08	235.08	8	11.74
Diheptyltin, ISTD	275.07	176.95	8	13.71
Diheptyltin, ISTD	277.07	178.95	8	13.71
Diheptyltin, ISTD	245.08	145.98	8	13.71
Dioctyltin	261.03	148.98	8	14.94
Dioctyltin	263.03	150.98	8	14.94
Dioctyltin	375.17	263.03	8	14.94
Triphenyltin	196.94	119.90	18	16.21
Triphenyltin	349.15	194.98	22	16.21
Triphenyltin	351.15	196.94	22	16.21
Tricyclohexyltin	233.08	150.98	8	16.28
Tricyclohexyltin	287.09	205.03	8	16.28
Tricyclohexyltin	315.04	150.98	8	16.28

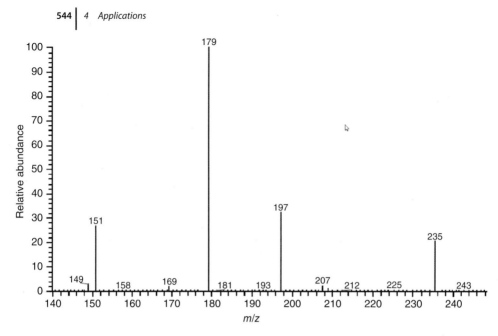

Figure 4.36 Product ion spectrum of monobutyltin (MBT) precursor ion *m/z* 235.

(*Continued*)

GC Method	System		Thermo Scientific TRACE GC Ultra
	Injector		PTV, temperature programmable injector
	Column	Type	TG-5MS, 5% phenyl silicone
		Dimensions	30 m length × 0.25 mm ID × 0.25 µm film thickness
	Guard column		DMTPS deactivated, 2 m × 0.53 mm ID pre-column
	Carrier gas		Helium
	Flow		constant flow, 1.4 mL/min
	PTV Injector	Injection mode	Splitless
		Inlet liner	Straight glass liner, SurfaSil treated
		Injection volume	3 µL
		Base temperature	50 °C, 0.1 min
		Transfer rate	8 °C/s
		Transfer temperature	280 °C, 1 min
		Cleaning step	350 °C, 11 min
	Split	Closed until	2 min
		Open	2 min – end of run
		Split flow	50 mL/min
	Oven program	Start	45 °C, 2 min
		Ramp 1	55 °C/min to 175 °C
		Ramp 2	35 °C/min to 300 °C
		Final temperature	300 °C, 2 min
	Transferline	Temperature	300 °C

(Continued)

MS Method	System		Thermo Scientific TSQ Quantum XLS
	Analyzer type		Triple quadrupole MS
	Ionization		EI
	Electron energy		70 eV
	Ion source	Temperature	250 °C
	Acquisition	Mode	Timed-SRM
			2 SRMs per analyte
	Collision gas		Ar, 1.0 mTorr
	Scan rate		0.2 s (5 spectra/s)
	Resolution	Setting	0.7 Da Q1 and Q3
Calibration	Internal standards		Tripropyltin, monoheptyltin, diheptyltin
	Calibration		Single-point, 10 ng/L spiked to a water sample

Results

The described method demonstrates the ability of GC-MS/MS analysis to comply and even exceed the requirements of the European Water Framework Directive for the detection and quantitation of organotin compounds. The used t-SRM functionality gives the instrument the ability to automatically determine optimum SRM dwell times for high sensitivity detection, even allowing for partially overlapping SRM transitions.

This method demonstrates detection and quantitation levels to 0.05 ng/L in Figure 4.37, exceeding the EU Directive's Annual Allowable Average of 0.2 ng/L.

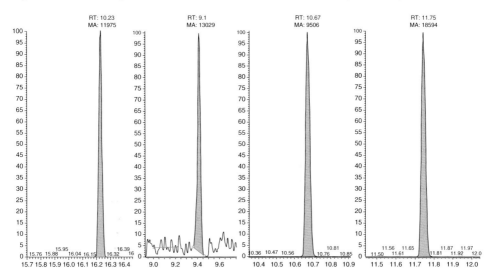

Figure 4.37 Mono-, di-, tri- and tetra-butyltin signals from a spiked water sample at 0.05 ng/L.

4.10
Multi-Method for Pesticides by Single Quadrupole MS

Keywords: multi-residue method; pesticides; fruits; vegetables; SPE cleanup; single quadrupole MS.

Introduction

The effective monitoring of pesticide residues and its degradation products require fast and comprehensive multi-methods. The referenced method fulfils the requirement for monitoring a wide scope of pesticide compounds and degradation products in fruit and vegetable samples.

The described internal standard method is suitable for the quantification of more than 200 components, using a single-quadrupole GC-MS. The extraction process with acetonitrile is followed by a clean-up including a salting-out step and solid phase extraction. The compound detection is done by SIM using the retention time and selected ion abundance ratios as qualifiers for identification. Two runs are required to determine the complete set of pesticides given in Table 4.10 according to the required SIM acquisition windows described as FV-1 and FV-2.

The quantitation using Aldrin as internal standard is based on blank matrix spikes to compensate for matrix effects.

In addition, the published method also includes the parallel determination of carbamates by HPLC with fluorescence detection (Health and Welfare Canada, 2000; Fillion, Sauvé and Selwyn, 2000).

Analysis Conditions

Sample material		Fruits, vegetables
Sample preparation	Liquid/liquid extraction	Homogenize 50 g sample with 100 mL ACN for 5 min, add 10 g $NaSO_4$, homogenize 5 min
	SPE Clean-up	Condition a C18 SPE tube with 2 mL ACN extract, transfer 15 mL ACN extract onto cartridge and elute 13 mL, add $NaSO_4$ to 15 mL, shake and centrifuge evaporate 10 mL aliquot (equivalent of 5 g sample) to 0.5 mL transfer extract to carbon SPE tube, elute with 20 mL ACN/toluene (3 + 1), evaporate, exchange solvent by adding 2 × 10 mL acetone, add 50 μL ISTD (Aldrin 1.0 ng/μL), add acetone to 2.5 mL, transfer 0.5 mL for analysis

Table 4.10 List of compounds included in the multi-method with retention times, target and qualifier ions (Q) and ratios (Q/tgt).

Compound	Method	RT (min)	Ions monitored (*m/z*)				Abundance ratio of qualifier ion/target		
			Target	Q1	Q2	Q3	Q/tgt	Q2/tgt	Q3/tgt
Dichlorvos-naled	FV-1	7.24	185	109	220	—	4.61	0.19	—
EPTC	FV-1	7.89	189	132	—	—	1.40	—	—
Bendiocarb degr.	FV-2	7.99	166	151	126	—	1.86	1.82	—
Allidochlor	FV-2	8.28	138	132	173	—	1.12	0.14	—
Methamidophos	FV-2	8.88	141	94	95	—	2.31	1.60	—
Butylate	FV-1	9.06	217	174	—	—	5.20	—	—
Promecarb degr.	FV-1	9.09	135	150	—	—	0.35	—	—
Chlorthiamid degr.	FV-2	9.36	171	173	136	—	0.64	0.18	—
Dichlobenil	FV-1	9.37	171	173	—	—	0.62	—	—
Dichlormid	FV-1	9.51	168	166	—	—	1.53	—	—
Vernolate	FV-2	9.59	86	128	—	—	1.14	—	—
Pebulate	FV-2	9.97	128	161	203	—	0.12	0.06	—
Aminocarb degr.	FV-1	10.06	150	151	136	—	1.27	0.79	—
Etridiazole	FV-2	10.12	211	183	—	—	0.97	—	—
Chlormephos	FV-1	10.46	234	121	—	—	3.19	—	—
Nitrapyrin	FV-2	10.85	194	196	198	—	0.97	0.31	—
Mexacarbate degr.	FV-2	11.18	165	150	134	—	0.81	0.34	—
Bufencarb degr.	FV-1	11.24	121	122	107	—	0.37	0.22	—
cis-Mevinphos	FV-1	11.82	127	192	164	109	0.20	0.06	0.23
Propham	FV-2	11.85	179	137	—	—	1.28	—	—
trans-Mevinphos	FV-1	12.17	127	192	164	109	0.22	0.06	0.23
Chloroneb	FV-2	12.43	191	193	206	—	0.64	0.53	—
o-Phenylphenol	FV-2	13.57	170	169	141	115	0.66	0.29	0.21
Tecnazene	FV-1	14.96	261	215	—	—	1.54	—	—
Cycloate	FV-2	15.09	154	83	215	—	2.26	0.07	—
Captafol degr.	FV-2	15.56	151	79	80	—	1.31	0.73	—
Captan degr.	FV-1	15.60	151	79	80	—	1.26	0.71	—
Heptenophos	FV-2	15.89	124	215	250	—	0.09	0.07	—
Acephate	FV-1	15.89	136	94	—	—	0.54	—	—
Demeton-*S*	FV-1	16.04	171	88	143	—	6.27	0.55	—
Hexachlorobenzene	FV-1	16.51	284	286	282	—	0.77	0.56	—
Ethoprophos	FV-2	16.98	158	242	139	—	0.08	0.47	—
Diphenylamine	FV-1	17.09	169	167	168	—	0.34	0.63	—
Di-allate 1	FV-1	17.46	234	236		—	0.36	—	—
Chlordimeform	FV-2	17.67	196	181		—	0.75	—	—
Propachlor	FV-1	17.68	120	176	211	—	0.23	0.05	—
Demeton-*S*-methyl	FV-2	18.08	88	109	142	—	0.26	0.17	—
Di-allate 2	FV-1	18.48	234	236	—	—	0.36	—	—
Ethalfluralin	FV-1	18.61	276	316	—	—	0.50	—	—
Phorate	FV-1	18.86	260	231	—	—	0.73	—	—
Trifluralin	FV-1	19.27	306	264	—	—	1.18	—	—
Sulfallate	FV-2	19.31	188	116	148	—	0.09	0.06	—
Chlorpropham	FV-2	19.39	213	127	—	—	3.21	—	—

(continued overleaf)

Table 4.10 (*Continued*)

Compound	Method	RT (min)	Ions monitored (*m/z*)				Abundance ratio of qualifier ion/target		
			Target	Q1	Q2	Q3	Q/tgt	Q2/tgt	Q3/tgt
Benfluralin	FV-1	19.45	292	264	—	—	0.23	—	—
Sulfotep	FV-1	19.54	322	202	—	—	1.15	—	—
α-BHC	FV-1	19.83	219	183	—	—	1.16	—	—
Bendiocarb	FV-2	20.74	151	166	223	—	0.41	0.08	—
Quintozene	FV-1	20.82	295	237	—	—	1.79	—	—
Promecarb degr.	FV-1	20.93	135	150	—	—	0.62	—	—
Omethoate	FV-1	21.15	156	110	—	—	1.10	—	—
Terbufos	FV-2	21.71	231	153	—	—	0.38	—	—
Demeton-*O*	FV-1	21.74	88	114	170	—	0.15	0.12	—
Desethylatrazine	FV-1	21.86	172	174	—	—	0.31	—	—
Clomazone	FV-2	21.92	204	125	—	—	2.14	—	—
Prometon	FV-2	22.01	225	210	168	—	1.44	1.17	—
Tri-allate	FV-1	22.32	268	270	—	—	0.67	—	—
Fonofos	FV-1	22.40	246	109	—	—	4.50	—	—
Diazinon	FV-1	22.67	304	179	—	—	3.58	—	—
Terbumeton	FV-2	22.77	210	169	225	—	0.94	0.25	—
Dicrotophos	FV-1	22.77	127	193	—	—	0.07	—	—
Lindane	FV-2	22.98	219	183	—	—	1.28	—	—
Dioxathion	FV-1	23.18	125	153	270	—	0.29	0.35	—
Disulfoton	FV-2	23.6	142	274	—	—	0.29	—	—
Profluralin	FV-1	23.68	318	330	—	—	0.3	—	—
Dicloran	FV-2	23.78	206	176	—	—	1.37	—	—
Propazine	FV-1	24.04	229	214	—	—	1.85	—	—
Atrazine	FV-2	24.12	215	200	—	—	1.88	—	—
Etrimfos	FV-2	24.14	292	277	—	—	0.44	—	—
Simazine	FV-1	24.21	201	186	—	—	0.70	—	—
Heptachlor	FV-1	24.5	272	274	—	—	0.79	—	—
Chlorbufam	FV-1	24.59	223	164	—	—	1.00	—	—
Schradan	FV-2	24.59	199	135	—	—	1.74	—	—
Aminocarb	FV-1	24.65	151	208	150	—	0.12	0.73	—
Terbuthylazine	FV-1	24.78	214	173	—	—	0.55	—	—
Monolinuron	FV-2	24.8	214	126	—	—	1.43	—	—
Secbumeton	FV-2	25.33	196	210	225	—	0.2	0.14	—
Dichlofenthion	FV-1	25.82	279	223	—	—	1.36	—	—
Cyanophos	FV-1	26.09	243	125	—	—	1.34	—	—
Isazofos	FV-2	26.1	161	257	313	—	0.28	0.04	—
Mexacarbate	FV-2	26.28	165	222	150	—	0.21	0.86	—
Propyzamide	FV-1	26.45	173	175	—	—	0.63	—	—
Aldrin ISTD	FV-1	26.48	263	265	—	—	0.66	—	—
Pirimicarb	FV-2	26.5	166	238	—	—	0.16	—	—
Dimethoate	FV-1	26.6	87	229	143	—	0.04	0.09	—
Monocrotophos	FV-2	26.68	127	192	109	—	0.09	0.11	—
Chlorpyrifos-methyl	FV-1	26.93	286	288	—	—	0.65	—	—

Table 4.10   (*Continued*)

| Compound | Method | RT (min) | Ions monitored (*m/z*) | | | | Abundance ratio of qualifier ion/target | | |
			Target	Q1	Q2	Q3	Q/tgt	Q2/tgt	Q3/tgt
Fluchloralin	FV-2	27.56	306	326	264	—	0.82	0.83	—
Fenchlorphos	FV-1	27.75	285	287	—	—	0.68	—	—
Desmetryn	FV-1	27.91	198	213	—	—	1.49	—	—
Dinitramine	FV-2	28.00	305	307	261	—	0.37	0.41	—
Dimetachlor	FV-2	28.23	210	134	197	—	8.78	3.13	—
Chlorothalonil	FV-1	28.75	266	264	—	—	0.78	—	—
Alachlor	FV-1	28.80	188	160	—	—	1.27	—	—
Prometryn	FV-1	29.22	241	226	—	—	0.68	—	—
Metobromuron	FV-2	29.23	258	61	—	—	16.26	—	—
Cyprazine	FV-2	29.33	212	227	229	—	0.89	0.12	—
Ametryn	FV-2	29.34	227	212	—	—	1.12	—	—
Simetryn	FV-1	29.45	170	155	—	—	0.93	—	—
Pirimiphos-methyl	FV-1	29.53	290	305	—	—	0.52	—	—
Vinclozolin	FV-1	29.55	285	287	—	—	0.62	—	—
Thiobencarb	FV-2	29.79	100	257	125	—	0.09	0.25	—
Metribuzin	FV-2	29.82	198	199	—	—	0.29	—	—
β-BHC	FV-1	29.90	219	183	—	—	1.17	—	—
Terbutryn	FV-1	30.21	226	241	—	—	0.51	—	—
Metalaxyl	FV-2	30.37	206	249	—	—	0.46	—	—
Parathionmethyl	FV-2	30.42	263	125	—	—	2.15	—	—
Chlorpyrifos	FV-1	30.61	314	199	—	—	3.18	—	—
Aspon	FV-1	30.97	211	253	—	—	0.27	—	—
Dicofol	FV-1	31.14	250	139	—	—	5.88	—	—
Oxychlordane	FV-2	31.21	115	185	149	—	0.62	0.51	—
Malaoxon	FV-1	31.25	268	195	—	—	1.85	—	—
Chlorthal-dimethyl	FV-2	31.26	301	299	—	—	0.81	—	—
Phosphamidon	FV-1	31.31	264	193	—	—	0.28	—	—
δ-HCH	FV-2	31.35	183	219	217	—	0.86	0.68	—
Metolachlor	FV-1	31.63	238	162	—	—	2.14	—	—
Terbacil	FV-2	31.91	160	161	216	—	1.33	0.02	—
Fenthion	FV-1	32.01	278	169	—	—	0.33	—	—
Bromophos	FV-1	32.15	331	329	—	—	0.77	—	—
Dichlofluanid	FV-1	32.39	226	123	—	—	5.26	—	—
Fenitrothion	FV-1	32.46	277	260	—	—	0.59	—	—
Pirimiphos-ethyl	FV-2	32.67	333	304	—	—	1.26	—	—
Malathion	FV-1	32.68	158	125	—	—	2.23	—	—
Paraoxon	FV-1	32.71	275	109	—	—	11.06	—	—
Heptachlor epoxide	FV-2	32.76	237	183	217	—	2.02	1.21	—
Nitrothal-isopropyl	FV-2	33.21	236	212	—	—	0.75	—	—
Butralin	FV-2	33.42	266	250	—	—	0.14	—	—
Ethofumesate	FV-1	33.48	161	286	207	—	0.25	1.04	—
Triadimefon	FV-1	33.99	208	210	—	—	0.30	—	—
Parathion	FV-2	34.05	291	139	—	—	1.20	—	—
Isopropalin	FV-1	34.15	280	238	—	—	0.60	—	—

(*continued overleaf*)

Table 4.10 (*Continued*)

Compound	Method	RT (min)	Target	Q1	Q2	Q3	Q/tgt	Q2/tgt	Q3/tgt
			— Ions monitored (m/z)				Abundance ratio of qualifier ion/target		
Pendimethalin	FV-2	34.34	252	281	—	—	0.08	—	—
Fenson	FV-2	34.36	268	141	—	—	4.41	—	—
Linuron	FV-2	34.40	248	160	250	—	1.44	0.62	—
α-Endosulfan	FV-1	34.42	277	339	243	—	0.57	1.37	—
Chlorthiamid	FV-2	34.45	205	170	—	—	2.36	—	—
Chlorbenside	FV-1	34.54	125	268	—	—	0.07	—	—
Allethrin	FV-2	34.82	123	136	—	—	0.28	—	—
Chlorfenvinphos	FV-1	34.97	323	267	—	—	2.51	—	—
trans-Chlordane	FV-1	35.22	373	375	—	—	0.93	—	—
Bromophos-ethyl	FV-1	35.35	359	303	—	—	1.31	—	—
Chlorthion	FV-2	35.56	297	125	—	—	4.38	—	—
Quinalphos	FV-1	35.59	146	298	—	—	0.09	—	—
Propanil	FV-2	35.69	161	163	—	—	0.66	—	—
Diphenamid	FV-2	35.71	167	239	—	—	0.16	—	—
Crufomate	FV-2	35.79	256	182	—	—	0.83	—	—
cis-Chlordane	FV-1	35.91	373	375	—	—	0.93	—	—
Isofenphos	FV-1	35.92	213	255	—	—	0.29	—	—
Metazachlor	FV-2	36.11	209	133	—	—	1.82	—	—
Phenthoate	FV-2	36.19	246	274	—	—	3.83	—	—
Chlorfenvinphos	FV-1	36.26	323	267	—	—	2.51	—	—
Penconazole	FV-1	36.33	248	159	—	—	1.85	—	—
Tolylfluanid	FV-1	36.61	238	137	—	—	3.56	—	—
p,p'-DDE	FV-1	37.06	318	246	—	—	1.75	—	—
Folpet	FV-2	37.21	260	262	—	—	0.69	—	—
Prothiofos	FV-2	37.50	309	267	—	—	1.34	—	—
Dieldrin	FV-1	37.55	277	263	—	—	1.47	—	—
Butachlor	FV-1	37.75	176	160	—	—	0.84	—	—
Chlorflurecol-methyl	FV-2	37.77	215	217	152	—	0.32	0.44	—
Captan	FV-1	38.05	149	79	—	—	6.28	—	—
Iodofenphos	FV-1	38.08	377	379	—	—	0.44	—	—
Tribufos	FV-2	38.16	169	202	—	—	0.5	—	—
Chlozolinate	FV-2	38.23	259	331	—	—	0.43	—	—
Crotoxyphos	FV-2	38.56	193	127	—	—	3.65	—	—
Methidathion	FV-1	38.70	145	85	—	—	0.75	—	—
Tetrachlorvinphos	FV-1	38.83	329	331	—	—	0.95	—	—
Chlorbromuron	FV-2	38-84	61	294	—	—	0.06	—	—
Procymidone	FV-1	38.85	283	285	—	—	0.69	—	—
Flumetralin	FV-2	38.98	143	157	—	—	0.14	—	—
Endrin	FV-1	39.09	263	281	—	—	0.44	—	—
Bromacil	FV-2	39.17	205	207	—	—	0.97	—	—
Flurochloridone 1	FV-2	39.19	311	187	—	—	2.48	—	—
Triadimenol	FV-2	39.22	112	168	—	—	0.40	—	—
Profenofos	FV-1	39.28	339	337	—	—	1.06	—	—

Table 4.10 (*Continued*)

Compound	Method	RT (min)	Ions monitored (*m/z*)				Abundance ratio of qualifier ion/target		
			Target	Q1	Q2	Q3	Q/tgt	Q2/tgt	Q3/tgt
o,p'-DDD	FV-1	39.67	235	237	—	—	0.64	—	—
Flurochloridone 2	FV-2	39.67	311	187	—	—	1.85	—	—
Ethylan	FV-2	39.78	223	165	—	—	0.10	—	—
Cyanazine	FV-1	40.04	225	240	—	—	0.49	—	—
Chlorfenson	FV-1	40.09	302	175	—	—	6.83	—	—
TCMTB	FV-2	40.40	238	180	—	—	6.23	—	—
o,p'-DDT	FV-1	40.71	235	237	—	—	0.63	—	—
Oxadiazon	FV-2	40.89	258	175	—	—	3.48	—	—
Carbetamide	FV-2	41.06	119	120	236	—	0.22	0.04	—
Tetrasul	FV-2	41.13	252	324	—	—	0.52	—	—
Imazalil	FV-1	41.15	215	173	217	—	0.98	0.61	—
Aramite 1	FV-1	41.17	185	319	—	—	0.18	—	—
Fenamiphos	FV-2	41.48	303	217	—	—	0.59	—	—
Erbon	FV-1	42.08	169	171	—	—	0.64	—	—
Aramite 2	FV-1	42.13	185	319	—	—	0.08	—	—
Methoprotryne	FV-2	42.18	256	213	—	—	0.37	—	—
Chloropropylate	FV-1	42.24	251	139	253	—	1.14	0.63	—
Methyl trithion	FV-2	42.40	157	314	—	—	0.10	—	—
Nitrofen	FV-1	42.42	283	202	—	—	0.73	—	—
Chlorobenzilate	FV-2	42.55	251	139	—	—	1.26	—	—
Carboxin	FV-2	42.81	143	235	—	—	0.29	—	—
Flamprop-methyl	FV-1	42.89	105	77	276	—	0.29	0.04	—
Bupirimate	FV-2	43.17	273	316	208	—	0.16	0.94	—
β-Endosulfan	FV-1	43.36	241	237	—	—	1.06	—	—
p,p-DDD	FV-2	43.62	235	237	—	—	0.64	—	—
Oxyfluorfen	FV-2	43.82	252	300	—	—	0.27	—	—
Chlorthiophos	FV-1	43.92	325	360	—	—	0.46	—	—
Ethion	FV-1	44.42	231	153	384	—	0.76	0.04	—
Etaconazole 1	FV-1	44.42	245	173	—	—	1.51	—	—
Sulprofos	FV-2	44.47	322	156	140	—	1.83	2.14	—
Etaconazole 2	FV-1	44.60	245	173	—	—	1.70	—	—
Flamprop-isopropyl	FV-2	44.85	105	276	363	—	0.07	0.01	—
p,p'-DDT	FV-1	44.96	235	237	—	—	0.64	—	—
Carbophenothion	FV-1	45.17	157	342	121	—	0.13	0.54	—
Fluorodifen	FV-2	45.19	190	328	162	—	—	0.15	—
Myclobutanil	FV-2	46.02	179	288	—	—	0.06	—	—
Benalaxyl	FV-1	46.15	148	206	325	—	0.16	0.02	—
Edifenphos	FV-1	46.84	173	310	201	—	0.27	0.27	—
Propiconazole 1	FV-2	47.25	259	261	—	—	0.61	—	—
Fensulfothion	FV-2	47.52	293	308	—	—	0.16	—	—
Mirex	FV-1	47.61	272	274	237	—	0.80	0.62	—
Propargite	FV-2	47.63	135	350	150	201	0.02	0.12	0.03
Propiconazole 2	FV-2	47.63	259	261	—	—	0.64	—	—

(continued overleaf)

Table 4.10 (Continued)

Compound	Method	RT (min)	Ions monitored (m/z)				Abundance ratio of qualifier ion/target		
			Target	Q1	Q2	Q3	Q/tgt	Q2/tgt	Q3/tgt
Diclofop-methyl	FV-1	47.64	253	340	281	—	0.40	0.34	—
Propetamphos	FV-2	48.22	124	208	—	—	2.24	—	—
Triazophos	FV-2	48.40	162	161	—	—	1.54	—	—
Benodanil	FV-1	48.92	231	323	203	—	0.17	0.21	—
Nuarimol	FV-1	49.15	314	235	203	—	3.39	3.03	—
Bifenthrin	FV-2	49.19	181	165	166	—	0.27	0.27	—
Endosulfan sulfate	FV-1	49.43	272	387	—	—	0.27	—	—
Bromopropylate	FV-2	50.26	341	183	—	—	1.11	—	—
Oxadixyl	FV-2	50.29	163	132	278	—	0.79	0.06	—
Methoxychlor	FV-1	50.41	227	228	—	—	0.16	—	—
Benzoylprop-ethyl	FV-2	50.84	105	77	292	—	0.26	0.04	—
Tetramethrin 1	FV-2	50.89	164	123	—	—	0.28	—	—
Fenpropathrin	FV-1	51.21	181	265	—	—	0.33	—	—
Tetramethrin 2	FV-2	51.25	164	123	—	—	0.29	—	—
Leptophos	FV-2	51.51	171	377	—	—	0.29	—	—
EPN	FV-1	51.70	169	157	—	—	2.46	—	—
Norflurazon	FV-2	51.87	303	145	—	—	2.48	—	—
Hexazinone	FV-1	52.08	171	128	—	—	0.13	v	—
Phosmet	FV-1	52.16	160	161	317	—	0.12	0.02	—
Iprodione	FV-2	52.30	314	316	187	—	0.64	1.31	—
Tetradifon	FV-1	52.47	229	356	—	—	0.40	—	—
Bifenox	FV-2	52.65	341	173	—	—	0.54	—	—
Oxycarboxin	FV-2	53.16	175	267	—	—	0.26	—	—
Phosalone	FV-1	53.43	182	367	—	—	0.08	—	—
Chloridazon	FV-2	53.48	221	220	—	—	0.51	—	—
Azinphos-methyl	FV-2	53.49	160	132	—	—	0.79	—	—
Nitralin	FV-2	53.59	274	316	—	—	0.95	—	—
cis-Permethrin	FV-1	53.69	183	163	165	—	0.22	0.18	—
Fenarimol	FV-2	53.79	219	139	—	—	2.18	—	—
trans-Permethrin	FV-1	54.05	183	163	165	—	0.30	0.23	—
Pyrazophos	FV-1	54.06	232	221	373	—	3.77	0.20	—
Azinphos-ethyl	FV-1	54.37	160	132	—	—	1.18	—	—
Dialifos	FV-2	54.40	210	208	—	—	3.13	—	—
Cyfluthrin 1	FV-2	55.79	226	206	—	—	1.08	—	—
Prochloraz	FV-2	55.80	180	308	—	—	0.33	—	—
Coumaphos	FV-2	56.02	362	210	—	—	1.20	—	—
Cypermethrin 1	FV-1	56.03	181	163	—	—	1.59	—	—
Cyfluthrin 4	FV-2	56.33	226	206	—	—	1.24	—	—
Cypermethrin 4	FV-1	56.58	181	163	—	—	1.91	—	—
Fenvalerate 1	FV-2	57.67	167	225	419	—	0.44	0.06	—
Fenvalerate 2	FV-2	58.16	167	225	419	—	0.44	0.06	—
Deltamethrin	FV-2	59.50	181	251	—	—	0.51	—	—

degr. = degradation product (Fillion, Sauvé and Selwyn, 2000).

(Continued)

GC Method	System		Agilent 5890 Series
	Autosampler		Agilent liquid autosampler 7673A
	Column	Type	DB-1701
		Dimensions	30 m length × 0.25 mm
			ID × 0.15 µm film thickness
	Pre-column		30 cm of same type is used as a retention gap
	Carrier gas		Helium
	Flow		Constant flow
	SSL Injector	Injection mode	Splitless
		Injection temperature	250 °C
		Injection volume	2 µL
	Oven program	Start	70 °C, 2 min
		Ramp 1	25 °C/min to 130 °C
		Ramp 2	2 °C/min to 220 °C
		Ramp 3	10 °C/min to 280 °C
		Final temperature	280 °C, 6.6 min
MS Method	System		Agilent MSD 5972A
	Analyzer type		Single quadrupole MS
	Ionization		EI
	Acquisition	Mode	SIM
	Scan rate		0.2 s (5 spectra/s)
Calibration	Internal standard		Aldrin, 1.0 ng/µL

Results

For most of the components the achieved recovery has been between 70 and 120%. For mirex, EPTC, butylate, HCB (hexachlorobenzene), folpet, oxycarboxin and chlorthiamid with recoveries below 50% the method is considered a screening procedure, for individual recoveries and discussion see the original reference (Fillion, Sauvé and Selwyn, 2000). The published LODs range between 0.02 and 1.0 mg/kg with 80% of the compounds having LODs below 0.04 mg/kg.

Additional compounds have been tested but not included to the method. Benzoximate gave four peaks; chloroxuron, metoxuron and oxydemeton-methyl gave three peaks of degradation products; fluazinam and flualinate showed many degradation peaks and poor sensitivity; diclone showed poor sensitivity; ditalimfos was not recovered; and vamidothion degraded in solution.

Data analyses have been done by macro driven automation with spreadsheet report and printouts of the selected ion chromatograms. It is reported that this method typically allows a sample throughput of 42 samples including blanks and spikes per week with results for priority samples within 1 day.

4.11
Analysis of Dithiocarbamate Pesticides

Keywords: food safety; dithiocarbamate fungicides; DTCs; hydrolysis; Thiram; GC-MS; SIM

Introduction

Dithiocarbamate fungicides (DTCs) are widely used in agriculture. They are non-systemic and typically remain at the site of application. DTCs are characterized by a broad spectrum of activity against various plant pathogens, low acute mammal toxicity, and low production costs (Crnogorac and Schwack, 2009).

 DTCs are not stable and cannot be extracted or analyzed directly. Contact with acidic plant juices degrades DTCs rapidly and they decompose into carbon disulfide (CS_2) and the respective amine (Crnogorac and Schwack, 2009). DTCs cannot be extracted by organic solvents from homogenize plant samples, as it is the QuEChERS standard procedure in pesticide-residue analyses. The described method is a non-specific DTC sum method that does not distinguish between the different species of DTCs in the sample. Interferences are known from natural precursors, for example from crops or brassica, that can produce CS_2 as well during hydrolysis (Reynolds, 2006; Crnogorac and Schwack, 2009).

 Dithiocarbamates can be quantitatively converted to carbon disulfide by reaction with tin(II)chloride in aqueous HCl (1 : 1) in a closed bottle at 80 °C. The CS_2 gas produced is absorbed into iso-octane and measured by GC-MS. The analysis of DTCs for this application follows the acid-hydrolysis method using $SnCl_2$/HCl (Reynolds, 2006). For method validation of the DTC pesticides, Thiram (99.5% purity) was used as representative DTC compound considering its simple structure (1 mol of Thiram = 2 mol of CS_2 => >1 mg of Thiram theoretically generates 0.6333 mg CS_2, 1 mL of 100 ppm Thiram in 25 g of grapes = 2.5 ppm of CS_2), see Figure 4.38. The residues were estimated by analysis of CS_2 as the DTC hydrolysis products by GC-MS.

Sample Preparation

The earlier published $SnCl_2$/HCl acid-hydrolysis method was employed for sample preparation (CRL, 2005). The described method follows the established methods applied in the EU reference laboratories and European commercial testing laboratories for CS_2 analysis. From the homogenized sample 25 g are taken in a 250 mL glass bottle, 75 mL of the reaction mixture is added, followed by 25 mL iso-octane. The bottle is closed gastight immediately and placed in a water bath at 80 °C for 1 h with intermittent shaking and inverting the bottle after every 20 min.

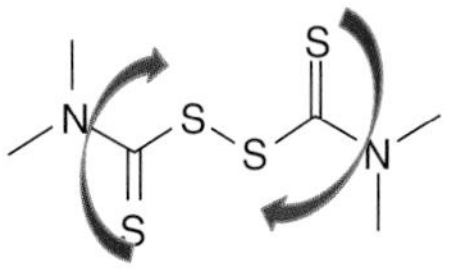

Figure 4.38 Thiram – 1 mol of Thiram generates 2 mol of CS_2.

After cooling the bottle to <20 °C by ice water, a 1–2 mL aliquot of the upper iso-octane layer is transferred into a micro centrifuge tube, and centrifuged at 5000 rpm for 5 min at 10 °C. The supernatant is then transferred into GC vials and the residues of DTCs are estimated by determining the CS_2 concentration by GC-MS. The sample preparation procedure depending on the type of food used takes ~1–2 h.

Preparation of Standard Solutions and Reaction Mixture

For method validation Thiram (99.5% purity) was used as representative DTC compound considering its simple structure (1 mol of Thiram = 2 mol of CS_2).

Carbon Disulfide Standard Solution

A stock solution of CS_2 (2000 µg/mL) was prepared by accurately pipetting 79 µL of CS_2 into a volumetric flask (certified A class, 50 mL) containing approximately 45 mL of iso-octane and made up to 50 mL with iso-octane. The CS_2 stock solution was kept in refrigerator at −20 °C and used within 2 days of preparation. CS_2 working standard solutions of 200 and 20 µg/mL concentrations (10 mL each) were prepared by serial dilution of stock solution with iso-octane.

Standard Solution of Thiram

Thiram 10 (±0.05) mg was weighed into a 10 mL volumetric flask (certified A class) and dissolved in ethyl acetate up to the mark to get a stock solution of 1000 µg/mL. A 100 µg/mL Thiram working standard was prepared from stock solution by dilution.

Preparation of Reaction Mixture

An amount of 30 g of tin(II)chloride was accurately weighed in the 1000 mL volumetric flask (certified A class) to which 1000 mL of concentrated HCL (35 %) was added. Then the solution was gradually added to 1000 mL water with continuous stirring to get clear solution.

Calibration standards

Calibration standard solutions of CS_2 at six different concentration levels (0.04, 0.08, 0.16, 0.32, 0.64 and 1.3 µg/mL) were prepared by appropriate dilutions of 20 µg/mL CS_2 working standard in iso-octane.

Matrix matched standards at the same concentrations were prepared by spiking the iso-octane extract of fresh control grapes, potato, tomato, green chilli and eggplant (all organically grown) using the following formula derived from above conversion of Thiram to CS_2:

$$\text{Spike quantity} = \frac{\text{Concentration to be achieved} * \text{weight of the sample}}{0.6333 * \text{concentration of the stock solution}}$$

Before preparation of matrix matched standards, the control samples were carefully monitored for absence of DTCs (in terms of CS_2) (Figure 4.39).

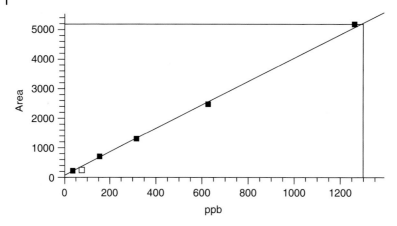

Figure 4.39 Calibration curve, range 0.04–1.300 µg/mL Thiram matrix spike, $R^2 = 0.9990$.

Experimental Conditions

Two GC columns of different polarity, stationary phase and film thickness have been evaluated. The first column was a medium polar cyanopropylphenyl phase (6% cyanopropylphenyl/94% dimethylpolysiloxane, 30 m × 0.32 mm ID, 1.8 µm film thickness, e.g. TraceGOLD TG-624) and as a second column a low polar 5%-phenyl stationary phase (5% diphenyl/95% dimethylpolysiloxane, 30 m × 0.25 mm ID, 0.25 µm film thickness, e.g. TraceGOLD TG-5MS). The TG-624 column type is a mid-polarity column ideally suited for the analysis of volatile analytes, whereas the TG-5MS column is more commonly used especially for pesticide analysis and mostly available with all laboratories. Both columns were thus tested for the applicability of the method. Either column can be used for DTC analysis (Figure 4.40).

Sample Measurements

A typical GC-MS batch consisted of matrix-matched calibration standards, samples, one matrix blank and one recovery sample for performance check after a set of every six samples.

The data acquisition was carried out in the SIM mode with compound-specific ions m/z 76 and 78 (the ^{34}S isotope, ion ratio 10 : 1) for a selective identification of CS_2 (Figure 4.41).

Results

Sensitivity
The sensitivity of the method was evaluated in terms of the LOD and limit of quantification (LOQ) which were respectively 0.005 and 0.04 µg/mL. The LOD is the

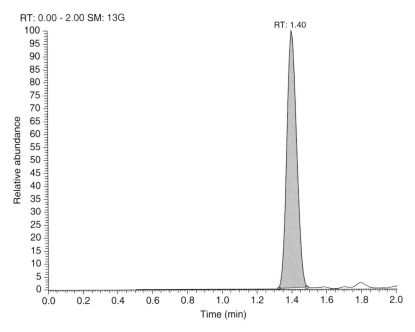

Figure 4.40 CS$_2$ chromatogram, 5 ppb matrix spike calibration.

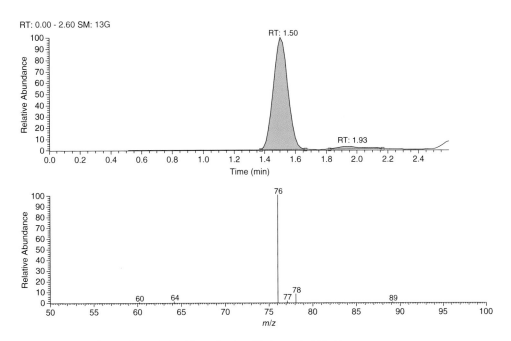

Figure 4.41 Chilli sample analysis with confirming CS$_2$ ion ratio 100 : 10.

concentration at which the signal to noise ratio (S/N) for the quantifier ion is >3, whereas LOQ is the concentration for which the S/N is >10.

Recovery

The recovery experiments were carried out on fresh untreated potato, tomato, eggplant, green chilli and grapes by fortifying 25 g of the samples with Thiram solution at 0.04, 0.16 and 1.30 µg/g levels in six replicates. The control samples of each of the tested commodities were obtained from an organic farm near Pune, India, and screened for absence of DTC residues before spiking. The spiked samples were extracted using sample preparation method as described above. The quantitation of the residues was performed using matrix matched standards (Table 4.11).

Accuracy

The precision of repeatability was determined by three analysts preparing six samples each on a single day. The intermediate precision was determined by the same analysts with six samples each on six different days. The method accuracy was determined with 0.04 mg/kg.

General Guidelines for DTC Analysis

The analysis of cruciferous crops, including brassica samples, may not be unequivocal, because they contain naturally occurring compounds that may generate carbon disulfide.

It is necessary to avoid the use of rubber material (natural/synthetic), for example gloves, when performing DTC analyses as they contain dithiocarbamates and this could lead to contamination problems. Silicone rubber and polythene do not contain dithiocarbamate.

Samples, other than fresh foodstuffs, will be comminuted by cryogenic milling. Fresh samples should be sub-sampled prior to extraction by removing segments from fresh samples following current Codex Alimentarius guidelines.

The samples should be analyzed within 4 weeks of cryogenic milling. If the storage of fresh produce is necessary it should be in a cool place (<−10 °C) keeping condensation at minimum (Amvrazi, 2005).

Table 4.11 Recoveries from different food commodities.

Spike level (ppb)	Grapes (%)	Chilli (%)	Potato (%)	Egg plant (%)	Tomato (%)
1300	96 (±4)	81 (±10)	90 (±9)	90 (±5)	81 (±4)
160	94 (±10)	80 (±13)	94 (±10)	92 (±8)	85 (±10)
40	104 (±15)	79 (±9)	104 (±15)	86 (±10)	96 (±15)

Analysis Conditions

Sample material			Fruit, vegetables, spices
Sample preparation			SnCl$_2$/HCl acid-hydrolysis method (Crnogorac and Schwack, 2009) The homogenized sample (25 g) was taken in a 250 mL glass bottle, 75 mL of the reaction mixture was added, followed by 25 mL isooctane. The bottle was closed immediately and placed in a water bath at 80 °C for 1 h with intermittent shaking and inverting the bottle after every 20 min. After cooling the bottle to <20 °C by ice water a 1–2 mL aliquot of the upper isooctane layer gets transferred into a micro centrifuge tube, and centrifuged at 5000 rpm for 5 min at 10 °C. The supernatant was transferred into GC autosampler vial
GC Method	System		Thermo Scientific™ Trace GC Ultra
	Autosampler		Thermo Scientific™ TriPlus
	Column 1	Type	TraceGOLD TG-624
		Dimensions	30 m length × 0.32 mm ID × 1.8 µm film thickness
	Column 2	Type	TraceGOLD TG-5MS
		Dimensions	30 m length × 0.25 mm ID × 0.25 µm film thickness
	Carrier gas		Helium
	Flow		Constant flow, 1 mL/min
	PTV Injector	Injection mode	PTV-LVI
		Injection volume	4 µL
		Base temperature	40 °C, 0.1 min at 100 kPa
		Evaporation	10 °C/s to 80 °C, 0.3 min at 200 kPa
		Transfer temperature	10 °C/s to 110 °C
		Transfer time	0.5 min at 200 kPa
		Cleaning phase	14.5 °C/s to 290 °C
		Final temperature	290 °C
	Solvent vent	Open until	0.17 min
	Split	Closed until	4 min
		Open	4 min to end of run
		Split flow	20 mL/min
	Oven program	Start	40 °C, 5 min
		Ramp 1	40 °C/min to 200 °C
		Final temperature	200 °C, 5 min
	Transferline	Temperature	295 °C Direct coupling

(*Continued*)			
MS Method	System		Thermo Scientific™ ITQ 900
	Analyzer type		Ion trap MS with external ionization
	Ionization		EI, 70 eV
	Ion source	Temperature	200 °C
	Acquisition	Mode	SIM
	Mass range	SIM mode	*m/z* 76,78
	Scan rate		0.5 s/scan
Calibration	External standardization		CS$_2$, Thiram
	Data points		6
	Calibration range		0.04 – 1.3 µg/mL

Conclusions

A reliable routine method for the analysis of dithiocarbamates with high precision in different vegetable and fruit commodities has been described (Dasgupta *et al.*, 2013). The extraction uses a SnCl$_2$/HCl acid-hydrolysis with iso-octane as solvent to form CS$_2$ which finally gets quantified by GC-MS. The recovery from different food commodities has been shown to be very high with 79 – 104%. The method allows a wide calibration range of 0.04 – 1.30 µg/mL Thiram. The LOQ has been determined as 0.04 µg/mL.

The GC injection method and column separation has been optimized for the injection of 4 µL of extract, using GC columns of standard film and dimensions, typically used for other types of residue analysis as well, so that a column change to a specific column for CS$_2$ determination only is not required.

4.12
GC-MS/MS Target Compound Analysis of Pesticide Residues in Difficult Matrices

Keywords: pesticides; residue analysis; selectivity; MS/MS; confirmation; quantitation

Introduction

The analysis of biological materials, such as plant and animal matter, for pesticide residue compounds using GC-MS has traditionally been difficult due to the very complex sample matrix. Approaches such as ECD (electron capture detector) and NPD (nitrogen/phosphorus detector) detection, SIM, large volume injection or sample concentration have been used for increasing the sensitivity of the method. But in matrices such as these, selectivity is often the limiting factor. Simply injecting more sample necessarily means injecting more sample matrix and chemical background. The SIM approach, while enhancing the sensitivity at the expense of full scan mass spectral information, increases the monitored signal of ions common to both the analyte and the background within the retention time window. Real improvements in trace level complex matrix analyses depend on improving both the sensitivity and the selectivity (Edwards *et al.*, 1998; Sheridan and Meola, 1999).

The use of MS/MS in GC detection introduces a structure related selectivity. While single quadrupole and TOF MS cannot distinguish ions of the same mass (isobaric) but different structure (pesticide vs. matrix), triple quadrupole instruments fragment the isobaric precursor ions in the collision cell, and detect only the fragments of the target structure. The resultant product ion spectrum at the target retention time is in presence and abundance due entirely to the m/z value of the selected pesticide precursor ion and not due to the chemical background. For quantitation typically the MRM mode is applied. Through this MS/MS process, the chemical background and matrix can be eliminated and low-level detection is enhanced.

The samples used for the assay were fish tissue and river water from an industrial region and prepared using liquid/liquid extraction (LLE). Aliquots of the homogenized fish tissue (typically 10 g) are treated with 100 g anhydrous sodium sulfate and extracted three times with 100 mL methylene chloride. The extract portions are combined, concentrated and purified using gel chromatography. Extracts are further cleaned using either a deactivated alumina/silica gel column or a Maxi Clean Florisil PR Cartridge (Alltech) and solvent elution. River water samples have been filtered and pre-concentrated using C_{18}-SPE (solid phase extraction) cartridges with standard methods described elsewhere.

For data acquisition, multiple MS/MS scan events are time programmed as known from SIM acquisitions throughout the chromatographic run. Individual scan events are automatically set at the appropriate times so that as different target analytes elute, the conditions necessary for their optimal determination are activated. Precursor ions are usually chosen using the known SIM masses. MS/MS fragmentation is achieved using initial standard values with little individual response optimization of the activation voltage (Gummersbach, 2011).

Analysis Conditions

Sample material		Food, fruit, vegetable, tea, water
Sample preparation	Liquid/liquid extraction	10 g of sample, 100 mL methylene chloride Clean-up by gel chromatography, alumina/silica, or Florisil column

(Continued)

GC Method	System		Thermo Scientific Trace GC Ultra
	Autosampler		Thermo Scientific AS 3000
	Column	Type	DB-XLB
		Dimensions	30 m length × 0.25 mm ID × 0.25 µm film thickness
	Carrier gas		Helium
	Flow		He, constant flow, 1.5 mL/min (about 40 cm/s)
	SSL Injector	Injection mode	Splitless, hot needle
		Inlet liner	4 mm ID deactivated glass liner with a 1 cm glass wool plug 3 cm from top
		Injection temperature	300 °C
		Hold time before injection	3 s
		Hold time after injection	0 s
		Injection volume	1 µL, 2 µL air plug
	Split	Closed until	1 min
		Open	1 min to end of run
		Split flow	50 mL/min
		Gas saver	2 min, 10 mL/min
		Purge flow	5 mL/min
	Oven program	Start	60 °C, 1 min
		Ramp 1	40 °C/min to 150 °C, 1 min
		Ramp 2	5 °C/min to 300 °C
		Final temperature	300 °C, 4.75 min
	Transferline	Temperature	300 °C
MS Method	System		Thermo Scientific Polaris Q
	Analyzer type		Ion trap MS with external ionization
	Ionization		EI
	Electron energy		70 eV
	Ion source	Temperature	200 °C
	Acquisition	Mode	MRM 2 SRMs per analyte
		Product ion scan	Full scan, 40–440 Da
	Collision gas		He buffer gas, about 1.5 mL/min
	Mass range		50–550 Da, range or SIM masses here
Calibration	External standard	Range	0.01–10 ng/µL
	Data points		6

RT, retention time.

Results

Several commercially available mixes of pesticide compounds were used for calibration of the system by serial dilution and external standard techniques. A least squares regression was applied to curve fit the calibration data. Calibration curves for organochlorine pesticides (OCPs) were linear ranging from below 0.01 to 10 ng/μL with correlation coefficients ranging from 0.994 to 0.999 (Tables 4.12 and 4.13).

The first example illustrates in Figures 4.42–4.44 the determination of two chlordane isomers. As can be seen in Figure 4.42, the single stage technique, whether monitoring the TIC or three characteristic masses (m/z 373 + 375 + 266), can barely distinguish between the analyte signal and the complex matrix at this concentration level (550 ppb α-chlordane in fish). MS/MS eliminates the matrix difficulties and allows much clearer detection of the analyte (Figure 4.43) with an additional full product ion mass spectrum for confirmation (Figure 4.44). Other compounds detected and confirmed are labelled. An example of a calibration curve is given in Figure 4.45 for α-chlordane showing a correlation factor of 0.998 over seven calibration points.

Another example shows the determination of captan (Figure 4.46) representative of nitrogen and phosphorus pesticides in difficult matrices. The single-stage GC-MS analysis of a river water sample shows a susceptible peak at retention

Table 4.12 MS/MS acquisition parameters for organochlorine pesticides.

Compound name	Acquisition mode	Precursor ion (m/z)	Excitation voltage (V)	q value	Product scan range (m/z)
4,4′-DDD	SRM MS/MS	235	0.80	0.225	40–440
4,4′-DDE	SRM MS/MS	318	0.80	0.225	40–440
4,4′-DDT	SRM MS/MS	235	0.80	0.225	40–440
Alachlor	SRM MS/MS	188	0.80	0.225	40–440
Aldrin	SRM MS/MS	293	0.80	0.225	40–440
α-HCH	SRM MS/MS	181	0.80	0.225	40–440
α-Chlordane	SRM MS/MS	373	0.80	0.225	40–440
β-HCH	SRM MS/MS	181	0.80	0.225	40–440
cis-Nonachlor	SRM MS/MS	407	0.80	0.225	40–440
Dieldrin	SRM MS/MS	279	0.80	0.225	40–440
Endosulfan I	SRM MS/MS	339	0.60	0.225	40–440
Endosulfan II	SRM MS/MS	339	0.60	0.225	40–440
Endosulfan sulfate	SRM MS/MS	387	0.65	0.225	40–440
Endrin	SRM MS/MS	281	0.80	0.225	40–440
Endrin aldehyde	SRM MS/MS	345	0.65	0.225	40–440
γ-HCH (Lindane)	SRM MS/MS	181	0.80	0.225	40–440
γ-Chlordane	SRM MS/MS	373	0.80	0.225	40–440
Heptachlor	SRM MS/MS	272	0.80	0.225	40–440
Heptachlor epoxide	SRM MS/MS	353	0.70	0.225	40–440
Methoxychlor	SRM MS/MS	227	0.90	0.225	40–440

Table 4.13 MS/MS acquisition parameters for nitrogen/phosphorus pesticides.

Compound name	Retention time (min : s)	Acquisition mode	Precursor ion (*m/z*)	Excitation voltage (V)	*q* value	Product scan range (*m/z*)
Dichlorvos	9 : 10	SRM MS/MS	185.00	1.20	0.33	70–200
Mevinphos	11 : 18	SRM MS/MS	192.00	0.80	0.33	100–200
Heptenophos	13 : 31	SRM MS/MS	215.00	1.10	0.33	80–220
Propoxur	13 : 84	SRM MS/MS	152.00	1.00	0.33	60–155
Demeton-*S*-methyl	13 : 99	SRM MS/MS	142.00	0.90	0.33	50–150
Ethoprophos	14 : 17	SRM MS/MS	200.00	0.90	0.33	100–210
Desethylatrazine	14 : 65	SRM MS/MS	172.00	1.10	0.33	80–180
Bendiocarb	14 : 86	SRM MS/MS	166.00	0.80	0.33	80–170
Tebutan	14 : 95	SRM MS/MS	190.00	0.90	0.33	80–200
Dimethoate	15 : 73	SRM MS/MS	125.00	0.90	0.33	50–130
Simazine	15 : 96	SRM MS/MS	201.00	0.90	0.33	100–210
Atrazine	16 : 14	SRM MS/MS	200.00	1.30	0.33	100–210
Terbumeton	16 : 35	SRM MS/MS	169.00	1.00	0.33	70–175
Terbuthylazine	16 : 62	SRM MS/MS	214.00	1.30	0.33	100–220
Dimpylate	17 : 03	SRM MS/MS	304.00	0.90	0.33	130–310
Terbasil	17 : 29	SRM MS/MS	161.00	1.20	0.33	80–170
Etrimfos	17 : 55	SRM MS/MS	292.00	0.90	0.33	150–300
Pirimicarb	17 : 95	SRM MS/MS	238.00	0.80	0.33	100–240
Desmethryn	18 : 23	SRM MS/MS	213.00	0.95	0.33	100–220
Ametryn	18 : 98	SRM MS/MS	227.00	0.90	0.33	100–230
Prometryn	19 : 10	SRM MS/MS	241.00	0.90	0.33	150–250
Terbutryn	19 : 56	SRM MS/MS	185.00	1.00	0.33	150–310
Pirimiphos-methyl	19 : 71	SRM MS/MS	305.00	0.75	0.33	60–190
Cyanazine	20 : 48	SRM MS/MS	225.00	1.10	0.33	60–230
Penconazole	21 : 62	SRM MS/MS	248.00	1.10	0.33	130–250
Triadimenol	22 : 05	SRM MS/MS	128.00	1.00	0.33	60–130
Methidathion	22 : 46	SRM MS/MS	145.00	0.80	0.33	60–150
Fluazifop-butyl	24 : 63	SRM MS/MS	383.00	1.00	0.33	150–390
Oxadixyl	25 : 41	SRM MS/MS	163.00	0.80	0.33	60–170
Benalaxyl	26 : 19	SRM MS/MS	266.00	0.90	0.33	85–270
Hexazinone	26 : 95	SRM MS/MS	171.00	1.20	0.33	80–180
Azinphos-methyl	29 : 61	SRM MS/MS	132.00	1.10	0.33	60–140
Fenarimol	30 : 55	SRM MS/MS	139.00	1.00	0.33	60–145
Pyrazophos	30 : 83	SRM MS/MS	265.00	0.90	0.33	100–280

time 16.12 min indicating the occurrence of captan (Figure 4.47). The corresponding mass spectrum is dominated by chemical background ions (Figure 4.48). The specific ions of captan can be found within at lower levels. Mass *m/z* 149 is not diagnostic as there may be a contribution from ubiquitously occurring phthalate plasticizers. The high ion intensity of *m/z* 299 cannot be explained from the molecular ion of captan in this case.

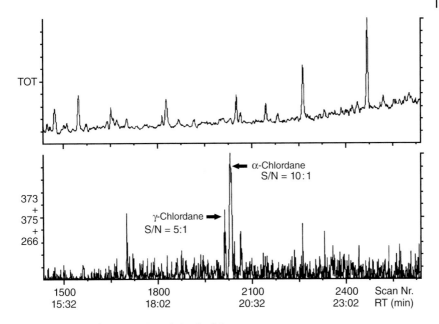

Figure 4.42   Single stage MS analysis of a fish extract.

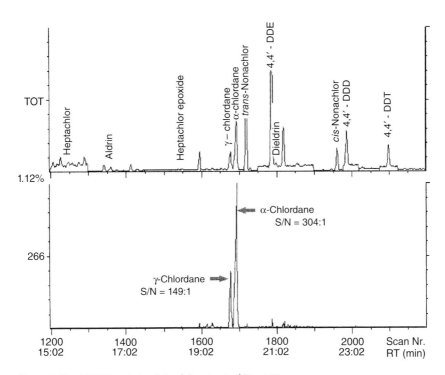

Figure 4.43   MS/MS analysis of the fish extract of Fig. 4.42.

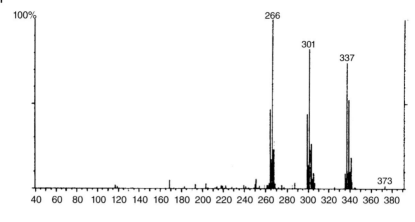

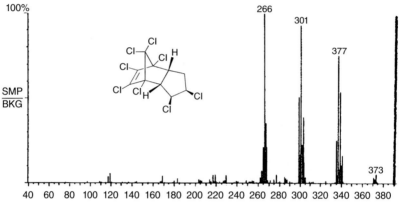

Figure 4.44 Confirmation of α-Chlordane by product ion spectrum (the top spectrum represents the MS/MS library spectrum for reference).

Using m/z 149 as the precursor ion for the MS/MS determination, the unambiguous confirmation of captan in the sample at high S/N (signal/noise) levels is achieved (Figure 4.49). The resulting product ion spectrum gives clear evidence of the structure-specific ions of captan as proof (Figure 4.50).

The retention and acquisition parameters of common nitrogen- and phosphorus-containing pesticides (phosphoric acid esters, atrazines, etc.) are listed in Table 4.13. It is recommended that the excitation parameters are used initially with equal standard values (0.9 V), which may be slightly adjusted in the case of response optimization. The quality of the product ion spectra is not affected (Figure 4.51). A typical standard run in MS/MS acquisition mode is shown in Figures 4.52 and 4.53. Even at the low 10 pg/μL level, the compounds are detected with high S/N values and full MS/MS product spectra for confirmation (Figure 4.51).

The use of MS/MS techniques can provide a very reliable and indeed essential means of analysis in the determination of target compounds in complex matrices.

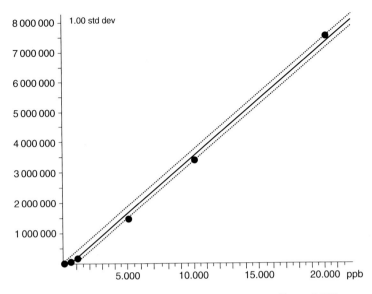

Figure 4.45   α-Chlordane calibration curve (correlation coefficient 0.998).

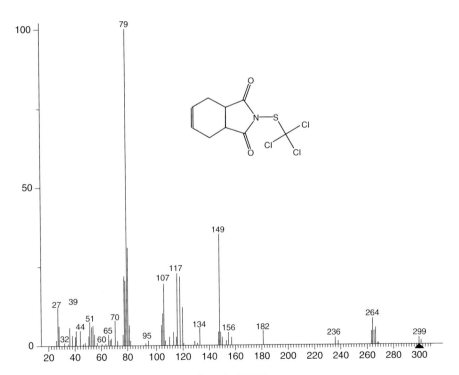

Figure 4.46   Captan – EI spectrum with formula (M 299).

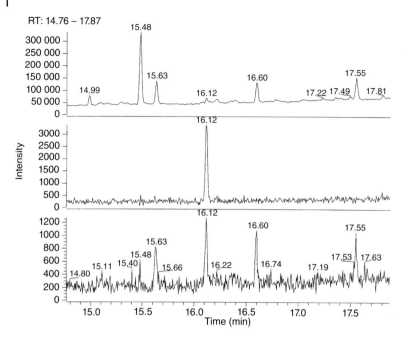

Figure 4.47 Single stage MS analysis of river water sample with top TIC and captan mass chromatograms *m/z* 79 and 149 (bottom).

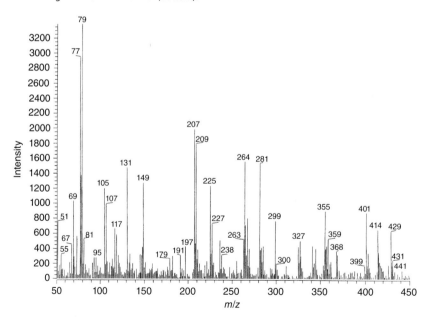

Figure 4.48 Mass spectrum of GC peak eluting 16.12 min – Captan? Specific ions *m/z* 79, 107 and 149 are within chemical background (*m/z* 149 may result from phthalate plasticizer).

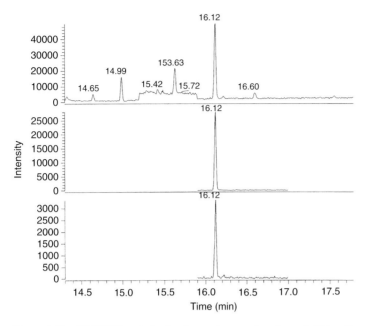

Figure 4.49 GC-MS/MS analysis of a river water sample using *m/z* 149 as the precursor ion (top TIC) with Captan product ion chromatograms *m/z* 105 and 79 (bottom).

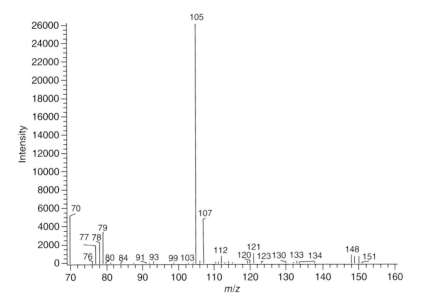

Figure 4.50 MS/MS spectrum of the GC peak eluting at 16.12 min (*m/z* 149 as precursor ion) – Confirmation for captan with structure-specific ions *m/z* 79, 105 and 107 free of chemical background.

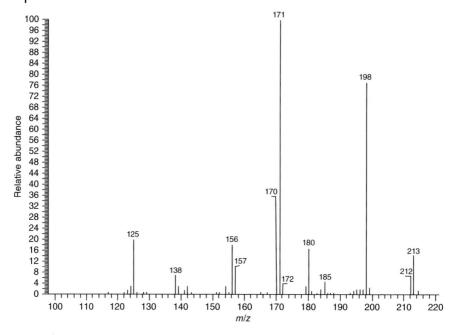

Figure 4.51 Product ion spectrum of desmetryn (precursor ion *m/z* 213).

The improved selectivity of the analysis from the MS/MS process, along with the inherent sensitivity in this particular example, of the ion trap mass spectrometer can often result in an order of magnitude or more of improvement in the signal to noise performance with related strong improvements in LOD, over single-stage full-scan or SIM methodologies.

4.13
Multi-Component Pesticide Analysis by MS/MS

Keywords: pesticides; food; multi-residue method; high throughput; QuEChERS; SRM (selected reaction monitoring); H-SRM (enhanced resolution selected reaction monitoring); MS/MS; confirmation; QED; data-dependent acquisition

Introduction

Food safety concerns are on the rise among consumers worldwide. In 2006, sweeping changes were made to the Food Hygiene Law in Japan regarding residual agricultural chemicals, including pesticides, in foods (The Japanese Ministry of Health, Labour, and Welfare, 2006). As a result, residue standards were created for all pesticides, and standard residue values were established for more than 800 pesticides. Because each type has different physicochemical properties, there are limitations on simultaneous analysis. Among the pesticides

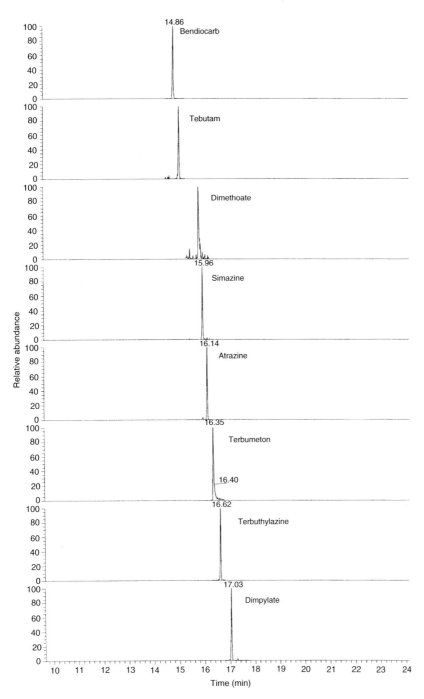

Figure 4.52   Selective GC-MS/MS mass chromatograms of a pesticide standard run at the 10 pg/μL level (see Table 4.14).

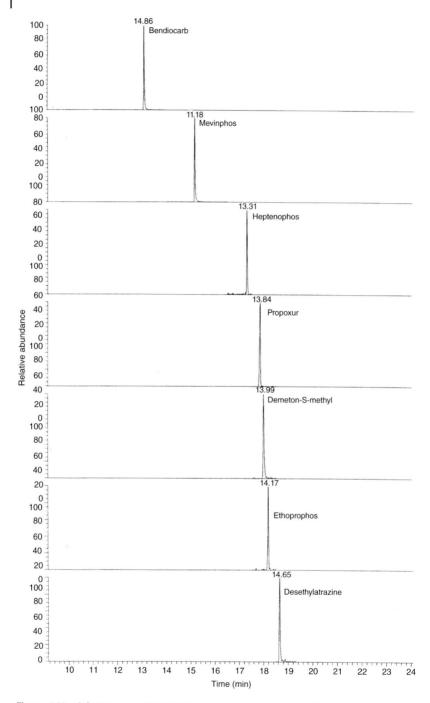

Figure 4.53 Selective mass GC-MS/MS chromatograms of a pesticide standard run at the 10 pg/μL level (see Table 4.14).

Table 4.14 Retention time, SRM conditions, calibration range, linearity and reproducibility of pesticide components investigated.

	RT (min)	Precursor ion (*m/z*)	Product ion (*m/z*)	Collision energy (V)	R^2	Calibration range (ppb)	CV (%) $n = 5$
Mevinphos	6.44	192	127	10	0.9999	0.1–100	4.03
XMC	7.52	122	107	10	0.9999	0.1–100	2.55
Tecnazene	8.03	261	203	15	0.9996	0.1–100	5.41
Ethoprophos	8.22	200	114	10	0.9981	0.1–100	7.91
Ethalfluralin	8.42	316	276	10	0.9997	0.1–100	4.14
Benfluralin	8.62	292	264	10	0.9989	0.1–100	1.86
Monocrotophos	8.62	192	127	10	0.9754	5–100	19.47
α-HCH	9.03	219	183	15	0.9999	0.1–100	4.51
Dicloran	9.25	206	176	10	0.9994	0.1–100	2.30
Simazine	9.30	201	172	10	0.9999	0.1–100	4.33
Propazine	9.50	214	172	10	0.9998	0.1–100	1.99
β-HCH	9.57	219	183	15	1.0000	0.1–100	3.51
γ-HCH	9.73	219	183	15	0.9998	0.1–100	6.57
Cyanophos	9.78	243	109	10	0.9996	0.1–100	3.56
Pyroquilon	9.90	173	130	20	0.9996	0.1–100	2.95
Diazinon	4.94	304	179	15	0.9995	0.1–100	4.40
Phosphamidon-1	10.06	264	127	10	0.9989	0.1–100	10.31
Prohydrojasmon-1	10.12	184	83	20	0.9992	0.1–100	7.39
δ-HCH	10.26	219	183	15	0.9994	0.1–100	5.17
Prohydrojasmon-2	10.66	264	127	10	0.9972	0.1–100	17.11
Benoxacor	10.7	259	120	15	0.9999	0.1–100	3.30
Propanil	10.95	262	202	10	0.9993	0.1–100	3.65
Phosphamidon-2	10.97	264	127	10	0.9970	0.1–100	8.77
Dichlofenthion	10.99	279	223	15	0.9994	0.1–100	2.21
Dimethenamid	11.06	230	154	10	0.9996	0.1–100	2.51
Bromobutide	11.09	232	176	10	0.9990	0.1–100	5.91
Parathionmethyl	11.24	263	109	10	0.9982	0.1–100	3.74
Tolclofos-methyl	11.38	265	250	15	0.9998	0.1–100	2.52
Ametryn	11.43	227	170	10	0.9999	0.1–100	0.90
Mefenoxam	11.57	249	190	10	0.9995	0.1–100	5.81
Bromacil	11.98	205	188	15	0.9988	0.1–100	3.87
Pirimiphos-methyl	12.00	305	276	10	0.9995	0.1–100	4.08
Quinoclamine	12.18	207	172	10	0.9989	0.1–100	4.24
Diethofencarb	12.34	225	125	15	0.9985	0.1–100	4.64
Cyanazine	12.52	225	189	10	0.9994	0.1–100	3.41
Chlorpyrifos	12.57	314	258	15	0.9991	0.1–100	3.37
Parathion	12.59	291	109	15	0.9962	0.1–100	9.76
Triadimefon	12.67	208	111	25	0.9986	0.1–100	6.10
Chlorthal-dimethyl	12.73	301	223	20	1.0000	0.1–100	1.23
Nitrothal-isopropyl	12.78	236	148	15	0.9974	0.1–100	5.53
Phthalide	13.04	272	243	10	0.9993	0.1–100	4.32
Fosthiazate	13.05[a)]	195	103	10	0.9956	5–100	6.29
Diphenamid	13.10	239	167	10	0.9997	0.1–100	4.67

(continued overleaf)

Table 4.14 (Continued)

	RT (min)	Precursor ion (m/z)	Product ion (m/z)	Collision energy (V)	R^2	Calibration range (ppb)	CV (%) $n = 5$
Pyrifenox-Z	13.64	262	200	15	0.9979	0.2–100	4.54
Fipronil	13.79	123	81	10	0.9991	0.1–100	3.49
Allethrin	13.67	136	93	10	0.9991	5–100	3.79
Dimepiperate	13.87	145	112	10	0.9987	0.1–100	3.74
Quinalphos	13.87	274	121	10	0.9987	0.1–100	1.82
Phenthoate	13.88	146	118	10	0.9984	0.1–100	1.96
Paclobutrazol	14.45	236	125	15	0.9961	0.1–100	7.41
Endosulfan	14.67	241	206	15	0.9996	0.1–100	4.54
Butachlor	14.73	237	160	10	0.9998	0.1–100	5.26
Imazamethabenz- methyl	14.81	256	144	20	0.9932	2–100	12.09
Butamifos	15.00	286	202	15	0.9958	0.1–100	4.66
Flutlanil	15.06	173	145	15	0.9986	0.1–100	1.93
Hexaconazole	15.06	214	172	15	0.9924	0.1–100	8.98
Profenofos	15.28	337	267	15	0.9968	0.1–100	6.61
Uniconazole-P	15.38	234	137	15	0.9966	0.1–100	11.37
Pretilachlor	15.37	162	132	15	0.9982	0.1–100	6.72
Flamprop-methyl	15.66	276	105	10	0.9986	0.1–100	3.93
Oxyfluorfen	15.69	361	300	10	0.9980	0.5–100	6.07
Azaconazole	15.79	217	173	15	0.9981	0.1–100	7.07
Bupirimate	15.82	316	208	10	0.9982	0.1–100	4.65
Thifluzamide	15.84	449	429	10	0.9972	0.1–100	2.75
Fenoxanil	16.25	293	155	20	0.9989	0.1–100	3.73
Chlorobenzilate	16.43	251	139	15	0.9976	0.1–100	0.81
Pyriminobac-methyl-Z	16.76	302	256	15	0.9986	0.1–100	2.70
Oxadixyl	16.86	163	132	10	0.9998	0.1–100	3.72
Triazophos	17.30	257	162	10	0.9941	0.2–100	6.72
Fluacrypyrim	17.38	189	129	10	0.9988	0.1–100	2.15
Edifenphos	17.72	310	173	10	0.9927	0.1–100	7.95
Quinoxyfen	17.74	272	237	10	0.9993	0.1–100	4.50
Lenacil	17.78	153	136	15	0.9979	0.1–100	5.19
Trifloxystrobin	18.01	222	162	10	0.9966	0.1–100	8.47
Pyriminobac-methyl-E	18.19	302	256	15	0.9982	0.1–100	2.12
Tebuconazole	18.39	250	125	20	0.9907	0.2–100	13.03
Diclofop-methyl	18.51	253	162	15	0.9991	0.1–100	2.14
Mefenpyr-diethyl	19.15	253	189	20	0.9992	0.1–100	3.35
Pyributicarb	19.24	165	108	10	0.9973	0.1–100	2.00
Pyridafenthion	19.46	152	116	20	0.9940	0.2–100	4.71
Acetamiprid	19.39	340	199	10	1.0000	50–100	–
Bromopropylate	19.64	341	185	15	0.9956	0.1–100	3.72
Piperophos	19.84	320	122	10	0.9939	0.2–100	7.51
Fenpropathrin	19.98	265	210	10	0.9973	0.1–100	6.87
Etoxazole	20.06	300	270	20	0.9969	0.1–100	8.84
Tebufenpyrad	20.10	333	171	20	0.9978	0.5–100	13.35
Anilofos	20.31	226	157	15	0.9948	0.2–100	5.56

Table 4.14 (*Continued*)

	RT (min)	Precursor ion (*m/z*)	Product ion (*m/z*)	Collision energy (V)	R^2	Calibration range (ppb)	CV (%) $n = 5$
Phenothrin-1	20.49	183	165	10	0.9967	5–100	16.13
Tetradifon	20.54	356	229	10	0.9998	0.2–100	4.17
Phenothrin-2	20.66	183	165	10	0.9968	0.1–100	3.79
Mefenacet	21.22	192	136	15	0.9955	0.1–100	4.90
Cyhalofop-buthyl	21.23	357	229	10	0.9967	0.1–100	5.52
Cyhalothrin-1	21.30	181	152	20	0.9975	0.2–100	3.21
Cyhalothrin-2	21.66	181	152	20	0.9984	0.2–100	6.67
Pyrazophos	22.06	373	232	10	0.9963	0.1–100	10.46
Bitertanol	22.80[b]	170	141	20	0.9873	0.1–100	6.76
Pyridaben	23.18	147	117	20	0.9958	0.1–100	1.29
Cafenstrole	24.03	100	72	5	0.9958	0.1–100	9.77
Cypermethrin-1	24.72	181	152	20	0.9983	2–100	9.29
Halfenprox	24.79	263	235	15	0.9979	0.1–100	10.25
Cypermethrin-2	24.92	181	152	20	0.9982	2–100	6.91
Cypermethrin-3	25.06	181	152	20	0.9985	2–100	16.27
Cypermethrin-4	25.13	181	152	20	0.9948	2–100	13.79
Fenvalerate-1	26.47	167	125	10	0.9977	0.1–100	3.11
Flumioxazin	26.50	354	176	20	0.9937	0.1–100	9.66
Fenvarelate-2	26.91	167	125	10	0.9979	0.1–100	3.26
Deltamethrin+ Tralomethrin	28.15	181	152	20	0.9967	0.2–100	8.20
Tolfenpyrad	29.11	383	171	20	0.9968	2–100	4.84
Imibenconazole	30.35	375	260	15	1.0000	50–100	—

a)   and 13.12 min.
b)   and 22.97 min.

for which standard values are currently set, GC-MS/MS can analyse more than 300 of them (Okihashi *et al.* 2005; Takatori *et al.*, 2011).

The superior selectivity and high-speed acquisition rate of the GC-MS/MS technique allows interference-free quantification, even with peak co-elution and provides positive confirmation in a single analytical run. To accurately monitor pesticide residues, a high throughput multi-residue method that can quantitate a large number of pesticide residues during a single analysis is described.

Sample Preparation

The QuEChERS sample preparation method is especially suited for low fat containing samples. Green pepper, carrot, grapefruit and banana samples were prepared for analysis using the QuEChERS method (Anastassiades *et al.*, 2003; QuEChERS 2014). A 10 g sample of food was homogenized in a food processor and placed in a polypropylene centrifuge tube. The sample was extracted with 20 mL of acetonitrile in a homogenizer. Then, 4 g of anhydrous magnesium sulfate and 1 g of sodium chloride were added and the resulting mixture was centrifuged.

After centrifugation, the supernatant was loaded onto a graphite carbon/PSA (primary secondary amine) dual-layer, solid-phase extraction column and eluted with 50 mL of acetonitrile/toluene (3 : 1). After the eluate was concentrated under reduced pressure, it was dissolved (1 g/mL) in 10 mL of acetonitrile/*n*-hexane to give the test solution (Hollosi *et al.*, 2013).

Analysis Conditions

Sample material			Food up to 15% fat content
Sample preparation			QuEChERS
GC Method	System		Thermo Scientific Trace GC Ultra
	Autosampler		Thermo Scientific TriPlus
	Column	Type	Rti-5MS
		Dimensions	30 m length × 0.25 mm
			ID × 0.25 µm film thickness
	Carrier gas		Helium
	Flow		Constant flow, 1.2 mL/min
	SSL Injector	Injection mode	Splitless, hot needle
		Injection Temperature	240 °C
		Injection Volume	1 µL
	Split	Closed until	1 min
		Open	1 min to end of run
		Split flow	50 mL/min
		Purge flow	5 mL/min
	Surge	Pressure, time	200 kPa, 0.5 min
	Oven program	Start	80 °C, 1 min
		Ramp 1	20 °C/min to 180 °C
		Ramp 2	5 °C/min to 280 °C
		Final temperature	280 °C, 10 min
	Transferline	Temperature	280 °C
MS Method	System		Thermo Scientific TSQ Quantum XLS
	Analyzer type		Triple quadrupole MS
	Ionization		EI
	Electron energy		70 eV
	Ion source	Temperature	280 °C
	Acquisition	Mode	Timed-SRM, H-SRM
			2 SRMs per analyte Table 4.14)
	Collision gas		Ar, 1.2 mTorr
	Scan rate		0.5 s/scan
	Resolution	Precursor peak width for H-SRM	Q1, 0.4 Da
		Product ion peak width	Q3, 0.7 Da
		Scan width	0.002 Da
Calibration	External standard	Range	0.1 – 100 pg/µL (ppb)
	Data points		8

Results

Simultaneous analysis was carried out on multi-component pesticide residues in various food products. Results obtained indicated excellent reproducibility (10% at 5 ppb) and linearity with R^2 better than 0.995 in the range of 0.1 – 100 ppb (0.1 – 100 pg/µL injected). Figure 4.54 shows as example the calibration curve for propazine at 0.1 – 100 ppb with a corresponding chromatogram at 1 ppb, showing excellent reproducibility ($R^2 = 0.9998$).

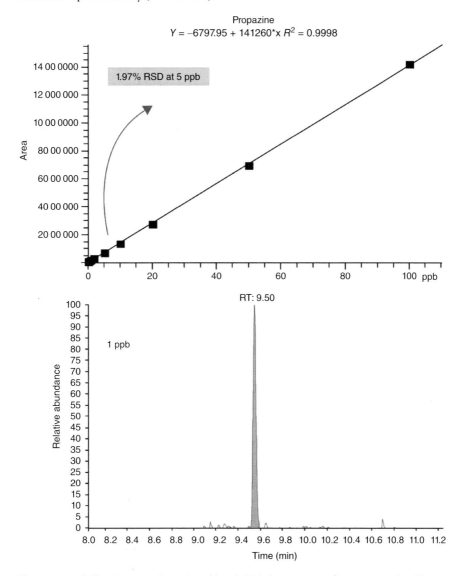

Figure 4.54   Calibration curve (0.1 – 100 ppb) and SRM chromatogram for propazine (1 ppb).

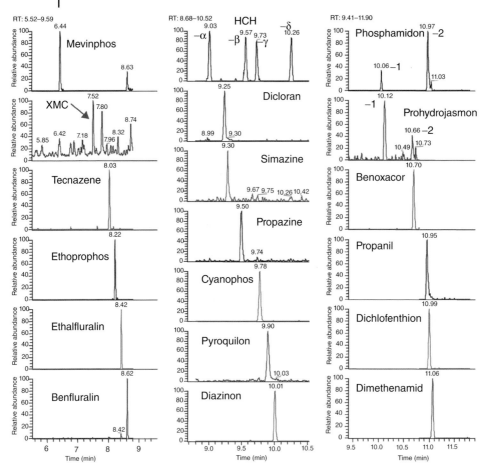

Figure 4.55 GC-MS/MS chromatograms of various pesticides at 1 ppb in green pepper samples.

Figure 4.55 shows examples of GC-MS/MS chromatograms of various pesticides in which 1 ppb of each pesticide was added to green pepper. Even at this low concentration (1/10 of the regulated MRL level for pesticides), it is possible to provide reliable quantitation measurements with remarkably high sensitivity using the highly resolved parent ions for MS/MS analysis. No cross talk was observed for the analysis of closely eluting multi-component mixtures. Using H-SRM, interferences from the sample matrix background were substantially reduced, leading to very low-level LOQs. Figure 4.56 illustrates the gain in selectivity when using the highly resolved Q1 precursor ion for SRM quantitation.

In addition, structural confirmation of the analytes during the quantification run have been provided using the data dependant acquisition of a MS/MS product ion spectrum (see Figures 4.57 and 4.58).

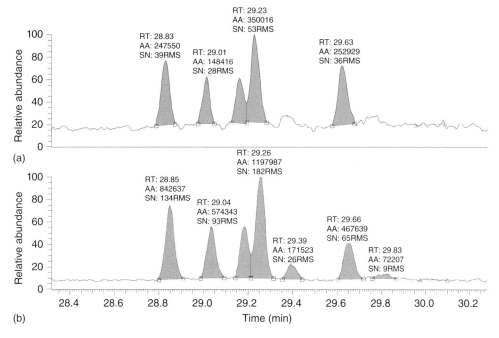

(a)

(b)

Figure 4.56   SRM and H-SRM analysis of pyrethroids in cabbage matrix at a 5 ppb level, TIC chromatograms. (a) 2 µL injection, Q1 set to 0.7 Da peak width. (b) 10 µL injection, Q1 set to 0.4 Da peak with shows increased selectivity with reduced background intensity even a fivefold increased injection volume.

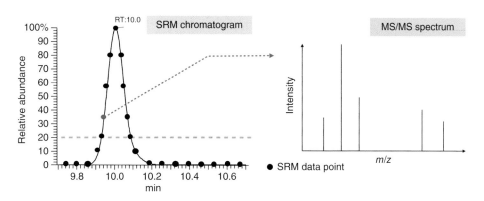

Figure 4.57   Data dependent data acquisition scheme. The first SRM scan intensity above a user defined threshold, (a) dotted line, is acquired as MS/MS product ion spectrum (b).

In combination with the QuEChERS, sample preparation GC-MS/MS using H-SRM turned out to provide the required sensitivity and certainty for a multi-component quantitation of trace pesticides in food for a high throughput of samples.

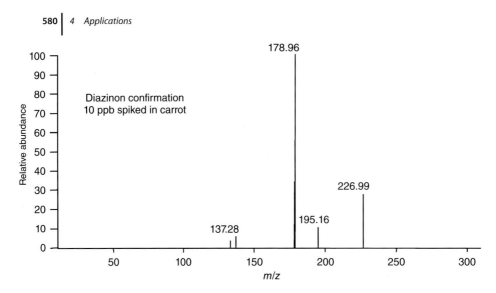

Figure 4.58 Confirmation of diazinon by the product ion spectrum at the 10 ppb level.

4.14
Multiresidue Pesticides Analysis in Ayurvedic Churna

Keywords: food safety; traditional herbal medicine; spices; liquid/liquid extraction; timed-SRM

Introduction

Ayurveda is a Sanskrit term, made up of the words 'ayus' and 'veda' meaning life and science, together translating to 'science of life'. A blend of several herbs and spices make up the powdered mixture known as 'churna'. Depending on its intended use for medicinal, beauty or culinary purpose, the recipe varies. Avipittakara 'churna' is a traditional Ayurvedic formula used widely and almost daily to control vitiated pitta dosha, remove heat in the digestive system, control indigestion, constipation, vomiting and anorexia. A major analytical challenge for these types of samples is mainly the addition of multiple herbs with sugar and the natural colour of herbs (Narayanaswamy, 1981; Lohar).

The dried leaves result in highly complex extracts from the sample preparation due to the rich content of active ingredients, essential oils and the typical high boiling natural polymer compounds. Owing to the use of pesticides in the fresh herbs the 'churna' may contain residual pesticides. Analysis of pesticide residues is governmentally regulated (The Pesticides Compound Database, 2014). Strict quality parameters have been implemented to preserve the quality and efficacy of these 'churnas'.

Sample Preparation

The sample preparation involves the extraction of 15 g powdered sample of Avipittakara 'churna' with 15 mL acetonitrile containing 1% acetic acid in the presence of 3 g magnesium sulfate, 1.5 g sodium acetate and 10 g NaCl. The supernatant (1 mL) was collected after centrifugation, and a dispersive clean-up was performed using 200 mg PSA and 10 mg GCB. The extract was centrifuged at 10 000 rpm for 5 min. Three microlitre of the supernatant was injected for analysis. For recovery and validation studies 15 g of a confirmed blank 'churna' matrix is fortified with appropriate quantities of a pesticide standard mixture.

Experimental Conditions

The analytical method comprises a triple quadrupole GC-MS/MS system with liquid autosampler. The gas chromatograph is equipped with a temperature programmable PTV injector for splitless injection. The separation is performed using a 5% phenyl phase column of standard dimensions. After a short dispersive SPE clean-up the acetonitrile extract is directly injected after centrifugation. The Thermo Scientific TraceFinder™ software with compound database of pesticides was used for method setup and data processing (The Pesticides Compound Database, 2014).

For all pesticide compounds two SRM transitions have been chosen for the MRM acquisition method. The first transition was used for quantitation, the second transition for confirmation by checking the ion intensity ratio during data processing, see Table 4.15.

The retention times had been synchronized between data processing of QC standard samples with the acquisition method for the timed-SRM protocol (see Figure 4.59) in order to update all compound retention times to keep track of potential RT variations due to a matrix impact from real life samples.

Analysis Conditions

Sample material		Ayurvedic churna, herbs, spices
Sample preparation		
	Liquid/liquid extraction	15 g powdered sample
		15 mL acetonitrile containing 1% acetic acid
		Vortex with 3 g MgSO$_4$, 1.5 g NaOAc, 10 g NaCl, centrifuge
		Shake 1 mL supernatant with 200 mg PSA and 10 mg GCB, centrifuge
		Inject from supernatant

Table 4.15 MRM transitions for quantitation and confirmation, optimized collision energy (CE) and the precision of the calibration (R^2) of matrix spike standards.

Number	Compound name	RT (min)	Quantitation (m/z)	CE (V)	Confirmation (m/z)	CE (V)	R^2
1.	Diflubenzuron (degr. i-cyanat)	5.24	153.02 > 90.01	20	153.02 > 125.01	20	0.9969
2.	Diflubenzuron (degr. aniline)	5.75	127.01 > 65.01	30	127.01 > 100.01	30	0.9949
3.	Methamidophos	5.87	141.00 > 95.00	10	141.00 > 126.00	5	0.9930
4.	Dichlorphos (DDVP)	5.94	184.95 > 92.98	17	219.95 > 184.95	10	0.9960
5.	Dichlobenil	6.82	135.97 > 99.98	10	170.96 > 135.97	15	0.9960
6.	Mevinphos	7.39	127.03 > 109.02	10	192.04 > 127.03	12	0.9964
7.	Acephate	7.50	136.01 > 42.00	10	136.01 > 94.01	15	0.9904
8.	Dichloraniline, 3,5-	7.61	160.98 > 89.99	25	160.98 > 98.99	25	0.9989
9.	Molinate (Ordram)	8.58	126.07 > 55.03	10	187.10 > 126.07	10	0.9941
10.	TEPP	8.60	263.06 > 179.04	15	263.06 > 235.06	5	0.9946
11.	Omethoate	9.00	110.01 > 79.01	15	156.02 > 110.01	10	0.9969
12.	Fenobucarb	9.11	121.07 > 77.05	15	150.09 > 121.07	10	0.9977
13.	Propoxur	9.13	110.06 > 64.03	10	152.08 > 110.06	10	0.9981
14.	Propachlor	9.16	176.06 > 120.04	10	196.07 > 120.04	10	0.9980
15.	Ethoprophos	9.38	158.00 > 80.90	15	158.00 > 114.00	5	0.9949
16.	Trifluralin	9.58	264.09 > 160.05	15	306.10 > 264.09	15	0.9944
17.	Chlorpropham	9.62	213.00 > 127.00	5	213.00 > 171.00	5	0.9981
18.	Benfluralin	9.63	292.10 > 160.05	21	292.10 > 264.09	10	0.9923
19.	Sulfotep	9.70	322.02 > 202.01	15	322.02 > 294.02	10	0.9943
20.	Bendiocarb	9.72	166.06 > 151.06	15	166.06 > 166.06	15	0.9996
21.	Monocrotophos	9.80	127.03 > 95.03	20	127.03 > 109.03	25	0.9971
22.	Methabenzthiazuron	9.82	164.05 > 136.04	12	164.05 > 164.05	10	0.9974
23.	HCH, alpha	10.15	180.91 > 144.93	15	218.89 > 182.91	15	0.9970
24.	Metamitron	10.36	202.09 > 174.07	5	202.09 > 186.08	10	0.9969
25.	Atrazine	10.54	215.09 > 173.08	10	215.09 > 200.09	10	0.9945
26.	Pencycuron	10.62	125.05 > 89.04	12	180.07 > 125.05	12	0.9914
27.	Dioxathion	10.72	125.00 > 97.00	15	125.00 > 141.00	15	0.9936
28.	HCH, beta	10.73	180.91 > 144.93	15	218.89 > 182.91	15	0.9933
29.	Propetamphos	10.74	236.07 > 166.05	15	236.07 > 194.06	5	0.9918
30.	HCH, gamma (Lindane)	10.81	180.91 > 144.93	15	218.89 > 180.91	5	0.9939
31.	Terbuthylazine	10.84	214.10 > 132.06	10	229.11 > 173.08	10	0.9935
32.	Diazinon	10.88	137.05 > 84.03	10	304.10 > 179.06	15	0.9987
33.	Propyzamide	10.93	173.01 > 145.01	15	175.02 > 147.01	15	0.9939
34.	Fluchloralin	10.95	264.04 > 206.03	10	306.05 > 264.04	10	0.9967
35.	Pyroquilon	11.07	173.08 > 130.06	20	173.08 > 145.07	20	0.9974
36.	Pyrimethanil	11.11	198.11 > 158.09	30	198.11 > 183.10	15	0.9953
37.	Tefluthrin	11.16	177.02 > 127.02	20	197.03 > 141.02	15	0.9991
38.	Etrimfos	11.29	292.06 > 153.03	10	292.06 > 181.04	10	0.9935
39.	Pirimicarb	11.50	166.10 > 96.06	10	238.14 > 166.10	15	0.9937
40.	HCH, delta	11.54	180.91 > 144.93	15	204.07 > 91.03	15	0.9949
41.	Iprobenfos	11.54	204.07 > 122.04	15	218.89 > 182.91	15	0.9997
42.	Formothion	11.74	126.00 > 93.00	8	172.00 > 93.00	5	0.9982
43.	Phosphamidon II	11.83	227.05 > 127.03	15	264.06 > 193.04	15	0.9977
44.	Dichlofenthion	11.90	222.98 > 204.98	10	278.97 > 222.98	15	0.9946
45.	Dimethachlor	11.94	197.08 > 148.06	10	199.08 > 148.06	10	0.9992

Table 4.15 (*Continued*)

Number	Compound name	RT (min)	Quantitation (*m/z*)	CE (V)	Confirmation (*m/z*)	CE (V)	R^2
46.	Dimethenamid	11.95	230.06 > 154.04	10	232.06 > 154.04	10	0.9953
47.	Propazine	12.02	214.09 > 172.08	12	214.09 > 214.09	10	0.9970
48.	Propanil	12.06	217.01 > 161.00	10	219.01 > 163.00	10	0.9934
49.	Malaoxon	12.07	127.02 > 99.02	10	127.02 > 109.02	20	0.9978
50.	Chlorpyrifos-methyl	12.08	124.96 > 78.97	10	285.91 > 92.97	20	0.9945
51.	Metribuzin	12.13	198.08 > 82.03	20	198.08 > 110.05	20	0.9997
52.	Spiroxamine I	12.15	100.09 > 58.05	15	100.09 > 72.06	15	0.9909
53.	Vinclozolin	12.16	212.00 > 172.00	15	285.00 > 212.00	15	0.9957
54.	Carbofuran, 3-Hydroxy	12.21	137.06 > 81.03	18	180.08 > 137.06	15	0.9974
55.	Parathion-methyl	12.22	263.00 > 109.00	15	263.00 > 246.00	15	0.9966
56.	Alachlor	12.23	161.07 > 146.06	12	188.08 > 160.07	10	0.9997
57.	Tolclofos-methyl	12.25	264.96 > 92.99	20	264.96 > 249.96	15	0.9932
58.	Propisochlor	12.31	162.08 > 144.07	10	223.11 > 147.07	10	0.9983
59.	Metalaxyl	12.37	249.13 > 190.10	10	249.13 > 249.13	5	0.9911
60.	Carbaryl	12.41	144.06 > 115.05	20	144.06 > 116.05	20	0.9919
61.	Fuberidazol	12.41	183.80 > 156.10	10	183.80 > 183.10	20	0.9902
62.	Fenchlorphos (Ronnel)	12.47	284.91 > 269.92	13	286.91 > 271.91	20	0.9994
63.	Prosulfocarb	12.63	100.00 > 72.00	10	128.00 > 43.10	5	0.9938
64.	Pirimiphos-methyl	12.66	290.09 > 233.07	10	305.10 > 290.09	15	0.9911
65.	Spiroxamine II	12.75	100.09 > 58.05	15	100.09 > 72.06	15	0.9916
66.	Ethofumesate	12.80	207.08 > 161.06	10	277.02 > 109.01	8	0.9907
67.	Fenitrothion Confirming 1	12.80	277.02 > 260.02	10	286.11 > 207.08	12	0.9997
68.	Methiocarb	12.84	168.06 > 109.04	15	168.06 > 153.06	15	0.9971
69.	Malathion	12.92	127.01 > 99.01	10	173.02 > 127.01	10	0.9951
70.	Dichlofluanid	12.95	223.97 > 122.99	15	225.97 > 122.99	15	0.9971
71.	Phorate sulfone	13.01	153.00 > 125.00	5	199.00 > 143.00	10	0.9942
72.	Dipropetryn	13.02	241.90 > 149.80	20	254.90 > 180.30	20	0.9906
73.	Chlorpyrifos (-ethyl)	13.12	198.96 > 170.96	15	313.93 > 285.94	12	0.9995
74.	Fenthionoxon	13.22	277.80 > 109.10	25	329.60 > 298.90	10	0.9927
75.	Chlorthal-dimethyl (DCPA)	13.24	300.91 > 300.91	15	331.90 > 300.91	15	0.9986
76.	Flufenacet	13.26	211.04 > 123.02	10	211.04 > 183.03	10	0.9959
77.	Endosulfan I (alpha)	13.43	240.89 > 205.91	20	264.88 > 192.91	22	0.9942
78.	Imazethapyr	13.49	201.90 > 133.00	15	252.00 > 145.90	20	0.9944
79.	Butralin	13.50	266.14 > 190.10	15	266.14 > 220.11	15	0.9996
80.	Pirimiphos (-ethyl)	13.54	304.12 > 168.06	15	333.13 > 318.12	15	0.9992
81.	Pendimethalin	13.86	252.12 > 162.08	12	252.12 > 191.09	12	0.9912
82.	Fipronil	13.87	212.97 > 177.98	16	366.95 > 212.97	25	0.9938
83.	Cyprodinil	13.91	224.13 > 208.12	20	225.13 > 210.12	18	0.9959
84.	Metazachlor	13.92	133.05 > 117.04	20	209.07 > 132.05	12	0.9939
85.	Penconazole	14.01	248.06 > 157.04	25	248.06 > 192.04	15	0.9977
86.	Tolylfluanid	14.05	137.05 > 91.03	20	238.09 > 137.05	15	0.9922
87.	Chlorfenvinphos-Z	14.05	266.98 > 158.99	15	322.97 > 266.98	15	0.9904
88.	Allethrin	14.06	123.08 > 81.05	10	136.08 > 93.06	10	0.9923
89.	Mecarbam	14.09	226.04 > 198.03	5	329.05 > 160.03	10	0.9979
90.	Phenthoate	14.18	146.01 > 118.01	10	274.03 > 246.02	10	0.9951

(*continued overleaf*)

Table 4.15 (*Continued*)

Number	Compound name	RT (min)	Quantitation (*m/z*)	CE (V)	Confirmation (*m/z*)	CE (V)	R^2
91.	Mephosfolan	14.20	196.02 > 140.02	15	196.02 > 168.02	10	0.9973
92.	Quinalphos	14.21	146.03 > 118.02	15	157.03 > 129.02	13	0.9943
93.	Triflumizole	14.31	179.04 > 144.04	15	206.05 > 179.04	15	0.9925
94.	Procymidone	14.31	283.02 > 96.01	15	283.02 > 255.02	10	0.9983
95.	Bromophos-ethyl	14.50	358.89 > 302.91	20	358.89 > 330.90	10	0.9985
96.	Methidathion	14.60	124.98 > 98.99	22	144.98 > 84.99	10	0.9945
97.	Chlordane, alpha (cis)	14.62	372.81 > 265.87	18	374.81 > 267.87	15	0.9967
98.	DDE, o,p	14.63	245.95 > 175.97	25	317.94 > 245.95	20	0.9946
99.	Sulfallate	14.68	188.02 > 132.02	22	188.02 > 160.02	16	0.9945
100.	Paclobutrazol	14.72	236.10 > 125.06	15	236.10 > 167.07	15	0.9926
101.	Disulfoton sulfone	14.74	213.01 > 125.01	10	213.01 > 153.01	5	0.9912
102.	Picoxystrobin	14.77	303.09 > 157.04	20	335.09 > 303.09	10	0.9937
103.	Endosulfan II (beta)	14.88	271.88 > 236.89	18	338.85 > 265.88	15	0.9973
104.	Mepanipyrim	14.89	222.11 > 207.10	15	223.11 > 208.10	15	0.9965
105.	Chlordane, gamma (trans)	14.89	372.81 > 265.87	18	374.81 > 267.87	15	0.9991
106.	Flutriafol	14.97	123.04 > 75.03	15	219.07 > 123.04	15	0.9915
107.	Napropamide	15.00	128.07 > 72.04	10	271.16 > 128.07	5	0.9972
108.	Flutolanil	15.03	173.06 > 145.05	15	173.06 > 173.06	15	0.9988
109.	Pretilachlor	15.13	162.09 > 147.08	15	216.05 > 174.04	20	0.9935
110.	Hexaconazole, confirming 1	15.13	231.06 > 175.04	10	262.14 > 202.11	15	0.9962
111.	Isoprothiolane	15.14	290.06 > 118.03	15	290.06 > 204.05	15	0.9961
112.	Profenofos	15.21	138.98 > 96.98	8	338.94 > 268.95	20	0.9939
113.	Oxadiazon	15.26	258.05 > 175.04	10	304.06 > 260.05	10	0.9927
114.	DDE, p,p	15.32	245.95 > 175.97	25	317.94 > 245.95	20	0.9964
115.	Myclobutanil	15.40	179.07 > 125.05	15	179.07 > 152.06	15	0.9912
116.	Buprofezin	15.43	172.09 > 57.03	10	249.13 > 193.10	10	0.9906
117.	Kresoxim-methyl	15.44	206.09 > 116.05	15	206.09 > 131.06	15	0.9921
118.	DDT, o,p′	15.47	234.94 > 164.96	15	234.97 > 164.98	20	0.9935
119.	DDT, o,p′, confirming 1	15.47	236.94 > 164.96	20	236.97 > 164.98	20	0.9963
120.	Aramite-1	15.48	185.06 > 63.02	15	319.10 > 185.06	15	0.9959
121.	Aramite-2	15.69	185.06 > 63.02	15	319.10 > 185.06	15	0.9971
122.	Carpropamid	15.78	139.00 > 103.10	10	222.00 > 125.00	18	0.9982
123.	Cyproconazole	15.79	222.09 > 125.05	20	224.09 > 127.05	20	0.9989
124.	Nitrofen	15.85	201.99 > 138.99	21	282.98 > 252.98	15	0.9997
125.	Chlorobenzilate	15.98	251.02 > 139.01	20	253.03 > 141.01	15	0.9978
126.	Oxadiargyl	15.99	149.90 > 122.90	15	285.00 > 255.00	14	0.9963
127.	Fenthion sulfoxide	16.05	279.01 > 153.01	15	294.02 > 279.01	8	0.9958
128.	Diniconazole	16.11	268.06 > 232.05	15	270.06 > 234.05	15	0.9949
129.	Ethion	16.12	230.99 > 202.99	15	383.99 > 230.99	10	0.9973
130.	Oxadixyl	16.16	132.06 > 117.05	15	163.07 > 132.06	10	0.9985
131.	DDT, p,p′	16.20	234.94 > 164.96	20	234.94 > 164.96	20	0.9979
132.	DDD, p,p′	16.20	234.97 > 164.98	20	236.97 > 164.98	20	0.9959
133.	Chlorthiophos1	16.20	324.96 > 268.97	15	324.96 > 296.97	10	0.9969
134.	Imiprothrin	16.36	123.00 > 81.00	5	324.90 > 269.20	14	0.9967
135.	Mepronil	16.45	269.14 > 119.06	10	269.14 > 210.11	10	0.9945
136.	Triazophos	16.46	161.03 > 134.03	10	257.05 > 162.03	10	0.9936

Table 4.15 *(Continued)*

Number	Compound name	RT (min)	Quantitation (*m/z*)	CE (V)	Confirmation (*m/z*)	CE (V)	R^2
137.	Ofurace	16.58	186.05 > 158.05	10	232.07 > 186.05	10	0.9973
138.	Carfentrazone-ethyl	16.59	330.03 > 310.03	20	340.03 > 312.03	10	0.9919
139.	Benalaxyl	16.63	234.12 > 174.09	10	266.14 > 148.08	10	0.9951
140.	Trifloxystrobin	16.65	116.04 > 89.03	15	190.06 > 130.04	10	0.9962
141.	Propiconazole, peak 1	16.77	259.02 > 69.01	20	259.02 > 173.02	20	0.9989
142.	Edifenphos	16.78	173.01 > 109.01	15	310.03 > 173.01	10	0.9904
143.	Quinoxyfen	16.84	272.00 > 237.00	20	307.00 > 272.00	10	0.9982
144.	Endosulfan sulfate	16.85	271.88 > 236.89	15	273.88 > 238.89	15	0.9929
145.	Clodinafop-propargyl	16.87	349.05 > 238.04	15	349.05 > 266.04	15	0.9991
146.	Fluopicolide	16.90	208.80 > 182.00	20	261.00 > 175.00	24	0.9988
147.	Hexazinone	17.02	171.00 > 71.00	10	171.00 > 85.00	10	0.9998
148.	Propargite	17.16	135.06 > 107.05	15	350.16 > 201.09	10	0.9991
149.	Diflufenican	17.21	266.05 > 246.05	10	394.07 > 266.05	10	0.9981
150.	Triphenyl phosphate (TPP)	17.26	325.07 > 169.04	25	326.07 > 325.07	10	0.9995
151.	Iprodione	17.65	187.02 > 124.01	20	187.02 > 159.02	40	0.9979
152.	Bifenthrin	17.77	181.05 > 153.05	6	181.05 > 166.05	15	0.9922
153.	Picolinafen	17.90	376.08 > 238.05	15	376.08 > 239.05	15	0.9981
154.	Bromopropylate	17.91	184.98 > 156.98	20	342.96 > 184.98	20	0.9967
155.	Fenoxycarb	17.93	186.08 > 186.08	10	255.11 > 186.08	10	0.9933
156.	Fenpropathrin	18.01	181.09 > 152.07	23	265.13 > 210.10	15	0.9956
157.	Fenamidone	18.10	238.08 > 237.08	20	268.09 > 180.06	20	0.9994
158.	Tebufenpyrad	18.11	276.13 > 171.08	15	333.16 > 276.13	10	0.9997
159.	Fenazaquin	18.23	145.08 > 117.07	15	160.09 > 117.07	20	0.9951
160.	Imazalil	18.25	173.03 > 145.02	20	215.04 > 173.03	15	0.9954
161.	Furathiocarb	18.27	163.07 > 107.04	10	325.13 > 194.08	10	0.9989
162.	Flurtamone	18.38	199.06 > 157.05	20	333.10 > 120.04	15	0.9945
163.	Tetradifon	18.46	226.93 > 198.94	18	353.88 > 158.95	15	0.9973
164.	Phosalone	18.54	181.99 > 111.00	15	181.99 > 138.00	10	0.9985
165.	Triticonazole	18.57	217.09 > 182.07	10	235.10 > 217.09	10	0.9945
166.	Pyriproxyfen	18.68	136.06 > 78.03	15	136.06 > 96.04	15	0.9941
167.	Cyhalofop butyl	18.70	256.10 > 120.05	10	256.10 > 256.10	10	0.9969
168.	Tralkoxydim	18.80	137.00 > 57.20	10	181.04 > 152.03	23	0.9995
169.	Cyhalothrin, lambda	18.80	197.04 > 141.03	15	234.90 > 217.20	15	0.9997
170.	Lactofen	18.83	344.04 > 223.02	15	344.04 > 300.03	15	0.9975
171.	Benfuracarb	19.03	164.08 > 149.07	10	190.09 > 144.07	10	0.9975
172.	Pyrazophos	19.05	221.05 > 193.04	10	232.05 > 204.05	10	0.9930
173.	Fenarimol	19.15	139.01 > 111.01	15	219.02 > 107.01	15	0.9993
174.	Azinphos-ethyl	19.20	132.01 > 77.01	20	160.02 > 132.01	5	0.9944
175.	Fenoxaprop-P	19.41	288.03 > 260.03	10	361.04 > 288.03	10	0.9998
176.	Bitertanol1	19.59	170.09 > 115.06	25	170.09 > 141.07	20	0.9993
177.	Permethrin, peak 1	19.68	183.04 > 165.03	15	183.04 > 168.03	15	0.9973
178.	Bitertanol2	19.71	170.09 > 115.06	25	170.09 > 141.07	20	0.9993
179.	Permethrin, peak 2	19.81	183.04 > 165.03	15	183.04 > 168.03	15	0.9909
180.	Prochloraz	19.88	180.01 > 138.01	15	310.03 > 268.02	10	0.9932
181.	Cafenstrole	20.21	100.04 > 72.03	15	188.08 > 119.05	15	0.9991

(continued overleaf)

Table 4.15 *(Continued)*

Number	Compound name	RT (min)	Quantitation (m/z)	CE (V)	Confirmation (m/z)	CE (V)	R^2
182.	Cyfluthrin, peak 1	20.26	163.02 > 91.01	12	163.02 > 127.02	10	0.9915
183.	Fenbuconazole	20.34	129.04 > 102.03	15	198.07 > 129.04	10	0.9996
184.	Cypermethrin I	20.65	163.03 > 127.02	10	181.03 > 152.03	25	0.9996
185.	Boscalid (Nicobifen)	20.84	342.03 > 140.01	15	344.03 > 142.01	15	0.9977
186.	Flucythrinate, peak 1	20.85	199.07 > 107.04	22	199.07 > 157.06	10	0.9958
187.	Quizalofop-Ethyl	20.92	299.07 > 255.06	20	372.09 > 299.07	15	0.9969
188.	Etofenprox	21.08	163.09 > 107.06	16	163.09 > 135.07	10	0.9987
189.	Flucythrinate, peak 2	21.12	199.07 > 107.04	22	199.07 > 157.06	10	0.9989
190.	Fenvalerate, peak 1	21.94	167.05 > 125.04	10	419.13 > 225.07	10	0.9978
191.	Fluvalinate, peak 1	22.09	250.06 > 200.05	20	252.06 > 200.05	20	0.9973
192.	Pyraclostrobin	22.17	132.03 > 77.02	15	325.08 > 132.03	20	0.9936
193.	Fluvalinate, peak 2	22.20	250.06 > 200.05	20	252.06 > 200.05	20	0.9977
194.	Fenvalerate, peak 2	22.28	167.05 > 125.04	10	419.13 > 225.07	10	0.9996
195.	Difenoconazole, peak 1	22.76	323.05 > 265.04	15	325.05 > 267.04	20	0.9995
196.	Indoxacarb	22.95	203.03 > 106.01	20	203.03 > 134.02	20	0.9996
197.	Deltamethrin II	23.28	252.99 > 93.00	18	252.99 > 173.99	18	0.9987
198.	Azoxystrobin	23.63	344.10 > 329.10	20	388.11 > 345.10	15	0.9991
199.	Dimethomorph-1	23.91	301.10 > 165.05	10	387.12 > 301.10	12	0.9992
200.	Dimethomorph-2	24.60	301.10 > 165.05	10	387.12 > 301.10	12	0.9990

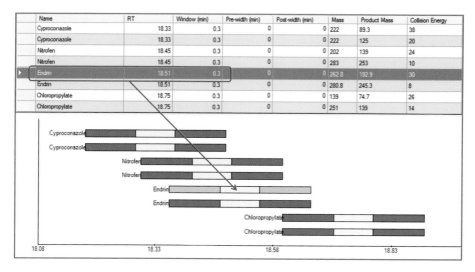

Figure 4.59 Principle of the timed-SRM acquisition method in the retention time range of 10.29–11.04 min. The white centre parts of the individual compound acquisition windows stand for the pesticide peak width centred to the compound retention time, the grey areas before and after the peak give the full SRM acquisition window. Yellow highlighted the beta HCH t-SRM window.

(*Continued*)

GC Method	System		Thermo Scientific™ Trace 1300 GC
	Autosampler		Thermo Scientific™ TriPlus RSH
	Column	Type	TraceGOLD™ TG-5SilMS
		Dimensions	30 m length × 0.25 mm ID × 0.25 μm film thickness
	Carrier gas		Helium
	Flow		Constant flow, 1.2 mL/min
	PTV Injector	Inlet liner	Baffled, Siltek™ deactivated
		Injection mode	Splitless
		Injection volume	3 μL
		Base temperature	87 °C, 0.3 min
		Transfer rate	14.5 °C/s
		Transfer temperature	285 °C
		Transfer time	2.5 min
		Cleaning ramp	14.5 °C/s
		Cleaning phase	20 min
		Final temperature	290 °C
	Split	Closed until	3 min
		Open	3 min to end of run
		Split flow	30 mL/min
		Purge flow	5 mL/min
	Oven program	Start	70 °C, 2 min
		Ramp 1	10 °C/min to 200 °C, 1 min
		Ramp 2	10 °C/min to 285 °C
		Final temperature	285 °C, 8.5 min
	Transferline	Temperature	285 °C
			Direct coupling
MS Method	System		Thermo Scientific TSQ 8000
	Analyzer type		Triple quadrupole MS
	Ionization		EI
	Electron energy		70 eV
	Ion source	Temperature	230 °C
	Acquisition	Mode	Timed-SRM
			2 SRMs per analyte
	Collision gas		Ar, 1.5 mL/min
	MRM Method		See Table 4.15
	Scan rate		0.2 s (5 spectra/s)
	Resolution	Setting	Normal (0.7 Da)
Calibration	External standard	Range	2.5–50 μg/L
	Data points		5
Data Processing			Thermo Scientific TraceFinder™ software

Results

All the 200 compounds included in the described method showed a precise calibration with good calibration correlation coefficients of >0.99 in the concentration range of 2.5–50 ng/g, as shown Figure 4.60. The obtained recoveries from spiked blank churna samples are reported high within 70–120% with <20% associated RSDs.

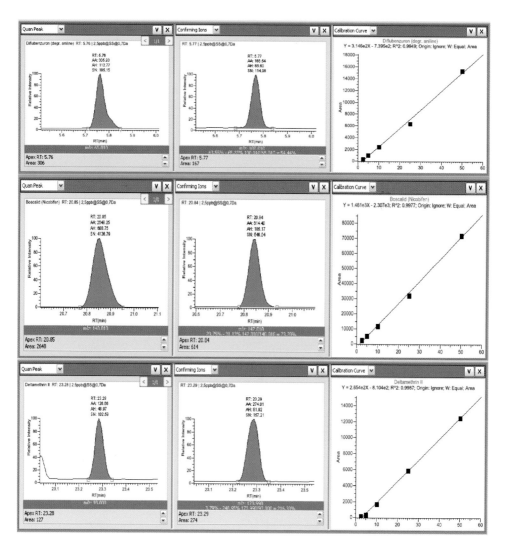

Figure 4.60 Selected pesticide chromatograms at 2.5 ng/g with quantitation peaks, confirming ions and calibration curves.

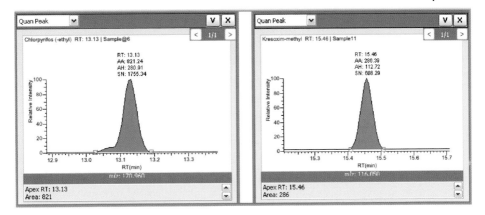

Figure 4.61 Traces of chlorpyrifos ethyl and kresoxim methyl were detected at 2.3 and 2.7 µg/kg respectively in regional market samples.

The method setup as describe above was applied for samples bought from the regional market. The results from analysis of market samples are presented in Figure 4.61 (Surwade, *et al.*, 2013).

Conclusions

The described method offers the routine screening for a large number of pesticides in herbs, spikes and similar products using a quick and fast sample preparation with a relatively short total analysis time of ~28 min. This method can be utilized for detection and confirmation of trace amounts of pesticides in difficult matrices such as herbal churnas or traditional medicine preparations.

A good linearity, specificity, recovery and repeatability of the method were established with minimal sample preparation time. GC-MS/MS as detection method provided a very high selectivity for the sensitive detection and quantitation of the pesticides even from samples with a high matrix load from the short sample preparation.

4.15
Determination of Polar Aromatic Amines by SPME

Keywords: nitroaromatic compounds; NAC; aromatic amines; *in situ* derivatization; iodination; SPME; drinking water; EU regulation; full-scan confirmation

Introduction

Aromatic amines are known to react with DNA and hence are suspected to cause a carcinogenic risk to humans. Six aromatic amines are classified as carcinogenic or probably carcinogenic by the International Agency for the

Research of Cancer (IARC), and the source of these compounds is manifold from several industries. The major use is in the poly-urethane production with possible emissions during production, use and disposal of such materials. A significant contribution to environmental distribution is the microbial reduction of nitroaromatic compounds (NACs) of which more than 70 compounds are mass produced with more than 1000 mt per year. Until today, more than 30 aromatic amines have been identified as metabolites, also deriving from applied pesticides.

The described quantification method provides a highly sensitive and robust method for monitoring aromatic amines at the low nanogram per litre level in water samples. The SPME method is particularly useful for the quantification of aromatic amines in drinking water. In contrast to the time-consuming SPE methodologies requiring high sample volumes, this SPME method utilizes a novel *in situ* derivatization of the polar compounds in aqueous solution (Pan and Pawliszyn, 1997) forming iodinated apolar derivatives. The automated SPME GC-MS procedure allows the routine quantitation and full-scan confirmation of a large number of water samples.

Analysis Conditions

Sample material			Drinking water, ground water
Sample preparation	In-situ derivatization		10 mL water sample, acidified, shaken with 0.5 mL NaNO$_3$ solution (10 g/L), 20 min, add 1 mL amidosulfonic acid solution (50 g/l), shake for 45 min, heat the solution to 100 °C, 5 min, cool to RT, add 0.25 mL NaSO$_3$ solution (sat.), adjust pH to approximately 8 with 0.25 mL K$_2$HPO$_4$ (0.25 mol/L) and 0.4 mL NaOH (5 mol/L), fill solution into 13 mL crimp top vials without headspace for analysis
	Extraction	SPME Immersion	Supelco PDMS/DVB fibre, 65 µm 30 min into sample solution using fibre vibration, use 'prep ahead' mode during GC runtime
	Desorption	Desorption time	5 min
		Temperature program	50 °C initial temperature 250 °C final temperature after 1 min splitless
		Split ratio	100 : 1 after 3 min

(*Continued*)

GC Method	System		Varian 3800
	Injector		Varian split/splitless, temperature programmable 1079 type
	Inlet liner		Siltek deactivated SPME liner, 0.8 mm ID
	Column	Type	Rtx-CLPesticides
		Dimensions	30 m length × 0.25 mm ID × 0.25 μm film thickness with 1.5 m retention gap and transfer capillary
	Carrier gas		Helium
	Flow		Constant flow, 2 mL/min
	Split	Closed until	3 min
		Open	3 min to end of run
		Split flow	200 mL/min
	Oven program	Start	40 °C, 3 min
		Ramp 1	15 °C/min to 130 °C
		Ramp 2	30 °C/min to 160 °C
		Isothermal	160 °C, 5 min
		Ramp 3	30 °C/min to 200 °C
		Final temperature	20 °C/min to 250 °C
	Transferline	Temperature	280 °C
MS Method	System		Varian Saturn 2000
	Analyzer type		Ion trap MS with internal ionization
	Ionization		EI, emission current 40 μA,
	Electron energy		70 eV
	Ion trap	Temperature	200 °C
	Acquisition	Mode	Full scan
	Mass range		60–450 Da
	Scan rate		0.35 s/scan
Calibration	Internal standard	Range	Aniline-d_5, 1 mg/L

Results

SPME has been successfully applied for the extraction of non polar components from aqueous samples. The *in situ* derivatization step lowers the polarity of the polar amines in an easy 'one pot' reaction with extraction efficiencies of the aromatic iodine derivatives of better than 95%, with the only exception of 2-amino-4,6-dinitrotoluene (2A4,6DNT) with only 77%. Spike solutions have been prepared with water and may not be diluted in alcohols, as the alcohol component will react as nucleophile (Figure 4.62).

SPME allows the complete transfer of the on the fibre accumulated analyte amount to the GC-MS analysis, also small sample sizes can be analyzed with the

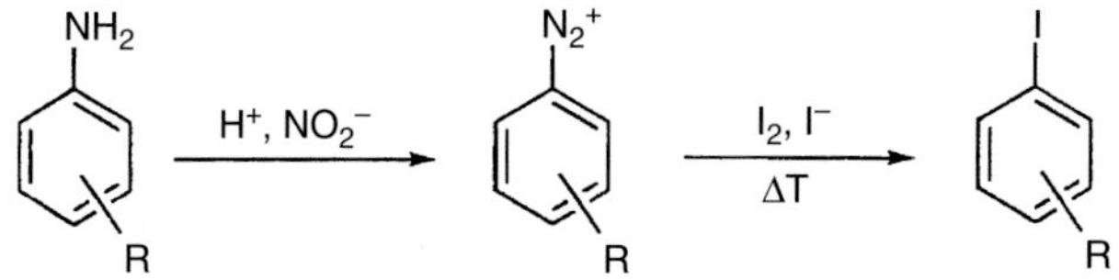

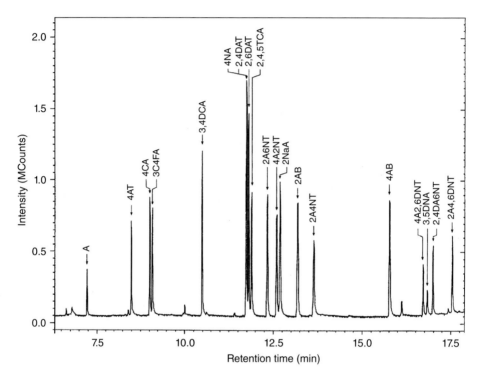

Figure 4.62 Derivatization reaction with diazotation followed by iodination.

Figure 4.63 Standard chromatogram of the derivatized aromatic amines (abbreviations see Table 4.16). (Zimmermann, Ensigner and Schmidt, 2004; reprinted with permission of Analytical Chemistry, Copyright 2004 American Chemical Society.)

given performance. The thermal stress to the fibre has been minimized by the temperature profile during injection resulting in the repeated use of one fibre bundle of up to 80 times.

For optimization of the chromatographic conditions, a standard sample diluted in ethyl acetate was applied using identical conditions. Figure 4.63 shows the separation of a standard mixture with almost baseline separation of all components. The MS was operated in full-scan mode to allow the additional detection of non-target analytes.

An overview of the components included in the described method is given in Table 4.16. Listed with the names and used abbreviations of the compounds are the chromatographic retention times, molecular weight of the iodine derivative

Table 4.16 List of aromatic amine compounds[a].

Compound	Abbreviation	Derivative	RT (min)	M	Detected ions[a] (m/z)	LOD (ng/L)	RSD % at 0.5 μg/L
Aniline	A	Iodobenzene	8.36	204	**204**, 77, 127	4	6.0
4-Aminotoluene	4AT	4-Iodotoluene	9.54	218	**218**, 91, 65	12	4.3
4-Chloroaniline	4CA	1-Chloro-4-iodobenzene	9.98	238	**238**, 111, 75	2	3.8
3,4-Dichloroaniline	3,4DCA	1,2-Dichloro-4-iodobenzene	11.55	272	**272**, 145, 109	3	6.7
2,4,5-Trichloroaniline	2,4,5TCA	1,2,4-Trichloro-5-iodobenzene	13.41	306	**306**, 179, 143	6	11
3-Chloro-4-fluoroaniline	3C4FA	2-Chloro-4-iodo-1-fluorobenzene	10.04	256	**256**, 129, 109	3	4.3
2,4-Diaminotoluene	2,4DAT	2,4-Diiodotoluene	13.22	344	**344**, 217, 90	13	11
2,6-Diaminotoluene	2,6DAT	2,6-Diiodotoluene	13.31	344	**344**, 217, 90	7	9.2
2-Naphthylamine	2NaA	2-Iodonaphthaline	14.59	254	**254**, 127, 74	11	7.4
2-Aminobiphenyl	2ABP	2-Iodobiphenyl	15.20	280	**280**, 152, 127	5	14
4-Aminobiphenyl	4ABP	4-Iodobiphenyl	16.98	280	**280**, 152, 127	9	11
4-Nitroaniline	4NA	1-Iodo-4-nitrobenzene	13.22	249	**249**, 219, 203	5	20
2-Amino-4-nitrotoluene	2A4NT	2-Iodo-4-nitrotoluene	15.69	263	**263**, 90, 105	8	16
2-Amino-6-nitrotoluene	2A6NT	2-Iodo-6-nitrotoluene	14.05	263	**246**, 89, 119	2	14
4-Amino-2-nitrotoluene	4A2NT	4-Iodo-2-nitrotoluene	14.44	263	**246**, 89, 119	3	16
2-Amino-4,6-dinitrotoluene	2A4,6DNT	2-Iodo-4,6 dinitrotoluene	18.49	308	**291**, 164, 89	38	20
4-Amino-2,6-dinitrotoluene	4A2,6DNT	4-Iodo-2,6-dinitrotoluene	17.76	308	**291**, 89, 63	27	13
2,4-Diamino-6-nitrotoluene	2,4DA6NT	2,4-Diiodo-6-nitrotoluene	18.00	389	**372**, 344, 216	30	16
Aniline-d5 (ISTD)	A-d5	Iodobenzene-d5	8.35	209	**209**, 82, 127	–	–

a) Quantitation mass in bold.
Zimmermann, Ensigner and Schmidt, 2004; reprinted with permission of Analytical Chemistry, Copyright 2004 American Chemical Society.

and its characteristic mass peaks. The base peaks used for compound selective quantitation has been typically the molecular ion. The observed $(M - 17)^+$ base peaks are due to the ortho-effect fragmentation (see also Section 3.2.6.9). With the iodinated derivatives the quantitation mass has been shifted by 111 Da per amino group to the higher mass range. This effect in particular allows the very sensitive detection of the derivatives with high S/N values as unspecific background typically appears in the lower mass range.

The calibration range has been selected in accordance with the European regulation for toxic organic pollutants in drinking water, with the regulated level of 0.1 μg/L for all target compounds. The calibration using aniline-d_5 as internal standard is linear from 0.05 μg/L over 2 orders of magnitude. A saturation of the SPME fibre has not been observed in this range. LOD have been calculated between 2

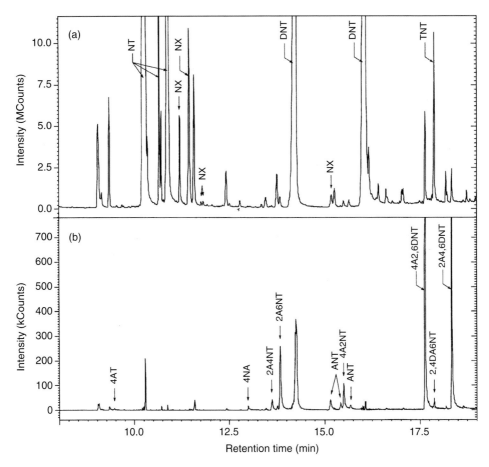

Figure 4.64 Analysis of a contaminated ground water from the area of a former ammunition plant. (Zimmermann, Ensigner and Schmidt, 2004; reprinted with permission of Analytical Chemistry, Copyright 2004 American Chemical Society.)

and 13 ng/L except for 2A4,6DNT, 4A2,6DNT and 2,4DA6NT with 27–38 ng/L, meeting excellently the EU regulatory levels.

Real water samples have been analyzed from different sources and demonstrated the applicability of the method for waste water, groundwater, surface water and drinking water. Figure 4.64 shows the analysis of a contaminated ground water sample from the area of a former ammunition plant. Repeated analyses gave RSDs in range of 3–10% (Zimmermann, Ensinger and Schmidt, 2004).

4.16
Analysis of Nitrosamines in Beer

Keywords: food safety; nitrosamines; beer; GC-MS/MS; MRM; quantitation; confirmation

Introduction

'Nitrosamine' is the common term used for compounds of the class of N-nitrosodialkylamines, with the general formula as shown in Figure 4.65. A large variety of compounds are known and described with different alkyl moieties (Boyd, Basic and Bethem, 2008). The simplest N-nitrosodialkylamine with two methyl groups is the N-nitrosodimethylamine (NDMA). Nitrosamines are in common highly toxic compounds with high cancerogenicity for humans and animals, in higher doses leading to severe liver damage with internal bleeding[1] (Agency for Toxic Substances & Disease Registry, 1989).

The 'classical' nitrosamine analysis was performed for many years by gas chromatography using a thermal energy analyzer (TEA) as detector. This special TEA detector was used due to its selectivity for nitrosamines based on the specific chemiluminescent reaction of ozone with the detector generated NO from nitrosamines. Today, with increased sensitivity requirements, the detection limits of the TEA, and also its complex operation, do not comply any more with the required needs for low detection limits and sample throughput. Mass spectrometric methods have increasingly replaced the TEA.

The EPA method 521 by Munch and Bassett from 2004 provided at that time a suitable GC-MS method based on chemical ionization (CI) using an ion trap mass spectrometer with internal ionization (Munch and Bassett, 2004; March and Hughes, 1989), in contrast to standard quadrupole or ion trap mass spectrometers

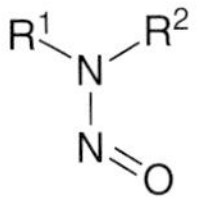

$$R^1{\diagdown}\underset{\underset{\displaystyle N{\diagup}\diagup O}{|}}{N}{\diagup}R^2$$

Figure 4.65 General formula of nitrosamines.

---

1) Material Safety Data Sheet NDMA.

using a dedicated (external) ion source design. Current developments in GC-MS triple quadrupole technology deliver today very high sensitivity and selectivity also in the small molecule mass range and allow the detection of nitrosamines at very low concentration levels even in complex matrix samples. This is made possible by using a much simpler and standard approach with the regular electron ionization (EI) for a very straightforward method for low level nitrosamine analysis.

This application describes a turn-key GC-MS/MS method for routine detection and quantitation of food borne nitrosamine compounds. The food matrix used here has been different malt beer products and as a final food product regular commercial beer.

The sample preparation is adapted and slightly modified from AOAC Official Method (2000), 982.11 (AOAC, 2000). An SPE column extraction method using a celite column and elution with DCM (dichloromethane) to isolate the nitrosamines from the beer samples was developed.

Analysis Conditions

Sample material			Mash, beer
Sample preparation			AOAC Official Method (2000) 982.11
	SPE	Material	Celite columns
		Elution	DCM
GC Method	System		Thermo Scientific™ Trace 1300 GC
	Autosampler		Thermo Scientific™ TriPlus RSH
	Carrier gas		Helium
	Flow		Constant flow, 1 mL/min
	Column	Type	TR-WAX MS
		Dimensions	30 m length × 0.25 mm ID × 0.5 µm film thickness
	Injection volume		1 µL
	SSL Injector	Injection mode	Splitless
		Injection Temperature	250 °C
		Injection Volume	
	Split	Closed	1 min
		Open	1 min to end of run
		Split flow	
	Surge	Pressure, time	300 kPa, 1 min
	Oven program	Start	45 °C for 3 min
		Ramp 1	25 °C/min to 130 °C
		Ramp 2	12 °C/min to 230 °C
		Final temperature	230 °C, 1 min hold

(Continued)

	Transferline	Temperature	250 °C Direct coupling
MS Method	System		Thermo Scientific™ TSQ 8000
	Analyzer type		Triple quadrupole MS
	Ionization	Mode	EI
	Electron energy		70 eV
	Ion source temperature		220 °C
	Acquisition	Mode	MRM using timed SRM
	Resolution	Setting	Normal (0.7 Da)
Calibration	Internal standard		NDPA added to each calibration level at 50 ppb
	Data points		7
	Calibration Range		1 to 500 ppb

MRM Method Setup

The triple quadrupole MS method setup was performed by using the AutoSRM software which is part of the employed TSQ 8000 software suite. The generated MRM method is shown in Table 4.17 using a short acquisition window of 60 s around the compound retention time for the retention time base data acquisition.

Table 4.17 MRM method setup using two transition for each compound, one for quantitation, the second one for confirmation by checking the calibrated ion ratio.

	Precursor ion (m/z)	Product ion (m/z)	Collision energy (eV)	Retention time (min)	Acquisition window (s)
NDMA	74.1	42.1	15	7.89	60
	74.1	43.8	5	7.89	60
NDEA	102.1	85.1	5	8.56	60
	102.1	44.1	10	8.56	60
NDPA (ISTD)	130.2	113.1	5	9.76	60
	130.2	42.9	10	9.76	60
NDBA	158.2	99.1	5	11.35	60
	158.2	141.1	5	11.35	60
NPIP	114.1	83.9	5	11.80	60
	114.1	41.5	15	11.80	60
NPYR	100.1	55.1	5	12.06	60
	100.1	43.0	10	12.06	60
NMOR	116.1	86.1	5	12.47	60
	116.1	56.1	10	12.47	60

Sample Measurements

From the large variety of potential nitrosamines the compounds that had been included in this method are those that are reported to be of relevance in the germinated malt drying process. Samples analyzed included malt beer as unspiked samples with 4% ethanol as sample blanks. In case of the analysis of other food matrices, additional compounds can be added to this method as described in the method setup by AutoSRM (Thermo Fisher Scientific, 2012a,b; Chen *et al.*, 2013).

Results

The chromatograms of the nitrosamines included in this method show a quick elution of the compounds from 7.87 min for NDMA to 12.47 min for NMOR allowing a short cycle time for a high sample throughput. High peak intensities as shown in Figure 4.66 are achieved at the lowest calibration level of 1 ppb. The nitrosamine NDMA can be detected with good S/N values.

The quantitative calibration has been performed in a wide concentration range from 1 to 500 ppb. Figure 4.67 shows the chromatogram peaks of NDMA from all the calibration runs with the S/N value at each dilution level. In all cases the

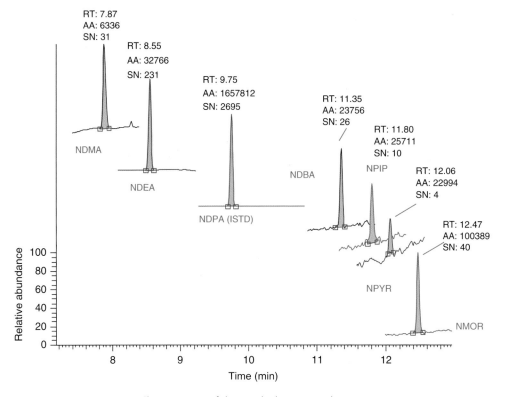

Figure 4.66 Chromatogram of the standard mix at 1 ppb.

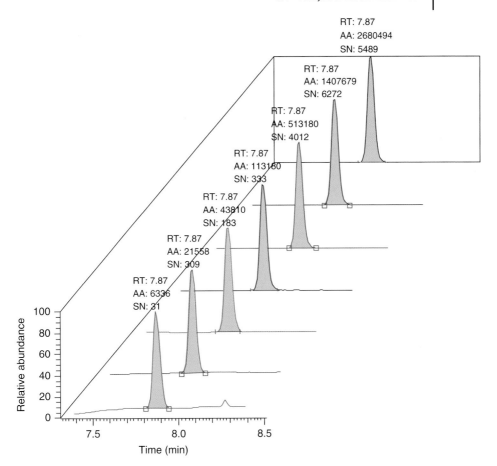

Figure 4.67 Eluting peaks with S/N values of the NDMA calibration runs from 1 (bottom) to 500 ppb (top).

NDMA peak shape is perfectly symmetrical, no tailing occurs and the peak area integration provides very reliable values without the need for any further manual corrections. The linear calibration of NDMA used to quantify the samples is shown in Figure 4.68 with very good correlation of R^2 better than 0.99. The same good calibration precision can be achieved for all nitrosamines.

LOQ Determination

The calculation of the LOQ and LOD was based on the S/N achieved for a chromatographic peak. For the LOD a minimum S/N value of 3 was requested, for the LOQ a minimum S/N of 10 (Table 4.18).

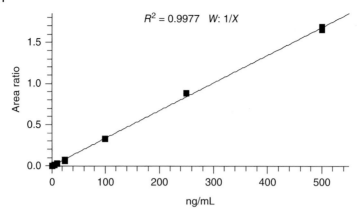

$R^2 = 0.9977$ W: 1/X

Figure 4.68 Linear calibration function for NMDA from 1 to 500 ppb.

Table 4.18 Calculation of the method LOQ and LOD.

Compound	S/N at 1 ppb	Calculated LOQ (ppb)	Calculated LOD (ppb)
NDMA	13	1.0	0.25
NDEA	231	0.05	0.02
NDBA	23	0.5	0.20
NPIP	10	1.0	0.50
NMOR	40	0.3	0.10
NPYR	24	3.0	1.0

Confirmation

For compound confirmation the ion ratio check provided by the quantitation software was used by comparing the ion intensity of the second acquired SRM transition with the first SRM used for quantitation. The precision for the ion ratio was calculated using the three replicate standard runs over the complete concentration range from 1 to 500 ppb and is shown Table 4.19. Although the detected ions

Table 4.19 Precision of the confirming ion ratio intensities from 1 to 500 ppb.

Concentration (ppb)	1	5	10	25	100	250	500	AVG	RSD (%)
NDMA	70.7	67.9	68.0	69.8	69.1	71.9	69.6	69.6	2.01
NDEA	20.8	22.1	22.5	22.4	22.5	22.5	22.5	22.2	2.84
NDBA	102.4	102.4	98.2	98.6	96.1	93.4	99.2	98.6	3.28
NPIP	6.1	5.5	6.2	5.9	6.0	6.1	6.2	6.0	3.88
NPYR	—	64.6	62.4	66.2	66.9	68.1	66.7	65.8	3.06
NMOR	67.8	70.4	69.7	69.6	69.8	69.1	69.7	69.4	1.18

all are in the low mass range and potentially subject to many interferences the precision of the product ion ratio is very good in the range of $1-4\%$.

For quality control purposes in sample analyses the confirmation of a positive result is done by the ion ratio check during the quantitation data processing. The ion ratio of the two acquired product ions is required to stay within $\pm5\%$ (10%) for all compounds, compared to the calibrated value from the standard runs. This provides a solid safety margin for routine sample measurements. Table 4.19 indicates the used average value (AVG) of the ion ratio for all nitrosamines investigated.

Sample Measurements

A number of samples have been measured, including blanks and spiked beer samples. The results of a blank sample are shown in Table 4.20. The found low NDMA concentration in this sample has been calculated below the calibration, and also below LOQ. The blank sample could be confirmed to be free from nitrosamine compounds at the given LOQ.

Another sample was prepared from beer that has been spiked with different amounts of nitrosamines. All nitrosamine compounds have been detected and quantified in a low concentration range of $9-13$ ppb, see Table 4.21. Each quantified peak passed the ion ratio quality control and could be positively confirmed at this low level by calculating the product ion ratios for each of the compounds.

Table 4.20 Results of a blank sample.

Compound	Area	ISTD Area	Area ratio	Ion ratio confirmation	Calculated amount (ppb)
NDMA	2591.4	2028129.8	0.001	Pass (65.1%)	0.74[a]
NDEA	1875.4	2028129.8	0.001	Fail (0%)	N/A
NDBA	6807.0	2028129.8	0.003	Fail (81.1%)	N/A
NPIP	N/A	2028129.8	N/A	N/A	N/A
NPYR	N/A	2028129.8	N/A	N/A	N/A
NMOR	4415.782	2028129.8	0.002	Fail (0%)	N/A

a) Below LOQ.

Table 4.21 Results from a spiked beer sample.

Compound	Area	ISTD area	Area ratio	Ion ratio	Calculated amount (ppb)
NDMA	91318.1	2282168.0	0.040	Pass (68.3%)	12.0
NDEA	480955.5	2282168.0	0.211	Pass (22.0%)	9.4
NDBA	402754.6	2282168.0	0.176	Pass (96.8%)	13.2
NPIP	280162.1	2282168.0	0.123	Pass (5.9%)	10.1
NPYR	318081.3	2282168.0	0.139	Pass (68.9%)	13.3
NMOR	1145719.1	2282168.0	0.502	Pass (67.9%)	10.1

Conclusions

With the described GC-MS/MS method using EI ionization all nitrosamine compounds under investigation could be safely detected and precisely quantified at the required low levels for a safe food control.

The LODs of all compounds have been determined to be below 1 ppb, using 1 ppb as the lowest concentration for the quantitative calibration. The linearity in the range of 1–500 ppb shows good quantitative precision. All calibration curves have been shown to be strictly linear with R^2 better than 0.99.

The described EI ionization GC-MS/MS method for food nitrosamines can serve as a turnkey method for routine use in food safety control. The presented method is fast, allows high sample throughput, and provides results with very high sensitivity and precision. With this standard EI ionization method setup this presented method for low level nitrosamine quantitation is recommended to be employed as a productive alternative to the earlier described chemical ionization ion trap procedure using liquid CI reagents.

4.17
Phthalates in Liquors

Keywords: food safety; spirits; plasticizer; phthalate acid ester; PAE; dinonyl-phthalate; DNP

Introduction

Phthalates (Phthalate Acid Esters, PAEs) have widespread use in the polymer industry as plasticizers and softeners to increase the elasticity of polymer materials. They are chemically inert, have high density, low to medium volatility, a high solubility in organic solvents, and are easily released during the ageing of the polymer materials. Phthalates had been reported as functional solvents in the aromatic, essential oil and even beverage industries. Phthalate plasticizers also migrate from plastic containers or closures into soft drinks and alcoholic beverages.

Phthalate residues in food and beverages are regulated internationally. In China the Ministry of Health issued a public notice on 1st June 2011, that phthalate esters are prohibited as non-food substances for use in food. PAEs are introduced into the food chain primarily through food packaging materials. Alcoholic beverages in plastic containers are of potential risk, since the containing ethanol provides a very good solubility for leaching the PAEs into the beverages from the plastic contact materials. The contamination risk increases with liquors having high ethanol content.

Sample Preparation

The beverage sample used for this application was a white spirit, bought from a local liquor store. Chinese liquor typically contains 30–60 vol% of ethanol. As phthalate esters are highly soluble in ethanol the extraction of phthalate esters using *n*-hexane as solvent is less effective (Dongliang, 2010). The removal of the major part of ethanol from the liquor before *n*-hexane extraction is necessary to avoid low recoveries. For optimization of the extraction procedure and recovery determination, one liquor sample was spiked with 4 mg/L concentration of a commercial phthalate standard.

An accurate amount of 5.0 mL sample was transferred in a glass centrifuge tube and then heated in a boiling water bath to remove the ethanol (Dongliang, 2010). The heating time depends on the alcoholic strength of the spirit sample. Usually the tube gets removed from the water bath with a residual volume of 2–3 mL. After cooling to room temperature, 2.0 mL of *n*-hexane was added. The glass tube was then shaken for extraction and left standing 5 min for phase separation. The supernatant was transferred to autosampler vials for analysis.

For the determination of the recoveries the standard solution was added to the sample to obtain a spiked solution at 0.80 mg/L concentration level. The results were compared with and without ethanol removal, shown in Table 4.22. After removal of ethanol before the extraction with *n*-hexane, good and consistent recoveries of the phthalate compounds in the range of 89–112% were obtained.

Table 4.22 Comparison of recovery of phthalates from liquor without and with prior removal of ethanol before extraction.

Compound	CAS #	Abbreviation	Without ethanol removal recovery (%)	With ethanol removal recovery (%)
Dimethyl phthalate	131-11-3	DMP	60.0	102.0
Diethyl phthalate	84-66-2	DEP	35.4	107.0
Di-isobutyl phthalate	84-69-5	DIBP	99.5	94.4
Di-n-butyl phthalate	84-74-2	DBP	106.0	104.0
Di-(4-methyl-2-pentyl) phthalate	146-50-9	DMPP	99.7	95.1
Di-(2-methoxy)-ethyl phthalate	117-82-8	DMEP	3.4	88.8
Diamyl phthalate	131-18-0	DPP	109.0	108.0
Di-(2-ethoxy)-ethyl phthalate	605-54-9	DEEP	13.6	103.0
Dihexyl phthalate	685-15-50-4	DHP	104.0	101.0
Butylbenzyl phthalate	85-68-7	BBP	88.4	108.0
Di-(2-ethylhexyl) phthalate	117-81-7	DEHP	106.0	108.0
Di-(2-butoxy)-ethyl phthalate	117-83-9	DBEP	83.1	104.0
Dicyclohexyl phthalate	84-61-7	DCHP	94.8	102.0
Di-*n*-octyl phthalate	117-84-0	DNOP	103.0	106.0
Diphenyl phthalate	84-62-8	DPhP	77.1	112.0
Dinonyl phthalate	84-76-4	DNP	110.0	109.0

Experimental Conditions

All measurements have been carried out using a single quadrupole GC-MS system equipped with split/splitless injector and liquid autosampler.

Sample Measurements

The elution order of the phthalate compounds was determined by analysing a standard mixture at medium concentration. The analyses had been run with full scan data acquisition. The spectra observed were compared with the NIST data base for identification and retention time determination. Although the full scan chromatogram as shown in Figure 4.69 gives high background and includes the elution of many other compounds dissolved in the spirit, the selective mass traces of the major phthalate ions allow a very good selectivity for a reliable peak area integration.

The compound quantitation was performed by selecting the most intense and unique ions of the PAE compounds providing selective mass chromatograms for individual peak integration. Finally, eight commercial liquor samples were prepared by the described sample preparation method for determining possible contamination by phthalate esters.

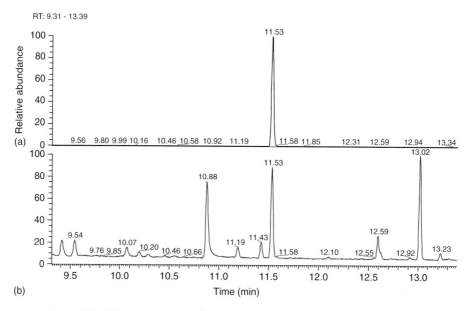

Figure 4.69 Dimethyl-phthalate chromatograms from spiked sample with the selective mass chromatogram *m/z* 163 (a) and the full scan trace (b) allowing the interference free peak area integration of the PAE compound.

Results

The full scan detection of three selected PAE components dimethyl-phthalate, di-isobutyl-phthalate and di-(2-ethylhexyl) phthalate are shown as examples in Figures 4.70–4.72. The mass spectra are taken for comparison with the NIST library to confirm the compound identity.

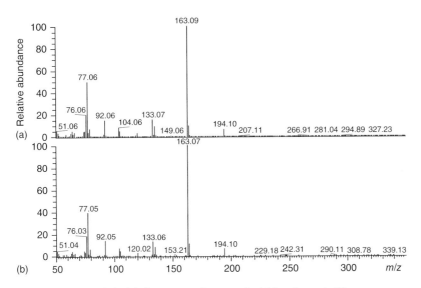

Figure 4.70   Dimethyl-phthalate spectra from standard (a) and sample (b).

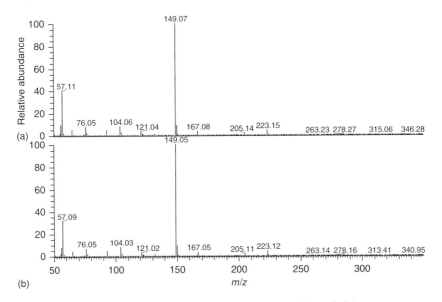

Figure 4.71   Di-isobutyl-phthalate spectra from standard (a) and sample (b).

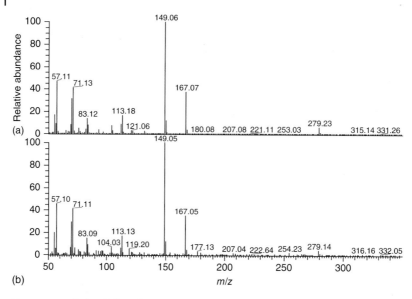

Figure 4.72 Di-(2-ethylhexyl) phthalate spectra from standard (a) and sample (b).

Quantitation

A series of matrix spiked samples with five different concentrations was prepared in the range of 0.10–4.00 mg/L of the standard solution. The samples were injected in sequence from low to high concentration. The peak areas were calculated for the calibration curve with linear regression with very good precision and R^2 value of 0.999 for all PAE compounds. The results for 15 phthalate esters show a very good linear relationship in the full calibration range of 0.1–4.0 mg/L.

The dinonyl-phthalates (DNP) create a special analytical challenge. The DNPs typically consist of a technical mixture of C9-isomers. Hence the response of DNP is distributed to individual isomers. The integration of the unresolved DNP peaks needs to be performed over a wider but constant retention-time range from the data processing software as shown in Figure 4.73. A linear calibration range for DNP of 0.4–4.0 mg/L could be achieved.

Sensitivity

The determination of the LOD and LOQ was based on the characteristic extracted ion mass chromatograms with a peak signal-to-noise ratio $S/N \geq 3$ for LOD, and $S/N \geq 10$ for LOQ, as given in Table 4.23 for the individual phthalate compounds. Figure 4.74 shows the calibration curves of 16 PAE compounds.

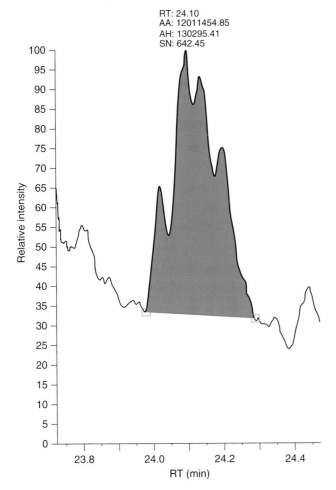

RT: 24.10
AA: 12011454.85
AH: 130295.41
SN: 642.45

Figure 4.73 Quantitation peaks of the unresolved DNP isomers over a set retention time range.

Method Precision and Determination of Recovery at Trace Level

The liquor samples were spiked by two low concentration levels at 0.1 and 0.3 mg/L, and measured five times at each level. The results show that the average recovery even at trace level was 83.2–110%, and the RSD range ($n = 5$) was 1.3–8.4%. The recovery and precision results are shown in Table 4.24.

Eight samples of commercially available liquor brands were analyzed using the described method. The concentrations of phthalate ester residues found are shown in Table 4.25. The samples tested showed that DEHP was found in low concentration in all samples, and DIBP, DBP, DEHP in many of the analyzed liquor samples.

Table 4.23 Phthalate quantitation – Linear dynamic range with limit of detection (LOD) and limit of quantification (LOQ), average R^2 0.9990.

Compound name	Retention time (min)	Quantitation ion (*m/z*)	Linear range (mg L)	Correlation coefficient R^2	LOD (µg L)	LOQ (µg L)
DMP	11.53	163	0.1–4.0	0.9994	0.1	0.3
DEP	13.02	149	0.1–4.0	0.9999	0.1	0.3
DIBP	15.64	149	0.1–4.0	0.9981	0.1	0.3
DBP	16.72	149	0.1–4.0	0.9986	0.1	0.3
DMPP	17.33/17.36	149	0.1–4.0	0.9993	0.2	0.6
DMEP	17.74	59	0.1–4.0	0.9984	0.2	0.6
DPP	18.43	149	0.1–4.0	0.9996	0.1	0.3
DEEP	18.59	72	0.1–4.0	0.9996	0.1	0.3
DHP	20.02	149	0.1–4.0	0.9990	0.1	0.3
BBP	20.94	149	0.1–4.0	0.9998	0.2	0.6
DEHP	21.37	149	0.1–4.0	0.9969	0.2	0.6
DBEP	21.45	149	0.1–4.0	0.9993	0.5	1.5
DCHP	22.50	149	0.1–4.0	0.9985	0.2	0.6
DOP	23.43	149	0.1–4.0	0.9998	0.5	1.5
DPhP	23.70	225	0.1–4.0	0.9988	0.2	0.6
DNP	24.0–24.4	149	0.4–4.0	0.9983	50	150

Analysis Conditions

Sample material			Liquors, alcoholic beverages
GC Method	System		Thermo Scientific™ Trace 1310 GC
	Autosampler		AS 1310 liquid autosampler
	Column	Type	Thermo Scientific™ TRACE™ TR-35MS
		Dimensions	30 m length × 0.25 mm ID × 0.25 µm film thickness
	Carrier gas		Helium
	Flow		Constant flow, 1 mL/min
	SSL Injector	Injection mode	Splitless
		Injection Temperature	280 °C
		Injection Volume	1 µL
	Oven program	Start	80 °C, 1 min
		Ramp 1	10 °C/min to 280 °C
		Final temperature	280 °C, 10 min
	Transferline	Temperature	280 °C
			Direct coupling

(*Continued*)

MS Method	System		Thermo Scientific™ ISQ
	Analyzer type		Single quadrupole MS
	Ionization		EI
	Electron Energy		70 eV
	Ion source	Temperature	280 °C
	Acquisition	Mode	Full scan
	Mass range		50–350 Da
	Scan rate		0.2 s (5 spectra/s)
	Resolution	Setting	Normal (0.7 Da)
Calibration	External standard	Range	0.10–4.00 mg/L, matrix spike

Conclusions

The described application follows the China regulation GB/T 21911-2008 for the determining of phthalates in food (Standardization Administration of China, 2008). The used sample preparation procedure was optimized from GB/T 21911-2008 with the ethanol removal from liquor beverages followed by an n-hexane extraction and GC-MS detection. The method is sensitive, rapid, and accurate and covers a wide linear concentration range to meet the need for trace level detection of phthalate esters in different types of beverages.

4.18
Analysis of the Natural Spice Ingredients Capsaicin, Piperine, Thymol and Cinnamaldehyde

Keywords: food safety; product safety; residue analysis; spices; active ingredients; traditional Chinese medicine, TCM; medical applications; personal defence products; pepper spray; MRM; analyte protectants

Introduction

Pungent spices are common ingredients for food preparations in all cooking traditions. Spices have been used as well for a long time in the traditional Chinese medicine (TCM). Beyond that there is a modern use of the active ingredients of spices in a variety of personal defence and law enforcement products, such as pepper spray, due to their immediate physiological irritation effects.

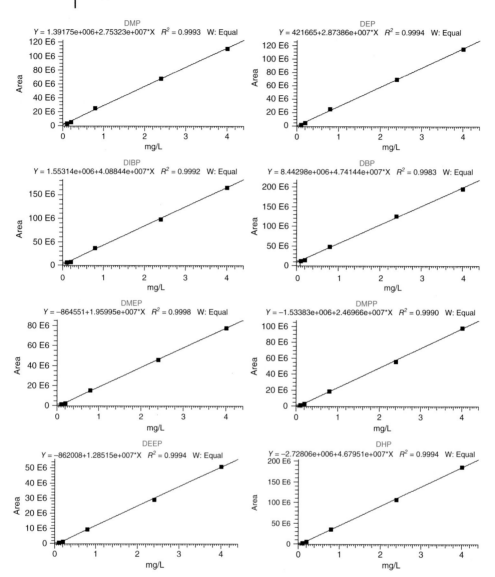

Figure 4.74 The linear calibration curves of all 16 PAEs.

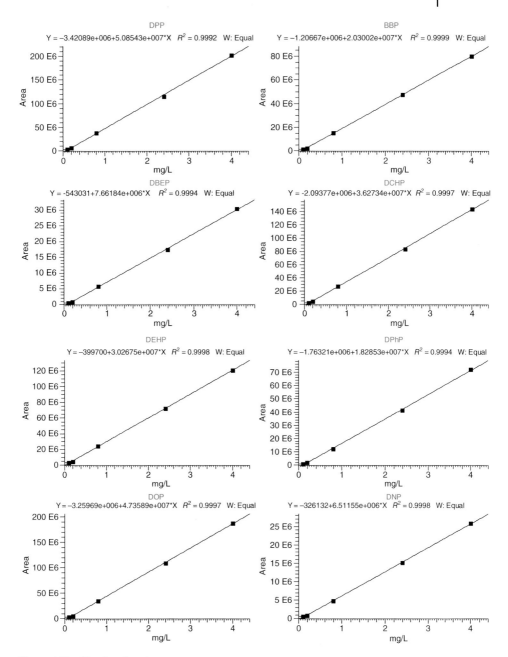

Figure 4.74 (*Continued*)

Table 4.24 Method recovery and precision data at trace level (average recovery 103%).

Compound name	Spike level 0.1 mg/L		Spike level 0.3 mg/L	
	Recovery %	RSD %	Recovery %	RSD %
DMP	95.0	5.4	99.0	4.7
DEP	103.0	5.5	108.0	2.2
DIBP	101.0	2.0	101.0	3.2
DBP	107.0	6.6	101.0	1.3
DMPP	105.0	3.3	107.0	5.7
DMEP	86.3	5.3	83.2	3.4
DPP	109.0	6.0	104.0	1.6
DEEP	103.0	4.1	104.0	3.2
DHP	104.0	4.6	109.0	3.7
BBP	110.0	3.6	103.0	3.7
DEHP	102.0	1.4	105.0	4.1
DBEP	104.0	5.0	108.0	4.6
DCHP	103.0	4.1	103.0	3.6
DOP	105.0	5.8	104.0	2.6
DPhP	108.0	4.2	109.0	1.8
DNP	107.0	8.4	101.0	5.4

Table 4.25 The phthalate ester concentration in eight commercial liquor samples (mg/L).

Compound	Sample 1	Sample 2	Sample 3	Sample 4	Sample 5	Sample 6	Sample 7	Sample 8
DMP	ND	0.303	ND	ND	0.005	ND	ND	0.025
DEP	ND	ND	ND	ND	0.011	ND	ND	ND
DIBP	ND	1.526	ND	1.373	0.106	ND	ND	ND
DBP	ND	1.024	0.045	0.656	0.133	ND	0.469	0.064
DMPP	ND	ND	ND	ND	ND	ND	ND	ND
DMEP	ND	ND	ND	ND	ND	ND	ND	ND
DPP	ND	ND	ND	ND	ND	ND	ND	ND
DEEP	ND	ND	ND	ND	ND	ND	ND	ND
DHP	ND	ND	ND	ND	ND	ND	ND	ND
BBP	ND	ND	ND	ND	ND	ND	ND	ND
DEHP	0.086	0.029	0.010	0.236	0.014	0.006	0.017	0.016
DBEP	ND	ND	ND	ND	ND	ND	ND	ND
DCHP	ND	ND	ND	ND	ND	ND	ND	ND
DOP	ND	ND	ND	ND	ND	ND	ND	ND
DPhP	ND	ND	ND	ND	ND	ND	ND	ND
DNP	ND	ND	ND	ND	ND	ND	ND	ND

Note: ND = not detected.

Piperine

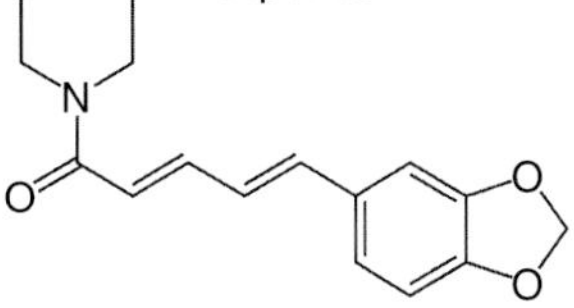

Capsaicin

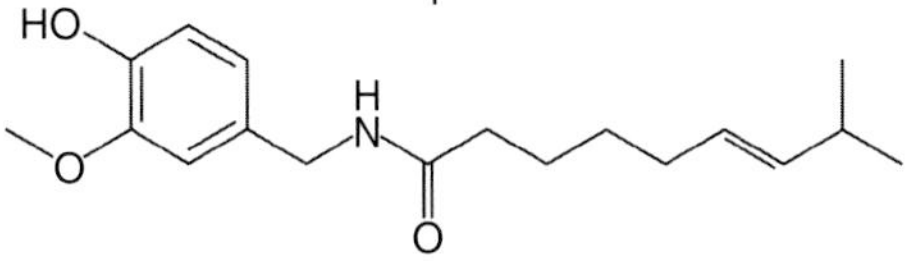

Thymol

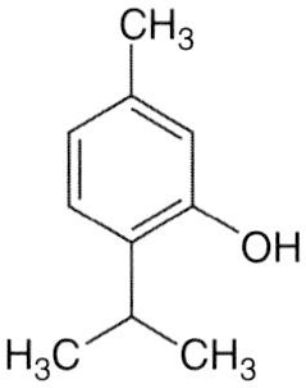

α-Methyl-t-Cinnamaldehyde

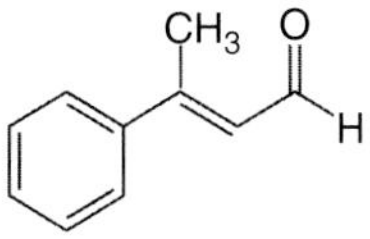

Experimental Conditions

This application describes the analysis of extracts from spices using GC-MS/MS as a highly selective tool for the quantitative trace determination of the representative ingredients of natural active spice ingredients capsaicin, piperine, thymol, and also cinnamaldehyde as a flavouring side component.

Sample Measurements

The active compounds capsaicin and dihydrocapsaicin elute with a short retention time difference. A good separation free from peak tailing is necessary for the reliable peak integration for low RSD values at low concentration levels. It was found with different types of GC columns that the quality of the column deactivation, age of the column and matrix deposits have a detrimental effect on the capsaicin and dihydrocapsaicin peak shape and quantitative reproducibility. Also, piperine and cinnamaldehyde were affected while thymol always showed symmetrical peak shapes, apparently being unaffected by the increasingly active column film conditions.

An analyte protectant was co-injected with the extract of active analytes to preserve inert conditions with the inlet liner and analytical column for high quantitative precision and reproducible results. In this case a concentration of 2 ppm of sorbitol was added to the extracts in the shown experiments.

Results

With the co-injection of sorbitol as analyte protectant symmetrical peak shapes for all described compounds could be achieved, including the critical pair capsaicin and dihydrocapsaicin, as shown in Figure 4.75. It is important to note here that it is not only the immediately visible peak shape and peak separation that is positively affected, but due to a significantly reduced tailing the integrated peak areas and S/N ratios are outstandingly increased. The individual peaks for selected compounds capsaicin and piperine of the calibration runs up to 1000 ppb, normalized to 100% each, are given in Figures 4.76 and 4.77. The linear quantitative calibrations with a zoom into the low concentration range of 10–200 ppb are shown in Figure 4.78.

Conclusions

The reproducibility for a series of measurements using a spiked real life spice sample with the application of above mentioned analyte protectant was determined on three consecutive days. The precision of the area results has been calculated as RSD (%). The peak area data in Table 4.26 indicate the good precision of the spiked spice sample analysis a low level spike of below 10 ppb. The reproducibility over three days for all compounds tested is in the range of 1–3%.

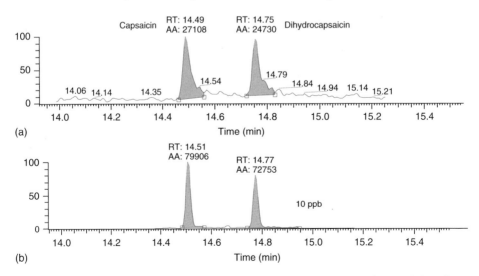

Figure 4.75 GC peak shapes of capsaicin and dihydrocapsaicin, (a) without and (b) with analyte protectant, both runs are at 10 ppb concentration using a 30 m 5%-phenyl column.

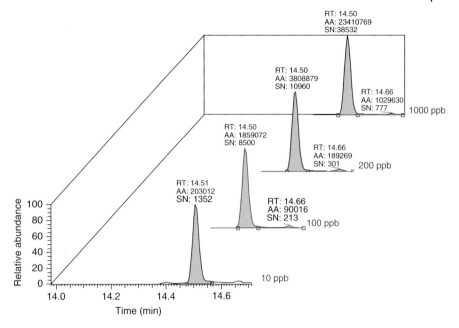

Figure 4.76 Capsaicin calibration peaks 10–1000 ppb.

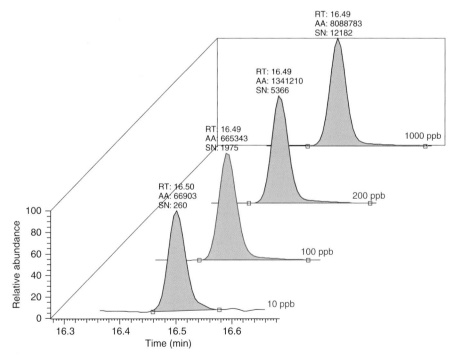

Figure 4.77 Piperine calibration peaks 10–1000 ppb.

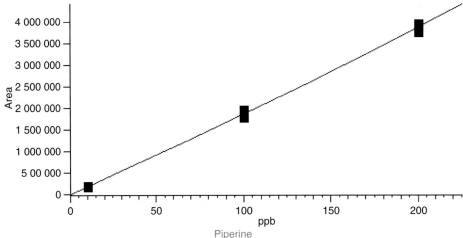

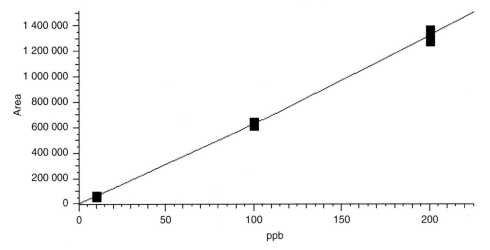

Figure 4.78 Quantitative calibrations on low concentration side for capsaicin and piperine.

Table 4.26 Precision of a spiked spice sample analysis.

Compound Name	Day 1 (area cts)	Day 2 (area cts)	Day 3 (area cts)	RSD (%)
Thymol	96.513	94.128	91.462	2.7
α-Methyl-*trans*-cinnamaldehyde	97.665	93.579	92.918	2.7
Capsaicin	100.669	105.363	99.392	3.1
Dihydrocapsaicin	102.752	103.852	101.089	1.4
Piperine	104.307	106.685	103.274	1.7

Analysis Conditions

Sample material			Spices, fruit, medical applications, personal defense products
Sample preparation			Dried material Solvent extraction with acetonitrile, methanol, or ethanol Filtration, concentration
GC Method	System		Thermo Scientific™ Trace 1300 GC
	Autosampler		Autosampler Thermo Scientific™ Triplus RSH
	Column	Type	Rtx-5Sil MS
		Dimensions	15 m length × 0.25 mm ID × 0.25 µm film thickness
	Carrier gas		Helium
	Flow		Constant flow, 1.2 mL/min
	SSL Injector	Injection mode	Splitless
		Injection temperature	300 °C
		Injection volume	1 µL
	Split	Closed until	1 min
		Open	1 min to end of run
		Split flow	50 mL
		Gas saver	2 min, 20 mL/min
		Purge flow	5 mL/min
	Oven program	Start	50 °C, 2 min
		Ramp 1	20 °C/min to 300 °C, 2 min
	Transferline	Temperature	280 °C Direct coupling
MS Method	System		Thermo Scientific™ TSQ 8000
	Analyzer type		Triple quadrupole MS
	Ionization		EI
	Electron energy		70 eV
	Ion source	Temperature	250 °C
	Acquisition	Mode	Timed-SRM
	Mass range		See MRM table (Table 4.27)
	Scan rate		200 ms
	Resolution	Setting	Normal (0.7 Da)
Calibration			External standardization
	Calibration range		10–1000 ppb

Table 4.27 MRM data acquisition method.

Compound name	CAS Number	RT (min)	Precursor mass (*m/z*)	Product mass (*m/z*)	Collision energy (V)	Peak width (min)
Thymol	89-83-8	6.24	135.1	91.1	15	5
Thymol	89-83-8	6.24	150.1	135.1	10	5
α-Methyl-*trans*-cinnamaldehyde	101-39-3	6.51	145.1	91.1	25	5
α-Methyl-*trans*-cinnamaldehyde	101-39-3	6.51	145.1	115.1	20	5
Capsaicin	404-86-4	12.64	137.0	94.0	20	5
Capsaicin	404-86-4	12.64	137.0	122.0	15	5
Dihydrocapsaicin	19408-84-5	12.89	137.0	94.0	20	5
Dihydrocapsaicin	19408-84-5	12.89	137.0	122.0	15	5
Piperine	94-62-2	14.08	200.8	115.1	20	5
Piperine	94-62-2	14.08	285.0	172.7	10	5

4.19
Aroma Profiling of Cheese by Thermal Extraction

Keywords: food quality; aroma profiling; off-odours; cheese; thermal extraction; dynamic headspace; TOF MS

Introduction

The aroma impression is important for consumers in spontaneous grocery food checking. Aroma profiling on a professional aspect is a key component in the industrial control amongst others of constant quality, shelf life, composition or maturity. In case of cheese a complex blend of low and high level compounds combine to create their distinctive odours, the subject of further analytical investigation.

The nature of cheese varies considerably, and the subtle differences in aroma, flavour and texture all add to its appeal. Various factors are involved in the nature of the final product, including the type of animal, its diet, processing of the milk, the cheese culture, and the ageing conditions. Manufacturers spend a great deal of effort in ensuring the products are of a consistently high quality.

Extraction of the diversity of aroma components from cheese at the typical temperature of consumption, trapping on sorbent tubes followed by thermal desorption to a GC-MS system allows the evaluation of the aroma profile and the identification and quantitation of individual compounds.

Sample Preparation

Samples of the following cheeses were analyzed:

- Full-fat Welsh Cheddar (extra-mature)
- Low-fat Cheddar (mild)

- Comte (a hard cheese made with unpasteurized cow's milk)
- Emmental (a medium-hard cheese)
- Channel Island Brie (a soft cheese).

The cheese samples were cut and placed in equal amounts into inert-coated heated extraction chambers for release and trapping of the volatile compounds. Nitrogen as an inert gas is passed through the chambers at a constant flow rate while sweeping the headspace vapours onto individual sorbent packed tubes. The used micro-chamber thermal extractor allows up to six samples to be extracted in parallel under identical conditions for comparable results.

Experimental Conditions

Dynamic Headspace Sampling
The sample chamber is set to a temperature of 40 °C for pre-heating and extraction. This temperature was chosen for a good representation of the released VOCs of the cheeses as a mock-up of the typical mouth feeling. In this dynamic headspace sampling step the volatile headspace is flushed from the cheese samples by using the inert gas nitrogen for collection onto the sorbent packed tubes. After collection the tubes had been transferred to a thermal desorption unit.

Thermal Desorption
Thermal desorption concentrates the collected cheese aroma constituents into a narrow elution band for GC-MS analysis. Each sorbent tube of the individual cheese samples was thermally desorbed by evaporating them inverse to the sampling direction into the GC carrier gas stream using an automated thermal desorber.

During the automated desorption process the released components were transferred to a Peltier cooled smaller internal 'focusing' trap integrated within the TD system. After completion of the slower primary sampling tube desorption, the focusing trap was itself very quickly desorbed by heating it rapidly in reverse flow for injection of the compounds into the GC column. This two-stage desorption process optimizes the concentration enhancement and produces narrow chromatographic peaks for optimum separation and sensitivity.

The choice for the INNOWax column with a polyethylene glycol (PEG) stationary phase was taken due to the high polarity and high upper temperature limits to provide optimum separation for the containing polar aroma compounds expected to be present in these samples.

The MS detection has been carried out using the BenchTOF-dx time-of-flight mass spectrometer for its sensitivity in the full scan acquisition mode. This allows recording of the complete mass spectra at similar levels to that of quadrupole instruments run in SIM mode. The full scan mass spectra had been identified by matching against commercial spectra libraries NIST and Wiley.

Sample Measurements

The chromatograms shown in Figure 4.79 are taken from four types of cheeses, collected to individual sorbent tubes simultaneously from adjacent chambers. The intensities are normalized to the chromatogram base peaks. All samples had been analyzed with identical TD–GC-MS conditions. The pattern of the collected volatile compounds can be compared easily by overlaying the chromatograms. Even without a quantitative calibration the peak areas give a good approximation of the relative abundances of the individual components, estimating comparable response factors for these organic compounds of similar structure.

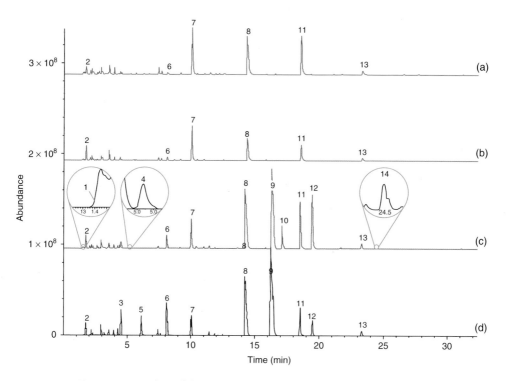

Figure 4.79 Analysis of the VOC profiles of (a) full-fat (extra-mature) Cheddar, (b) low-fat Cheddar, (c) Comte and (d) Emmental, with major compounds and some sulfur compounds labelled (insets). (1) Hydrogen sulfide; (2) carbon disulfide; (3) propan-1-ol 4 Dimethyl disulfide; (5) pentan-2-ol; (6) 3-methylbutan-1-ol; (7) 3-hydroxybutan-2-one (Acetoin); (8) acetic acid; (9) propanoic acid; (10) 2-methylpropanoic acid; (11) butanoic acid; (12) 3-methylbutanoic acid; (13) hexanoic acid; (14) dimethyl sulfone.

Analysis Conditions

Sample material			Food, cheese
Sample preparation			
	Thermal extraction	Markes Ltd.	Micro-Chamber/Thermal Extractor
		Purge gas	Nitrogen
		Trap	Quartz wool – Tenax®
			TA – Carbograph™ 5TD
		Sample volume	5 g cheese, grated
		Chamber temperature	40 °C
		Equilibration time	20 min
		Purge flow	50 mL/min
		Sampling time	20 min
	Thermo desorber	Markes Ltd.	UNITY 2
		Flow path temperature	160 °C
		Transfer line	200 °C
		Focusing trap	Material emissions
		Dry-purge	2 min, 20 mL/min flow to split
		Primary (tube) first stage	150 °C for 5 min
		Desorption	50 mL/min trap flow, no split flow
		Second stage	300 °C for 5 min
			50 mL/min trap flow, no split flow
		Secondary (trap)	Trap low: 30 °C, trap high: 300 °C
		Desorption	Heating rate: 24 °C/s
		Hold time	5.0 min, split flow
			50 mL/min (high split)
			5 mL/min (low split)
		Pre-trap fire purge	2 min, 50 mL/min trap flow
			50 mL/min split flow (high split)
			5 mL/min split flow (low split)
		Overall TD split	51 : 1 (high split), 6 : 1 (low split)
GC Method	System	Thermo Scientific	TRACE GC Ultra
	Autosampler	Markes Ltd.	TD-100
	Column	Type	HP-Innowax
		Dimensions	30 m length × 0.25 mm ID × 0.25 μm film thickness
	Carrier gas		Helium
	Flow		Constant flow, 1 mL/min (initial $P = 6.4$ psi)
	SSL Injector	Injection mode	Splitless
		Injection temperature	160 °C
	Oven program	Start	40 °C, 2.0 min
		Ramp 1	5 °C/min to 180 °C
		Ramp 2	20 °C/min to 260 °C
		Final temperature	260 °C, 6.0 min
	Transferline	Temperature	265 °C

(*Continued*)

MS Method	System		ALMSCO International BenchTOF-dx™
	Analyzer type		Time-of-flight MS with axial ion acceleration
	Ionization		EI
	Ion source	Temperature	260 °C
	Acquisition	Mode	Full scan
	Mass range		33–350 Da
	Scan rate		2 Hz

Results

Similar aroma profiles had been acquired of the low-fat and full-fat Cheddar cheese with the main differences being in peak intensities.

Looking at the individual compounds identified, the presence of carbon disulfide is seen in all four samples, along with trace levels of hydrogen sulfide, dimethyl disulfide and dimethyl sulfone in the Comte.

Fatty acids, both straight-chain and branched, are key contributors to the aromas of various cheeses. The presence of large amounts of propanoic acid and 3-methylbutanoic acid had been found in the odour profiles of the Emmental and Comte, with the added presence of 2-methylpropanoic acid in the latter.

The presence of 3-methylbutan-1-ol has been identified in all four samples. This compound, which is known to confer a pleasant aroma of fresh cheese, was most abundant in the Comte and Emmental. In the Emmental relatively large amounts of propan-1-ol and pentan-2-ol has been determined. The respective 'sweet' and 'fresh' notes that have been reported for these components may contribute to the distinctive aroma of this cheese.

Conclusions

Thermal extraction of foods in combination with thermal desorption of the collected volatile headspace constituents, in this reported case of different cheeses varieties, is a viable method for the reproducible aroma characterization.

The described technology was used to evaluate the aroma profile from a range of cheese samples. Besides the chromatographic pattern, both the identification and relative quantitation of desirable compounds as well as off-odours can be performed. The method also enables the identification of changes in the composition over time, providing useful data relating to shelf life and product safety (Markes Int. Ltd., 2010, 2012).

4.20
48 Allergens

Keywords: product safety; consumer protection; toys; fragrance; allergens; extraction; ion trap; MS/MS; EU regulation; liquid-liquid extraction (LLE)

Introduction

More than 2500 fragrance ingredients are used in perfumes and perfumed consumer goods such as cosmetics, detergents, fabric softeners and other household products to give them a specific, usually pleasant smell. They can sometimes cause skin irritations or allergic reactions (EU Scientific Committee on Consumer Safety, 2011). The safety of children toys is regulated by the EU Directive 2009/48/EC, also including a list of allergens that shall not be contained. The presence of traces of these fragrances shall be allowed provided that such presence is technically unavoidable under good manufacturing practice and does not exceed 100 mg/kg (European Commission, 2009).

The referenced publication provides the general methodology for the determination of 48 out of the 66 restricted fragrance allergens by the EU regulation. Eight of the remaining 18 fragrance allergens are natural extracts that are not amenable to a GC method, and another 10 are reported to be commercially unavailable (Lv *et al.*, 2013).

Sample Preparation

Extraction
The extraction method uses a complete dissolution of plastic toys, then the liquid extraction by an immiscible polar solvent, in this case methanol. Clays, paper and other not dissolvable materials are extracted directly after chopping into small pieces of less than 2 mm size. Also cryo milling to preserve the volatile fragrances is recommended (Hopfe, 2009).

Plastic toys: 1 g of the sample is dissolved with 10 mL of an appropriate solvent, for instance acetone for acrylonitrile butadiene styrene (ABS), dichloromethane for polystyrene (PS) and tetrahydrofuran for polyvinylchloride (PVC) in an ultrasonic bath. 10 mL of methanol is added to the solution and shaken. The methanol phase is centrifuged, applied to SPE clean-up and concentrated.

Play clays: 1 g of the sample is extracted with 10 mL acetone and methanol in an ultrasonic bath. The combined solutions get centrifuged, applied to SPE clean-up and concentrated.

Plush and paper toys: The samples are cut into pieces of less than 5 mm size and extracted with 20 mL of acetone in an ultrasonic bath. Prior to GC-MS analysis the extract is filtered using a 0.45 µm PTFE filter.

For clean-up and concentration an SPE step is used. The eluate gets concentrated to 5 mL volume for injection to GC-MS.

Experimental Conditions

As the MS detection method the MS/MS mode was chosen. The single ion detection (SIM/SIS) of the ion trap analyser has been compared using matrix samples and did not deliver the required compound selectivity for trace level quantitation in matrix, see Figure 4.80.

For MS/MS method development the individual standards are analysed in full scan mode (40–400 Da) for selection of the precursor ions. Two MS/MS transition has been used for each compound, one for quantitation and the second as a qualifier for confirmation. The ion trap MS/MS excitation voltage was optimized for the precursor ions of each individual compound for best isolation efficiencies and maximum response of the characteristic product ions. The complete setup for MS/MS acquisition is shown in Table 4.28. For data acquisition the chromatogram was divided into 13 retention time segments with the parallel detection of the assigned compounds.

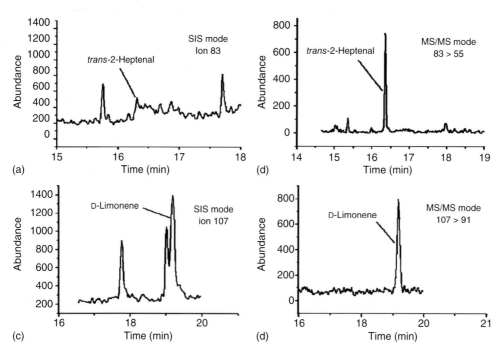

Figure 4.80 Comparison of SIM and MS/MS mode, (a and b) trans-2-heptenal, 2.6 mg/kg and (c and d) D-limonene, 18.2 mg/kg. (Lv *et al.* 2013, reproduced with permission of John Wiley & Sons.)

Table 4.28 Fragrance allergens with MS/MS acquisition and quantitation details. (Lv *et al.* 2013, reproduced with permission of John Wiley & Sons.)

No.	Compound name	EU Regulation	Retention time (min)	CAS Number	Quantitation Transition m/z > m/z	Excitation voltage (V)	isol. wind. (Da)	Qualifier transition m/z > m/z	Excitation voltage (V)	isol. wind. (Da)	Linear ranges (mg/L)	Correlation coefficient R^2
1	Ethyl acrylate	a)	8.17	140-88-5	99 > 77	0.4	4	99 > 81	0.4	4	0.2–50	0.9990
2	Methyl-*trans*-2-butenoate	a)	9.97	623-43-8	85 > 57	0.3	3	100 > 69	0.4	3	0.05–20	0.9995
3	5-Methyl-2,3-hexanedione	a)	12.30	13706-86-0	85 > 57	0.5	3	85 > 41	0.5	3	0.02–10	0.9996
4	*trans*-2-Heptenal	a)	16.34	18829-55-5	83 > 55	0.4	4	95 > 79	0.4	5	0.05–10	0.9993
5	*trans*-2-Hexenal-dimethyl-acetal	a)	17.58	18318-83-7	113 > 71	0.6	3	113 > 97	0.6	3	0.2–20	0.9968
6	Benzyl alcohol	a)	18.93	100-51-6	108 > 79	0.6	3	91 > 65	0.4	4	0.005–5	0.9995
7	D-Limonene	b)	19.65	5989-27-5	107 > 91	0.6	3	136 > 94	0.5	3	0.05–10	0.9992
8	Dimethyl citraconate	a)	20.53	617-54-9	127 > 99	0.4	4	127 > 69	0.4	4	0.005–5	0.9993
9	*trans*-2-Hexenal-diethyl-acetal	a)	21.33	67746-30-9	127 > 85	0.7	3	127 > 98	0.8	3	0.5–10	0.9943
10	Linalool	b)	21.51	78-70-6	93 > 77	0.6	4	121 > 93	0.4	3	0.02–10	0.9994
11	Benzyl cyanide	a)	21.87	140-29-4	117 > 90	0.4	4	90 > 63	0.9	4	0.002–5	0.9985
12	Diethyl maleate	a)	22.93	141-05-9	127 > 99	0.4	4	127 > 82	0.5	4	0.01–10	0.9995
13	Methyl heptine carbonate	b)	24.29	111-12-6	123 > 93	0.7	4	123 > 67	0.5	4	0.05–20	0.9996
14	4-Methoxyphenol	a)	24.70	150-76-5	124 > 109	0.6	4	109 > 81	0.6	3	0.01–20	0.9988
15	Citronellol	b)	25.98	106-22-9	95 > 67	0.4	4	128 > 81	0.7	3	0.05–20	0.9991
16a	Citral, Isomer 1	a)	26.41	5392-40-5	137 > 95	0.5	3	137 > 109	0.4	3	0.2–20	0.9996
16b	Citral, Isomer 2	a)	27.62	5392-40-5	137 > 95	0.5	3	137 > 109	0.4	3	0.2–20	0.9996
17	Geraniol	a)	27.06	106-24-1	93 > 65	1.2	4	123 > 81	0.7	5	0.5–20	0.9986
18	Cinnamal	a)	27.36	104-55-2	131 > 103	0.7	4	131 > 77	0.8	4	0.01–20	0.9999
19	4-Ethoxy-phenol	a)	27.71	622-62-8	138 > 110	0.6	3	110 > 82	0.6	3	0.02–20	0.9995
20	Anisyl alcohol	b)	27.82	105-13-5	138 > 109	0.5	4	109 > 94	0.7	5	0.02–20	0.9990

(continued overleaf)

Table 4.28 (Continued)

No.	Compound name	EU Regulation	Retention time (min)	CAS Number	Quantitation Transition m/z > m/z	Excitation voltage (V)	isol. wind. (Da)	Qualifier transition m/z > m/z	Excitation voltage (V)	isol. wind. (Da)	Linear ranges (mg/L)	Correlation coefficient R^2
21	Hydroxy-citronellal	a)	28.10	107-75-5	95 > 67	0.4	4	121 > 93	0.7	3	0.5–50	0.9991
22	4-*tert*-Butylphenol	a)	28.66	98-54-4	135 > 107	0.6	3	107 > 77	0.6	3	0.005–5	0.9995
23	Cinnamyl alcohol	a)	29.12	104-54-1	92 > 65	1.0	4	115 > 89	0.7	4	0.1–20	0.9986
24	4-Phenyl-3-buten-2-one	a)	31.64	122-57-6	103 > 77	0.5	4	131 > 103	0.4	3	0.02–10	0.9996
25	Eugenol	a)	31.91	97-53-0	164 > 149	0.6	3	164 > 131	0.7	3	0.2–20	0.9992
26	Dihydrocoumarin	a)	32.51	119-84-6	120 > 91	0.6	4	148 > 120	0.8	4	0.005–5	0.9991
27	Coumarin	a)	35.17	91-64-5	118 > 90	0.4	3	146 > 118	0.8	3	0.005–5	0.9987
28	Isoeugenol	a)	36.23	97-54-1	164 > 149	0.6	4	164 > 131	0.7	4	0.2–20	0.9985
29	2,4-Dihydroxy-3-methylbenzaldehyde	a)	36.83	6248-20-0	151 > 67	1.2	4	151 > 95	1.2	4	0.2–10	0.9768
30	alpha-iso-Methylionone	b)	38.54	127-51-5	135 > 79	0.8	4	135 > 107	0.4	4	0.01–10	0.9999
31	Lilial	b)	39.95	80-54-6	189 > 131	1.0	3	147 > 129	0.8	4	0.01–20	0.9999
32a	Pseudoionone, Isomer 1	a)	40.16	141-10-6	109 > 79	0.6	4	149 > 93	0.7	4	0.2–20	0.9995
32b	Pseudoionone, Isomer 2	a)	42.34	141-10-6	109 > 79	0.6	4	149 > 93	0.7	4	0.2–20	0.9995
33	6-Methylcoumarin	a)	40.88	92-48-8	160 > 132	0.4	4	132 > 103	1.6	4	0.02–20	0.9993
34	7-Methylcoumarin	a)	40.88	2445-83-2	160 > 132	0.4	4	132 > 103	1.6	4	0.02–20	0.9993
35	Diphenylamine	a)	43.62	122-39-4	169 > 140	3.2	3	169 > 115	3.2	3	0.005–5	0.9998
36	4-(*p*-Methoxyphenyl)-3-butene-2-one	a)	44.50	943-88-4	161 > 133	0.6	3	176 > 145	0.6	4	0.02–10	0.9983
37	Amylcinnamal	a)	45.09	122-40-7	203 > 145	1.2	4	203 > 129	1.6	4	0.05–20	0.9991
38	Lyral	a)	45.47	31906-04-4	136 > 107	0.4	3	136 > 79	0.7	3	0.5–50	0.9996
39	Amylcinnamyl alcohol	a)	46.57	101-85-9	133 > 115	0.7	4	187 > 130	0.7	3	0.5–50	0.9997

(continued overleaf)

Table 4.28 (Continued)

No.	Compound name	EU Regulation	Retention time (min)	CAS Number	Quantitation Transition m/z > m/z	Excitation voltage (V)	isol. wind. (Da)	Qualifier transition m/z > m/z	Excitation voltage (V)	isol. wind. (Da)	Linear ranges (mg/L)	Correlation coefficient R^2
40a	Farnesol, Isomer 1	b)	47.28	4602-84-0	93 > 77	0.4	4	107 > 91	0.5	4	1.0–50	0.9963
40b	Farnesol, Isomer 2	b)	47.81	4602-84-0	93 > 77	0.4	4	107 > 91	0.5	4	1.0–50	0.9963
40c	Farnesol, Isomer 3	b)	48.24	4602-84-0	93 > 77	0.4	4	107 > 91	0.5	4	1.0–50	0.9963
41	7-Methoxycoumarin	a)	47.37	531-59-9	148 > 133	0.7	3	176 > 148	0.8	3	0.05–20	0.9989
42	1-(p-Methoxyphenyl)-1-penten-3-one	a)	48.64	104-27-8	161 > 133	0.6	4	190 > 161	1.4	4	0.01–10	0.9997
43	Hexyl-cinnamaldehyde	b)	49.08	101-86-0	145 > 117	0.7	4	129 > 102	1.7	5	0.05–20	0.9999
44	Benzyl benzoate	b)	49.46	120-51-4	194 > 165	1.7	3	105 > 77	0.6	3	0.02–20	0.9992
45	4-tert-Butyl-3-methoxy-2,6-dinitrotoluene	a)	52.23	83-66-9	253 > 219	0.6	5	253 > 121	1.3	5	0.2–50	0.9985
46	Benzyl salicylate	a)	53.47	118-58-1	91 > 65	0.4	3	228 > 210	0.8	3	0.2–10	0.9850
47	7-Ethoxy-4-methylcoumarin	a)	56.21	87-05-8	204 > 148	1.9	4	148 > 91	1.1	4	0.1–10	0.9988
48	Benzylcinnamate	b)	60.61	103-41-3	131 > 103	0.7	3	192 > 115	1.1	4	0.05–20	0.9984

a) Banned compound as of EU Directive 2009/48/EC, technically unavoidable traces may not exceed 100 mg/kg.
b) Requires declaration if concentration exceeds 100 mg/kg.

Analysis Conditions

Sample material			Plastic, plush and paper toys, clays
Sample preparation	Liquid extraction		Dissolution of the plastic material, methanol liquid extraction
			Liquid extraction with acetone (clays, plush, paper toys)
	SPE	Material	ENVI-Carb™ cartridges
		Conditioning	5 mL methanol
		Sample application	Flow 3 mL/min, effluent collected
		Elution	With 15 mL dichloromethane, effluent collected
		Concentration	Combined effluents to 5 mL volume
		Filter	0.45 µm PTFE membrane prior to injection
GC Method	System		Varian 450
	Autosampler		Varian CP-8400
	Column	Type	HP-1MS
		Dimensions	50 m length × 0.20 mm ID × 0.50 µm film thickness
	Carrier gas		Helium, 99.999%
	Flow		Constant flow, 0.7 mL/min
	SSL Injector	Injection mode	Splitless
		Injection temperature	280 °C
		Injection volume	1 µL
	Split	Closed until	1 min
		Open	1 min to end of run
	Oven program	Start	50 °C, 1 min
		Ramp 1	5 °C/min to 155 °C, 6 min
		Ramp 2	3 °C/min to 260 °C
		Final temperature	260 °C
	Transferline	Temperature	280 °C
			Direct coupling
MS Method	System		Varian 240
	Analyzer type		Ion trap MS with internal ionization
	Ionization		EI
	Filament/multiplier delay time		8 min
	Electron energy		70 eV
	Ion source	Temperature	280 °C
	Manifold temperature		50 °C
	Ion trap temperature		220 °C
	Acquisition	Mode	MS/MS, resonant waveform CID
	Precursor ion width		3–5 Da
Calibration	External standard		Target compounds spiked into blank samples
	Range		0.005 to >100 mg/kg

Results

The separation of the 48 fragrance allergens could be achieved with good performance using a 50 m unpolar thick film column. The chromatogram of the total ion current of a standard compound mix is shown in Figure 4.81, with a zoom into the congested peak region in the centre of the chromatogram in Figure 4.82.

The LOD and LOQ values of the method have been determined based on the attained S/N ratio with three respectively 10 times the S/N. The achieved LODs in toy materials ranged from 0.005 to 5.0 mg/kg and the LOQs from 0.02 to 20 mg/kg, being significantly lower than the required limits in the EU regulation of 100 mg/kg. The obtained recoveries ranged from 79.5 to 109.1%.

Commercial toys were analyzed by the authors in order to demonstrate the applicability of the method (Lv *et al.*, 2013).

Conclusions

The described method provides a general approach for the accurate and effective determination of 48 fragrance allergens in toy samples. The MS/MS acquisition mode instead of single ion detection allows the identification and quantitation of the fragrance allergens at levels compliant with the current EU regulations in the complex matrix of the toy samples.

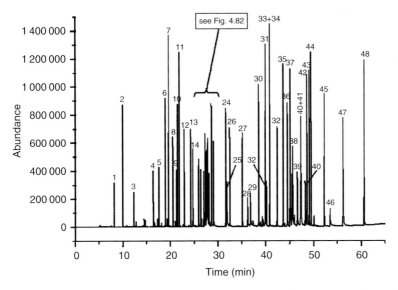

Figure 4.81   Separation of the 48 fragrance allergens, standard at 40 mg/L, peak numbers as of Table 4.28. (Lv *et al.* 2013, reproduced with permission of John Wiley & Sons.)

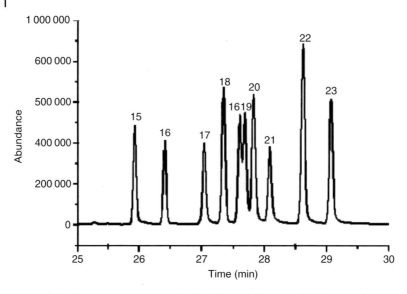

Figure 4.82 Compound separation detail of Figure 4.81. (Lv *et al.* 2013, reproduced with permission of John Wiley & Sons.)

4.21
Analysis of Azo Dyes in Leather and Textiles

Keywords: product safety; textiles; leather; azo dyes; cancerogenic amines; EN ISO standard method; fast full scan detection; quantitation; confirmation; library search

Introduction

Azo dyes are compounds characterized by their vivid colours and provide excellent colouring properties. They are important and widely used as colouring agents in the textile and leather industries. The risk in the use of azo dyes arises mainly from the breakdown products that can be created *in vivo* by reductive cleavage of the azo group into aromatic amines. Due to the toxicity, carcinogenicity and potential mutagenicity of thus formed aromatic amines the use of certain azo dyes as textile and leather colorants, and the exposure of consumers using the textile and leather coloured with azo compounds causes a serious health concern (European Commission – Health and Consumers, 1999).

The EU Commission classified 22 amines as proven or suspected human carcinogens. Azo dyes which, by reductive cleavage of one or more azo groups, may release one or more of these aromatic amines in detectable concentrations, that is above 30 ppm in the finished articles or in the dyed parts thereof may not be used in textile and leather articles which may come into direct and prolonged contact with the human skin or oral cavity (European Commission, 2002). The EU Directive 2002/61/EC has banned the use of dangerous azo colorants, placing textiles and leather articles coloured with such substances on the market, and requested the development of a validated analytical methodology for control. The described application is compliant with the requirements of the EN ISO 14362-1 standard procedure for the analysis of certain azo dyes in cotton and silk textiles. Following appropriate sample preparation methods the described analytical setup can be used for the analysis of azo dyes in leather and synthetic fabric as well.

The MS was set to run in full scan mode, providing the complete mass spectra of the detected compounds for identification, confirmation and quantitation. The complete list of azo compounds and internal standards is given in Table 4.29 with retention times for the given analytical conditions and the masses for the selective quantitation (Purwanto and Chen, 2013).

Table 4.29 Amine compounds included in the method[a] .

Amine compound	CAS #	RT (min)	Quan (m/z)	Comment
Aniline	62-53-3	5.42	93	2
o-Toluidine	95-53-4	6.43	106	
2,4-Xylidine	95-68-1	7.35	121	2
2,6-Xylidine	87-62-7	7.39	121	2
d-Naphthalene	1146-65-2	7.50	136	ISTD
2-Methoxyaniline	90-04-0	7.61	108	
p-Chloroaniline	106-47-8	7.97	127	
m-Anisidine	536-90-3	8.40	123	2
p-Cresidine	120-71-8	8.50	122	
2,4,5-Trimethylaniline	137-17-7	8.57	120	
4-Chor-o-toluidine	95-69-2	8.90	106	
1,4-Phenylenediamine (1,4-Benzendiamine)	106-50-3	8.91	108	1, 2
2,4-Toluenediamine	95-80-7	10.09	122	1
2,4-Diaminoanisole	615-05-4	10.91	123	1
2,4,5-Trichloroaniline	636-30-6	11.08	195	ISTD
2-Napthylamine	91-59-8	11.44	143	
5-Nitro-o-toluidine	99-55-8	11.85	152	
4-Aminodiphenyl	92-67-1	12.66	169	
p-Aminoazobenzene	60-09-3	14.36	92	
4,4-Oxydianiline	101-80-4	14.62	200	
4,4-Diaminodiphenylmethane	101-77-9	14.66	198	
Benzidine	92-87-5	14.71	184	
o-Aminoazobenzene	2835-58-7	15.01	106	2
3,3-Dimethyl-4,4-diaminodiphenylmethane	838-88-0	15.31	226	
3,3′-Dimethylbenzidine	119-93-7	15.47	212	
4,4′-Thiodianiline	139-65-1	16.04	216	
4,4-Methylene-bis-2-chloroaniline	101-14-4	16.23	231	
3,3′-Dimethoxybenzidine	119-90-4	16.24	244	
3,3-Dichlorobenzidine	91-94-1	16.26	252	

a) CAS # is CAS Registry Number, RT is the expected retention time, Quan is the quantitation ion.
Comments:
1. Compounds are unstable and tend to deteriorate at temperatures above 20°C
2. Additional compounds included in the assay, not part of EN ISO 17234-1

Analysis Conditions

Sample material	Textiles, leather
Sample preparation	Textiles made of cellulose and protein fibres, for example cotton, viscose, wool or silk (ISO, 2014) make the azo dyes accessible to a reducing agent without prior extraction. The EN ISO 17234-1 standard method for the analysis of such textiles is based on the chemical reduction of azo dyes followed by solid phase extraction (SPE) with ethyl acetate Synthetic fibres like polyester, polyamide, polypropylene, acrylic or polyurethane materials requires prior extraction of the azo dyes and is described in the EN 14362-2 standard method The analysis of leather samples follows the EN ISO 17234 standard method The azo group of most azo dyes can be reduced in the presence of sodium dithionite ($Na_2S_2O_4$) under mild conditions (pH = 6, T = 70 °C), resulting in the cleavage of the diazo group and forming two aromatic amines as the reaction products. The amines are extracted by liquid–liquid extraction with methyl t-butyl ether (MTBE), concentrated, adjusted to a certain volume with MTBE, then analyzed by GC-MS. The quantitation is performed with the internal standards d-naphthalene and 2,4,5-trichloroaniline. In the EN ISO 17234-1 standard method the directly reduced amines are isolated by SPE

GC Method	System	Thermo Scientific™	Trace 1300 GC
	Carrier gas		Helium
	Flow		Constant flow, 1.0 mL/min
	Column		TG-35MS or equivalent polarity column 30 m length × 0.25 mm ID × 0.25 μm film thickness
	Injection		Autosampler, AS 1310
	SSL Injector	Injection mode	Splitless
		Injection temperature	200 °C
		Injection volume	1.0 μL
	Split	Closed	1 min
		Open	1 min to end of run
		Split flow	50 mL/min
		Gas Saver	10 mL/min at 3 min

(*Continued*)

		Start	60 °C, 1 min
	Oven program	Ramp 1	15 °C/min to 200 °C
		Ramp 2	25 °C/min to 310 °C
		Final temperature	310 °C, 5 min
	Transferline	Temperature	295 °C
			Direct coupling
MS Method	System		Thermo Scientific™ ISQ
	Analyzer type		Single quadrupole MS
	Ionization		EI
	Electron energy		70 eV
	Ion source temperature		220 °C
	Acquisition mode	Full scan	50–350 Da
	Scan rate		0.075 s/scan
Calibration	Internal standards		d-Naphthalene, 2,4,5-trichloroaniline
	Calibration range		Representing 5, 30 and 100 ppm in the textile

Results

The scan rate of the MS run in full scan mode has been set to a very fast scan speed of only 75 ms/scan. This allows a very high chromatographic resolution especially of unresolved chromatographic peaks in the applied complex mixture of amines. Short analysis cycle times can be achieved for an increased sample throughput.

Isomeric amine compounds need to be separated on the chromatographic time scale for individual component quantitation. The mass spectra of isomers are very similar and typically do not offer unique quantitation ions for the independent quantitation. This is the case for the isomeric xylidine compounds 2,4-xylidine and 2,6-xylidine. The applied method separated both compounds very well with 7:35 and 7:39 min retention time, see Figure 4.83.

All other target amine analytes produce distinct mass spectra with unique ions available for selective quantitation, even if the chromatographic peaks in the total ion chromatogram are not resolved from each other. In these situations when several analytes coelute, individual mass spectra can still be isolated and identified by a library search. The intensities of the unique ions allow the interference-free quantitation based on the separated mass traces. In case of the coelution of the three compounds 4,4-methylene-bis-2-chloroaniline (RT 16:23 min), 3,3'-dimethoxybenzidine (RT 16:24 min), and 3,3-dichlorobenzidine (RT 16:24 min) shown in Figure 4.84 only a partial chromatographic resolution with the given conditions of the method is achieved, but due to specific fragment ions a safe peak integration and quantitation is accomplished.

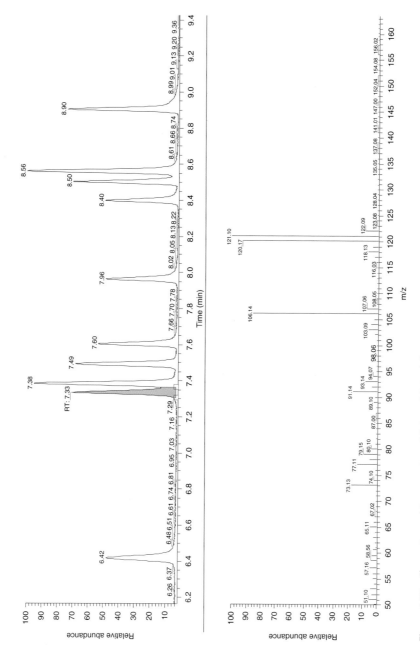

Figure 4.83 Chromatographic separation of 2,4- and 2,6-xylidine at retention times 7:33 and 7:38 min, with mass spectrum of the integrated 2,4-Xylidine peak below.

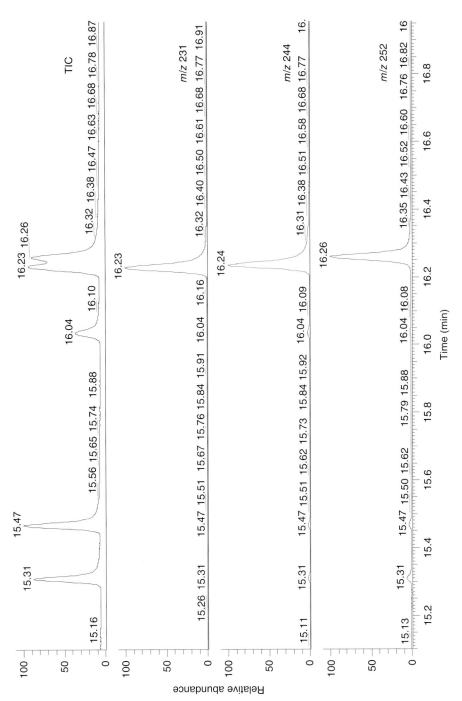

Figure 4.84   Coelution of three different amines, separated by individual mass traces.

Although coeluting the quantitation is performed separately by means of individual fragment ions. The fast scanning of the mass spectrometer allows the detection of small retention time differences so that even the clean spectra of each of the compounds can be isolated for library search, see Figure 4.85a – c. All three compounds are safely identified by searching the NIST library. The spectra have been taken from a spiked calibration file at the 3 ppm level. The high sensitivity of the mass spectrometer provides even at the required low detection level and fast

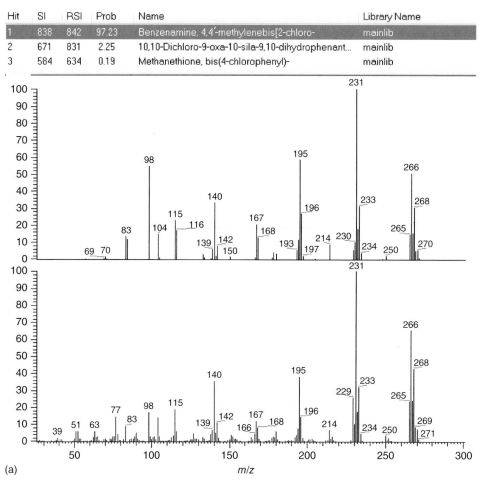

Hit	SI	RSI	Prob	Name	Library Name
1	838	842	97.23	Benzenamine, 4,4'-methylenebis[2-chloro-	mainlib
2	671	831	2.25	10,10-Dichloro-9-oxa-10-sila-9,10-dihydrophenant...	mainlib
3	584	634	0.19	Methanethione, bis(4-chlorophenyl)-	mainlib

(a)

Figure 4.85 Acquired mass spectra and library search results for the coeluting compounds of Figure 4.86 top acquired spectrum from data file in Figure 4.86 bottom spectrum from NIST library) (a) 4,4-Methylene-bis-2-chloroaniline (CAS 101-14-4). (b) Dimethoxybenzidine (CAS 119-90-4) (c) 3,3-Dichlorobenzidine (CAS 91-94-1).

Hit	SI	RSI	Prob	Name	Library Name
1	823	845	55.18	2,2'-Dimthoxy-biphenyl-4,4'-diamine	mainlib
2	817	818	43.37	Benzidine, 3,3'-dimethoxy-	mainlib
3	650	658	1.00	Propanedinitrile, [(3,4,5-trimethoxyphenyl)methyle...	mainlib

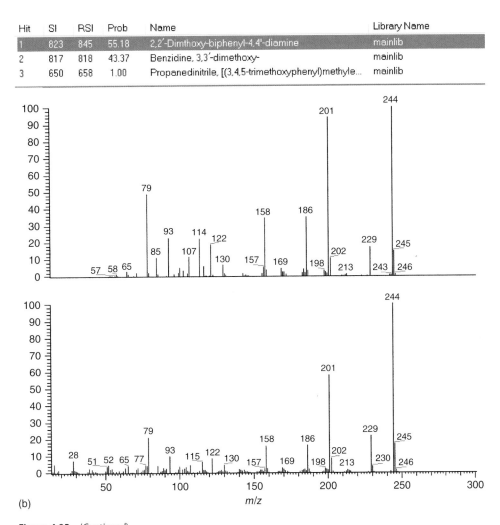

(b)

m/z

Figure 4.85   *(Continued)*

Hit	SI	RSI	Prob	Name	Library Name
1	867	885	97.31	[1,1′-Biphenyl]-4,4′-diamine, 3,3′-dichloro-	mainlib
2	673	727	1.38	1-Chloro-11H-indolo[3,2-c]quinoline	mainlib
3	626	711	0.28	8-Chloro-11H-indolo[3,2-c]quinoline	mainlib

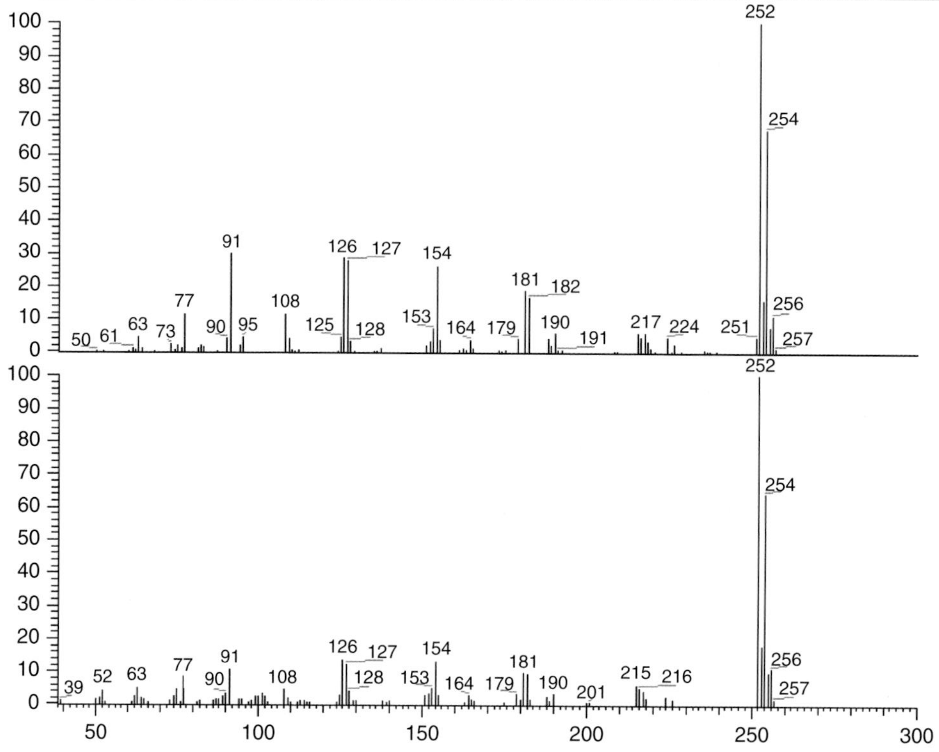

(c)

Figure 4.85   (*Continued*)

scanning speed very complete mass spectra for a solid compound confirmation, in parallel to the quantitation.

The quantitation was done by using the two internal standards d-naphthalene RT 7:50 min and 2,4,5-trichloroaniline RT 11:08 min. A linear calibration with $R^2 > 0.9998$ was created using standards over three levels at 0.5, 3.0 and 10.0 ppm, representing 5, 30 and 100 ppm in the textile. An example of a routine QC report for the decision level of 30 ppm is shown in Figure 4.86. The list of compounds and the chromatogram of the QC sample is shown. All compounds were detected, integrated and passed the confirmation check based on the compound mass spectrum.

Component name	Response	Expected RT	RT	Calculated amount	Ion ratio status
Aniline	155538574.30	5.42	5.40	0.000	Passed
o-Toluidine	160660665.78	6.43	6.42	0.000	Passed
2,4-Xylidine	118418341.17	7.35	7.33	0.000	Passed
2,6-Xylidine	160179028.35	7.39	7.39	0.000	Passed
d-Naphthalene	249479939.19	7.50	7.49	N/A	Passed
2-Methoxyaniline	97026223.41	7.61	7.60	0.000	Passed
p-Chloroaniline(b)	121883651.37	7.97	7.97	0.000	Passed
m-Anisidine(b)	96035123.15	8.40	8.40	0.000	Passed
p-Cresidine(b)	137451748.57	8.50	8.50	0.000	Passed
2,4,5-Trimetilaniline(b)	176109539.11	8.57	8.56	0.000	Passed
4-Chlor-o-toluidine(b)	133205514.66	8.90	8.90	0.000	Passed
1,4-Phenylenediamine(b)	52487228.40	8.91	8.93	0.000	Passed
2,4-Toluenediamine(b)	60852176.48	10.09	10.07	0.000	Passed
2,4-Diaminoanisole(b)	42982997.18	10.91	10.88	0.000	Passed
2-Napthylamine(b)	223788762.28	11.44	11.44	0.000	Passed
5-Nitro-o-toluidine(b)	38296755.97	11.85	11.85	0.000	Passed
4-Aminodiphenyl(b)	226914065.95	12.66	12.66	0.000	Passed
p-Aminobenzene	117133397.82	14.36	14.35	0.000	Passed
4,4-Oxydianiline(b)	37224262.23	14.62	14.62	0.000	Passed
4,4-Diaminodiphenylmethane(b)	53234915.02	14.66	14.66	0.000	Passed
Benzidine(b)	160850923.04	14.71	14.71	0.000	Passed
o-Aminoazobenzene(b)	144094875.03	15.01	15.00	0.000	Passed
3,3-Dimethyl-4,4-diaminodiphenylmethane(b)	57071128.20	15.31	15.31	0.000	Passed
3,3′-Dimethylbenzidine(b)	145275874.72	15.47	15.47	0.000	Passed
4,4′-Thiodianiline(b)	36700000.54	16.04	16.04	0.000	Passed
4,4-Methylene-bis-2-chloroaniline(b)	43603905.21	16.23	16.23	0.000	Passed
3,3′-Dimethoxybenzidine	43367566.07	16.24	16.24	0.000	Passed
3,3-Dichlorobenzidine	85388633.63	16.26	16.26	0.000	Passed

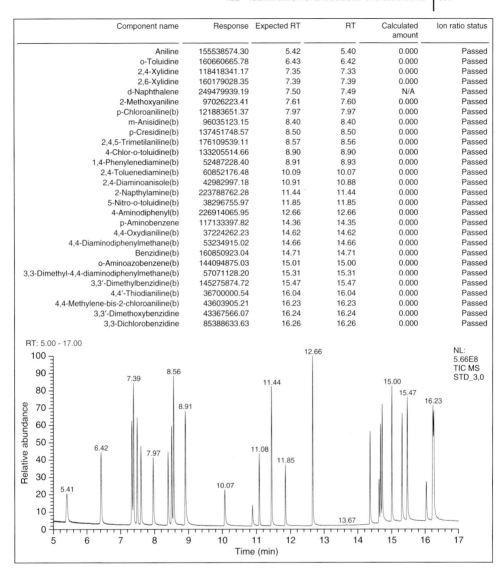

Figure 4.86 Report of a routine QC sample at the decision level of 30 ppm.

4.22
Identification of Extractables and Leachables

Keywords: product safety; pharmaceutical product; packaging; quality control; AMDIS deconvolution; unknown identification; Mass Frontier spectrum interpretation

Introduction

Leachables in pharmaceutical products are 'trace amounts of chemicals originating from containers, medical devices or process equipment that end up as contaminants in medicinal products resulting in exposure to patients' (Moffat, 2011).

The US Food and Drug Administration (FDA) defines extractables as compounds that can be extracted from a container material when in the presence of a solvent, and leachables as those compounds that leach into the drug product formulation as a result of direct contact with the formulation (Lewis, 2011).

The European Agency for the evaluation of Medical Products Guideline (European Medicines Agency, EMA) states that 'it should be determined whether any of the extractables are also leachables present in the formulation at the end of the shelf life of the product or to the point equilibrium is reached if sooner'.

The identification of leachables, and attribution to the contact component from which they originate, is important because such species may react with the drug product or formulation ingredients, compromise the efficacy of the drug product or interfere with dosage consistency, finally may pose a negative health effect.

GC-MS analysis is mostly applied for volatile components using headspace analysis, or after a solvent extraction step for semi volatile compounds. This application describes the analysis of a polymer plunger material using different extraction techniques, derivatization and headspace analysis by single quadrupole GC-MS. A parallel classical FID detection channel was configured which shows the chromatogram similarity and can be used for a future simplified GC-FID routine method.

Sample Preparation

The elastomeric plunger material was examined in different ways. The volatiles were determined via direct headspace analysis (HS). For the headspace injection 10 plungers had been placed in a 20 mL headspace vial.

The extractables of the sample were analysed by preparing different liquid extracts using three extraction procedures and derivatization:

1) Aqueous extraction of the elastomer material, followed by a dichloromethane (DCM) extraction of the aqueous phase, no derivatization.
2) The above DCM extract has been derivatized using BSTFA.
3) Isopropanol (IPA) extraction, no derivatization.

Experimental Conditions

The analyses was performed using GC with dual detection by FID and single quadrupole MS. The dual detection was accomplished by using a SilFlow™ microfluidic connection device, also allowing a no-vent option for easy column change without a venting of the mass spectrometer. The GC was equipped with an autosampler for both liquid and headspace injections.

Data Processing and Results

The mass spectrometer detection in full scan mode was used for identification of the unknown compounds. The parallel FID detection was checked for compliance with the MS total ion chromatogram.

The chromatogram of the headspace analysis is shown in Figure 4.87. The analyses of the different liquid extraction and derivatization procedures follow in Figures 4.88–4.90, all of them with the total ion (TIC) and FID traces. All chromatograms demonstrate the dual FID plus MS detection in very good agreement of the eluted compound pattern.

AMDIS was used for a deconvolution of the complex chromatograms extracting the 'clean' background and coelution corrected single compound mass spectra. For search and spectral comparison the NIST library was used. AMDIS associates the found retention time and mass spectrum for an improved identification. All results can optionally be transferred to Excel™ for further investigation and documentation.

Mass spectra not satisfactorily identified by library search were analyzed in the fragmentation pattern by using Mass Frontier resulting in structural proposals.

AMDIS Chromatogram Deconvolution

The AMDIS deconvolution program works in three steps (Mallard and Reed, 1997):

– *Step 1*: AMDIS analyses the chromatogram. It counts the number of eluted compounds based on a minimum of ions that show a common retention time maximum (maximizing masses peak finder). The corresponding mass spectrum is extracted and cleared from potential contributions from baseline and coeluting compound mass intensities.
– *Step 2*: AMDIS checks if target compounds from a user library are present by matching simultaneously retention time (or retention index, if available) and mass spectrum.
– *Step 3*: All detected compound spectra are compared with the spectra of the linked libraries allowing a filter with different criteria.

The list of AMDIS identified compounds of the headspace analysis and liquid extractions are given in the following Tables 4.30–4.32 with compound name, CAS numbers as well as the peak quality parameters retention time and the measured peak width and tailing information. The result of the mass spectrum library comparison is given in the most right columns. The 'Reverse' fit column informs about the match quality in % of the spectrum pattern of the proposed library entry with the unknown spectrum (reverse search).

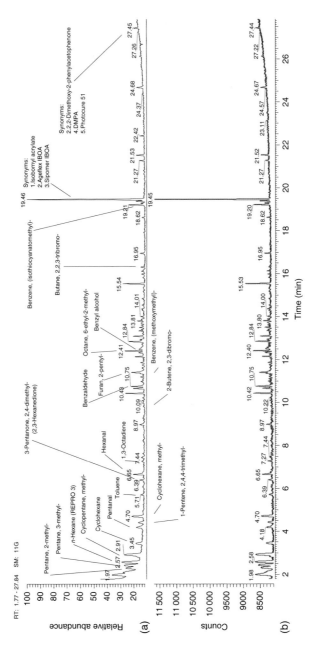

Figure 4.87 Chromatogram of the elastomeric plunger material by headspace analysis, (a) MS TIC, (b) FID.

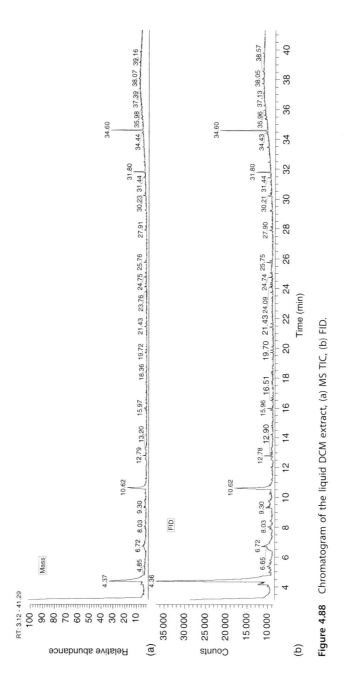

Figure 4.88 Chromatogram of the liquid DCM extract, (a) MS TIC, (b) FID.

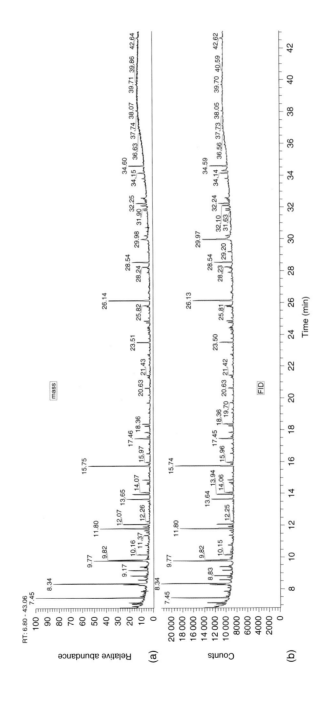

Figure 4.89 Chromatogram of the liquid DCM extract, derivatized with BSTFA, (a) MS TIC, (b) FID.

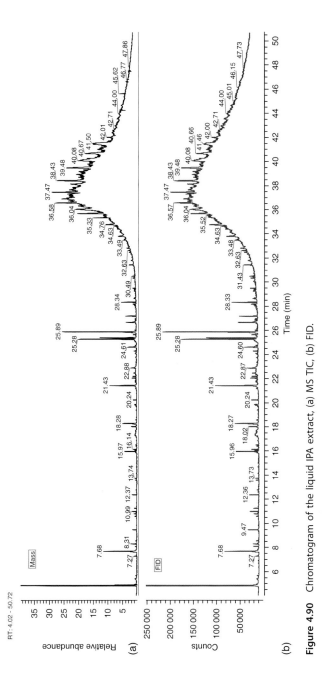

Figure 4.90 Chromatogram of the liquid IPA extract, (a) MS TIC, (b) FID.

Table 4.30 Compounds identified by headspace analysis.

Name	RT (min)	Reverse fit	Area (cts)	Area (%)	Height (cts)	Height (%)
Pentane, 2-methyl-	2.28	85	1917	4.7	368	2.4
Pentane, 3-methyl-	2.40	87	2308	5.6	442	2.8
Cyclopentane, methyl-	2.93	79	2847	7.0	515	3.3
Cyclohexane, methyl-	4.72	82	1333	3.3	346	2.2
2,3-Hexanedione	6.40	83	238	0.6	106	0.7
Hexanal	6.65	82	825	2.0	300	1.9
Propanal, 2,2-dimethyl-	10.43	88	950	2.3	538	3.5
Octane, 2,2,6-trimethyl-	10.66	88	445	1.1	259	1.7
Benzaldehyde	10.75	92	1031	2.5	461	3.0
Octane, 2,6,6-trimethyl-	13.34	79	164	0.4	115	0.7
2-Propenoic acid	19.46	95	6465	15.8	3438	22.1
Ethanone, 2,2-dimethyl	27.44	82	217	0.5	125	0.8

Mass Frontier Spectrum Interpretation

Some of the acquired spectra are not included in the commercial libraries, some matches show structural similarities. The Mass Frontier software analyses the unknown mass spectrum and associates fragmentation pathways and ion structures to the unknown spectral pattern calculated from the included knowledge base of known fragmentation rules (HighChem, 2014). Two examples of spectrum interpretation of unknown compound spectra are given in Figure 4.91 from the DCM extract and in Figure 4.92 from the IPA extract. The generated proposals of the Mass Frontier expert system show a good plausibility and explain the mass spectrum pattern well (Table 4.33).

Analysis Conditions

Sample material		Elastomer material, polymers
Sample preparation	Purge and Trap	Tekmar purge and trap system LSC 2000
	Sparge vessel	Frit sparge, 25 mL volume
	Purge gas	Nitrogen
	Trap	Tenax
	Sample volume	5 mL
	Sample temperature	40 °C
	Standby	30 °C
	Sample preheat	2.50 min
	Purge duration	10 min
	Purge flow	40 mL/min
	Desorb preheat	175 °C
	Desorb temperature	180 °C

Table 4.31 Compounds identified by liquid IPA extraction.

Name	RT (min)	Reverse fit	Area (cts)	Area (%)	Height (cts)	Height (%)
Isopropyl alcohol	4.85	78	1737928	31.0	626333	28.9
Tricyclo[3.1.0.0(2,4)]hex-3-ene-3-carbonitrile	8.52	78	11506	0.2	5183	0.2
Benzyl alcohol	9.48	94	43713	0.8	23090	1.1
Benzene, (bromomethyl)-	10.80	90	23050	0.4	10571	0.5
Benzyl isopentyl ether	10.99	89	31562	0.6	18724	0.9
Benzyl isocyanate	11.25	94	21095	0.4	10942	0.5
Heptanoic acid, propyl ester	11.41	86	3680	0.1	2310	0.1
Decanone-2	12.60	79	3716	0.1	2230	0.1
Dodecane	12.77	92	5686	0.1	3521	0.2
Butane, 1,2,2-tribromo-	13.12	79	1611	0.0	1096	0.1
Tridecane	14.69	81	3736	0.1	2055	0.1
Benzene, (isothiocyanatomethyl)-	15.97	91	114760	2.1	47845	2.2
2-Dodecanone	16.40	84	7773	0.1	3757	0.2
Tetradecane	16.52	94	11436	0.2	5989	0.3
1-Bromo-3-(2-bromoethyl)heptane	18.02	65	57416	1.0	19643	0.9
1-Bromo-3-(2-bromoethyl)heptane	18.31	66	143599	2.6	50332	2.3
Pentadecane, 3-methyl-	19.39	88	5468	0.1	2763	0.1
2-Tetradecanone	19.82	89	16697	0.3	7514	0.4
Hexadecane	19.87	90	18647	0.3	8612	0.4
N-Benzylidenebenzylamine	22.32	94	17660	0.3	5632	0.3
Tetradecanenitrile	24.29	81	17783	0.3	8962	0.4
Hexadecanoic acid, methyl ester	24.61	97	61160	1.1	28869	1.3
Phthalic acid, butyl cyclobutyl ester	25.06	88	10308	0.2	4055	0.2
Decane, 5,6-bis(2,2-dimethylpropylidene)-, (*E,Z*)-	25.29	72	403320	7.2	163977	7.6
Isopropyl palmitate	25.89	76	572407	10.2	204258	9.4
Oleanitrile	26.72	80	90096	1.6	41750	1.9
Heptadecanenitrile	26.97	76	8891	0.2	4527	0.2
Octadecanoic acid, methyl ester	27.19	92	73043	1.3	38750	1.8
Isopropyl stearate	28.34	90	99362	1.8	53712	2.5
Diisooctylphthalate	31.81	82	22826	0.4	6778	0.3

(*Continued*)

Desorb time	4 min
MCM Desorb	5 °C
Bake	10 min at 200 °C
MCM bake	90 °C
BGB	Off
Mount	60 °C
Valve	220 °C
Transfer line	200 °C
Connection to GC	LSC 2000, inserted into the carrier gas supply of the injector

Table 4.32 Compounds identified by liquid DCM extraction.

Name	RT (min)	Reverse fit	Area (cts)	Area (%)	Height (cts)	Height (%)
Toluene	4.17	96	17526	3.7	2914	3.2
Benzene, 1-fluoro-4-methyl-	4.30	88	6223	1.3	2039	2.2
Benzene, 1-fluoro-2-methyl-	4.37	93	244339	51.0	27516	29.7
Benzyl alcohol	7.89	90	800	0.2	307	0.3
Benzylalcohol	9.38	88	3632	0.8	1081	1.2
Cyclopropyl carbinol	10.63	77	55086	11.5	9086	9.8
Decanal	12.79	88	4070	0.9	2030	2.2
Diethylphthalate	19.72	92	1195	0.3	682	0.7
Dibutyl phthalate	25.05	90	1253	0.3	559	0.6
Ethylhexyl phthalate	31.81	93	6673	1.4	3399	3.7

(Continued)

GC Method				
	System		Thermo Scientific™ TRACE 1300 GC	
	Autosampler		Thermo Scientific™ TriPlus RSH	
	Column	Type	TG-5MS	
		Dimensions	30 m length × 0.25 mm ID × 0.25 µm film thickness	
	Transfer capillaries	To FID	0.2 m × 0.2 mm ID	
		To MS	2.0 m × 0.15 mm ID	
	Carrier gas		Helium	
	Flow		Constant pressure, 125 kPa	
	Injection		Autosampler Thermo Scientific™ Triplus RSH/Manual	
	SSL Injector	Inlet liner	4 mm ID with glass wool	
		Injection mode	Splitless	
		Injection temperature	320 °C	
		Injection volume	1 µL of liquid extracts 1 mL headspace volume	
	Split	Closed until	1 min for liquid extracts	
		Open	For headspace injections	
		Open	1 min to end of run	
		Split flow	20 mL/min	
		Purge flow	5 mL/min	
	Oven program		For liquid injections	For headspace injections
		Start	40 °C, 1 min	30 °C, 3 min
		Ramp 1	8 °C/min to 325 °C	8 °C/min to 280 °C
		Final temperature	325 °C, 14 min	280 °C, 10 min
	Transferline	Temperature	300 °C Direct coupling	

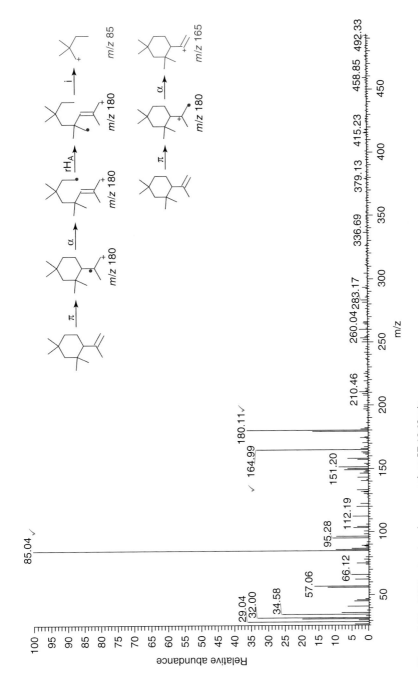

Figure 4.91   DCM extract, unknown peak at RT 12.68 min.

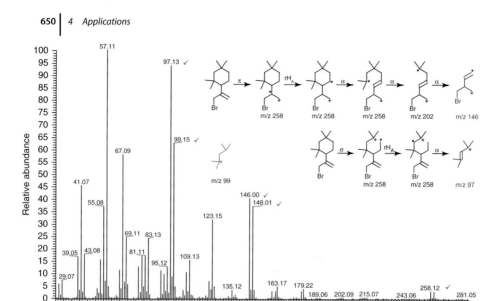

Figure 4.92 IPA extract, unknown peak at RT 18.31 min.

(Continued)			
	Detectors		FID
		Temperature	300 °C
		Air	350 mL/min
		Hydrogen	35 mL/min
		Nitrogen	40 mL/min
MS Method	System		Thermo Scientific™ ISQ
	Analyzer type		Single quadrupole MS
	Ionization		EI
	Electron energy		70 eV
	Ion source	Temperature	220 °C
		Emission current	50 µA
	Acquisition	Mode	Full scan
	Mass range		25–700 Da
	Scan rate		4 scans/s (250 ms/Scan)
	Resolution	Setting	Normal (0.7 Da)

Conclusions

The parallel detection using full scan MS and FID shows very good compliance in the detected compound pattern. After identification of the typical major components using the mass spectrometer the routine quality control for such compounds can be run reliably using a simplified GC-FID method.

Table 4.33 Compounds identified by liquid DCM extraction and BSTFA derivatization.

Name	RT (min)	Reverse fit	Area (cts)	Area (%)	Height (cts)	Height (%)
Disiloxane, hexamethyl-	6.83	91	1426	0.5	1588	1.0
Dimethyl sulfone	7.14	92	5078	1.7	3467	2.1
Trifluoromethyl-bis-(trimethylsilyl)methyl ketone	7.45	94	14274	4.8	10694	6.6
Octane, 4-ethyl-	7.85	94	1932	0.6	1158	0.7
1,2-Bis(trimethylsiloxy)ethane	8.34	93	20903	7.0	12757	7.8
Cyclopropane, 1-heptyl-2-methyl-	8.58	87	3614	1.2	2127	1.3
Silane, (cyclohexyloxy)trimethyl-	8.83	89	4655	1.6	3102	1.9
Tetrasiloxane, decamethyl-	9.34	86	3844	1.3	2285	1.4
Silane, (1-cyclohexen-1-yloxy)trimethyl-	9.77	90	18016	6.0	10881	6.7
Propanoic acid, 2-[(trimethylsilyl)oxy]-, trimethylsilyl ester	9.82	95	8888	3.0	4868	3.0
Glycolic acid	10.17	88	3744	1.3	1516	0.9
Silane, trimethyl(phenylmethoxy)-	11.80	94	16044	5.3	9758	6.0
3,6,9-Trioxa-2-silaundecane, 2,2-dimethyl-	11.93	89	1501	0.5	974	0.6
Benzoic acid trimethylsilyl ester	13.65	94	7195	2.4	4126	2.5
Octanoic acid, trimethylsilyl ester	13.95	87	5175	1.7	3037	1.9
Octane, 2,4,6-trimethyl-	14.70	85	1338	0.5	790	0.5
Butanedioic acid, bis(trimethylsilyl) ester	14.86	89	1936	0.6	1089	0.7
Nonanoic acid, trimethylsilyl ester	15.75	91	19185	6.4	10747	6.6
Benzene, (isothiocyanatomethyl)-	15.97	86	2663	0.9	1295	0.8
Decanoic acid	17.46	90	5016	1.7	2737	1.7
Lauric acid TMS	20.63	91	1867	0.6	1006	0.6
Tetradecanoic acid, trimethylsilyl ester	23.51	86	4824	1.6	2576	1.6
Phthalic acid, butyl cyclobutyl ester	25.05	89	655	0.2	348	0.2
Hexadecanoic acid, trimethylsilyl ester	26.14	89	14483	4.8	7421	4.6
Octadecanoic acid, trimethylsilyl ester	28.55	86	6429	2.1	3375	2.1
Phthalic acid, ethy hexyl ester	31.81	89	1724	0.6	869	0.5
4-Methyl-2,4-bis(4'-trimethylsilyl-oxyphenyl)pentene-1	33.43	91	652	0.2	380	0.2

The deconvolution using the AMDIS software allows a precise isolation of the mass spectra even from coeluting compounds. The possibility to use an individual library of target compounds, combining retention time and mass spectrum makes it a powerful tool for analytical control.

For unknown mass spectra the Mass Frontier software is a unique tool for spectrum interpretation. Structure proposals and fragmentation pathways are provided for mass spectra allowing a deeper sample and unknown elucidation.

4.23

Metabolite Profiling of Natural Volatiles and Extracts

Keywords: metabolomics; phenotype; genotype; SPME; GC-MS; GC-MS/MS; AMDIS; deconvolution; biomarker; identification; library search; MRM; quantitation

Introduction

Analyses focusing on a group of metabolites or watching as many metabolites as possible at specified environmental or developmental stages is called 'metabolite profiling'. The analysis of metabolites is very complex due to the chemical diversity of these small biological molecules. On the other hand metabolite target analyses may be restricted to specific metabolites of interest, which can be selectively monitored and quantified (Weckwerth, Wenzel and Fiehn, 2004).

The analytical challenges are the very complex chromatographic coelution of a large number of compounds, hence requiring a mass spectral deconvolution for identification. An integrated workflow approach of two analytical phases for discovery and quantitation is applied. This two stage concept combines the advantages of the full scan measurements for deconvolution and library search compound identification with the MS/MS selectivity for chemical structures and allows for absolute quantification in a targeted analysis using multiple reaction monitoring (MRM) as shown in Figure 4.93 (Hübschmann *et al.*, 2012). With the described 2-tier workflow scientists are able to conduct both discovery phase with identification and the quantification analysis with a single instrument using the triple stage quadrupole GC-MS/MS MRM technology.

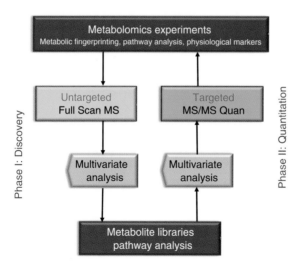

Figure 4.93 Metabolite Profiling GC-MS/MS Workflow.

Workflow Phase I: Discovery

The discovery phase provides the identification of as many metabolites as possible by GC-MS analysis in Full Scan mode. This first phase is dominated by the deconvolution of the chromatograms to extract the full and representative mass spectra as well as the compound retention time for subsequent identification. Figure 4.94 shows the elution of the different metabolite compound classes using the described analytical method (Fragner, Weckwerth and Hübschmann, 2010).

The deconvolution step is greatly facilitated by using the AMDIS program and the NIST mass spectral library search program (Mallard and Reed, 1997). Unique compound mass spectra are extracted by the analysis of all the transient ion signals allocating ion masses and relative intensities for each of the eluting compounds.

The identification of metabolites is based on the characteristic EI fragmentation patterns as well as on retention time (RT). The mass spectrum identification is facilitated by searching large data bases with NIST, Wiley or dedicated collections of mass spectra like the Fiehn metabolite spectrum library, or already available in-house metabolite databases, which are linked seamlessly into the search procedure.

Workflow Phase II: Targeted Quantitation

For the targeted quantitation phase suitable MS/MS transitions for each compound are applied using the identical chromatography setup and the retention

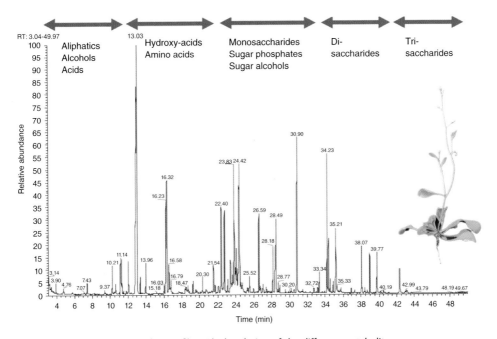

Figure 4.94 General metabolite profile with the elution of the different metabolite compound classes (Arabidopsis Thaliana).

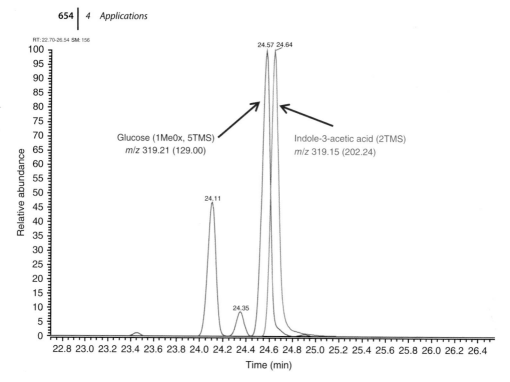

Figure 4.95 Separation of coeluting compounds with same precursor ion by MS/MS, indole-3-acetic acid (IAA): *m/z* 319.15 > 202.24, glucose: *m/z* 319.21 > 129.00.

times defined in phase I. Using a triple quadrupole MS the selected precursor ions get fragmented to structure specific product ions in the collision cell. With the applied TSQ 8000 in the described experiments the AutoSRM optimization procedure can be used at this stage for a large number the selected metabolites. The integration of the detected product ion peaks provides the selective quantitation of all target metabolites. Figure 4.95 shows the chromatograms of indole-3-acetic acid and glucose compounds at the level of 50 pmol injected amount. Both compounds although coeluting can be integrated independently from each other due to the different SRM transitions used.

Sample Preparation

For the analysis of volatile compounds the sample material, for example rice or leaves are weight in equal amounts into headspace vials and capped. For the analysis of the extracted metabolites plant material (leaves) gets homogenized under liquid nitrogen. About 50 mg are applied to extraction with a water/chloroform/methanol mixture to extract water soluble metabolites. The polar phase is dried in a vacuum centrifuge. A 2-step derivatization can be applied: First a methoxyamination (methoxyamine hydrochloride in pyridine) can be used to suppress keto-enol tautomerism, followed by a regular silylation using BSTFA or MSTFA (*N*-methyl-*N*-trimethylsilyl-trifluoroacetamide) to derivatize polar functional groups. The final derivatization volume is 100 μL. Standards get dissolved in methanol or water, diluted into various concentrations, dried and derivatized using the same procedure (Weckwerth, Wenzel and Fiehn, 2004; Fragner, Weckwerth and Hübschmann, 2010).

Analysis Conditions

Sample material			Seeds, grains, plant material
Sample preparation	SPME extraction		Fibre type DVB/CAR/PDMS
			4 g of rice into a 20 mL headspace vial, capped
			Incubation temperature 80 °C, 30 min
			Extraction time 30 min
			Injector desorption time 5 min
	Derivatization		Liquid extraction with methanol/water/DCM
			Dried extracts derivatized with BSTFA or MSTFA
GC Method	System		Trace 1310 GC
	Injector		Split/splitless
	Column	Type	TG-5MS
		Dimensions	30 m length × 0.25 mm ID × 0.25 μm film thickness
	Carrier gas		Helium
	Flow		Constant flow, 1.2 mL/min
	SSL Injector	Injection mode	Split 1 : 10
		Inlet liner	Split liner with glass wool
		Injection temperature	285 °C
		Injection volume	1 μL
	Oven program	Start	60 °C, 4 min
		Ramp 1	8 °C/min to 170 °C
		Ramp 2	4 °C/min to 300 °C
		Final temperature	300 °C, 15 min
	Transferline	Temperature	285 °C
MS Method	System		Thermo Scientific™ TSQ 8000
	Analyzer type		Triple quadrupole MS
	Ionization		EI
	Electron energy		70 eV
	Ion source	Temperature	300 °C
	Acquisition	Mode	Full scan, timed-SRM
			MRM, 2 SRMs per compound
	Collision gas		Ar, 1.5 mL/min
	Mass range		29-350 Da (SPME), 44-600 Da (derivatized extracts)
	Scan rate		0.2 s (5 spectra/s)
	Resolution	Setting	Normal (0.7 Da)

Results

The SPME measurements presented here had been carried out to distinguish different varieties of rice. The achieved full scan chromatograms in Figure 4.96 show a significantly different pattern of the collected volatile components. Based on the recorded complete mass spectra the AMDIS deconvolution with peak identification can be performed. Selected compounds that had been identified with their relative signal intensity to the first sample are given in Table 4.34.

The chromatograms of the derivatized extracts are shown in Figure 4.97. Besides major common components differentiating compounds can be identified on a large number at low level intensities. A detailed data analysis for differentiators can be performed by using the SIEVE software. After the retention time alignment of the chromatograms in Figure 4.98 the quantitative analysis of the deconvoluted peaks is performed and presented in tabular from and for visual inspection in Figure 4.99 with the overlaid peak profiles from the chromatograms under investigation, as well as bar graphs of the peak areas as quantitative result.

Conclusion

The ability to use a single instrument when performing both the screening for discovery and the target quantitation of metabolites not only accelerates the profiling process, thereby increasing throughput, but also provides optimized reproducibility in fragmentation pattern, peak intensities and mass accuracies. In addition,

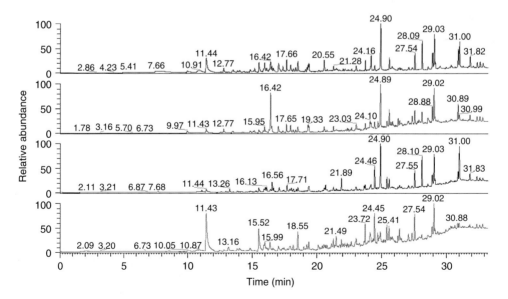

Figure 4.96 Phenotype analysis by SPME GC-MS (four rice samples)

Table 4.34 Selected compounds detected by SPME in samples of four different rice varieties with relative intensity to sample 1.

Compound name	RT (min)	Sample 1	Sample 2	Sample 3	Sample 4
Toluene	10.03	+++	++	+	+
Hexanal	11.42	+++	+	+	++
Xylene	12.77	+++	++	+	+
Styrene	13.51	+++	++	+	+
2-Butylfuran	13.14	+++	+	+	++
2-Heptanone	13.83	+++	+	+	++
Heptanal	14.01	+++	+	+	++
2-Pentylfuran	15.50	+++	+	+	++
1-Octen-3-ol	15.91	+++	+	++	++
Benzaldehyde	15.95	+++	++	+	+
2-Ethyl-1-hexanol	17.05	+++	++	+	+
Undecane	17.36	+++	++	++	+
Phenol	17.66	+++	++	++	+
Nonanal	18.54	+++	+	+	++
1-Hexadecanol	19.33	+++	++	+	-
Dodecane	19.39	+++	++	+	+
1-Methylene-1H-indene	20.56	+++	++	++	+
2-Butyl-2-octenal	23.71	+++	+	+	++
4-Octadecylmorpholine	24.90	+++	++	++	+

compounds which typically cannot be separated through gas chromatography can be separated using compound-specific MRM transitions in a single run for individual quantitation. The high sensitivity, dynamic range and selectivity of the triple quadrupole method enable scientists to effectively perform metabolite profiling.

GC-MS/MS provides both metabolomics workflow phases on one instrument platform:

Phase I: Discovery phase analysis
- Marker identification in Full Scan mode
- Fast full scan analysis with deconvolution
- Access to the largest mass spectral know-how bases
- Reference library building

Phase II: Target compound quantitation
- Selective and accurate compound quantification using the MS/MS mode
- High dynamic range for interesting metabolite biomarkers
- Complex mixture analysis using the compound-specific MRM mode
- Coeluting compounds get separated and individually quantified by MRM

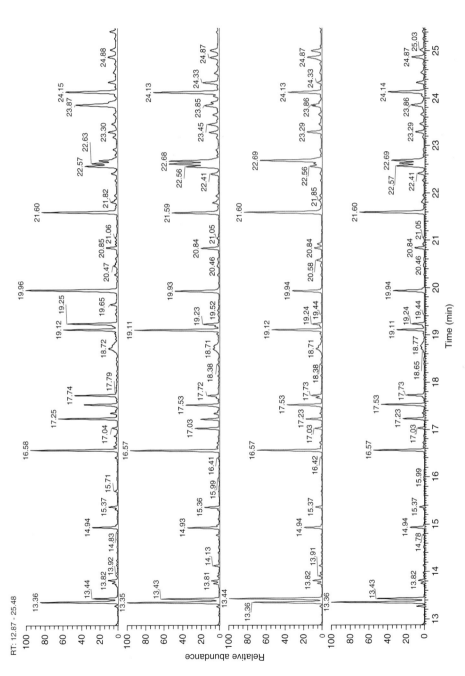

Figure 4.97 Extract analyses after BSTFA derivatization (four difference rice samples).

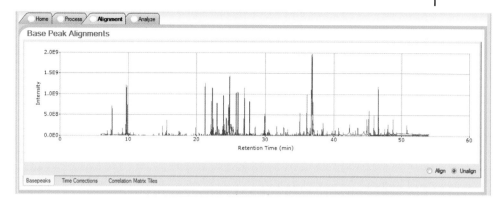

Figure 4.98 Retention time alignment of the sample chromatogram (SIEVE software).

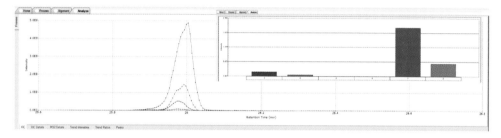

Figure 4.99 A discriminator compound identified in four of six samples, three times with low intensity, one sample with high intensity (SIEVE software).

4.24
Fast GC Quantification of 16 EC Priority PAH Components

Keywords: PAH; benzo[*a*]pyrene; EU regulation; Fast GC; high resolution; HRMS; food; meat; flavour; smoke; flavouring

Introduction

The European Commission Regulation (EC) No 208/2005 of 4 February 2005, which came into force on 1 April 2005, and the Commission Recommendation of 4 February 2005 on the further investigation into the level of polycyclic aromatic hydrocarbons in food, and the Commission Directive 10/2005/EC of 4 February 2005 laying down the sampling methods and the methods of analysis for the official control of the levels of benzo(a)pyrene in foodstuffs, regulate the maximum levels for benzo[*a*]pyrene in different groups of food of which have the strongest ruling foods for infants and young children with maximum 1.0 µg/kg. The maximum level for smoked meats and smoked meat products is 5.0 µg/kg.

The best known carcinogenic compound – benzo[*a*]pyrene – is used as leading substance out of about 250 different compounds which belong to the PAH (polycyclic aromatic hydrocarbon) group. The German revision of the flavour directive (Aromenverordnung) of 2 May 2006 in § 2 (4) regulates the maximum level for benzo[*a*]pyrene at 0.03 µg/kg for all types of food with added smoke flavourings.

The Commission Recommendation of 4 February 2005 on the further investigation on the levels of PAHs in certain types of food is directed to analyse the levels of 15 PAH compounds, which are classified as priority (Figure 4.100) and to check the suitability of benzo[*a*]pyrene as a marker.

In addition, the Joint FAO/WHO Experts Committee on Food Additives (JECFA) identified the PAH compound benzo[*c*]fluorene as to be monitored as well (Figure 4.101) (JECFA, 2005).

During GC-MS method setup, it turned out that quadrupole desktop MS instruments could not provide the required selectivity and sensitivity at the low decision level for the list of 16 PAH for a reliable detection in real-life samples. High-resolution mass spectrometry with its particularly high selectivity provided the requested robustness and safety in the quantitative determination (Ziegenhals and Jira, 2006 and 2007; Kleinhenz *et al.*, 2006).

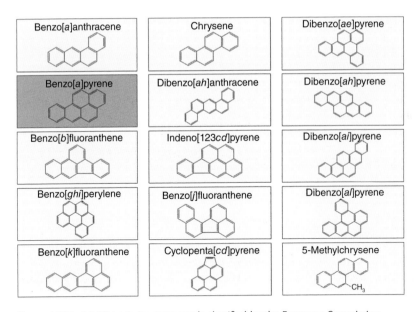

Figure 4.100   15 PAH priority compounds classified by the European Commission Regulation.

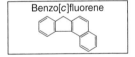

Figure 4.101   Additional PAH priority compound to be monitored according to JECFA 2005.

Table 4.35 Exact masses of PAH and labelled internal standards.

PAH native	Abbreviations	Exact mass native (Da)	PAH ISTD labelled	Exact mass labelled (Da)
Benzo[c]fluorine	BcL	216.0934	5-F-BcL	234.0839
Benzo[a]anthracene	BaA	228.0934	$^{13}C_6$-BaA	234.1135
Chrysene	CHR	228.0934	$^{13}C_6$-CHR	234.1135
Cyclopenta[cd]pyrene	CPP	226.0777	—	—
5-Methylchrysene	5MC	242.1090	d_3-5MC	245.1278
Benzo[b]fluoranthene	BbF	252.0934	$^{13}C_6$-BbF	258.1135
Benzo[j]fluoranthene	BjF	252.0934	—	—
Benzo[k]fluoranthene	BkF	252.0934	$^{13}C_6$-BkF	258.1135
Benzo[a]pyrene	BaP	252.0934	$^{13}C_4$-BaP	256.1068
Indeno[123cd]pyrene	IcP	276.0934	d12-IcP	288.1687
Dibenzo[ah]anthracene	DhA	278.1090	d_{14}-DhA	292.1969
Benzo[ghi]perylene	BgP	276.0934	$^{13}C_{12}$-BgP	288.1336
Dibenzo[al]pyrene	DlP	302.1090	13-F-DlP	320.0996
Dibenzo[ae]pyrene	DeP	302.1090	$^{13}C_6$-DeP	308.1291
Dibenzo[ai]pyrene	DiP	302.1090	$^{13}C_{12}$-DiP	314.1493
Dibenzo[ah]pyrene	DhP	302.1090	—	—
PAH recovery standards deuterated				
D_{12}-Benzo[a]anthracene			d_{12}-BaA	240.1687
D_{12}-Benzo[a]pyrene			d_{12}-BaP	264.1687
D_{12}-Benzo[ghi]perylene			d_{12}-BgP	288.1687

The quantitation was done using isotope dilution technique by the addition of isotope-labelled and fluorinated standards before extraction, as well as for the determination of the response factors of all PAH (Ziegenhals, 2008). Recovery values have been determined by the addition of three deuterium-labelled compounds (Table 4.35).

Analysis Conditions

Sample material		Food, meat, meat products, herbs, spices
Sample preparation	Extraction	Pressurized solvent extraction (ASE)
	Clean-up	Size exclusion chromatography Solid phase extraction
GC Method	System	Thermo Scientific TRACE GC Ultra
	Autosampler	Thermo Scientific TriPlus

(Continued)

	Column	Type	TR-50MS
		Dimensions	10 m length × 0.1 mm ID × 0.1 µm film thickness
	Carrier gas		Helium
	Flow		Constant flow, 0.6 mL/min
	SSL Injector	Injection mode	Splitless, hot needle
		Injection temperature	320 °C
		Injection volume	1.5 µL
	Split	Closed until	1 min
		Open	1 min to end of run
	Oven program	Start	140 °C, 1 min
		Ramp 1	10 °C/min to 240 °C
		Ramp 2	5 °C/min to 270 °C
		Ramp 3	30 °C/min to 280 °C
		Ramp 4	4 °C/min to 290 °C
		Ramp 5	30 °C/min to 315 °C
		Ramp 6	3 °C/min to 330 °C
		Final temperature	330 °C, 0 min
	Transferline	Temperature	300 °C
MS Method	System		Thermo Scientific DFS
	Analyzer type		Magnetic sector high resolution MS
	Ionization		EI
	Electron energy		45 eV
	Ion source	Temperature	280 °C
	Acquisition	Mode	SIM (MID) 2 ions per analyte (see Table 4.36)
	Scan rate		0.7 s
	Resolution	Setting	R 8000 at 5% peak height, R 16000 FWHM
Calibration	Internal standard	Range	^{13}C labelled internal standards

Results

Although an initial use of a 50% phenyl capillary column of 60 m length (60 m × 0.25 mm × 0.25 µm, at constant pressure) provided the required chromatographic resolution of the various isomers, the retention time of more than 90 min turned out to be not appropriate for a control method with high productivity. The application of fast GC column technology reduced the required retention by more than 75% to only 25 min maintaining the necessary chromatographic

Table 4.36 MID descriptor for PAH fast-GC/HRMS data acquisition.

RT (min)	Exact mass (Da)	Function	Dwell time (ms)
8:50	216.09375	Native	82
	218.98508	Lock	2
	226.07830	Native	82
	228.09383	Native	82
	234.08450	Native	82
	234.11400	Native	82
	240.16920	Native	82
	263.98656	Cali	6
13:00	218.98508	Lock	2
	242.10960	Native	82
	245.12840	Native	82
	252.09390	Native	82
	256.10730	Native	82
	258.11400	Native	82
	263.98656	Cali	6
	264.16920	Native	82
19:00	263.98656	Lock	2
	276.09390	Native	74
	278.10960	Native	74
	288.13410	Native	74
	292.19740	Native	74
	313.98340	Cali	6
22:00	263.98656	Lock	2
	302.10960	Native	120
	308.12970	Native	120
	313.98340	Cali	6
	314.13980	Native	120
	320.10010	Native	120

resolution (see Figure 4.102). The critical separation components are shown in Figures 4.103 and 4.104. For all components the fast GC method provides a robust peak separation and quantitative peak integration.

Applicability for different matrices has been shown for many critical matrices. Figure 4.105 shows the analysis of the extract from caraway seeds with a determined concentration of benzo[*a*]pyrene of 0.02 μg/kg. An LOD of 0.005 μg/kg and an LOQ of 0.015 μg/kg can be estimated for the analysis of spices, when the sample weight is 1–1.5 g. The recovery values achieved with the described sample preparation has been between 50 and 120% [6].

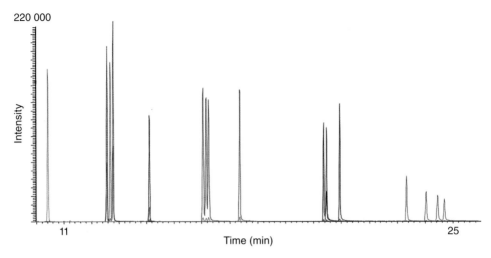

Figure 4.102 Fast-GC-HRMS separation of 16 EC priority PAHs in only 25 min. (Data acquisition MID descriptor as of Table 3.36)

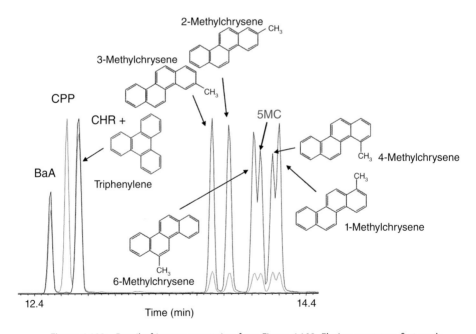

Figure 4.103 Detail of isomer separation from Figure 4.102. Elution sequence first peak cluster BaA, CPP, CHR, second peak cluster 3MC, 2MC, 6MC, 5MC, 4MC, 1MC.

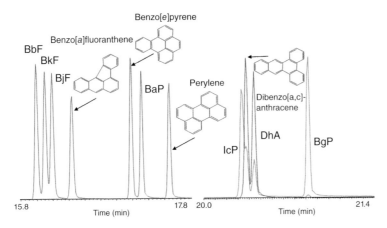

Figure 4.104 Detail of isomer separation from Figure 4.102. Elution sequence left peak cluster 15.8–17.8 min is BbF, BkF, BjF, BaF, BeP, BaP, PER and right peak cluster 20.0–21.4 min is IcP, DcA, DhA, BgP.

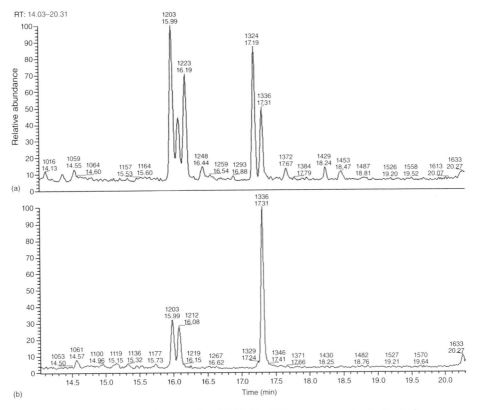

Figure 4.105 Benzo[a]pyrene determination (RT 17:31 min) in caraway seeds at a level of 0.02 µg/kg (a. native PAH and b. ^{13}C-BaP, elution sequence BbF, BkF, BjF, BeP, BaP).

4.25
Multiclass Environmental Contaminants in Fish

Keywords: food safety; micro pollutants; environment; POPs; OCPs; PCBs; PAHs; BFRs; fish feed; fish farming; liquid extraction

Introduction

Fish with high fat contents is an important source of long-chain n-3 polyunsaturated fatty acids (PUFAs) in the human diet, in particular for docosahexaenoic acid (DHA) and eicosapentaenoic acid (EPA) (EFSA, 2005). On the other side, a wide range of environmental contaminants have been reported to be accumulated in fish that can pose a potential human health hazard (Leonard, 2011). Major contaminations include the different OCPs, PAHs or the widely distributed persistent organic pollutants (POPs) with the polychlorinated biphenyls (PCBs) and the flame retardants compound class of the polybrominated diphenylethers (PBDEs). Also farmed fish can significantly contribute to dietary exposure to various contaminants due to the use of land sourced fish feed. A multi method for efficient control of fish and fish feed for various groups of contaminants from PCBs, OCPs, BFRs (brominated flame retardants) and PAHs is outlined in the following analytical procedure published by Kamila Kalachova *et al.* (Kalachova *et al.*, 2013).

Sample Preparation

The multiclass sample preparation for PCBs, OCPs, BFRs and PAHs is based on the liquid extraction of the hydrophobic target analytes from an aqueous sample

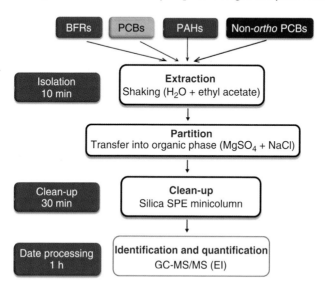

Figure 4.106 Multiclass sample preparation workflow.

suspension into ethyl acetate. The clean-up step of the ethyl acetate phase provides the removal of the co-extracted lipids on a silica minicolumn, see Figure 4.106 (Kalachova *et al.*, 2011).

Ten grams of fish tissue is homogenized and spiked with 10 ng of the internal recovery standards BDE 37 (brominated diphenyl ether) and a ^{13}C-labelled PCB 77, mixed with 5 mL distilled water and shaken with 10 mL ethyl acetate for 1 min. Four grams anhydrous $MgSO_4$ and 2 g of NaCl are added to the mixture and shaken for another 1 min, and centrifuged. Finally 5 mL of the organic layer are withdrawn and dried under nitrogen. The residue is dissolved in 1 mL of *n*-hexane and purified using a silica minicolumn.

As internal quantitation standards ^{13}C-labelled BDE 209 (50 ng/mL), ^{13}C-labelled PCB 101 (40 ng/mL) and a 16 EPA-PAHs mix as ^{13}C-labelled compounds (2 ng/mL) is spiked to the final extract.

Sample Measurements

Special attention in this multicomponent setup is necessary to the GC inlet system. A temperature programmable injector (PTV) is used, offering a cleaning step at an elevated temperature after each extract injection. The choice of liner and deactivation turned out to be critical for the long-term robustness. This is caused by the short sample preparation generating extracts with a high matrix load, as it is shown in Figure 4.107.

A mid-polarity phase GC column dedicated for the separation of PAH components was selected (Thomas *et al.*, 2010). The GC oven temperature program was optimized to obtain the best separation for all target analytes BFRs, PCBs, OCPs and PAHs. Particular attention is required for the coeluting isomeric PAHs compounds (i) BbFA (benzo[b]fluoranthene), BjFA (benzo[j]fluoranthene), BkFA (benzo[k]fluoranthene), (ii) IP, BghiP, DBahA and (iii) BaA, CHR (chrysene), CPP (cyclopenta[cd]pyrene) which are using the same SRM transition for detection (for the acronyms used see Table 4.37). Each target compound was detected using two MS/MS transitions to fulfil the EU SANCO identification criteria (European Commission, 2012). The quantitation ions, confirmation transitions and collision energies had been optimized for each analyte as given in Tables 4.38–4.41.

The chosen target analytes are present in fish typically in a wide concentration range. Trace levels of dl-PCBs or PAHs can occur together with high concentrations of major PCB congeners. For this reason the standard solutions for the quantitative calibration covers three orders of magnitude with a concentration range from 0.05 to 100 ng/mL.

For quality assurance the standard reference material Lake Michigan Fish Tissue, SRM 1947, mussel tissue, SRM 1974b, fresh fish from the retail market, and fish feed were used. The fish feed used was composed of fish meal (48.8%), fish oil (5.7%), wheat (17.4%), wheat by-products (8.8%), soya (13.4%) and other components (5.9%). Typical chromatograms of the target contaminants in retail market samples are shown in Figures 4.108–4.110.

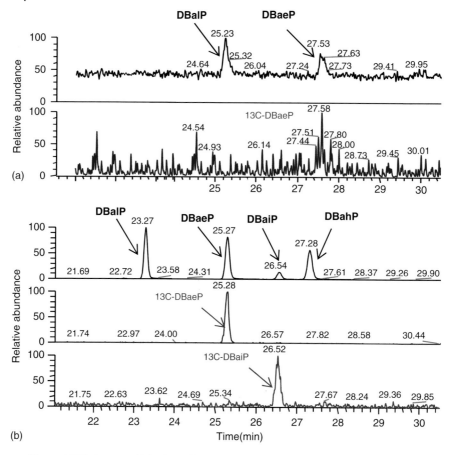

Figure 4.107 Impact of inlet liner choice on GC performance after months of operation, (a) Silcosteel liner and (b) Siltek deactivated baffle liner.

Analysis conditions

Sample material		Fish, fish feed
Sample preparation	Liquid extraction	10 g fish tissue, homogenized, (1 g for fish feed) Add recovery ISTDs BDE 37 (3,4,4′-TriBDE), $^{13}C_{12}$-PCB 77 (3,3′,4,4′-TetraCB, 77 L) Mix with 5 mL distilled water (14 mL distilled water for fish feed) Shake with 10 mL ethyl acetate Add 4 g anhydr. $MgSO_4$, 2 g NaCl, shake, centrifuge

Table 4.37 Compound acronyms used.

Acronym	Compound name
PBDE	Polybrominated diphenyl ether
BDE	Brominated diphenyl ether
HBB	Hexabromobenzene
PBT	Pentabromotoluene
PBEB	Pentabromoethylbenzene
BTBPE	bis(2,4,6-Tribromophenoxy) ethane
OBIND	Octabromo-1-phenyl-1,3,3-trimethylindane
DBDPE	Decabromodiphenyl ethane
cis-HEPO	*cis*-Heptachloroepoxide
trans-HEPO	*trans*-Heptachloroepoxide
AC	Acenaphthene
ACL	Acenaphthylene
AN	Anthracene
BaA	Benz[*a*]anthracene
BaP	Benzo[*a*]pyrene
BbFA	Benzo[*b*]fluoranthene
BcFL	Benzo[*c*]fluorene
BeP	Benzo[*e*]pyrene
BjFA	Benzo[*j*]fluoranthene
BkFA	Benzo[*k*]fluoranthene
BghiP	Benzo[*ghi*]perylene
CHR	Chrysene
CPP	Cyclopenta[*cd*]pyrene
DBahA	Dibenz[*ah*]anthracene
DBaeP	Dibenzo[*ae*]pyrene
DBahP	Dibenzo[*ah*]pyrene
DBaiP	Dibenzo[*ai*]pyrene
DBalP	Dibenzo[*al*]pyrene
dl-PCBs	Dioxin-like PCBs
FA	Fluoranthene
FL	Fluorene
IP	Indeno[1,2,3-*cd*]pyrene
NA	Naphthalene
PHE	Phenanthrene
PY	Pyrene
TRI	Triphenylene
1MC	1-Methylchrysene
1MN	1-Methylnaphthalene
1MPH	1-Methylphenanthrene
1MP	1-Methylpyrene
2MA	2-Methylanthracene
2MN	2-Methylnaphthalene
3MC	3-Methylchrysene
5MC	5-Methylchrysene

Table 4.38 MRM transitions of target BFRs.

Analyte	M (g/mol)	RT (min)	Precursor ion (*m/z*)	Product ion (*m/z*)	CE (V)
BDE 28	406.9	10.5	405.8	245.9	15
			407.9	247.9	18
BDE 37	406.9	10.9	405.8	245.9	15
			407.9	247.9	18
PBT	486.6	10.9	406.7	246.7	20
			485.5	324.8	30
PBEB	500.7	11.1	499.7	485.1	10
			501.5	487.0	15
BDE 49	485.8	12.0	485.8	325.8	18
			487.8	327.8	18
BDE 47	485.8	12.3	485.8	325.8	18
			487.8	327.8	18
HBB	551.5	12.5	551.6	470.6	25
			551.6	389.6	30
BDE 66	485.8	12.6	485.8	325.8	18
			487.8	327.8	18
BDE 77	485.8	13.0	485.8	325.8	18
			487.8	327.8	18
BDE 100	564.7	13.8	561.8	401.8	18
			565.8	405.8	18
BDE 99	564.7	14.2	561.8	401.8	18
			565.8	405.8	18
BDE 85	564.7	15.4	561.8	401.8	18
			565.8	405.8	18
BDE 154	643.6	15.5	641.7	481.7	18
			645.7	485.7	18
BDE 153	643.6	16.3	641.7	481.7	18
			645.7	485.7	18
BDE 183	722.5	20.2	721.8	561.8	20
			723.7	563.8	20
BTBPE	687.6	21.0	356.7	277.4	15
			356.7	328.4	15

(Continued)

Clean-up	Silica minicolumn	5 mL organic layer, dried under nitrogen
		Redissolve in 1 mL hexane
		Apply to clean-up
		1 g sorbent (for up to 0.1 g fat)
		5 g sorbent (for up to 0.8 g fat)
		Elute with 20 mL hexane-DCM (3 : 1, v/v)
		Evaporate solvents
		Redissolve in 0.5 mL i-octane
		Add quantitation ISTDs

Table 4.39 MRM transitions of target PCBs.

Analyte	*M* (g/mol)	RT (min)	Precursor ion (*m/z*)	Product ion (*m/z*)	CE (V)
PCB 28	257.5	8.2	255.9	219.9	20
			257.9	150.9	35
PCB 52	291.9	8.4	291.9	221.9	22
			291.9	256.9	15
PCB 101	324.6	9.2	323.8	253.8	30
			325.8	290.8	14
$^{13}C_{12}$-PCB 101	336.6	9.2	338.8	265.8	30
			336.8	302.8	14
PCB 81	291.9	9.7	291.9	221.9	22
			291.9	256.9	15
PCB 77	291.9	9.9	291.9	221.9	22
			291.9	256.9	15
$^{13}C_{12}$-PCB 77	303.9	9.9	302.9	233.9	22
			302.9	268.9	15
PCB 123	324.6	10.0	323.8	253.8	30
			325.8	290.8	14
PCB 118	324.6	10.1	323.8	253.8	30
			325.8	290.8	14
PCB 153	360.9	10.2	357.8	287.9	25
			359.8	289.9	25
PCB 114	324.6	10.3	323.8	253.8	30
			325.8	290.8	14
PCB 105	324.6	10.6	323.8	253.8	30
			325.8	290.8	14
PCB 138	360.9	10.8	357.8	287.9	25
			359.8	289.9	25
PCB 126	324.6	10.9	323.8	253.8	30
			325.8	290.8	14
PCB 167	360.9	11.0	357.8	287.9	25
			359.8	289.9	25
PCB 156	360.9	11.5	357.8	287.9	25
			359.8	289.9	25
PCB 157	360.9	11.6	357.8	287.9	25
			359.8	289.9	25
PCB 180	395.3	11.6	391.8	321.8	25
			393.8	323.8	25
PCB 169	360.9	12.0	357.8	287.9	25
			359.8	289.9	25
PCB 189	395.3	12.5	391.8	321.8	25
			393.8	323.8	25

Table 4.40 MRM transitions of target OCPs.

Analyte	M (g/mol)	RT (min)	Precursor ion (*m/z*)	Product ion (*m/z*)	CE (V)
HCB	284.8	7.4	248.8	213.9	20
			283.8	213.9	20
HCH-α	290.8	7.5	216.9	180.9	15
			218.9	182.9	15
HCH-γ	290.8	7.9	216.9	180.9	15
			218.9	182.9	15
HCH-β	290.8	8	216.9	180.9	15
			218.9	182.9	15
Heptachlor	373.3	8.1	271.9	236.9	15
			273.9	238.9	12
Aldrin	364.9	8.5	262.9	192.9	32
			262.9	227.9	32
HEPO-*cis*	389.3	9	352.8	262.9	15
			354.8	264.9	15
HEPO-*trans*	389.3	9.1	288.9	218.9	15
			352.8	252.9	15
Chlordane-trans	409.8	9.3	276.9	203.9	16
			372.8	265.9	15
o,p'-DDE	318.0	9.3	246	176	25
			317.9	245.9	20
Chlordane-*cis*	409.8	9.4	372.8	265.9	18
			409.8	374.8	5
Endosulfan-α	406.9	9.5	240.9	205.9	20
			264.9	192.9	22
p,p'-DDE	318.0	9.6	246	176	25
			317.9	245.9	20
Dieldrin	380.9	9.9	262.9	192.9	26
			262.9	227.9	5
o,p'-DDD	320.0	10	235	165	20
			237	165	20
Endrin	380.9	10.3	262.9	190.9	25
			280.9	244.9	12
p,p'-DDD	320.0	10.3	235	165	20
			237	165	20
o,p'-DDT	354.5	10.4	234.9	165	15
			236.9	165	20
Endosulfan-β	406.9	10.7	240.9	205.9	20
			271.9	236.9	18
p,p'-DDT	354.5	10.8	234.9	165	20
			236.9	165	20
Endosulfan sulfate	422.9	11.3	271.9	236.9	15
			273.9	238.9	15

HEPO; heptachloroepoxide.

Table 4.41 MRM transitions of target PAHs.

Analyte	M (g/mol)	RT (min)	Precursor ion (*m/z*)	Product ion (*m/z*)	CE (V)
$^{13}C_6$-NA	134.2	5.0	134.0	83.0	30
			134.0	107.0	20
NA	128.2	5.0	128.0	77.0	30
			128.0	102.0	20
1MN	142.2	5.6	141.0	115.0	15
			142.0	115.0	25
2MN	142.2	5.7	141.0	115.0	15
			142.0	115.0	25
$^{13}C_6$-ACL	158.2	6.5	158.0	130.0	30
			158.0	156.0	20
ACL	152.2	6.5	152.0	102.0	30
			152.0	126.0	20
$^{13}C_6$-AC	160.2	6.6	159.0	158.0	20
			160.0	159.0	20
AC	154.2	6.6	153.0	126.0	40
			153.0	151.0	40
$^{13}C_6$-FL	172.2	7.1	171.0	145.0	30
			171.0	169.0	30
FL	166.2	7.1	165.0	139.0	30
			165.0	163.0	30
$^{13}C_6$-PHE	184.2	8.1	184.0	156.0	30
			184.0	182.0	30
PHE	178.2	8.1	178.0	152.0	20
			178.0	176.0	20
$^{13}C_6$-AN	184.2	8.1	184.0	156.0	30
			184.0	182.0	30
AN	178.2	8.1	178.0	152.0	20
			178.0	176.0	20
1MPH	192.3	8.6	192.0	165.0	30
			192.0	189.0	30
2MA	192.3	8.8	192.0	165.0	30
			192.0	189.0	30
$^{13}C_6$-FA	208.3	9.7	208.0	206.0	30
			208.0	180.0	30
FA	202.3	9.7	202.0	176.0	30
			202.0	200.0	30
$^{13}C_3$-PY	205.3	10.1	205.0	203.0	30
			205.0	204.0	30
PY	202.3	10.1	202.0	176.0	30
			202.0	200.0	30
BcFL	216.3	10.7	216.0	189.0	30
			216.0	215.0	20
1MP	216.3	11.0	216.0	189.0	30
			216.0	215.0	20
$^{13}C_6$-BaA	234.3	12.3	234.0	208.0	30

(continued overleaf)

Table 4.41 (*Continued*)

Analyte	M (g/mol)	RT (min)	Precursor ion (*m/z*)	Product ion (*m/z*)	CE (V)
			234.0	232.0	30
BaA	228.3	12.3	228.0	202.0	30
			228.0	226.0	30
$^{13}C_6$-CHR	234.3	12.5	234.0	208.0	30
			234.0	232.0	30
CPP	226.3	12.5	226.0	200.0	30
			226.0	224.0	30
CHR	228.3	12.5	228.0	202.0	30
			228.0	226.0	30
1MC	242.3	13.1	242.0	226.0	30
			242.0	240.0	30
5MC	242.3	13.3	242.0	226.0	30
			242.0	240.0	30
3MC	242.3	13.4	242.0	226.0	30
			242.0	240.0	30
$^{13}C_6$-BbFA	258.3	14.6	258.0	230.0	30
			258.0	256.0	30
BbFA	252.3	14.6	252.0	226.0	30
			252.0	250.0	30
$^{13}C_6$-BkFA	258.3	14.7	258.0	230.0	30
			258.0	256.0	30
BkFA	252.3	14.7	252.0	226.0	30
			252.0	250.0	30
BjFA	252.3	14.8	252.0	226.0	30
			252.0	250.0	30
$^{13}C_4$-BaP	256.3	15.7	256.0	230.0	30
			256.0	256.0	30
BaP	252.3	15.7	252.0	226.0	30
			252.0	250.0	30
$^{13}C_6$-IP	282.3	19.3	282.0	254.0	30
			282.0	280.0	30
IP	276.3	19.3	276.0	248.0	40
			276.0	274.0	40
$^{13}C_6$-DBahA	284.3	19.3	284.0	256.0	30
			284.0	282.0	30
DBahA	278.3	19.3	278.0	252.0	30
			278.0	276.0	30
13C$_{12}$-BghiP	288.3	20.8	288.0	260.0	30
			288.0	286.0	30
BghiP	276.3	20.8	276.0	248.0	40
			276.0	274.0	40
DBalP	302.4	27.9	302.0	276.0	40
			302.0	300.0	30
$^{13}C_6$-DBaeP	308.4	30.7	308.0	282.0	40
			308.0	306.0	30

Table 4.41 (*Continued*)

Analyte	M (g/mol)	RT (min)	Precursor ion (*m/z*)	Product ion (*m/z*)	CE (V)
DBaeP	302.4	30.7	302.0	276.0	40
			302.0	300.0	30
$^{13}C_{12}$-DBaiP	314.4	32.5	314.0	288.0	40
			314.0	312.0	30
DBaiP	302.4	32.5	302.0	276.0	40
			302.0	300.0	30
DBahP	302.4	33.5	302.0	276.0	40
			302.0	300.0	30

BcFL; benzo[*c*]fluorene.

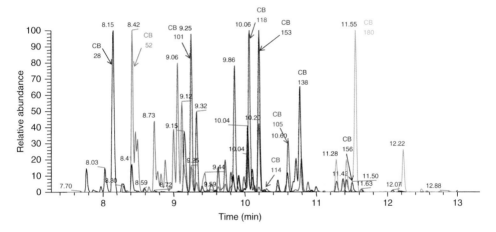

Figure 4.108 Herring fresh fish from the Baltic Sea (PCBs).

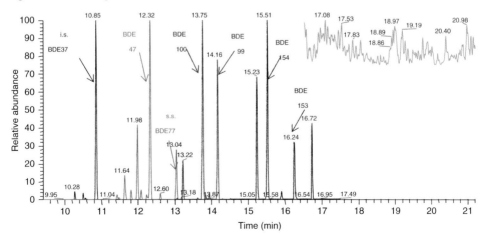

Figure 4.109 Herring fresh fish from the Baltic Sea (PBDEs).

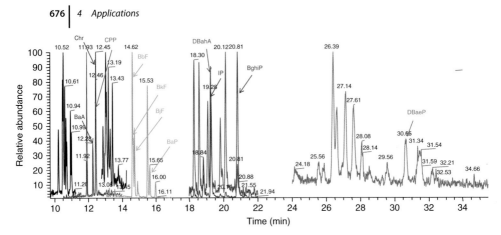

Figure 4.110 Bivalves fresh from Northeast Atlantic Ocean (PAHs).

(*Continued*)			
GC Method	System		Thermo Scientific™ Trace GC Ultra
	Autosampler		Thermo Scientific™ Triplus
	Column	Type	Rxi-17Sil MS
		Dimensions	30 m length × 0.25 mm ID × 0.25 μm film thickness
	Carrier gas		Helium
	Flow		Constant flow, 1.3 mL/min
	PTV Injector	Injection mode	Splitless, 2 min
		Injection volume	1 μL
		Base temperature	95 °C, 0.05 min
		Transfer rate	14.5 °C/s
		Transfer temperature	200 °C
		Transfer time	1 min
		Cleaning phase	4.5 °C/s to 320 °C
		Final temperature	320 °C, 3 min
		Inlet liner	Baffle liner, 2 mm ID, Siltek™ deactivated
	Split	Closed until	2 min
		Open	2 min until end of run
	Oven program	Start	80 °C, 2 min
		Ramp 1	30 °C/min to 240 °C
		Ramp 2	10 °C/min to 340 °C
		Final temperature	340 °C, 20 min
	Transferline	Temperature	320 °C
			Direct coupling
MS Method	System		Thermo Scientific™ TSQ Quantum GC
	Analyzer type		Triple quadrupole MS
	Ionization		EI
	Electron energy		70 eV
	Ion source	Temperature	270 °C
		Emission current	50 μA

(*Continued*)

	Acquisition	Mode	Timed-SRM
			2 SRMs per analyte
	Collision gas		Ar, 1.5 mL/min
	Scan rate		0.3 s
	Resolution	Setting	0.7 Da for Q1, Q3
Calibration	Internal standard		Recovery and quantitation ISTDs
	Calibration range		0.05–100 ng/mL

Results

The described sample preparation method provides recoveries for all target analytes in the range from 70% to 121%, except for the OCPs dieldrin, endrin, endosulfan sulfate, α- and β-endosulfan, and the volatile PAHs AC, ACL (acenaphthylene), PHE (phenanthrene), NA, 1MN and 2MN. For these analytes, recoveries close to zero are reported. The overall recoveries and the RSDs for both matrices fish and fish feed were in the ranges for PCBs 74–119% (RSD 1–19%), OCPs 72–120% (RSD 3–20%), BFRs 73–116% (RSD 3–19%) and for PAHs 70–119% (RSD 1–20%), see also Table 4.42.

For the quantitative calibrations a precision with R^2 values higher than 0.99 are reported for all calibration ranges tested. The quantitation of PAHs was highly influenced by matrix effects that are explained by active sites of the injector inlet liner, as can be seen in Figure 4.107. Injections effects have been minimized by the use of a deactivated inlet liner. The result calculation corrected recoveries with the internal syringe standard.

The achieved method quantitation limits (MQLs) allow the determination of dl-PCBs and the major PCBs well below the maximum limits set by the EU legislation (European Commission, 2011a,b). Also for BaP and the sum of BaP, BaA, BbFA and CHR, low MQLs of 0.025 and 0.1 μg/kg could be achieved. The reported MQLs were defined by a S/N > 6 for the quantitative transition and S/N > 3 for the confirmation transition (European Commission, 2012).

Method limitations

The labile highly brominated compounds BDE 196, 197, 203, 206, 207 and 209, OBIND (octabromo-1-phenyl-1,3,3-trimethylindane) and DBPDE cannot

Table 4.42 Method validation results.

$n = 6$	PAHs	PCBs	PBDEs
Recovery (%)	76–90	90–115	92–116
RSD (%)	5–14	2–11	9–16
LOQ (μg/kg)	0.025–0.5	0.005–0.01	0.025–0.5
Linearity R^2	0.993–1	0.993–0.999	0.991–0.999

be included in the method due to an unsatisfactory sensitivity. For the trace analysis of these compounds a dedicated shorter capillary column is required (Krumwiede, Griep-Raming and Muenster, 2005).

With the chosen analytical column a coelution of several PCBs is observed. PCBs 28/31, 84/101, 138/163 and PCBs 118/123 are not baseline separated. Coelution also occurs for the PAHs CHR/TRI (triphenylene) which are detected on the same SRM transition. These coeluting compounds need to be reported as a sum value.

The described method cannot be used for the OCPs dieldrin, endrin, endosulfan sulfate and α- and β-endosulfan, as well as the PAHs AC, ACL, PHE, NA, 1MN and 2MN. These volatile PAHs get lost during the evaporation step in the clean-up.

Conclusions

This method describes the trace determination of multiclass environmental contaminants in fish and fish feed. The analytical performance meets the EU SANCO criteria for the control of food and feed contaminants with high recoveries and good repeatability.

A total number of 73 target compounds comprising 18 PCBs, 16 OCPs, 14 BFRs and 25 PAHs can be determined using a common liquid extraction sample preparation with a short silica minicolumn clean-up. Very low MQLs are achieved for all target analytes in the range from 0.005 to 1 µg/kg for fish muscle tissue and 0.05 – 10 µg/kg for fish feed. The recoveries in both matrices are in the range of 70 – 120%.

The analytical relevance of those compounds which were excluded from the method in real fish or fish feed samples is seen of minor significance with respect to their frequency of presence and contamination levels. The volatile PAHs are typical air pollutants, and their presence in aquatic organisms indicates acute environmental contamination. Also the occurrence of the five excluded OCPs dieldrin, endrin, endosulfan sulfate, α- and β-endosulfan in fish are not seen as a significant source of human exposure (Kalachova *et al.*, 2013).

4.26
Fast GC of PCBs

Keywords: Fast GC; PCB; dl-PCBs; WHO-PCBs; sewage sludge; HT-8; separation power; carborane

Introduction

Due to the widespread use of PCBs as dielectric and heat-transferring fluids in power transformers, hydraulic fluids, intense use as plasticizers and flame retardants and due to their stable molecular structure, PCBs with its 209 possible congeners, are still today subject to be monitored in environmental analysis (US EPA,

2003). In particular, the coplanar 'dioxin like' dl-PCBs, or non-ortho-substituted PCBs, are of increasing analytical importance because of their toxicity similar to 2,3,7,8-TCDD (tetrachlorodibenzodioxin) with a significant contribution to the sample TEQ value.

The 12 so-called dl-PCBs or WHO-PCBs of toxicologic importance are the coplanar non-ortho substituted congeners of a total number of 68 coplanar congeners of which 20 in total are non-ortho substituted:

Tetra Cl-PCB	77, 81
Penta Cl-PCB	105, 114, 118, 123, 126
Hexa Cl-PCB	156, 157, 167, 169
Hepta Cl-PCB	189

The GC separation of these PCB congeners has been based for a long time on low and intermediate-polarity phases, such as 5% phenylsilicone (typically DB-5MS, TR-5MS, etc.). Problems with co-elution, poor detection at low levels and the need for better separation of the highly toxic coplanar PCBs demand columns with increased selectivity.

The HT8 carborane phase provides good selectivity for the above dl-PCBs as well as the seven indicator congeners (IUPAC 28, 52, 101, 118, 153, 138 and 180) and separates them from most of the potential co-eluting congeners (Bøwadt and Larsen, 2005; SGE, 2005). The analysis is usually performed with a 50 m and 0.25 mm ID capillary column and takes up to 60 min, depending on the used length of the column.

The analysis time for PCBs can be significantly reduced with Fast GC conditions by using a short 10 m column with 0.1 mm ID. The HT8 FAST PCB column has been successfully used as an excellent screening capillary column for MS detection (Gummersbach, 2011).

Analysis Conditions

Sample material			sewage sludge
GC Method	System		Thermo Scientific TRACE GC Ultra
	Injector		Split/splitless
	Column	Type	HT-8
		Dimensions	10 m length × 0.1 mm ID × 0.1 µm film thickness
	Carrier gas		Helium
	Flow		Constant flow, 0.6 mL/min
	SSL Injector	Injection mode	Splitless
		Inj. Temperature	250 °C
		Inj. Volume	1 µL
	Split	closed until	1 min
		open	1 min – end of run

(Continued)

	Oven program	Start	90 °C, 1 min
		Ramp 1	30 °C/min to 220 °C
		Ramp 2	15 °C/min to 330 °C
		Final temp.	300 °C, 5 min
	Transferline	Temperature	280 °C
MS Method	System		Thermo Scientific DSQ II
	Analyzer type		Single quadrupole MS
	Ionization		EI
	Electron energy		70 eV
	Ion source	Temperature	280 °C
	Acquisition	Mode	Full scan
	Mass range		100–500 Da
	Scan rate		0.1 s (10 spectra/s)

Results

The Fast GC separation of PCBs still maintains excellent congener separation within a total analysis time of only 9 min including the internal standard decachlorobiphenyl (PCB 209). Laboratories with a high sample throughput benefit in particular from the increased productivity. The chromatogram of the test mix shows a very good separation of some of the critical pairs, for example congeners 31/28 (about 80%) and 163/138 (about 50%). This excellent separation can be achieved in a total analysis time of less than 10 min, see Figure 4.111. Due to the very high scan speed of the employed MS detector a full-scan mass spectrum of full integrity for substance confirmation is provided as given for HCB in Figure 4.112. The achieved separation power (Figure 4.113) shows that the FAST PCB capillary column in combination with a very fast scanning quadrupole mass spectrometer is an excellent choice for the screening and quantitation for PCBs.

Further potential for increased detection sensitivity could be achieved by running the MS in SIM detection mode. The analysis of a contaminated sewage sludge sample acquired in the full-scan mode given in Figure 4.114 shows the well-known mass chromatogram patterns of the PCB chlorination degrees, but in a very short analysis runtime of less than 10 min including the elution of the internal quantification standard PCB 209.

The unique selectivity of the HT8 column is attributed to the presence of the carborane unit having an affinity towards chlorinated biphenyls with the least number of ortho-substitutions. Fewer substitutions in the ortho position increase freedom of rotation, allowing the chlorinated biphenyl moiety to have greater interaction with the carborane unit (de Boer, Dao and van Dortmond, 1992). This phenomenon causes non-ortho substituted PCBs to have increased elution times compared to their ortho substituted congeners. This allows detection and quantitation of the important congeners used to monitor PCB occurrence and distribution in the environment.

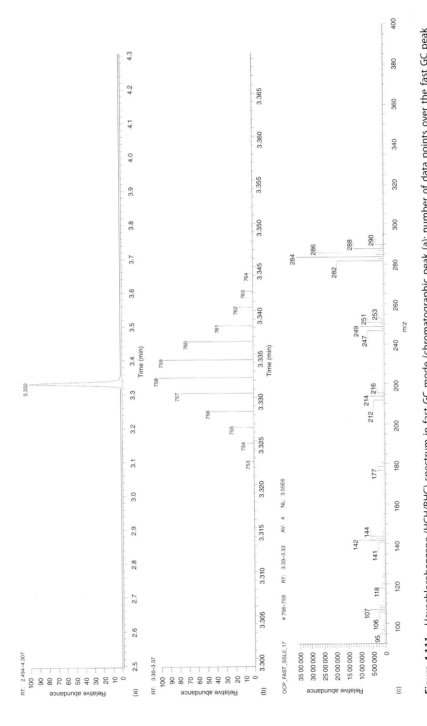

Figure 4.111 Hexachlorobenzene (HCH/BHC) spectrum in fast GC mode (chromatographic peak (a); number of data points over the fast GC peak (b) and full scan spectrum (c).

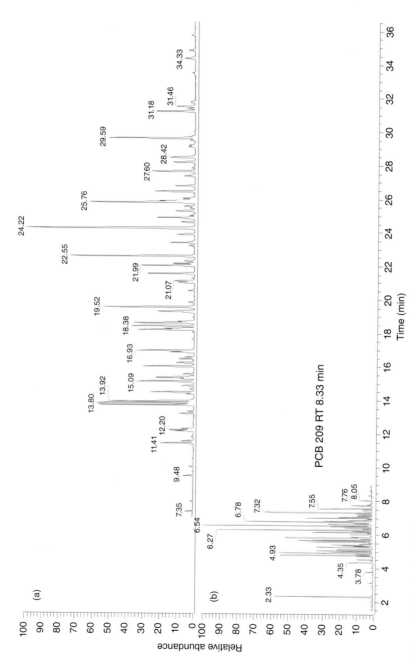

Figure 4.112 Aroclor Mix A30, A40, A60: Comparison of Fast GC. (b) (10 m column length, 0.1 μm ID) to (a) normal chromatography (30 m column length, 0.25 μm ID).

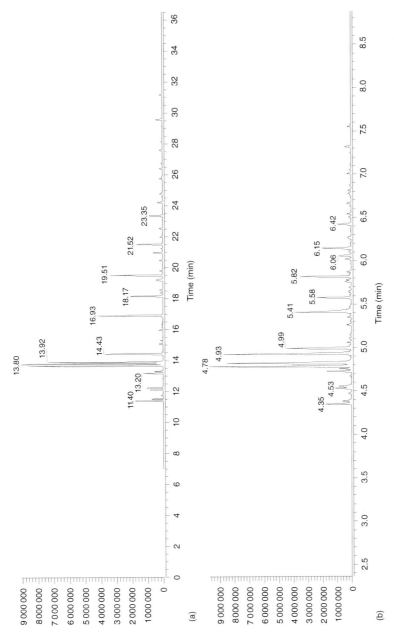

Figure 4.113 Aroclor Mix A30, A40 and A60: separation power comparison of Fast GC with 10 m column length (b) to normal chromatography with 30 m column length (a).

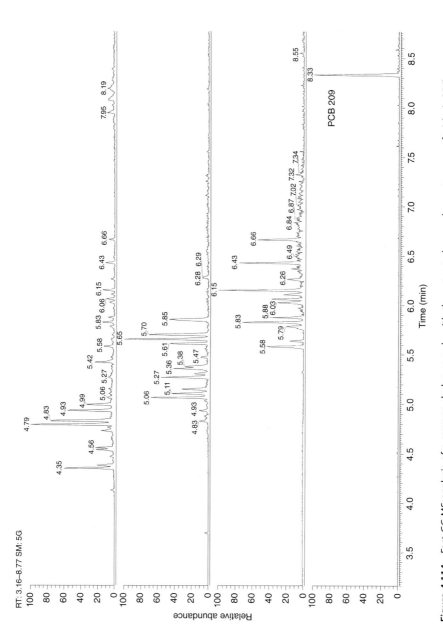

Figure 4.114 Fast GC-MS analysis of a sewage sludge sample with the extracted mass chromatograms of tri-(m/z 256), tetra-(m/z 293), penta-(m/z 326) and deca-chlorbiphenyl (m/z 498, ISTD).

4.27
Congener Specific Isotope Analysis of Technical PCB Mixtures

Keywords: PCB; congener-specific δ^{13}C values; source; fate; environment; 2DGC;
MCSS; IRMS; CSIA

Introduction

Stable carbon isotope analysis is increasingly applied for the understanding of the
sources and fate of anthropogenic organic contaminants in the environment. In
particular, the compound specific isotope analysis (CSIA) of complex mixtures
provides a powerful analytical tool to trace the origin and conversion of organic
compounds in the environment. Although traditional approaches such as finger-
printing, which involves matching of the isomer profiles in samples with that in
technical preparations, have been used to determine sources of anthropogenic
chemicals, for example PCBs, dioxins (Horii *et al.*, 2005; Horii, 2008). GC-IRMS
complements the existing methods to understand the sources and environmental
destiny of anthropogenic chemicals.

   In this application, 2DGC-C-IRMS technique is used for the isotopic analysis of carbon, and to determine congener-specific δ^{13}C values of PCB and
PCN congeners in several technical mixtures produced in the United States,
Japan, Germany, France, former USSR, Poland and former Czechoslovakia
(see Figure 4.115).

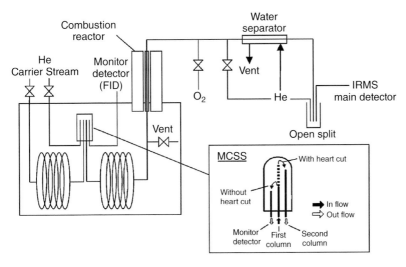

Figure 4.115   Schematics of the 2DGC-C-IRMS instrument setup.   (Reprinted with permission from Environ. Sci. Technology, Copyright 2005 American Chemical Society.)

Analysis Conditions

Sample material			Technical PCB mixtures
GC Method	System		Thermo Scientific TRACE GC 2000 with monitor detector FID
	Injector	Type	Split/splitless
	Column	Type	DB5-MS
		Dimensions	30 m length × 0.25 mm ID × 0.25 μm film thickness and as second dimension column
		Type	Rtx-200
		Dimensions	15 m length × 0.32 mm ID × 0.25 μm film thickness
		Column switching	MCSS, moving capillary switching system
	Carrier gas		Helium
	Flow		Constant flow, 1 mL/min
	SSL Injector	Injection mode	Splitless
		Injection temperature	260 °C
		Injection volume	1 μL
	Split	Closed until	1 min
		Open	1 min to end of run
	Oven program	Start	70 °C, 1 min
		Ramp 1	15 °C/min to 180 °C
		Ramp 2	2 °C/min to 260 °C
		Final temperature	260 °C, 5 min
	Oxidation reactor	Temperature	940 °C
MS Method	System		Thermo Scientific MAT 252
	Analyzer type		Magnetic sector IRMS
	Ionization		EI
	Electron energy		70 eV
	Acquisition	Mode	Parallel ion detection
	Collision gas		Ar, 1.5 mL/min
	Mass range		m/z 44, 45, 46, continuous monitoring

Results

Two-dimensional gas chromatography separation by using heart-cutting (2DGC-C-IRMS by MCSS) enabled significant improvement in the resolution and sensitivity of individual PCB isomers by at least an order of magnitude better than the traditional one-dimensional GC-IRMS method.

Only selected target compounds, which have been identified using the FID monitor detector, have been transferred onto a second column via heart cutting for online conversion with the oxidation furnace (Figure 4.116). The background signal of *m/z* 44 was significantly reduced to 10 mV even at a high temperature

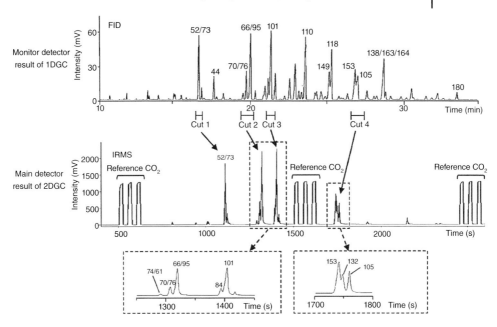

Figure 4.116 Chromatograms of the monitor detector and IRMS for a mixture of Kanechlor 500. The CO_2 reference gas injections occur at the beginning, between cuts and at the end of the separation (Horii *et al.*, 2005). (Reprinted with permission from Environ. Sci. Technology, Copyright 2005 American Chemical Society.)

(260 °C) due to minimized column bleed. The estimated sensitivity of carbon using 2DGC-C-IRMS was <7 ng, which corresponds to 10–20 ng of individual PCB congeners injected. Thirty-one PCB congeners were selected in 18 technical PCB preparations for the determination of $\delta^{13}C$ values of individual PCB congeners (Table 4.43) (Horii *et al.*, 2005).

It has been observed that lower chlorinated PCB congeners showed higher $\delta^{13}C$ values in each technical PCB mixture, which might be influenced by isomer-specific isotopic partitioning. Geographical differences in the $\delta^{13}C$ values among PCB preparations, particularly those of Delors, Sovol, Trichlorodiphenyl and Chlorofen, indicate possible differences in the raw materials used during the production processes (Figure 4.117).

The $\delta^{13}C$ values determined for PCB and also for separately investigated PCN congeners were similar to those from petroleum and terrestrial plants. However, these values are apparently different from those of carbonates and marine plants. It could be proven that GC-IRMS by accurate instrumental methods such as 2DGC-C-IRMS provides the necessary analytical tools to reveal isotopic partitioning by unknown environmental and geochemical processes.

Table 4.43 δ^{13}C values of selected PCB congeners from technical mixtures (extracted, for the full table refer to Horii et al., 2005).

		United States 1930–1975 Aroclor			Germany 1930–1982 Clophen			Czechoslovakia 1959–1984 Delor		Soviet Union 1939–1993	
Congener	Structure Cl-PCB	21	40–42	52–54	40–42	48	52–54	NA		41	53
		A 1221	A 1242	A 1254	A 30	A 40	A 50	D 103	D 106	Sovol	TCP
Biphenyl	no Cl										
1	2-	−25.2									
3	4-	−27.2									
10/4	2,6-/2,2'-	−24.7	−25.9		−26.5			−31.3			−22.7
8/5	2,4'-/2,3-		−26.2		−27.3			−31.4			−23.1
18	2,2',5-		−26.9		−28.2		−25.5	−33.2			−23.9
16/32	2,2',3-/2,4',6-		−26.7		−28.0		−25.8	−33.0			−24.7
31/28	2,4',5-/2,4,4'-		−26.7		−27.7		−26.9	−32.8			−24.8
33/20/53	2,3',4'-		−26.3		−28.1		−26.7	−33.1			−24.6
52/73	2,2',5,5'-/2,3',5',6-		−27.7	−26.3	−28.8	−29.2	−27.1	−34.0		−23.4	−25.1
43/49	2,2',3,5-/2,4,5'-		−27.0	−25.0	−28.1	−28.1	−26.0	−32.8			−24.6
47/75/48	2,2',4,4'-/2,4,4',6-/2,2',4,5-		−27.2				−26.6				
44	2,2',3,5'-		−27.3	−25.3	−29.0	−28.6	−26.6	−33.9			−24.6
59/42	2,3,3',6-/2,2',3,4'-		−27.6		−27.2		−28.5	−31.7			−25.9
74/61	2,4,4',5-/2,3,4,5-		−26.7				−26.5				
70/76	2,3',4',5-/2,3',4',5'-		−26.7	−24.0	−28.5	−27.6	−26.1	−33.1		−22.0	
66/95	2,3',4,4'-/2,2',3,5',6-		−27.0	−26.5	−28.4	−28.4	−27.2	−33.1	−31.6	−23.6	−25.2
89/101/90	2,2',3,4,6'-/2,2',4,5,5'-/2,2',3,4',5-			−27.4		−31.2	−28.3		−31.3	−27.1	
110	2,3,3',4',6-			−27.3		−29.9	−28.1		−31.2	−25.3	
149/139	2,2',3,4',5',6-/2,2',3,4,4',6-			−27.8			−29.7		−32.6		

Table 4.43 *(Continued)*

		United States 1930–1975 Aroclor			Germany 1930–1982 Clophen			Czechoslovakia 1959–1984 Delor		Soviet Union 1939–1993	
	Product	21 A 1221	40–42 A 1242	52–54 A 1254	40–42 A30	48 A 40	52–54 A 50	NA D 103	D 106	41 Sovol	53 TCP
Cl content by weight (%)											
118/106	2,3',4,4',5-/2,3,3',4,5-			−26.4		−27.8	−27.4		−30.9	−25.1	
153/132/168	2,2',4,4',5,5'-/2,2',3,3',4,6'-/2,3',4,4',5',6-			−27.2			−28.4		−32.0	−25.1	
105	2,3,3',4,4'-			−27.1			−28.2			−25.0	
164/163/138	2,3,3',4',5',6-/2,3,3',4',5,6-/2,2',3,4,4',5'-			−27.5			−30.5		−32.5	−26.4	
182/187	2,2',3,4,4',5,6-/2,2',3,4',5,5',6-								−34.2		
174	2,2',3,3',4,5,6'-								−34.2		
180	2,2',3,4,4',5,5'-								−34.4		
170/190	2,2',3,3',4,4',5-/2,3,3',4,4',5,6-								−33.2		
199	2,2',3,3',4,5,5',6'-										
203/196	2,2',3,4,4',5,5',6-/2,2',3,3',4,4',5,6'-										
194	2,2',3,3',4,4',5,5'-										
206	2,2',3,3',4,4',5,5',6-										
Mean		−25.7	−26.9	−26.5	−28.0	−27.8	−27.8	−32.8	−32.5	−24.8	−24.5
Maximum		−24.7	−25.9	−24.0	−26.5	−25.5	−26.0	−31.3	−30.9	−22.0	−22.7
Minimum		−27.2	−27.7	−27.8	−29.0	−31.2	−30.5	−34.0	−34.4	−27.1	−25.9

Reprinted with permission from Environ. Sci. Technology, Copyright 2005 American Chemical Society.

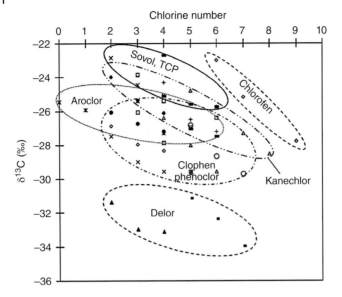

Figure 4.117 PCB production characteristics by $\delta^{13}C$ value plot ag. chlorine number of the selected congeners from Table 4.49 (Horii *et al.*, 2005). (Reprinted with permission from Environ. Sci. Technology, Copyright 2005 American Chemical Society.)

4.28
Dioxin Screening in Food and Feed

Keywords: food safety; POPs; PCDDs; PCDFs; dl-PCBs; official control; EU regulations; triple quadrupole; GC-MS/MS; GC-HRMS; ion ratio; confirmation

Introduction

The monitoring of food and feed stuff for the presence of polychlorinated dibenzo-*p*-dioxins (PCDDs), polychlorinated dibenzofurans (PCDFs) and the dioxin-like polychlorinated biphenyls (dl-PCBs) can be performed with screening and confirmatory methods.

The methods of sampling and analysis for the official control of levels of PCDD/PCDFs and dl-PCBs for food and feeding stuffs are defined in the European Commission Regulation (EC) No 152/2009 and Commission Regulation (EC) No 1883/2006. For the analysis of PCDD/PCDFs and dl-PCBs for food (European Commission, 2012a) and feed (European Commission, 2012b) screening and confirmatory methods can be applied. Screening methods comprise GC-MS, GC-MS/MS or bioanalytical methods, confirmatory methods are defined as GC-HRMS (US EPA, 1994) and GC-MS/MS methods (European Commission Regulation No. 709/2014 of 20 June 2014 amending Regulation

Table 4.44 Criteria for screening and confirmatory methods (European Commission, 2012a,b).

	Screening with bioanalytical or physicochemical methods	Confirmatory methods
False-compliant rate	<5%	
Trueness		−20 to + 20%
Repeatability (RSD$_r$)	<20%	
Within-laboratory reproducibility (RSD$_R$)	<25%	<15%

(EC) No 152/2009 as regards the determination of the levels of dioxins and polychlorinated biphenyls).

For the control of the regulated maximum levels screening methods should allow a cost-effective high sample-throughput with the goal to avoid false negative results. Here GC-MS methods as well as bioanalytical methods are used. The confirmation of results in samples with significant levels requires a confirmatory method. The quality criteria for the screening and confirmatory methods in the EU regulations are oriented towards high reproducibility and a low false compliant rate (European Commission, 2012a, 2012b) (Table 4.44).

For GC-MS methods the method performance need to be demonstrated to cover the level of interest with an acceptable precision (coefficient of variation, CV). The required LOQ needs to be at 20% of the level of interest, which requires in absolute amounts detectable quantities of PCDD/PCDFs in the upper femtogram range, non-ortho-PCBs in the low pictogram range, and other dl-PCBs in the nanogram range. The limit of quantitation of an individual congener is defined in this context as the concentration of an analyte in the extract of a sample which produces an instrumental response on two different ions with a S/N ratio of 3 : 1 for the less sensitive signal.

Also, GC-MS screening methods require the control of recoveries by the addition of ^{13}C-labelled internal standards at the first sample preparation step. Screening methods require at least one congener as internal standard for any homologous group or mass spectrometric recording function. In confirmatory methods all ^{13}C-labelled internal standards shall be used. The range of recoveries shall cover 30–140% for screening methods and 60–120% for confirmatory methods.

For the clean-up and fractionation as well as the chromatographic separation of isomers the same requirements apply for screening and confirmation.

Due to the high performance of modern triple quadrupole mass spectrometers for PCDD/PCDF and dl-PCB analysis (Kotz *et al.*, 2012) further amendments of the criteria for confirmatory methods for inclusion of MS/MS (tandem mass spectrometry) have been set in effect with the European Commission Regulation No 709/2014 of 20 June 2014. "Technical progress and developments have shown that the use of gas chromatography/tandem mass spectrometry (GC-MS/MS) should be allowed for use as a confirmatory method for checking compliance with the maximum level, in addition to gas chromatography/high resolution mass spectrometry (GC-HRMS)" (European Commission (2014)).

- The calculation of LOQ needs to be based on signal-to-noise ratio or lowest concentration point on calibration curve.
- The data acquisition requires the monitoring of at least two specific precursor ions, each with one specific corresponding transition product ion for all labelled and unlabelled analytes (comparable to GC-HRMS).
- The mass resolution setting for each quadrupole needs to be set equal to or better than unit mass resolution in order to minimize possible interferences on the analytes of interest.
- The maximum permitted tolerance of relative ion intensities needs to be demonstrated to be in the range of ±15% for selected SRM product ions in comparison to calculated or measured values (average from calibration standards).

Sample Preparation

The extraction and clean-up process for food and feed samples was performed according Figure 4.118 for GC-MS/MS and GC-HRMS. The identical extracts were measured on both instrument types.

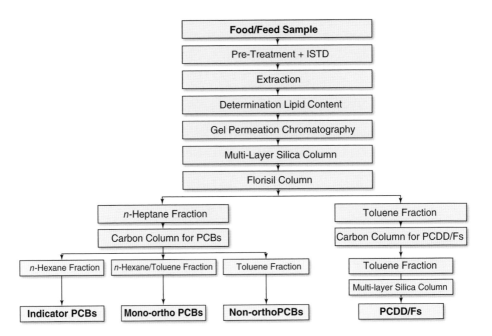

Figure 4.118 Extraction and clean-up process for food and feed samples (Kotz *et al.*, 2011).

Experimental Conditions

For GC-MS/MS analysis the same GC conditions with PTV injection are used as they are standard for GC-HRMS measurements.

Special attention is required concerning the analytical column choice, carrier gas flow and the setup of the GC temperature program. The EPA method 1613 sets certain requirements for the TCDD and TCDF (tetrachlorodibenzofuran) congener separation (US EPA, 1994). The valley between 2,3,7,8-TCDD and the other tetra-dioxin isomers with m/z 319.8965, and between 2,3,7,8-TCDF and the other tetra-furan isomers at m/z 303.9016 shall not exceed 25%, as shown in Figure 4.119 using a 60 m TR-5MS column type. The column length of 60 m is also required to separate the 1,2,3,4,7,8-HxCDF and 1,2,3,6,7,8-HxCDF isomers with better than 25% valley.

Other known interferences observed on the most used 5% phenyl columns can be caused by the content of PCBs in the sample. PCB 87 can interfere with PCB 81, and PCB 129, 178 with PCB 126. Also PCB 77 and 110 can interfere on the column and need be separated in clean-up step. Other interferences are reported

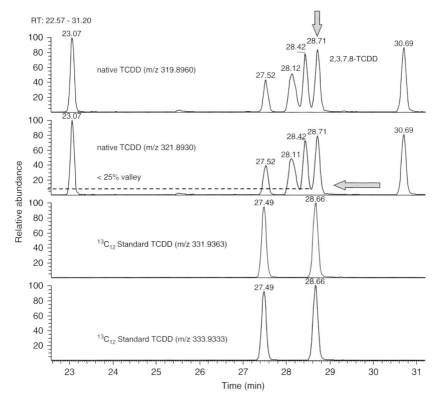

Figure 4.119 Separation performance of <25% valley between 2,3,7,8-TCDD and the other tetra-dioxin congeners.

Table 4.45 SRM transitions for tetra to octa PCDFs/PCDDs and labelled internal standards.

PCDD/PCDF Type	Precursor Ion 1 m/z	Product Ion 1 m/z	Precursor Ion 2 m/z	Product Ion 2 m/z
TCDF	303.90	240.94	305.90	242.94
[$^{13}C_{12}$]TCDF ISTD	315.94	251.97	317.94	253.97
TCDD	319.90	256.90	321.89	258.89
[$^{13}C_{12}$]TCDD ISTD	331.94	267.97	333.93	269.97
PeCDF	339.86	276.90	341.86	278.89
[$^{13}C_{12}$]PeCDF ISTD	351.90	287.93	353.90	289.93
PeCDD	355.85	292.85	357.85	294.85
[$^{13}C_{12}$]PeCDD ISTD	367.90	303.90	369.89	305.89
HeCDF	371.82	308.86	373.82	310.86
[$^{13}C_{12}$]HeCDF ISTD	383.86	319.90	385.86	321.89
HeCDD	387.82	324.82	389.82	326.82
[$^{13}C_{12}$]HeCDD ISTD	399.86	335.86	401.86	337.86
HpCDF	407.78	344.82	409.78	346.82
[$^{13}C_{12}$]HpCDF ISTD	419.82	355.86	421.82	357.85
HpCDD	423.78	360.78	425.77	362.77
[$^{13}C_{12}$]HpCDD ISTD	435.82	371.82	437.81	373.81
OCDF	441.76	378.80	443.76	380.79
[$^{13}C_{12}$]OCDF ISTD	453.78	389.82	455.78	391.81
OCDD	457.74	394.74	459.74	396.74
[$^{13}C_{12}$]OCDD ISTD	469.78	405.78	471.78	407.78

to occur on the masses of the HxCBs 169 and 1,2,3,7,8-PeCDD which are very similar and coelute on 30 and 60 m DB-5MS. Other newly developed column phases which are specialized for the dioxin/PCB congener separation should be considered for improvements of particular congener coelutions. For instance the BPX-DXN column is able to separate between HxCB 169 and 1,2,3,7,8-PeCDD, but at lower sensitivity of the 60 m BPX-DXN compared to a 30 m DB-5MS column. The Rtx-Dioxin2 and TG-Dioxin columns offer a different selectivity and separates well 2,3,7,8-TCDD and -TCDF from other congeners, also improves the 2,3,4,7,8-PeCDF and 1,2,3,7,8-PeCDF separation.

All target congeners are monitored using two precursor ions with each one product ion. As precursor ions usually the intense M^+ ions are used. For the PCDD/PCDFs the product ions typically generated by COCl loss are selected, see Table 4.45. With PCBs the elimination of chlorine is observed, see Table 4.46.

For calibration the concentrations as given in Table 4.47 were applied. The chromatogram peaks are displayed in Figure 4.120. For checking of the performance of the GC-MS/MS system in the low concentration range additionally 1:2 and 1:5 dilutions of the lowest calibration point were measured. The calculated WHO-PCDD/PCDF-TEQ for these calibrations are based on the analysis equivalent of 3 g of fat and an injection of 5 µL out of 20 µL of the concentrated final extract volume, ranged between 0.12 and 24 pg/g fat.

Table 4.46 SRM transitions for PCBs and labelled internal standards.

PCB Type	Precursor Ion 1 m/z	Product Ion 1 m/z	Precursor Ion 2 m/z	Product Ion 2 m/z
MoCB	188.04	153.04	190.04	153.04
[$^{13}C_{12}$] MoCB ISTD	200.08	165.10	202.08	165.10
DiCB	222.00	152.06	224.00	152.06
[$^{13}C_{12}$] DiCB ISTD	234.04	164.10	236.04	164.10
TrCB	255.96	186.02	257.96	186.02
[$^{13}C_{12}$] TrCB ISTD	268.00	198.02	270.00	198.02
TeCB	289.92	219.98	291.92	219.98
[$^{13}C_{12}$] TeCB ISTD	301.96	232.02	303.96	232.02
PeCB	323.90	253.95	325.90	255.95
[$^{13}C_{12}$] PeCB ISTD	335.92	265.99	337.92	267.99
HxCB	357.80	287.90	359.80	289.95
[$^{13}C_{12}$] HxCB ISTD	369.90	299.51	371.90	301.95
HpCB	391.80	321.90	393.80	323.90
[$^{13}C_{12}$] HpCB ISTD	403.80	333.90	405.80	335.90
OcCB	427.80	357.80	429.80	357.80
[$^{13}C_{12}$] OcCB ISTD	439.80	369.90	441.80	369.90
NoCB	461.70	391.80	463.70	393.80
[$^{13}C_{12}$] NoCB ISTD	473.80	403.80	475.80	405.80
DeCB	495.70	425.80	497.70	427.80
[$^{13}C_{12}$] DeCB ISTD	507.70	437.80	509.70	439.80

Table 4.47 Concentrations of individual congeners in calibration solutions.

	Cal 1 (pg/µL)	Cal 2 (pg/µL)	Cal 3 (pg/µL)	Cal 4 (pg/µL)	Cal 5 (pg/µL)
2,3,7,8-TCDD	0.0125	0.025	0.05	0.2	0.5
1,2,3,7,8-PeCDD	0.0250	0.050	0.10	0.4	1.0
1,2,3,4,7,8-HxCDD	0.0250	0.050	0.10	0.4	1.0
1,2,3,6,7,8-HxCDD	0.0625	0.125	0.25	1.0	2.5
1,2,3,7,8,9-HxCDD	0.0250	0.050	0.10	0.4	1.0
1,2,3,4,6,7,8-HpCDD	0.1250	0.250	0.50	2.0	5.0
OCDD	0.2500	0.500	1.00	4.0	10.0
2,3,7,8-TCDF	0.0125	0.025	0.05	0.2	0.5
1,2,3,7,8-PeCDF	0.0125	0.025	0.05	0.2	0.5
2,3,4,7,8-PeCDF	0.0625	0.125	0.25	1.0	2.5
1,2,3,4,7,8-HxCDF	0.0250	0.050	0.10	0.4	1.0
1,2,3,6,7,8-HxCDF	0.0250	0.050	0.10	0.4	1.0
1,2,3,7,8,9-HxCDF	0.0125	0.025	0.05	0.2	0.5
2,3,4,6,7,8-HxCDF	0.0125	0.025	0.05	0.2	0.5
1,2,3,4,6,7,8-HpCDF	0.0250	0.050	0.10	0.4	1.0
1,2,3,4,7,8,9-HpCDF	0.0125	0.025	0.05	0.2	0.5

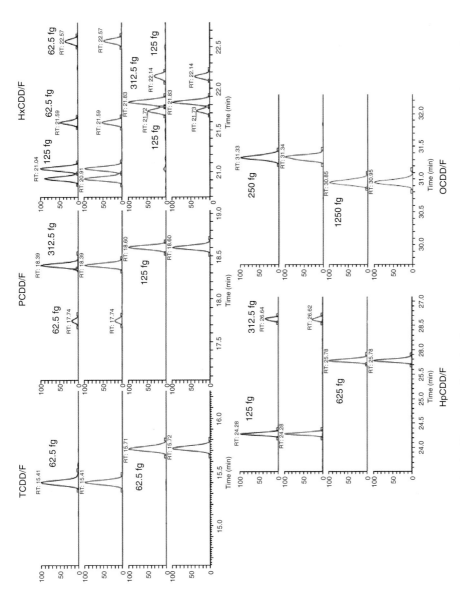

Figure 4.120 Lowest level calibrations peaks (Cal 1) for tetra- to octa-CDD/F congeners, top traces PCDDs, bottom PCDFs with injected amounts.

Analysis conditions

Sample material			Food, feeding stuff
GC Method	System		Thermo Scientific™ TRACE GC Ultra
	Autosampler		Thermo Scientific™ TriPlus RSH
	Column	Type	TR-Dioxin, TG-5MS, or DB-5MS
		Dimensions	60 m length × 0.25 mm ID × 0.25 µm film thickness
	Carrier gas		Helium
	Flow		Constant flow, 1 mL/min
	PTV Injector	Injection mode	Splitless
		Injection volume	5 µL, extract in toluene
		Injection speed	5 µL/s
		Base temperature	100 °C
		Transfer rate	13.3 °C/s
		Transfer temperature	340 °C
		Final temperature	340 °C
	Split	Closed until	1 min
		Open	1 min to end of run
		Flow	20 mL/min
	Oven program	Start	120 °C
		Ramp 1	17 °C/min to 250 °C
		Ramp 2	2.5 °C/min to 285 °C
		Final temperature	285 °C, 13 min
	Transferline	Temperature	280 °C
			Direct coupling
MS Method	System		Thermo Scientific™ TSQ Quantum XLS Ultra
	Analyzer type		Triple quadrupole MS
	Ionization		EI
	Electron energy		40 eV
	Ion source	Temperature	250 °C
		Emission current	50 µA
	Acquisition	Mode	Timed-SRM
			2 SRMs per analyte
	Collision gas		Ar, 1.5 mTorr
	Collision energy		22 V (all congeners)
	Resolution	Setting	Normal (0.7 Da) for Q1 and Q3
Calibration	Internal standard		Isotope dilution method, ^{13}C-labelled standards
	Calibration range		0.0125–10.0 pg/µL, congener dependent
	Data points		5

Results

One important criterion for quality control in the identification of the measured PCDD/PCDF congeners is the ion ratio between the two monitored product ions for each congener. For quality control the ion abundance ratios can be compared with calculated or measured values as demonstrated in Figure 4.121. The ratios are comparable with the measured ratios, if identical collision energy and collision gas pressure is applied for both transitions. The measured ion abundance ratios in the calibration run match the calculated theoretical values within the QC limits of $\pm 15\%$. The calculated ratio depends on the relative abundance of the two selected precursor ions of the molecular ion and the probability of the loss of the two chlorine isotopes with the possible leaving groups $CO^{35}Cl$ or $CO^{37}Cl$ leading to the formation of two product ions of different mass. Figures 4.122 and 4.123 shows the principle for TCDD. The ion abundance ratio 1.04 for TCDD between quantitation ion and confirming ion for TCDD is calculated by:

$$\frac{\text{Relative abundance (quantitation ion)} \times \text{Probability (quantitation ion)}}{\text{Relative abundance (confirming ion)} \times \text{Probability (confirming ion)}} = 1.04$$

The results of GC-MS/MS have been compared with GC-HRMS measurements. The deviation of the GC-MS/MS results from GC-HRMS for different food samples and human milk, covering a concentration range between 0.1

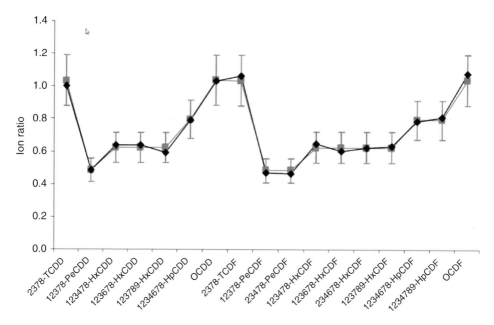

Figure 4.121 Comparison of the calculated (square) and measured (diamond) ion ratio on the TSQ Quantum XLS Ultra GC-MS/MS system.

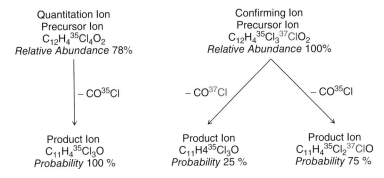

Figure 4.122 Calculation of theoretical ion abundance ratio for TCDD.

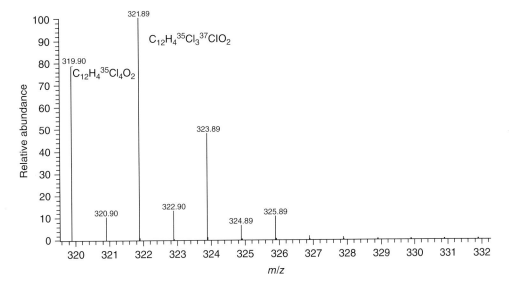

Figure 4.123 TCDD molecular ion cluster with m/z 319.90 $C_{12}H_4{}^{35}Cl_4O_2$ and m/z 321.89 $C_{12}H_4{}^{35}Cl_3{}^{37}ClO_2$ due to the statistical distribution of the chlorine isotopes.

and 10 pg WHO-PCDD/PCDF-TEQ/g fat (WHO-TEF 1998), are below 20% for most of the samples. The applicable maximum levels for butter, eggs, meat and fat (European Commission, 2006) together with difference GC-MS/MS vs. GC-HRMS are shown in Figure 4.124. For all sample types the maximum levels can be controlled by triple quadrupole GC-MS/MS with good compliance compared to the GC-HRMS method. As examples a spiked fat sample is shown in Figure 4.125 at a concentration of 62.5 fg 2,3,7,8-TCDD, and a real life sample of a contaminated fat sample in Figure 4.126.

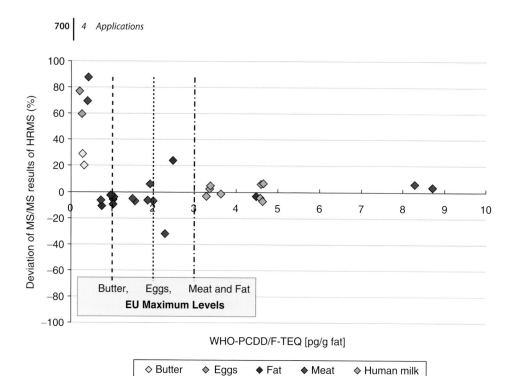

Figure 4.124 Compliance of GC-MS/MS with the required EU PCDD/PCDF maximum levels for butter, eggs, meat and fat in pg/g fat and % difference to the GC-HRMS method (European Commission, 2006; Kotz *et al.*, 2011).

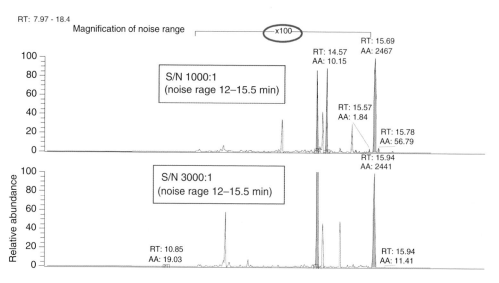

Figure 4.125 2,3,7,8-TCDD detection at RT 15:69 min in a spiked fat sample by GC-MS/MS, determined as 62.5 fg, top quantitation, bottom ion ratio transition.

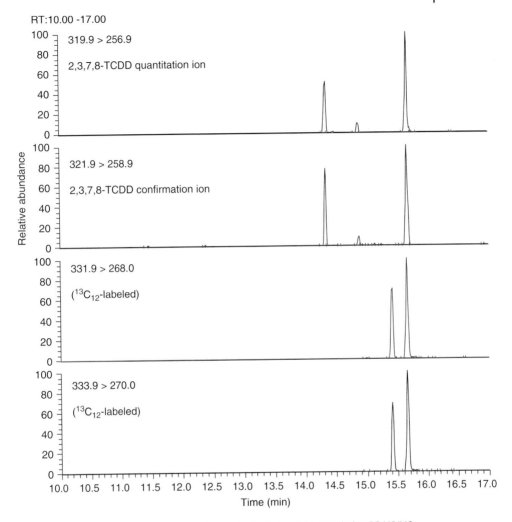

Figure 4.126 2,3,7,8-TCDD mass traces of a contaminated real fat sample by GC-MS/MS, determined as 150 fg, from top quantitation, ion ratio confirmation, bottom labelled internal standard transitions.

Conclusions

Latest developments in triple quadrupole technology make GC-MS/MS applicable for routine analysis of PCDD/PCDFs and dioxin-like PCBs in food and feed at the levels of regulatory importance. These GC-MS/MS systems have the potential to serve as a cost efficient alternative confirmatory method for determination of PCDD/PCDFs and dioxin-like PCBs in food and feed.

The sample preparation and clean-up workflow, as well as the chromatographic separation are kept unchanged as of the standard procedures established in GC-HRMS protocols.

A good correlation between the results of GC-MS/MS and GC-HRMS could be observed for the concentration range above 0.5 pg WHO-PCDD/PCDF-TEQ/g fat for food and human milk samples. Also for fish and feed samples acceptable deviations were observed in the concentration range considerably below the established legal limits (Kotz *et al.*, 2011).

4.29
Confirmation Analysis of Dioxins and Dioxin-like PCBs

Keywords: dioxin; PCDD; PCDF; dioxin-like PCB; TEF; TEQ; WHO; screening; confirmation; HRGC/HRMS; accurate mass; MID

Introduction

Over the past 30 years, dioxin TEQ levels and body-burden levels in the general population have been on the decline and continue to decrease (Lorber, 2002). But, more than 90% of human exposure to dioxins and dioxin-like substances is through food, mainly meat and dairy products, fish and shellfish, and hence lead to continuous control (WHO 2014). With increasingly lower dioxin levels in food, feed and tissues, more demanding LOD, selectivity, sensitivity and QC checks are required to trace their presence at these further decreasing levels.

Recent studies document the declines in exposure to and body burdens of dioxins in the United States (Centers for Disease Control and Prevention, 2009, 2014). Data available on body burdens in the general population have been provided by Patterson. The data and analyses highlight the importance of taking the age group into account. While the mean dioxin TEQ increases as age increases (Figure 4.127), the sharpest increase is observed among the age group 60+ (Aylward and Hays, 2002; Lorber, 2002; Hays and Aylward, 2003; Patterson *et al.*, 2004, 2006; Turner *et al.*, 2006). As the dioxin and dl-PCB concentrations are related to the low fat level in blood, significantly improved sensitivities of the analytical instrumentation applied are required to further establish an efficient control at further decreasing levels, especially in the younger population. With children and toddlers, the small available blood sample sizes become the limiting factor (Turner *et al.*, 2004).

The described isotope dilution method for the quantification of polychlorinated dioxins, furans and dioxin-like polychlorinated biphenyls (dl-PCBs) follows the EPA 1613 Rev.B method (US EPA, 1994, 1998a/b). The MID scheme on the accurate target masses typically uses isotope ratio qualifiers besides the specific retention time for all native dioxin/furan congeners, as well as for their specific ^{13}C labelled internal standards, one quantification mass and one ratio mass. The analytical setup for the high resolution GC-MS is given with the MID descriptor, as

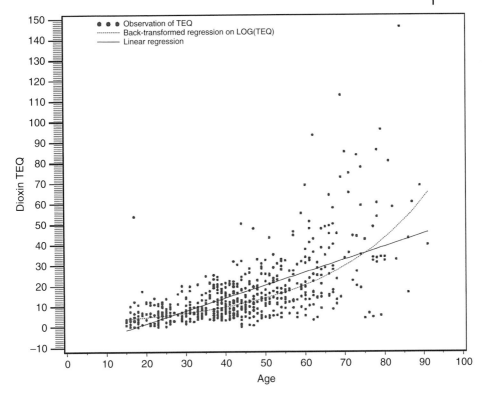

Figure 4.127 Dioxin body burden in TEQ versus age (Patterson *et al.*, 2004).

shown in Table 4.48 listing the exact masses of PCDD/PCDF and PCB including the ^{13}C labelled standards. A typical MID setup for the data acquisition of the individual groups of chlorinated compounds is given in Figure 4.128 (Krumwiede and Hübschmann, 2007, 2008). The mass spectrometer is operated at a mass resolution power of equal or better than 10 000 as required by the EPA 1613 b method. The effective resolution achieved during the analysis of real-life samples has to be constantly monitored on the reference masses and documented in the data file for each MID window.

Analysis Conditions

Sample material			Food, meat eggs, milk, blood, and so on
GC Method	System		Thermo Scientific TRACE GC Ultra
	Injector		Split/splitless
	Column	Type	TG-5MS
		Dimensions	60 m length × 0.25 mm ID × 0.1 μm film thickness
	Carrier gas		Helium

Table 4.48 Accurate mass MID set up for PCDD and PCDF analysis in MID lock-and-cali mode (width first lock: 0.3 Da and voltage settling time delays: 10 ms).

MID window no. (time window)	Reference masses (FC43) m/z lock mass (L), cali mass (C)	Target masses m/z native (n) ^{13}C internal standard (is)	MID cycle time (intensity, dwell time ms)
1. Tetra-PCDD/PCDF (9.00 – 19.93 min)	313.98336 (L), 363.98017 (C)	303.90088(n), 305.89813(n), 315.94133(is), 317.93838(is), 319.89651(n), 321.89371(n), 331.93680(is), 333.93381(is)	0.75 s (L/C: 30, 4 ms; n: 1, 137 ms; is: 7, 19 ms)
2. Penta-PCDD/PCDF (19.93 – 23.52 min)	313.98336 (L), 363.98017(C)	339.85889(n), 341.85620(n), 351.89941(is), 353.85702(n), 353.89646(is!), 355.85400(n), 365.89728(is), 367.89433(is)	0.80 s (L/C: 30, 4 ms; n: 1, 147 ms; is: 7, 21 ms)
3. Hexa-PCDD/PCDF (23.52 – 26.98 min)	375.97974 (L), 413.97698 (C)	371.82300(n), 373.82007(n), 385.86044(is), 387.85749(is), 389.81494(n), 391.81215(n), 401.85535(is), 403.85240(is)	0.80 s (L/C: 30, 4 ms; n: 1, 147 ms; is: 7, 21 ms)
4. Hepta-PCDD/PCDF (26.98 – 32.06 min)	413.97698 (L), 463.97378 (C)	407.78101(n), 409.77826(n), 419.82147(is), 421.81852(is), 423.77588(n), 425.77317(n), 435.81638(is), 437.81343(is)	0.90 s (L/C: 35, 4 ms; n: 1, 169 ms; is: 7, 24 ms)
5. Octa-PCDD/PCDF (32.06 – 36.00 min)	425.97681 (L), 463.97378 (C)	441.74219(n), 443.73929(n), (453.78250(is)), (455.77955(is)), 457.73706(n), 459.73420(n), 469.77741(is), 471.77446(is)	0.95 s (L/C: 40, 4 ms; n: 1, 183 ms; is: 7, 22 ms)

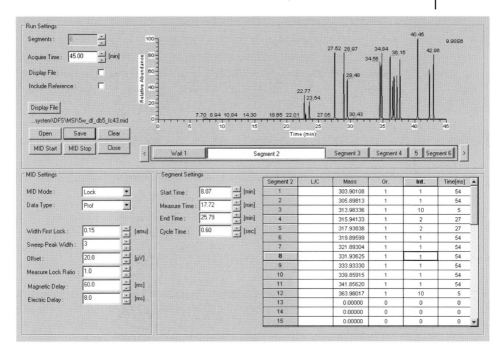

Figure 4.128   Typical MID setup with the individual acquisition window (segment) for each eluting congener group of different chlorination degree.

(Continued)

	Flow		Constant flow, 0.8 mL/min
	SSL Injector	Injection mode	Splitless
		Injection temperature	260 °C
		Injection volume	2 μL
	Split	Closed until	1 min
		Open	1 min to end of run
		Gas saver	2 min, 10 mL/min
	Oven program	Start	120 °C, 3 min
		Ramp 1	19 °C/min to 210 °C
		Ramp 2	3 °C/min to 275 °C
		Final temperature	275 °C, 12 min
	Transferline	Temperature	280 °C
MS Method	System		Thermo Scientific DFS
	Analyzer type		Double focussing magnetic sector MS
	Ionization		EI
	Electron energy		48 eV
	Ion source	Temperature	270 °C
	Resolution	Setting	R 10 000 (at 5% peak height)
Calibration	Internal standard	Range	^{13}C-labelled congeners

Results

A typical GC separation using the parameters from Table 4.48 of an EPA 1613 CS1 dioxin standard at 50 fg/µL of TCDD and TCDF is shown in Figure 4.129 (Fishman *et al.*, 2007). These GC parameters were also employed for the analysis of blood samples. For a quantitative analysis with the calculation of the TEQ values, a summary of the WHO 1998 and the WHO 2005 TEF values is given in Table 4.49 (van den Berg *et al.*, 1998, 2005).

The confirmation ratios (relative areas of quantification and ratio masses) for all dioxins/furans in repeated injections of a 17 fg/µL were evaluated for a blood pool sample (see Figure 4.130) (Krumwiede and Hübschmann, 2006). All 2,3,7,8-TCDD results of a sample measurement series over several days of the blood sample gave excellent results within the required ±15% window at the lowest detection levels and provided the confirmation ion ratios in compliance with EPA 1613 requirements.

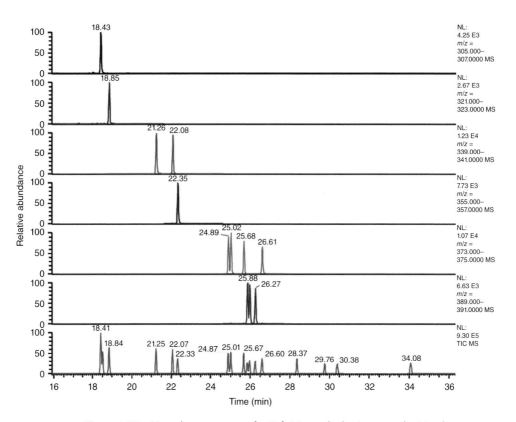

Figure 4.129 Mass chromatograms of a 50 fg/µL standard using a regular 30 m long column.

Table 4.49 Summary of WHO 1998 and 2005 TEF values.

Compound	WHO 1998 TEF	WHO 2005 TEF[a]
Chlorinated dibenzo-p-dioxins		
2,3,7,8-TCDD	1	1
1,2,3,7,8-PeCDD	1	1
1,2,3,4,7,8-HxCDD	0.1	0.1
1.2.3.6.7.8- HxCDD	0.1	0.1
1.2.3.7.8.9- HxCDD	0.1	0.1
1,2,3,4,6,7,8-HpCDD	0.01	0.01
OCDD	0.0001	**0.0003**
Chlorinated dibenzofurans		
2,3,7,8-TCDF	0.1	0.1
1,2,3,7,8-PeCDF	0.05	**0.03**
2,3,4,7,8-PeCDF	0.5	**0.3**
1,2,3,4,7,8-HxCDF	0.1	0.1
1.2.3.6.7.8- HxCDF	0.1	0.1
1.2.3.7.8.9- HxCDF	0.1	0.1
2,3,4,6,7,8-HxCDF	0.1	0.1
1.2.3.4.6.7.8-HpCDF	0.01	0.01
1.2.3.6.7.8.9-HpCDF	0.01	0.01
OCDF	0.0001	**0.0003**
Non-ortho substituted PCBs		
3,3′,4,4′-tetraCB (PCB 77)	0.0001	0.0001
3,4,4′,5-tetraCB (PCB 81)	0.0001	**0.0003**
3,3′,4,4′,5-pentaCB (PCB 126)	0.1	0.1
3,3′,4,4′,5,5′-hexaCB (PCB 169)	0.01	**0.03**
Mono-ortho substituted PCBs		
2,3,3′,4,4′-pentaCB (PCB 105)	0.0001	**0.00003**
2,3,4,4′,5-pentaCB (PCB 114)	0.0005	**0.00003**
2,3′,4,4′,5-pentaCB (PCB 118)	0.0001	**0.00003**
2′,3,4,4′,5-pentaCB (PCB 123)	0.0001	**0.00003**
2,3,3′,4,4′,5-hexaCB (PCB 156)	0.0005	**0.00003**
2,3,3′,4,4′,5′-hexaCB (PCB 157)	0.0005	**0.00003**
2,3′,4,4′,5,5′-hexaCB (PCB 167)	0.00001	**0.00003**
2,3,3′,4,4′,5,5′-heptaCB (PCB 189)	0.0001	**0.00003**

a) Bold values indicate a change in TEF value.
Reference: Van den Berg *et al.* (2006).

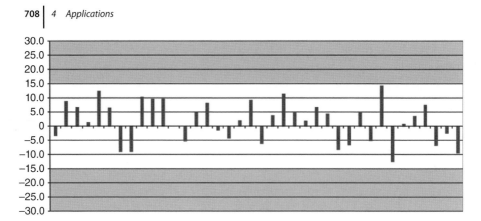

Figure 4.130 Confirmation of ion ratios *m/z* 320/322 in % for repeated injections of a blood sample extract at 17 fg/μL, ±15% window complies with EPA 1613 method.

4.30
Analysis of Brominated Flame Retardants PBDE

Keywords: BFRs; PBDE; deca-BDE; MID; PTV; on-column; EPA 1614

Introduction

BDEs are among the most important and most widely used flame retardants in a range of different industrial and consumer products. They are found worldwide in practically all types of matrices moving them into the focus of legislation resulting in a ban for certain BDE congeners (European Parliament and of the Council of the European Union, 2003). The EU directive 2003/11/EC prohibits the use of penta-BDE and octa-BDE for the member states of the European community (European Parliament, 2003).

As a result, BDEs have received rising interest in recent years by the analytical community (Stapleton *et al.*, 2008). The by far most efficient analysis technique is high resolution GC-MS using isotope dilution technique according to EPA 1614 for highest precision quantification with highest significance (Krumwiede and Hübschmann, 2006).

Analysis Conditions

Sample material			Industrial and consumer products
GC Method	System		Thermo Scientific TRACE GC Ultra
	Injector		Split/splitless
	Column	Type	TG-5MS
		Dimensions	15 m length × 0.25 mm ID × 0.1 μm film thickness

Table 4.50 Accurate mass MID setup: MID lock-and-cali mode (target masses in brackets: optional second ratio mass for native BDE).

MID window no.	Reference masses (PFK) L = lock mass, C = cali mass	Target masses (second ratio mass native)	MID cycle time (s)
1. Tri-BDE	392.9753 (L), 430.9723 (C)	(403.8041), 405.8021, 407.8001, 417.8424, 419.8403	0.55
2. Tetra-BDE	480.9688 (L), 492.9691 (C)	(483.7126), 485.7106, 487.7085, 495.7529, 497.7508	0.55
3. Penta-BDE	554.9644 (L), 592.9627 (C)	(561.6231), 563.6211, 565.6190, 575.6613, 577.6593	0.60
4. Hexa-BDE	480.9688 (L), 504.9691 (C)	481.6976, 483.6956, (485.6937), 493.7372, 495.7352	0.60
5. Hepta-BDE	554.9644 (L), 592.9627 (C)	(559.6082), 561.6062, 563.6042, 573.6457, 575.6436	0.70
6. Deca-BDE	754.9531 (L), 766.9531 (C)	797.3349, 799.3329, (801.3308), 809.3752, 811.3731	0.90

(Continued)

	Carrier gas		Helium
	Flow		Constant flow, 1.0 mL/min
	SSL Injector	Injection mode	Splitless
		Injection temperature	280 °C
		Injection volume	2 µL
	Split	Closed until	1 min
		Open	1 min to end of run
		Slit flow	50 mL/min
	Oven program	Start	120 °C, 2 min
		Ramp 1	15 °C/min to 230 °C
		Ramp 2	5 °C/min to 270 °C
		Ramp 3	10 °C/min to 330 °C
		Final temperature	330 °C, 5 min
	Transferline	Temperature	280 °C
MS Method	System		Thermo Scientific DFS
	Analyzer type		Double focussing magnetic sector MS
	Ionization		EI
	Electron energy		40 eV
	Ion source	Temperature	270 °C
	Acquisition	Mode	MID (Table 4.50)
	Mass resolution		R 10 000 (at 5% peak height)

Results

The full-scan results proved that for all bromination degrees, either the molecular ion or the fragment ion showing the loss of 2 Br atoms are the most

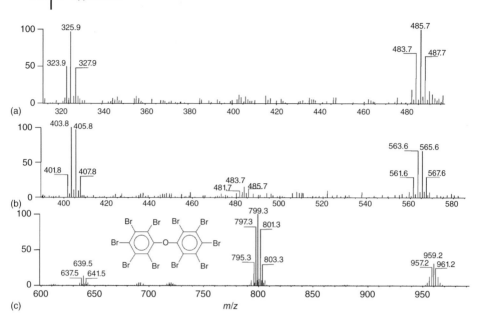

Figure 4.131 Mass spectra of PBDE congeners; (a) tetra-BDE, (b) penta-BDE and (c) deca-BDE and structure of deca-BDE.

abundant ions (see Figure 4.131). The change of the most intense ion from M^+ to $[M-2Br]^+$ is typically observed with higher bromination degrees and starting to become significant already with penta- and hexa-BDEs, depending on the ion source conditions. The loss of Br is slightly temperature dependent and varies with GC elution and ion source temperatures. Therefore, with different instrument conditions the transition of the most abundant ion from M^+ to $[M-2Br]^+$ might be shifted to tetra/penta- or hexa/hepta-BDE. In general, the relative intensity $[M-2Br]^+/M^+$ is increasing with the degree of bromination. For deca-BDE, the intensity gain when using the $[M-2Br]^+$ mass peak for MID detection is at least a factor of 4 compared to M^+ (Krumwiede and Hübschmann, 2007, 2008). For a list of exact masses including relative intensities, see Tables 4.51 and 4.52.

Mass spectrometer tuning parameters similar to those typically used for dioxin/PCB analysis were found to give optimum sensitivity for BDE analysis as well. A slightly higher ion source temperature of 270 °C is recommended, taking the high boiling characteristics of the BFRs into account (although highly brominated compounds appear at lower elution temperatures as same molecular weight hydrocarbons). The use of PFK (perfluorokerosene) as internal mass reference for accurate mass measurements is mandatory, because lock and cali masses in the high mass range are needed (e.g. for deca-BDE). Autotuning for gaining highest sensitivity was carried out on PFK mass m/z 480.9688.

Table 4.51 PBDE exact mass references (1) for natives and internal standards (molecular ions).

#Br	Native		$^{13}C_{12}$ Standard		$^{13}C_6$ Standard	
	Exact mass M$^+$ (Da)	Relative intensity (%)	Exact mass M$^+$ (Da)	Relative intensity (%)	Exact mass M$^+$ (Da)	Relative intensity (%)
Br1	247.98313	100.0	260.02339	100.0	254.00326	100.0
	249.98108	97.3	262.02134	97.3	256.00121	97.3
Br2	325.89364	51.4	337.93390	51.4	331.91377	51.4
	327.89159	100.0	339.93185	100.0	333.91172	100.0
	329.88955	48.6	341.92981	48.6	335.90968	48.6
Br3	403.80415	34.3	415.84441	34.3	409.82428	34.3
	405.80211	100.0	417.84237	100.0	411.82224	100.0
	407.80006	97.3	419.84032	97.3	413.82019	97.3
	409.79801	31.5	421.83827	31.5	415.81814	31.5
Br4	481.71467	17.6	493.75492	17.6	487.73480	17.6
	483.71262	68.5	495.75288	68.5	489.73275	68.5
	485.71057	100.0	497.75083	100.0	491.73070	100.0
	487.70853	64.9	499.74878	64.9	493.72866	64.9
	489.70648	15.8	501.74674	15.8	495.72661	15.8
Br5	559.62518	10.6	571.66544	10.6	565.64531	10.6
	561.62313	51.4	573.66339	51.4	567.64326	51.4
	563.62109	100.0	575.66134	100.0	569.64122	100.0
	565.61904	97.3	577.65930	97.3	571.63917	97.3
	567.61699	47.3	579.65725	47.3	573.63712	47.3
	569.61495	9.2	581.65520	9.2	575.63508	9.2
Br6	637.53569	5.4	649.57595	5.4	643.55582	5.4
	639.53365	31.7	651.57390	31.7	645.55377	31.7
	641.53160	77.1	653.57186	77.1	647.55173	77.1
	643.52955	100.0	655.56981	100.0	649.54968	100.0
	645.52751	73.0	657.56776	73.0	651.54763	73.0
	647.52546	28.4	659.56572	28.4	653.54559	28.4
	649.52341	4.6	661.56367	4.6	655.54354	4.6
Br7	715.44620	3.1	727.48646	3.1	721.46633	3.1
	717.44416	21.1	729.48442	21.1	723.46429	21.1
	719.44211	61.7	731.48237	61.7	725.46224	61.7
	721.44006	100.0	733.48032	100.0	727.46019	100.0
	723.43802	97.3	735.47828	97.3	729.45815	97.3
	725.43597	56.8	737.47623	56.8	731.45610	56.8
	727.43392	18.4	739.47418	18.4	733.45405	18.4
	729.43188	2.6	741.47214	2.6	735.45201	2.6
Br8	793.35672	1.6	805.39697	1.6	799.37685	1.6
	795.35467	12.4	807.39493	12.4	801.37480	12.4
	797.35262	42.3	809.39288	42.3	803.37275	42.3
	799.35058	82.2	811.39084	82.2	805.37071	82.2
	801.34853	100.0	813.38879	100.0	807.36866	100.0

Table 4.51 (Continued)

#Br	Native		$^{13}C_{12}$ Standard		$^{13}C_6$ Standard	
	Exact mass M⁺ (Da)	Relative intensity (%)	Exact mass M⁺ (Da)	Relative intensity (%)	Exact mass M⁺ (Da)	Relative intensity (%)
	803.34648	77.8	815.38674	77.8	809.36661	77.8
	805.34444	37.9	817.38470	37.8	811.36457	37.8
	807.34239	10.5	819.38265	10.5	813.36252	10.5
	—	—	—	—	815.36047	1.3
Br9	871.26723	0.9	883.30749	0.9	877.28736	0.9
	873.26518	7.8	885.30544	7.8	879.28531	7.8
	875.26314	30.2	887.30339	30.2	881.28327	30.1
	877.26109	68.5	889.30135	68.5	883.28122	68.5
	879.25904	100.0	891.29930	100.0	885.27917	100.0
	881.25700	97.3	893.29725	97.3	887.27713	97.3
	883.25495	63.1	895.29521	63.1	889.27508	63.1
	885.25290	26.3	897.29316	26.3	891.27303	26.3
	887.25086	6.4	899.29111	6.4	893.27099	6.4
Br10	949.17774	0.5	961.21800	0.5	955.19787	0.5
	951.17570	4.4	963.21595	4.4	957.19582	4.4
	953.17365	19.4	965.21391	19.4	959.19378	19.4
	955.17160	50.3	967.21186	50.3	961.19173	50.3
	957.16956	85.7	969.20981	85.7	963.18968	85.7
	959.16751	100.0	971.20777	100.0	965.18764	100.0
	961.16546	81.1	973.20572	81.1	967.18559	81.1
	963.16342	45.1	975.20367	45.1	969.18354	45.1
	965.16137	16.4	977.20163	16.4	971.18150	16.4
	967.15932	3.6	979.19958	3.6	973.17945	3.6

The calculated reference masses are based on the following values for isotopic masses: ^1H 1.0078250321 Da, ^{12}C 12.0000000000 Da, ^{13}C 13.0033548378 Da, ^{16}O 15.9949146221 Da, ^{79}Br 78.9183376 Da and ^{81}Br 80.9162910 Da. All listed masses refer to singly positively charged ions. Masses for isotope peaks have been calculated for a resolving power of 10 000 (10% valley definition). The mass of the electron (0.000548579911 Da) was taken into account for the calculation of the ionic masses.
Reference: Audi and Wapstra (1995) and Peter and Taylor (1999).

All congeners in the employed BDE standard could be separated on the 15 m column (see Figure 4.132). Similar to dioxins, the BDE congeners are separated on unpolar columns groupwise in the order of their bromination degree. The use of a short 10–15 m column with a thin film is recommended to analyse the thermolabile deca-BDE more efficiently.

The LOQs achieved had been similar to those known for dioxin and PCB analysis. They can also be achieved for the analysis of the far higher boiling BDEs in the low femtogram range (see Figure 4.133). Also the quantitation linearity proved to fulfil highest standards as shown in Figure 4.134.

Table 4.52 PBDE exact mass references (2) for natives and internal standards (M − 2Br ions).

# Br of M	Ratio [%] M⁺/ (M − 2Br)⁺	Native		¹³C$_{n2}$ Standard		¹³C$_6$ Standard	
		Exact mass (Da) (M − 2Br)⁺	Relative intensity (%)	Exact mass (Da) (M − 2Br)⁺	Relative intensity (%)	Exact mass (Da) (M − 2Br)⁺	Relative intensity (%)
Br1	100	n/a		n/a		n/a	
Br2	100	168.056966	100.0	180.097224	100.0	174.077095	100.0
Br3	100	245.967479	89.8	258.007737	100.0	251.987608	95.8
		247.965432	100.0	260.005690	97.7	253.985561	100.0
Br4	100	323.877991	47.5	335.918249	51.0	329.898120	49.4
		325.875945	100.0	337.916203	100.0	331.896074	100.0
		327.873898	54.4	339.914156	48.5	333.894027	51.4
Br5	85	401.788504	31.0	413.828762	33.9	407.808633	33.0
		403.786457	96.4	415.826715	100.0	409.806586	99.6
		405.784411	100.0	417.824669	97.4	411.804540	100.0
		407.782364	35.3	419.822622	31.4	413.802493	34.2
Br6	60	479.699017	16.0	491.739275	17.4	485.719146	16.7
		481.696970	65.7	493.737228	68.2	487.717099	66.9
		483.694923	100.0	495.735181	100.0	489.715052	100.0
		485.692877	69.5	497.733135	64.7	491.713006	66.9
		487.690830	17.9	499.731088	15.7	493.710959	17.0
Br7	55	557.609529	2.8	569.649787	3.0	563.629658	2.9
		559.607483	19.8	571.647741	21.0	565.627612	20.5
		561.605436	59.7	573.645694	61.7	567.625565	60.5
		563.603389	99.7	575.643647	100.0	569.623518	100.0
		565.601343	100.0	577.641601	97.4	571.621472	98.5
		567.599296	60.9	579.639554	56.9	573.619425	59.0
Br8	50	635.520042	4.8	647.560300	5.3	641.540171	5.1
		637.517995	29.5	649.558253	31.6	643.538124	30.6
		639.515949	74.2	651.556207	77.2	645.536078	75.5
		641.513902	100.0	653.554160	100.0	647.534031	100.0
		643.511855	75.6	655.552113	73.1	649.531984	74.6
		645.509809	31.2	657.550067	28.4	651.529938	29.8
		647.507762	5.1	659.548020	4.6	653.527891	5.0
Br9	40	713.430554	2.8	725.470812	3.0	719.450683	2.9
		715.428508	19.8	727.468766	21.0	721.448637	20.5
		717.426461	59.7	729.466719	61.7	723.446590	60.5
		719.424414	99.7	731.464673	100.0	725.444544	100.0
		721.422368	100.0	733.462626	97.3	727.442497	98.5
		723.420321	60.9	735.460579	56.9	729.440450	58.9
		725.418275	20.7	737.458533	18.4	731.438404	19.6
		727.416228	2.8	739.456486	2.6	733.436357	2.8
Br10	25	791.341067	1.4	803.381325	1.6	797.361196	1.5
		793.339020	11.5	805.379278	12.3	799.359149	11.9

(continued overleaf)

Table 4.52 (*Continued*)

# Br of M	Ratio [%] M⁺/ (M − 2Br)⁺	Native		¹³C_{n2} Standard		¹³C₆ Standard	
		Exact mass (Da) (M − 2Br)⁺	Relative intensity (%)	Exact mass (Da) (M − 2Br)⁺	Relative intensity (%)	Exact mass (Da) (M − 2Br)⁺	Relative intensity (%)
		795.336974	40.1	807.377232	42.2	801.357103	41.3
		797.334927	80.5	809.375185	82.1	803.355056	81.2
		799.332880	100.0	811.373138	100.0	805.353009	100.0
		801.330834	80.5	813.371092	77.8	807.350963	79.0
		803.328787	40.8	815.369045	37.8	809.348916	39.4
		805.326741	11.9	817.366999	10.5	811.346870	11.2
		807.324694	1.4	819.364952	1.3	813.344823	1.4

The calculated reference masses are based on the following values for isotopic masses: ^{1}H 1.0078250321 Da, ^{12}C 12.0000000000 Da, ^{13}C 13.0033548378 Da, ^{16}O 15.9949146221 Da, ^{79}Br 78.9183376 Da and ^{81}Br 80.9162910 Da. All listed masses refer to singly positively charged ions. Masses for isotope peaks have been calculated for a resolving power of 10 000 (10% valley definition). The mass of the electron (0.000548579911 Da) was taken into account for the calculation of the ionic masses.
Reference: Audi and Wapstra (1995) and Peter and Taylor (1999). The given ratio M+/(M − 2Br)+ provides typical values, depends on actual ion source conditions.

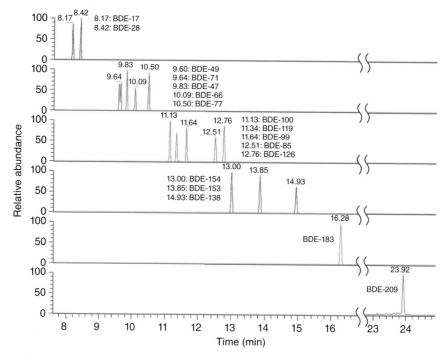

Figure 4.132 Mass chromatograms of tri- to hepta- and deca-BDE showing the separation according to the bromination degree on a 5% phenyl phase column, length 15 m.

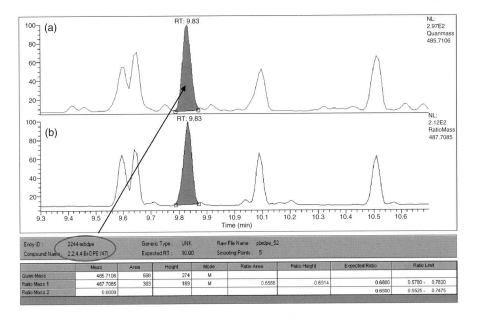

Figure 4.133 25 fg BDE 47 (tetra-BDE); (a) quan mass and (b) ratio mass.

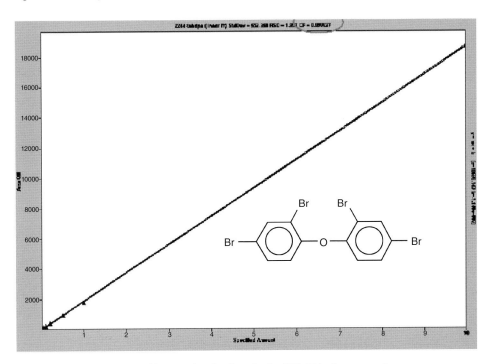

Figure 4.134 Quantitative calibration showing linearity for BDE 47 in the range of 25–10 000 fg/μL.

4.31

SPME Analysis of PBBs

Keywords: PBBs; PBDEs; headspace; SPME; GC-MS/MS; quantitation

Introduction

Polybrominated biphenyls (PBBs) and PBDEs are among the most widely used flame retardants with a widely distributed contamination in the environment causing a high risk or subsequent accumulation in food and feedstuff. Although typically the highly brominated congeners are applied in flame retardants, the lower brominated species are found more often in environmental samples (de Boer *et al.*, 2001).

According to the focus of potential accumulation from environmental sources, this application covers the congener range of up to the hexa brominated species of both PBBs and PBDEs. A fast and reliable automated online extraction method for BFR from water samples is described. For the first time, SPME is applied for the trace determination of BFRs. In contrast to conventional methods using the Soxhlet extraction with an extended extract cleanup, this method uses the fast and straightforward extraction of the analytes from the sample headspace. This is possible due to the high water vapour volatility of the PBBs and PBDEs and the low partition coefficient of these BFR compounds in aqueous media. The SPME method can be integrated in regular GC-MS autosamplers, is fast and prevents the chromatographic system from the typical background load of Soxhlet extracted samples and provides clean chromatograms. Tap water and urban waste water has been analyzed (Polo *et al.*, 2004).

The mass spectrometric detection in this application is achieved by MS/MS using an ion trap mass spectrometer providing excellent selectivity and sensitivity for trace-level quantitation in environmental water samples. Alternatively a triple quadrupole system can be employed using a similar setup. Highest selectivity and lowest determination levels have been achieved also by using high-resolution mass spectrometer (De Boer *et al.*, 2001; Krumwiede, 2006).

Analysis Conditions

Sample material			Environmental samples, water
Sample preparation	SPME		10 mL of a filtered aqueous sample are filled in 22 mL headspace vials, sealed with Teflon faced septum, equilibrated at 100 °C for 5 min
	Extraction	Fibre	Supelco PDMS fibre, 65 µm
		Time	30 min, stir bar agitation while sampling
	Desorption	Temperature	300 °C
GC Method	System		Varian 3800
	Injector		Split/splitless

(*Continued*)

	Column	Type	CP-Sil 8 CB
		Dimensions	25 m length × 0.25 mm ID × 0.25 µm film thickness
	Carrier gas		Helium
	Flow		Constant flow, 1.2 mL/min
	SSL Injector	Injection mode	Splitless
		Injection temperature	300 °C
	Split	Closed until	2 min
		Open	2 min to end of run
		Split flow	50 mL/min
	Oven program	Start	60 °C, 2 min
		Ramp 1	30 °C/min to 250 °C
		Ramp 2	5 °C/min to 280 °C
		Final temperature	280 °C, 8 min
	Transferline	Temperature	280 °C
MS Method	System		Varian Saturn 2000
	Analyzer type		Ion trap MS with internal ionization (with MS/MS waveboard)
	Ionization		EI
	Electron energy		70 eV
	Ion Trap	Temperature	250 °C
	Acquisition	Mode	Full scan, MS/MS resonant waveform
	Mass range		50–650 Da
	Scan rate		1 s (1 spectrum/s)
Calibration	External standard	Range	0.12–500 pg mL

Results

Despite of the high molecular weight of the BFRs, the investigated compounds have been extracted with higher recoveries from the headspace than from direct immersion into the aqueous sample. This is due to an obviously high water vapour volatility of the brominated aromatics, which is known from PCBs as well. Using headspace SPME, the transfer of high boiling matrix components to the chromatographic system is prevented usually causing high background levels during MS detection. At the same time, the useful lifetime of the fibre is significantly extended. Extractions are supported by stir bar agitation in the headspace vial. The method parameters are based on a comprehensive and systematic optimization study (Polo *et al.*, 2004).

The described method shows excellent sensitivity, linearity and quantitative precision up to the hexabromo congeners. The detection limits are in the low picogram per litre range (7.5–190 pg/L) with a calibration range by spiking tap water between 120 fg/mL and 500 pg/mL, as given in Table 4.53. The achieved correlation values are excellent in the range of 0.9977–1.0. Method precision and

Table 4.53 Linearity and LODs.

Compounds	Concentration range (pg/mL)	Correlation factor (R^2)	LOD S/N 3 (pg/L)
BDE-3	1.00–498	0.9977	190
PBB-15	1.01–503	1.0000	9.0
PBB-49	0.95–476	1.0000	7.5
BDE-47	0.41–205	0.9998	20
BDE-100	0.12–60	1.0000	60
BDE-99	0.41–205	0.9999	47
BDE-154	0.34–17	1.0000	150
BDE-153	0.23–12	0.9995	100

Table 4.54 Repeatability at two different concentration levels ($n = 3$).

Compounds	Concentration 1 (pg/mL)	Repeatability (RSD)	Concentration 2 (pg/mL)	Repeatability (RSD)
BDE-3	10.0	4.4	199	6.3
PBB-15	10.1	3.8	201	12
PBB-49	9.5	2.8	190	1.7
BDE-47	4.1	17	82	1.2
BDE-100	1.2	15	24	10
BDE-99	4.1	20	82	1.2
BDE-154	0.34	24	6.8	9.3
BDE-153	0.23	26	4.6	8.8

LOD have been determined with triplicates measurements as given in Table 4.54. The recovery experiments have been performed with three different types of blank matrix samples including tap water, effluent and influent waste waters from an urban sewage plant. The results are given in Table 4.55 and Figure 4.135.

Table 4.55 Recovery in real water matrix samples.

Compounds	Tap water spiked conc. (pg/mL)	Recovery (%)	Effluent water spiked conc. (pg/mL)	Recovery (%)	Influent water[a] spiked conc. (pg/mL)	Recovery ± RSD (%)
BDE-3	1.00	99 ± 1	10.0	100 ± 4	199	106 ± 10
PBB-15	1.01	97 ± 4	10.1	92 ± 2	201	90 ± 6
PBB-49	0.95	90 ± 21	9.5	94 ± 1	190	93 ± 5
BDE-47	0.41	91 ± 8	4.1	97 ± 7	82	87 ± 7
BDE-100	0.12	100 ± 4	1.2	83 ± 17	24	95 ± 4
BDE-99	0.41	87 ± 19	4.1	90 ± 12	82	92 ± 8
BDE-154	0.034	nd	0.34	100 ± 25	6.8	74 ± 11
BDE-153	0.023	nd	0.23	117 ± 13	4.6	82 ± 11

a) Chromatogram see Figure 4.135.

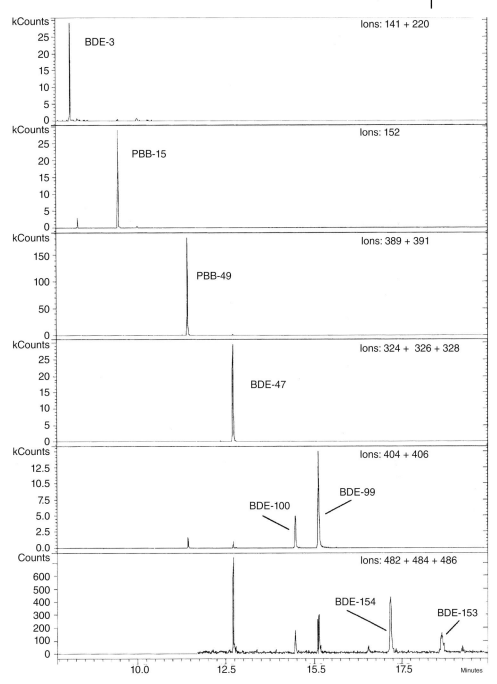

Figure 4.135 Analysis of a spiked influent waste water sample as of Table 4.55. (Polo *et al.*, 2004; reprinted with permission of Analytical Chemistry, Copyright 2004, American Chemical Society.)

4.32
Analysis of Military Waste

Keywords: military waste; nitroaromatics; TNT; degradation products; carcinogenicity; Soxhlet; choice of column; ECD; PID; chemical ionization; PAHs; water Cl

Introduction

In the past, little attention was paid to the ecological consequences of rearmament and disarmament. Only frequently has the problem of military waste become a topic in residue analysis. The combined consequences of two world wars have not been considered for a long time and current political changes are leading to new types of contamination which will require analytical solutions within the framework of demilitarization. This overall area will also therefore increase in importance in the future in the field of analysis.

There is no absolute definition of military waste. The term *encompasses* disused military sites where munitions or chemical weapons were manufactured, processed, stored or deposited. Soil and ground water are contaminated by the substances concerned and also by synthesis products, by-products, degradation products or by problematic fuels. Some of these compounds are liable to explode giving off toxic products. Explosives (nitroaromatics) and chemical warfare agents are of particular importance. The large quantity of water required for the manufacture of nitroaromatics led to the setting up of production plants in areas with large water supplies which are often still also used for public consumption. Many substances which are recovered today as metabolites or stabilizers are carcinogenic.

Besides HPLC for routine control, GC with ECD and PID (photo ionization detector) has been the normal procedure for residue analysis up till now (Kuitunen *et al.*, 1991). GC-MS is increasing in importance as the number of substances in the trace region as well as in high concentrations which need to be determined is increasing. As the spectrum of expected contaminants is usually difficult to estimate, the mass spectrometer is indispensable as the universal and specific detector (Yinon, 1993).

The analysis of nitroaromatics and their metabolites is of particular importance in the assessment of military waste. Fourteen by-products and degradation products of 2,4,6-TNT (trinitrotoluene) alone are definitely classified or suspected as carcinogens (Table 4.56).

The analysis of TNT degradation products involves Soxhlet extraction or pressurized liquid extraction (PLE) for soil, whereby the differing polarity of the substances to be extracted needs to be taken into consideration when choosing the extraction agent (Figure 4.136). Nitroaromatics can be separated by HPLC or GC. GC-MS is now the method of choice for determining them together with their metabolites. Classical GC detection using NPD, PID or ECD has limitations (Figure 4.137 and Table 4.57). ECD is frequently used and exhibits high sensitivity for all compounds with two or more nitro groups. However, it can therefore only be used for the detection of certain metabolites (Figure 4.138).

Table 4.56 By-products and degradation products of TNT with assessment of the carcinogenic potential.

Compound	Assessment	Compound	Assessment
2-Nitrotoluene		2,6-Diaminotoluene	
3-Nitrotoluene		2,4,6-Triaminotoluene	
4-Nitrotoluene		2-Amino-6-nitrotoluene	
2,3-Dinitrotoluene	III A2	2-Amino-4-nitrotoluene	
2,4-Dinitrotoluene	III A2	6-Amino-2,4-dinitrotoluene	
2,6-Dinitrotoluene	III A2	4-Amino-2,6-dinitrotoluene	
3,4-Dinitrotoluene	III A2	2-Amino-4,6-dinitrotoluene	
2,4,6-Trinitrotoluene		3-Nitrobiphenyl	
2,4,5-Trinitrotoluene	III B	4-Nitrobiphenyl	III A2
2,3,4-Trinitrotoluene		3-Nitrobiphenyl	
1,2-Dinitrobenzene	III B	4-Nitrobiphenyl	III A2
1,3-Dinitrobenzene	III B	2,2′-Dinitrobiphenyl	
1,4-Dinitrobenzene	III B	2-Nitronaphthalene	III A2
2,3-Diaminotoluene		4-Aminobiphenyl	III A1
2,4-Diaminotoluene		1,3-Dinitronaphthalene	III B

The substances are assessed according to the ordinance on hazardous materials and the German MAK value list (Maximum Workplace Concentrations, Deutsche Forschungsgemeinschaft, 2014): III A1 = definitely carcinogenic in humans; III A2 = definitely carcinogenic in animals and III B = suspicion of potential carcinogenicity.

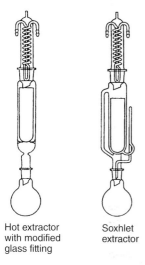

Hot extractor with modified glass fitting

Soxhlet extractor

Compound/ extraction procedure	Methanol soxhlet	Methanol flow extraction	Diethyl ether extraction
2-Nitrotoluene	85%	100%	92%
3-Nitrotoluene	82%	95%	90%
4-Nitrotoluene	73%	90%	85%
2,6-Dinitrotoluene	91%	95%	95%
2,4-Dinitrotoluene	72%	85%	90%
2,3-Dinitrotoluene	74%	90%	90%
3,4-Dinitrotoluene	34%	70%	80%
2,4,6-Trinitrotoluene	6%	70%	80%
1,3-Dinitrobenzene	66%	70%	85%
1,4-Dinitrobenzene	1%	70%	85%
4-Amino-2,6-dinitrotoluene	47%	45%	90%
2-Amino-4,6-dinitrotoluene	41%	45%	90%
3-Nitrobiphenyl	96%	95%	90%
2,2-Dinitrobiphenyl	92%	95%	95%
1,3-Dinitronaphthalene	56%	80%	95%
2,3-Diaminotoluene	–	35%	75%
2,4-Diaminotoluene	–	45%	80%
2,6-Diaminotoluene	–	50%	80%
2-Amino-4-nitrotoluene	–	65%	80%
2-Amino-6-nitrotoluene	–	65%	80%
2,6-Diamino-4-nitrotoluene	–	45%	80%

– Not determined

Figure 4.136 Comparison of liquid extraction procedures for the analysis of nitroaromatics.

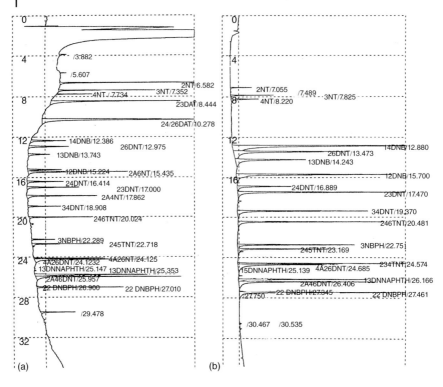

Figure 4.137 Comparison of the PID (a) and ECD detector (b) in the analysis of nitroaromatics and their metabolites (column HT8, 400 pg/component, for conditions see text).

Analysis Conditions

Sample material			Soil, environmental armament contaminations, military waste
Sample preparation			Continuous flow extraction with methanol or diethyl ether
GC Method	System (1)		Agilent 5890
	Injector (1)		Split/splitless
	System (1)		Finnigan Magnum GC
	Injector (1)		PTV cold injection system
	Column (1)	Type	Rtx-200
		Dimensions	30 m length × 0.25 mm ID × 0.25 μm film thickness
	Column (2)	Type	HT8
		Dimensions	25 m length × 0.22 mm ID × 0.25 μm film thickness
	Carrier gas		Hydrogen for GC-ECD
			Helium for GC-MS

Table 4.57 Detectors for the analysis of nitroaromatics and their metabolites.

Detector	Advantages	Disadvantages
FID	High linearity	Low selectivity Low sensitivity
ECD	Good sensitivity	Low linearity Contamination when using highly concentrated samples, low sensitivity for substances with less than two nitro groups, correspondingly difficult detection of metabolites
NPD	Good selectivity for nitro compounds and metabolites	Low sensitivity
PID	The same response for nitro and amino compounds, field tests	Low sensitivity
ELD	N-specific, high linearity, simple calibration	Low sensitivity
GC-MS/EI	High sensitivity, good identification even of other accompanying substances	—
GC-MS/CI	High selectivity, additional confirmation from the molecular mass, PCI for nitroaromatics and metabolites, NCI only for nitro compounds	—

(Continued)

	Flow		2 bar, constant pressure
	SSL Injector (1)	Injection mode	Split
		Injection temperature	270 °C
	PTV Injector	Injection mode	Slit
		Base temperature	40 °C, 1 min
		Evaporation	250 °C/min to 300 °C
		Final temperature	300 °C, 15 min

MS Method	System (1)		Agilent MSD 5971A
	Analyzer type		Single quadrupole MS
	Ionization		EI
	Electron energy		70 eV
	Acquisition	Mode	Full scan
	Mass range		29–250 Da
	Scan rate		1 s (1 spectrum/s)
	System (2)		Finnigan Magnum
	Analyzer type		Ion trap MS with internal ionization
	Ionization		EI/CI
	Reagent gas		Water
	Electron energy		70 eV
	Acquisition	Mode	Full scan
	Mass range		100–400 Da
	Scan rate		1 s (1 spectrum/s)

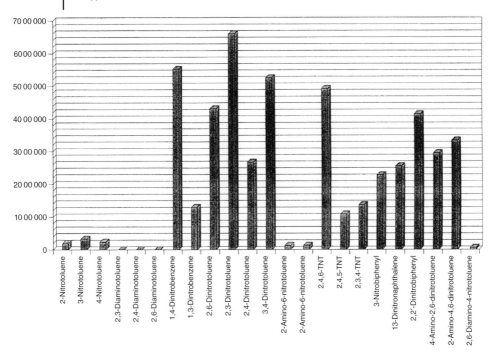

Figure 4.138 Differing responses of nitroaromatics and their metabolites on ECD (400 pg samples).

Results

An analysis of nitroaromatics with ECD as the detector and hydrogen as the carrier gas is shown in Figure 4.139. This separation is designed to be rapid and does not take metabolites into consideration. Rapid separation requires the appropriate choice of GC column. The critical pair 2,4- and 2,6-diaminotoluene can be separated well on the Rtx-200 column (trifluoro-propylmethylsilicone phase) (Figure 4.140), which is not possible with columns of the same length containing pure methylsilicone or carborane phases (HT8). As the isomers cannot be differentiated by mass spectrometry, gas chromatographic separation is a prerequisite for separate determination (Figure 4.141). Other compounds from TNT analysis which can sometimes co-elute with them can be differentiated by mass spectrometry.

The EI spectra of nitroaromatics frequently show an intense fragment at m/z 30 which results from rearrangement and fragmentation of the nitro group. The molecular ion signals are small and can therefore be completely undetectable in matrix-rich samples. With chemical ionization, all the substances involved form intense (quasi)molecular ions, which can be detected with high S/N values because of their higher masses (Figure 4.142). A series of typical CI spectra measured using water as the reagent gas is shown in Figure 4.143 for various

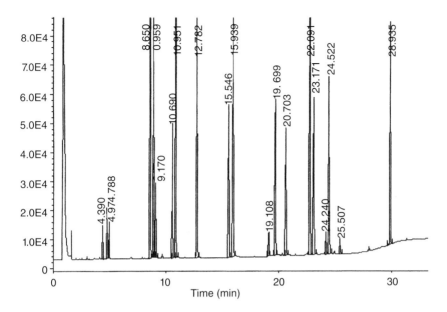

```
========================================================
                   External standard report
========================================================
```

Sample name   : 2 ppm mixture   Injection number: 1

RT (min)	Area (cts)	Integration	Compound name
4.390	22076	BB	2MNT
4.768	30355	BB	3MNT
4.974	17798	BV	4MNT
8.658	26792	BB	14DNB
8.959	25766	BB	26DNT
9.178	78678	BB	13DNB
10.640	151300	BB	24DNT
10.951	374679	BB	23DNT
12.762	297095	BB	34DNT
15.546	213138	BB	3-NO$_2$biphenyl
15.939	331069	BB	246TNT
19.108	30392	BB	245TNT
19.699	213808	BB	234TNT
20.703	188891	BB	1.3-DiNO$_2$naphthalene
22.831	396779	BB	2.2′-DiNO$_2$biphenyl
23.171	220906	BB	4NH$_2$-2.6-DiNO$_2$toluene
24.522	251604	BB	2NH$_2$-4.6-DiNO$_2$toluene
29.935	220999	BB	2.4-DiNO$_2$Diphenylamine

Figure 4.139 Separation of nitroaromatics on a Restek Rtx-1701 phase
(60 m × 0.32 mm × 0.25 µm, H$_2$, HP 5890 GC-ECD, 80 °C, 2 min, 25 – 130 °C/min, 4 °C/min
to 220 °C, 10 °C/min to 260 °C, 4.5 min).

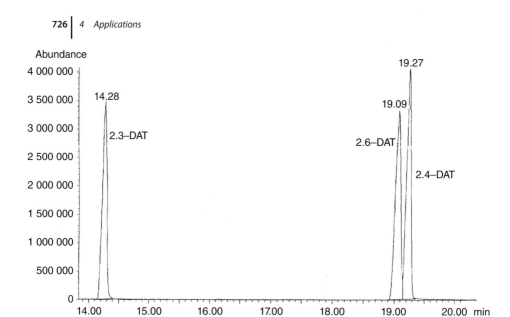

Figure 4.140 Chromatographic separation of the critical pair 2,4- and 2,6-diaminotoluene on the Rtx-200 capillary column.

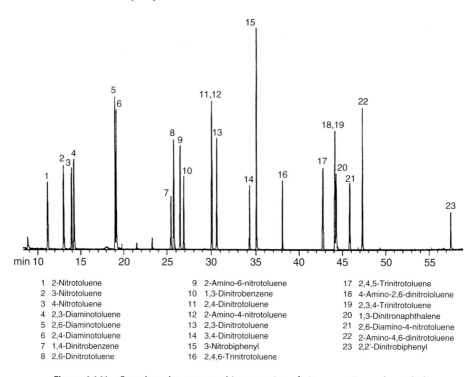

1 2-Nitrotoluene	9 2-Amino-6-nitrotoluene	17 2,4,5-Trinitrotoluene
2 3-Nitrotoluene	10 1,3-Dinitrobenzene	18 4-Amino-2,6-dinitrololuene
3 4-Nitrotoluene	11 2,4-Dinitrotoluene	19 2,3,4-Trinitrotoluene
4 2,3-Diaminotoluene	12 2-Amino-4-nitrotoluene	20 1,3-Dinitronaphthalene
5 2,6-Diaminotoluene	13 2,3-Dinitrotoluene	21 2,6-Diamino-4-nitrotoluene
6 2,4-Diaminotoluene	14 3,4-Dinitrotoluene	22 2-Amino-4,6-dinitrotoluene
7 1,4-Dinitrobenzene	15 3-Nitrobiphenyl	23 2,2'-Dinitrobiphenyl
8 2,6-Dinitrotoluene	16 2,4,6-Trinitrotoluene	

Figure 4.141 Complete chromatographic separation of nitroaromatics and metabolites on the Rtx-200 capillary column.

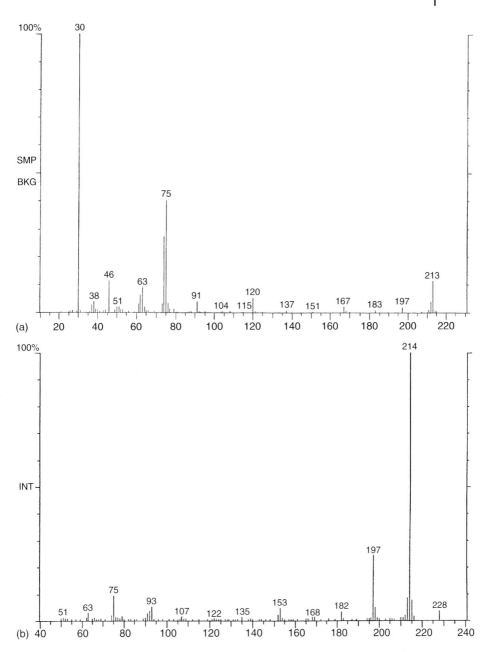

Figure 4.142 Mass spectra of trinitrobenzenes, M 213 (a) EI spectrum with extensive fragmentation down to *m/z* 30 (NO⁺) from the nitro group. (b) CI spectrum (methane) with (M + H)⁺ as the base peak.

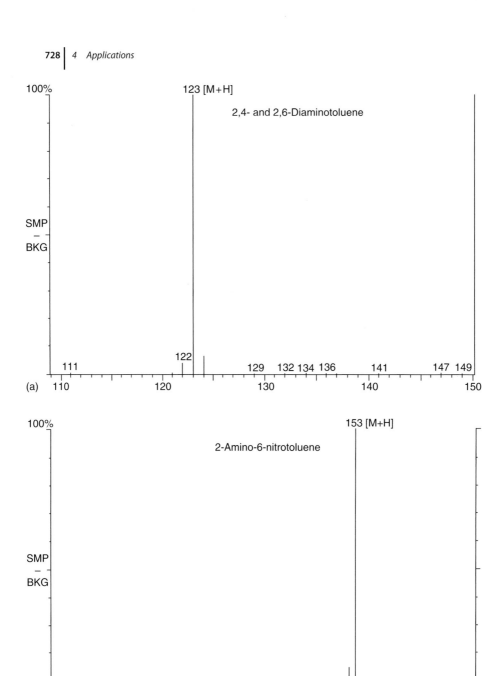

Figure 4.143 Typical CI spectra with different amino and nitro substitutions (a) diamino-toluene, (b) aminonitrotoluene, (c) aminodinitrotoluene and (d) diaminonitrotoluene.

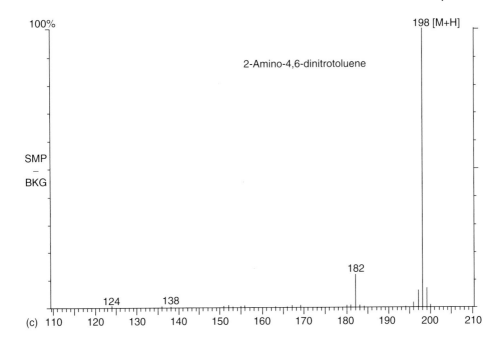

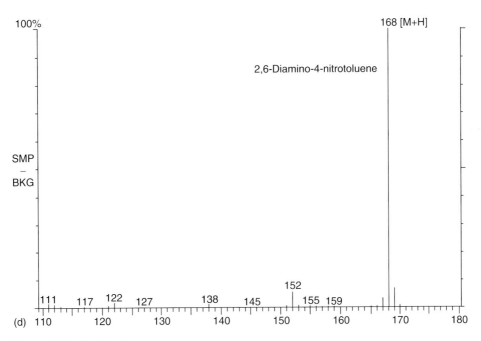

Figure 4.143 *(Continued)*

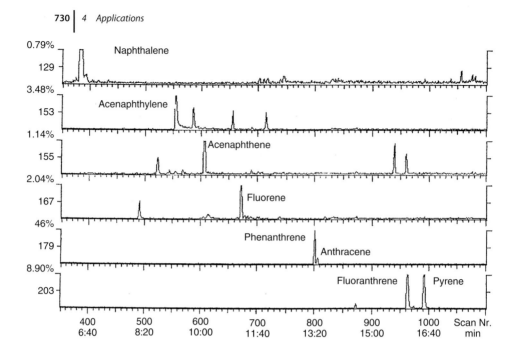

Figure 4.144 Analysis of a TNT soil sample for PAH components (CI mode).

nitro-and amino-substitutions. The outstanding selectivity of chemical ionization allows the determination of the equally important concentrations of PAHs in the same analysis. As the PAHs are very accessible by CI, it is only necessary to assign the corresponding $(M + H)^+$ masses in order to be able to determine their concentrations (Figure 4.144). The evaluation of the analysis is thus only directed to a further group of compounds. A second analysis run on the same sample is no longer necessary.

4.33
Comprehensive Drug Screening and Quantitation

Keywords: emergency toxicology; poisoning; full scan screening; targeted; non-targeted analysis; liquid-liquid extraction; one-point calibration

Introduction

The analytical tasks in emergency toxicology laboratories are manifold and comprise for instance the support for poisonings, detoxication or diagnosis. The described analytical workflow covers the full scan screening, spectrum identification and a fast one-point calibration for quantitation (Maurer, 2007, 2012; Meyer, Weber and Maurer, 2014).

Sample Preparation

The liquid–liquid extraction (LLE) is the standard extraction method in clinical and forensic toxicology for a broad range of drugs (Maurer, Pfleger and Weber, 2011). The available sample volume allows the application of the extracts for GC-MS and LC-MS analysis. For the sample preparation 1 mL of plasma is mixed with internal standard (IS), 100 μL of methanolic trimipramine-d3 (1.0 mg/L) and 2.0 mL of saturated aqueous sodium sulfate solution. The LLE is performed with a mixture of diethyl ether and ethyl acetate (1 : 1; v/v, 5 mL, 30 s). The organic layer is taken after centrifugation and evaporated to dryness at 60 °C. The remaining aqueous layer is then basified with 0.5 mL of aqueous sodium hydroxide (1 M) and extracted again. The organic layer is transferred to the same flask and evaporated to dryness. The dry residue is reconstituted in 100 μL of methanol and transferred into autosampler vials for GC injection (Meyer, Weber and Maurer, 2014).

Experimental Conditions

The extracts are analysed using a single quadrupole GC-MS system. As analytical column a 12 m short non polar capillary column with 0.2 mm ID and a thick film of 0.33 μm has been used. One microlitre of the extract has been injected in splitless mode.

For the GC-MS quality control check a methanolic solution of typical drugs is injected routinely. The ratio of peak areas between morphine and codeine (both at 50 ng/μL concentration) should be at least 1 : 10 as minimum QC requirement. Otherwise the system requires preventive maintenance.

Sample Measurements

The set of drugs comprising the described screening solution shown in Table 4.58 was based on the experience of the most often quantified compounds by GC-MS in the referenced laboratory and selected for a fast and reliable quantification in emergency toxicology (Meyer, Weber and Maurer, 2014).

The total number of compounds has been split in two one-point calibrator groups (OPC A and B, shown in Table 4.58) to avoid chromatographic overlapping. The identification is based on the complete mass spectra using AMDIS (Meyer, Peters and Maurer, 2010). The compound quantitation is performed on the specific ion chromatograms for each compound using the quantifier ions (QIs), given in Table 4.58, in ratio to the applied internal standard. The QI had been selected according to selectivity and sensitivity, but could show potential

Table 4.58 Drugs, quantifier ions, one-point calibrator group, therapeutic and toxic plasma reference concentrations, and plasma concentrations in OPCs, QC low and high levels. (Meyer *et al.* 2014, reproduced with permission of John Wiley & Sons.)

Drug name	Quantifier Ion (*m/z*)	OPC group	Therapeutic plasma concentration (mg/L)	Toxic plasma concentration higher than (mg/L)	Plasma concentration		
					OPC (mg/L)	QC low (mg/L)	QC high (mg/L)
Amitriptyline	58	A	0.05−0.2	0.5	1	0.2	4
Biperiden	98	A	0.005−0.1	0.08	1	0.1	4
Bromazepam	236	B	0.08−0.17	0.3	1	0.5	4
Citalopram	58	B	0.02−0.2	0.5	1	0.2	4
Clobazam	300	B	0.1−0.4	0.5	1	0.4	4
Clomipramine	269	B	0.1−0.2	0.4	1	0.2	4
Clozapine	243	B	0.35−0.6	0.8	1	0.6	4
Codeine	299	A	0.01−0.05	0.3	1	0.1	4
Diazepam	256	A	0.1−1	1.5	1	1	4
Diphenhydramine	58	A	0.1−1.0	1	1	1	4
Dihydrocodeine	301	B	0.03−0.25	1	1	0.25	4
Diltiazem	58	B	0.05−0.4	0.8	1	0.4	4
Doxepin	58	A	0.02−0.15	0.5	1	0.3	4
Doxylamine	58	A	0.05−0.2	1	1	0.5	4
Imipramine	58	B	0.05−0.15	0.4	1	0.2	4
Lamotrigine	185	B	4−10	15	5	10	30
Levetiracetam	126	A	10−37	400	10	40	400
Levomepromazine	185	A	0.02−0.1	0.4	1	0.2	4
Melperone	112	B	0.05−1	1	1	1	4
Methadone	72	A	0.05−0.8	0.5	0.5	1	4
Metoclopramide	86	B	0.04−0.15	0.2	1	0.1	4
Midazolam	310	B	0.08−0.25	1	0.5	0.25	4
Mianserin	193	B	0.015−0.07	0.5	1	0.1	4
Mirtazapine	195	A	0.04−0.3	1	1	0.4	4
Moclobemide	100	B	0.3−1	5	1	1	4
Nordazepam	242	A	0.2−0.8	1.5	1	1	4
Olanzapine	242	B	0.02−0.08	0.2	2	0.1	4
Perazine	70	B	0.02−0.35	0.5	0.5	0.3	4
Pethidine	247	A	0.1−0.8	1	1	1	4
Phenprocoumon	251	B	1−3	5	5	5	10
Promethazine	72	A	0.1−0.4	1	1	0.25	4
Prothipendyl	58	A	0.05−0.2	0.5	1	0.2	4
Quetiapine	210	B	0.025−0.9	1.8	2	1	4
Sertraline	274	A	0.05−0.25	0.3	1	0.2	4
Temazepam	271	A	0.3−0.9	1	1	1	4
Tramadol	58	A	0.01−0.25	1	1	0.5	4
Venlafaxine	58	A	0.2−0.75	1	1	0.5	4
Verapamil	303	B	0.05−0.35	0.9	1	0.4	4
Zolpidem	235	A	0.08−0.3	0.5	1	0.3	4
Zotepine	58	A	0.01−0.12	0.2	1	0.1	4

interferences from matrix or contaminants at high concentrations: tramadol on m/z 58 with palmitic acid (alternatively use m/z 263 at lower sensitivity), methadone on m/z 72 with stearic acid, and nordazepam on m/z 242 with diisoctylphthalate.

Analysis Conditions

Sample material			Plasma
Sample preparation	Liquid-liquid extraction		1 mL plasma Add ISTD 100 µL of methanolic trimipramine-d3 (1.0 mg/L) Add 2.0 mL saturated aqueous Na$_2$SO$_4$ solution Add diethyl ether, ethyl acetate (1 : 1; v/v, 5 mL, shaking 30 s) Centrifuge Evaporate organic layer to dryness Add to aqueous layer 0.5 mL of aqueous sodium hydroxide (1 M) Extract again with solvent mixture Evaporate organic layer to dryness Reconstitute residue in 100 µL of methanol Transfer into an autosampler vial for injection
GC Method	System		Agilent 6890 GC
	Column	Type	HP-1
		Dimensions	12 m length × 0.2 mm ID × 0.33 µm film thickness
	Carrier gas		Helium
	Flow		Constant flow, 1 mL/min
	SSL Injector	Injection mode	Splitless
		Injection temperature	200 °C
		Injection volume	1 µL
	Oven program	Start	100 °C
		Ramp 1	30 °C/min to 310 °C
		Final temperature	310 °C, 5 min
	Transferline	Temperature	280 °C
			Direct coupling
MS Method	System		Agilent 5973 MSD
	Analyzer type		Single quadrupole MS
	Ionization		EI
	Electron Energy		70 eV
	Ion source	Temperature	220 °C
	Acquisition	Mode	Full scan
	Mass range		50–550 Da
	Scan rate		1 scan/min

(Continued)

Calibration	Standard	Onepoint calibrator (OPC), 0.5–10.0 mg/L
	Internal standard	Trimipramine-d3, 1.0 mg/L
	Data points	1 per compound

Results

The recovery for the drug compounds screened was achieved for the most analytes between 72 and 97%. Low recovery exceptions are reported for levetiracetam 23%, phenprocoumon 37% and olanzapine with 66%.

The linearity of the quantitation was tested for the practical application of a fast OPC approach with levels in the practical range from high therapeutic up to toxic plasma concentrations, typically comprising 0.1, 0.5, 2, 4 and 6 mg/L. Exceptions are the highly dosed drugs such as levetiracetam, lamotrigine, phenprocoumon or quetiapine (see Table 4.58).

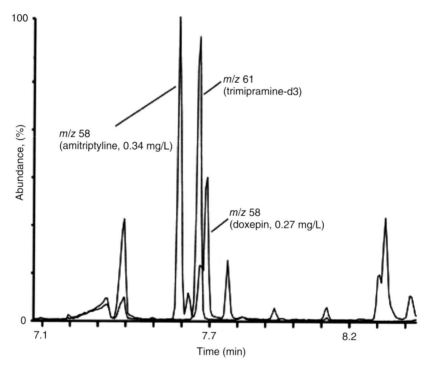

Figure 4.145 Mass chromatograms from a plasma proficiency test for quantitation: amitriptyline (m/z 58, 0.34 mg/L), doxepin (m/z 58, 0.27 mg/L) and IS timipramine-d3 (m/z 61). (Meyer *et al.* 2014, reproduced with permission of John Wiley & Sons.)

The proof of method applicability has been achieved by analysing proficiency test samples separately validated by reference methods. See the typical reconstructed compound specific mass chromatograms in Figure 4.145.

Conclusions

The described GC-MS screening method allows the fast, accurate, and reliable quantitation of 36 drugs relevant to emergency toxicology in the range of therapeutic to toxic concentrations.

4.34
Determination of THC-Carbonic Acid in Urine by NCI

Keywords: forensics; THC-A; Cannabis; Marinol™; Dronabinol; metabolite; derivatization; negative chemical ionization

Introduction

It is a well-documented fact that the widespread use of marijuana products commonly named like grass, weed, dope, hashish, bhang, pot, shit or ganja continues to make Cannabis sativa one of the most used drug plants of our time (Kapusta *et al.*, 2006; Beck, Legleye and Spilka, 2007; Jones *et al.*, 2009). Furthermore, recent research has led to the use of THC (tetrahydrocannabinol), which is the main psychoactive ingredient of the plant, as a FDA approved therapeutically drug under the brand name Marinol in the USA, marketed under the generic international non-proprietary name (INN) Dronabinol in other countries (Levin and Kleber, 2008).

Often urine is the matrix of choice in epidemiological studies dealing with addictive drugs in physiological or forensic research, workplace drug testing as well as in doping control (Musshoff and Madea, 2006). This application was developed at the Robert Koch-Institute in Berlin, Germany, and presents a highly reliable routine method for an easy and effective sample pre-treatment with high recovery. The selective quantitative determination of the THC acid metabolite (THC-A, THC-COOH) was achieved with a THC-COOH-PFPOH-HFBA derivative in the low pico- to middle femtogram-range using the negative chemical ionization (NCI) mode (Melchert, Hübschmann and Pabel, 2009).

Sample Preparation

Hydrolysis Procedure
Because the THC metabolite THC-COOH is excreted in urine in conjugated form, a hydrolysis step is necessary to receive the free THC-COOH before extraction. Hydrolysis is done for a 1 mL urine sample with 100 µL 12 N KOH solution after addition of 20 µL of the internal standard, equivalent to 100 ng of deuterated THC-COOH. Hydrolysis is achieved by heating the mixture in a crimp-cap septum

closed N 11-1 vial to 60 °C for 20 min. After cooling to room temperature, 350 μL of acetic acid anhydride are added to set the pH < 4 for the reaction mixture.

Extraction Procedure

The hydrolysed sample is transferred quantitatively on an Extrelut™ NT1 extraction column. After waiting for 5 min, the sample is first extracted with 4 mL isooctane and then with additional 2 mL isooctane. The combined extracts are collected in glass vials (volume 10 mL) with a conical bottom, evaporated to dryness and concentrated in the conical part of the vial.

Derivatization Conditions

The extracted THC-COOH is transferred completely into N 11 glass vials. As derivatization agents 50 μL pentafluoropropanol and 80 μL heptafluorobutyric acid anhydride are added and the reaction mixture is heated to 65 °C for 20 min in the crimp-closed vial. The remaining derivatization reagents are then completely evaporated and the remainder is carefully transferred to conical autosampler microvials (ND 8, 1.1 mL) using a microliter syringe and toluene as solvent, so that an end volume of 200 μL is achieved for the injection of 1 μL aliquots of the derivatized extract.

Experimental Conditions

A single quadrupole DSQ II GC-MS system equipped with a CTC PAL liquid autosampler was used. The GC-MS measurements have been done in the electron ionization (EI) mode for method development and the NCI mode for routine quantitation.

Sample Measurements

Reproducibility of Retention Times

Under the described GC-conditions the reproducibility of the retention times for the THC-COOH-derivatives was < ±0.5% within one day and < ±3% within a total working time of 6 months, covering more than 1000 sample injections.

Limit of Detection (LOD), Limit of Quantification (LOQ)

For measurements of THC-COOH as PFPOH-HFBA-derivatives from spiked urine samples in the NCI mode a LOD value of 300 fg with a signal-to-noise ratio S/N > 3 could be achieved. As LOQ value 1 pg with a signal-to-noise ratio S/N > 10 was determined.

Recovery and Calibration

For urine samples spiked with THC-COOH between 15 and 65 ng/mL recovery rates of 85–95% were found. The fragment ions m/z 483 (deuterated derivative) and m/z 474 (native derivative) have been used for quantification.

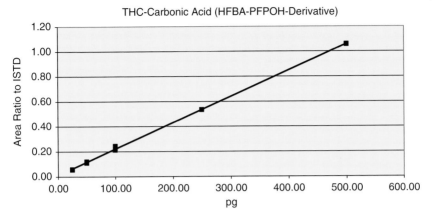

Figure 4.146 Internal standard calibration for the THC-COOH-PFPOH-HFBA derivative, range 25–500 pg, R^2 0.9989.

For standard analytical work the internal standard calibration method was used in a range of 25–500 pg for the THC-COOH-PFPOH-HFBA derivatives as shown in Figure 4.146. For urine samples of low level exposure samples the calibration was done in the 1–10 pg range.

Analysis Conditions

Sample material			Urine
Sample preparation	SPE		
		Material	Extrelut™ NT1
		Sample application	Wait 5 min
		Elution	1.4 mL isooctane
			2.2 mL isooctane
		Concentration	Evaporated close to dryness
	Derivatization	Reagents	50 µL pentafluoropropanol
			80 µL heptafluorobutyric acid anhydride
		Reaction	Heated to 65 °C for 20 min
			Remaining reagents evaporated
		Solvent	200 µL toluene
GC Method	System		Thermo Scientific™ TRACE GC Ultra
	Autosampler		CTC PAL
	Column	Type	Restek Rtx 5Sil MS with 10 m Integra-Guard
		Dimensions	30 m length × 0.25 mm ID × 0.25 µm film thickness
	Carrier gas		Helium
	Flow		Constant flow, 30 cm/s
	PTV Injector	Injection mode	Splitless
		Injection Volume	1 µL

(Continued)

		Base temperature	90 °C
		Transfer rate	10 °C/s
		Transfer temperature	200 °C
	Split	Closed until	1 min
		Open	1 min to end of run
		Purge flow	5 mL/min
	Oven program	Start	70 °C, 1 min
		Ramp 1	40 °C/min to 200 °C
		Ramp 2	20 °C/min to 265 °C
		Final temperature	265 °C, 3.5 min
	Transferline	Temperature	285 °C
			Direct coupling
MS Method	System		Thermo Scientific™ DSQ II
	Analyzer type		Single quadrupole MS
	Ionization	EI	For method development
		CI	For routine quantitation
	CI Gas		Methane, 1.2 mL/min
	Electron Energy		70 eV
	Ion source	Temperature	230 °C
	Acquisition	Mode	SIM
	Mass range		*m/z* 197, 474 and 483
Calibration	Internal standard		THC-COOH-d9-HFBA-PFPOH
	Calibration range		1–10 pg, 25–500 pg
			THC-COOH-d9-HFBA

Results

Mass Spectra and GC Separation

Figure 4.147 shows the mass spectrum of the THC-COOH-PFPOH-HFBA derivative measured in the EI-mode. Figure 4.148 shows the mass spectrum in the NCI mode. Both spectra had been submitted for inclusion into the NIST

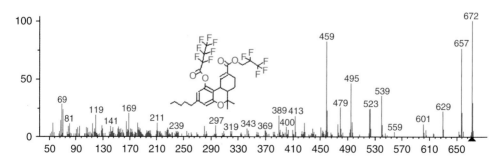

Figure 4.147 EI mass spectrum of the THC-COOH-PFPOH-HFBA derivative.

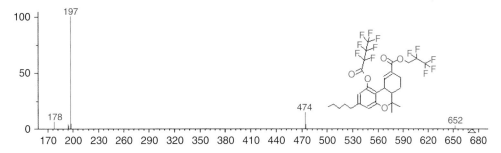

Figure 4.148   NCI mass spectrum of the THC-COOH-PFPOH-HFBA derivative.

mass spectral library. For both spectra the fragmentation patterns are given. The following Figure 4.149 presents the mass chromatogram of the native and deuterated THC-COOH derivative from a hydrolyzed, extracted and derivatized urine sample. The three traces show the total ion current trace (top), the mass trace m/z 474 for the native THC-COOH derivative (middle) and m/z 483 for the internal standard d_9-THC-COOH derivative (bottom). Although the ion intensity of m/z 474 appears to be low in the NCI spectrum (Figure 4.148) the higher mass delivers less matrix interference leading to higher S/N values in urine samples than the use of the lower mass fragment m/z 197.

Measurement of Quality Control Samples

Multiple measurements ($n = 10$) of the Medidrug BTM U-Screen Level 2 quality control urine samples with a target value of 65 ng/mL has been used for verification. The results have been well within the expected range ($\pm 4\%$). CEDIA control urine samples of the THC 25 Control Set 'low' and 'high' (Cat.-Nr. 1661086) were measured as additional control. The expected target ranges for the 'low' samples (18.75 µg/L) and for the 'high' samples (31.25 µg/L) could be verified with $\pm 6\%$ for the 'low' and $\pm 4\%$ for the 'high' urine samples. Spiked native urine samples of cannabis naive persons did show that the above mentioned recovery rates between 85 and 95% could be achieved continuously. In none of the analysed urines of cannabis naive persons as well as in commercial drug free urine control samples THC-COOH was detected.

Conclusions

THC-COOH, as the main metabolite of THC in urine, can be detected for many days after the last consumption of cannabis products. The THC-COOH concentration depends not only on the amount of drugs used, but also on the THC-content of the consumed cannabis product. Measurements with the method described above have shown that none of the commonly prescribed or OTC drugs interfered with the quantification of THC-carbonic acid in urine samples. This is due to the effective Extrelut extraction procedure. In addition,

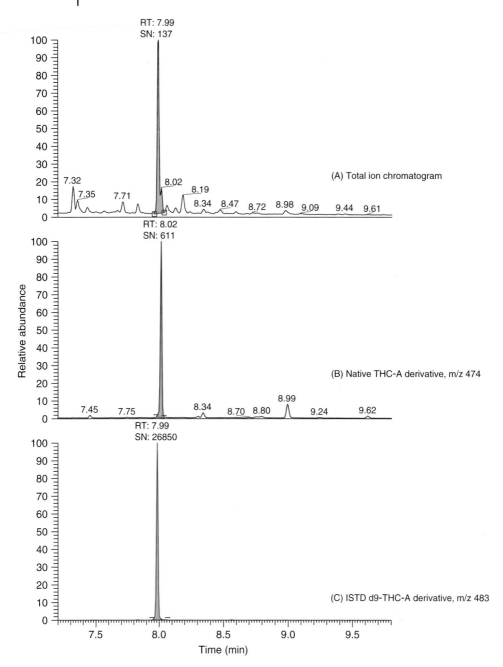

Figure 4.149 Mass chromatograms of the total ion (A), native (B) and deuterated (C) THC-COOH derivative from a hydrolyzed, extracted and derivatized urine sample.

the derivatization of the isolated THC-COOH leads to very selective and 'clean' chromatograms. The NCI mass spectral fragmentation pattern is unique to the THC-COOH-PFPOH-HFBA derivative. So the high resolution power of capillary GC, coupled with mass spectrometry in the NCI mode, can safely overcome interferences which are often observed in immunological methods from present drugs.

4.35
Detection of Drugs in Hair

Keywords: SFE; opiates; pharmaceuticals; cocaine; methadone; fortifying; MS/MS

Introduction

Analysis of hair allows information to be obtained concerning a period extending further back than that which can be examined using blood and urine samples. Testing of hair can be used to detect repeated and chronic drug misuse. Narcotics and other addictive drugs and, for example medicines can be detected unambiguously using the method described and this satisfies the current legal requirements (Martz *et al.*, 1991; Schwinn, 1992; Sachs and Raff, 1993). With hair analysis the chronic misuse of these substances can be traced back over weeks and months (hair growth amounts to about 1 cm/month). Amphetamines, cannabis, cocaine, methadone and opiates, for example can be detected. As this is a stepwise investigation, the development of drug use over a period can be visualized. Monitoring drug dependence in the methadone programme is one application of this type of analysis, as is doping detection in competitive sport (Möller *et al.*, 1993). Hair analysis cannot be used to detect an acute drug problem, as a blood or urine test can. The quantity examined should be at least 50 mg. Sometimes, a very small quantity such as the quantity of hair found daily in a shaver can be sufficient.

SFE is used for extraction in this application, whereby it is assumed that all the active substances which can be determined by GC can be extracted with supercritical CO_2 (Sachs and Uhl, 1992). SFE has the advantage of a higher extraction rate, as, unlike other procedures, it takes only a few minutes. In addition, the extraction unit can be coupled directly to the GC-MS allowing the method to be automated and ever smaller quantities to be analyzed.

Analysis Conditions

Sample material	Hair
Sample preparation	A bundle of hair of pencil thickness, fixed that movement of the individual hairs over one another is inhibited, sections of 1 cm length are cut and ground

(Continued)

	Extraction	SFE	20 mg hair powder, treated with 20 µL ethyl acetate in the extraction cartridge (volume 150 µL)
		Oven temperature	60 °C
		Static extraction	15 min, 200–300 bar
		Dynamic extraction	5 min, 200–300 bar
		Extract isolation	Transfer of the eluate to a sample bottle with about 1 mL ethyl acetate.
	Derivatization		With pentafluoropropionic anhydride (PFPA).
GC Method	System		Varian 3400
	Injector		Gerstel CIS, cold injection system, Split/splitless
	Column	Type	DB-5
		Dimensions	30 m length × 0.31 mm ID × 0.25 µm, film thickness
	Carrier gas		Helium
	Flow		1 mL/min
	CIS Injector	Injection mode	Splitless
	Oven program	Start	60 °C, 2 min
		Ramp 1	25 °C/min to 320 °C
		Isothermal	230 °C, 5 min
		Ramp 2	25 °C/min to 280 °C
		Final temperature	280 °C, 5 min
	Transferline	Temperature	280 °C
MS Method	System		Finnigan TSQ 700
	Analyzer type		Triple quadrupole MS
	Ionization		EI
	Electron energy		70 eV
	Acquisition	Mode	Product ion scan
	Collision gas, energy		Ar, 2 mbar, 25 V collision energy

Results

In the method described, the SFE was not coupled directly to the GC-MS (on-line SFE-GC-MS). The extraction, derivatization and analysis were carried out separately. The use of different pressures (100, 200, 300 bar) shows the different extraction possibilities. At 200 bar, the extract produced showed the best signal/noise ratio for heroin (Figure 4.150). Using derivatization, morphine is detected as morphine-2-PFP and monoacetylmorphine (MAM) as MAM-PFP together with underivatized heroin.

As it is known that the more lipophilic substances, such as heroin, cocaine and THC carboxylic acid can be deposited directly in hair and do not need to be derivatized for chromatography, SFE can in fact be coupled directly with GC-MS/MS (Traldi *et al.*, 1991).

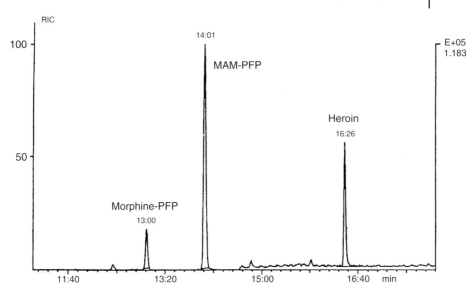

Figure 4.150   Chromatogram of a hair sample extracted by SFE (200 bar), taken in product ion scan mode (MS/MS).

4.36
Screening for Drugs of Abuse

Keywords: drug screening; heroin; codeine; morphine; hydrolysis; derivatization

Introduction

Detection of drug taking by the investigation of blood and urine samples is one of the tasks of a forensic toxicology laboratory in workplace drug testing. The routine analysis of drug screening can be carried out by TLC, HPLC or by immunological methods. Positive results require confirmation. Determination using GC-MS is recognized as a reference method. Also GC-MS/MS methods providing increased selectivity for a list of target compounds are increasingly applied (Smith *et al.*, 2007; Weller *et al.*, 2000). The full-scan method is preferred over the SIM procedure for differentiating between different drugs in a comprehensive screening procedure because of the higher specificity and universality (Pfleger *et al.*, 1992). The required decision limits (Table 4.59) are laid down by the US Ministry of Health (HHS) and the US Ministry of Defense (DOD).

Heroin (3,6-diacetylmorphine) is usually not determined directly (Musshoff *et al.*, 2004). The unambiguous detection of heroin consumption is carried out by determining 6-MAM, which is formed from heroin as a metabolite (Figure 4.151). If morphine is difficult or impossible to detect when clean-up is carried out without hydrolysis, then the latter should be used. It should be noted that nonspecific hydrolases can also effect the degradation of 6-MAM to morphine. If the substitute drugs methadone and dihydrocodeine are present, these are also determined

Table 4.59 Decision limits in drug screening (data in ng/mL).

Active substance	Screening		Confirmation	
	HHS	DOD	HHS	DOD
THC	100[a]	50	15	15
BZE	300	150	150	100
Opiates	300	300	300	300
AMPs	1000	500	500	500
PCP	25	25	25	25
BARBs	n.s.	200	n.s.	200
LSD	n.s.	0.5	n.s.	0.2

a) HHS suggestion 50 ng/mL.
THC, 11-nor-A-9-tetrahydrocannabinol-9-carboxylic acid; PCP, Phencyclidine; BZE, benzoylecgonine; BARB, barbiturates; AMP, amphetamines; LSD, lysergic acid diethylamide.

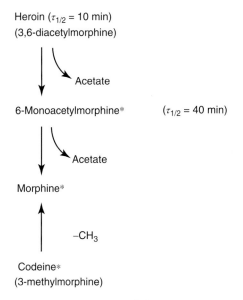

Heroin ($\tau_{1/2}$ = 10 min)
(3,6-diacetylmorphine)

Acetate

6-Monoacetylmorphine*   ($\tau_{1/2}$ = 40 min)

Acetate

Morphine*

–CH₃

Codeine*
(3-methylmorphine)

Figure 4.151 Opiate metabolism (* determination by GC-MS).

using the procedure described. As blood from corpses also has to be processed during routine operations to some extent, a comparatively time-consuming extraction including a reextraction is carried out. For the gas chromatography of morphine derivatives, derivatization of the extract by silylation, acetylation or pentafluoropropionylation, for example is basically necessary (Maurer 1990, 1993). Mass spectrometric detection using NCI is possible by using fluorinated derivatives. In the procedure described, silylation with MSTFA and detection in the EI mode has been chosen (Donike, 1992).

Analysis Conditions

Sample material			Urine samples, blood
Sample preparation	Serum (blood)		1 mL sample, add internal standard (e.g. 100 ng morphine-d_3 or codeine-d_3) Dilute to 7 mL with phosphate buffer (pH 6)
	SPE extraction		Condition with 2 mL methanol and 2 mL phosphate buffer pH 6 (Bond-Elut-Certify 130 mg) Elute with 2×1 mL chloroform/isopropanol/25% ammonia (70 : 30 : 4)
	Derivatization		With 50 µL MSTFA at 80 °C, 30 min
	Urine		Hydrolyse 2 mL sample with 20 µL enzyme solution (ß-glucuronidase), 60 min, 60 °C Adjust sample pH to 8–9 with 0.1 N sodium carbonate solution
	SPE extraction		Activate a C18-cartridge with methanol, condition with 0.1 N sodium carbonate solution Apply sample and washing the cartridge with 0.1 N sodium carbonate solution Elute with 1 mL acetone/chloroform (50 : 50) Evaporate eluate to dryness
	Derivatization		With MSTFA at 80 °C, 30 min
GC Method	System		Finnigan GCQ GC
	Injector		Split/splitless
	Column	Type	DB-1
		Dimensions	30 m length $\times$ 0.25 mm ID $\times$ 0.25 µm film thickness
	Carrier gas		Helium
	Flow		constant flow, 40 cm/s
	SSL Injector	Injection mode	Splitless
		Injection temperature	275 °C
		Injection volume	1 µL
	Split	Closed until	0.1 min
		Open	0.1 min to end of run
	Oven program	Start	100 °C
		Ramp 1	10 °C/min to 310 °C
		Final temperature	310 °C
	Transferline	Temperature	280 °C
MS Method	System		Finnigan GCQ
	Analyzer type		Ion trap MS with internal ionization

(Continued)

Ionization		EI
Acquisition	Mode	Full scan
Mass range		70–440 Da
Scan rate		0.5 s (2 spectra/s)

Results

The chromatogram and mass spectra after clean-up of a typical serum sample are shown in Figure 4.152. The mass traces for codeine, morphine-d_3 and morphine and the TIC are shown. For morphine a concentration of 60 ng/L is calculated with reference to morphine-d_3 using a calibration curve. The resulting mass spectra

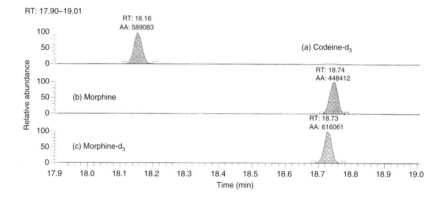

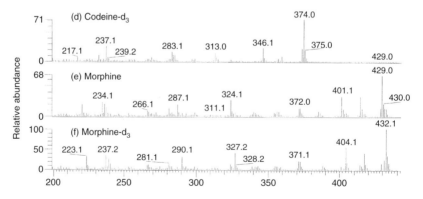

Figure 4.152 Analysis of an authentic serum sample fortified with 100 ng morphine-d_3 and 100 ng codeine-d_3 Above: (a) Mass trace for codeine-d_3 (*m/z* 374), (b) mass trace for morphine (*m/z* 429) and (c) Mass trace for morphine-d_3 (*m/z* 432). Below: (d) spectrum of codeine-d_3, (e) spectrum of morphine and (f) spectrum of morphine-d_3.

Table 4.60 Selective masses of the opiates as TMS derivatives.

Opiate	Retention index	*m/z* values
Levorphanol-TMS	2188	329, 314
Pentazocine-TMS	2262	357, 342
Levallorphan-TMS	2318	355, 328
Dihydrocodeine-TMS	2365	373, 315
Codeine-TMS	2445	371, 343
Hydrocodone-TMS	2447	371, 356
Dihydromorphine-TMS	2459	431, 416
Hydromorphone-TMS	2499	357
Oxycodone-TMS	2503	459, 444
Morphine-bis-TMS	2513	429, 414
Norcodeine-bis-TMS	2524	429
Monoacetylmorphine-TMS	2563	399, 340
Nalorphine-bis-TMS	2656	455, 440

Note: All TMS derivatives show an intense signal for the trimethylsilyl fragment at *m/z* 73, which does not appear when the scan is begun at *m/z* 100 and is unnecessary for substance confirmation. The retention indices were determined on DB-1, 30 m × 0.25 mm × 0.25 µm with an *n*-alkane mixture of up to C_{32} (Weller and Wolf, 1990).

allow identification, for example on comparison with relevant toxicologically orientated spectral libraries. The clean-up and analysis methods can be used for other basic drugs, such as amphetamine derivatives, methadone and cocaine and their metabolites.

A list of morphine derivatives and synthetic opiates, which have been found by the procedure described after clean-up of serum samples, is shown in Table 4.60. For the corresponding trimethylsilyl derivatives, the specific search masses are given which allow identification in combination with the retention index.

4.37
Structural Elucidation by Chemical Ionization and MS/MS

Keywords: chemical analysis; identification; chemical ionization; molecular ion; MS/MS; ion trap; amines; Mass Frontier; fragmentation pathways

Introduction

Structural elucidation is important in any synthetic work to check the identity of final products and by-products. Mass spectrometry in this context is the tool of choice for the confirmation of expected compound structures or the identification of new components found, especially after a chromatographic separation of mixtures.

The mass spectra for structure elucidation are typically generated by electron impact ionization (EI) providing rich information from the fragmentation to the structure of an unknown compound. The reliable interpretation requires within the first steps the allocation of the molecular ion of the compound in the spectrum, giving access to the possible fragmentation pathways from the chemical structure of the molecule. Many compound classes only show a molecular ion of low intensity, or even do not show the molecular ion in the spectrum at all. A final structure confirmation using the EI spectrum only is not concluding in these cases. The vast majority of members of the large compound classes of hydrocarbons, alcohols, ketones, acids, esters or amines are known to not reveal the molecular information in their respective EI mass spectra.

The CI mode in combination with EI ionization can be the solution for a routine structure elucidation in industrial chemical analysis. Most useful is the positive chemical ionization (PCI) using different protonating reagent gases like methane, iso-butane or ammonia. PCI leads to the formation of the quasimolecular ion $(M + H)^+$ in good abundance, which then can be also used for MS/MS fragmentation revealing the required structural details. Only for special compound classes with high abundance of electronegative groups like nitrates or halogens the NCI is used.

This example of structural elucidation deals with the analysis of long chain alkyl amides applying the reagent gases methane and ammonia. MS/MS fragmentation of the molecular precursor ion is used for structural confirmation. The data system supported interpretation using the Mass Frontier program is used to confirm the fragmentation pathways.

Experimental Conditions

As a model compound representing the general MS behaviour of alkyl amides the compound 'N,N-dimethyldecanamide', CAS No. 14433-76-2, is used for the development of the methodology. The GC injection and separation parameters were optimized for the separation of the amides on the applied amine column.

Sample Measurements

First the regular 70 eV EI mass spectrum of the amide has been acquired (Vale, 2011). The search of the EI spectrum against the NIST library gave a match for 'N,N-dimethyldecanamide', which has a molecular weight of 199 g/mol, see Figure 4.153. The abundance of the molecular ion m/z 199 with ~2% is very low.

PCI with methane and ammonia was used to form the molecular ion, which in the next steps was fragmented for structural elucidation by MS/MS. The Thermo Scientific Mass Frontier software was then used to predict the theoretical fragments and fragmentation pathways of the detected product ions to confirm their identity.

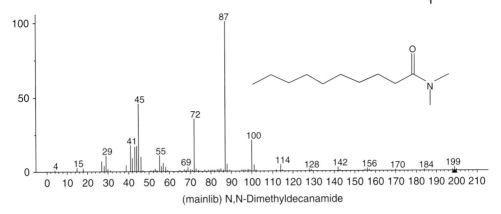

Figure 4.153 NIST library entry 'N,N-dimethyldecanamide', M 199.

Analysis Conditions

Sample material			Industrial amines, cationic surfactant products
Sample preparation			Aliquot weighed out, dissolved in i-propanol, diluted in acetone to final concentration of about 2000 µg/mL
GC Method	System		Thermo Scientific™ TRACE GC Ultra™
	Autosampler		Thermo Scientific AS 3000 II
	Column	Type	TG5-Amine
		Dimensions	30 m length × 0.25 mm ID × 0.50 µm film thickness
	Carrier gas		Helium
	Flow		Constant flow, 1 mL/min
	PTV Injector	Injection mode	Splitless
		Injection Volume	1 µL
		Base temperature	50 °C
		Transfer rate	14 °C/s
		Transfer temperature	300 °C
		Transfer time	10 min
	Split	Closed until	0.05 min
		Open	0.05 min to end of run
		Gas saver	2 min, 10 mL/min
		Purge flow	5 mL/min
	Surge	Pressure, time	120 kPa, 2.55 min
	Oven Program	Start	40 °C, 2.5 min
		Ramp	30 °C/min to 300 °C
		Final Temperature	300 °C, 15 min
MS Method	System		Thermo Scientific™ ITQ
	Analyzer type		Ion trap MS with external ionization

(Continued)

Ionization		EI/PCI
CI Gas		Methane, ammonia, 1 mL/min at 10 psi
Electron energy		70 eV
Ion source	Temperature	200 °C
	Buffer gas	0.3 mL He/min
	Emission current	250 µA
Acquisition	Mode	Full scan
	Mass range	40–450 Da
Acquisition	Mode	MS/MS
	Precursor ion	*m/z* 200.3
	Product ions	*m/z* 72.1, 85.0, 116.1, 130.2
	Coll. Energies	0.54, 1.07, 1.93 (ACE optimized)
	Scan time	0.17 s (6 spectra/s)

Results

The chromatographic separation was optimized by a cold injection on the PTV and a capillary column specialized for the analysis of amines.

The ITQ GC-MS/MS system provided the required structure elucidation and identification of *N,N*-dimethyldecanamide, see Figure 4.154. By choosing the proper reagent gas, PCI techniques can selectively protonate molecules and provide a high intensity of the quasimolecular ion for the MS/MS process. The

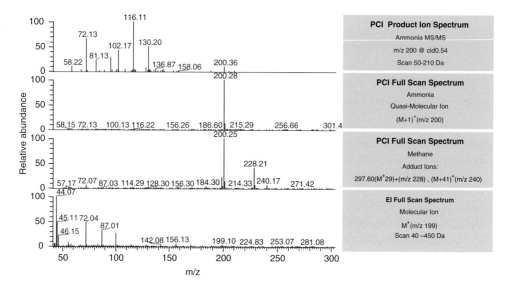

Figure 4.154 PCI MS/MS product ion and full scan spectra with ammonia and methane, EI full scan spectrum.

use of methane as reagent provides the confirmation of the molecular weight 199 with the protonated molecular ion m/z 200 and the expected adduct formation of M + 29 to *m/z* 228 and M + 41 to *m/z* 240.

Ammonia does not form adducts and provides the quasi-molecular $(M + 1)^+$ ion unfragmented with high intensity due to its high proton affinity of 854.2 kJ/mol. The high intensity quasi-molecular ion is the ideal precursor for MS/MS. The product ion spectrum with the fragment ion structures predicted by Mass Frontier software for the protonated molecular ion is given with Figure 4.155. The mechanism for confirmation of the *m/z* 116 ion by Mass Frontier is shown as a fragmentation pathway in Figure 4.156. In conclusion, the generation of the

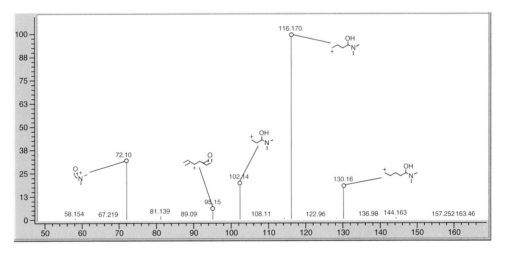

Figure 4.155 Structure Annotated Product Ion Spectrum (Mass Frontier).

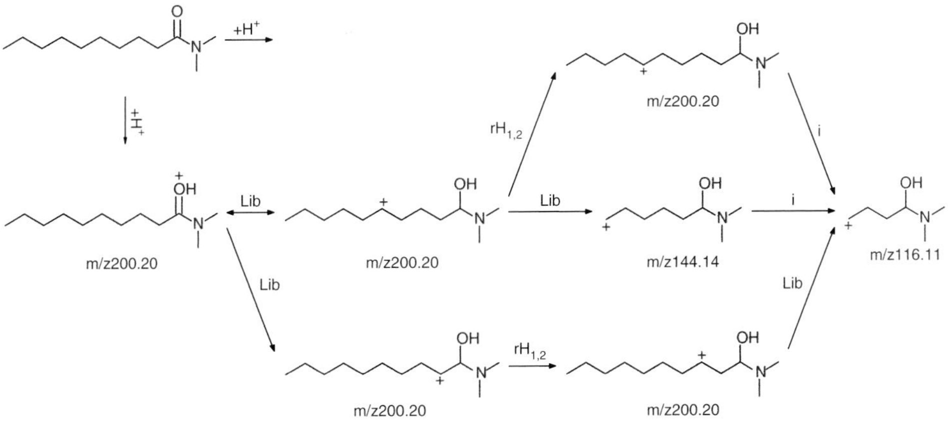

Figure 4.156 Product Ion m/z 116 Pathway Confirmation (Mass Frontier).

molecular ion by PCI and Mass Frontier interpretation assisted in the final structural identification of N,N-dimethyldecanamide.

4.38
Volatile Compounds in Car Interior Materials

Keywords: automobile industry; thermal desorption; degassing; automotive interior materials; VOCs; SVOCs; VDA 278; ISO 11890-2

Introduction

The smell of new cars is for many the enchanting scent of a long planned and eagerly awaited delight. This typical odour is generated by a variety of polymer materials of the interior of new automobiles. But owners become increasingly concerned about the air quality inside of a new car as the emitted chemicals are supposed to create a potential health risk.

Studies show that the indoor air of new vehicles carries a high amount of VOCs released from new vehicle interiors. The total volatile organic compound (TVOC) concentration within the interior of a minivan was determined as high as $7500 \, \mu g/m^3$ of inside air on the second day after delivery, which is approximately two orders of magnitude higher than regular outdoor TVOC concentrations. Over 60 chemicals had been identified in this study inside the interior of vehicles released from different materials such as carpets, pedals, seat covers, door linings and so on (Grabbs, Corsi and Torres, 2000).

The European so called 'solvents emissions' directive provides the currently most stringent regulations on the release of odours from VOCs or fogging of the interior windscreen by semi volatile emissions (SVOCs (semi-volatile organic compounds), FOG) with the maximum limit values for vehicle refinishing products (European Commission, 2004, 2010).

The standards for the sampling and gas chromatographic testing of paints, varnishes and related products used in the automobile industry are set by the ISO norm 11890-2 in the new revision from 2013 (ISO, 2013). Other methods ISO 11890-1 and ASTM D2369 are using gravimetric weight difference methods (ASTM, 2010). In the United States the California Standard Section 01350 specification is the relevant standard for evaluating and restricting VOC emissions in indoor air, which is also applied to volatile emissions in automobiles (California Department of Public Health, 2010). In China the overall automotive interior air quality discussion has attracted the government departments, agencies, car inspection and decorative materials manufacturers with the first release of the 'passenger air quality assessment guidelines' as of 1 March 2012 (Lee, 2013).

On the background of these new international regulations the organic materials used for automobile interiors need to be screened at the manufacturer and the raw material suppliers for VOC and SVOC release to ensure the air quality inside the car. The reference method for the determination of VOCs and FOG in automotive interior materials is the VDA 278 standard (or GMW 15634) using a thermal desorption GC-MS method (GM Engineering Standards, 2008; VDA 278, 2011). VDA stands for the German Quality Management System (QMS) for the automobile industry (Verband der Automobilindustrie, Germany). The VDA 278 is part of the delivery specs of the car manufacturers Daimler, BMW, Porsche and VW. GM/Opel uses the corresponding GM Engineering Standards GMW 15634. The VDA 278 analysis procedure serves for the determination of emissions from non-metallic materials which are used for interior parts in motor vehicles like textiles, carpets, adhesives, scaling compounds, foam materials, leathers, plastic parts, foils, lacquers or combinations of different materials (VDA 278, 2011). It provides semi-quantitative values of the emission from these materials of VOCs and the semivolatile condensable substances (SVOCs, FOG). The term *'fog'* is used here as these less volatile substances can condense at ambient temperatures and contribute to the fogging of the windshield. The suggested analysis method uses thermal desorption GC-MS for both of the VOCs and SVOC/FOG analysis.

In a first step the VOC analysis are determined with a thermal desorption at 90 °C for 30 min. The emitted compounds are analyzed and calibrated using a toluene standard. The VOC concentration is expressed as toluene equivalent. The SVOC analysis (FOG) is run from the same sample in a second desorption at 120 °C for 60 min. The emitted substances are calibrated using a C16 alkane standard. The FOG result is expressed as hexadecane equivalent (VDA 278, 2011; Lee, 2013).

Sample Measurements

Sample Preparation

Samples are taken directly into a glass adsorption tube. Specific sampling requirements apply according to the investigated materials. For ABS, PVC, leather and other plastic parts about $30 \, mg \pm 5 \, mg$ are used, cut into pieces of about 4 cm length and 3 mm width.

VOC testing

A standard series with concentrations of 10, 50, 100, 200, 500 and 1000 ng/μL in methanol has been prepared as the working standards used for calibration. The calibration solutions have been applied directly into Tenax filled desorption tubes and analyzed using the above described method for TD-GC-MS measurements. The chromatogram of the VOC compounds as the total ion current is shown in Figure 4.157. In Table 4.61 the BTEX compounds analyzed are listed with retention times and the specific ions used for selective quantification, also showing the resulting R^2 values giving the precision of the quantitative calibration.

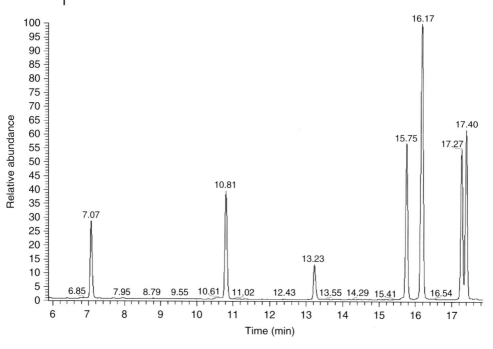

Figure 4.157 TD-GC-MS total ion chromatogram (TIC) of the volatile organic standard.

Table 4.61 Volatile organic compounds of the BTEX test with retention times and quantitative precision.

Compounds	Retention time (min)	Quantitation ion (*m/z*)	Linearity R^2
Benzene	7.07	78	0.9991
Toluene	10.81	91	0.9999
Ethylbenzene	15.75	91	0.9990
p/m-Xylene	16.17	91	0.9998
Styrene	17.27	91	0.9993
o-Xylene	17.40	91	0.9994

SVOC testing

The SVOC testing provides the semivolatile emission value (TVOC) of the material analyzed. It is determined according to VDA 278 by the peak area integration of the sample chromatogram after thermal desorption at the higher temperature of 120 °C. The result calculation is based and expressed as the area comparison with a 100 ng toluene analysis as standard. The calibration solution dissolved in methanol has been applied directly into Tenax filled desorption tubes and analyzed using the above described method for TD-GC-MS measurements.

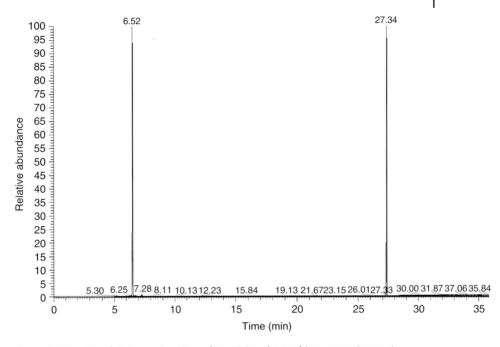

Figure 4.158 C6 and C16 retention times determining the total ion current integration limits.

The TVOC value of a sample is determined by the integration of the chromatographic peak area between C6 and C16 with 100 ng total integrated area toluene peak comparison calculated. The two standards hexane and *n*-hexadecane determine the retention time position of the C6 and C16 peak (see Figure 4.158). For the test samples the peak area between the calibrated retention times of n-hexane and n-hexadecane is determined as a total peak area. Representative chromatograms of a leather and a sponge sample are shown with the total ion current in Figures 4.159 and 4.160.

The TVOC concentration of the sample is calculated according to the following formula:

$$C_s = \frac{100 \times A}{A_T \times m_s}$$

A	:	Sample, C6–C16 chromatographic total peak area integration (area cts)
A_T	:	Toluene reference, 100 ng injection, chromatographic peak area integration (area cts/100 ng)
m_s	:	Sample volume (mg)
C_s	:	TVOC concentration (FOG value) in the sample (ng/mg).

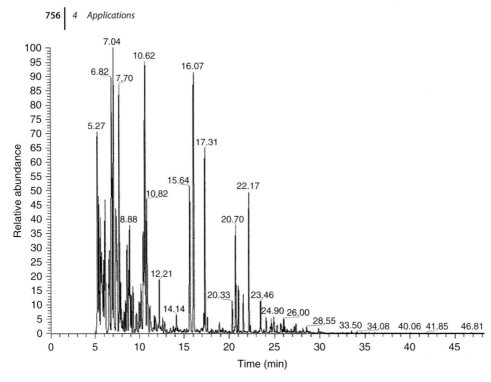

Figure 4.159 A leather sample TD-GC-MS analysis (total ion current).

Analysis Conditions

Sample material			All vehicle interior materials, polymers, leather, rubber
Sample preparation	Thermo Desorber	Type	Markes Ltd. UNITY 2
		Desorption tubes	Empty glass tube ID 4 mm, length 9 cm, with glass wool plug for calibration
		Carrier gas	Helium
		Transfer line	200 °C
		Focusing trap	Material emissions
	Desorption	Pre-purge	1 min
		Desorption temperature/time	90 °C for 30 min (VOCs) 120 °C for 60 min (SVOCs) 300 °C for 10 min (calibration, standards)
		Desorption flow	50 mL/min trap flow, splitless
	Cold trap focussing	Initial temperature	−30 °C
		Heating rate	100 °C/s
		Desorption temperature	300 °C

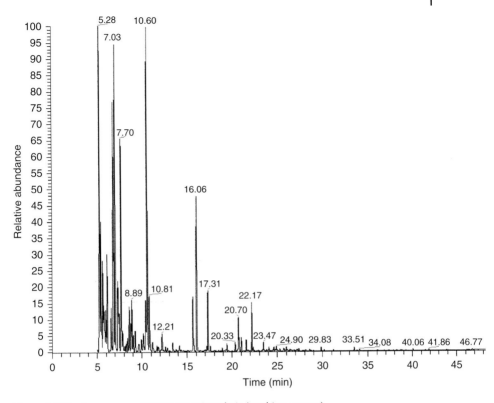

Figure 4.160 A sponge sample TD-GC-MS analysis (total ion current).

(*Continued*)

		Hold time	3 min (VOCs)
			5 min (SVOCs)
			10 min (standards)
		Pre-purge	3 min, bypass 20 mL/min
		Cold trap desorption flow	30 mL/min trap flow
		Split flow	20 mL/min
GC method	System	Thermo Scientific	TRACE 1300 GC
	Column	Type	TG-5MS, 5% phenyl methyl siloxane
		Dimensions	60 m length × 0.25 mm ID × 0.25 μm film thickness
	Carrier gas		Helium
	Flow		Constant flow, 1.3 mL/min
	SSL Injector	Injection mode	Splitless
		Injection Temperature	280 °C
	Oven program	Start	40 °C, 2.0 min
		Ramp 1	20 °C/min to 80 °C, 2 min

(*Continued*)

		Ramp 2	10 °C/min to 160 °C, 5 min
		Ramp 3	20 °C/min to 320 °C
		Final temperature	310 °C, 15.0 min
	Transferline	Temperature	280 °C
MS Method	System		Thermo Scientific ISQ
	Analyzer type		Single quadrupole MS
	Ionization		EI, 70 eV
	Ion source	Temperature	280 °C
	Acquisition	Mode	Full scan
	Mass range		33–350 Da
	Scan rate		300 ms/scan

Conclusion

This application demonstrates the analysis of VOCs and SVOCs in automotive interior materials thermal desorption GC-MS using a single quadrupole MS. The described analytical method follows the international recognized method VDA 278 for the analysis of volatiles for the automotive industry. The sample preparation using thermal desorption glass tubes is simple. The measurements can be run fully automated providing high sensitivity, precision and a wide linear range. This method setup provides the standard analytical solution for testing automotive interior materials for VOC and SVOC/FOG data.

References

Section 4.1 Air Analysis According to EPA Method TO-14

Madden, A.L. (1994) Analysis of Air Samples for the Polar and Non-Polar VOCs Using a Modified Method TO-14. Tekmar Application Report Vol. 5.3, Cincinnati, OH.

Schnute, B. and McMillan, J. (1993) TO-14 Air Analysis Using the Finnigan MAT Magnum Air System. Application Report No. 230.

Section 4.2 BTEX in Surface Water as of EPA Method 8260

Butler, J. (2013) *Analysis of BTEX in Wastewater by ISQ GC-MS*, Thermo Fisher Scientific, Austin, TX, personal communication.

EPA (2006) Method 8260C, Volatile Organic Compounds by Gas Chromatography/Mass Spectrometry (GC-MS), Revision 3 August 2006.

OI Analytical (2010) Proper Trap Selection for the OI Analytical Model 4560 and 4660 Purge-and-Trap Sample Concentrators. Application Note 12861111, OI Analytical, College Station, TX.

Section 4.4 Static Headspace Analysis of Volatile Priority Pollutants

Belouschek, P., Brand, H., and Lönz, P. (1992) Bestimmung von chlorierten Kohlenwasserstoffen mit kombinierter Headspace- und GC-MS-Technik. *Vom Wasser*, **79**, 3–8.

Rinne, D. (1986) Direkter Vergleich der Headspace-GC- Technik zum Extraktionsverfahren von leichtflüchtigen

Halogen-Kohlenwasserstoffen, Gewässer-
schutz, Wasser, Abwasser 88, 291–325.

Section 4.6 irm-GC/MS of Volatile Organic Compounds Using Purge and Trap Extraction

Schmidt, T.C., Zwank, L., Elsner, M., Berg,
M., Meckenstock, R. U., and Haderlein, S.
B. (2004) Compound-Specific Stable Iso-
tope Analysis of Organic Contaminants in
Natural Environments: A Critical Review
of the State of the Art, Prospects, and
Future Challenges, *Anal. Bioanal. Chem.*,
378, 283–300.
Zwank, L. and Berg, M. (2013) Enhanced
Method Detection Limits for irm-GC-MS
of Volatile Organic Compounds Using
Purge and Trap Extraction., Application
Note: 30053, Thermo Electron.
Zwank, L. *et al.* (2003) Compound specific
carbon isotope analysis of volatile organic
compounds in the low)μg/L-range. *Anal.
Chem.*, **75**, 5575–5583.

Section 4.7 Geosmin and Methylisobor-neol in Drinking Water

Chang, J., Biniakewitz, R. and Harkey, G.
(2007) Determination of Geosmin and
2-MIB in Drinking Water by SPME-PTV-
GC-MS. Application Note AN10213,
Thermo Fisher Scientific, San Jose, CA.
McMillan, J. (1994) Analysis of 2-
Methylisoborneol and Geosmin Using
Purge and Trap on the Finnigan MAT
MAGNUM GC-MS. Finnigan MAT Environmental Analysis Application Report No.
232, Finnigan MAT, San Jose.
Preti, G., Gittelman, T.S. *et al.* (1993) Letting
the nose lead the way – malodorous components in drinking water. *Anal. Chem.*,
65, 699A–702A.

Section 4.8 Polycyclic Musks in Waste Water

Eschke, H.-D., Traud, J., and Dibowski,
H.-J. (1994a) Analytik und Befunde
künstlicher Nitromoschus-Substanzen in
Oberflächen- und Abwässern sowie Fis-
chen aus dem Einzugsgebiet der Ruhr.
Vom Wasser, **83**, 373–383.

Eschke, H.-D., Traud, J., and Dibowski, H.-J.
(1994b) Untersuchungen zum Vorkommen
polycyclischer Moschus-Duftstoffe in ver-
schiedenen Umweltkompartimenten. *Z.
Umweltchem. Ökotox.*, **6**, 183–189.
Rimkus, G. and Wolf, M. (1994) Analysis and
bioaccumulation of nitro musks in aquatic
and marine biota, in *16th International
Symposium on Capillary Chromatography,
Riva del Garda, September 27–30, 1994*
(eds P. Sandra and G. Devos), Hüthig,
Heidelberg, pp. 433–445.

Section 4.9 Organotin Compounds in Water

Butler, J. and Phillips, E. (2009) Analysis of
Organotins by LVI GC-MS SIM, Application Note 10305, Thermo Fisher Scientific,
Austin, TX.
European Commission (2008) DIRECTIVE
2008/105/EC OF THE EUROPEAN PAR-
LIAMENT AND OF THE COUNCIL
of 16 December 2008 on environmental
quality standards in the field of water
policy, amending and subsequently repealing Council Directives 82/176/EEC,
83/513/EEC, 84/156/EEC, 84/491/EEC,
86/280/EEC and amending Directive
2000/60/EC of the European Parliament
and of the Council. *Off. J. Eur. Union*, **L
348**, 84.
Morabito, R. and Quevauviller, P. (2002)
Spectrosc. Eur., **4**, 18.
OSPAR Commission (2011) Background
Document on Organic Tin Compounds,
OSPAR Commission, Publication Number:
535/2011, ISBN: 978-1-907390-76-0.
Ostrakhovitch, E. and Cherian, M. (2007)
Tin, in Handbook on the Toxicology of
Metals, 3rd edn (eds G.F. Nordberg, B.A.
Fowler, M. Nordberg, and L. Friberg),
Elsevier, 839–854.
Takeuchi, M., Mizuishi, K., and Hobo, T.
(2000) *Anal. Sci.*, **16**, 349.
Thermo Scientific (2014) Hydrocar-
bon Soluble Siliconizing Fluid,
*http://www.thermoscientific.com/en/
product/hydrocarbon-soluble-siliconizing-
fluid.html* (accessed December 2013).

Section 4.10 Multi-Method for Pesticides by Single Quadrupole MS

Audi, G. and Wapstra, A.H. (1995) The 1995 update to the atomic mass evaluation. *Nuclear Phys. A*, **595**, 409–480.

Fillion, J., Sauvé, F., and Selwyn, J. (2000) Multiresidue method for the determination of residues of 251 pesticides in fruits and vegetables by gas Chromatography/mass spectrometry and liquid chromatography with fluorescence detection. *J. AOAC Int.*, **83**, 698–713.

Health and Welfare Canada (2000) *Analytical Methods for Pesticide Residues in Food*, Health Canada, Ottawa.

Peter, J. and Taylor, B.N. (1999) CODATA recommended values of the fundamental physical constants. *J. Phys. Chem. Ref. Data*, **28** (6), 1713–1852, and references cited therein.

Section 4.11 Analysis of Dithiocarbamate Pesticides

Amvrazi, E.G. (2005) in Fate of Pesticide Residues on Raw Agricultural Crops after Postharvest Storage and Food Processing to Edible Portions, Pesticides – Formulations, Effects, Fate (ed M. Stoytcheva), ISBN: 978-953-307-532-7.

CRL (2005) Analysis of Dithiocarbamate Residues in Foods of Plant Origin involving Cleavage into Carbon Disulfide, Partitioning into Isooctane and Determinative Analysis by GC-ECD, *http://www.crl-pesticides.eu/library/docs/srm/meth_DithiocarbamatesCs2_EurlSrm.PDF* (accessed 2 November 2014).

Crnogorac, G. and Schwack, W. (2009) Residue analysis of dithiocarbamate fungicides. *Trends Anal. Chem.*, **28** (1).

Dasgupta, S., Mujawar, S. *et al.* (2013) Analysis of Dithiocarbamate Pesticides by GC-MS, Thermo Fisher Scientific Application Note AN10333.

Reynolds, S. (2006) Analysis of dithiocarbamates. Presented at the SELAMAT Workshop, Bangkok, Thailand, 2006.

Section 4.12 GC-MS/MS Target Compound Analysis of Pesticide Residues in Difficult Matrices

Edwards, J., Fannin, S.T., Klein, D. and Steinmetz, G. (1998) The Analysis of Pesticide Residue Compounds in Biological Matrices by GC-MS. TR 9137 Technical Report, Finnigan Corp., Austin, TX, 7/1998.

Gummersbach, J. (2011) Thermo Fisher Scientific Applications Laboratory, Dreieich, Germany, personal communication.

Sheridan, R.S. and Meola, J.R. (1999) Analysis of pesticide residue in fruits, vegetables and milk by gas chromatography/tandem mass spectrometry. *AOAC Int.*, **82** (4), 982–990.

Van den Berg, M. *et al.* (2006) The 2005 World Health Organization Re-evaluation of human and mammalian toxic equivalency factors for dioxins and dioxin-like compounds, 2005 WHO Re-evaluation of TEFs. *Tox. Sci.*, **93** (2), 223–241.

Section 4.13 Multi-component Pesticide Analysis by MS/MS

Anastassiades, M., Lehotay, S.J., Stajnbaher, D., and Schenck, F.J. (2003) A fast and easy multiresidue method employing acetonitrile extraction/partitioning with dispersive solid-phase extraction for the determination of pesticide residues in produce. *J. AOAC Int.*, **86**, 412.

Hollosi, L., Bousova, K., and Godula, M. (2013) *Validation of the Method for Determination of Pesticide Residues by Gas Chromatography – Triple-Stage Quadrupole Mass Spectrometry, Thermo Fisher Scientific Method 63899*. Food Safety Response Center, Dreieich, Germany.

Okihashi, M., Kitagawa, Y., Akutsu, K., Obana, H., and Tanaka, Y. (2005) Rapid method for the determination of 180 pesticide residues in foods by gas chromatography/mass spectrometry and flame photometric detection. *J. Pestic. Sci.*, **30** (4), 368–377.

QuEChERS *www.quechers.com* (accessed 12 May 2014).

The Japanese Ministry of
Health, Labour and Welfare
*http://www.mhlw.go.jp/english/topics/
foodsafety/positivelist060228/index.html*
(English), *http://www.mhlw.go.jp/topics/
bukyoku/iyaku/syoku-anzen/zanryu2/
index.html* (Japanese) (accessed 12 May
2014).

Takatori, S., Okihashi, M., Kitagawa, Y.,
Fukui, N., Kakimoto-Okamoto, Y., Obana,
H. (2011) Pesticides – Strategies for Pesticides Analysis. Ed. M Stoytcheva, Intech,
197–214.

Section 4.14 Multiresidue Pesticides Analysis in Ayurvedic Churna

Lohar, D.R. Protocol for Testing Guideline for Ayurvedic, Siddha and Unani
Medicines, Chapter 2.5.1, Government of India, Department of AYUSH,
Ministry of Health & Family Welfare, Pharmacopoeial Laboratory
for Indian Medicines, Ghaziabad,
*http://www.plimism.nic.in/Protocol_For_
Testing.pdf* (accessed 2 November 2014).

Narayanaswamy, V. (1981) Origin and development of ayurveda (a brief history). *Anc.
Sci. Life*, **1** (1), 1–7.

Surwade, M., Kumar, S.T., Karkhanis,
A., Kumar, M., Dasgupta, S. and
Hübschmann, H.-J. (2013) Analysis
of Multiresidue Pesticides Present in
Ayurvedic Churna by GC-MS/MS, Thermo
Fisher Scientific Application Note AN
10361.

The Pesticides Compound Database (2014)
Thermo Fisher Scientific, Austin, TX.

Section 4.15 Determination of Polar Aromatic Amines by SPME

Pan, L. and Pawliszyn, J. (1997)
Derivatization/solid-phase microextrac-
tion: New approach to polar analytes, *J.
Anal. Chem.*, **69**, 196–205.

Zimmermann, T., Ensinger, W.J., and
Schmidt, T.C. (2004) In situ
derivatisation/solid-phase microextraction: Determination of polar aromatic
amines. *Anal. Chem.*, **76**, 1028–1038.

Section 4.16 Analysis of Nitrosamines in Beer

Agency for Toxic Substances & Disease
Registry (1989) Public Health Statement for n-Nitrosodimethylamine,
*http://www.atsdr.cdc.gov/toxprofiles/
phs141.html* (accessed 2 November 2014).

AOAC (2000) AOAC Official Method 982.11.

Boyd, R.K., Basic, C., and Bethem, R.A.
(2008) Trace Quantitative Analysis by Mass
Spectrometry, John Wiley & Sons, Ltd.

Chen, A., Hübschmann, H.-J., Chan, S.H.,
Li, F. and Chew, Y.F. (2013) High Sen-
sitivity Analysis of Nitrosamines Using
GC-MS/MS, Thermo Scientific Application
Note AN10315.

March, R.E. and Hughes, R.J. (1989)
Quadrupole Storage Mass Spectrometry,
2nd edn, John Wiley & Sons, Ltd.

Munch, J.W. and Bassett, M.V. (2004)
Method 521: Determination of
Nitrosamines in Drinking Water by Solid
Phase Extraction and Capillary Column
Gas Chromatography with Large Volume
Injection and Chemical Ionization Tandem
Mass Spectrometry (MS/MS) (Version 1.0),
U.S. Environmental Protection Agency.

Thermo Fisher Scientific (2012a) Introducing
AutoSRM, Thermo Fisher Scientific, Technical Brief No. AB52298.

Thermo Fisher Scientific (2012b) Thermo
Scientific TSQ 8000 Triple Quadrupole
GC-MS/MS Instrument Method, Thermo
Fisher Scientific, Technical Brief No.
AB52299.

Section 4.17 Phthalates in Liquors

Dongliang, S. (2010) Determination of
phthalate ester residues in white spirit by
GC-MS. *Chemical Analysis and Meterage*,
19 (6), 33–35.

Lv, J., Liang, L. and Hübschmann, H.-J.
(2013) Determination of Phthalates in
Liquor Beverages by Single Quadrupole
GC-MS. Application Note AN10339,
Thermo Fisher Scientific.

Standardization Administration of China
China method GB/T 21911-2008 (2008)
Determination of Phthalates in Food,
Standardization Administration of China.

Section 4.19 Aroma Profiling of Cheese by Thermal Extraction

Markes Int. Ltd. (2010) Food Decomposition Analysis Using the Micro-Chamber/Thermal Extractor and TD-GC-MS, Application Note TDTS 95, Markes Int. Ltd.

Markes Int. Ltd. (2012) Rapid Aroma Profiling of Cheese Using a Micro-Chamber/Thermal Extractor with TD–GC-MS Analysis, Application Note TDTS 101, Markes Int. Ltd.

Section 4.20 48 Allergens

European Commission (2009) Directive 2009/48/EC of the European Parliament and of the Council of 18 June 2009 on the safety of toys. *Off. J. Eur. Union*, **L170**, 1–37.

EU Scientific Committee on Consumer Safety (2011) Opinion on Fragrance Allergens in Cosmetic Products, SCCS/1459/11, ISSN: 1831-4767.

Hopfe, W. (2009) Scented Toys – Analysing Allergenic Substances, Fritsch GmbH Application Report.

Lv, Q., Zhang, Q. *et al.* (2013) Determination of 48 fragrance allergens in toys using GC with ion-trap MS/MS. *J. Sep. Sci.* (online publication ahead of print).

Section 4.21 Analysis of Azo Dyes in Leather and Textiles

European Commission (2002) Directive 2002/61/EC of the European Parliament and of the Council of 19 July 2002 amending for the nineteenth time Council Directive 76/769/EEC relating to restrictions on the marketing and use of certain dangerous substances and preparations (azocolourants) of 19 July 2002 amending for the nineteenth time Council Directive 76/769/EEC relating to restrictions on the marketing and use of certain dangerous substances and preparations (azocolourants). *Off. J.*, **L243**, 0015–0018.

European Commission – Health and Consumers Scientific Committees Opinion (1999) Brussels, *http://ec.europa.eu/health/scientific_committees/environmental_risks/opinions/sctee/sct_out27_en.htm* (accessed 18 January 1999).

ISO EN ISO 14362-1. (2014) *Textiles – Methods for the Determination of Certain Aromatic Amines Derived from Azo Colorants – Part 1: Detection of the Use of Certain Azo Colorants Accessible without Extraction*, ISO.

Purwanto, A. and Chen, A. (2013) Detection, Identification and Quantitation of Azo Dyes in Leather and Textiles by GC-MS, Thermo Fisher Scientific Application Note.

Section 4.22 Identification of Extractables and Leachables

HighChem (2014) Mass Frontier Operation, *http://www.highchem.com/* (accessed 2 November 2014).

Lewis, D.B. (2011) Current FDA perspective on leachable impurities in parenteral and ophthalmic drug products. Presentation at the AAPS Workshop on Pharmaceutical Stability, Washington, DC, 2011.

Mallard, W.G. and Reed, J. (1997) AMDIS – User Guide, U.S. Department of Commerce, Technology Administration, National Institute of Standards and Technology (NIST), Standard Reference Data Program, Gaithersburg, MD.

Moffat, F. (2011) Extractables and Leachables in Pharma – A Serious Issue. Solvias Whitepaper, Solvias AG, Kaiseraugst.

Section 4.23 Metabolite Profiling of Natural Volatiles and Extracts

Fragner, L., Weckwerth, W. and Hübschmann, H.-J. (2010) Metabolomics Strategies Using GC-MS/MS Technology, Application Note AN 51999, Thermo Fisher Scientific.

Hübschmann, H.-J., Fragner, L., Weckwerth, W., Cardona, D. (2012) Metabolomics strategies using GC-MS/MS technology. Poster Metabolomics Conference, 2012.

Mallard, W.G. and Reed, J. (1997) *AMDIS User Guide*, National Institute of Standards and Technology (NIST).

Weckwerth, W., Wenzel, K., and Fiehn, O. (2004) Process for the integrated extraction identification and quantification of metabolites, proteins and RNA to reveal their co-regulation in biochemical networks. *Proteomics*, **4** (1), 78–83.

Section 4.24 Fast GC Quantification of 16 EC Priority PAH Components

Kleinhenz, S., Jira, W., and Schwind, K.-H. (2006) Dioxin and polychlorinated biphenyl analysis: automation and improvement of clean-up established by example of spices. *Mol. Nutr. Food Res.*, **50** (4–5), 362–367.

European Commission (2005) COMMISSION REGULATION (EC) No 208/2005 of 4 February 2005 amending Regulation (EC) No 466/2001 as regards polycyclic aromatic hydrocarbons. **L 34**, 3–5.

European Commission (2005) COMMISSION DIRECTIVE 10/2005/EC of 4 February 2005 laying down the sampling methods and the methods of analysis for the official control of the levels of benzo(a)pyrene in foodstuffs. L34, 15–34.

The joint FAO/WHO Expert Committee (2005) Summary and Conclusion of the Joint FAO/WHO Expert Committee on Food Additives, Sixty-Fourth Meeting, Rome, February 8 – 17, 2005, JCEFA/64/SC.

Ziegenhals, K. and Jira, W. (2006) Bestimmung der von der EU als prioritär eingestuften polyzyklischen aromatischen Kohlenwasserstoffe (PAK) in Lebensmitteln, Kulmbach Kolloquium, September, 2006.

Ziegenhals, K. and Jira, W. (2007) High sensitive PAH method to comply with the new EU directives. Presentation at the European High Resolution GC-MS Users Meeting, Venice, Italy, March 23–24, 2007.

Ziegenhals, K., Speer, K., Hübschmann, H.-J., and Jira, W. (2008) Fast GC/HRMS to quantify the EU priority PAH. *J. Sep. Sci.* **31** (10): 1779–86.

Section 4.25 Multiclass Environmental Contaminants in Fish

EFSA (2005) Opinion of the scientific panel on contaminants in the food chain on a request from the European parliament related to the safety assessment of wild and farmed fish. *EFSA J.*, **236**, 1–118.

European Commission (2011a) Commission regulation (EU) No 1259/2011 of 2 December 2011 amending Regulation No 1881/2006 as regards maximum levels for dioxins, dioxin-like PCBs and non dioxin-like PCBs in foodstuffs. *Off. J. Eur. Union*, **320**, 18–23.

European Commission (2011b) Commission regulation (EU) No 835/2011 of 19 August 2011 amending Regulation (EC) No 1881/2006 as regards maximum levels of polycyclic aromatic hydrocarbons in foodstuffs. *Off. J. Eur. Union*, **215**, 4–8.

European Commission (2012) Document no SANCO/12495/2011: Method Validation and Quality Control Procedures for Pesticides Residues Analysis in Food and Feed, Implemented by 01/01/2012, *http://ec.europa.eu/food/plant/protection/pesticides/docs/qualcontrol_en.pdf* (accessed 2 November 2014).

Kalachova, K., Pulkrabova, J., Drabova, L., Cajka, T., Kocourek, V., and Hajslova, J. (2011) Simplified and rapid determination of polychlorinated biphenyls, polybrominated diphenyl ethers and polycyclic aromatic hydrocarbons in fish and shrimps integrated into a single method. *Anal. Chim. Acta*, **707**, 84–91.

Kalachova, K., Pulkrabova, J., Cajka, T., Drabova, L., Stupak, M., and Hajslova, J. (2013) Gas chromatography–triple quadrupole tandem mass spectrometry: a powerful tool for the (ultra)trace analysis of multiclass environmental contaminants in fish and fish feed. *Anal. Bioanal. Chem.*, **405**, 7803–7815.

Krumwiede, D., Griep-Raming, J. and Muenster, H. (2005) Comparative studies of PTV on-column like injection for improved sensitivity in GC-MS analysis of thermolabile high boiling brominated flame retardants, Thermo Electron, Bremen, Germany, Poster, ASMS Conference, 2005.

Leonard, B. (ed) (2011) Chapter 9: Environmental chemical contaminants and pesticides, in Fish and Fishery Products: Hazards and Controls Guidance, 4th edn, DIANE Publishing, pp. 155–180.

Thomas, J., Dorman, F., Cochran, J., Wittrig, M., Rhoads, D., de Zeeuw, J. and Stidsen, G. (2010) Analysis of Polyaromatic Hydrocarbons Using Next Generation Highly Selective Stationary Phases, Restek Corporation, Bellefonte PA, Poster, ISCC Riva del Garda.

Section 4.26 Fast GC of PCBs

de Boer, J., Dao, Q., and van Dortmond, R. (1992) Retention times of fifty one chlorobiphenyl congeners on seven narrow bore capillary columns coated with different stationary phases. *J. High Resolut. Chromatogr.*, **15**, 249–255.

Bøwadt, S. and Larsen, B. (2005) Rapid screening of chlorobiphenyl congeners by GC-ECD on a carborane – polydimethylsiloxane copolymer. *J. High Resolut. Chromatogr.*, **15** (5), 350–351.

Gummersbach, J. (2011), Thermo Fisher Scientific, Application Laboratory Dreieich, Germany, personal communication.

SGE (2005) HT8: The perfect PCB Column, Publication No. AP-0040-C Rev:04 08/05.

Section 4.27 Congener Specific Isotope Analysis of Technical PCB Mixtures

Horii, Y., Kannan, K., Petrick, G., Gamo, T., Falandysz, J., and Yamashita, N. (2005) Congener-specific carbon isotopic analysis of technical PCB and PCN mixtures using two-dimensional gas chromatography-isotope ratio mass spectrometry. *Environ. Sci. Technol.*, **39**, 4206–4212.

Horii, Y., van Bavel, B., Kannan, K., Petrick, G., Nachtigall, K., and Yamashita, N. (2008) Novel Evidence for Natural Formation of Dioxins in Ball Clay. *Chemosphere*, **70**, 1280–1289.

Section 4.28 Dioxin Screening in Food and Feed

European Commission (2006) COMMISSION REGULATION (EC) No 1881/2006 of 19 December 2006 setting maximum levels for certain contaminants in foodstuffs. *Off. J.*, **L 364**, 5–24.

European Commission (2012a) COMMISSION REGULATION (EC) No 278/2012 of 28 March 2012 amending Regulation (EC) No 152/2009 as regards the determination of the levels of dioxins and polychlorinated biphenyls. *Off. J.*, **L 91**, 8–22.

European Commission (2012b) COMMISSION REGULATION (EC) No 252/2012 of 21 March 2012 laying down methods of sampling and analysis for the official control of levels of dioxins and dioxin-like PCBs in certain foodstuffs and repealing Regulation (EC) No 1883/2006. *Off. J.*, **L 84**, 1–22.

European Commission (2014) COMMISSION REGULATION (EU) No 709/2014 of 20 June 2014 amending Regulation (EC) No 152/2009 as regards the determination of the levels of dioxins and polychlorinated biphenyls. *Off. J.*, **L 888**, 1–18.

Kotz, A., Malisch, R., Wahl, K., Bitomsky, N., Adamovic, K., Gerteisen, I., Leswal, S., Schächtele, J., Tritschler, R., and Winterhalter, H. (2011) *Organohalogen Compd.*, **73**, 688–691.

Kotz, A., Malisch, R., Focant, F., Eppe, G., Cederberg, T.L., Rantakokko, P., Fürst, P., Bernsmann, T., Leondiadis, L., Lovász, C., Scortichini, G., Diletti, G., di Domenico, A., Ingelido, A.M., Traag, W., Smith, F., and Fernandes, A. (2012) Analytical criteria for use of MS/MS for determination of dioxins and dioxin-like PCBs in feed and food. *Organohalogen Compd.*, **74**, 156–159.

US EPA (1994) Tetra- Through Octa-Chlorinated Dioxins and Furans by Isotope Dilution HRGC-HRMS, Method 1613, Rev. B, U.S. Environmental Protection Agency, Washington, DC.

Section 4.29 Confirmation Analysis of Dioxins and Dioxin-like PCBs

Aylward, L.L. and Hays, S.M. (2002) Temporal trends in human TCDD body burden: Decreases over three decades and implications for exposure levels. *J. Expo. Anal. Environ. Epidemiol.*, **12**, 319–328.

Centers for Disease Control and Prevention. Fourth Report on Human Exposure to Environmental Chemicals (2009) U.S. Department of Health and Human Services, Centers for Disease Control and Prevention: Atlanta, GA. *http://www.cdc.gov/exposurereport/*

Centers for Disease Control and Prevention. Fourth Report on Human Exposure to Environmental Chemicals, Updated Tables (2014) U.S. Department of Health and Human Services, Centers for Disease Control and Prevention: Atlanta, GA. *http://www.cdc.gov/exposurereport/*

Fishman, V.N., Martin, G.D., and Lamparski, L.L. (2007) Comparison of a variety of gas chromatographic columns with different polarities for the separation of chlorinated dibenzo-p-dioxins and dibenzofurans by high-resolution mass spectrometry, *J. Chromatogr. A*, **1139**, 285–300.

Hays, S.M. and Aylward, L.L. (2003) Dioxin risks in perspective: Past, present and future. *Regul. Toxicol. Pharmacol.*, **37** (2), 202–217.

Krumwiede, D. and Hübschmann, H.-J. (2006) Confirmation of Low Level Dioxins and Furans in Dirty Matrix Samples using High Resolution GC-MS. Application Note AN30112, Thermo Fisher Scientific, Bremen.

Lorber, M. (2002) A pharmacokinetic model for estimating exposure of Americans to dioxin-like compounds in the past, present and future. *Sci. Total Environ.*, **288**, 81–95.

Patterson, D.G., Canady, R., Wong, L.-Y., Lee, R., Turner, W., Caudill, S., Needham, L., and Henderson, A. (2004) Age specific dioxin TEQ reference range. *Organohalogen Compd.*, **66**, 2878–2883.

Patterson, D.G., Jr., Welch, S.M., Focant, J.-F. and Turner, W.E. (2006) The use of various gas chromatography and mass spectrometry techniques for human biomonitoring studies. Lecture at the 26th International Symposium on Halogenated Persistent Organic Pollutants, Oslo, Norway, August 21–25, 2006.

Turner, W., Welch, S., DiPietro, E., Cash, T., McClure, C., Needham, L., and Patterson, D. (2004) The phantom menace – determination of the true Method Detection Limit (MDL) for background

levels of PCDDs, PCDFs and cPCBs in human serum by high-resolution mass spectrometry. *Organohalogen Compd.*, **66**, 264–271.

Turner, W.E., Welch, S.M., DiPietro, E.S., Whitfield, W.E., Cash, T.P., McClure, P.C., Needham, L.L. and Patterson, D.G. Jr, (2006) Instrumental approaches for improving the detection limit for selected PCDD congeners in samples from the general U.S. Population as background levels continue to decline. Poster at the 26th International Symposium on Halogenated Persistent Organic Pollutants, Oslo, Norway, August 21–25, 2006.

US EPA (1994) *Tetra-through Octa-Chlorinated Dioxins and Furans by Isotope Dilution HRGC-HRMS*, U.S. Environmental Protection Agency Office of Water Engineering and Analysis Division, Washington, DC, EPA Method 1613 Rev.B, October 1994.

US EPA (1998a) Polychlorinated Dibenzo-p-dioxins and Polychlorinated Dibenzofurans by High Resolution Gas Chromatography-Low Resolution Mass Spectrometry (HRGC-LRMS), US EPA Method 8280B, Rev. 2, January 1998.

US EPA (1998b) Polychlorinated Diebnzo-p-dioxins and Polychlorinated Dibenzofurans by High Resolution Gas Chromatography-High Resolution Mass Spectrometry (HRGC-HRMS), US EPA Method 8290B, Rev. 1, January 1998.

US EPA (2003) Table of PCB Species by Congener Number, online, *www.epa.gov/osw/hazard/tsd/pcbs/pubs/congenertable.pdf* (accessed Oct 2014).

Van den Berg, M.L. *et al.* (1998) Toxic equivalent factors (TEFs) for PCBs, PCDDs, PCDFs for humans and wildlife. *Environ. Health Perspect.*, **106** (12), 775–792.

Van den Berg, M. *et al.* (2006) The 2005 World Health Organization re-evaluation of human and mammalian toxic equivalency factors for dioxins and dioxin-like compounds, 2005 WHO re-evaluation of TEFs. *Tox. Sci.*, **93** (2), 223–241.

WHO (2014) Dioxins and Their Effects on Human Health, Fact sheet, **255**, 1–4 *http://www.who.int/mediacentre/factsheets/fs225/en/* (accessed Oct 2014).

Section 4.30 Analysis of Brominated Flame Retardants PBDE

EU (2003) Directive 2002/95/EC of the European Parliament and of the Council of the European Union, January 27, 2003

EU (2003) Directive 2003/11/EC of The European Parliament and of The Council of 6 February 2003 amending for the 24th time Council Directive 76/769/EEC relating to restrictions on the marketing and use of certain dangerous substances and preparations (pentabromodiphenyl ether, octabromodiphenyl ether)

Krumwiede, D. and Hübschmann, H.-J. (2006) Trace Analysis of Brominated Flame Retardants with High Resolution Mass Spectrometry, LCGC Europe Supplement, July 2006.

Krumwiede, D. and Hübschmann, H.-J. (2007) High resolution GC-MS as a viable solution for conducting environmental analyses. *Peak*, **11**, 7–15.

Krumwiede, D. and Hübschmann, H.-J. (2008) DFS – Analysis of Brominated Flame Retardants with High Resolution Mass Spectrometry. Application Note 30098, Thermo Fisher Scientific, Bremen, Germany.

Stapleton, H.M., Kelly, S.M., Allen, J.G., McClean, M.D., and Webster, T.F. (2008) Measurement of polybrominated diphenyl ethers on hand wipes: Estimating exposure from hand-to-mouth contact. *Environ. Sci. Technol.*, **42** (9), 3329–3334.

Section 4.31 SPME Analysis of PBBs

De Boer, J., Allchin, C., Law, R.J., Zegers, B., and Boon, J.P. (2001) Methods for the analysis of polybrominated diphenylethers in sediments and biota, *Trends Anal. Chem.*, **10**, 591–599.

Krumwiede, D. and Hübschmann, H.-J. (2006) Analysis of Brominated Flame Retardants by High Resolution GC-MS. Thermo Scientific Application Note 30098, Thermo Scientific, Bremen.

Polo, M., Gomez-Noya, G., Quintana, J.B., Llompart, M., Garcia-Jares, C., and Cela, R. (2004) Development of a solid-phase microextraction gas chromatography/ tandem mass spectrometry method for polybrominated diphenyl ethers and polybrominated biphenyls in water samples. *Anal. Chem.*, **76**, 1054–1062.

Section 4.32 Analysis of Military Waste

Deutsche Forschungsgemeinschaft (2014) *MAK- und BAT-Werte-Liste 2014: Maximale Arbeitsplatzkonzentrationen und Biologische Arbeitsstofftoleranzwerte*, Wiley-VCH, Weinheim, Germany.

Kuitunen, M.L., Hartonen, K., and Riekkola, M.L. (1991) Analysis of chemical warfare agents from soil samples using off-line supercritical fluid extraction and capillary gas chromatography, in *13th International Symposium on Capillary Chromatography, Riva del Garda May 1991* (ed P. Sandra), Huethig, Heidelberg, pp. 479–488.

Yinon, J. and Zitrin, S. (1993) *Modern Methods and Applications of Explosives*, John Wiley & Sons, Ltd, Chichester.

Section 4.33 Comprehensive Drug Screening and Quantitation

Maurer, H.H. (2007) Demands on scientific studies in clinical toxicology. *Forensic Sci. Int.*, **165**, 194.

Maurer, H.H. (2012) How can analytical diagnostics in clinical toxicology be successfully performed today? *Ther. Drug Monit.*, **34**, 561.

Maurer, H.H., Pfleger, K., and Weber, A.A. (2011) Mass Spectral and GC Data of Drugs, Poisons, Pesticides, Pollutants and their Metabolites, Wiley-VCH Verlag GmbH, Weinheim.

Meyer, M.R., Peters, F.T., and Maurer, H.H. (2010) Automated mass spectral deconvolution and identification system for GC-MS screening for drugs, poisons and metabolites in urine. *Clin. Chem.*, **56**, 575.

Meyer, G.M.J., Weber, A.A., and Maurer, H.A. (2014) Development and validation of a fast and simple multi-analyte procedure for quantification of 40 drugs relevant to emergency toxicology using GC-MS and one-point calibration. *Drug Test. Anal.*, **6** (5), 472–81. doi: 10.1002/dta.1555.

Section 4.34 Determination of THC-Carbonic Acid in Urine by NCI

Beck, F., Legleye, S., and Spilka, S. (2007) *Sante Publique*, **19**, 481.

Jones, A.W., Kugelberg, F.C., Holmgren, A., and Ahlner, J. (2009) *Forensic Sci. Int.*, **181**, 40.

Kapusta, N.D., Ramskogler, K., Hertling, I., Schmid, R., Dvorak, A., Walter, H., and Lesch, O.M. (2006) *Alcohol Alcohol.*, **41**, 188.

Levin, F.R. and Kleber, H.D. (2008) *Am. J. Addict.*, **17**, 161.

MARINOL® is a Registered Trademark of Unimed Pharmaceuticals, Inc.

Melchert, H.-U., Hübschmann, H.-J., and Pabel, E. (2009) Analytik der THC-Carbonsäure – Spezifische Detektion und hochsensitive Quantifizierung im Harn durch NCI-GC-MS. *Labo* Heft 1, S. 8–S. 12, *https://de.wikipedia.org/wiki/Datei:THC-COOH-Analytik.pdf* (accessed 2 November 2014).

Musshoff, F. and Madea, B. (2006) *Ther. Drug Monit.*, **28**, 155.

Section 4.35 Detection of Drugs in Hair

Martz, R., Donelly, B. *et al.* (1991) The use of hair analysis to document a cocaine overdose following a sustained survival period before death. *J. Anal. Toxicol.*, **15**, 279–281.

Möller, M.R., Fey, P., and Wennig, R. (1993) Simultaneous determination of drugs of abuse (opiates, cocaine and amphetamine) in human hair by GC-MS and its application to the methadon treatment program, in *Special Issue: Hair Analysis as a Diagnostic Tool for Drugs of Abuse Investigation*, Forensic Science International, vol. **63** (ed P. Saukko), Elsevier, Amsterdam, pp. 185–206.

Sachs, H. and Raff, I. (1993) Comparison of quantitative results of drugs in human hair by GC-MS, in *Forensic Science International*, vol. **63**, Special Issue: Hair Analysis as a Diagnostic Tool for Drugs of Abuse Investigation (ed P. Saukko), Elsevier, Amsterdam, pp. 207–216.

Sachs, H. and Uhl, M. (1992) Opiat-Nachweis in Haar-Extrakten mit Hilfe von GC-MS/MS und Supercritical Fluid Extraktion (SFE). *Toxichem + Krimichem*, **59**, 114–120.

Schwinn, W. (1992) Drogennachweis in Haaren. *Kriminalist*, **11**, 491–495.

Traldi, P., Favretto, D., and Tagliaro, F. (1991) Ion trap mass spectrometry, a new tool in the investigation of drugs of abuse in hair. *Forensic Sci. Int.*, **63**, 239–252.

Section 4.36 Screening for Drugs of Abuse

Donike, M. (1969) N-Methyl-N-Trimethylsilyl-Trifluoracetamid ein neues Silylierungsmittel aus der Reihe der silylierten Amide, *J. Chromatogr.*, **42**, 103–104.

Maurer, H.H. (1990) Identifizierung unbekannter Giftstoffe und ihrer Metaboliten in biologischem Material. *GIT Suppl.*, **1**, 3–10.

Maurer, H.H. (1993) GC-MS contra Immunoassay? Symposium Aktuelle Aspekte des Drogennachweises, Mosbach, Germany, April 15, 1993, Abstracts.

Musshoff, F., Trafkowski, J., and Madea, B. (2004) Validated assay for the determination of markers of illicit heroin in urine samples for the control of patients in a heroin prescription program. *J. Chromatogr. B*, **811**, 47–52.

Pfleger, K., Maurer, H.H., and Weber, A. (1992) *Mass Spectra and GC Data of Drugs, Poisons, Pesticides, Pollutants and Their Metabolites, Parts 1,2,3*, 2. erw. Aufl, Weinheim, VCH Publishers.

Smith, M., Vorce, S.P., Holler, J.M., Shimomura, E., Magluilo, J., Jacobs, A.J. (2007) Modern Instrumental Methods in Forensic Toxicology. *J. Anal. Tox.*, **315**, 237–53–8A–9A.

Weller, J.P., Wolf, M., and Szidat, S. (2000) Enhanced selectivity in the determination of Δ9-tetrahydrocannabinol and two major metabolites in serum using ion trap GC-MS/MS. *J. Anal. Toxicol.*, **24**, 1–6.

Section 4.37 Structural Elucidation by Chemical Ionization and MS/MS

Vale, G., Butler, J., Herbold, R., Hübschmann, H.-J. and O'Brian, P. (2011) Structural Elucidation of Alkylamines in Process Streams by PCI Ammonia GC-MS/MS, Application Note AN52255, Thermo Fisher Scientific.

Section 4.38 Volatile Compounds in Car Interior Materials

ASTM D2369 (2010) *Standard Test Method for Volatile Content of Coatings*, ASTM International, West Conshohocken, PA, doi: 10.1520/D2369-10E01, *www.astm.org* (accessed 2 November 2014).

California Department of Public Health (2010) Standard Method for the Testing and Evaluation of Volatile Organic Chemical Emissions from Indoor Sources Using Environmental Chambers, Version 1.1, February 2010.

European Commission (2010) Commission Directive 2010/79/EU of 19 November 2010 on the adaptation to technical progress of Annex III to Directive 2004/42/EC of the European Parliament and of the Council on the limitation of emissions of volatile organic compounds. *Off. J. Eur. Union*, **L304**, 18.

European Commission (2004) Directive 2004/42/CE of the European Parliament and of the Council of 21 April 2004 on the limitation of emissions of volatile organic compounds due to the use of organic solvents in certain paints and varnishes and vehicle refinishing products and amending Directive 1999/13/EC. *Off. J.*, **L143**, 87–96.

GM Engineering Standards GM15634. (2008) *Determination of Volatile and Semi-Volatile Organic Compounds from Vehicle Interior Materials*, Jan. 2008, General Motors Corporation.

Grabbs, J., Corsi, R., and Torres, V. (2000) Volatile organic compounds in new automobiles: screening assessment. *J. Environ. Eng.*, **126** (10), 974–977.

ISO ISO 11890-2:2013. (2013) *Paints and Varnishes – Determination of Volatile Organic Compound (VOC) Content – Part 2: Gas-Chromatographic Method*, ISO.

Lee, C. (2013) New Chinese Guidelines for Car Interior Air Quality to Come into Effect Tomorrow, *www.gasgoo.com* (accessed 28 February 2013).

VDA 278 (2011) Thermal Desorption Analysis of Organic Emissions for the Characterization of Non-Metallic Materials for Automobiles, October 2011, VDA Verband der Automobilindustrie, Germany, *www.vda.de* (accessed 2 November 2014).

Glossary

This glossary of chromatographic and mass spectrometric terms also contains selected terms of the third draft of recommendations for nomenclature, definitions of terms and acronyms in mass spectrometry, currently undergoing review for publication in the IUPAC Journal Pure and Applied Chemistry.

A

α	Separation factor of 2 adjacent peaks; $\alpha = k_2/k_1$.
α-Cleavage	Homolytic cleavage where the bond fission occurs between at the atom adjacent to the atom at the apparent charge site and an atom removed from the apparent charge site by two bonds.
A_p	Peak area.
a-Ion	Fragment ion containing the N-terminus formed upon dissociation of a protonated peptide at a backbone $C-C$ bond.
AC	Alternating current.
Accelerating voltage	Electrical potential used to impart translational energy to ions in a mass spectrometer.
Accelerator mass spectrometry (AMS)	Mass spectrometry technique in which atoms extracted from a sample are ionized, accelerated to megaelectron volt energies and separated according to their momentum, charge and energy.
Acceptable quality range	The interval between specified upper and lower limits of a sequence of values within which the values are considered to be satisfactory.
Acceptable value	An observed or corrected value that falls within the acceptable range.

Handbook of GC-MS: Fundamentals and Applications, Third Edition. Hans-Joachim Hübschmann.
© 2015 Wiley-VCH Verlag GmbH & Co. KGaA. Published 2015 by Wiley-VCH Verlag GmbH & Co. KGaA.

Accreditation	A formal recognition that a laboratory is competent to carry out specific tests or specific types of tests.
Accuracy	The closeness of agreement between a test result and the accepted reference value, it is determined by determining trueness and precision.
Accurate mass	Experimentally determined mass of an ion that is used to determine an elemental formula. Note: accurate mass and exact mass are not synonymous. The former refers to a measured mass and the latter to a calculated mass.
ACN	Acetonitrile.
Adiabatic ionization	Process whereby an electron is removed from an atom, ion or molecule in its lowest energy state to produce an ion in its lowest energy state.
Adduct ion	Ion formed by the interaction of an ion with one or more atoms or molecules to form an ion containing all the constituent atoms of the precursor ion as well as the additional atoms from the associated atoms or molecules.
AFS	Amperes full scale.
AGC	Automatic gain control, controls the variable ionization time in ion trap mass spectrometers depending from the total ion current in the selected mass range by a quick pre-scan. This control provides the high inherent full scan sensitivity of ion trap mass spectrometers at low ion streams by storing ions until the full trap capacity is reached.
Alumina	A gas–solid adsorbent stationary phase.
Aliquot	A subsample derived by a divisor that divides a sample into a number of equal parts and leaves no remainder; a subsample resulting from such a division. In analytical chemistry, the term *aliquot* is generally used to define any representative portion of the sample, regardless of whether a remainder is left or not.
AMDIS	Automated mass spectra deconvolution and identification system, analyses the individual ion signals and extracts and identifies the spectrum of each component in co-eluting peaks analysed by GC-MS → deconvolution.

Analogue ion	Ions that have similar chemical valence, for example, the acetyl cation CH_3CO^+ and the thioacetyl cation CH_3CS^+.
Analyte	The substance that has to be detected, identified and/or quantified and derivatives emerging during its analysis.
Analytical scan	The part of the ion trap MS scan function that produces the mass spectrum.
Analyte spike recovery	Recovery of an analyte spike added to a sample prior to sample preparation. Determination of spike recovery is based on results provided by spiked and unspiked sample. Used to estimate matrix effects and sample preparation losses → surrogate standard.
APCI	Atmospheric pressure chemical ionization, ionization mode in LC-MS.
APE	Atom percent excess, commonly used expression in tracer experiment employing labelled substances for the degree of enrichment above the natural isotope content: $APE = at.\% - at.\%_{nat}$ Deprecated term, replace SI conform with 'atom fraction' expressed as percent.
APGC	GC coupling method to an APCI interface.
Appearance energy (AE)	Minimum energy that must be imparted to an atom or molecule to produce a detectable amount of a specified ion. In mass spectrometry it is the voltage, which corresponds to the minimum electron energy necessary for the production of a given fragment ion. The term *appearance potential* (AP) is deprecated.
ARC	Automatic reaction control, variable ionization and reaction time control in internal ionization ion trap mass spectrometers for chemical ionization. A built-up of CI ions in the ion trap is facilitated by the AGC control until the maximum capacity or preset maximum reaction time is reached. This results in the inherent high sensitivity of the ion trap analyser for full scan data acquisition in full scan mode.
Aromagram	A chromatogram representative for the odour activity of a sample, → GC-olfactometry.

Array detector	Detector comprising several ion collection elements, arranged in a line or grid where each element is an individual detector.
Associative ion/ molecule reaction	Reaction of an ion with a neutral species in which the reactants combine to form a single ion.
Associative ionization	Ionization process in which two atoms or molecules, one or both of which is in an excited state, react to form a single positive ion and an electron.
Atmospheric pressure chemical ionization (APCI)	Chemical ionization in LC-MS that takes place using a nebulized liquid and atmospheric pressure corona discharge at atmospheric pressure before ions enter the vacuum of the MS.
Atmospheric pressure ionization (API)	Ionization process in which ions are formed in the gas phase at atmospheric pressure before ions enter the vacuum of the MS.
Atmospheric pressure matrix-assisted laser desorption/ionization (APMALDI)	Matrix-assisted laser desorption/ionization in which the sample target is at atmospheric pressure.
Atmospheric pressure photoionization (APPI)	Atmospheric pressure photo ionization in which the reactant ions are generated by photoionization before they enter the vacuum of the MS.
Atom% or at.%	Unit commonly used for the expression of isotope ratios, for example, tracer experiments. $$\text{at.\%} = \frac{n}{n_i} \cdot 100 = \frac{1}{(1+1/R)} \cdot 100 \ (\%)$$ with n = number of isotope atoms, $n_i = R = {}^{13}C/{}^{12}C$.
Autodetachment	Formation of a neutral species when a negative ion in a discrete state with an energy greater than the detachment threshold loses an electron spontaneously without further interaction with an energy source.
Autoionization	Formation of an ion when an atom or molecule in a discrete state with an internal energy greater than the ionization threshold loses an electron spontaneously without further interaction with an energy source.
Average mass	Mass of an ion or molecule calculated using the average mass of each element weighted for its natural isotopic abundance.

B

ß	Phase ratio. The ratio of mobile to stationary-phase volumes. Thicker stationary-phase films yield longer retention times and higher peak capacities. For open-tubular columns, $\beta = V_G/V_L \sim d_c/4d_f$.
ß-Cleavage	Homolytic cleavage where the bond fission occurs between at an atom removed from the apparent charge site atom by two bonds and an atom adjacent to that atom and removed from the apparent charge site by three bonds.
b-Ion	Fragment ion containing the N-terminus formed upon dissociation of a protonated peptide at a backbone $C-N$ bond.
Backflush	Occurs when compounds (often high boiling matrix) in a pre-column or at the end of a chromatogram are flushed from the column to vent or to another column by flow reversal.
Bake out	Generally a thermal cleaning step in various applications: Gas chromatography – the process of removing contaminants from a column by operation at elevated temperatures, which should not exceed a column's maximum operating temperature (MAOT). Purge and trap methodology – the purification step of the adsorption trap. Mass spectrometry – the cleaning of the analyser from dissolved gases and contaminants by heating the steel manifold for an extended time.
Balanced pressure injection	Headspace injection technique whereby the equilibrated sample vial is depressurized to its maximum against the column pre-pressure.
Band broadening	Several processes that cause solute profiles to broaden as they migrate through a column.
Base peak (BP)	The peak in a mass spectrum that has the greatest intensity. Note: This term may be applied to the spectra of pure substances or mixtures.
Beam instruments	Type of mass spectrometers in which beams of ions are continuously formed from an ion source, passed

through ion optics and are resolved for mass analysis, typically magnetic sector and quadrupole instruments.

Benzyl cleavage — A fragmentation reaction of alkylaromatics forming the benzylic carbenium ion, which appears as the tropylium ion with m/z 91 in many spectra of aromatics.

Bias — The difference between the mean measured value and the true, accepted reference value.

Blank — 1. Material (a sample or a portion or extract of a sample) known not to contain detectable levels of the analyte(s) sought. Also known as a *matrix blank*
2. A complete analysis conducted using the solvents and reagents only; in the absence of any sample (water may be substituted for the sample, to make the analysis realistic). Also known as a *reagent blank* or *procedural blank.*

Blank value — Many analysis methods require the determination of blank values in order to be able to compensate for nonspecific analyte/matrix interaction. A differentiation is made between reagent blank samples and sample blank samples.

Bleed — The loss of material from a column or septum caused by high-temperature operation. Bleed can result in ghost peaks and increased detector baseline offset and noise.

Blank sample — A clean sample or sample of matrix processed so as to measure artefacts in the measurement (sampling and analysis) process, providing the blank value.

Bonded phase — A stationary phase that is chemically bonded to the inner column wall, → also Cross-linked phase.

Bracketing calibration — Organization of a batch of determinations such that the detection system is calibrated immediately before and after the analysis of the samples. For example, calibrant 1, calibrant 2, sample 1, sample n, calibrant 1, calibrant 2.

BSIA — Bulk sample isotope analysis, the isotope ratio MS analysis of a sample after bulk conversion in a crucible of an elemental analyser, contrary to an → CSIA.

BSTFA	Silylating agent for derivatization reactions, Bis (trimethylsilyl) trifluoroacetamide.
BTEX	Abbreviation for the analysis of the benzene, toluene, ethylbenzene and xylene isomers group of aromatics, mostly by headspace sample analysis.
BTV	Breakthrough volume, for example, of an adsorption trap.
BTX	→ BTEX.
Buffer gas	Inert gas used for collisional deactivation of internally excited ions or of the translational energies of ions confined in an ion trap.

C

CAD	Collision activated decomposition in MS/MS experiments, → CID.
Calibrant	→ Calibration Standard.
Calibration	The set of operations that establish, under specified conditions, the relationship between values indicated by a measuring instrument or measuring system, and the corresponding known values. The result of a calibration is sometimes expressed as a calibration factor or as a series of calibration factors in the form of a calibration curve.
Calibration check standard	A standard independently prepared (different source, different analyst) from the calibration standards and run after the original calibration to verify the original calibration. There is usually at least one calibration check standard per batch.
Calibration curve	Defines the relation between analyte concentration and analytical response. Normally at least three to five appropriately placed calibration standards are needed to adequately define the curve. The curve should incorporate a low standard not exceeding 10 times the detection limit. Analytical response, where appropriate, is zeroed using a reagent blank. Either a linear or other curve fit, as appropriate, may be used. Standards and samples must have equivalent reagent backgrounds (e.g. matrix, solvent, acid content, etc.) at the point of analysis.

Calibration drift	The difference between the instrument response and a reference value after a period of operation without recalibration.
Calibration standard	A solution (or dilution) of the analyte (and internal standard, if used) or reference material used to calibrate an instrument.
Capacity factor	The 'k' value of a column describes the molar ratio of a substance in the stationary phase to that in the mobile phase from the relationship of the net retention time to the dead volume.
Carbosieve	Carbon molecular sieve used as an adsorption material for air analysis, $\rightarrow$ also VOCARB (Supelco).
Carboxen	Carbon molecular sieve used as an adsorption material for air analysis, $\rightarrow$ also VOCARB (Supelco).
Carry over	Taking the analyte to the next analysis, also known as memory effect.
CAS No.	The unique registration number of a chemical compound or substance assigned by the Chemical Abstract Service, a division of the American Chemical Society. The intention is to make database searches more convenient, as chemicals often have many names. Almost all molecule databases today allow searching by CAS number. The CAS no. usually is given in the substance entries of many library mass spectra.
Cationized molecule	An ion formed by the association of a cation with a molecule M, for example, $[M + Na]^+$ and $[M + K]^+$. The terms *quasi-molecular ion* and *pseudo-molecular ion* are deprecated.
CCM	A scan-by-scan calibration correction method used, for example, in high resolution selected reaction monitoring, $\rightarrow$ H-SRM.
CDEM	$\rightarrow$ Continuous dynode electron multiplier.
CE	Coating efficiency. A metric for evaluating column quality. The minimum theoretical plate height divided by the observed plate height; $CE = H_{min}/H$.

Centroid	The calculated centre of a mass peak acquired in scan mode. The centroid value can be calculated precisely independent of the resolution power of the mass spectrometer in use. Values displayed in the spectrum with three or more digits are often misleadingly associated with the resolving power of the instrument. In LRMS special care has to be taken as the centroid value gives the centre of gravity of the mass peak composed of many compounds falling in the wide mass window. In HRMS centroid mass values are used to calculate a possible sum formula within a deviation of <2 ppm mass precision.
Centroid acquisition	Procedure of recording mass spectra in which an automated system detects mass peaks, calculates their centroids and assigns m/z values based on a mass calibration file. Only the centroid m/z and the peak intensity values are stored; $\rightarrow$ Continuum acquisition.
Certification	A formal recognition that a laboratory is competent to carry out specific tests or specific types of tests.
Certified reference material (CRM)	A reference material having one or more property values that are certified by a technically valid procedure, accompanied by or traceable to a certificate or other documentation that is issued by a certifying authority.
Charge exchange ionization	Type of CI reaction (PCI), interaction of an ion with an atom or molecule in which the charge on the ion is transferred to the neutral without the dissociation of either $\rightarrow$ charge transfer ionization.
Charge transfer reaction	Type of CI reaction (NCI), action of an ion with a neutral species in which some or all of the charge of the reactant ion is transferred to the neutral species.
Chemical ionization	Formation of a new ion in the gas phase by the reaction of a neutral analyte species with an ion. The process may involve transfer of an electron, a proton or other charged species between the reactants. Note 1: When a positive ion results from chemical ionization the term may be used without qualification. When a negative ion results

the term *negative ion chemical ionization* should be used.
Note 2: this term is not synonymous with chemi-ionization.
Through chemical ionization usually a soft ionization is achieved providing information on the (quasi-) molecular weight of a substance. The selectivity of the reaction and extent of fragmentation are controlled by the choice of the reagent gas.

Chemi-ionization	Reaction of an atom or molecule with an internally excited atom or molecule to form an ion. Note that this term is not synonymous with chemical ionization.
CI	→ Chemical ionization.
CID	Collision induced dissociation, leads in MS/MS to the formation of a product ion spectrum from a selected precursor ion.
C-ion	Fragment ion containing the N-terminus formed upon dissociation of a protonated peptide at a backbone $N-C$ bond.
Clean-up	Generally, the sample preparation procedure involving removal of the matrix and concentration of the analyte.
Cluster ion	Ion formed by a multicomponent atomic or molecular assembly of one or more ions with atoms or molecules, such as $[(H_2O)_n H]^+$, $[(NaCl)_n Na]^+$ and $[(H_3PO_3)_n HPO_3]^-$.
Co-chromatography	A procedure in which the extract prior to the chromatographic step(s) is divided into two parts. Part one is chromatographed as such. Part two is mixed with the standard analyte that is to be measured and chromatographed. The amount of added standard analyte has to be similar to the estimated amount of the analyte in the extract. This method is designed to improve the identification of an analyte, especially when no suitable internal standard can be used, → also Standard addition method.
Cold injection	An GC injection that occurs at temperatures lower than the final oven temperature, usually at or below the solvent boiling point.

Cold trapping	A chromatographic technique for focussing volatile compounds at the beginning of the GC column by cooling a section of the column, for example, with ice or liquid CO_2 below the boiling point of the compounds or solvent, → also Cryofocussing.
Collision gas	Inert gas used for collisional excitation and ion/molecule reactions. The term *target gas* is deprecated.
Collision-induced dissociation (CID)	Dissociation of an ion after collisional excitation. The term *collisionally activated dissociation* (CAD) is deprecated.
Collision quadrupole	Transmission quadrupole to which an oscillating radio frequency potential is applied so as to focus a beam of ions through a collision gas with no *m/z* separation. Note: a collision quadrupole is often indicated by a lower case q as in QqQ.
Collisional activation (CA)	→ Collisional excitation.
Collisional excitation	Reaction of an ion with a neutral species in which all or part of the translational energy of the collision is converted into internal energy of the ion, for example, for the CID process in MS/MS.
Collision cell	A transmission hexapole, octapole or square quadrupole collision cell to which an oscillating radio frequency potential is applied that is filled with a collision gas at low pressure and used to generate collision induced dissociation (CID) of ions to form a product ion spectrum. The collision cell has no mass separating capabilities. Axial fields accelerate product ions to leave the collision cell for a fast switch of → SRM reactions.
Comminution	The process of reducing a solid sample to small fragments.
Comprehensive GC (GC × GC)	Two-dimensional GC technique in which all compounds experience the selectivity of two columns connected in series by a retention modulation device, thereby generating much higher resolution than that attainable with any single column.

Compressibility correction factor (*j*)	This factor compensates for the expansion of a carrier gas as it moves along the column from the entrance, at the inlet pressure (p_i), to the column exit, at the outlet pressure (p_o).
Concurrent solvent recondensation (CSR)	Large volume injection technique with splitless injectors, due to recondensation of solvent inside a retention gap the resulting pressure difference speeds up significantly the sample vapour transfer from the injector allowing larger samples volumes to be injected.
Confidence interval	The set of possible values within which the true value will reside with a stated probability (i.e. confidence level).
Confidence level	The probability, usually expressed as a percentage, that a confidence interval will include a specific population parameter; confidence levels usually range from 90% to 99%.
Confidence limit	Upper or lower boundary values delineating the confidence interval.
Confirmation	Confirmation is the combination of two or more analyses that are in agreement with each other (ideally, using methods of orthogonal selectivity), at least one of which meets identification criteria). It is impossible to confirm the complete absence of residues. Adoption of a 'reporting limit' at the LCL avoids the unjustifiably high cost of confirming the presence or absence, of residues at unnecessarily low levels. The nature and extent of confirmation required for a positive result depends upon importance of the result and the frequency with which similar residues are found. Assays based on an ECD tend to demand confirmation, because of their lack of specificity. Mass spectrometric techniques are often the most practical and least equivocal approach to confirmation. AQC procedures for confirmation should be rigorous.
Confirmatory method	A method that provides full or complementary information enabling the substance to be unequivocally identified and if necessary quantified at the level of interest.

Consecutive reaction monitoring (CRM)	MS^n experiment with three or more stages of m/z separation and in which a particular multistep reaction path is monitored, typical with ion trap mass analyzers.
Constant neutral loss scan	A scan procedure for a tandem mass spectrometer designed to monitor a selected neutral loss mass difference from precursor ions by detection of the corresponding product ions produced by metastable ion fragmentation or collision-induced dissociation. Synonymous terms are constant neutral mass loss scan and fixed neutral fragment scan.
Constant neutral loss spectrum	Spectrum of all precursor ions that have undergone an operator-selected m/z decrement, obtained using a constant neutral loss scan; → Constant neutral mass loss spectrum and fixed neutral mass loss spectrum.
Constant neutral mass gain scan	Scan procedure for a tandem mass spectrometer designed to produce a constant neutral mass gain spectrum of different precursor ions by detection of the corresponding product ions of ion/molecule reactions with a gas in a collision cell.
Constant neutral mass gain spectrum	Spectrum formed of all product ions produced by gain of a pre-selected neutral mass following ion/molecule reactions with the gas in a collision cell, obtained using a constant neutral mass gain scan.
Contamination	Unintended introduction of the analyte into a sample, extract, internal standard solution and so on, by any route and at any stage during sampling or analysis.
Continuous dynode electron multiplier	An ion-to-electron detector in which the ion strikes the inner continuous resistance surface of the device and induces the production of secondary electrons that in turn impinge on the inner surfaces to produce more secondary electrons. This avalanche effect produces an increase in signal in the final measured current pulse, → also SEM.
Continuous injection	Process for the production of defined atmospheres for the calibration of thermodesorption tubes.
Control chart	A graph of measurements plotted over time or sequence of sampling, together with control

limit(s) and, usually, a central line. Control charts may be used to monitor ongoing performance as assessed by method blanks, verification standards, control standards, spike recoveries, duplicates and references samples.

Control limits

Specified boundaries on a control chart that, if exceeded, indicates a process is out of statistical control, and the process must be stopped, and corrective action taken before proceeding. These limits may be defined statistically or based on protocol requirements. Control limits may be assigned to method blanks, verification/control standards, spiked recoveries, duplicates and reference samples.

Control sample

A sample with predetermined characteristics which undergoes sample processing identical to that carried out for test samples and that is used as a basis for comparison with test samples. Examples of control samples include reference materials, spiked test samples, method blanks.

Conversion dynode

Surface that is held at high potential so that ions striking the surface produce electrons that are subsequently detected, → Post-acceleration detector.

CRM

Certified reference material, means a material that has a specified analyte content assigned to it.

Cross-linked phase

A stationary phase that includes cross-linked polymer chains. Usually, it is bonded to the column inner wall, → also Bonded phase.

Cross-talk

MS/MS signal artifacts in an SRM transition from the previous SRM scan, can potentially occur when fragment ions from one SRM transition remain in the collision cell while a second SRM transition takes place, can be the source of false positives when different SRM events have the same product ions formed from different precursor ions. Linear acceleration of the product ions inside of the collision cell prevent cross-talk effects.

Cryofocussing

Capillary GC injection technique for volatile compounds in headspace, purge and trap or thermodesorption systems. Instead of an evaporation injector, online coupling to the sample injection is set up in such a way that a defined region of the column is cooled by liquid CO_2 or liquid N_2 to focus a volatile sample. The chromatography starts with heating up of the focussing region.

CSIA	Compound specific isotope analysis, the isotope ratio MS analysis of individual compounds after chromatographic separation followed by the online conversion to simple gases, contrary to a → BSIA.
CSR	→ Concurrent solvent recondensation.
Cyclotron motion	Circular motion of a particle of charge q moving at velocity v in a magnetic flux density B that results from the Lorentz force $qv \times B$.

D

2DGC	Two-dimensional GC, abbreviation used for describing heart cutting as well as comprehensive $GC \times GC$ methods. → Comprehensive GC.
3D trap	Three-dimensional ion trap mass spectrometer, usually reflected with this term is the typical internal ionization capability; modern 3D traps also use external ion sources providing ion injection into the trap for storage and mass analysis.
δ notation	Notation in units per mil (‰), commonly used in isotope ratio mass spectrometry to express isotope abundance differences as a ‰ deviation between the sample against an international standard, for example, for ^{13}C

$$\delta^{13}C = \left[\left(\frac{R_{\text{Sample}}}{R_{\text{Standard}}} \right) - 1 \right] \times 10^3 \; [‰]$$

with $R = {}^{13}C/{}^{12}C$
0.01 ‰ equals 10^{-5} at.% (→ also atom%). The international standards are assigned a value of 0.0 ‰ on their respective δ scales.

d_c	Average column inner diameter.
d_f	Average stationary-phase film thickness.
D_G	Gaseous diffusion coefficient; ~0.05 for hydrocarbons in helium carrier gas and 0.1 for hydrogen carrier gas.

D_L	Liquid–liquid diffusion coefficient; $\sim 1 \times 10^{-5}$ for hydrocarbons in silicones.
d_p	Average particle diameter.
Dalton, Da	Non-SI unit of mass (symbol Da) that is identical to the unified atomic mass unit, based since 1961 on the mass definition of $^{12}C = 12.00000$ g/mol.
Daly detector	Detector consisting of a conversion dynode, scintillator and photomultiplier. A metal knob at high potential emits secondary electrons when ions impinge on the surface. The secondary electrons are accelerated onto the scintillator that produces light that is then detected by the photomultiplier detector.
DAT	Diaminotoluene isomers.
DC	Direct current.
DEA	The U.S. Drug Enforcement Agency.
Dead time	The time taken for a substance that is not retarded in a GC column to reach the detector, for example, air, methane for silicone phases, to pass through a chromatographic column. The carrier gas velocity is calculated from the dead time and the column length, → also HETP.
Dead volume	Extra volume experienced by analytes as they pass through a chromatographic system. Excessive dead volume causes additional peak broadening.
Decision limit	The limit at and above which it can be concluded with an error probability of that a sample is non-compliant.
Deconvolution	The term is used in AMDIS in the broad sense of extracting one signal from a complex mixture. The treatment of noise, the correction for base line drift, and the extraction of closely co-eluting peaks from one another are part of the deconvolution process.
DEGS	Diethylene glycol succinate; used as a stationary phase.
Delayed extraction (DE)	Application of the accelerating voltage pulse after a time delay in desorption/ionization from a surface.

The extraction delay can produce energy focussing in a time-of-flight mass spectrometer.

Desorb preheat	Purge and trap technique step for heating up the adsorption trap. The effective desorption and transfer of analytes from the adsorption trap into the GC-MS system is effected by switching a six-way valve first after a set preheat temperature of approx. 5 °C below of the desorption temperature has been reached.
Desorption ionization (DI)	Formation of ions from a solid or liquid material by the rapid vaporization of that sample in the ion source.
Detection capability	The smallest content of the substance that may be detected, identified and/or quantified in a sample with an error probability of β.
Detection limit	It describes the smallest detectable signal of the analyte that can be clearly differentiated from a blank sample (assessed in the signal domain, detection criterion). It is calculated as the upper limit of the distribution range of the blank value.
Diagnostic ion	MS term for ions that are highly characteristic for the compound measured, especially for ions whose formation reveals structural or compositional information. For instance, the phenyl cation in an electron ionization mass spectrum is a diagnostic ion for benzene and derivatives. Used typically, for extracted ion chromatograms, → XIC.
DI-SPME	Direct immersion headspace solid phase micro extraction, the SPME fibre is exposed directly into an (aqueous) liquid sample for collection of analytes.
Dimeric ion	An ion formed by ionization of a dimer or by the association of an ion with its neutral counterpart such as $[M_2]^{+\bullet}$ or $[M - H + M]^+$.
Direct injection	The sample enters an inlet and is transferred in its entirety into a column by carrier gas flow. No sample splitting or venting occurs during or after the injection.
Direct insertion probe	Device for introducing a single sample of a solid or liquid, usually contained in a quartz or other

non-reactive sample holder (e.g. crucible), into a mass spectrometer ion source.

Dissociative ionization	Reaction of a gas-phase molecule in chemical ionization that results in its decomposition to form products, one of which is an ion.
Distonic ion	Radical cation or anion arising formally by ionization of diradicals or zwitterionic molecules (including ylides). In these ions the charge site and the unpaired electron spin cannot be both formally located in the same atom or group of atoms as it can be with a conventional ion.
dl-PCB	Dioxin-like PCBs, → PCB.
DMCS	Dimethylchlorosilane; used for silanizing glass GC parts.
DNB	Dinitrobenzene isomers.
DOD	The US Department of Defense.
Double-focussing mass spectrometer	Mass spectrometer that incorporates a magnetic sector and an electric sector connected in series in such a way that ions with the same m/z but with distributions in both the direction and the translational energy of their motion are brought to a focus at a point.
Dry purge	Purge and trap analysis step whereby moisture is removed from the trap by the carrier gas before desorption.
Duty cycle	Degree of the effective ion acquisition time of, for example, a mass spectrometer. Here the duty cycle is determined by the sum of the ion dwell times relative to the total scan cycle time including jump and stabilization times of the analyser voltages.
Dwell time	Effective ion acquisition time typically in milliseconds, maximized by using a selected ion monitoring scan (→ SIM, MID, SRM)

E

ECD	Electron-capture detection, a detector ionizes analytes by collision with metastable carrier gas molecules produced by ß-emission from a

radioactive source such as ^{63}Ni. The electron capture detector is one of the most sensitive detectors and it responds strongly to halogenated analytes and others with high electron-capture cross-sections ($\rightarrow$ also electron capture dissociation). Relative ECD-response to hydrocarbons:

10^1 Esters, ethers.
10^2 Monochlorides, alcohols, ketones, amines.
10^3 Dichlorides, monobromides.
10^4 Trichlorides, anhydrides.
$10^5 - 10^6$ Polyhalogenated, mono-, diiodides.

Eddy diffusion	Multipath effect in chromatography, the cause of peak broadening through diffusion processes.
Effective plates	The number of effective theoretical plates in a column, taking the dead volume into consideration ($\rightarrow$ also HETP).
Efficiency	The ability of a column to produce sharp, well-defined peaks. More-efficient columns have more theoretical plates (N) and smaller theoretical plate heights (H).
EI	Electron ionization, with modern GC-MS instruments ionization usually takes place at ionization energy of 70 eV. Positive ions are formed predominantly.
Einzel lens	Three element ion lens in which the first and third elements are held at the same voltage. Such a lens produces focussing without changing the translational energy of the particle.
ELCD	Electrolytic-conductivity detector, also Hall detector; gives a mass-flow dependent signal. The detector catalytically reacts halogen-containing analytes with hydrogen (reductive mode) to produce strong acid by-products that are dissolved in a working fluid. The acids dissociate and the detector measures increased electrolytic conductivity. Other operating modes modify the chemistry for response to nitrogen or sulfur-containing substances.
Electric sector	$\rightarrow$ Electrostatic energy analyser.

Electron affinity, E_{EA}	Electron affinity of a species M is the minimum energy required for the process $M^{-\bullet} \rightarrow M + e^-$ where $M^{-\bullet}$ and M are in their ground rotational, vibrational and electronic states and the electron has zero translational energy.
Electron attachment ionization	Ionization of a gaseous atom or molecule by attachment of an electron to form $M^{-\bullet}$ ions.
Electron capture	The capture of thermal electrons by electronegative compounds. This process forms the basis of the ECD (electron capture detector) and is also made use of in negative chemical ionization (NCI, ECD-MS).
Electron capture dissociation (ECD)	Process in which multiply protonated molecules interact with low energy electrons. Capture of the electron leads the liberation of energy and a reduction in charge state of the ion with the production of the $[M + nH]^{(n-1)+}$ odd electron ion, which readily fragments.
Electron energy	Magnitude of the electron charge multiplied by the potential difference through which electrons are accelerated, for example, from a filament in order to effect electron ionization.
Electron impact ionization	$\rightarrow$ Electron ionization.
Electron ionization	Ionization of an atom or molecule by electrons that are typically accelerated to energies between 10 and 150 eV in order to remove one or more electrons from the molecule. The term *electron impact* is deprecated.
Electron volt, eV	Non-SI unit of energy (symbol eV) defined as the energy acquired by a particle containing one unit of charge through a potential difference of 1 V. An electron volt is equal to $1.60217733(49) \times 10^{-19}$ J.
Electrospray ionization (ESI)	A process for LC-MS in which ionized species in the gas phase are produced from a solution via highly charged fine droplets, by means of spraying the solution from a narrow-bore needle tip at atmospheric pressure in the presence of a high electric field (1000–10 000 V potential). Note: When a pressurized gas is used to aid in the formation of a stable spray, the term *pneumatically*

assisted electrospray ionization is used. The term *ion spray* is deprecated.

Electrostatic energy analyser (ESA)	A device consisting of conducting parallel plates, concentric cylinders or concentric spheres that separates charged particles according to their ratio of translational energy to charge by means of a voltage difference applied between the pair.
Elution temperature	The temperature of the GC oven at which an analyte reaches the detector.

Energy

The SI unit is the Joule (J)

$$1\,J = 1\,Nm = 1\,Ws$$

In mass spectrometry the energy of ions is usually given in (eV) and is calculated from the elemental charge and the acceleration voltage in the ion source.
Conversion factors:

$$1\,eV = 23.0\,kcal = 96.14\,kJ$$

(with $1\,cal = 4.18\,J$).

E/2 mass spectrum	Mass spectrum obtained using a sector mass spectrometer in which the electric sector field *E* is set to half the value required to transmit the main ion-beam. This spectrum records the signal from doubly charged product ions of charge-stripping reactions.
EPA	The US Environmental Protection Agency.
ESI	Electrospray ionization, an LC-MS coupling technique, → Electrospray ionization.
ESTD	External standard, quantitation by using external standardization. The analyte itself is used for quantitative calibration as a clean standard or added to a blank standard matrix. The signal height for a known concentration of the analyte is used for the calibration procedure. The calibration runs are carried out separately (externally) from the analysis of the sample.
Even-electron ion	An ion containing no unpaired electrons in its ground electronic state.

Exact mass	Calculated mass of an ion or molecule containing a single isotope of each atom, most frequently the lightest isotope of each element, calculated from the masses of these isotopes using an appropriate degree of accuracy.
External ionization	Process for the production of ions for an ion storage mass spectrometer, for example, ion trap MS. The ionization does not take place inside of the ion trap analyser; instead, ions are formed in an attached ion source and transferred to the ion trap analyser. The special decoupling of ionization and mass analysis in ion trap instruments allows the independent use of GC parameters, the application of negative chemical ionization as well as several MS/MS scan techniques.
Extracted ion chromatogram (XIC)	Chromatogram created by plotting the intensity of the signal observed at a chosen m/z value or series of values in a series of mass spectra recorded as a function of retention time, → Reconstructed ion chromatogram.

F

F_a	The column outlet flow-rate corrected to room temperature and pressure; for example, the flow-rate as measured by a flow metre. F_a can be calculated from the average carrier gas linear velocity and the column dimensions.
F_s	The split-vent flow-rate, measured at room temperature and pressure.
False negative	A result wrongly indicating that the analyte concentration does not exceed a specified value.
False positive	A result wrongly indicating that the analyte concentration exceeds a specified value.
FAME	Fatty acid methyl ester.
Faraday cup	A conducting cup or chamber that intercepts a charged particle beam and is electrically connected to a current measuring device.
Fast atom bombardment (FAB)	Ionization of any species by the interaction of a focussed beam of neutral atoms having a translational energy of several thousand

electronvolts with a sample that is typically dissolved in a solvent matrix. Related term: *secondary ionization*.

Fast ion bombardment (FIB)

Ionization of any species by the interaction of a focussed beam of ions having a translational energy of several thousand electronvolts with a solid or liquid sample. For a liquid sample this is the same as liquid secondary ionization.

FC43

Perfluorotributylamine (PFTBA), a widely used reference substance for calibration of the mass scale, M 671.

FFAP

Free fatty-acid phase, a polar stationary phase for GC columns.

FID

Flame ionization detector, providing a mass flow dependent signal, the detector ionizes most classes of organic compounds. FID is a universal detection technique. Little or no response have noble gases, CO, CO_2, O_2, N_2, H_2O, CS_2, NO_x, NH_3, perhalogenated compounds, formic acid/aldehyde.

Field desorption (FD)

Formation of gas-phase ions in the presence of a high electric field from a material deposited on a solid surface. The term *field desorption/ionization* is deprecated.

Field-free region (FFR)

Section of a mass spectrometer in which there are no electric or magnetic fields.

Field ionization (FI)

Removal of electrons from any species, usually in the gas phase, by interaction with a high electric field.

Fixed product ion scan

In a sector instrument, either a high voltage scan or a linked scan at constant B^2/E. Both give a spectrum of all precursor ions that fragment to yield a pre-selected product ion.
Note: The term *daughter ion* is deprecated.

Fortified sample

A sample enriched with a known amount of the analyte to be detected.

Forward library search

A procedure of comparing a mass spectrum of an unknown compound with a mass spectral library so that the unknown spectrum is compared with the library spectra, considering only the mass signals and intensities of the unknown to be compared with the library spectra. Mass signals not belonging to the compound of interest, e.g. matrix, column bleed or coeluting compounds lead to lower match values, and require background subtraction. → reversed library search.

Fourier transform ion cyclotron resonance mass spectrometer (FT-ICR)	A mass spectrometer based on the principle of ion cyclotron resonance in which an ion in a magnetic field moves in a circular orbit at a frequency characteristic of its m/z value. Ions are coherently excited to a larger radius orbit using a pulse of radio frequency energy and their image charge is detected on receiver plates as a time domain signal. Fourier transformation of the time domain signal results in a frequency domain signal which is converted to a mass spectrum based in the inverse relationship between frequency and m/z. A characteristic of FT-ICR is the very high mass resolution power ($R > 10^6$) and mass precision (< 1 ppm).
FPD	Flame photometric detector, providing a mass flow dependent signal, the detector burns in a hydrogen-rich flame where analytes are reduced and excited. Upon decay of the excited species light is emitted of characteristic wavelengths. The visible-range atomic emission spectrum is filtered through an interference filter and detected with a photomultiplier tube. Different interference filters can be selected for sulfur, tin or phosphorus emission lines. The flame photometric detector is sensitive and selective.
Fragment ion	A product ion that results from the dissociation of a precursor ion. Note: The term *daughter ion* is deprecated.
Fringe field	Electric or magnetic field that extends from the edge of a sector, lens or other ion optics element.
Frit sparger	U-shaped tube for the purge and trap analysis of water samples with built-in frit for fine dispersion of the purge gas.
Fritless sparger	U-shaped tube without a frit for the purge and trap analysis of moderately foaming water or solid samples.
FS	Fused silica.

FSOT	Fused-silica open-tubular column.
Full scan	Acquisition mode for the recording of complete mass spectra over a specified mass range.
FWHM	*Full width at half peak maximum*, term used in the definition of peak resolution for the measurement of the peak width at half peak height.

G

GALP	Good automated laboratory practice.
Gas isotope ratio mass spectrometry	Common name for the area of isotope ratio mass spectrometry for the determination of the stable isotopes of H, N, C, O, S and Si. Compounds containing these elements can be quantitatively converted into simple gases for mass spectrometric isotope ratio analysis, that is, H_2, N_2, CO, CO_2, O_2, SO_2, SiF_4 fed by viscous flow or entrained into a continuous He flow into the ion source of a dedicated isotope ratio mass spectrometer.
GCB	Graphitized carbon black, → VOCARB (Supelco).
GC-O	Abbreviation used for → GC-olfactometry.
GC-olfactometry	Refers to the use of human assessors as a sensitive and selective detector for odour-active compounds. The aim of this technique is to determine the odour activity of volatile compounds in a sample extract, and assign a relative importance to each compound.
GC×GC	→ Comprehensive GC.
Ghost peaks	Peaks from compounds not present in the original sample. Ghost peaks can be caused by septum bleed, analyte decomposition or carrier gas contamination.
GIRMS	→ Gas isotope ratio mass spectrometry.
Glass cap cross	→ Werkhoff splitter.
GLC	Gas–liquid chromatography, using this technique analytes partition between a gaseous mobile phase and a liquid stationary phase. Selective interactions between the analytes and the liquid phase cause different retention times in the column. Most

current GC stationary phases are immobilized liquid phases at the specified operation temperature of the column.

GLP — Good laboratory practice.

GLPC — Gas–liquid phase chromatography, → GLC; gas–liquid chromatography.

GPC — Gel permeation chromatography, gel chromatography used for sample preparation, for example, to remove lipid matrix components in pesticide analysis clean-up.

Gridless reflectron — A reflectron in which ions do not pass through grids in their deceleration and turnaround thereby avoiding ion loss due to collisions with the grid.

GSC — Gas–solid chromatography, this technique, analytes partition between a gaseous mobile phase and a solid stationary phase. Selective interactions between the analytes and the solid phase cause different retention times in the column.

H

H — Height equivalent to one theoretical plate. The distance along the column occupied by one theoretical plate; $H = L/N$.

H_{meas} — Height equivalent to one theoretical plate as measured from a chromatogram

$$H_{meas} = \frac{L}{5.54 \left(\frac{t_R}{W_h} \right)^2}$$

H_{min} — Minimum theoretical plate height at the optimum linear velocity, ignoring stationary-phase contributions to band broadening. For open-tubular columns:

$$H_{min} = \left(\frac{d_c}{2} \right) \sqrt{\frac{1+6k+11k^2}{3(1+k)^2}}$$

h_p — Peak amplitude.

H_{theor} — Theoretical plate height. For open-tubular columns (Golay equation):

$$H_{theor} = \left(\frac{2D_G}{\bar{u}} \right) + \bar{u} \left\{ \left[\frac{(1+6k+11k^2)}{96(1+k)^2} \right] \left(\frac{d_c^2}{D_G} \right) + \left[\frac{2k}{3(1+k)^2} \right] \left(\frac{d_f^2}{D_L} \right) \right\}$$

Hall detector	→ ELCD.
Hard ionization	Formation of gas-phase ions accompanied by extensive fragmentation typically → EI ionization is considered a hard ionization technique.
Headspace GC	Static headspace GC.
Headspace sweep	Technique in purge and trap analysis for treating foaming samples whereby the purge gas is passed only over the surface of the sample instead of through it.
Heartcut	GC-GC technique in which two or more partially resolved peaks that are eluted from one column are directed onto another column of different polarity or at a different temperature for improved separation.
Heterolysis	→ Heterolytic cleavage.
Heterolytic cleavage	Fragmentation of a molecule or ion in which both electrons forming the single bond that is broken remain on one of the atoms that were originally bonded, → Heterolysis.
HETP	Height equivalent to one theoretical plate; discontinued term for plate height (H). The dependence of the plate high value on the carrier gas velocity determined from the van Deemter curve is used to optimize chromatographic separation.
HFBA	Heptafluorobutyric anhydride, derivatization agent for preparing volatile heptafluorobutyrates. It is frequently used for introducing halogens into compounds to increase the response in ECD or NCI.
High-energy collision-induced dissociation	Collision-induced dissociation process wherein the projectile ion has laboratory-frame translational energy higher than 1 keV.
High resolution MS	→ HRMS.
Homolysis	→ Homolytic cleavage.
Homolytic cleavage	Fragmentation of an ion or molecule in which the electrons forming the single bond that is broken are shared between the two atoms that were originally bonded. For an odd electron ion, fragmentation results from one of a pair of

	electrons that form a bond between two atoms moving to form a pair with the odd electron on the atom at the apparent charge site. Fragmentation results in the formation of an even electron ion and a radical. This reaction involves the movement of a single electron and is represented by a single-barbed arrow, → Homolysis.
HRGC	High resolution gas chromatography, the inherent meaning of this term is capillary gas chromatography in contrast to packed column chromatography.
HRMS	High resolution mass spectrometry, the separation of C, H, N and O multiples in mass spectrometry. The empirical formula of an ion is usually obtained through accurate mass determination as part of the structure elucidation. Meaningful accurate mass determinations require high mass resolution to separate close isobaric mass signals. In GC-MS target compound analysis this term demands for a resolution power of better than 10 000 at 10% valley providing the accurate ion mass (originating from the EPA method 1613). Mass spectrometers providing resolution power of $R > 10\,000$ (at 5% peak height) or $R > 20\,000$ (at FWHP) are generally termed high mass resolution instruments.
HS	Headspace sampling, gas-phase sampling technique in which the sample is taken from an enclosed space above a solid or liquid sample after attaining equilibrium conditions.
HSGC	Headspace gas chromatography.
H-SRM	Highly resolved selected reaction monitoring, MS/MS target compound scan technique using enhanced mass resolution at Q1 of a triple quadrupole analyzer for increased analyte selectivity.
HS-SPME	Headspace solid phase micro extraction, the SPME fibre is exposed to the headspace of a sample for collection of analytes.
HxCDD	Hexachlorodibenzodioxin isomers.
Hybrid mass spectrometer	A mass spectrometer that combines m/z analysers of different types to perform tandem mass spectrometry.

Hydrogen/deuterium exchange	Exchange of hydrogen atoms with deuterium atoms in a molecule or preformed ion in solution prior to introduction into a mass spectrometer, or by reaction of an ion with a deuterated collision gas inside a mass spectrometer.

I

ICR-MS	Ion cyclotron resonance mass spectrometer, an ion storage mass spectrometer providing very high resolution power $R > 10^6$ and accurate mass measurements by measuring the cycle frequency of ions in a strong magnetic field followed by Fourier transformation.
IDL	Instrument detection limit, defined statistically as a measure for the potential instrument sensitivity at the 99% confidence level from the area precision (RSD%) of a series of measurements, applied in cases a $\rightarrow$ S/N cannot be calculated due to the absence of a suitable noise band in modern digitally filtered MS systems. The IDL does not inform about the lowest detectable concentration $\rightarrow$ LOD.
INCOS	*Integrated computer systems*, the term refers to the search approach for mass spectra in former Finnigan GC-MS systems.
Indicator PCBs	$\rightarrow$ PCB.
Inductive cleavage	A heterolytic cleavage of an ion. For an odd electron ion, inductive cleavage results from the pair of electrons that forms a bond to the atom at the apparent charge site moving to that atom while the charge site moves to the adjacent atom. The movement of the electron pair is represented by a double-barbed arrow.
In-house validation	$\rightarrow$ Single laboratory study.
In-source collision-induced dissociation	The dissociation of an ion as a result of collisional excitation during ion transfer from an atmospheric pressure ion source and the mass spectrometer vacuum. This process is similar to ion desolvation but uses higher collision energy.
Interference	A positive or negative response produced by a compound(s) other than the analyte, contributing to the response measured for the analyte or making

integration of the analyte response less certain or accurate. Interference is also loosely referred to as *chemical noise* (as distinct from electronic noise, 'flame noise', etc.).→ Matrix effects are a subtle form of interference. Some forms of interference may be minimized by greater selectivity of the detector using GC-MS or GC-MS/MS. If the interference cannot be eliminated or compensated, its effects may be acceptable if there is no significant impact on accuracy.

Interlaboratory study

The organization, performance and evaluation of tests on the same sample by two or more laboratories in accordance with predetermined conditions to determine testing performance. According to the purpose the study can be classified as collaborative study or proficiency study.

Internal Standard (ISTD)

A compound for quantitative reference not contained in the sample with chemical characteristics similar to those of the analyte. Labeled analogues of the target analytes are typically applied, for example, in dioxin analysis. The ISTD provides an analytical response that is distinct from the analyte and not subject to interference. Internal standards are used to adjust for variations in the analytical response due to
− matrix effects,
− sample preparation losses,
− final sample volume,
− variable injection volumes, or
− instrumental effects.

One or more ISTDs are added to the sample immediately prior to sample preparation (surrogate standard) and/or the extract vial before injection (recovery standard). The ratio between surrogate and recovery standard is used to calculate the analyte recovery.

Ion

An atomic, molecular or radical species with an unbalanced electrical charge. The corresponding neutral species need not be stable.

Ion desolvation

The removal of solvent molecules clustered around a gas-phase ion by means of heating and/or collisions with gas molecules.

Ionic dissociation	The dissociation of an ion into another ion of lower mass and one or more neutral species or ions with a lower charge.
Ion injection	The transfer of ions formed in an external ion source to the analyser of an ion storage mass spectrometer for subsequent mass analysis.
Ion/ion reaction	The reaction between two ions, typically of opposite polarity. The term *ion/ion reaction* is deprecated.
Ion mobility spectrometry (IMS)	Separation of ions according to their velocity through a buffer gas under the influence of an electric field.
Ion/molecule reaction	Reaction of an ion with a molecule. Note: the term *ion-molecule reaction* is deprecated because the hyphen suggests a single species that is both an ion and a molecule.
Ion/neutral complex	A particular type of transition state that lies between precursor and product ions on the reaction coordinate of some ion reactions.
Ion/neutral reaction	Reaction of an ion with an atom or molecule.
Ion/neutral exchange reaction	Reaction of an ion with a neutral species to produce a different neutral species as the product.
Ion-pair formation	Reaction of a molecule to form both positive ion and negative ion fragments among the products.
Ion source	Region in a mass spectrometer where ions are produced.
Ion storage MS	Mass spectrometer equipped with internal or external ion generation and ion collection. The collection and analysis of ions take place discontinuously, for example, ion traps (LRMS) and ICR or Orbitrap MS (HRMS).
Ion-to-photon detector	Detector in which ions strike a conversion dynode to produce electrons that in turn strike a phosphor layer and the resulting photons are detected by a photomultiplier, → Daly detector.
Ion trap (IT)	Device for spatially confining ions using electric and magnetic fields alone or in combination.

Ionization cross section	A measure of the probability that a given ionization process will occur when an atom or molecule interacts with a photon, electron, atom or molecule.
Ionization efficiency	Ratio of the number of ions formed to the number of molecules consumed in the ion source.
Ionizing collision	Reaction of an ion with a neutral species in which one or more electrons are removed from either the ion or neutral.
IRMS	→ Isotope ratio mass spectrometry.
Isomers	Two substances with the same molecular sum formula, but different structural formula, or different spatial arrangement of the atoms (stereoisomers). Isomers differ chemically and physically. Frequently mass spectra of isomers cannot be differentiated, but because of different interactions with the stationary phase, they can be chromatographically separated.
Isotope	Although having the same nuclear charge (number of protons) most of the elements exist as atoms that have nuclides with varying numbers of neutrons, known as *isotopes*. They belong to the same chemical element but have different physical behaviour, for example, masses. While in chemical synthesis the natural isotope composition is not taken into consideration (use of the average atomic weight), in mass spectrometry the distribution of the isotopes over the different masses is visible and the isotope pattern is assessed, for example, dioxin analysis (molecular weight calculation based on the principle isotope). IRMS determines isotope ratios as a result of fractionation processes highly precise.
Isotope dilution mass spectrometry	A quantitative mass spectrometry technique based on the measurement of the isotopic abundance of a nuclide after isotope dilution with the test portion; an isotopically enriched compound is used as an internal standard.
Isotope effect	Alteration of either the equilibrium constant or the rate constant of a reaction if an atom in a reactant molecule is replaced by one of its isotopes, distinguished are kinetic isotope effect, equilibrium isotope effect, primary isotope effect, secondary isotope effect.

Isotope pattern	Characteristic intensity pattern in a mass spectrum derived from the different abundance of the isotopes of an element. From the isotope pattern of an element in the mass spectrum conclusions can be drawn on the number of atoms of this element in the elemental formula. Important isotope patterns in organic analysis are shown by Cl, Br, Si, C and S. Organometallic compounds show characteristic patterns rich in lines. Molecular ions show the isotope pattern of all the elements of the chemical sum formula.
Isotope ratio	Ratio of the number of atoms of one isotope to the number of atoms of another isotope of the same chemical element in the same system. The number ratio, R, is the number obtained by counting a specified entity (usually molecules, atoms or ions) divided by the number of another specified entity of the same kind in the same system.
Isotope ratio mass spectrometry (IRMS)	The precise measurement of the relative quantity of the different isotopes of an element in a material using a mass spectrometer. Isotope ratio differences are typically measured against international standards and expressed on a delta per mille scale (‰).
Isotopologue ions	Ions that differ only in the isotopic composition of one or more of the constituent atoms. For example, $CH_4^{+\bullet}$ and $CH_3D^{+\bullet}$ or $^{10}BF_3$ and $^{11}BF_3$ or the ions forming an isotope cluster. The term *isotopologue* is a shortening of isotopic homologue.
Isotopomeric ions	Isomeric ions having the same numbers of each isotopic atom but differing in their positions. Isotopomeric ions can be either configurational isomers in which two atomic isotopes exchange positions or isotopic stereoisomers. The term *isotopomer* is a shortening of isotopic isomer.
ISTD	→ Internal standard.
ITD	Ion-trap detector, a mass spectrometric (MS) detector that uses an ion-trap device to generate mass spectra.
ITEX	In-tube extraction, automated dynamic headspace extraction technique using a packed syringe needle, injection by thermal desorption in a GC injector.

J

j	Mobile phase compressibility correction factor, a factor, applying to a homogeneously filled column of uniform diameter, that corrects for the compressibility of the mobile phase in the column, also called *Compressibility Correction Factor*. In liquid chromatography the compressibility of the mobile phase is negligible. In gas chromatography, the correction factor can be calculated as:

$$j = \frac{3}{2}\frac{p^2-1}{p^3-1} = \frac{3}{2}\frac{(p_i/p_o)^2-1}{(p_i/p_o)^3-1}$$

Jet separator	Interface construction for the coupling of mass spectrometers to wide bore and packed chromatography columns. The quantity of the lighter carrier gas is reduced by dispersion. Loss of analyte cannot be prevented.

K

k	Retention factor, a measurement of the retention of a peak;

$$k = \frac{(t_R - t_M)}{t_M}$$

K	Partition coefficient. The relative concentration of an analyte in the mobile and stationary phases;

$$K = \beta k.$$

Kovats index	The most used index system in gas chromatography for describing the retention behaviour of substances. The retention values are based on a standard mixture of alkanes: Alkane index = number of C-atoms × 100.

L

L	Column length.
Laboratory sample	A sample prepared for sending to a laboratory and intended for inspection or testing.
Laser ionization (LI)	Formation of ions through the interaction of photons from a laser with a material or with gas-phase ions or molecules.

Level of interest	The concentration of substance or analyte in a sample that is significant to determine its compliance with legislation.
Line spectrum	Representation of a mass spectrum as a series of vertical lines indicating the ion abundance intensities across the mass sale of m/z values.
Linear ion trap (LIT)	A two-dimensional Paul ion trap in which ions are confined in the axial dimension by means of an electric field at the ends of the trap.
Linear range (LR)	Also called *linear dynamic range*. The range of analyte concentration or amount in which the detector response per amount is constant within a specified percentage.
Linear velocity (u)	The speed at which the carrier gas moves through the column, usually expressed as the average carrier gas linear velocity (u_{avg}).
Linked scan	A scan in a tandem mass spectrometer with two or more m/z analysers e.g a triple quadrupole or a sector mass spectrometer that incorporates at least one magnetic sector and one electric sector. Two or more of the analysers are scanned simultaneously so as to preserve a predetermined relationship between scan parameters to produce a product ion, precursor ion or constant neutral loss or gain spectrum.
Liquid phase	In GC, a stationary liquid layer coated or chemically bonded on the inner column wall (WCOT column) or on a support (packed, SCOT column) that selectively interacts with the analytes to produce different retention times.
LLOD	lower limit of detection, also called *detection limit* or *limit of detection*, → LOD.
LOD	Limit of detection, the lowest concentration of a substance that can still be detected unambiguously (assessed in the signal domain). The value is obtained from the decision limit (smallest detectable signal) using the calibration function or the distribution range of the blank value, typically expressed in a S/N value > 3.
LOQ	Limit of quantitation, unlike the limit of detection, the limit of quantitation is confirmed by the

calibration function. The value gives the lower limiting concentration, which differs significantly from a blank value and can be determined unambiguously and quantitatively with a given precision. The LOQ value is therefore dependent on the largest statistical error, which can be tolerated in the results. Often expressed using the signal domain for the determination of the lowest quantifiable concentration providing a S/N value > 10, here ignoring the calibration function with a significantly higher S/N requirement than applied for the LOD.

Low-energy collision-induced dissociation	A collision-induced dissociation process wherein the precursor ion has translational energy lower than 1 keV. This process typically requires multiple collisions for the transformation of kinetic into internal vibrational energy. The collisional excitation process is cumulative.
LRMS	Low-resolution mass spectrometry, covers all MS analyser technologies providing nominal mass resolution, in contrast to → HRMS.
LVSI	Large volume splitless injection, using the → concurrent solvent recondensation effect (CSR).

M

μ scan	The shortest scan unit in ion trap mass spectrometers, depending on the pre-selected scan rate of the chromatogram, several μ scans are accumulated to form the stored and displayed mass spectrum.
Magic 60	A laboratory term referring to the ~60 analytes of the combined volatile halogenated hydrocarbon/BTEX determination (VOC), which are analysed together in EPA methods, for example, by purge and trap GC-MS.
Magnetic sector	A device that produces a magnetic field perpendicular to a charged particle beam that deflects the beam to an extent that is proportional to the particle momentum per unit charge. For a monoenergetic beam, the deflection is proportional to m/z.

Magnetic sector MS	Single or double focussing mass spectrometer with magnetic (and electrostatic) analyser for the spatial separation of ions on individual flight paths and focussing on the exit slit at the detector, double focussing mass spectrometer are employed for high resolution MS, → also HRMS.
MAM	Monoacetylmorphine.
MAOT	Maximum allowable operating temperature, highest continuous column operating temperature that will not damage a column, if the carrier gas is free of oxygen and other contaminants. Slightly higher temperatures are permissible for short periods of time during temperature programed operation or intermediate column bakeouts.
Mass accuracy	The deviation of the measured accurate mass from the calculated exact mass of an ion. It can be expressed as an absolute value in milliDaltons (mDa) or as a relative value in parts-per-million (ppm) error. The absolute accuracy is calculated as accurate mass − exact mass. Example: The experimentally measured mass m/z = 239.15098, the theoretical exact mass of the ion m/z = 239.15028. The absolute mass accuracy = (239.15098 − 239.15028) = 7.0 mDa. In relative terms the mass accuracy is calculated as ((accurate mass − exact mass)/exact mass) × 10^6. The relative mass accuracy = (239.15098 − 239.15028)/239.15028 ×10^6 = 2.9 ppm.
Mass calibration	A means of determining m/z values in a scan from their times of detection relative to initiation of acquisition of a mass spectrum. Most commonly this is accomplished using a computer-based data system and a calibration file obtained from a mass spectrum of a compound that produces ions whose m/z values are known.
Mass defect	The difference between the (nominal) mass number and the exact monoisotopic mass of a molecule or atom.
Mass excess	The negative of the mass defect.
Mass filter	A quadrupole analyser works as a mass filter. From the large number of different ion species formed in

the ions source a quadrupol filters ions of specific *m/z* values only. A mass spectrum is acquired by cyclic ramping in the control (scan) by continuously moving the transmission window of 1 Da width over the selected mass range, typically upwards (full scan). Switching the filter characteristics to pre-selected masses only ions of distinct m/z values are acquired (SIM, selected ion monitoring).

Mass gate	A set of plates or grid of wires in a time-of-flight mass spectrometer that is used to apply a pulsed electric field with the purpose of deflecting charged particles in a given *m/z* range.
Mass number	Sum of the number of protons and neutrons in an atom, molecule or ion, → Nucleon number.
Mass range	Range of *m/z* over which a mass spectrometer can detect ions or is operated to record a mass spectrum.
Mass resolution	Smallest mass difference Δm between two equal magnitude peaks so that the valley between them is a specified fraction of the peak height.
Mass resolving power	The measure R describing the mass resolution setting or capability of a mass analyzer, expressed as the observed mass divided by the difference between two masses that can be separated: $R = m/\Delta m$. The procedure by which Δm was obtained (e.g. FWHM), e.g. at half peak height (FWHM), and the mass at which the measurement was made should be reported. For magnetic sector instruments another definition is used. The '10% valley', method measures the peak width at 5% peak height. The difference between the two definitions is a factor of 2 (i.e. 10 000 resolving power by the 10% valley method equals 20 000 resolving power by FWHM). Note: Mass resolving power is often confused or interchangeably used with → mass resolution.
Mass scale	A mass scale always implies the *m/z* scale (mass to charge value). Multiply charged ions with n charges appear in the spectrum at $m/n \cdot z$, e.g. doubly charged coronene (M 300) with a signal at *m/z* 150.

Mass selective axial ejection   Use of mass selective instability to eject ions of selected *m/z* values from an ion trap.

Mass selective instability   A phenomenon observed in a Paul ion trap whereby an appropriate combination of oscillating electric fields applied to the body and the end-caps of the trap leads to unstable trajectories for ions within a particular range of *m/z* values and thus to their ejection from the trap.

Mass spectral library   A collection of mass spectra of different compounds, usually represented as arrays of signal intensity vs. the *m/z* value rounded off to the integral mass number. Spectral libraries can carry additional compound information like exact molecular mass, Kovacz retention index, compound structure and a collection of synonym names. MS/MS and accurate mass libraries are the current developments.

Mass spectrometry/mass spectrometry (MS/MS)   The acquisition and study of the spectra of selected precursor ions, or of precursor ions of a selected neutral mass loss. Structure selective target analyte quantitations use → SRM/MRM data acquisition methods. MS/MS can be accomplished using beam instruments incorporating more than one analyser (tandem mass spectrometry in space, triple quadrupole analyzer) or in trap instruments (tandem mass spectrometry in time).
→ Tandem mass spectrometry.

Mass spectrum   A plot of the relative abundances of ions as a function of the their *m/z* values. The mass spectrum is the quantitative mass analysis of the ions generated in the ion source resp. collision cell.

Matrix   The medium (e.g. food, soil, seawater) in which the analyte of interest may be contained.

Matrix duplicate   An intralaboratory split sample that is used to document the precision of a method in a given sample matrix.

Matrix effect   The under- or overestimation of analytical results due to response differences between a clean standard solution and a matrix spike calibration. The matrix can be standardized by using → analyte protectants. → Analyte spike recovery.
→ Interference.

The response of some determination systems (e.g. GC-MS, LC-MS) to certain analytes may be affected by the presence of co-extractives from the sample matrix. Partition in headspace analyses and SPME is also frequently affected by components present in the samples. These matrix effects derive from various physical and chemical processes and may be difficult or impossible to eliminate. More reliable calibration may be obtained with → matrix-matched calibration when it is necessary to use techniques or equipment that are potentially prone to the effects.

Matrix-matched calibration

Calibration intended to compensate for matrix effects and acceptable interference, common use in pesticide residue analysis. The matrix blank (→ 'blank') should be prepared for analysis of samples. In practice, the pesticide is added to a blank extract (or a blank sample for headspace analysis) of a matrix similar to that analysed. The blank matrix used may differ from that of the samples if it is shown to compensate for the effects. However, for determination of residues approaching or exceeding the MRL, the same matrix (or standard addition) should be used. A matrix-matched calibration may compensate for matrix effects but does not eliminate the underlying cause. Because the underlying cause remains, the intensity of effect may differ from one matrix or sample to another, and also according to the 'concentration' of matrix. Isotope dilution or standard addition should be considered where matrix effects are sample dependent.

Matrix spike

A confirmed blank sample spiked with a known concentration of target analyte(s). The spiking occurs prior to sample preparation and analysis. A matrix spike is used to document the bias of a method in a given sample matrix (→ recovery, → surrogate standard). The term is also used in case of quantitative calibrations using a confirmed blank sample to prepare the calibration curve instead of clean solvent to prevent systematic errors by the → matrix effect, → matrix spike calibration.

| Matrix spike calibration | Calibration procedure using a confirmed blank matrix as basis for a quantitative calibration instead of clean solvents, → matrix effect. |
| Mass unit | The SI unit for the atomic mass m is given in kg. The additional unit used in chemistry is the atomic mass unit defined as |

$$1\,u = 1.660 \times 10^{-27}\,kg$$

	In mass spectrometry the Dalton [Da] is commonly used as a mass unit, defined as 1/12 of the mass of the carbon isotope ^{12}C, → Dalton. In earlier literature mass units are given also in amu (atomic mass units), and the mmu (millimass unit) used for 1/1000 amu (use deprecated by IUPAC).
Mathieu stability diagram	A graphical representation expressed in terms of reduced coordinates that describes the stability of charged particle motion in a quadrupole mass filter or quadrupole ion trap mass spectrometer, based on an appropriate form of the Mathieu differential equation.
Mattauch–Herzog geometry	An arrangement for a double-focussing mass spectrometer in which a deflection of $n/(4\sqrt{(2)})$ radians in a radial electrostatic field is followed by a magnetic deflection of $n/2$ radians.
Maximizing masses peak finder	A routine method of data handling in GC-MS analysis. The change in each in individual ion intensity with time is analysed. If several ions have a common peak maximum at the same retention time, the elution of an individual substance is recognized and noted even in case of a co-elution. The peak finder is used for automatic analysis of complex chromatograms, → also AMDIS.
McLafferty rearrangement	A dissociation reaction triggered by transfer of a hydrogen atom via a six-member transition state to the formal radical/charge site from a carbon atom four atoms removed from the charge/radical site (the α-carbon); subsequent rearrangement of electron density leads to expulsion of an olefin molecule. This term was originally applied to ketone ions where the charge/radical site is the carbonyl oxygen, but it is now more widely applied.

MCP	→ Microchannel plate detector, typically found as ion detector in TOF instruments, → Array detector.
MDL	→ Method detection limit.
MDQ	Minimum detectable quantity, the amount of analyte that produces a signal twofold that of the noise level.
MEPS	Micro-extraction by packed sorbent, miniaturized SPE sample preparation method using a packed syringe needle, and GC injection via liquid desorption in the GC injector.
Method blank	An analyte-free sample to which all reagents are added in the same volumes or proportions as used in sample processing. The method blank must be carried through the complete sample preparation and analytical procedure. The method blank is used to assess contamination resulting from the analytical process.
Method detection limit (MDL)	The minimum amount of analyte that can be analyzed within specified statistical limits of precision and accuracy, including sample preparation. The lower detection limit (→ LOD, LLOD) when applying the complete method from sample preparation, clean-up and analysis.
Merlin seal	Alternative injector septum solution 'Merlin Microseal' using a septum less seal formed like a duck bill and sealed by the inside carrier pressure and spring load.
Membrane inlet (MI)	A semi-permeable membrane separator that permits the passage of analytes directly from solutions or ambient air to the mass spectrometer ion source.
Metastable ion	An ion that is formed with internal energy higher than the threshold for dissociation but with a lifetime great enough to allow it to exit the ion source and enter the mass analyser where it dissociates before detection.
MHE	Multiple headspace extraction, a quantitation procedure used in static headspace involving multiple extraction and measurement from a single sample. The vial pressure is released between the

consecutive measurements using a dedicated MHE device on autosampler platforms.

Microchannel plate (MCP)	A thin plate that contains a closely spaced array of channels that each act as a continuous dynode particle multiplier. A charged particle, fast neutral particle or photon striking the plate causes a cascade of secondary electrons that ultimately exits the opposite side of the plate.
Microchannel devices	The microchannel or microfluidic devices are used for flow switching in GC, consisting of laser cut metal sheets (shims) with thicknesses from 20 to 500 µm. The resulting channel dimensions are similar to the conventional fused silica capillary columns. Shim stacks are bonded together with top and bottom plates for column connections, having low thermal mass for fast GC oven ramp rates.
Microfluidic devices	→ Microchannel devices.
MID	Multiple ion detection, the recording of several individual ions (m/z values) for increased dwell time and detection sensitivity in contrast to full scan, → also SIM.
Mixture search	Search mode of the PBM library search algorithm whereby a mixture is analysed by forming and searching difference spectra between sample and library.
MNT	Mononitrotoluene isomers.
Modifier	The addition of organic solvents in small amounts to the extraction agent in SFE. The modifier can be added directly to the sample (and is effective only in the static extraction step) or continuously by using a second pump. Typically up to 10% modifier is added to CO_2.
Molar mass	Mass of 1 mol of a compound:

$6.022\,1415(10) \times 10^{23}$ atoms or molecules.

Note: The term *molecular weight* is deprecated because 'weight' is the gravitational force on an object that varies with geographical location. Historically, the term has been used to denote the molar mass calculated using isotope averaged atomic masses for the constituent elements.

Molecular ion	The non-fragmented ion formed by the removal of one or more electrons to form a positive ion or the addition of one or more electrons to form a negative ion.
Molecular sieve	A stationary phase that retains analytes by molecular size interactions.
Molecular weight	The sum of the atomic weights of all the atoms present in a molecule. The term *molecular weight* is commonly used although actually masses are involved. The average molecular weight is calculated taking the natural isotopic distribution of the elements into account (stoichiometry), → Molar mass. Molecular weights in MS: The calculation is carried out exclusively using the atomic masses of the principle isotopes instead of averaged atomic masses. → Nominal mass.
Monoisotopic mass	Exact mass of an ion or molecule calculated using the mass of the principle (usually the lightest) isotope of each element.
MQL	Minimum quantitation limit, → also LOQ (limit of quantitation).
MRL	Maximum residue level.
MRM	Multiple reaction monitoring, MS/MS scan to monitor selected product ions only, essentially the same experiment as selected reaction monitoring, → SRM.
MRPL	Minimum required performance limit, means minimum concentration of an analyte in a sample, which at least has to be detected and confirmed. It is used to harmonize the analytical performance of methods for substances for which no permitted limited has been established.
MS/MS^n, MS^n	This symbol refers to multistage MS/MS experiments designed to record product ion spectra where n is the number of product ion stages (fragment ions). For ion traps, sequential MS/MS experiments can be undertaken where $n > 2$ whereas for a simple triple quadrupole system $n = 2$, → Multiple-stage mass spectrometry.

M-series	*n*-Alkyl-bis (trifluoromethyl) phosphine sulfides, a homologous series used to construct a retention time index system. The compounds can be used in gas chromatography with FID, ECD, ELCD, NPD, FPD, PID and MS detectors.
MSTFA	*N*-Methyl-*N*-trimethylsilyltrifluoroacetamide, a derivatization agent for silylation, typically used, for example, in anabolic steroid derivatization.
Multidimensional	Separations performed with two or more columns of different separation mechanism (orthogonal) in which peaks are selectively directed onto or removed from at least one of the columns by a timed valve system, → Backflush, → Heartcut, → Pre-cut.
Multiphoton ionization (MPI)	Photoionization of an atom or molecule in which in two or more photons are absorbed.
Multiple-stage mass spectrometry	Multiple stages of precursor ion *m/z* selection followed by product ion detection for successive fragment ions.
m/z	Symbol *m/z* is used to denote the dimensionless quantity formed by dividing the mass of an ion in unified atomic mass units by its charge number (regardless of sign). The symbol is written in italicized lower case letters with no spaces.
	Note 1: The term *mass-to-charge-ratio* is deprecated. Mass-to-charge-ratio has been used for the abscissa of a mass spectrum, although the quantity measured is not the quotient of the ion's mass to its electric charge. The symbol *m/z* is recommended for the dimensionless quantity that is the independent variable in a mass spectrum.
	Note 2: The proposed unit Thomson (Th) is deprecated.

N

η	Viscosity. Carrier gas viscosity increases with increasing temperature.
N	Number of theoretical plates:

$$N = 5.54\left(\frac{t_{\mathrm{R}}}{w_{\mathrm{h}}}\right)^2 \sim 16\left(\frac{t_{\mathrm{R}}}{w_{\mathrm{b}}}\right)^2$$

N_{eff}	The number of effective plates. This term is an alternate measurement of theoretical plate height that compensates for the non-partitioning nature of an unretained peak $$N = 16\left(\frac{t'_R}{w_b}\right)^2$$
N_{req}	The number of theoretical plates required to yield a particular resolution (R) at a specific peak separation (α) and retention factor (k): $$N_{\text{req}} = 16\,R^2\left(\frac{\alpha}{\alpha-1}\right)^2\left(\frac{k+1}{k}\right)^2$$
Nafion dryer	Device for drying analytical gas streams using capillary membrane tubes of polar Nafion material with outer counter flow of dry carrier gas.
NB	Nitrobenzene.
NBS	The U.S. National Bureau of Standards, now NIST.
ndl-PCB	Non dioxin-like PCBs, → PCB.
Needle sparger	Glass vessel used in purge and trap analysis of solids or foaming samples. The purge gas is passed into the sample via a needle perforated on the side or by means of a headspace sweep.
Negative ion	An atomic or molecular species having a net negative electric charge.
Negative ion chemical ionization (NCI, NICI)	Chemical ionization that results in the formation of negative ions.
Neutral loss	Loss of an uncharged species from an ion during either a rearrangement process or direct dissociation.
Neutral loss scan	MS/MS experiment for analysing precursor ions which undergo a common loss of neutral particles of the same mass. Analytes with functional groups in common are detected.
Nier–Johnson geometry	Arrangement for a double-focussing mass spectrometer in which a deflection of $n/2$ radians in a radial electrostatic field analyser is followed by a magnetic deflection of $n/3$ radians.
NIH	U.S. National Institute of Health.
NIST	U.S. National Institute of Standards and Technology, part of the U.S. Department of Commerce.

Nitrogen rule	An organic molecule containing the elements C, H, O, S, P, or a halogen has an odd nominal mass if it contains an odd number of nitrogen atoms.
Nominal mass	Mass of an ion or molecule calculated using the mass of principle, often most abundant isotope of each element rounded to the nearest integer value and equivalent to the sum of the mass numbers of all constituent atoms. According to this convention CH_4 and CD_4 have the same nominal mass!
Nominal mass resolution	The mass spectrometric resolution for the separation of mass signals with a uniform peak width of 1000 mDa (1 mass unit) in the quadrupole or ion trap analyser, $\rightarrow$ LRMS. The resulting mass numbers are generally given as nominal mass numbers or to one decimal place (unlike high resolution data).
NPD	Nitrogen–phosphorus detection, the nitrogen–phosphorus detector, providing a mass flow dependent signal, catalytically ionizes nitrogen or phosphorus-containing solutes on a heated rubidium or cesium surface in a reductive atmosphere. The nitrogen–phosphorus detector is highly selective and provides sensitivity that is better than that of a flame ionization detector.
NT	Nitrotoluene isomers.
Number of theoretical plates	Describes the separating capacity of a column.

O

OCDD	Octachlorodibenzodioxin.
OCI	On-column injection, sample enters the column directly from the syringe and does not contact other surfaces. On-column injection usually signifies cold injection for capillary columns.
Odd-electron ion	$\rightarrow$ Radical ion.
Odd-electron rule	Odd-electron ions may dissociate to form either odd or even-electron ions, whereas even-electron ions generally form even-electron fragment ions.
Olfactometry	$\rightarrow$ GC-olfactometry, GC-O.

On-column	Sample injection technique in GC whereby a diluted liquid extract is injected directly (without evaporation) on to a pre-column or on to the separation column itself. The temperature of the injection site at the beginning of the column is controlled by the oven temperature.
Onium ion	A positively charged hypervalent ion of the non-metallic elements. Examples are the methonium ion CH_5^+, the hydrogenonium ion H_3^+ and the hydronium ion H_3O^+. Other examples are the oxonium, sulfonium, nitronium, diazonium, phosphonium and halonium ions.
Orbitrap	The Kingdon trap, an ion trapping device that consists of an outer barrel-like electrode and a coaxial inner spindle-like electrode that form an electrostatic field with quadro-logarithmic potential distribution. The frequency of harmonic oscillations of the orbitally trapped ions along the axis of the electrostatic field is independent of the ion velocity and is inversely proportional to the square root of m/z so that the trap can be operated as a mass analyser using image current detection and Fourier transformation of the time domain signal.
	The trademarked term *Orbitrap* has been used to describe a Kingdon trap used as a mass spectrometer.
Orthogonal extraction	Pulsed acceleration of ions perpendicular to their direction of travel into a time-of-flight mass spectrometer. Ions may be extracted from a directional ion source, drift tube or m/z separation stage.

P

P	Relative pressure across the column; $P = p_i/p_o$.
Δp	Pressure drop across the column; $\Delta p = p_i - p_o$.
p_i	Absolute inlet pressure.
p_o	Absolute outlet pressure.
PAH	Polyaromatic hydrocarbons, the group of polycyclic hydrocarbons from naphthalene to and beyond coronene.

Partition coefficient	The partition coefficient k used in the headspace analysis technique is given by the partition of a compound between the liquid and the gaseous phases $c_{\text{liqu.}}/c_{\text{gas}}$.
PAT	Purge and trap technique, $\rightarrow$ P&T.
Paul ion trap	Ion trapping device that permits the ejection of ions with an m/z lower than a prescribed value and retention of those with higher mass. It depends on the application of radio frequency voltages between a ring electrode and two endcap electrodes to confine the ion motion to a cyclic path described by an appropriate form of the Mathieu equation. The choice of these voltages determines the m/z below which ions are ejected. The term *cylindrical ion trap* is deprecated.
	The name was given to the device developed by Prof. Paul, University of Bonn, Germany, and known as the *ion trap analyser*. Prof. Paul received the Nobel price in 1989 for his work on the QUISTOR development at the beginning of the 1950s.
PBB	Polybrominated biphenyls.
PBDE	Polybrominated diphenylether.
PBM	Probability based match, library search procedure for mass spectra developed by Prof. McLafferty.
PCB	Polychlorinated biphenyls with a total number of 209 congeners.
	Seven indicator PCBs are routinely monitored:
	tetra Cl-PCB: 28, 52
	penta Cl-PCB: 101, 118
	hexa Cl-PCB: 138, 153
	hepta Cl-PCB: 180.
	12 so-called dioxin-like PCBs or WHO-PCBs are the coplanar non-ortho substituted congeners of a total number of 68 coplanar congeners of which 20 in total are non-ortho substituted:
	tetra Cl-PCB: 77, 81.
	penta Cl-PCB: 105, 114, 118, 123, 126
	hexa Cl-PCB: 156, 157, 167, 169.
	hepta Cl-PCB: 189.
PCI	Positive chemical ionization.
PCN	Polychlorinated naphthalenes.

PCXE	Polychlorinated xanthenes.
PCXO	Polychlorinated xanthones.
Peak capacity	In quantitative GC: The amount of analyte that can be injected without a significant loss of column efficiency. In multidimensional GC: The maximum number of peaks that can be resolved.
Peak (in mass spectrometry)	Localized region of relatively large ion signal in a mass spectrum. Although peaks are often associated with particular ions, the terms *peak* and *ion* should not be used interchangeably.
Peak intensity	Height or area of a peak in a mass spectrum.
Peak matching	Procedure for measuring the accurate mass of an ion using scanning mass spectrometers, in which the peak corresponding to the unknown ion and that for a reference ion of known m/z are displayed alternately on a display screen and caused to overlap by adjusting the acceleration voltage.
Peak overload	If too much of analyte is injected, its peak can be distorted into a triangular shape exceeding the column capacity.
PEEK	Polyether ether ketone, hard plasticizer-free polymer with the general structure (with $x = 2$, $y = 1$):

	PEEK is used as a sealing and tubing material and for screw joints in HPLC, SFE, SFC and in high vacuum areas of MS, thermally stable up to 340 °C.
PEG	Polyethylene glycol.
Permitted limit	The maximum residue limit, maximum level or other maximum tolerance for substances established in legislation.
PFBA	Pentafluoropropionic anhydride, derivatization agent, it is also used for introducing halogens to increase the response in ECD and NCI.
PFK	Perfluorokerosene, calibrant in mass spectrometry, widely used with magnetic sector MS instruments.

PFPD	Pulsed flame photometric detector, provides two simultaneous signals for S and P by measurement of a fluorescence/time profile in the range of 2–25 ms with about 5 Hz cycle time (after Prof. Aviv Amirav, University Tel Aviv, Israel).
PFTBA	Perfluorotributylamine, calibrant in mass spectrometry, widely used with quadrupole, ion trap and magnetic sector MS instruments, → FC43.
Phase ratio	In the headspace analysis technique the phase ratio of $V_{gas}/V_{liqu.}$ gives the degree of filling the headspace bottle. In capillary GC the phase ratio describing the ratio of the internal volume of a column (volume of the mobile phase) to the volume of the stationary phase. High-performance columns are characterized by high-phase ratios. A table showing different combinations of internal diameters and film thicknesses can be used to optimize the choice of column with regard to analysis time, resolution and capacity.
Photodissociation	Process wherein the reactant ion is dissociated as a result of absorption of one or more photons.
Photoionization (PI)	Ionization of an atom or molecule by a photon, $M + h\nu \rightarrow M^{+\bullet} + e^-$. The term *photon impact* is deprecated.
PID	Photoionization detector, the photoionization detector ionizes analyte molecules with photons in the UV energy range, provides a concentration dependent signal. The photoionization detector is a selective detector that responds to aromatic compounds and olefins when operated in the 10.2 eV photon range, and it can respond to other materials with a more energetic light source.
PIONA	Paraffins, isoparaffins, olefins, napthenes and aromatic compounds.
PLOT	Porous-layer open-tubular column, a fused silica capillary column type with a modified inner wall that has been etched or otherwise treated to increase the inner surface area or to provide gas-solid chromatographic retention behaviour.
PONA	Paraffins, olefins, napthenes and aromatic compounds.

Porous polymer	A stationary-phase material that retains analytes by selective adsorption or by molecular size interaction.
Positive ion	Atomic or molecular species having a net positive electric charge.
Post-acceleration detector	Detector in which a high voltage is applied after m/z separation to accelerate the ions and produce an improved signal, $\rightarrow$ Conversion dynode.
Post-source decay (PSD)	Technique specific to reflectron time-of-flight mass spectrometers where product ions of metastable transitions or collision-induced dissociations generated in the flight tube prior to entering the reflectron are m/z separated to yield product ion spectra.
PPINICI	Pulsed positive ion negative ion chemical ionization, the alternating data acquisition of positive and negative ions formed during chemical ionization, patented by former Finnigan Corp., San Jose, CA, USA.
Precision	The closeness of agreement between independent test results obtained under stipulated (predetermined) conditions. The measure of precision usually is expressed in terms of imprecision and computed as standard deviation of the test result. Less precision is determined by a larger standard deviation.
Precursor ion	Ion that reacts to form structure related product ions. The reaction can be unimolecular dissociation, ion/molecule reaction, isomerization or change in charge state. The term *parent ion* is deprecated.
Precursor ion scan	MS/MS scan function or process that records a precursor ion spectrum. It is used to detect substances with related structures which give common fragments. The term *parent ion scan* is deprecated.
Precursor ion spectrum	Mass spectrum recorded from any spectrometer in which the appropriate m/z separation function can be set to record the precursor ion or ions of selected product ions. The term *parent ion spectrum* is deprecated.

Pre-cut	Peaks at the beginning of a chromatogram are removed to vent or directed onto another column of different polarity or at a different temperature for improved resolution.
Pre-purge	A preliminary step in purge and trap analysis, the atmospheric oxygen is removed by the purge gas before the sample is heated to avoid side reactions.
Pre-scan	The step in the ion trap scan function before the analytical scan. During the pre-scan the variable ionization time or ion collection time is adjusted to fill the trap to its optimum capacity with ions.
Pre-search	Part of library searching of mass spectra in which a small group of candidates is selected from the whole number for detailed comparison and ranking.
Press fit	Glass tube connectors for fused silica capillaries. The cross cut column end is simply pushed into the conical opening and seal is achieved with the external polyimide coating. Caution is necessary when applying high temperatures during oven ramping for weakening the polymer at the sealing site.
Pressure units	The SI unit is given in Pascal:

$$1\,Pa = 1\,Nm^{-2}$$
$$10^5\,Pa = 10^5\,Nm^{-2} = 1\,bar$$

Pressure values are often given in traditional units. In MS vacuum technologies pressures are given in (Torr) or (mTorr) and gas pressures of GC supplies frequently in (kPa), (bar) or (psi).

Pressure units conversion table:

	Pa	bar	Torr	psi	at	atm
Pa	1	$1\cdot10^{-5}$	$7.5\cdot10^{-3}$	$1.45\cdot10^{-4}$	$1.02\cdot10^{-5}$	$9.87\cdot10^{-6}$
bar	$1\cdot10^5$	1	750	14.514	1.02	0.987
Torr	133	$1.33\cdot10^{-3}$	1	$1.94\cdot10^{-2}$	$1.36\cdot10^{-3}$	$1.32\cdot10^{-3}$
psi	$6.89\cdot10^3$	$6.89\cdot10^{-2}$	51.67	1	$7.03\cdot10^{-2}$	$6.80\cdot10^{-2}$
at	$9.81\cdot10^4$	0.981	736	14.224	1	0.968
atm	$1.0133\cdot10^5$	1.0133	760	14.706	1.033	1

at: technical atmosphere $1\,kp\,cm^{-2}$.
atm: physical atmosphere $1.033\,kp\,cm^{-2}$.
psi: pound per square inch.

Primary reaction	Conversion of the reagent gas used for CI into the reagent ions by electron ionization, → also CI.
Principal ion	Most abundant ion of an element or molecular isotope cluster, such as the $^{11}B^{79}Br_2^{81}Br^{+\bullet}$ ion of m/z 250 of the cluster of isotopologue molecular ions of BBr_3. The term *principal ion* has also been used to describe ions that have been artificially isotopically enriched in one or more positions such as $CH_3-^{13}CH_3^{+\bullet}$ or $CH_2D_2^{+\bullet}$, but those are best defined as isotopologue ions.
Product ion	An ion formed as the product of a reaction involving a particular precursor ion. The reaction can be unimolecular dissociation to form fragment ions, an ion/molecule reaction or simply involve a change in the number of charges. The term *fragment ion* is deprecated. The term *daughter ion* is deprecated.
Product ion scan	Specific scan function or process that records a product ion spectrum. The terms *fragment ion scan* and *daughter ion scan* are deprecated.
Product ion spectrum	Mass spectrum recorded from any spectrometer in which the appropriate m/z separation scan function is set to record the product ion or ions of selected precursor ions. The terms *fragment ion scan* and *daughter ion scan* are deprecated. Note: The term *MS/MS spectrum* is deprecated; a scan specific term, for example, *precursor ion spectrum* or *second-generation product ion spectrum* should be used.
Proficiency study	Analysing the same sample, allowing laboratories to choose their own methods, provided these methods are used under routine conditions. The study has to be performed according to ISO guide 43-1 and 43-2 and can be used to assess the reproducibility of methods.
Proton abstraction	Type of NCI reaction, ionization is effected by transfer of a proton from the analyte (abstraction) to the reagent ion, for example, from analytes with phenolic OH groups.
Proton affinity	Proton affinity of a species M is the negative of the enthalpy change for the reaction $M + H^+ \rightarrow [M+H]^+$ at 298 K.

Protonated molecule	An ion formed by interaction of a molecule with a proton, and represented by the symbolism $[M + H]^+$.
	Note 1: The term *protonated molecular ion* is deprecated; this would correspond to a species carrying two charges.
	Note 2: The terms *pseudo-molecular ion* and *quasi-molecular ion* are deprecated; a specific term such as protonated molecule or a chemical description such as $[M + Na]^+$, $[M - H]^-$, and so on, should be used.
Protonation	Type of PCI reaction, ionization is effected by transfer of one ore more protons to the substance molecule. Protonating reagent gases include methane, methanol, water, isobutene and ammonia.
PTGC	Programmed-temperature GC, the column temperature changes in a controlled manner as peaks are eluted.
PTI	Programmed-temperature injection, a cold injection technique in which the inlet temperature is specifically programmed from the gas chromatograph.
PTV	Programmed-temperature vaporizer, an inlet system designed to perform a temperature programmed injection, a cold injection system for direct liquid injection for split or splitless injection, solvent split technique and cryo-enrichment.
P&T	Purge-and-trap sampling, dynamic headspace procedure, a concentration technique for volatile solutes. The sample is purged with an inert gas that entrains volatile components onto an adsorptive trap. The trap is then heated to desorb trapped components into a GC column.
Pure search	A mode of the PBM search procedure, which only uses the forward search capability of the library search, → also Forward library search.
PyGC	Pyrolysis GC, the sample is pyrolysed (decomposed) in the inlet before GC analysis.
PyMS	→ Pyrolysis mass spectrometry.

Pyrogram	Chromatogram received from the separation of pyrolysis products.
Pyrolysis mass spectrometry (PyMS)	A mass spectrometry technique in which the sample is heated to the point of decomposition and the gaseous decomposition products are introduced into the ion source.

Q

QED	Quantification enhanced by data dependant MS/MS, a QED scan on a triple quadrupole instrument delivers an information rich product ion mass spectrum that can be used to confirm the existence of compounds by an in-built MS/MS library while they are being quantified using the MRM scan mode.
QIT	Quadrupole ion trap, → Paul ion trap.
QMS	Quadrupole mass spectrometer, → Transmission → Quadrupole mass spectrometer.
QqQ	Triple quadrupole mass spectrometer. Note: The lower case q denotes the collision cell.
Quality assurance (QA)	An integrated system of activities involving quality planning, quality control, quality assessment, quality reporting and quality improvement to ensure that a product or service meets defined standards of quality with a stated level of confidence.
Quality control (QC)	The overall system of technical activities whose purpose is to measure and control the quality of a product or service so that it meets the needs of users. The aim is to provide quality that is satisfactory, adequate, dependable and economical. For analytical chemistry, QC is a set of procedures applied to an analytical methodology to demonstrate that the analysis is in control.
Quality control sample	A sample or standard used either singly or in replicate to monitor method performance characteristics.
Quasimolecular ion	An ion to which the molecular mass is assigned, which is formed by, for example, chemical ionization as $(M+H)^+$, $(M+NH_4)^+$, $(M-H)^+$,

$(M − H)^−$. The term *quasimolecular ion* is deprecated by the IUPAC to use 'cationized molecule' instead.

QuEChERS	Quick Easy Cheap Effective Rugged Safe, acronym for a 'Fast and easy multi-residue method employing acetonitrile extraction/partitioning' and 'dispersive solid-phase extraction for the determination of pesticide residues in produce', → *www.quechers.com*, M. Anastassiades, S.J. Lehotay, D. Stajnbaher and F.J. Schenck, J AOAC Int 86, (2003) 412.
QUISTOR	Quadrupole ion storage trap, → Paul ion trap.

R

r	Relative retention. For peak i relative to standard peak s; $r = k_i/k_s$.
R	Resolution, the quality of separation of two peaks. In GC for two closely eluted peaks using the formula:

$$R = (t_{R,2} − t_{R,1})/w_{b,2}$$

where the subscripts 1 and 2 refer to the first and second peaks.
From N, k_2 and α

$$R = \left(\frac{\sqrt{N}}{4} \right) \left(\frac{\alpha−1}{\alpha} \right) \left(\frac{k_2}{k_2+1} \right)$$

where k_2 is the retention factor of the second peak. A resolution of 1.5 is said to be baseline resolution. R incorporates both efficiency and separation.

Radical ion	An ion, either a cation or anion, containing unpaired electrons in its ground state. The unpaired electron is denoted by a superscript dot alongside the superscript symbol for charge, such as for the molecular ion of a molecule M, that is, $M^{+\bullet}$. Radical ions with more than one charge and/or more than one unpaired electron are denoted such as $M^{(2+)(2\bullet)}$. Unless the positions of the unpaired electron and charge can be associated with specific atoms, superscript charge designation should be placed before the superscript dot designation.

Reagent gas cluster	The spectrum of ions formed from the reagent gas for chemical ionization, monitored to adjust the correct reagent gas pressure in the ion source.
Reagent ion	An ion produced in large excess in a chemical ionization source that reacts with neutral sample molecules to produce an ionized form of the molecule through an ion/molecule reaction.
Reagent ion capture	Type of CI reaction in PCI/NCI. The ionization of the analyte is achieved by an addition reaction of the reagent ion.
Recombination energy	Energy released when an electron is added to an ionized molecule or atom, that is, the energy involved in the reverse process to that referred to in the definition of vertical ionization energy.
Recovery	The percentage of the true concentration of a substance recovered during the analytical procedure. It is determined during validation, if no certified reference material is available → surrogate standard.
Reference ion	Stable ion whose structure is known with certainty. These ions are usually formed by direct ionization of a molecule of known structure, and are used to verify by comparison the structure of an unknown ion.
Reference material	A material of which one or several properties have been confirmed by a validated method, so that it can be used to calibrate an apparatus or to verify a method of measurement.
Reference method	A sampling or measurement method that has been officially specified by an organization as meeting its data quality requirements.
Reflectron	Constituent of a time-of-flight mass spectrometer that uses a static electric field to reverse the direction of travel of the ions entering it. A reflectron improves mass resolution by assuring that ions of the same m/z but different translational kinetic energy arrive at the detector at the same time.
Relative response factor	Ratio of the analyte response to the response of the related internal (recovery) standard.

Repeatability	Precision under repeatability conditions, where independent test results are obtained with the same method on identical test items in the same laboratory by the same operator using the same equipment.
Reproducibility	Precision under reproducibility conditions, where test results are obtained with the same method on identical test items in different laboratories with different operators using different equipment.
RER	Reduced energy ramp, collision energy regime during the generation of a MS/MS product spectrum in which the applied collision energy is ramped down to produce a richer product ion spectrum for structure elucidation and confirmation, especially in the higher mass range part of the spectrum.
Residual gas analyser (RGA)	Mass spectrometer used to measure the composition and pressure of gases in an evacuated chamber.
Resonance-enhanced multiphoton ionization (REMPI)	Multiphoton ionization in which the ionization cross section is significantly enhanced because the energy of the incident photons is resonant with an intermediate excited state of the neutral species.
Resonance ion ejection	Mode of ion ejection in a Paul ion trap that relies on an auxiliary radio frequency voltage that is applied to the endcap electrodes. The voltage is tuned to the secular frequency of a particular ion to eject it.
Resonance ionization (RI)	→ Resonance-enhanced multiphoton ionization.
Response	The specific height of the detector signal, usually calculated as the ratio of the peak area or height to the quantity of the analyte.
Retention gap	A short piece of deactivated but uncoated column as a pre-column or guard column placed between the inlet and the analytical column. There is no retention of the analytes. A retention gap often helps relieve solvent flooding. It also retains non-volatile sample contaminants from on-column injection. After evaporation of the solvent the analytes are focussed at the beginning of the analytical column.

Retention index	A uniform system of retention classification according to a solute's relative location between a pair of homologous reference compounds on a specific column under specific conditions. It compares the time a compound is eluting from the column to the times of a set of standard compounds. The most common set of compounds used for retention indices are hydrocarbons typically from C5 to C30, separated in the order of boiling points, → Kovats index.
Retention time	The time that the compound is held up on the GC column. Usually a retention index is used as compound specific rather than a retention time which is method dependant.
Retention volume	The carrier gas volume required to elute a component.
Reverse library search	A procedure of comparing a mass spectrum of an unknown compound with a mass spectral library so that the unknown spectrum is compared in turn with the library spectra, considering only the m/z peaks observed to have significant intensity in the current library spectrum. This comparison procedure serves to expose mixed spectra or to search for a substance in a GC-MS chromatogram.
RF	Response factor; radio frequency.
RI	Retention index, for example, Kovats index.
RIC	Reconstructed ion chromatogram. The total ion current (TIC) is calculated from the sum of the intensities of all of the acquired mass signals in a mass spectrum. This value is generally stored with the mass spectrum and shown as the RIC or TIC value in the mass spectrum. Plotting the RIC or TIC value along the scan axis (retention time) gives the conventional chromatogram diagram (older magnetic sector instruments used a dedicated total ion current detector for this purpose). For this reason the shape of the chromatogram in GC-MS is dependent from the mass range scanned.
Robustness	A measure of the capacity an analytical method to obtain comparable and acceptable results when

perturbed by small but deliberate variations in procedural parameters. It provides an indication of the method's suitability and reliability during normal use. During a robustness study the GC-MS method parameters such as solvent amounts, carrier gas flow, heating ramps and detector settings are intentionally varied to study the effects on analytical results.

RSD

Relative standard deviation.

RT

Retention time.

Ruggedness

The susceptibility of an analytical method to changes in experimental conditions which can be expressed as a list of the sample materials, analytes, storage conditions, environmental and/or sample preparation conditions under which the method can be applied as presented or with specified minor modifications.

S

s

Split ratio. The ratio of the sample amount that is vented to the sample amount that enters the column during split injection. Higher split ratios place less sample on the column. *s* is usually measured as the ratio of total inlet flow to column flow; $s = (F_s + F_c)/F_c$.

Sandwich technique

A syringe injection technique in which a sample plug is placed after or between one or two solvent plugs in the syringe barrel, separated by a short air plug, to wash the syringe needle with solvent and obtain better sample transfer into the inlet, $\rightarrow$ also Solvent flush.

SBSE

Stir bar sorptive extraction, a glass coated magnetic stir bar is coated with PDMS which serves as high capacity sorption phase, extraction by thermal desorption in a dedicated injector system.

Scan function

The control of ion trap or quadrupole mass analysers by changing the applied voltages in time, represented by a diagram of voltage (U) against time *t* (ms).

SCD

Sulfur chemiluminescence detection, a sulfur chemiluminescence detector responds to

	sulfur-containing compounds by generating and measuring the light from chemiluminescence.
SCOT	Support-coated open-tubular column, a capillary column in which stationary phase is coated onto a support material that is distributed over the column inner wall. A SCOT column generally has a higher peak capacity than a WCOT column with the same average film thickness.
Screening method	Methods that are used to detect the presence of a substance or class of substances at the level of interest. These methods have the capability for a high sample throughput and are used to sift large numbers of samples for potential non-compliant results. They are specifically designed to avoid false compliant results.
Secondary reaction	The chemical reaction of analyte molecules with the reagent ions in a CI ion source to form stable product ions, → also CI.
Sector mass spectrometer	Mass spectrometer consisting of one or more magnetic sectors for m/z selection in a beam of ions. Such instruments may also have one or more electric sectors for energy focusing.
Selected ion monitoring (SIM)	Operation of a mass spectrometer for target compound analysis in which the abundances of several ions of specific m/z values are recorded rather than the entire mass spectrum.
Selected reaction monitoring (SRM)	MS/MS scan technique, whereby a product ion spectrum is produced by collision induced decomposition of a selected precursor ion, in which only one or more selected fragment ions are detected. In MS/MS instruments the SRM technique provides the highest possible selectivity together with the highest possible sensitivity (→ dwell time) for target compound quantitation. Data acquired from specific product ions corresponding to m/z selected precursor ions recorded via two or more stages of mass spectrometry. Selected reaction monitoring for target compound analysis can be performed as tandem mass spectrometry in time or tandem mass spectrometry in space. The term *multiple reaction monitoring* is deprecated.

Selectivity	In analytical chemistry it is the recommended term to express the extent to which a particular method can be used to determine analytes under given conditions in mixtures or matrices without interferences from other components of similar behaviour. Selectivity can be graded (in contrast to → specificity). In chromatographic methods the fundamental ability of a stationary phase to retain substances selectively based upon their chemical characteristics, including vapour pressure and polarity. In mass spectrometry the capability of a method or instrument to distinguish small mass differences, for example, between target ions and non-target matrix substances.
Selectivity tuning	Several techniques for adjusting the selectivity of separations that involve more than one column or stationary-phase type. Serially coupled columns and mixed-phase columns can be selectivity-tuned.
SEM	Secondary electron multiplier for high amplification factors, usually built from discrete dynodes which are electrically connected by resistors to provide a voltage ramp across the number of dynodes (contrary to a CDEM).
Sensitivity	The degree of detector response to a specified analyte amount per unit time or per unit volume.
Separation α	The degree of separation of two peaks in time, → α and R.
Septum	Silicone or other elastomeric material that isolates inlet carrier flow from the atmosphere and permits syringe penetration for injection (alternative septum solution → Merlin seal).
Septum purge	The carrier gas swept across the injector internal septum face to a separate vent so that material emitted from the septum does not enter the inlet.
SFC	Supercritical fluid chromatography, uses a supercritical fluid as the mobile phase, can be coupled directly to mass spectrometry by using capillary columns.
SFE	Supercritical fluid extraction, as extraction medium mostly CO_2 is used, modifiers, for

example, methanol or ethyl acetate can be added to optimize matrix specific extraction efficiencies.

Silica gel	A gas-solid adsorbent, a porous form of silicon dioxide with high affinity to water.
Significant figures	Those digits in a number that are known with certainty, plus the first uncertain digit. Example: three significant figures have 0.104, 1.04, 104, 1.04×10^4. The 1 and the middle 0 are certain, and the 4 is uncertain, but significant. Initial zeroes are never significant. Exponential number has no effect on the number of significant figures.
SIM	Selected ion monitoring, recording of individual pre-selected ion masses, as opposed to full scan, $\rightarrow$ also MID.
SIM descriptor	Data acquisition control file for SIM analyses, contains the selected specific masses of the analytes (m/z values), the individual dwell times and the retention times for timely switching the detection to other analytes in the same run.
SIMDIS	Simulated distillation, a boiling-point separation technique that simulates physical distillation of petroleum products.
Single laboratory study	An analytical study involving a single laboratory using one method to analyse the same or different test materials under different conditions over justified long time intervals (in-house validation)
SIS	Selected ion storage, describes the SIM measuring technique in ion trap MS, $\rightarrow$ SIM, $\rightarrow$ also Waveform ion isolation
SISCOM	Search for identical and similar compounds, search procedure for mass spectra in libraries developed by Henneberg, Max Planck Institute for Coal Research, Mühlheim, Germany.
Skewing	Reversal of the relative intensities in a mass spectrum caused by changing the substance concentration during a scan. Skewing occurs with beam instruments during slow mass scans in the steep rising or falling slopes of GC peaks.

SN	Separation number or Trennzahl (TZ). A measurement of the number of peaks that could be placed with baseline resolution between two sequential peaks, z and $z+1$, in a homologous series such as two hydrocarbons:

$$SN = \frac{t_{R(z+1)} t_{R(z)}}{w_{h(z+1)} + w_{h(z)}} - 1$$

S/N	Signal-to-noise ratio, the ratio of the peak height to the noise level. Peak height is measured from the average noise level to peak top. Noise is measured as the width of the noise band (typically 4 σ), excluding known signals.
Soft ionization	Formation of gas-phase ions without extensive fragmentation.
Solutes	Chemical substances that can be separated by chromatography.
Solvent effect	Focussing the analyte to a narrow band at the beginning of a capillary column by means of condensation of the solvent and directed evaporation.
Solvent flooding	A source of peak-shape distortion caused by excessive solvent condensation inside the column during and after splitless or on-column injection.
Solvent flush	GC injection technique, also known as *sandwich technique*, whereby a liquid (solvent, derivatization agent) is first drawn up into the syringe, followed by some air to act as a barrier and finally the sample, → Sandwich technique.
Solvent flushing	A column rinsing technique that can remove non-volatile sample residue and partially restore column performance.
Solvent split	GC injection technique using the PTV injector so that larger quantities of diluted extracts can be applied from more than 2 µL up to a LC-GC coupling. The excess of solvent is evaporated through the split line while analytes are concentrated in the inlet liner. A large difference between the boiling points (volatility) of the solvent and analytes is required.
Space charge effect	Result of mutual repulsion of particles of like charge that limits the current in a charged-particle

	beam or packet and causes some ion motion in addition to that caused by external fields.
SPE	Solid-phase extraction, a sample clean-up technique.
Specificity	The term *specificity* is considered as an absolute term, and thus cannot be graded. The IUPAC states that specificity is the ultimate of selectivity. A method is specific for one analyte or not. This suggests that no component other than the analyte contributes to the result. Hardly any method is that specific and, in general, the term should be avoided, → selectivity.
SPI	Septum equipped programmable injector, special design of a cold injection system, which can be used exclusively for total sample transfer.
Spiked sample	A sample prepared by adding the target analytes to a blank matrix sample. Used in an analytical method to determine the effect of the matrix on the recovery efficiency, → matrix effect.
Split injection	The sample size is adjusted to suit capillary column requirements by splitting off a major fraction of sample vapours in the inlet so that as little as 0.1% enters the column. The rest is vented.
Splitless injection	A derivative of split injection for total sample transfer. During the first minutes of the injection process, the sample is not split and enters only the analytical column. Splitting is restored afterwards to purge the sample remaining in the inlet. As much as 99% of the sample enters the column.
SPME	Solid phase micro extraction, a sample clean-up technique that uses a removable sorptive micro extraction device.
Spray and trap	Extraction procedure for foaming aqueous liquids, the liquid stream is sprayed into a purge gas stream.
SRM	→ Selected reaction monitoring.
SSL	Split-/splitless injector.
Stable ion	Ion with internal energy sufficiently low that it does not rearrange or dissociate prior to detection in a mass spectrometer.

Standard addition method	A procedure in which the test sample is divided in two (or more) test portions. One portion is analysed as such and known amounts of the standard analyte are added to the other test portions before analysis. The amount of the standard analyte added has to be between two and five times the estimated amount of the analyte in the sample. This procedure is designed to determine the content of an analyte in a sample, taking account of the recovery of the analytical procedure.
	Often used quantitative calibration method in headspace analysis for matrix independent calibrations.
Standard operating procedures (SOPs)	The established written and approved analytical procedures of a laboratory.
Static field	Electric or magnetic field that does not change in time.
Stationary phase	Liquid or solid materials coated inside an analytical column that selectively retain analytes.
SUMMA canister	Passivated stainless steel canisters for air analysis (e.g. EPA method TO 14/15) for collection of samples and standardization. The inner surface is deactivated by a patented procedure involving a Cr/Ni oxide layer.
Surrogate standard	The internal standard for quantification, which is added to a homogenized sample before starting the sample preparation. The ratio of the surrogate standard to other internal standards added during the course of the clean-up is used to calculate the recovery efficiency.
SWIFT	Stored waveform inverse Fourier transformation, technique to create excitation waveforms for ions in FT-ICR mass spectrometer or Paul ion traps. An excitation waveform in the time-domain is generated by taking the inverse Fourier transform of an appropriate frequency domain programmed excitation spectrum, in which the resonance frequencies of ions to be excited are included. This procedure may be used for selection of precursor ions in MS/MS experiments.

Systematic error	A consistent deviation in the results of sampling or analytical processes from the expected or known value. Such error can be caused by a faulty instrument operation or calibration, by human or methodological bias.

T

T_o	Room temperature.
T_c	Column temperature.
t_M	Unretained peak retention time. The time required for one column volume (V_G) of carrier gas to pass through a column.
t_R	Retention time. The time required for a peak to pass through a column.
t'_R	Adjusted retention time; $t'_R = t_R - t_M$.
Tandem-in-space	MS/MS analysis with beam mass spectrometers. Ion formation, selection, collision induced decomposition and acquisition of the product ion spectrum take place continuously in separate sections of the mass spectrometer, for example, triple quadrupole MS.
Tandem-in-time	MS/MS analysis with ion storage mass spectrometers. Ion formation, selection, collision-induced decomposition and acquisition of the product ion spectrum take place in the same location of the mass spectrometer but sequentially in time, for example, ion trap MS, ICR MS.
Tandem mass spectrometry	→ Mass spectrometry/mass spectrometry coupling (MS/MS).
Target compound	Analyte to be quantitatively determined, usually taken from a list of compounds from regulations or directives.
TCA	Target compound analysis, multicomponent analysis, multimethod. The analytical strategy with the setup and data evaluation for the quantitative monitoring of a selected group of target compounds. For data evaluation the analytes are grouped together in a target compound list for analysis and are searched for using automated

routines by spectrum comparison in a retention time window, identified and quantified when found.

TCD

Thermal-conductivity detection, a thermal conductivity detector measures the differential thermal conductivity of carrier and reference-gas flows. Solutes emerging from a column change the carrier gas thermal conductivity and produce a response. TCD is a universal detection technique with moderate sensitivity depending on the thermal conductivity of the analytes.

Analytes	Thermal conductivity (10^5 cal/cm s^{-1} °C)
H_2	49.9
He	39.9
N_2	7.2
Ethane	7.7
H_2O	5.5
Benzene	4.1
Acetone	4.0
CH_3Cl	2.3

TCDD

Tetrachlorodibenzo-*p*-dioxin isomers.

TCDF

Tetrachlorodibenzofuran isomers.

TCEP

Tris(cyanoethoxy)propane.

Tenax

Non-polar synthetic polymer based on 2,6-diphenyl-*p*-phenyleneoxide used as an adsorption material for the concentration of air samples or in purge and trap analysis.

TE/GC-MS

Thermal extraction GC-MS using a thermal extraction unit for solid sample material as inlet system of the gas chromatograph.

Theoretical plate

A hypothetical entity inside a column that exists by analogy to a multipleplate distillation column. As solutes migrate through a column, they partition between the stationary phase and the carrier gas. Although this process is continuous, chromatographers often visualize a step-wise model. One step corresponds approximately to a theoretical plate.

Thermal ionization (TI)

Ionization of a neutral species through contact with a high-temperature surface.

TIC	Total ion current, sum of all acquired intensities in a mass spectrum, → also RIC.
Time lag focussing	Energy focussing in a time-of-flight mass spectrometer that is accomplished by introducing a time delay between the formation of the ions and the application of the accelerating voltage pulse. Ion formation may be in the gas phase or at a sample surface. Related term: *delayed extraction*.
Time-of-flight mass spectrometer (TOF-MS)	Instrument that separates ions by m/z in a field-free region after acceleration to a fixed kinetic energy.
TMCS	Trimethychlorosilane, trimethylsilyl chloride, silylation agent (catalyst).
TMS derivative	Trimethylsilyl derivative, a chemical derivative to increase substance volatility to facilitate chromatographic application, typical diagnostic mass m/z 73.
TMSH	Trimethylsulfonium hydroxide, derivatization agent for esterification and methylation, for example, free fatty acids, phenylureas, and so on. TMSH was used successfully to save on extra steps for the derivatization in the insert of the PTV cold injection system.
TNT	Trinitrotoluene.
Total ion current (TIC)	Sum of all the separate ion currents carried by the ions of different m/z contributing to a complete mass spectrum or in a specified m/z range of a mass spectrum.
Total ion current chromatogram	Chromatogram obtained by plotting the total ion current detected in each of a series of mass spectra covering the acquired mass range recorded as a function of retention time.
Total sample transfer	GC injection technique whereby the entire injection volume reaches the column with or without evaporation, → splitless injection.
TPH	Total petroleum hydrocarbons.
Traceability	The property of a result of a measurement whereby it can be related to appropriate standards, generally international or national standards, through an unbroken chain of comparisons.

Transmission	The ratio of the number of ions leaving a region of a mass spectrometer to the number entering that region, e.g. ratio of ions leaving the ion source vs. reaching the detector.
Transmission quadrupole mass spectrometer	A mass spectrometer that consists of four parallel rods whose centres form the corners of a square and whose opposing poles are connected. The voltage applied to the rods is a superposition of a static potential and a sinusoidal radio frequency potential. The motion of an ion in the x and y dimensions is described by the Mathieu equation whose solutions show that ions in a particular m/z range can be transmitted by oscillation along the z-axis.
Triple quadrupole mass spectrometer	A tandem mass spectrometer comprising two transmission quadrupole mass spectrometers in series, with a non-resolving (RF-only) quadrupole between them to act as a collision cell.
Tropylium ion	The characteristic ion in the spectrum of alkylaromatics, $C_7H_7^+$ m/z 91, is formed by benzyl cleavage of the longest alky chain. The structure of the seven-membered ring (after internal rearrangement) is responsible for the high stability of the ion.
TRT	Temperature rise time, heating up period in pyrolysis until the set pyrolysis temperature is reached. A fast TRT is a quality feature of the pyrolyser.
TSD	Thermionic-specific detection, → NPD (nitrogen–phosphorus detection).
TSIM	N-Trimethysilyl-imidazole, a potent derivatization agent for silylation.
TSP	Thermospray, LC-MS coupling device.
TZ	Trennzahl, → Separation number.

U

$\bar{u}_{avrg}$	Average linear carrier gas velocity; $\bar{u}_{avrg} = L/t_M$.
u_o	Carrier gas velocity at the column outlet; $u_o = \bar{u}_{avrg}/j$.

u_{opt}	Optimum linear gas velocity. The carrier gas velocity corresponding to the $\bar{u}$ minimum theoretical plate height, ignoring stationary-phase contributions to band broadening: $$u_{opt} = 8 \left(\frac{D_G}{d_c} \right) \sqrt{\frac{3(1+k)^2}{1+6k+11k^2}}$$
u	The unified atomic mass unit u, a non-SI unit of mass defined as one twelfth of the mass of one atom of ^{12}C in its ground state and equal to $1.6605402(10) \times 10^{-27}$ kg. The term *atomic mass unit* is deprecated. $\rightarrow$ Dalton. Note: The term *atomic mass unit* (amu) is ambiguous as it has been used to denote atomic masses measured relative to a single atom of ^{16}O or to the isotope-averaged mass of an oxygen atom or to a single atom of ^{12}C.
UAR	*Unknown analytical response*, this term refers to GC peaks in a multicomponent target compound analysis (TCA), which are not included in the list of target compounds.
Unimolecular dissociation	Fragmentation reaction in which the molecularity is treated as one, irrespective of whether the dissociative state is that of a metastable ion produced in the ion source or results from collisional excitation of a stable ion.
Unstable ion	Ion with sufficient energy to dissociate within the ion source.
UTE, UTE%	Utilization of theoretical efficiency, $\rightarrow$ CE (coating efficiency).

V

V_G	The volume of carrier gas contained in a column. For open-tubular columns and ignoring the stationary-phase film thickness: $(d_f), V_G = L(d_c^2/4)$.
V_L	The volume of (liquid) stationary phase contained in a column.
Validation	The confirmation by examination and the provision of effective evidence that the particular requirements of a specific intended use are fulfilled.

Validated method	A method that has been determined to meet certain performance criteria for sampling or measurement operations.
van Deempter plot	→ HETP.
Verification	Confirmation by examination and provision of evidence that specified requirements have been met.
Viton	The brand name of synthetic rubber and fluoropolymer elastomer commonly used in O-rings and other moulded or extruded goods. The name is trademarked by DuPont Performance Elastomers L.L.C., → *http://en.wikipedia.org/wiki/Viton*.
VOC	Volatile organic compounds, a group of analytes, which are determined together using an EPA method and which contain both volatile halogenated hydrocarbons and BTEX.
VOCARB	Non-polar adsorbent filling for concentration of air samples, and in purge and trap analysis, multilayer filling based on graphitized carbon black (Carboxen) and carbon molecular sieves (Carbosieve), Supelco.
VOCOL	Volatile organic compounds column, Sigma-Aldrich.

W

w_b	The peak width at its base, measured in seconds. For a Gaussian peak, $w_b = 1.596\,(A_p/h_p)$.
w_h	The peak width at half height, measured in seconds. For a Gaussian peak, $w_h = 0.940\,(A_p/h_p)$.
Waveform ion isolation	Ion storage technique for the ion trap analyser. By using resonance frequencies the ion trap can exclude ions of several m/z values from storage (e.g. from the matrix) and collect selectively determined pre-selected analyte ions, also SIS.
WBOT	Wide-bore open-tubular column, open-tubular (capillary) column with a nominal inner diameter of 530 µm.

WCOT	Wall-coated open-tubular column, a capillary column in which the stationary phase is coated directly on the column wall.
Werkhoff splitter	Adjustable flow splitter for sample injection into two capillary columns, or a split located after the analytical column, also know as the *glass cap cross divider*.
Wiswesser line notation	Code for chemical structures using an alphanumerical system, for example, for Lindane, L6TJ AG BG CG DG EG FG *GAMMA, contained e.g. in the Wiley library of mass spectra.
Within-laboratory reproducibility	Precision obtained in the same laboratory under stipulated (predetermined) conditions (concerning, e.g. method, test materials, operators, environment) over justified long time intervals.
Working standard	A general term used to describe dilutions produced from the stock standard, which are used, for example, to spike for recovery determination or to prepare calibration standards.

X

XAD	Synthetic polymer (resin) used as an adsorption material in air or purge and trap analysis as well as in the clean-up for chlorinated compounds, for example, dioxins, furans, PCBs.
x-Ion	Fragment ion containing the C-terminus formed upon dissociation of a protonated peptide at a backbone C–C bond.

Y

y-Ion	Fragment ion containing the C-terminus formed upon dissociation of a protonated peptide at a backbone C–N bond.

Z

z-Ion	Fragment ion containing the C-terminus formed upon dissociation of a protonated peptide at a backbone N–C bond.

Further Reading

Coplen, T.B. (2008) Explanatory Glossary of Terms Used in Expression of Relative Isotope Ratios and Gas Ratios, International Union of Pure and Applied Chemistry Inorganic Chemistry Division, Commission on Isotopic Abundances and Atomic Weights, Peer Review, January 16, 2008, *http: //www.iupac.org* (accessed 14 May 2014).

Coplen, T.B. (2011) Guidelines and recommended terms for expression of stable isotope-ratio and gas-ratio measurement results. *Rapid Commun. Mass Spectrom.*, **25**, 2538–2560.

Hinshaw, J.V. (1992) A compendium of GC terms and techniques. *LCGC*, **10** (7), 516–522.

IUPAC (1993) Nomenclature for chromatography. *Pure Appl. Chem.*, **65**, 819–872.

IUPAC Task Group MS Terms, Murray, K.K., Boyd, R.K., Eberlin, M.N., Langeley, G.J., Li, L., and Naito, Y. (2013) Definitions of terms relating to mass spectrometry (IUPAC Recommendations 2013). *Pure Appl. Chem.*, **85** (7), 1515–1609,

http://iupac.org/publications/pac/asap/ PAC-REC-06-04-06/ (accessed 26 November 2013).

Library4Science Chromatography Topics, *http://www.chromatography-online.org/topics/* (accessed 16 May 2014).

Mass Spectrometry-Wikipedia Articles on Mass Spectrometry, Based on the Wikipedia Mass Spectrometry Category, *http://mass-spec.lsu.edu/ msterms/index.php/Main_Page*, (accessed 26 November 2013).

SANCO (2012) Method Validation and Quality Control Procedures for Pesticide Residue Analysis in Food and Feed, Document No. SANCO/12495/2011, implemented by 01/01/2012.

Schröder, E. (1991) *Massenspektroskopie*, Springer, Berlin.

Vessman, J., Stefan, B.I., van Staden, J.F., Danzer, K., Lindner, W., Burns, T.D., Fajgelj, A., and Müller, H. (2001) Selectivity in analytical chemistry (IUPAC Recommendations 2001). *Pure Appl. Chem.*, **73** (8), 1381–1386.

Index

Handbook of GC-MS: Fundamentals and Applications, Third Edition. Hans-Joachim Hübschmann.
© 2015 Wiley-VCH Verlag GmbH & Co. KGaA. Published 2015 by Wiley-VCH Verlag GmbH & Co. KGaA.